Fault Diagnosis for Electrical Machines, Power Electronics, and Drives

Fault Diagnosis for Electrical Machines, Power Electronics, and Drives

Guest Editors

Jose M. Guerrero
Araitz Iturregi Aio
Sergio Toscani

MDPI

Basel • Beijing • Wuhan • Barcelona • Belgrade • Novi Sad • Cluj • Manchester

Guest Editors
Jose M. Guerrero
Department of Electrical Engineering
Universidad del País Vasco/Euskal Herriko Unibertsitatea
Bilbao
Spain

Araitz Iturregi Aio
Department of Electrical Engineering
Universidad del País Vasco/Euskal Herriko Unibertsitatea
Bilbao
Spain

Sergio Toscani
Dipartimento di Elettronica, Informazione e Bioingegneria (DEIB)
Politecnico di Milano
Milano
Italy

Editorial Office
MDPI AG
Grosspeteranlage 5
4052 Basel, Switzerland

This is a reprint of the Special Issue, published open access by the journal *Applied Sciences* (ISSN 2076-3417), freely accessible at: https://www.mdpi.com/journal/applsci/special_issues/JJ7309YABX.

For citation purposes, cite each article independently as indicated on the article page online and as indicated below:

Lastname, A.A.; Lastname, B.B. Article Title. *Journal Name* **Year**, *Volume Number*, Page Range.

ISBN 978-3-7258-8001-0 (Hbk)
ISBN 978-3-7258-8002-7 (PDF)
https://doi.org/10.3390/books978-3-7258-8002-7

Contents

About the Editors

Jose M. Guerrero

Jose M. Guerrero is a Research Associate Professor at the Department of Electrical Engineering, University of the Basque Country (UPV/EHU). He received a B.Sc. in energy resources, fuels and explosive engineering, an M.Sc. in electrical engineering, and a Ph.D. degree in Electrical Engineering from Universidad Politécnica of Madrid, Spain, in 2018, 2019, and 2022 respectively. His main field of interest relates to the diagnosis and protection of electrical systems that involve power electronics, such as electric drives, DC links, energy storage systems or DC microgrids. He has authored and co-authored more than 100 scientific contributions and more than 25 patents in the field. His expertise mainly resides in ground fault diagnosis and frequency response analysis applied to drives and electrical machines for condition monitoring. As researcher, he has taken part in multiple interdisciplinary collaborative projects. The combined funding of these projects exceeds 1,750,000 € for the organizations in which he has participated. Furthermore, he is member of the technical committee for research project evaluation at the Agency for Scientific and University Quality of Andalusia, and of the local committee of the Developments in Power System Protection Conference.

Araitz Iturregi Aio

Araitz Iturregi Aio is an Associate Professor at the Department of Electrical Engineering, University of the Basque Country (UPV/EHU), where she also received her M.Sc. and PhD in 2006 and 2013, respectively. She began her professional career at SEW Eurodrive, specializing in frequency inverters and decentralized drive systems. In 2007, she joined the Department of Electrical Engineering at the Bilbao School of Engineering (UPV/EHU), a role she still holds today. Her main research interests include current interruption techniques, the simulation of electric arc behaviors in current interruption switches, and the design of power systems protection. Her primary interests also encompass HIL simulation techniques for design and validate protection algorithms in power systems. She has completed pre- and post-doctoral research stays at the Technische Universität Ilmenau in Germany. Additionally, she also collaborates with the Leibniz Institute for Plasma Science and Technology (INP Greifswald), Ostschweizer Fachhochschule, TU Delft or companies such as Iberdrola, ZIV or Tecnalia in several joint national and international research projects.

Sergio Toscani

Sergio Toscani is an Associate Professor with the Dipartimento di Elettronica, Informazione e Bioingegneria, Politecnico di Milano. In 2007, he received an M.Sc. degree (summa cum laude) in Electrical Engineering from Politecnico di Milano, where he also obtained his Ph.D. degree (summa cum laude) in 2011. His research activity primarily focuses on the characterization of electrical devices and measurements in power systems. His main research interests include the development and characterization of voltage and current transducers, advanced digital signal processing techniques for power system measurements, diagnostic methods for electrical devices, and the monitoring of photovoltaic systems. He serves as Technical Program Co-Chair of the IEEE Workshop on Applied Measurements for Power Systems and he is member of the Technical Committee TC39—Measurements in Power Systems of the IEEE Instrumentation and Measurement Society. He has been Guest Editor for several Special Sections published in leading journals in the field of electrical measurements and is member of the Editorial Board of the MDPI journal *Applied Sciences*.

Preface

The reprint "Fault Diagnosis for Electrical Machines, Power Electronics, and Drives" brings together a collection of review and research articles addressing one of the most significant challenges in modern electrical engineering.

This reprint aims to contribute to the current literature by presenting advanced fault diagnosis and protection methods for electrical drives incorporating power electronics. The increasing use of power electronic converters, while enabling higher efficiency, better performance, and greater flexibility, also intensifies stress on electrical machines, produced by modulation strategies and high-frequency switching. This increases the likelihood of electrical, mechanical and magnetic faults. Additionally, conventional protection systems often struggle to properly and timeously detect faults under non-sinusoidal and variable-frequency operating conditions for which they were not originally designed.

Despite their negative impact on system lifetime, power electronic converters also provide new opportunities, as they can actively take part in advanced fault diagnosis and protection techniques. The coexistence of these challenges and opportunities constitutes the main motivation behind this reprint.

On behalf of the editorial team and the contributing authors, we would like to recommend this reprint to researchers working on fault detection and protection of electrical machines supplied by power electronics, as well as to practicing engineers interested in implementing advanced diagnostic techniques in industrial environments.

This reprint includes an extensive review of general fault diagnosis methods for aerospace and satellite applications, together with research contributions addressing specific and relevant failures such as inter-turn short-circuits, broken rotor bars in induction machines, demagnetization in permanent magnet synchronous machines, and multiple fault identification in DC motors. Through these advanced diagnostic techniques, this reprint provides readers with relevant insights for the development and implementation of fault diagnosis strategies in electrical machines, power electronics, and drives.

Jose M. Guerrero, Araitz Iturregi Aio, and Sergio Toscani
Guest Editors

applied sciences

MDPI

Review

Fault Diagnosis and Prognosis of Satellites and Unmanned Aerial Vehicles: A Review

MohammadSaleh Hedayati †, Ailin Barzegar † and Afshin Rahimi *,†

Department of Mechanical, Automotive and Materials Engineering, University of Windsor, 401 Sunset Ave, Windsor, ON N9B 3P4, Canada; hedayat2@uwindsor.ca (M.H.); barzegaa@uwindsor.ca (A.B.)
* Correspondence: arahimi@uwindsor.ca
† These authors contributed equally to this work.

Abstract: This paper comprehensively analyzes advanced Fault Diagnosis and Prognosis (FDP) techniques employed in aerial and space agents such as satellites, spacecraft, and Unmanned Aerial Vehicles (UAVs). The critical engineering functions of fault diagnostics and prognosis, particularly the emerging field of fault prognosis, emphasize the necessity for further advancement. Integrating these methodologies enriches the system's capacity to diagnose faults in their early stages. Additionally, it enables the prediction of fault propagation and facilitates proactive maintenance to mitigate the risk of severe failure. This paper aims to introduce diverse FDP methods, followed by a discussion on their application and evolution within single and multisatellite/UAV systems. Throughout this review, eighty-five relevant works are analyzed and discussed and their evaluation metrics are expanded upon as well. Within the works analyzed in this review, it was found that data-driven methods constitute 54% and 7% of the methodologies utilized in single- and multiagent FDP, respectively, which underscores the rise of these methods in the field of single-agent FDP and their unexplored potential in multiagent condition monitoring. Finally, this review is brought to a close with a suggested classification scheme of the utilized methodologies in the field, a quantitative analysis of their contributions to the field, and remarks and mentions of the potential gaps in the area.

Keywords: fault detection; fault isolation; fault identification; fault prognosis; single/multisatellite/UAV

1. Introduction

The reliability of satellite and Unmanned Aerial Vehicle (UAV) missions has received significant attention in recent years owing to the cost and time needed to launch satellites [1]. The possible severe consequences in the case of satellite and UAV failures also contribute to this cause in addition to the manufacturing and operational costs. According to [2], 32% of on-orbit spacecraft failures are attributed to the attitude and orbit control system (AOCS) and 27% to the power subsystem. Consequently, addressing the AOCS and power subsystem health monitoring could significantly boost onboard spacecraft mission reliability. Moreover, 51% of the malfunctions are due to Reaction Wheel (RW), momentum wheels, Control Moment Gyro (CMG), thrusters, and gyroscopes, and 18% are due to unknown reasons, warranting the need to explore health monitoring in the mentioned system components. Regarding the time of failure after launch, 41% of the recorded failures in [2] occurred in the first year of operation, making Fault Diagnosis and Prognosis (FDP) strategies imperative to extending on-orbit spacecraft and satellites' lifetimes. On the other hand, for UAVs, malfunctions could be caused by intentional or unintentional factors. Among the various operation-endangering abnormalities, the following are the most prevalent: cyber-attacks, such as signal jamming, spoofing, measurement sensor, keylogging viruses, control system attacks, and exogenous physical factors, such as sudden aerodynamic disturbances, and damaged actuators [3]. The satellite and UAV actuator and sensor failures could also be attributed to intrinsic hardware wear-and-tear and degradation. To increase the mission reliability of these agents, all kinds of significant faults should be fixed or circumvented regardless of their source. There are two main approaches for increasing the mission reliability of satellites and UAVs [4]: hardware redundancy and analytical redundancy. Hardware redundancy involves incorporating redundant components for switching to a healthy component once the primary unit malfunctions. However, this poses a major issue. With the rise in the number of launched satellites and the decrease in their sizes, embedding redundant components becomes impractical [5]. The limited space available in UAVs also does not allow for hardware redundancy. In addition, switching from a faulty to a healthy component could disturb the operational continuity of missions [6]. Due to hardware redundancy's evident shortcomings, there is a need for advanced analytical redundancy methods. Analytical redundancy methods comprise vastly different means to monitor the health status of agents, namely, data-driven, model-based, and expert manpower-based methods. Compared with the methods reliant on expert personnel to go through the operational data of the agents, autonomous methods are distinctly superior, especially since they are more cost-effective and time-efficient while being less prone to errors. Hence, the advantages of autonomous analytical redundancy methods have inspired researchers to explore this area further. Among the host of analytical redundancy methods, data-driven and model-based methods are the most prevalent in the literature owing to their advantages. To address the health monitoring requirements of agents more precisely, analytical redundancy methods are divided into fault diagnosis and fault prognosis fields. The former accounts for informing the system of a fault in it, whereas the latter is responsible for estimating the system's Remaining Useful Life (RUL) given the diagnosed faults. From a mission reliability perspective, the insights derived from the FDP modules are used as input to the fault-tolerant control module. To increase the readability and understandability of this work among the myriad of works on general system reliability applications, we employ and explain the terminology in our work decided upon by the International Federation of Automatic Control (IFAC) Technical Committee SAFEPROCESS and outlined in [7]. Here, we briefly go over the most prevalent definitions presented in [7]: "Fault Detection" is defined as the determination of an existing system fault with an emphasis on the time of detection. Following this stage, "Fault Isolation" is described as the stage responsible for determining the location, kind, and time of detection. The subsequent step is labeled "Fault Identification", which involves determining the size details of the isolated fault and the time-variant behavior that it exhibits. Alternatively, "Fault Diagnosis" is defined as a step that follows "Fault Detection", contains "Fault Isolation and Identification" in itself, and entails measuring the size, kind, location of the fault, and the time of its detection. Finally, "Monitoring" means continuous online or real-time physical system condition evaluation through storing operational information and discerning and annunciating the emerging anomalies in its operation.

In the current study, we review the different FDP methods for satellites and UAVs that work individually or as members of a multisatellite/UAV network using analytical redundancy methods. The inclusion of both satellites and UAVs is due to the similarities in how health-monitoring approaches have been implemented for these agents. Investigating the methods employed in one agent could help provide valuable takeaways for utilizing them in the other agent. As a result of the prominence of data-driven and model-based methods in this area, this study has been divided into sections covering each category while also reviewing other methods that are fusions of the two mentioned categories (i.e., hybrid methods) or completely different methodologies.

The diagnostic and prognostic methods used for increasing the mission reliability of satellites, spacecraft, and UAVs in this work can be applied to more than one class of each of these agents. Many of the techniques addressed in this work are applicable to various actuators, sensors, and subsystems onboard satellites and UAVs. Satellites are usually categorized based on their mass, which can lie in ranges of 1 g to upwards of 1000 kg [8]. Some of the most prominent satellite classes include nanosatellites and small-, medium-, and large-sized satellites. However, the methods analyzed in this review are presented with the mindset of applying them to more than their original classes. Conversely, when a method is applicable to only one or a few scenarios, it is declared. Furthermore, different classes of UAVs are also considered. The main categories of UAVs include the two groups of Vertical Takeoff and Landing (VTOL) and Conventional Takeoff and Landing (CTOL) UAVs. Some examples of VTOLs covered in this work are Hex-rotor, quadrotor, and tilt-rotor UAVs and fixed-wing UAVs for the CTOL class. The same analysis logic described for the FDP of different satellite classes also holds for different UAV classes.

It is notable to mention that in this review, relevant quantitative validation results of the discussed studies are presented when possible. However, the test results of different studies are usually acquired under different circumstances unless compared directly. Therefore, deep analysis solely based on the quantitative results might prove erroneous, but they are provided for reference in this review and expanded upon when possible.

The rest of the article is structured as follows. Section 2 provides initial details, encompassing the problem's definitions and setting the groundwork for subsequent discussions. Section 3 goes over the review methodology that was adopted in this review and important considerations and assumptions for the readers of this review. Sections 4 and 5 explore and review FDP approaches, respectively. Finally, Section 6 summarizes this paper, presents a classification of the prominent methodologies in the field of satellite and UAV FDP, and proposes potential future research directions based on the identified gaps in the field.

2. Preliminaries and Definitions

Faults are an inevitable part of dynamic systems and must be accounted for and dealt with, particularly for applications where reliability is paramount. Faults are defined as unpredicted alterations of the system behavior occurring naturally or intentionally, prohibiting the system's subcomponents from continuing nominal operation These faults could leave a detrimental effect on the whole system and take some time to cause failure, or in some cases, they could cause instantaneous system failures. Consequently, searching for means of diagnosing and prognosing faults is warranted to deal with faults with lasting effects as part of a health monitoring strategy. General satellite, spacecraft, and UAV fault types could be broken down into four categories: abrupt, transient, intermittent, and incipient. Abrupt malfunctions are classified as spontaneously caused by hardware failure and will not be alleviated until fully repaired. In other words, they are characterized by system changes that are faster than the system's nominal dynamics. Transient malfunctions comprise deviations from the nominal operating conditions that may disappear without external intervention. Recurring transient faults are called intermittent faults. The likelihood of intermittent faults occurring is proportional to the system complexity [9]. Finally, incipient faults are defined as malfunctions caused by the deterioration of a system over time [10]. These four fault types are illustrated in Figure 1. Faults could also be modeled as additive and multiplicative faults depending on their effects on a system's dynamics. Multiplicative faults affect a system's inputs and outputs in a multiplicative way and stem from systematic parameter variations [11].

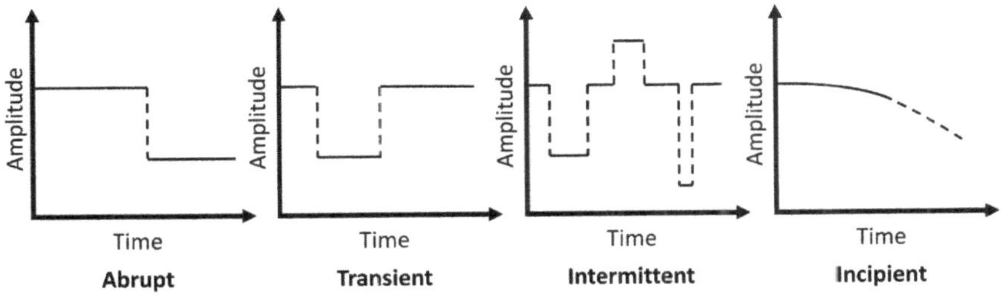

Figure 1. Behavior of different fault types over time.

The fault diagnosis process in nonlinear systems, including satellites and UAVs, involves multiple layers. The common classification used to delineate these stages of fault diagnosis includes detection, isolation, identification, estimation, and reconstruction. The term "malfunction detection" or "fault detection" refers to the capability of an analytically redundant methodology to recognize the presence of a fault. We have achieved the detection stage once our diagnostic schemes accurately identify when a fault originated. Isolation, conversely, involves determining the fault's location within the system. Therefore, a system's detection and isolation processes have temporal and spatial aspects, respectively. Detection always precedes or co-occurs with isolation but never follows it. Fault identification, estimation, and reconstruction are often used interchangeably to convey a similar concept. This stage of diagnosis aims to determine the characteristics of the malfunction. A malfunction's characteristics could include its severity, shape, depth, etc. Two main approaches for performing fault identification on a system are qualitative and quantitative. The two methods are subsets of identification techniques; however, their implementations tend to be vastly different and depend on the application. For instance, qualitative techniques necessitate a predefined set of malfunction types for classifying future occurrences as one of them or, conversely, for clustering them based on emerging malfunction types. In contrast, quantitative techniques rely on quantifying system performance degradation to achieve identification. The next step towards health monitoring is to predict the future course of faults and their effect on the RUL of a system. Prognosis becomes especially beneficial when a system is dealing with incipient faults, as its RUL could be evaluated using prognostics and lengthened by operators by taking appropriate remedying measures accordingly [12]. The results of the FDP schemes are then incorporated into fault-tolerance and recovery strategies to compensate for the malfunctions of single and multiagent systems. The visual representation of the mentioned process is depicted in Figure 2. Nevertheless, the intricate details of fault-tolerant methods and the literature related to them fall beyond the scope of this paper.

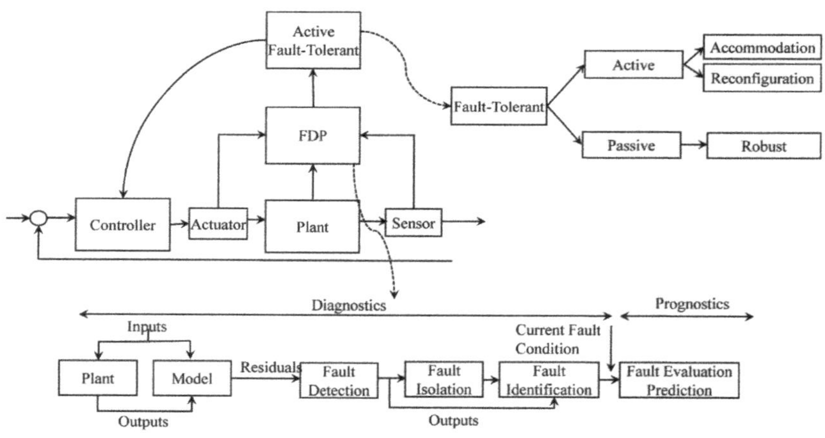

Figure 2. FDP and fault-tolerant control schemes.

FDP methodologies mostly monitor the health state of satellites, spacecraft, and UAVs through model-based and data-driven methods. Model-based techniques rely on high-fidelity models of systems and estimators to evaluate fault-sensitive states, analyze them, and compare the results with the model of the system operating nominally. Model-based methods comprise deterministic and stochastic strategies. Stochastic estimation schemes account for the added process and measurement noise while estimating states. Among stochastic schemes, Kalman Filter (KF) are utilized for linear models with added Gaussian noise, and Unscented Kalman Filter (UKF) and Extended Kalman Filter (EKF) for nonlinear models with added Gaussian noise. When several faults are involved, interacting multiple model (IMM) methods could be used which employ several interacting filters to classify different fault modes [13]. Meanwhile, the Particle Filter (PF) class is reliable for estimation in nonlinear models with non-Gaussian noise. All in all, the main drawback of using model-based methods is their requirement for precise models that are not always available and are difficult to derive for every application. Essentially, the family of KFs can estimate the internal states of a linear or nonlinear dynamic system in the presence of known parameters and noisy processes and measurements. They could also be used to estimate the parameters instead of states, considering only states, and the process and measurement noises are known. Keep in mind that in this text, states refer to the dynamical representation of systems that tend to exhibit rapid changes. In contrast, parameters are virtually time-invariant elements closely tied to the system's characteristics. System state estimation is used to study the deviations of the system's dynamical behavior from nominal conditions. However, the system's parameters, as opposed to states, could also indicate a system's health, especially when it is undergoing degradation. There are also approaches for simultaneously estimating the states and parameters of a system, namely, dual and joint estimation. On the other hand, parameter estimation is usually, but not necessarily, associated with time delays, restricting their usage to offline health monitoring and not accommodating online monitoring requirements [14]. Model-based methods usually utilize residual-based algorithms to diagnose faults in control systems. To clarify the practicality of residual-based algorithms in fault diagnosis, an example of their use in a general discrete linear system along with a linear KF was outlined in [15]. Accordingly, the linearized system model in state-space form is represented by

$$\begin{cases} X_k = A_{k-1}X_{k-1} + \Gamma_{k-1}w_{k-1} \\ Z_k = C_kX_k + v_k + f_{k,\varphi}\gamma \end{cases} \tag{1}$$

where $X_k \in \mathbb{R}^n$ denotes the system state, $Z_k \in \mathbb{R}^m$ denotes the measurement output, $A_{k-1} \in \mathbb{R}^{n\times n}$ is a constant matrix, $\Gamma_{k-1} \in \mathbb{R}^{n\times r}$ denotes the noise matrix, $w_k \in \mathbb{R}^r$ denotes the process zero-mean Gaussian white noise sequence, and $v_k \in \mathbb{R}^m$ denotes the measurement zero-mean Gaussian white noise sequence. The process and measurement noises are independent. In addition, the control input is assumed to be zero. γ denotes a random vector representing the fault size, while $f_{k,\phi}$ is set to be one when the system is faulty and zero when the system is nominal.

Subsequently, we aim to update the a priori state estimate with a gain of weighted time duration and the new measurement in each iteration of running the following iterative equation:

$$\begin{cases} \hat{X}_k^- = A_{k-1}\hat{X}_{k-1}^+ \\ P_{k/k-1} = A_{k-1}P_{k-1}A_{k-1}^T + \Gamma_{k-1}Q_{k-1}\Gamma_{k-1}^T \\ L_k = P_{k/k-1}C_k^T\left[C_kP_{k/k-1}C_k^T + R_k\right]^{-1} \\ \hat{X}_k^+ = [I - L_kC_k]A_{k-1}\hat{X}_{k-1}^+ + L_kZ_k \\ P_k = [I - L_kC_k]P_{k/k-1} \end{cases} \tag{2}$$

where $\hat{X}_k^+$ denotes the estimated state using the measurement update vector of Z_k, $\hat{X}_k^-$ denotes the predicted state using the last iteration's estimated state only. $P_{k/k-1}$ and P_k denote the error covariance time and measurement updates, respectively. Finally, the state residual is defined as the difference between the estimated and the predicted states and is evaluated as follows:

$$\beta_k = \hat{X}_k^- - \hat{X}_k^+ \tag{3}$$

The calculated residual is then continuously compared with a threshold. The residual can also be created using differences between nominal and estimated measurement outputs instead of predicted and estimated outputs. Figure 3 represents the process of generating output residuals and delivering them to the fault diagnosis block.

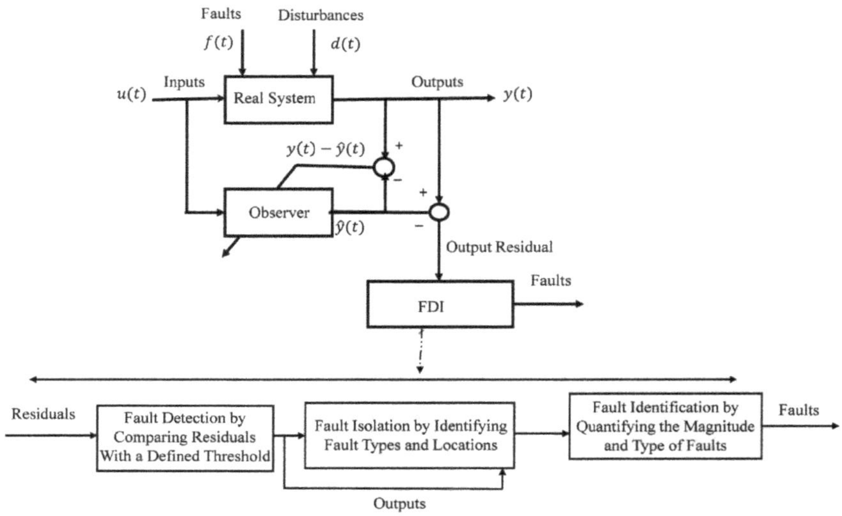

Figure 3. Fault diagnosis strategy.

When a threshold is surpassed, a fault is identified, and depending on the context, it may be isolated and described based on where and the extent to which the threshold has been surpassed. Residual-based algorithms have the flexibility to employ different residual definitions, but they fundamentally share the same concept as exemplified in [15]. However, using constant thresholds could yield unsatisfactory results or cause divergence, especially when they are poorly defined and when the nonmeasurable parameters of the system undergo abrupt changes [16]. To avoid this problem, adaptive thresholding methods have been utilized in the literature. Data-driven methods are another popular approach for health monitoring in nonlinear systems. Data-driven techniques encompass a broad spectrum of Machine Learning (ML) and Artificial Intelligence (AI) algorithms that can learn and extract patterns from datasets. The major advantages of using these techniques are analyzing complex nonlinear systems through learning the patterns in their output and accommodating applications where either labeled or unlabeled datasets are available, eliminating manual work. Data-driven methods develop an understanding of the studied linear or nonlinear dynamic systems and essentially model them as black boxes capable of analyzing real systems' performance. However, this imposes two major challenges: (i) large amounts of data are needed for data-driven methods to become reliable, and (ii) they lack interpretability compared with other methods. For instance, a broad range of neural networks are utilized in literature for learning the features of nonlinear systems; however, generally, we do not have meaningful and effective access to the inner workings of neurons and how they reach a specific output. Meanwhile, model-based method outputs are easily justifiable since high-fidelity models are accessible and can be examined. Nevertheless, the advantages of data-driven methods outweigh their disadvantages, and although they must be extensively addressed, data-driven techniques remain one of the most viable approaches for health monitoring. Furthermore, data-driven approaches for health monitoring of satellites and UAVs mainly focus on examining time series datasets, as is warranted by the nature of these agents' operations. An underlying assumption in data-driven approaches is that operational time series datasets exhibit similar characteristics, which is needed to justify their use in such applications [17]. Since time series datasets constitute a substantial portion of the available training data for satellites, spacecraft, and UAVs, neural networks, particularly Recurrent Neural Network (RNN), have emerged as attractive approaches for conducting FDP on such datasets. Another option for FDP in satellites and UAVs is signal processing and frequency analysis.

They consist of the two categories of parametric and nonparametric methods. Parametric techniques operate under the assumption that the analyzed signals conform to a specific mathematical model, such as autoregression and state-space models [18]. Parametric methods are most effective when the assumed model matches closely with the signal. On the other hand, nonparametric methods are those where the analyzed signals are not associated with a known model or their attributes are not well-known. For instance, the Fourier transform, fast Fourier transform (FFT), and wavelet transform. Nonparametric methods are more robust to deviations and more flexible in application. However, these methods suffer from spectrum leakage and the picket fence effect, causing inaccuracies in frequency estimation. The drawbacks of nonparametric methods have been addressed in the literature; however, some barriers still exist in utilizing them [18]. Signal processing methods can also be combined with data-driven methods. In the literature, they provide time–frequency spectrums from the system and train Convolutional Neural Network (CNN) models with them, leveraging the advantages of data-driven methods. One of the drawbacks of signal processing methods that discourage their use in the health monitoring of satellites and UAVs is their predisposition toward making false alarms [19].

A visual representation of the discussed monitoring modes and methodologies is provided in Figure 4.

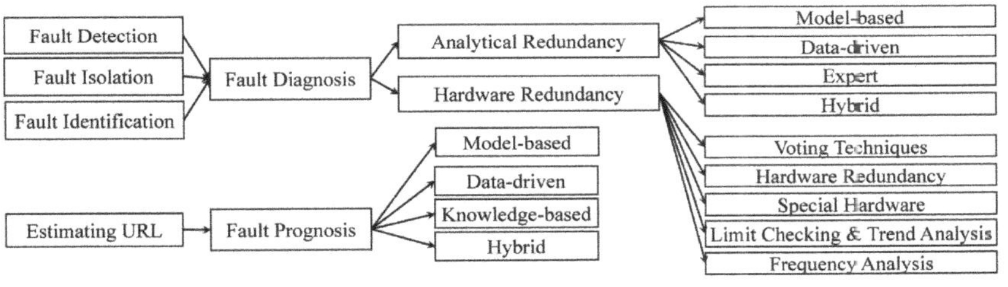

Figure 4. Fault Diagnosis and Prognosis approaches.

On the other hand, if a fault is to be addressed through a network of satellites or UAVs, we must be aware of the communication between these agents in all fault diagnosis methods. By evaluating the communication topology, we can determine the required quantity of observers/filters and the designed architectures to diagnose faults across the entire network. A communication graph/topology consists of nodes interconnected by edges. Directed edges create connections between groups of nodes in a single specified direction. Conversely, undirected graphs do not restrict the direction of information flow. The fault diagnosis observers for a network of agents can have three main architectures based on an analysis provided in [20]:

1. A centralized architecture involves a single master observer or filter that gathers information and manages inputs throughout the entire network. This approach results in a significant communication load because the master filter/observer must provide state information to all agents in the team for state and fault estimation.
2. A distributed architecture utilizes a hierarchical system of filters/observers. Each spacecraft operates a local filter that offers estimates to the master filter. This design effectively distributes the computational load among the local agents, simplifying the process.
3. A decentralized architecture operates without a hierarchy, eliminating the need for central supervision. Within this structure, each spacecraft processes its measurement data simultaneously with other spacecraft in the network, enabling efficient state and fault estimation without a centralized control mechanism. Clear schematic diagrams of all three fault diagnosis architectures are presented in Figure 5.

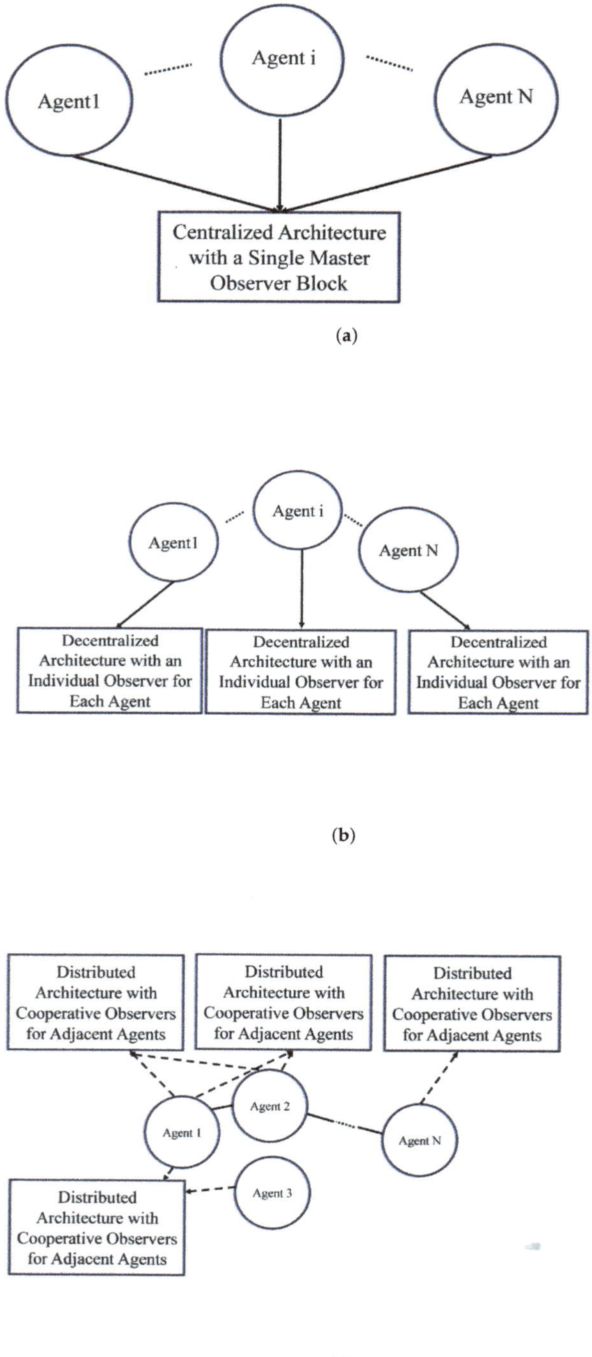

Figure 5. Fault diagnosis architectures: (**a**) centralized; (**b**) decentralized; (**c**) distributed.

The communication between agents can be either continuous or event-triggered. The event-triggered communication method allows communication when necessary to reduce the overall communication load. This flexibility ensures efficient utilization of resources and minimizes unnecessary data transmission.

In Sections 4 and 5, we review the existing publications concerning the FDP in both single satellite/UAV systems and networks comprising satellites and UAVs.

3. Review Methodology

In this review paper, a systematic search and screening procedure (depicted by Figure 6) for relevant works in the field was carried out. First, an initial database of all the work that is reliable and relevant to the authors' knowledge pertaining to satellite, spacecraft, and UAV condition monitoring published by relatively reputable sources was curated. In this step, the main keywords and features of the gathered works were extracted. Then, a subsequent search procedure for relevant works was carried out with the most prominent scholarly search engines by using the main keywords. Even so, there is always the possibility of human oversight in this type of search process. Therefore, after conducting a comprehensive exploration using academic search engines, we studied the works cited by the already identified works and added the relevant citations to our database or pool of accumulated studies. The same process was repeated until all the relevant citations in every work in our database were examined.

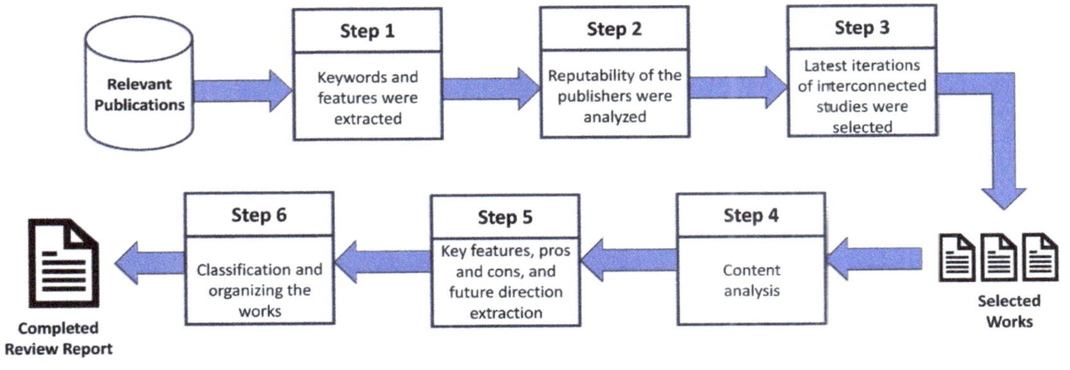

Figure 6. The review's systematic search, screening, and analysis approach.

Furthermore, having a reputable publisher was an important screening criterion while gathering our database of studies. By reputable sources, we do not necessarily mean a journal that has a high impact factor or a high-ranking conference. Especially in the field of satellite FDP, with the limited work that has been published in this field, it might prove fallacious to consider a study publisher's ranking as the definitive deciding factor. Therefore, there were more nuanced approaches involved in judging a source's credibility. The challenge in this step arose from the lesser known sources. The papers published by the distinguished sources in the field were automatically entered into the next screening step, which is step #3 in Figure 6. The credibility of the other sources was gauged through an examination of the works published by them. If the publications of a source displayed glaring issues and mishaps, that source and its corresponding studies in our pool were disqualified. If the source exhibited patterns of bulk research papers with minimal contribution to their field, which was assessed by inspecting their content, year-of-publication, and number of citations, they were disqualified. The other factor that caused disqualifications was works that built upon prior methodologies but with insignificant and superficial additions and innovations.

Through the reviewing process, for the sake of brevity, the works that were determined to be continuations of each other were discarded, except for the latest variation of the research conducted in them, while conserving the main takeaways of each series of papers. This does not imply that the studies that shaped the current state of the literature of the field were not suitable but that they were so developed throughout different researchers' work that repeating them did not add much value to this review. Furthermore, in this review, emphasis was placed on promising works with high potential for improvement or real-life implementation. Therefore, as long as an article was scientifically sound, sophisticated, and with promising potential, they were included in the review regardless of its publication date. However, since the latest iterations of ongoing research movements were included in this review, many of the works are quite recent. The reviewing procedure is illustrated in Figure 6 for more clarity. By this step in the reviewing process, a pool of promising works had been compiled. The initial works contained about 130 works, of which 85 were selected for analysis in this review through the screening process. After this stage, a comprehensive content analysis was carried out, and the results are provided in Sections 4 and 5.

Different criteria tables were also curated through the reviewing process with a summary table outlining the focus of each in Table 1, under Tables 2–8, that outlined the pros and cons of different categories of work, as well as the finer details of their research. This was performed to provide the reader with a comprehensive overview of the important strong suits and weaknesses of the works and their assumptions for ease in reimplementing and reproducing them in the future. At times, some details we thought would be useful for reproducing or enhancing the existing works were left out in their respective manuscripts or not greatly and definitively expanded on. In such cases, we left those specifications blank in our tables. The layout of the criteria tables also proved to be greatly helpful in determining whether a scientific work is relevant to our review's scope.

There is a need to further elaborate on the advantages and disadvantages provided in Tables 2–4. An advantage or pro was defined as a feature that served to solve a persisting problem in the area or gave the proposed strategy an edge over its competitors to be applied in real life. Conversely, a disadvantage or con was defined as an inherent shortcoming that deterred their implementation in real-life applications. The most commonly identified advantage across the works was their capability to be applied to more than only fault detection, isolation, identification, estimation, prognosis, etc. In contrast, the most prominent disadvantage was observed to be the computational burden of the proposed strategies. Generally, the more fault diagnosis modes a strategy can support, the better they are at being suited for more real-life problems. However, higher generalizability often comes with higher computational intensity, making them less suitable for solving other problems requiring less consideration. Therefore, many of the strengths and limitations listed in Tables 2–4 are due to their respective works' problem statements or focus. Even so, there are other types of strengths and limitations that are specific to each study's methodology, such as their assumptions, which should not be neglected if one decides to adopt them in real life. The summary and strengths and limitations tables also aim to provide a bite-sized summary of each work that facilitates the study of this review by readers.

The defining criteria and key characteristics have been compiled in Tables 5–8. Table 5 contains the works on single-agent fault diagnosis with the criteria being focused on diagnosis mode, the specific approach, the strategies' application, the fault types considered, and the measurement parameters that are used for monitoring. The diagnosis mode is imperative since it is one of the deciding factors as to whether a strategy is applicable to a problem or not. The strategies' application was included as a criterion to highlight the areas where they have been employed and which applications they have not been utilized in yet. This serves to reduce the risk of oversight on the researchers' part in failing to consider all the possible solutions for a specific application. The fault type details are also extremely useful since different types might need dissimilar monitoring methods or some are more prevalent in certain applications which helps in identifying which works are beneficial in addressing a condition monitoring problem. The studied measurement outputs are also compiled to highlight what information is needed from a system for the readers to be able to incorporate the methodologies in this review into their desired systems. Table 6 includes the works on single-agent fault prognosis with its focus being on the same criteria. Moving on, Table 7 consists of diagnosis studies on multiagent systems with their main criteria being the type of diagnosis, approach, application, and fault type. In the field of multiagent fault diagnosis, there is more focus placed on the relationship between satellites and UAVs. Therefore, detailed information about the measurements that are used for monitoring is not substantial and not provided in Table 7. Table 8 presents the secondary criteria that are useful for identifying the contributions of the studied works in multiagent fault diagnosis. These criteria include their architecture, communication topology and types, agent dynamic modeling, and control approaches. The architecture plays a crucial role in reducing computational complexity and system burden, as explained in Figure 5 and Section 2. In a centralized architecture, a single master observer receives information from all agents to diagnose and estimate agents' faults. While this approach involves high computational complexity, it is highly accurate. In contrast, the decentralized architecture reduces computational complexity by allowing each agent to run its own observer. However, since there is no exchange of relative information between agents, this architecture sacrifices some precision. Finally, the distributed architecture offers a trade-off between computational complexity and accuracy by assigning observers to groups of neighboring agents. Regarding communication, agents can transmit information either unidirectionally (directed) or bidirectionally (undirected). In practice, the communication within a graph can be a combination of both. Additionally, communication can either be continuous over time or event-based, where information is transmitted only when necessary. This event-based approach helps reduce the overall communication load. Moving on, we have the homogeneous and heterogeneous agent dynamic modeling and control logic criteria. The agent dynamics focus on whether the agents in the system have identical (homogeneous) or different (heterogeneous) dynamic behaviors. The control approaches in multiagent systems are divided into four main categories: consensus, formation, coverage, and flocking, depending on the mission goals. Literature shows that formation and consensus are more commonly used than other methods. It is also important to consider that, in real-world applications, the dynamics of agents in multiagent systems may vary due to operational conditions or environmental factors. This variability should be considered when working on control strategies and fault diagnosis in multiagent systems to ensure a more realistic and robust approach. An overview of all of the descriptive tables is provided in Table 1. These tables serve to guide the reader to know the defining features of the studies included in this review. For example, suppose the reader refers to this review to find a prospective methodology to compare their own with. In that case, these criteria are crucial in determining which methodologies are eligible for comparison. Or if the reader aims to find the gaps within the applications and strategies employed in the said studies, the tables serve to highlight the works that are promising in that regard.

Finally, classification and organization of the utilized methodologies and studies in the literature were achieved through the content analysis, and the results are provided in Section 6.1. The identified gaps and shortcomings of the existing works are also compiled in Section 6.2 for readers' consideration.

For the works in which the final quantitative results are explicitly provided, they are included in this review as well. In some works, detailed context is needed to understand the quantitative results. For these works, the final quantitative results are not included in this review to minimize the risk of misconstruing the results by the reader and inflating the text; however, their main takeaways are provided in Sections 4 and 5. Some works proposed solutions to solving persisting issues in the field and quantitative results do not highlight their contributions. Even so, among the works that do have explicit and easily understandable final quantitative results, direct comparison of these results might prove erroneous. For example, comparing the results of fault classifiers proposed by different works is not easily and intuitively possible. Different classifiers are usually trained on different datasets with different assumptions in the absence of a significant and rigorous dataset in the field. Furthermore, the faults inserted into the training and testing datasets might not be identical either. A classifier with a high efficiency and accuracy rate that is trained and tested on a specific dataset might have a much lower efficiency and accuracy rate if applied to a dataset used by another classifier. In such cases, if explicit comparisons are provided by the authors of the studies, they are provided in this review, and if not, they are not included to conserve the scientific integrity of this review, since comparisons prove fallacious when they are conducted between different methodologies with different assumptions and different main features, as described by Tables 5–8. In this way, Tables 5–8 also help the reader identify which methodologies are eligible for comparison with their own prospective strategies. Consequently, to avoid the risk of presenting wrong comparisons to readers, we refrained from providing a detailed quantitative results comparison between all the works.

Table 1. Reference tables provided in this review for summary, extra, and detailed information about the outlined studies.

Table Ref.	Description
Table 2	strengths and limitations of diagnosis studies on single-agent systems
Table 3	strengths and limitations of diagnosis studies on multiagent systems
Table 4	strengths and limitations of prognosis studies on single-agent systems
Table 5	Criteria and key features of diagnosis studies on single-agent systems
Table 6	Criteria and key features of prognosis studies on single-agent systems
Table 7	Main criteria and key features of diagnosis studies on multiagent systems
Table 8	Secondary important criteria of diagnosis studies on multiagent systems

Table 2. Papers on single-agent fault diagnostics and their strengths and limitations.

Refs.	Strengths (+)/Limitations (−)
1. Generalizability and Data Efficiency	
[10,16,21–23]	(+) Generalizable frameworks applicable to various motors and systems (+) Reduced requirements for historical data and training samples (+) Capable of diagnosing multiple faults and unexpected/untrained faults (−) Limitations in accuracy under conditions of scarce training data
2. Fault Detection and Sensitivity	
[13,15,24–26]	(+) Enhanced sensitivity to fault amplitudes and shapes (+) Improved accuracy and interpretability through advanced techniques (−) Higher computational requirements in complex scenarios
3. Computational Complexity and Model Robustness	
[27–31]	(+) Robustness to noise and disturbances with efficient algorithms (+) Enhanced interpretability and efficiency through various models (−) Increased computational burden due to model complexity (−) Potential for false positives and negatives in certain methodologies
4. Fault Diagnosis Methods and Techniques	
[32–36]	(+) Application of diverse techniques (e.g., ML, NN) improves diagnosis accuracy (+) Flexibility in diagnosing both actuator and sensor faults (−) Assumptions regarding system behavior may lead to inaccuracies
5. Limitations and Challenges	
[31,37–40]	(+) Addressing key challenges like fault detection and model interpretability (−) Various limitations including the need for specific data types and assumptions that may not hold in real-world applications
6. Specific Applications and Innovations	
[36,41–44]	(+) Innovative applications in digital twins and communication resources (+) Enhanced methodologies for fault diagnosis in specific systems (−) Complexity in adapting these methods across different scenarios

Table 3. Papers on multiagent fault diagnostics and their strengths and limitations.

Refs.	Strengths (+)/Limitations (−)
Fault Detection and Isolation Techniques	
[45]	(+) Considers simultaneous faults (–) Limited by over-actuated systems for isolating multiple faults
[46]	(+) Instant fault detection via local observers (+) Reliable detection regardless of fault magnitude
[47]	(+) Hybrid approach isolates low-severity faults
[48]	(+) Structural perturbation minimizes observer errors
[49]	(+) Identifies multiple faults in linear systems with correlated signatures
[50]	(+) Operates with inaccurate information
[51]	(+) Distributed FDI using local measurements
[52]	(+) Decentralized control strategy satisfies communication constraints
[53]	(+) Decentralized strategy for communication limitations
[54]	(+) Switching method contains errors in specific parts
Comm. Strategies	
[55]	(+) Decentralized control meets deep space communication needs (–) Healthy satellites must allocate more resources
[56]	(+) Expanded framework for fault diagnosis in broader contexts
[57]	(+) Nonlinear system capability in adverse conditions
[58]	(+) Uses primary satellite data to diagnose faults without extra sensors
[59]	(+) Develops decentralized UIO for actuator faults
[60]	(+) Heterogeneous clusters reduce computational load
[61]	(+) Similar advantages as previous entry
[62]	(+) Robust against disturbances in varying conditions
Resource Management and Performance	
[63]	(+) Reduces operational costs while enhancing reliability
[64]	(+) MIMO structure analysis for optimality
[65]	(+) CKF achieves superior estimation performance (–) Increases communication cost
[66]	(+) Quaternion representation simplifies complex systems (–) Computational load managed by rule-based systems
[67]	(+) Robustness through H-infinity method
[68]	(+) Multiple-leader structure considers communication data noise

Table 4. Papers on prognostics and their strengths and limitations.

Ref.	Approach	Strengths (+) / Limitations (−)
[5,12]	**D-D**	(+) Could operate even with limited historical data (+) Employs regressors for prognosis, which adds to the interpretability and quantitativeness of the predictions
[69]	**D-D**	(+) Eliminates the need for component-level measurements (+) Significantly diminishes the need for historical operational data
[70]	**M-B**	(+) The methodology is robust to Gaussian and non-Gaussian noises (+) Does not require historical data due to being a M-B strategy (–) Implementation complexity and computational burden incurred by using CAUKF and PFs
[71]	**M-B**	(+) Accounts for non-Gaussian noise (+) More accurate than classic PFs (+) Investigates the bearing lubrication faults in more detail using the lubrication temperature (–) Only accounts for lubrication faults as a result of temperature gradients
[72]	**M-B**	(+) Real-time prognosis analysis of the UAV (+) Multiactuator RUL prediction (–) Nonlinear but stationary degradation models are employed for brushless Direct Current (DC) motors (–) Only addresses the hovering state of the UAV

Table 5. An overview of papers on single-agent fault diagnosis.

Ref.	Monitor Type	Approach	Application	Fault Type	Measurement Parameters
		Data-Driven Approaches			
[3,15,73]	FD, FId	**D-D** Various Neural Networks (CNN, LSTM)	UAV Sensors	Operational, Bias, Drift, Stuck Faults	Sensor Data (IMU, Gyroscope, GPS)
		Model-Based Approaches			
[6,10,16,45,74]	FD, FId	**M-B (Det.)** Observer-Based Methods	Satellite, UAV, and Other Systems	Multiplicative, Additive	Attitude Rates, Angular Velocities, Position Data
		Hybrid Approaches			
[14,21,24,36,37]	FD, FIs, FId	Combination of Various Algorithms (e.g., SVM, Ensemble Learning)	Multiple Applications Including UAVs and Power Systems	Anomalies, Failures	Multiple Sensor Outputs Including Voltage, Current, and Attitude Data
		Fault Types			
1. Operational and Physical Attacks					
[3,15,73]	FD, FId	**Various Data-Driven Methods**	UAV Sensors	Operational Faults, Physical Attacks	Sensor Data
2. Sensor and Actuator Faults					
[27,38]	FD, FIs, FId	**Model-Based and Hybrid**	Various UAV Systems	Stuck, Bias, and Intermittent Faults	Sensor and Actuator Data
3. Systematic Anomalies and Failures					
[29,33]	FD	**Data-Driven Approaches**	Telemetry Data	Anomalies	Telemetry Sensor Outputs
4. Comprehensive Fault Detection and Isolation					
[23,41,75]	FD, FId	**Model-Based and Data-Driven**	Satellite and UAV Systems	Various Fault Types	Multiple Measurement Parameters

Table 6. An overview of papers on single-agent fault prognosis.

Ref.	Approach	Application	Fault Type	Measurement Parameters
[5]	**D-D** LSTM	Single RW	Multiplicative	Motor current, RW rates
[12]	**D-D** ARIMA, LSTM	Single RW	Multiplicative	RW rates
[69]	**D-D** Chebyshev NN	Pyramid CMG	Multiplicative	Satellite rates
[70]	**M-B (Stoch.)** CAUKF and PF	Single RW	Multiplicative	Motor current, RW rates
[71]	**M-B (Stoch.)** PF	Single RW	Lubrication loss over time	Lubrication temperature, Lubrication degradation model, Lubricant's injected amount
[72]	**M-B (Stoch.)** TSKF, Entropy weight method, Laguerre function-based model predictive control	Quadrotor UAV Actuators	Degradation	—

Table 7. An overview of papers on multiagent fault diagnosis and their main features.

Ref.	Monitor Type	Approach	Application	Fault Type
[45]	FD, FIs, Recon.	**M-B** Observer-based Sliding mode	RW	loss-of-effectiveness
[46]	FD	**D-D** DNN	RW	Bus voltage, Motor current fault, Viscous friction
[63]	FD, FIs	**D-D** NNs	PPT actuators	Change in the Amount of Produced Thrust
[66]	FD	**M-B** CDM and fuzzy rule-based reasoning	RW, Battery	Intermittent bus voltage, RW current, Viscous Friction
[52,55]	FA	**M-B** KF	Actuator	loss-of-effectiveness
[47]	FD, FIs	**M-B** Hybrid and consists of both low-level (agent/team level) and high-level	Actuator	loss-of-effectiveness
[48]	FD, FIs	**M-B** Hybrid consists of a bank of residual generators and a DES-based fault diagnose	Actuator, Sensor	loss-of-effectiveness
[53]	FA	**M-B** Conventional Recovery Method	Actuator	loss-of-effectiveness
[54]	FE, FA	**M-B** KF	Actuator	loss-of-effectiveness
[76]	FD, FE	**M-B** KF	Actuator	Fault signal
[56]	FA	**M-B** KF	Actuator	loss-of-effectiveness
[57]	FD, FIs	**M-B** Hybrid consists of a bank of residual generators and a DES-based fault diagnoser	Actuator	loss-of-effectiveness
[65]	FE	**M-B** CKF	Actuator	loss-of-effectiveness
[64]	FD, FIs	**M-B** KF	Actuator	Change in the Amount of Produced Thrust
[77]	FD, FIs	**M-B** EKF	Actuator	loss-of-effectiveness
[78]	FE	**M-B** Observer-based $H\infty$	Actuator	intermittent fault
[58]	FD, FIs	**M-B** Decision-making algorithm based on relative dynamic	Gyroscopes	Bias faults, Scale factor faults
[67]	FD, FIs, FE	**M-B** Observer-based	Actuator	Additive Abrupt Fault, Slow-varying Fault
[59]	FE	**M-B** Observer-Based UIO	Actuator	loss-of-effectiveness
[60]	FE	**M-B** Observer-based UIO	Actuator	loss-of-effectiveness
[61]	FD, FE	**M-B** Observer-Based UIO	Actuator	loss-of-effectiveness
[62]	FD, FE	**M-B** Observer-based UIO	Actuator	loss-of-effectiveness
[51]	FD, FIs	**M-B** Observer-Based UIO	Actuator	Cyber-attack/additive fault
[49]	FD, FIs	**M-B** KF	Actuator	loss-of-effectiveness
[50]	FE	**M-B** KF	Actuator	loss-of-effectiveness
[68]	FD, FIs	**M-B** KF	Actuator	loss-of-effectiveness
[79]	FD, FIs	**M-B and Knowledge-based** Fuzzy reasoning	RW	loss-of-effectiveness

Table 8. Overview of fault diagnosis techniques in multiagent systems.

Ref.	Architecture	Comm. Topology	Homogeneous/ Heterogeneous	Control Logic
[46]	Decentralized	Directed	Homogeneous	Formation
[63]	Leader–Follower	Undirected	Homogeneous	Formation
[66]	Leader–Follower	Directed	Homogeneous	Formation
[52]	Decentralized	Undirected	Homogeneous	Formation
[55]	Distributed	Undirected	Homogeneous	Formation
[47]	Centralized	Undirected	Homogeneous	Formation

Table 8. *Cont.*

Ref.	Architecture	Comm. Topology	Homogeneous/ Heterogeneous	Control Logic
[48]	Distributed	Undirected	Homogeneous	Formation
[53]	Distributed	Undirected	Homogeneous	Formation
[54]	Distributed	Directed	Homogeneous	Formation
[76]	Distributed	Undirected	Homogeneous	Formation
[56]	Distributed	Directed	Homogeneous	Formation
[57]	Centralized, Decentralized	Continuous	Homogeneous	Formation
[65]	Distributed	Directed	Homogeneous	Formation
[64]	Decentralized-MIMO	Undirected	Homogeneous	Formation
[77]	Centralized, Decentralized, and Distributed	Undirected	Homogeneous	Formation
[78]	Distributed	Undirected	Homogeneous	Formation
[58]	Decentralized	Directed	Homogeneous	Formation
[67]	Distributed	Directed	Homogeneous	Formation
[59]	Leader–Follower	Directed	Homogeneous	Formation
[60]	Distributed	Directed	Heterogeneous	Consensus
[61]	Distributed	Directed	Heterogeneous	Formation
[51]	Distributed	Undirected	Homogeneous	Formation
[49]	Distributed	Undirected	Homogeneous	Formation
[50]	Distributed	Undirected	Homogeneous	Formation
[68]	Distributed /Leader–Follower	Directed	Homogeneous	Formation
[79]	Leader–Follower	Undirected	Homogeneous	Formation

4. Diagnostics

The domain of diagnostics plays a crucial role in the health assessment of satellites, spacecraft, and UAVs. Detecting, localizing, and identifying faults are essential to rescuing a malfunctioning agent's mission. Furthermore, successful prognosis relies on the awareness of a system's fault and the availability of relevant information. As a result, fault diagnosis for these agents has garnered significant interest, spurring extensive research in this field. Due to the increasing attention that the field of fault diagnosis has received in monitoring a broad range of agents, scientists have expanded its implementation to include satellite, spacecraft, and UAV fault diagnosis. The predominant strategies employed in the current body of literature for performing fault diagnosis within Aerospace applications, with a specific focus on satellites and UAVs, encompass data-driven and model-based techniques. While there have been occasional references to their fusion and alternative methods in published works, these approaches have not been subjected to the same rigorous scrutiny in the literature as the previously mentioned strategies. This section comprehensively lists and reviews the applications of data-driven, model-driven, hybrid, and other methods on diagnostics of a single and a group of satellites, spacecraft, and UAVs in the current literature. To provide a detailed and yet concise overview of each paper's strong suits and shortcomings, Table 2 presents the advantages and disadvantages of each study on single-agent fault diagnostics. Table 3 offers a similar analysis for a team of aerial and space agents. In order to provide a broad yet comprehensive overview of the innovations featured in the studies reviewed in this research, each study has been scrutinized and an extensive list of their characteristics is presented in Tables 5 and 7 for single-agent and multiagent fault diagnosis, respectively. Table 5 is structured in a way to highlight the exact architectures, health monitoring modes, considered fault types and variables, and applications of each study on single-agent systems. Consequently, facilitating the process of deriving insights from each paper. In addition, Table 5 serves as a comparative standpoint over the methodologies utilized in this field. Furthermore, Table 7 offers an overview of the attributes in research related to fault diagnosis in multisatellite/UAV systems.

4.1. Data-Driven Methods

Data-driven methods have become one of the most prevalent methods for diagnostics in satellites and UAVs, especially in recent years. The literature on diagnostics has utilized these methods to a great extent to capitalize on their capability of assessing the health status of systems without having rigid mathematical models. Deep learning approaches as well as other supervised, and unsupervised approaches constitute most of the body of work in this area. Their ability to accommodate applications where autonomous diagnosis is required or when manually defined features are not available has made data-driven techniques a popular method to perform diagnostics. This section reviews how the existing literature has employed data-driven techniques for fault diagnostics.

Talebi et al. [27] proposed a robust data-driven methodology for fault diagnosis of actuators and sensors onboard a broad range of nonlinear systems. Their methodology incorporates two recurrent neural network-based observers for Fault Detection, Isolation, and Identification (FDII) and operates based on the studied nonlinear system's state-space model. More specifically, FDII is achieved simultaneously as the diagnosis scheme gets alerted. This is performed by indicating where the fault has formed using fault-sensitive thresholds and obtaining its intensity. The proposed methodology is also strengthened against unmodeled dynamics, unknown faults, noises, and uncertainties. The methodology was tested using a case study on magnetorquer actuators and magnetometer sensors onboard the Attitude Control System (ACS) subsystem of Low Earth Orbit (LEO) satellites, and the test results verified the methodology's capability to be used in real-life situations. While actuator fault diagnosis is of paramount significance to increase the reliability of control systems, sensor FDII is also influential in affecting an agent's lifetime. Sensor faults could emerge due to external disturbances, aging, and power fluctuations [6]. The work of Talebi et al. addressed the reliability requirements for actuators and sensors.

Rahimi et al. [80] advanced the idea of utilizing data-driven approaches for Fault Detection and Isolation (FDI) using an ensemble of Adaboost decision tree, Adaboost Random Forest (RF), MultiLayer Perceptron (MLP), and K-Nearest Neighbors (KNN) ML algorithms applied to a four-wheel configuration of RWs. Principal Component Analysis (PCA) was also incorporated into the scheme to perform dimension reduction. Their method in [80] underperformed the required expectations for an FDI system with the highest classifier precision of 58.78%. Rahimi et al. claimed that implementing more complicated datasets for running the algorithms and utilizing optimization methods for hyperparameter evaluation could lead to more accurate results. Expanding the scheme to address different malfunction intensities, symmetrical RW arrangements, and computation overload would be the next step in this scheme's progression.

Varvani Farahani et al. [21] discussed data-driven fault isolation for locating multiple concurrently existing malfunctions in a four-CMG pyramid assembly. The proposed data-driven scheme consists of a feature extraction method, a feature reduction method, and a classification algorithm, which, through testing, are chosen to be Correlation Analysis, PCA, and an optimized Support Vector Machine (SVM), respectively. A sensitivity analysis was carried out to demonstrate the proposed method's robustness against added noise signals, unavailable sensors, and skipped measurements. The results of the sensitivity analysis displayed suitable error margins for practical implementations. The simulations exhibit 100% and 99% accuracies for healthy and single-malfunctioning-CMG assemblies, respectively, and 97.8%, 91.4%, and 77% accuracies which correspond to two, three, and four in-phase malfunctions coexisting in a system. The problem of diagnosing simultaneous faults is best countered using data-driven methods. As Jiang and Khorasani have demonstrated in [45], where concurrent faults can occur, e.g., RW assemblies, the model-based methods may struggle with diagnosing all of the faults depending on the system and if more than two are present.

Rahimi et al. [32] explored detecting and isolating in-phase malfunctions using automated data-driven methods applied on three-orthogonal, standard four-wheel, and pyramid RW configurations. Their proposed approach preprocesses the time series input data by automatically extracting features from the temporal, spectral, and statistical domains and then passing it through a feature reduction stage. Subsequently, several classification methods were considered: Gradient Boosting, RF, Decision Tree, and MLP. It was shown that Gradient Boosting and RF performed the best among the methods. However, both operated poorly under complex fault scenarios. The FDI methodology fared relatively better for the three-orthogonal case than the standard four-wheel, and pyramid arrangements, especially due to the symmetry in the pyramid assembly. Expanding upon the presented scheme by improving the accuracy of FDI for more complicated malfunction scenarios would be the next step in developing a more robust FDI scheme. Moreover, Rahimi et al. mentioned that for RW configurations that have each RW manipulating torques in more than one axis (e.g., pyramid and standard four-wheel), fault complexities (including inception and duration) could have a detrimental effect on the FDI scheme's accuracy.

Vaz Carneiro et al. [81] formulated a data-driven approach to fault detection in RW assemblies employing supervised and unsupervised learning techniques, including Gaussian Mixture Model (GMM), SVM, Artificial Neural Networks, and Long Short-Term Memory (LSTM) Neural Networks. The learners receive a combination of RW attitude, temperature, control torques, and power consumption to detect the occurrence of faults. However, temperature proved to be the most sensitive variable to faults. The models are trained with the simulated datasets provided by the Basilisk simulation framework [82], a novel approach to training data-driven models for satellite fault diagnosis applications. Producing and using large amounts of synthetic data could be conducive to achieving adeptly trained models due to not having a limitation on the amount of data produced and knowing the exact details of the occurring faults beforehand as opposed to using historical data for training models [81]. Conversely, to avoid overfitting, amply-sized synthetic and historical datasets could also be created using generative models. The datasets acquired using generative AI could become a promising approach to alleviate the problems associated with the training phase of data-driven methods.

Suo et al. [83] developed a feature selection algorithm based on the fuzzy Bayesian risk theory along with a heuristic forward greedy algorithm for fault diagnosis of power systems of in-orbit satellites. The proposed algorithm works with an SVM for classifying faulty or nominal states. While the proposed feature selection method is observed to obtain more accurate results than other prevalent methods, it is computationally intensive, and overcoming this shortcoming could be a favorable approach to progressing this research.

Cui et al. [17] provided a solution to overcome the problem of faulty dataset scarcity for training data-driven satellite fault detection models. The solution incorporates redistributing the healthy and faulty datasets by oversampling the faulty samples using Dynamic Time Warping (DTW). Their proposed fault detection scheme is then completed by employing the fast DTW method for assessing correlations between samples and a KNN model for detecting malfunctions. Cui et al. tested the methodology, and DTW oversampling significantly increased the fault detection accuracy. The proposed method by Cui et al. could also help mitigate the training difficulties of data-driven methods for applications where several distinct fault types could be present in a system. Data augmentation is a technique that has been widely employed in the field of ML to make classification models more robust to diverse test cases. Data augmentation is especially common in the field of computer vision. To demonstrate how data augmentation works, an example is provided in Figure 7. Let us assume that we want to train a data-driven model to be able to discern dog images among other classes. If our dataset only consists of images like the original image in Figure 7, our dataset would be scarce and the data-driven model would be prone to overfitting. For example, the data-driven model through the training process might come to the conclusion that dogs are only dogs if the ear-shaped features are towards the top of the image or the paw-shaped features are toward the bottom of the image. To us humans, this conclusion appears clearly erroneous; however, this must be clarified to data-driven models by curating a comprehensive dataset or using a more robust architecture. In this case, a myriad of image modifications are imposed on the original dataset to make the dataset more diverse as shown in Figure 7. This process is referred to as data augmentation, and it is applicable to more than just images. For example, in the work of Cui et al. [17], DTW oversampling was used that in some way operates as a deterministic way of finding an in-between time series between two original time series data points to make the training dataset more diverse. Therefore, data augmentation is also a viable way of enhancing data-driven methods' performance aside from feeding them synthetically generated datasets.

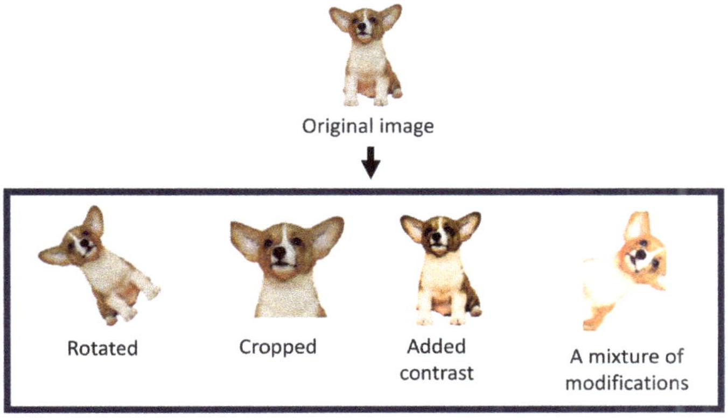

Figure 7. Data augmentation for a sample image of a dog (the original image is available https://unsplash.com/photos/short-coated-brown-and-white-puppy-eoqnr8ikwFE here) (accessed on 10 July 2024).

Hedayati et al. [84] proposed a data-generative AI model based on the Wasserstein Generative Adversarial Network (WGAN) architecture to balance scarce satellite RW datasets for use in training data-driven models. The utilized WGAN model comprises 1D convolutional layers within its generator and critic (discriminator). First, to train the WGAN model, multiple datasets for each fault type are curated. Each dataset contains very few data points as compared with the nominal RW dataset to reflect a real-life scenario. Then, identical instances of their proposed WGAN model are trained on each RW faulty class dataset. In this way, the overall training dataset becomes balanced. The final datasets augmented by the proposed framework of Hedayati et al. were tested quantitatively using an LSTM architecture and qualitatively as well. The LSTM model was trained separately on the WGAN-augmented, naturally balanced, imbalanced, artificially duplicated (inflated), and DTW-oversampled datasets and obtained accuracy rates of 84.17%, 93.25%, 48.35%, 97.51%, 89.32% as tested on a natural test dataset. The reason for WGAN-augmented dataset's underperformance as compared with other augmentation methods was attributed to WGAN adding too much diversity to faulty datasets and some resultant classes resembling each other. However, qualitatively it was observed that the WGAN added immense diversity and variability to the base dataset while retaining its most defining patterns whereas the other data augmentation methods added little to no diversity to the base dataset.

Pan et al. [85] used a novel data-driven method to detect faults in satellite power systems. Their presented method consists of forming correlations between time series datasets of several sensor measurements employing association rules between sensor datasets and incorporating them into a Kernel Principal Component Analysis (KPCA) model for detection. Association rule learning is especially effective since the sensor relationships change when an anomaly occurs. The novelty of the study of Pan et al. warrants the need for it to be examined further, but scant work has been carried out on this approach and application.

Ganesan et al. [28] presented a method employing a One Dimensional (1D)-CNN to address the fault detection problem in the power systems of satellites. The input data are preprocessed using the Stockwell transform before being fed into the CNN. When tested using the Advanced Diagnostics and Prognostics Testbed [86], the univariate detection methodology achieved an accuracy of 96.7%. Furthermore, Ganesan et al. claimed that the methodology could be extended to multivariate detecting, locating, and classifying malfunctions in the power systems.

Muthusamy and Kumar [10] presented a data-driven methodology for FDI in CMGs. Their methodology relies on the following elements for fault diagnosis: a predictive data-driven model reducing the dependency of the method on *a priori* operational data by 93.75%, a Chebyshev neural network for detecting malfunctions in CMGs, and an optimization-based scheme using a Genetic algorithm for locating faults. The input data are limited to attitude rate measurements to accommodate satellite measurement limitations, but more comprehensive input data could be incorporated into the method. Simulated results showed a 93.25% accuracy rate in isolating different types of faults. Muthusamy and Kumar's proposed method is suitable for applications where sensor redundancy poses problems or when historical flight records are unavailable.

Nozari et al. [24] proposed a mixed-learning approach to FDI of a tetrahedral RW assembly and performed a comparative study of the mixed and individual models. Nozari et al. incorporated the RF, SVM, partial least square, and Naïve Bayes (NB) algorithms into their mixed-learning strategy. The proposed model consists of training several local classifiers whose outputs are used to train a meta-level classifier. The proposed mixed strategy mostly outperformed its constituent learners under noisy, noiseless, and mixed circumstances. Additionally, training using mixed noisy and noiseless datasets yielded more accurate results. However, one of the main shortcomings of mixed-learning strategies is that some mediocre learners within the scheme could negatively affect the overall mixed-learning strategy's performance [32].

The research conducted by Abdelghafar et al. [29] presented a data-driven predictive method employing an Extreme Learning Machine (ELM) for detecting faults by analyzing satellite telemetry data. The proposed methodology of Abdelghafar et al. first predicts the nominal operating data of the satellite. Then, it applies a static confidence interval to obtain a range containing the values corresponding to a normal operating state. ELMs are especially useful for time-sensitive applications such as satellite system fault diagnosis due to their learning speed and generalizability. They also require less expert manpower utilization [87]. The gray wolf optimization algorithm is also utilized to enhance the biases and input weights of the learner. Testing using the NASA shuttle valve benchmark dataset [88] achieved 98.5% to 99.6% accuracy in detecting deviations from nominal operating states, outperforming both unoptimized ELM and SVM.

A fault detection method for satellite telemetry data is presented by Xie et al. [33]. The detection method employs a graph neural network that considers the correlations between extracted features. Based on the cyclical operation schedule of satellites, malfunction thresholds are dynamically evaluated for detecting instances where the thresholds are exceeded. The proposed method demonstrated an accuracy of 98.30% for fault detection during testing on a satellite power system telemetry information dataset. Contrary to most ML methods, graph neural networks do not require the assumption of independent features as within graph-like data, nodes are interconnected, and there are dependencies between the nodes [89].

Luo et al. [34] developed a ResNeXt-based slice residual attention network for the health monitoring of a CMG onboard spacecraft. The proposed data-driven method utilizes random slicing and an attention mechanism stage. As part of the data preprocessing phase of the proposed method, the short-time Fourier transform (STFT) is employed to generate spectrogram images to boost the diagnosis capability. The model identifies the CMG's fault states but falls short in accurately measuring each sub-component's fault severity. Luo et al. expressed the prospects of converting the approach of distinct fault classification into a regression model for more rigorous fault diagnosis or using reinforcement learning to address more fault types.

The research carried out by Zhao et al. [35] presented a CNN-based spacecraft and mobile robot CMG fault diagnosis scheme that comprises two attention-enhanced convolutional blocks. The input data are preprocessed into time–frequency spectrums using the STFT to enhance the fault diagnostic scheme. The scheme successfully classifies the overall CMG fault states; however, it fails to quantify each subcomponent's state of failure accurately. A demonstration of time–frequency transformation and the final images that get fed into the data-driven models is provided in Figure 8.

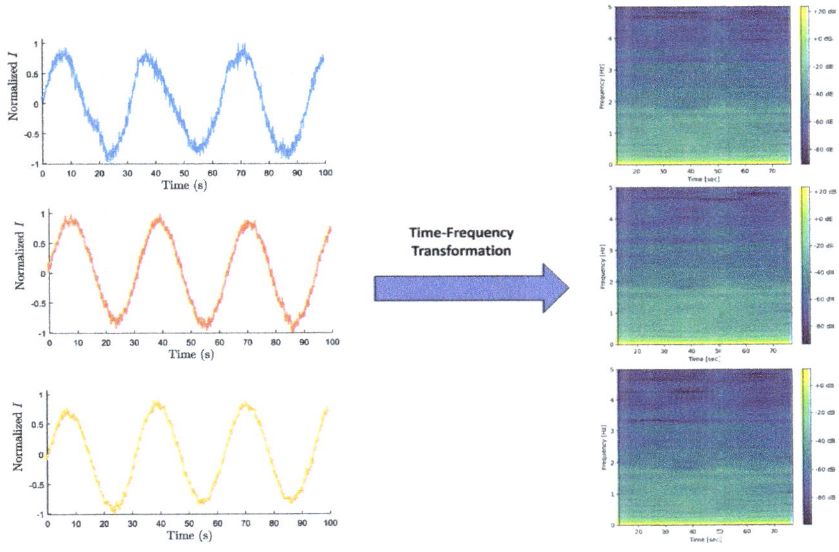

Figure 8. An example of generating spectrograms from RW time series signals for use in image-based data-driven models.

Liu et al. [90] developed two data-driven models to identify fault scenarios in a CMG onboard spacecraft. Both models employ the K-means algorithm to carry out diagnostics. However, one of the models utilizes the PCA for feature extraction, whereas the other uses the t-distribution random neighborhood embedding method based on the t-distributed Stochastic Neighbor Embedding (t-SNE) technique. Simulation results indicate the latter model provides significantly better performance. Liu et al. also indicated three ways of approaching the fault diagnosis of CMGs, namely, from the perspective of the system, assembly, and component. Their investigation into CMGs was conducted from a component-centric standpoint, considering both the physical and digital variables associated with CMGs for diagnostic purposes.

Jado and Moncayo [25] proposed a data-driven framework utilizing an multiple-model adaptive estimation (MMAE) comprised of multiple autoencoders to perform fault diagnosis on spacecraft. Each autoencoder is trained on an operational mode of the spacecraft and a Bayesian probability framework is used to gauge the probabilities of each model correctly representing the real system. The proposed methodology was numerically and experimentally tested and demonstrated to be able to timely diagnose faults resembling the training dataset. However, there lies an issue with this approach. The issue is that due to each autoencoder being trained on a singular operational mode, the in-between modes and relatively unknown failures could cause uncertainties in the model performance even though it is capable of detection, isolation, and identification to some extent. Their model could be bolstered by training more autoencoders to account for new failure modes.

Xiao and Yin [37] introduced a data-driven approach for FDI in satellite thrusters. They transformed the input data from the thrusters into image representations and established distinct fault categories based on these images. Subsequently, a CNN was employed for binary image classification. Their testing results revealed accuracy rates ranging from 98.30% to 98.71%, varying according to the thrusters' axis directions.

Sadhu et al. [3] developed a data-driven approach to detect and identify malfunctions in UAVs using sensor data from the Inertial Measurement Unit (IMU). Their proposed model combines a convolutional and bidirectional LSTM autoencoder for fault detection and a fault identification system based on CNN and LSTM. The identification component is activated only upon detecting a malfunction. During testing with empirical data, the detection module achieved an accuracy rate exceeding 90%, while the identifier module reached 85%. Sadhu et al. suggested that future work should involve applying their techniques to diverse UAV fleets and evaluating their method's performance in such scenarios.

An LSTM-based multivariate regressor using residual filtering is proposed by Wang et al. [73] for fault detection and recovery of UAVs. The proposed data-driven model automatically derives spatial and temporal correlations as features and utilizes a filter to cancel out the effect of random noise. Fault detection is accomplished by applying thresholds to the estimated parameters, and recovery is achieved by using the model's regressive nature to reconstruct the detected faulty data. The model's accuracy rate in detecting faults was determined by tests to be 99% and 93% for the gyroscope sensor's bias and drift faults, respectively, which outperforms the least-squares SVM and the LSTM model without residual filtering. Wang et al. also indicated that the possible future steps in developing this model are considering different fault magnitudes and flight modes' interference and addressing the online monitoring requirements by applying the model to an embedded monitoring unit.

Du et al. [26] proposed a data-driven approach to the mechanical fault diagnosis of damaged and cracked UAV rotors. The proposed methodology employs a 1D CNN to estimate faults using rotors' surface vibration acceleration signal as input during hovering and rising. Du et al. found that the preprocessing of input data significantly affects the model's performance in identifying small faults. Consequently, they used interval sampling instead of sequential sampling to reconstruct the vibration signals to obtain better results in diagnosing minor faults. The developed model's accuracy during the hovering and rising phases reached about 100% and 98%, respectively, while using the proposed sampling method and outdid the results yielded by sequential sampling. The model's accuracy rate was also indifferent to various loading conditions.

Taimoor et al. [91] proposed an adaptive radial basis functions (ARBF) approach based on the Lyapunov function theory and sliding-mode concept-based methodology to tune the weight parameters of neural network-based observers. The Neural Network (NN) observers are then utilized for fault estimation of sensors onboard quadrotor UAVs. The ARBF method's performance is compared with conventional radial basis functions (CRBF), adaptive multilayer perceptron (AMLP), conventional multilayer perceptron (CMLP), and Extended State Observer (ESO) methods and was shown to be more accurate and efficient than them.

Park et al. [22] developed a fault detection scheme based on unsupervised learning for diagnosing cyber-attacks and physical damages in the control systems of UAVs. This approach was incentivized by the shortcomings of supervised learning-based methods in requiring labeled data and failing to detect unexpected faults. The proposed model consists of a stacked autoencoder and uses thresholds for detecting faults. The input data features comprise UAV coordinates, attitude, IMU and sensors' data, and control inputs to the actuators. The next steps in this research could be adding to the features or incorporating sequential flight features, sequential autoencoders, or RNNs for enhanced detection performance. The presented model performed better in detecting Global Positioning System (GPS) spoofing attacks, denial-of-service (DoS) attacks, and rudder failure than in detecting elevator, aileron, and engine failures.

Li et al. [23] developed a novel Siamese hybrid neural network (SHNN) scheme based on Few-Shot Learning (FSL) for fault diagnosis of fixed-wing UAVs. FSL is an ML approach specializing in learning from a limited dataset for mainly supervised learning-based applications and exhibits high generalizability [92]. The FSL strategies' advantages have been leveraged to compensate for the lack of sufficient faulty state data for training the data-driven models, considering that the available faulty data for fixed-wing UAVs are scarce. Li et al. carried out comprehensive experimental testing to validate the proposed scheme under the following conditions: different training data sizes, different flights, and unknown faults. The tests also demonstrate that the SHNN framework performs better than SVM and 1D CNN models for fewer training sample sizes. The future scope of this research includes utilizing attention mechanisms and alleviating the computational burden of the algorithm.

The work by Huang et al. [31] proposes an ensemble data-driven model for actuator fault diagnosis of UAVs, specifically the ailerons. The proposed ensemble method uses a weighted combination of three different hybrid models. The difference between ensemble and hybrid methods in this context is that ensemble methods take the final diagnosis output of different models and make a weighted decision based on each individual model's output and are more akin to a parallelized structure between the individual models. Whereas a hybrid data-driven model combines different architectures within one holistic model and their combination only outputs one diagnosis. Weighted ensemble architectures could be used to alleviate the dominance within lopsided models and increase the generalizability of the model. On the other hand, in a hybrid model, each segment is usually responsible for one duty, e.g., feature extraction. Huang et al. used an overarching ensemble model of three hybrid models as follows: (1) A CNN-bidirectional LSTM (CNN-BiLSTM), (2) A CNN-bidirectional Gated Recurrent Unit (CNN-BiGRU), and (3) A CNN-bidirectional Gated Recurrent Unit-LSTM (CNN-BiGRU-LSTM). The CNN model that is shared across the three models is responsible for feature extraction whereas the subsequent structures are responsible for further processing and diagnostics. A combination of random search with grid search is used for pinpointing the optimal weights of the ensemble method. Through testing and comparing the model to different state-of-the-art models for this application, the proposed ensemble method mostly outperformed the other models at the cost of increased computational intensity.

Li et al. [75] proposed a transformer and local interpretable model-agnostic explanations (LIME) combination to perform fault diagnosis on a fixed-wing UAV's elevator. The LIME strategy was used to increase the interpretability of the transformer model by locally approximating the output of the proposed transformer and examining the effect of different features in the final diagnosis. The LIME algorithm also gives insights about how to simplify the transformer further by specifying the features that do not contribute extensively to the diagnosis and could be dropped. Furthermore, Li et al. implemented a loss function to mitigate the problem of faulty categories' imbalance. The proposed methodology displayed high accuracy in diagnosing faults and efficient convergences owing to the utilized loss function.

Huang and Ferguson [93] conducted work on satellite RW fault detection by proposing a simple One-Class Linear Regression (OC-LR) algorithm. The proposed OC-LR framework works by only training on nominal RW operational datasets which is curated using a simple polynomial representation of RW motor current's relationship with its angular velocity and acceleration. This simplified approach to monitoring the RW health state enables the algorithm to perform without the need for a rigorous mathematical model. Then, three detection metrics are used as follows: if a faulty data point falls within one, two, and three standard deviations of the training dataset's mean, there is a 68%, 95%, and 99.7% chance that it would be detected as a faulty data point, respectively. Through simulated and real-life tests, it was found that depending on how obvious the emerged fault is, any of the three detection metrics could surpass the other two, with their accuracy rates ranging from 76.7% to 99.1%. Huang and Ferguson's proposed algorithm is relatively simple, readily applicable, and not demanding, especially when dynamic models and faulty operational datasets are not available. However, this approach carries a relatively higher risk of false positives and negatives that could be catastrophic in a space system. Moreover, solely performing anomaly detection while only outputting the anomaly's degree of deviation from the nominal operation may not be enough for a fault-tolerant subsystem to save the whole satellite system, especially considering that different faults with different dynamics that require dissimilar remedying measures might exhibit the same levels of deviation. One of the possible avenues for enhancing this methodology is to implement a voting strategy between the different detection metrics.

Mousavi and Khorasani [46] introduced a Dynamic Neural Network (DNN)-based system for FDI in RWs of satellite formation flying (SFF) missions with a decentralized architecture. DNNs are trained with extended backpropagation using input/output data from the attitude control subsystem to model the nonlinear dynamics of each spacecraft. A fine-tuned set of DNN parameters is used to minimize estimation errors and to meet performance criteria. The methodology effectively detects low-severity actuator faults by employing local and neighboring spacecraft-based fault detectors. In the study, multilayer DNNs are trained based on relative attitude measurements and used to represent SFF dynamics.

Using DNNs, Valdes and Khorasani [63] developed an effective FDI system for Pulsed Plasma Thruster (PPT) used in ACS for SFF missions. The proposed approach involves three FDI strategies: a basic FDI scheme, an advanced FDI scheme, and an integrated FDI scheme. Basic schemes accurately detect and isolate faults in PPT actuators but have low precision and high error rates. The advanced scheme analyzes relative attitude data from formation flying, resulting in excellent detection but lacking isolation capability. With the integrated scheme, accuracy, precision, and minimal misclassification rates can be achieved, as well as insights into thrust production levels during faults can be obtained. Compared with traditional attitude control actuators, fault diagnosis for PPTs has been less explored because of challenges with force measurement and mathematical modeling.

4.2. Model-Based Methods

Model-based methodologies have been a staple of fault diagnostic strategies, specifically ACS diagnostics, for many years. Their capacity to carry out precise and timely FDII using high-fidelity models has made them reliable techniques for time-sensitive missions. On top of that, the interpretability present in the outputs provided by model-based methods is conducive to making appropriate judgments and taking remedying fault-tolerant measures. This section discusses their applicability in a broad spectrum of satellite and UAV diagnostic case studies.

Jiang and Khorasani [45] utilized a second-order nonlinear sliding mode observer for FDII in a tetrahedron RW assembly. Jiang and Khorasani indicated that due to the over-actuatedness of the RW assembly and redundant RWs, no more than two in-phase faults could be fully diagnosed, and postprocessing is needed to perform isolation and identification. However, even with two concurrently occurring faults, isolating their locations proved challenging since several permutations of faulty RWs could have the same effect on the generated residuals. Jiang and Khorasani worked around this problem by taking measures to make distinctions between the different possibilities.

Chen and Liu [41] utilized a two-stage EKF to estimate satellite actuator and sensor faults. The focal point of their study is estimating both multiplicative and additive faults while modeling actuator faults as multiplicative and additive and sensor faults as additive. Chen and Liu represented multiplicative faults as control effectiveness parameters and additive faults as added amplitudes. Their proposed model was also tested using the telemetry data of an on-orbit satellite, and the faults were successfully estimated.

Another work addressing the problem of diagnosing multiplicative faults is the study by Shahriari-kahkeshi et al. [11], which presents an adaptive model-based methodology applicable to Lipschitz nonlinear systems. The proposed methodology performs fault detection and identification simultaneously. It involves an adaptive state observer leveraging a robust adaptive law effective against unknown faults, measurement, and modeling uncertainties. The scheme was tested on a single-link flexible joint robot arm and displayed rapid convergence and better accuracy rates than a similar existing fault estimation scheme.

To accommodate the adaptivity requirements in satellite ACS diagnosis applications, Rahimi et al. [19] used Adaptive Unscented Kalman Filter (AUKF) techniques for fault detection of RWs. They expanded upon them by combining them with the Particle Swarm Optimization (PSO) method. The PSO method is used to circumvent the complicated procedure of setting up the AUKF for parameter estimation and making it more precise and efficient. The model-based methodology is based on a high-fidelity RW model developed by Bialke [94]. It is indicated by Rahimi et al. that for more rigorous results, a more sophisticated implementation of the PSO and AI manipulation is required.

Furthermore, Rahimi et al. addressed the fault diagnosis of RW assemblies in [95] by introducing a model-based hierarchical approach for FDII of RWs onboard satellites. Their three-step proposed methodology consists of (1) an AUKF for detection, (2) multiple UKFs along with Bayes' probability theorem and probability distributions for isolation, and (3) dual-state and parameter estimation using UKFs for identification. The methodology's accuracy rate in detecting, isolating, and identifying faults was determined by simulations to be 89.5% on average.

During their research on enhancing AUKFs for single RW usages, Rahimi et al. [16] also presented the Covariance-based Adaptive Unscented Kalman Filter (CAUKF), a variant of the AUKF with the aim of refining parameter estimation. The model-based method incorporates adaptive covariance matrices of states and parameters to be robust to sudden fluctuations in nonmeasurable parameters of the system. The proposed CAUKF generally takes more time to diagnose faults compared with the AUKF while displaying 4%, 14%, 42%, and 90% less mean square errors for residuals of the RW model parameters under abrupt, transient, intermittent, and incipient malfunction scenarios, respectively.

Moreover, Rahimi et al. [14] built upon their work in [16] by proposing the binary Binary Grid Covariance Adaptive Unscented Kalman Filter (GAUKF) methodology and a two-step hierarchical method for FDII of CMGs. The first step employs the adaptive thresholding approach, and the second step encapsulates fault isolation and identification by introducing an adaptive covariance-based binary grid search procedure. The GAUKF method was developed to address the high computational cost of the CAUKF and its suboptimal operation when the control system is closed-loop. GAUKF's fault isolation and identification performance surpasses both CAUKF and UKF methodologies in terms of precision. Additionally, the average execution time for the GAUKF is shorter than that of CAUKF and UKF when utilized in a closed-loop control system. However, the symmetry of the pyramid configuration poses problems in isolating the faults in RWs across from each other. Also, another drawback is the inaccurate estimation of one of the four wheels' fault parameters due to the system's lower dynamical sensitivity to the fourth RW's output [14,95].

Later on, Rahimi [96] proposed the Simplified Binary Grid Covariance Adaptive Unscented Kalman Filter (SGAUKF) methodology and further developed the isolation and identification modules presented in [14,16]. The SGAUKF method updates only the even or odd rows or columns of the posterior estimate covariance matrix's diagonal elements, cutting the order of computations to its square root. The method also features either an increase or no change, as the name binary implies. The Monte Carlo simulation results show that SGAUKF adds to the GAUKF isolation accuracy by 1% while speeding up the process by about 27%, making it an even more viable approach to online monitoring. The SGAUKF also suffers from the issue of imprecise identification performance for the fourth flywheel, which could be addressed in future research.

Nasrolahi and Abdollahi [6] designed a nonlinear observer for fault detection of attitude and rate sensors onboard satellites and presented a fault recovery strategy. Fault detection is achieved using the satellite's measured angular velocities and attitude parameters. However, attitude parameters are described using Modified Rodriguez Parameters (MRP), which enables the fault detection of different attitude sensor types with different configurations using the same observer.

Carvalho et al. [74] presented a fault detection filter that incorporates both H_∞ and H_2 simultaneously for Markovian Jump Linear System (MJLS). Carvalho et al.'s proposed approach aims to make the filter robust to disturbances, noises, and applied inputs and sensitive to malfunction signals. The filter is then applied to a CMG to demonstrate its performance. A brief description of MJLSs is included below. A discretized dynamical Markovian jump linear system's state-space model is defined as follows [97]:

$$X_{k+1} = A_{\theta(k)} X_k + B_{\theta(k)} u_k \tag{4}$$

where $X_k \in \mathbb{R}^d$ denotes the system's state at discrete time step k, the initial state X_0 follows a distribution of D, $u_k \in \mathbb{R}^r$ denotes the control command at time step k, $\theta(k)$ takes a value corresponding to a finite set of $N = 1, \ldots, n$ for each time step k, where $\theta(k)$ forms a discretized Markov process from N. The probability of transition between two states for $\theta(k)$ is defined as

$$p_{ij} = P(\theta(k+1) = j \mid \theta(k) = i) \tag{5}$$

where for each i the sum of all p_{ij} amounts to one. The initial distribution of $\theta(k)$ is also represented as follows:

$$\omega = [\omega_1 \ldots \omega_n]^T \tag{6}$$

where the sum of all ω_i is one. From Equation (5), Markov processes are described by the quality of their transition probability from one state to another solely depending on the current state of the system, irrespective of how it arrived at its current state. For demonstration, Figure 9 depicts a Markov chain of a fault-prone system with a transition probability matrix of $P = \begin{bmatrix} p_{11} & p_{12} \\ p_{21} & p_{22} \end{bmatrix}$.

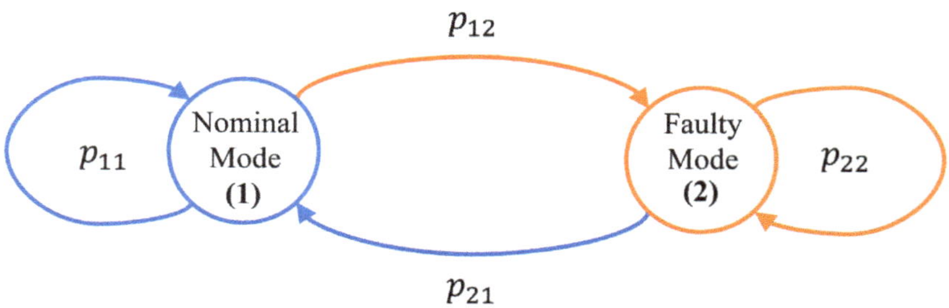

Figure 9. A schematic inspired by [98] of how a dynamic system abiding by a simple Markov process transitions between nominal and faulty modes.

Iglesis et al. [13] utilized the idea of jump Markov processes by presenting a Jump Markov Regularized Particle Filter (JMRPF) for nonlinear Inertial Navigation Sensor (INS) fault identification used in longitudinal control of a fixed-wing UAV. The jump Markov approach was introduced into a regularized PF to address the transitions of the system between faulty and nominal modes based on the Markov property. The proposed scheme of Iglesis et al. does not require predefined fault models to operate. Instead, estimation is carried out for additive abrupt and incipient faults with unexpected dynamics and amplitudes. A Kalman correction strategy is also incorporated into the proposed scheme, which distributes the particle states in the more probable state-space areas to enhance the state and malfunction estimation further. Under numerical simulations, JMRPF showed 77% less root-mean-square error in estimation than a regularized PF while exhibiting shorter convergence times and more robustness to faults with larger amplitudes.

Wang et al. [99] developed a model-based method for actuator FDII of a Hex-rotor UAV. Faults are classified as total actuator failures and gain faults attributed to the actuator lift factor's deviations from the nominal condition. The diagnosis model includes a set of EKF-based fault observers whose output is employed to achieve fault reconstruction. The proposed methodology of Wang et al. surpasses the accuracy of sliding mode observers in attitude angle tracking.

Maqsood et al. [42] developed an enhanced high gain observer for FDII of angular rate sensors in quadrotor UAVs to further address UAV sensor health monitoring. The proposed model is tested under incipient, oscillatory, and intermittent fault conditions, and its results are compared with the integral chain differentiator and basic high-gain observer techniques. The accuracy rate of the proposed approach significantly surpasses that of traditional techniques while being less computationally intensive and providing fast diagnostics.

Gai et al. [43] proposed a novel dynamic Event-Triggered Mechanism (ETM) for fault detection in UAV actuators to achieve minimal communication resource usage and remove correlations between generated residuals and dynamic event-triggered transmission errors while circumventing the Zeno phenomenon. Zeno phenomenon refers to the existence of an infinite number of events in a finite time interval, making ETMs prone to errors. The dynamic ETM approach employed by Gai et al. is based on the H_i/H_∞ optimization problem, which is solved using the Riccati recursion method. Their fault detection scheme also utilized a new residual evaluation and thresholding method. The simulation results indicate that the proposed dynamic ETM detection scheme outperforms the static variant in terms of accuracy, requiring at least 13% less event transmission information. Gai et al. also expressed that the next steps in this research would be to evaluate the effect of the dynamic ETM on quantitative fault inspection and examine the effect of closed-loop control systems on faults' noticeability.

Gao et al. [36] used a stochastic-model-based method for FDII of actuators onboard tilt-rotor UAVs. Their employed methodology consists of several EKF observers, each assigned to a single corresponding actuator. Moreover, it utilizes the MMAE method to use each observer's residuals and the state error covariance matrix for obtaining the fault conditional probabilities, which are then employed to achieve fault diagnosis. The proposed methodology of Gao et al. also enhances the efficiency of the MMAE method. It does not require additional sensors to monitor actuator deflections and change the flight controller.

Guzmán-Rabasa et al. [38] designed an H_∞ observer for FDI of actuators onboard a quadrotor UAV where the rotational dynamics are modeled as a reduced quasi-linear parameter-varying system. The numerical test results demonstrate that the proposed approach performs on the same accuracy level as similar methods. However, it is claimed to be more efficient because it only accounts for the system's rotational dynamics. On the other hand, it falls short in carrying out fault diagnosis while considering translational dynamics. The next progressions for this research could include validating the proposed approach with an experimental setup, developing a mixed H/H_∞ scheme for fault diagnosis or using it with a fault-tolerant controller.

To also address surface icing, Rotondo et al. [39] presented a model-based methodology for UAV icing and actuator fault diagnosis. Their approach is structured around combining the concepts of internal and Unknown Input Observers (UIO) and exploiting their fusion to develop a linear parameter-varying observer to circumvent the issues surrounding model uncertainties and unknown inputs. Rotondo et al. also suggested the following additions to enhance the proposed methodology: Alleviating the methodology's conservativeness, improving its performance, and generalizing it for more comprehensive operating conditions.

An innovative hierarchical approach for fault diagnosis in satellite components or subsystems is presented in [66]. This method emphasizes dividing the complex system into smaller pieces for structured diagnostic reasoning based on rules specific to each element. These modules are interconnected through a Component Dependency Model (CDM) using Bayesian networks. A leader–follower SFF is used to demonstrate the method's functionality and prove its potential to enhance satellite health monitoring. The approach will be extended to dynamic Bayesian network-based CDMs and systematically generated, providing promising directions for enhancing fault diagnosis capabilities.

In [52,55], Azizi and Khorasani explored strategies for managing actuator malfunctions in a decentralized manner for SFF. This is facilitated through a cooperative framework comprising three layers. Initially, established recovery methodologies are employed at the bottom-layer fault recovery based on assessments of fault severity. Nevertheless, inaccurate fault assessments can result in violations of mission error protocols. The higher layer supervising identifies faults and initiates the formation layer of fault recovery. This high-level supervision endeavors to adjust for the satellite partially restored at the lower level. In another paper, Tousi and Khorasani [47] developed a hybrid method that combined FDI stages such as [52,55] for a team of cooperating UAVs. Their hybrid methodology incorporates both bottom-layer and top-layer FDI modules. Simulation results involving a team of five UAVs are provided, demonstrating the effectiveness of this approach. In [48], Tousi et al. expanded the work in [47] and used Discrete-Event Systems (DES) at the top level to diagnose faults, while traditional diagnostic methods are used at the lower level. A broader spectrum of faults can be detected and isolated using this method than has been reported in the literature.

In another study [53], Azizi and Khorasani explored a novel distributed KF approach designed to estimate actuator faults for SFF. When addressing a complicated hierarchical system, they converted the representation of the system from an Overlapping Block-Diagonal State Space (OBDSS) to their innovative Constrained-State Block-Diagonal State Space (CSBDSS). The proposed approach can simplify the implementation of KFs in a distributed manner. During Kalman filtering iterations, the constrained-state condition must be met, equivalent to solving local constrained optimization cost functions. Various systems, including power systems and sensor networks, can be addressed with this technique.

A hierarchical fault estimation and accommodation approach is proposed in [54]. To minimize the adverse effects of unmodeled dynamics, uncertainty, and disturbances within the SFF, Azizi, and Khorasani seek to encourage cooperative interactions across levels and modules. Additionally, they highlighted a crucial finding: centralized estimation schemes exhibit notable drawbacks when faced with unmodeled dynamics, uncertainties, and disturbances. As a result, the cooperative estimation procedure proposed in [54] proves to be a highly applicable method for overcoming the mentioned challenges.

An advanced approach to cooperative actuator fault estimation in deep space is presented in [76] that integrates hybrid and switching techniques. Every operational mode represents a distinct cooperative estimation strategy and communication pattern between the localized filters responsible for detecting and estimating the SFF's status. With the help of this approach, fault estimation accuracy is likely to be enhanced in a cooperative and dynamically changing space environment.

A cooperative fault accommodation challenge within SFF is explored in [56]. It used absolute rather than relative measurements, like other studies by Khorasani's team [45,52,55,57], making the approach applicable to a wide range of SFFs operating in planetary orbits. In this study, collaboration between controllers maximizes the efficiency of supporting a faulty satellite. Therefore, the scope of this study is expanded to include a broader spectrum of SFF based on the cooperative fault accommodation framework.

In another study [65], Azizi and Khorasani introduced an innovative, collaborative framework designed to estimate the states of SFF. This framework employed the concept of sub-observers, each dedicated to estimating specific states based on provided input, output, and state information. To maintain estimation errors at a manageable level, they proposed a directed graph representing the interdependencies among subobservers. By carefully selecting the optimal path within these subobservers, a higher-level supervisor gains the ability to thoughtfully choose and configure a set of sub-observers, ensuring accurate estimation of all system states. In cases where unreliable information is introduced due to significant disturbances, noise, or actuator faults, specific subobservers may lose their efficacy. In such scenarios, the supervisor dynamically adjusts the subobserver set by selecting a new path within the subobserver's digraph. This adaptive approach effectively manages and restricts the impact of these uncertainties, ensuring they only influence local estimates of states and faults. Consequently, the spread of uncertainties throughout the estimation process is minimized, preventing extensive performance degradation across the entire SFF [65].

In [64], a particular structure is described by Meskin and Khorasani. The SFF is considered to be a Multiple-Input, Multiple-Output (MIMO) system in this study. By employing this structural approach, the overall formation is optimally managed, and a viable stability analysis can be conducted. One problem, however, is that the MIMO structure is not resilient to local faults. As a result, it is imperative to develop FDI filters capable of detecting such localized faults. FDI filters operating in a local or decentralized capacity were devised and presented to mitigate this vulnerability.

Ghasemi and Khorasani proposed an innovative FDI strategy for the ACS in SFF [77]. This method can use three different setups: decentralized, centralized, and semi-decentralized (distributed). As a result of these three configurations, a fault diagnostic system can be created for multisatellite systems. Compared with the other two FDI architectures, the centralized FDI architecture is more effective in missions incorporating angular velocity sensors. However, it also generates more false alarms. On the other hand, FDI decentralized architectures announce fault occurrences more frequently and produce fewer false alarms. The centralized Fault Detection (FD) architecture performs better in FDI, resulting in fewer false alarms in missions when formation flight integrates attitude-measuring sensors than the other two systems. Furthermore, the decentralized FD architecture communicates fault occurrences less frequently while producing fewer false alarms compared with the distributed FD architecture.

An H_∞-based robust distributed observer to synchronize the orientations of several satellites is investigated in [78]. A powerful distributed H_∞ is established at each satellite to detect intermittent faults. To calculate the gains for observers, they designed two sets of Linear Matrix Inequality (LMI) criteria. Numerous satellite numerical simulations show that the resilient H_∞ observer effectively estimates time-varying and intermittent faults when the satellite's communication layout is an undirected graph. Additionally, the designed observer is shown to be effective when several satellites have problems at once.

In [58], Shakouri and Assadian introduced a novel approach involving intersatellite measurements, considering relative positional data and orbital parameters. Their focus was on FDI within spacecraft rate gyroscopes. This innovative technique identifies a consistent motion characteristic, specifically restricting dynamic states during relative motion. Consequently, the primary satellite's angular velocity vector aligns with a quadratic surface. This ascertained motion characteristic aids in diagnosing faults within the gyroscope system and providing a rough estimation of the relevant scale factor or bias for the rate gyroscopes associated with the primary satellite. The proposed approach eliminates the need for additional subsystems within the SFF framework.

An exploration concerns a comprehensive approach for FDI within a specific class of nonlinear SFF systems [67]. The methodology considers model uncertainties, input variations, and environmental disturbances throughout the diagnostic process. To address fault detection, a nonlinear observer is carefully engineered to minimize uncertainty within the robust H_∞ framework. The observer gain matrices are computed using an LMI formulation. Robust UIO are developed to pinpoint the faulty actuator and facilitate fault isolation. This isolation is executed through a strategic implementation of generalized observer techniques. The novel observer architecture can concurrently estimate faults and states while effectively mitigating the impact of unknown input disturbances, model uncertainties, and external disturbances. This is achieved by employing Lipschitz formulations and the Linear Parameter Varying (LPV) method across all proposed observers, resulting in less conservative LMI conditions. It is essential to mention that this approach allows each satellite to diagnose its faults and those of its adjacent satellites.

Gao and Wang [59] introduced an approach to estimate faults and ensure fault tolerance in the control system for nonlinear SFF systems, explicitly addressing actuator faults. A decentralized UIO is designed to estimate the actuator fault factors. The estimated fault values obtained are then utilized to design a distributed fault-tolerant controller for SFF. Subsequently, an algorithm for fault-tolerant formation control is developed, using Adaptive Terminal Sliding Mode Control (ATSMC) techniques to improve the synchronization of each follower spacecraft with the leader spacecraft, even when faced with actuator faults.

In another research, Barzegar and Rahimi [60] undertook to design a robust UIO specifically crafted to estimate faults within clusters of small satellites. Operating within clusters, these satellites present heterogeneous dynamics characterized by significant diversity owing to environmental conditions and inherent nonlinearities. The study extends its focus beyond fault estimation, investigating that the designed observer and controller logic not only estimate faults accurately but also arrange synchronized behavior among satellites within each cluster. This synchronization shows as a state of consensus, steering the satellites towards a collective achievement of predefined objectives and trajectories designed for their mission. In a separate study, Barzegar and Rahimi [61] investigated the problem of fault diagnosis within clusters of small satellites. These satellites operate during diverse environmental conditions, presenting a difficult challenge in mitigating both disturbances and faults while maintaining the desired formation state among the satellites. The study aimed to alleviate the adverse effects of disturbances through an elaborate fault diagnosis approach, ultimately ensuring a consistent and reliable formation state among the small satellites, even in dynamically changing environments. To fortify the observer's resilience against external disturbances as outlined in Barzegar and Rahimi's work [60,61], they employed the H_∞ approach. In a separate study conducted by the same authors, a dissipativity-based UIO is implemented for the Lipschitz nonlinear multisatellite systems. This strategy is designed to mitigate the impact of disturbances across two scenarios: one where matching conditions are met and another where they are not [62].

Negash et al. [51] aimed to detect cyber-threats during UAV Formation Flight (UFF). Negash et al. have adopted a decentralized approach using UIO for FDI, which is particularly suited to the distinctive attributes of UFF and their associated control algorithms, which may involve unknown parameters. This approach facilitates the identification of system faults and extends its utility to detecting cyber intrusions and precisely locating compromised UAVs. The primary objective of this study is to introduce an algorithm capable of identifying compromised UAVs without compromising the overall performance of the formation. Numerical simulations conducted by the authors provide evidence that diagnosing cyber intrusions within a hexagonal UAV formation is highly effective [51].

Meskin et al. [49] proposed a unique solution to the problem of implementing UAVs in environments subject to significant disturbances. The FDI method was developed to integrate a continuous-time residual generator with a fault diagnosis system using a DES. Specifically, their hybrid FDI algorithm is applied to detect and isolate actuator faults in a quad-rotor network to demonstrate its efficacy.

Meskin and Khorasani [50] explored investigating, designing, and examining FDI filters for actuator systems within a network of aerial and space unmanned vehicles. The study emphasizes the actuator fault patterns within the team of agents, characterizing it as an excessively actuated system. An insolubility measure is introduced to evaluate fault patterns and design a novel set of structured residuals. This set is engineered to perform detection selectively and accurately, as well as isolate multiple faults that have been identified, particularly those with dependent fault patterns, such as excessively actuated systems. Based on these concepts, their algorithm is implemented to tackle the actuator FDI challenge in unmanned vehicle networks operating under different architectures, including centralized, distributed, and decentralized. Additionally, in [50], a comparative analysis was conducted, evaluating each architecture's advantages and limitations.

Zaeri Amirani et al. [68] focused on managing formations that change with time, especially when leaders move and followers lack information about their actions. The study also investigated altering the number of leaders and followers in the formation. Arrays of KFs were developed to handle noise reduction and integrate data using the state vector. A χ^2-test was utilized for FDI in this study. The coefficients in the formation control logic were set in advance, considering changes in the model, eliminating the need for real-time adjustments. The formation control law utilized integral state feedback based on the relative position integral of the agents, allowing followers to track leaders without their input.

4.3. Hybrid Methods

Hybrid methods are characterized by their ability to integrate data-driven and model-based methods. Using hybrid methodologies, the idea is to cover both main approaches' shortcomings using each approach's specializations. The term "Hybrid" implies the fusion of data-driven and model-driven methods; however, it does not restrict how the two approaches are combined. In other words, the word hybrid mostly serves as an umbrella term. This is also because hybrid methods in satellite, spacecraft, and UAV health monitoring have not been thoroughly explored in the literature, leaving hybrid methods a promising solution to further boost the existing FDII methods. This section reviews the few existing hybrid methodologies applicable to fault diagnosis applications.

Abbasi Nozari et al. [100] presented an enhanced FDI methodology for two RWs of a four-RW assembly, including both model-based and data-driven approaches used in residual generation and evaluation, respectively. The residual generation is conducted using the nonlinear geometric approach, and evaluation is carried out using a blended learning strategy with two separate hierarchically activated detection and isolation modules. The strategy incorporates robustness against external disturbances and model parametric uncertainties through the residual generation and evaluation modules. Consequently, the advantages of both the model-based and data-driven techniques are leveraged for the extra robustness of the strategy. The learning algorithm used in Abbasi Nozari et al.'s methodology employs a two-step classification approach by training several classifiers and using their test outputs to train the second-level learner. They adjusted their strategy to perform diagnostics using parallel and series formations. However, the series formation mostly surpassed the parallel formation. The simulations verify the hierarchical residual evaluation module's capability to provide satisfactory predictions using different validation datasets under real-life circumstances.

An NN-based observer is presented by Li et al. in [40] to minimize the impact of external interference on malfunction isolation and identification of single gimbal CMGs onboard spacecraft in periodic orbits. In addition, Li et al. employed multiple local fault adaptive estimators for fault isolation and basic identification for spacecraft with extra actuators. Then, their results are used to fuse gimbal and attitude data via neural networks. Finally, these NNs are used to isolate and identify malfunctions in the system.

Chen et al. [30] proposed a hybrid method for fault detection, isolation, and recovery of satellite IMU sensor faults. The proposed scheme works in two steps that get activated in succession: (1) a preliminary diagnosis and (2) a more precise diagnosis. The first step uses an adaptive unscented KF and the QUaternion ESTmiator (QUEST) algorithm in conjunction with each other to be able to examine different sensors and make the fault isolation problem deterministic. The second step consists of two Radial Basis Function (RBF) neural networks and an adaptive complementary filter. During normal operation, the two RBF neural networks get trained on the quaternion outputs of the AUKF and QUEST, and their training is halted when the first step is begun. After stopping their training, they predict the system quaternion outputs and contribute to achieving a higher level of accuracy in fault diagnosis and recovery of IMU faults. Furthermore, Chen et al. incorporated hypothesis testing, Parity equation, and Chi-Square approaches into their residual generators to address the problem of the residual thresholding being sensitive to factors beyond only faults and false alarms. Their proposed scheme was tested using an air-bearing table and demonstrated to have consistently better accuracies than the AUKF and QUEST algorithms used individually.

Another implementation of hybrid methods is conducted by Zhang et al. in [15] by devising a hybrid fault detection methodology for sensor health monitoring of the integrated navigation system of UAVs. Their proposed methodology uses a KF to estimate the residual sequences of sensor data. Then, a fuzzy logic system is employed to detect and determine the fault type where the membership functions are updated using a neural network.

Baura and Khorasani proposed a structured hierarchical framework for FDI within SFF [79]. In this framework, different diagnostic techniques are incorporated in a hierarchical structure, facilitating the integration of model- and knowledge-based diagnostic techniques. As part of our investigation, they used fuzzy reasoning to isolate faults in a leader–follower formation control system, focusing on ACS faults. The method is also considerably more efficient than traditional diagnostic approaches, which rely on individual satellite telemetry data trend analysis.

4.4. Other Methods

This section covers methodologies that do not fall into either the data-driven or model-driven spectrum. The literature addressing methods other than the focal points of this paper could provide us with useful insight, especially on combining extra methods with the existing model-based and data-driven solutions, for example, the methods incorporating time–frequency spectrums into CNN models.

Shangguan et al. [44] suggested a digital twinning approach to the fault diagnosis of satellites based on the Modelica language [101] for accommodating precise and online monitoring requirements. The model obtains valuable information for online monitoring by fusing the real satellite's telemetry data and simulation data. The proposed method was tested for the power system diagnosis of a satellite and provided accuracies of 83.33% to 93.33%, depending on the preset fault detection threshold.

Zhang et al. [18] proposed a time-domain frequency estimation methodology utilizing the subspace identification method. The presented frequency spectrum analysis approach is used with a gap-metric-oriented indicator and thresholding to achieve structural fault detection and identification in the rotor blades of UAVs. Test results expressed that the proposed methodology surpasses the signal reconstruction accuracy of the FFT and an interpolating FFT method for longer data lengths. It is also observed that blade faults are detected within one second of receiving damage. Future research could include devising more rapid recursive estimation methods and incorporating the malfunction detection scheme into a fault-tolerant control system.

5. Prognostics

Earlier, we covered the various ways the literature has achieved fault prognosis in satellites, spacecraft, and UAVs. The next step to advancing the health monitoring of the studied agents is to forecast the effect of the diagnosed malfunctions and obtain their RULs, namely, conducting fault prognosis. Performing prognostics for satellite and spacecraft applications is captivating, especially since their launch and operation are costly. Therefore, fault prognosis is a favorable approach to lengthening the lifetime or optimizing the use of satellites by knowing their RUL and time-to-failure. For UAVs, prognostics is not as pressing an issue since their flight time is usually limited and they can undergo maintenance between flights. Nonetheless, predicting their RUL and prolonging it is favorable. In addition, embedded analytical redundancy techniques for performing prognostics could incur high computational burdens, impeding the UAVs' main functionalities. As a result, UAVs have not received extensive attention in the field of prognostics from researchers. This section reviews the work that has currently been conducted on these agents' fault prognoses. Tables 2–4 list the strengths and limitations of each work on fault prognostics of single-agent systems. Table 6 outlines the main features of the fault prognosis works in a manner similar to Table 5. The reviewed literature indicates that, to the best of the authors' knowledge, very few published works exist on the actuator and sensor fault prognosis of a team of satellites or UAVs.

5.1. Data-Driven Methods

With the ability of data-driven methods to obtain predictive information on systems' time series operational data with access to prior historical data, performing fault prognosis on the studied agents becomes feasible. However, given the unrealized potential of data-driven methods in prognosis and unexplored areas of satellite and UAV prognostics, not many works are available in the literature about these agents' fault prognoses using data-driven methods. The following outlines the studies at hand regarding this area.

Islam and Rahimi [12] presented the first part in a double-module RW prognosis scheme using a data-driven approach. The first module is responsible for estimating the future values of the accessible system states. An AutoRegressive Integrated Moving Average (ARIMA) model and an LSTM network are tested for time series prediction, and the latter is shown to fare better in the first module on the input datasets. Later, Islam and Rahimi [5] expanded upon this scheme by augmenting it with the second module. The second module uses the first module's estimated states to predict the RW's inaccessible parameters. Another LSTM network is employed to evaluate the nonmeasurable system parameters and their future values that could give us insight into the health state of the RW. Relating these parameters to the RW's RUL is left for future research by the authors.

Muthusamy and Kumar discussed fault prognosis in CMGs in [69] using a predictive data-driven method that solely involves attitude rate sensor responses to circumvent the problem of unavailable measurement data for component-level variables. Component-level information includes temperature, vibrational and rotational measurements, and lubricant specifications. Muthusamy and Kumar's method consists of a Chebyshev neural network to simulate healthy system dynamics, along with a general path model, historical failures, and real-time measurements to estimate the remaining operating life of the system. According to performance results, the proposed method provides 96.25% accuracy, assuming 70% of the real-time satellite data are unavailable.

5.2. Model-Based Methods

Model-based strategies enable researchers to perform prognostics on a relatively straightforward but basic level when high-fidelity models are available compared with data-driven techniques. This is because data-driven techniques can accept numerous features that could affect the RUL of the studied agents, thus yielding more precise and descriptive estimations of agents' RUL. Incorporating the said features into models and accounting for coupling between different parameters could prove challenging and sometimes infeasible. Despite the established usage of model-driven techniques, their implementation in the fault prognosis area is scarce in the current literature. Some of the works in this area are reviewed in the following.

Rahimi et al. [70] suggested a model-based approach to the fault prognosis of RWs that is robust to Gaussian and non-Gaussian noises. After malfunction detection, the prognosis methodology activates by obtaining the fault-sensitive and nonmeasurable system parameters using the CAUKF scheme presented in [16] and outlined in Section 4.2. A PF is then used along with a health-decay model to estimate the damage model parameters and, through propagation in time, yields the RUL of the RW. The simulation test results demonstrate that implementing this approach would be conducive to acquiring accurate RW RUL estimations.

Alidadi and Rahimi [71] designed a stochastic model-based methodology for RUL estimation of RWs onboard satellites that accounts for the bearing lubrication faults. The proposed methodology consists of three key steps: First, lubricant temperature state estimation using an adjusted PF; second, model parameter estimation via a two-step PF accepting the lubricant's injected volume, estimated temperature, and degradation model as inputs, and third, RW RUL estimation under two lubrication malfunction cases. Alidadi and Rahimi also presented a PF resampling method for enhancing the accuracy and efficiency of the filters. The adaptive resampling method used in conjunction with the PFs demonstrated more accurate results for higher measurement noises as compared with the adjusted PF. The proposed methodology's performance in estimating the RWs' RUL exhibited errors between 0.1 and 4%. Future research could involve developing hybrid model-based and data-driven methods for more confident results or enhancing the proposed methodology by monitoring more RW parameters.

In the relatively unexplored field of UAV actuator fault prognosis, Shen et al. [72] proposed a prognostic and predictive maintenance strategy to increase the mission lifetime of quadrotor UAV actuators. Shen et al. first use a Two-Stage Kalman Filter (TSKF) to estimate the UAV state and actuator degradations in real time. To more easily analyze the degradation of multiple actuators, a comprehensive degradation variable was defined using the entropy weight method that assigns weights to actuators based on their degradation processes. Then, a health index is defined employing the Mahalanobis distance and three health stages are described based on the deviation of the UAV's health index from a nominal health index. Stage 1 encompasses the health conditions that are more degraded than the nominal state but do not influence the UAV performance heavily. Stage 2 is representative of health conditions that have lessened the UAV performance due to degradation and the inferred RUL may not realize the mission's requirements. Finally, the third stage describes the health states that are not able to meet mission stipulations. Then the results of these steps are used to autonomously extend the mission lifetime of the system. The proposed method of Shen et al. is applicable to hovering UAVs, and future research could address extending their method to other flight states.

6. Conclusions and Future Work

In this literature review, we have explored FDP within space and aerial vehicles, primarily focusing on satellites and Unmanned Aerial Vehicles. The insights from this study pave the way for future recommendations and research directions in this critical field of study. Tables 5–7 delineate a comprehensive overview of each study indicating the approaches and applications that have already been addressed in the literature.

6.1. Literature Analysis

According to the review strategy outlined in Section 1, the main details of the analyzed works are compiled in Tables 5–8. Each work's significance, advantages and disadvantages, and key features have been discussed in detail in the closing tables. However, in this section, we are going to carry out an analysis of the state of the literature in this field.

One of the key characteristics of each work in the literature was the methodology that was employed in them. Through compiling all these works, we arrived at a methodology/framework classification that explains the relationships between the employed strategies. This classification is illustrated by Figure 10.

In Figure 10, the three main classes that are presented include data-driven, model-based, and other methods. In the existing literature of this field, great emphasis was placed on data-driven and model-based methods with limited attention extended to other miscellaneous methods. Therefore, the same outlook is kept in this review to represent the current state of the existing works while shedding light on other methods, as well as pronounce their potential. The two main subcategories of data-driven methods are ML algorithms and deep learning, which are comprehensively discussed in Sections 4.1 and 5.1, as well as the commonly used dimensionality reduction techniques. Statistical models such as ARIMA are a simpler set of data-driven methods that share some aspects with ML techniques. Moving on to model-based methods, deterministic and stochastic dynamic system observers or state and parameter estimators are prevalent in this field. Deterministic methods are usually used when robustness to noise is not a strict requirement. Some works within this category also employed rule-based systems such as fuzzy logic-based works, and other probabilistic analysis methods which resemble the stochastic state and parameter estimators. The intersection of data-driven and model-based methods are referred to as hybrid methods and they have been covered in Section 4.3. The research into hybrid methods is not as prolific as its parent categories. These methods mostly include

NN observer-based strategies or hierarchical structures of data-driven and model-based frameworks that get activated hierarchically and work in conjunction with each other. Moreover, the vibration analysis-based methods as well as human-operated condition monitoring were classified as other methods since they were not integral to the research trends of this field but influential nonetheless. The attention that data-driven and model-based methods have been met in recent years is also a testament to the effects of the Fourth Industrial Revolution (4IR) or Industry 4.0 that has popularized the use of AI-based methodologies, big data analytics, and Internet of Things (IoT) (in autonomous FDP techniques and interconnected sensors and other components).

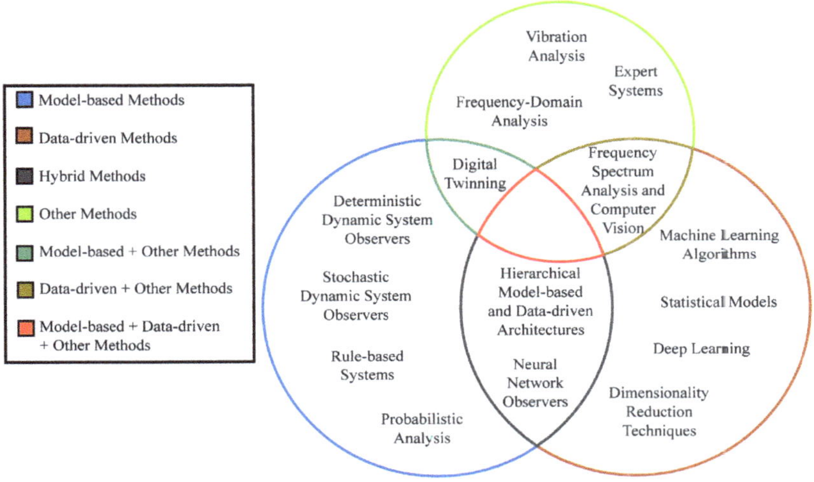

Figure 10. Compilation and classification of the methodologies utilized in the literature on satellite/UAV Fault Diagnosis and Prognosis.

The intersection of other methods and data-driven methods results in strategies that combine spectrograms with computer vision capabilities enabled by data-driven methods. In the same vein, digital twinning frameworks were identified as part of the intersection between other methods and model-based methods. Strictly speaking, digital twinning could be classified as a solely model-based methodology; however, these techniques usually require more insight into the other unrelated subsystems of the studied system than the unit FDP is carried out on. For example, many of the model-based methods in this field only require the models of the power subsystems to monitor their condition, whereas a digital twinning approach might need more context into the interactions of all the other subsystems with the power subsystem. For this reason, digital twinning approaches have been recognized to be a combination of model-based and other methods. To the authors' knowledge, the combination of all three categories of data-driven, model-based, and other methods does not exist in the literature. This combination on the surface appears to be too complicated and does not have any obvious advantages. Possible implementations within the red area in Figure 10 could include digital twin models with embedded AI capabilities, and finite element method-based models that account for satellites' and UAVs' surface vibrations that use data-driven models for FDP. These methodologies have not been explored in the literature; therefore, their respective area is left empty in Figure 10

Furthermore, to obtain more insight into how often these methods are used in the current literature, pie charts of the number of works designed around each of the prominent method categories are demonstrated by Figures 11 and 12 for single-agent and multiagent literature, respectively.

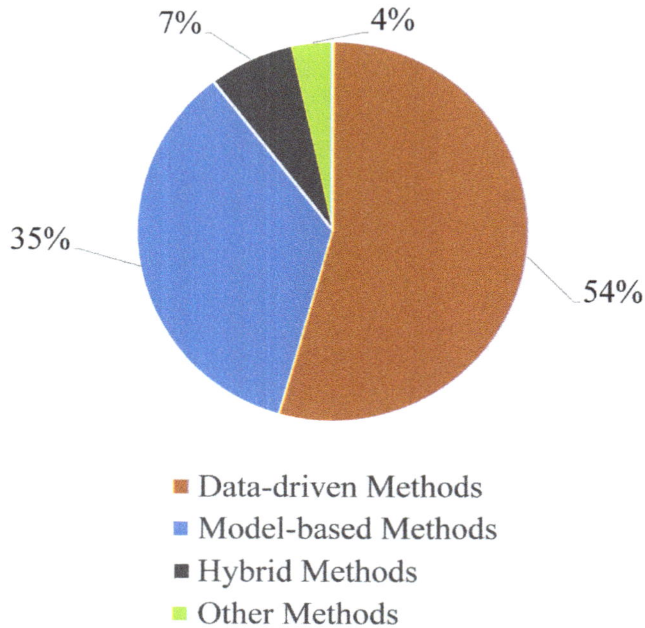

Figure 11. Breakdown of the methodologies used in the single-agent FDP literature.

It is noticeable from Figure 11 that data-driven methods are being heavily employed in the area of single satellite and UAV FDP. Consequently, with these methods now composing a significant portion of the literature (54%), it is imperative to address some of the underlying issues that inhibit their real-life implementations such as real-time monitoring shortcomings, faulty dataset dependencies, and poor integration with other subsystems such as the fault-tolerant module. Model-based methods also constitute an extensive share of the research on single-agent FDP especially because of their added robustness to noise. Hybrid methods are shown to be on the rise, with only 7% of the works focused on them, underscoring the need to develop them further. Finally, other methods are rarely used on their own in this field only appearing in 4% of the works, but it is beneficial to be familiarized with their potential combinations with other method categories.

On the other hand, the same pie chart depicted by Figure 12 for multiagent FDP shows a complete detachment from hybrid and other methods which could be due to the works on multiagent systems being fewer in numbers. It can also be seen that model-based methods are significantly more popular in this area appearing in 93% of the works. This signifies the potential for data-driven methods to become increasingly involved with multiagent FDP.

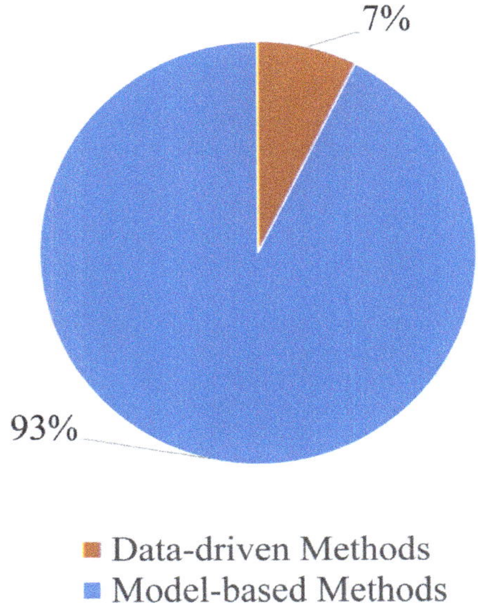

Figure 12. Breakdown of the methodologies used in the multiagent FDP literature.

6.2. Future Directions

To provide promising avenues to further the works presented in this review, below, we outline the potential areas for future exploration and improvements based on the gaps that have been identified:

Data-Driven FDP: Due to the scarcity of available historical operational datasets, generating artificial datasets that resemble real-life scenarios using AI-based methods is a promising approach for more rigorous exploration. Artificial datasets can address the limitations of mathematical models and provide a more comprehensive basis for FDP.

Increasing user trust in data-driven methods: Due to the low interpretability of many data-driven methods, user trust in these methods could suffer in spite of their effectiveness given how sensitive satellites, spacecraft, and UAVs could be. Methods such as LIME, which is utilized in [75], could alleviate this issue, although with an added trade-off with computational burden.

Dynamical Variables: Many works in the literature have considered different dynamical variables to achieve reliable fault diagnosis in single-agent systems according to their intended use. Experimenting with all possible dynamical variables in a system to train data-driven methods could yield intriguing results. Incorporating new features into data-driven methods allows for flexibility, albeit with additional computational demands.

Hybrid Methods: Leveraging a hybrid approach combining model-based and data-driven methods could help overcome the shortcomings of each individual method, leading to superior analytical techniques in the field of FDP.

Prognostics: While diagnostics for single-agent systems have been explored, there is room for further research in prognostics, particularly for satellites and UAVs Comprehensive research on fault prognosis for multisatellite/UAV systems is currently lacking and warrants further exploration.

Markovian Jump Systems: The integration of Markovian Jump Systems (MJS) and multiple models into fault diagnosis methods such as [13,74,102] for both single and multiagent systems presents an attractive opportunity, as they can handle abrupt system changes efficiently.

Association Rule Learning: As demonstrated by Pan et al. [85], association rule learning could serve as a not widely explored but viable means of analyzing the relationships between a bank of sensors or actuators. This approach could benefit the mission reliability of an agent even more if hardware redundancy is also available.

Graph Neural Networks: These networks differ from most traditional ML methods in the sense that they do not require feature independence [89]. Therefore, they have the capability to analyze a broader range of variables within the space and aerial agents which could render examining certain fault-sensitive parameters possible. These networks have also been utilized in the pertinent literature [33].

Multiagent Systems: There is a significant gap in research concerning fault diagnosis methods for multisatellite/UAV systems that prioritize consensus, coverage, or multiagent system control logic. Addressing this gap is a valuable direction for future research. Two of the few examples in this field include [60,62].

Heterogeneous Dynamics: Existing research mainly focuses on multiagent systems with homogeneous dynamics, while real-world systems exhibit heterogeneous dynamics influenced by environmental factors and mission objectives. Investigating FDP strategies that consider dynamic variability is essential for future research on fault diagnosis in multiagent systems. Examples of such work include [60–62].

Event-Triggered Communication: Exploring event-triggered communication can optimize information flow and reduce network workload, aligning research with practical applications and enhancing system efficiency. A summary is provided in Table 8.

Fractional-Order Systems: Future research should address fault diagnosis for fractional-order or reduced-order multisatellite systems. Balancing model accuracy and computational load is crucial when simplifying system complexity.

The recommendations and insights derived from this study will guide and inform future research endeavors in the critical field of Fault Diagnosis and Prognosis for satellites, spacecraft, and UAVs, ensuring the reliability and fault-tolerance of these systems.

Author Contributions: Conceptualization, methodology, software, validation, data curation, formal analysis, resources, and writing—original draft preparation, M.H.; conceptualization, methodology, software, validation, data curation, formal analysis, resources, and writing—original draft preparation, A.B.; funding acquisition, project administration, software, resources, formal analysis, supervision, visualization, and writing—review and editing, A.R. All authors have read and agreed to the published version of the manuscript.

Funding: This study was supported by Natural Sciences and Engineering Research Council of Canada (NSERC) through RGPIN-2020-05513, Canada. The APC was funded by RGPIN-2020-05513.

Institutional Review Board Statement: Not applicable as this study did not involve humans or animals.

Informed Consent Statement: Not applicable as this study did not involve humans or animals.

Data Availability Statement: No new data were created or analyzed in this study.

Acknowledgments: This study was supported by Natural Sciences and Engineering Research Council of Canada (NSERC) through RGPIN-2020-05513, and the University of Windsor, Canada.

Conflicts of Interest: The authors declare that they have no known competing financial interests or personal relationships that could have appeared to influence the work reported in this paper.

References

1. Barzegar, A.; Rahimi, A. Fault Diagnosis and Prognosis For Satellite Formation Flying: A Survey. *IEEE Access* **2022**, *10*, 26426–26442. [CrossRef]
2. Tafazoli, M. A study of on-orbit spacecraft failures. *Acta Astronaut.* **2009**, *64*, 195–205. [CrossRef]
3. Sadhu, V.; Zonouz, S.; Pompili, D. On-board Deep-learning-based Unmanned Aerial Vehicle Fault Cause Detection and Identification. In Proceedings of the 2020 IEEE International Conference on Robotics and Automation (ICRA), Paris, France, 31 May–31 August 2020; pp. 5255–5261. [CrossRef]
4. Pourtakdoust, S.; Fakhari Mehrjardi, M.; Hajkarim, M.; Nasihati Gourabi, F. Advanced fault detection and diagnosis in spacecraft attitude control systems: Current state and challenges. *Proc. Inst. Mech. Eng. Part J. Aerosp. Eng.* **2023**, *237*, 2679–2699. [CrossRef]

5. Sirajul Islam, M.; Rahimi, A. Fault Prognosis of Satellite Reaction Wheels Using a Two-Step LSTM Network. In Proceedings of the 2021 IEEE International Conference on Prognostics and Health Management (ICPHM), Detroit (Romulus), MI, USA, 6–8 June 2021; pp. 1–7. [CrossRef]
6. Nasrolahi, S.; Abdollahi, F. Sensor fault detection and recovery in satellite attitude control. *Acta Astronaut.* **2018**, *145*, 275–283. [CrossRef]
7. Isermann, R.; Ballé, P. Trends in the application of model-based fault detection and diagnosis of technical processes. *Control Eng. Pract.* **1997**, *5*, 709–719. [CrossRef]
8. Barnhart, D.J.; Vladimirova, T.; Baker, A.M.; Sweeting, M.N. A low-cost femtosatellite to enable distributed space missions. *Acta Astronaut.* **2009**, *64*, 1123–1143. [CrossRef]
9. Abbaspour, A.; Mokhtari, S.; Sargolzaei, A.; Yen, K. A Survey on Active Fault-Tolerant Control Systems. *Electronics* **2020**, *9*, 1513. [CrossRef]
10. Muthusamy, V.; Kumar, K. A novel data-driven method for fault detection and isolation of control moment gyroscopes onboard satellites. *Acta Astronaut.* **2021**, *180*, 604–621. [CrossRef]
11. Shahriari-kahkeshi, M.; Sheikholeslam, F.; Askari, J. Adaptive fault detection and estimation scheme for a class of uncertain nonlinear systems. *Nonlinear Dyn.* **2015**, *79*, 2623–2637. [CrossRef]
12. Sirajul Islam, M.; Rahimi, A. Use of A Data-Driven Approach for Time Series Prediction in Fault Prognosis of Satellite Reaction Wheel. In Proceedings of the 2020 IEEE International Conference on Systems, Man, and Cybernetics (SMC), Toronto, ON, Canada, 11–14 October 2020; pp. 3624–3628. [CrossRef]
13. Iglesis, E.; Horri, N.; Dahia, K.; Brusey, J.; Piet-Lahanier, H. Nonlinear Estimation of Sensor Faults With Unknown Dynamics for a Fixed Wing Unmanned Aerial Vehicle. In Proceedings of the 2021 International Conference on Unmanned Aircraft Systems (ICUAS), Athens, Greece, 15–18 June 2021; pp. 404–412. [CrossRef]
14. Rahimi, A.; Dev Kumar, K.; Alighanbari, H. Fault detection and isolation of control moment gyros for satellite attitude control subsystem. *Mech. Syst. Signal Process.* **2020**, *135*, 106419. [CrossRef]
15. Zhang, Q.; Wang, X.; Xiao, X.; Pei, C. Design of a fault detection and diagnose system for intelligent unmanned aerial vehicle navigation system. *Proc. Inst. Mech. Eng. Part J. Mech. Eng. Sci.* **2019**, *233*, 2170–2176. [CrossRef]
16. Rahimi, A.; Kumar, K.; Alighanbari, H. Fault estimation of satellite reaction wheels using covariance based adaptive unscented Kalman filter. *Acta Astronaut.* **2017**, *134*, 159–169. [CrossRef]
17. Cui, L.; Zhang, Q.; Shi, Y.; Yang, L.; Wang, Y.; Wang, J.; Bai, C. A Method For Satellite Time Series Anomaly Detection Based on Fast-Dtw and Improved-Knn. *Chin. J. Aeronaut.* **2023**, *36*, 149–159. [CrossRef]
18. Zhang, X.; Luo, H.; Li, K.; Kaynak, O. Time-Domain Frequency Estimation with Application To Fault Diagnosis Of The Unmanned Aerial Vehicles' Blade Damage. *IEEE Trans. Ind. Electron.* **2022**, *69*, 5257–5266. [CrossRef]
19. Rahimi, A.; Kumar, K.; Alighanbari, H. Enhanced Adaptive Unscented Kalman Filter For Reaction Wheels. *IEEE Trans. Aerosp. Electron. Syst.* **2015**, *51*, 1568–1575. [CrossRef]
20. Kang, B.H.; Hadaegh, F.Y.; Scharf, D.P.; Ke, N.P. Decentralized and self-centered estimation architecture for formation flying of spacecraft. In Proceedings of the 16th International Symposium on Space Flight Dynamics, Pasadena, CA, USA, 3–7 December 2001.
21. Varvani Farahani, H.; Rahimi, A. Data-Driven Fault Diagnosis For Satellite Control Moment Gyro Assembly with Multiple In-Phase Faults. *Electronics* **2021**, *10*, 1537. [CrossRef]
22. Park, K.; Park, E.; Kim, H. Unsupervised Fault Detection on Unmanned Aerial Vehicles: Encoding and Thresholding Approach. *Sensors* **2021**, *21*, 2208. [CrossRef]
23. Li, C.; Li, S.; Zhang, A.; Yang, L.; Zio, E.; Pecht, M.; Gryllias, K. A Siamese Hybrid Neural Network Framework For Few-Shot Fault Diagnosis Of Fixed-Wing Unmanned Aerial Vehicles. *J. Comput. Des. Eng.* **2022**, *9*, 1511–1524. [CrossRef]
24. Nozari, H.; Castaldi, P.; Banadaki, H.; Simani, S. Novel Non-Model-Based Fault Detection and Isolation Of Satellite Reaction Wheels Based on A Mixed-Learning Fusion Framework. *Ifac Pap. Online* **2019**, *52*, 194–199. [CrossRef]
25. Jado, R.; Moncayo, H. Deep-Learning Based Multiple-Model Bayesian Architecture for Spacecraft Fault Estimation. In Proceedings of the AIAA SCITECH 2024 Forum, Orlando, FL, USA, 8–12 January 2024. [CrossRef]
26. Du, C.; Zhang, X.; Zhong, R.; Li, F.; Yu, F.; Rong, Y.; Gong, Y. Unmanned Aerial Vehicle Rotor Fault Diagnosis Based on Interval Sampling Reconstruction of Vibration Signals and A One-Dimensional Convolutional Neural Network Deep Learning Method. *Meas. Sci. Technol.* **2022**, *33*, 065003. [CrossRef]
27. Talebi, H.; Khorasani, K.; Tafazoli, S. A Recurrent Neural-Network-Based Sensor and Actuator Fault Detection and Isolation For Nonlinear Systems with Application To The Satellite's Attitude Control Subsystem. *IEEE Trans. Neural Netw.* **2009**, *20*, 45–60. [CrossRef]
28. Ganesan, M.; Lavanya, R.; Nirmala Devi, M. Fault Detection in Satellite Power System Using Convolutional Neural Network. *Telecommun. Syst.* **2020**, *76*, 505–511. [CrossRef]
29. Abdelghafar, S.; Darwish, A.; Hassanien, A.; Yahia, M.; Zaghrout, A. Anomaly Detection Of Satellite Telemetry Based on Optimized Extreme Learning Machine. *J. Space Saf. Eng.* **2019**, *6*, 291–298. [CrossRef]
30. Chen, X.; Bettens, A.; Xie, Z.; Wang, Z.; Wu, X. Kalman Filter and Neural Network Fusion for Fault Detection and Recovery in Satellite Attitude Estimation. *Acta Astronaut.* **2024**, *217*, 48–61. [CrossRef]

31. Huang, Q.; Liang, B.; Dai, X.; Su, S.; Zhang, E. Unmanned Aerial Vehicle Fault Diagnosis Based on Ensemble Deep Learning Model. *Meas. Sci. Technol.* **2024**, *35*, 046205. [CrossRef]
32. Rahimi, A.; Folami, M. Reaction Wheels Fault Isolation Onboard 3-Axis Controlled Satel-Lite Using Enhanced Random Forest with Multidomain Features. *Int. J. Progn. Health Manag.* **2021**, *12*. [CrossRef]
33. Xie, L.; Pi, D.; Zhang, X.; Chen, J.; Luo, Y.; Yu, W. Graph Neural Network Approach For Anomaly Detection. *Measurement* **2021**, *180*, 109546. [CrossRef]
34. Luo, T.; Liu, M.; Zhao, H.; Duan, G.; Cao, X. Data-Driven Fault Monitoring For Spacecraft Control Moment Gyro with Slice Residual Attention Network. *J. Frankl. Inst.* **2022**, *359*, 9313–9333. [CrossRef]
35. Zhao, H.; Liu, M.; Sun, Y.; Chen, Z.; Duan, G.; Cao, X. Fault Diagnosis Of Control Moment Gyroscope Based on A New Cnn Scheme Using Attention-Enhanced Convolutional Block. *Sci. China Technol. Sci.* **2022**, *65*, 2605–2616. [CrossRef]
36. Gao, J.; Zhang, Q.; Chen, J. Ekf-Based Actuator Fault Detection and Diagnosis Method for Tilt-Rotor Unmanned Aerial Vehicles. *Math. Probl. Eng.* **2020**, *2020*, 1–12. [CrossRef]
37. Xiao, B.; Yin, S. A Deep Learning Based Data-Driven Thruster Fault Diagnosis Approach For Satellite Attitude Control System. *IEEE Trans. Ind. Electron.* **2021**, *68*, 10162–10170. [CrossRef]
38. Guzmán-Rabasa, J.; López-Estrada, F.; González-Contreras, B.; Valencia-Palomo, G.; Chadli, M.; Pérez-Patricio, M. Actuator Fault Detection and Isolation on A Quadrotor Unmanned Aerial Vehicle Modeled As A Linear Parameter-Varying System. *Meas. Control* **2019**, *52*, 1228–1239. [CrossRef]
39. Rotondo, D.; Cristofaro, A.; Johansen, T.; Nejjari, F.; Puig, V. Robust Fault and Icing Diagnosis in Unmanned Aerial Vehicles Using Lpv Interval Observers. *Int. J. Robust Nonlinear Control* **2018**, *29*, 5456–5480. [CrossRef]
40. Li, Y.; Hu, Q.; Shao, X. Neural Network-Based Fault Diagnosis for Spacecraft with Single-Gimbal Control Moment Gyros. *Chin. J. Aeronaut.* **2022**, *35*, 261–273. [CrossRef]
41. Chen, X.; Liu, M. A Two-Stage Extended Kalman Filter Method for Fault Estimation of Satellite Attitude Control Systems. *J. Frankl. Inst.* **2017**, *354*, 872–886. [CrossRef]
42. Maqsood, H.; Taimoor, M.; Ullah, Z.; Ali, N.; Sohail, M. Novel Sensor Fault Detection and Isolation for An Unmanned Aerial Vehicle. In Proceedings of the 2021 International Bhurban Conference on Applied Sciences and Technologies (IBCAST), Islamabad, Pakistan, 12–16 January 2021; pp. 486–493. [CrossRef]
43. Gai, W.; Li, S.; Zhang, J.; Zheng, Y.; Zhong, M. Dynamic Event-Triggered H_i/H_∞ Optimization Approach To Fault Detection for Unmanned Aerial Vehicles. *IEEE Trans. Instrum. Meas.* **2022**, *71*, 1–11. [CrossRef]
44. Shangguan, D.; Chen, L.; Ding, J. A Digital Twin-Based Approach for The Fault Diagnosis and Health Monitoring of A Complex Satellite System. *Symmetry* **2020**, *12*, 1307. [CrossRef]
45. Tao, J.; Khorasani, K. A Fault Detection, Isolation and Reconstruction Strategy for A Satellite's Attitude Control Subsystem with Redundant Reaction Wheels. In Proceedings of the 2007 IEEE International Conference on Systems, Man and Cybernetics, Montreal, QC, Canada, 7–10 October 2007; pp. 3146–3152. [CrossRef]
46. Mousavi, S.; Khorasani, K. Fault Detection of Reaction Wheels in Attitude Control Subsystem of Formation Flying Satellites. *Int. J. Intell. Unmanned Syst.* **2014**, *2*, 2–26. [CrossRef]
47. Tousi, M.; Khorasani, K. A Hybrid Fault Detection and Isolation Strategy for a Team of Cooperating Unmanned Vehicles. *Int. J. Control* **2014**, *88*, 90–103. [CrossRef]
48. Tousi, M.M.; Aghdam, A.G.; Khorasani, K. A Hybrid Fault Diagnosis for A Team of Unmanned Aerial Vehicles. In Proceedings of the 2009 IEEE International Conference on System of Systems Engineering (SoSE), Albuquerque, NM, USA, 30 May–3 June 2009; pp. 1–6.
49. Meskin, N.; Khorasani, K.; Rabbath, C. A Hybrid Fault Detection and Isolation Strategy for a Network of Unmanned Vehicles in Presence of Large Environmental Disturbances. *IEEE Trans. Control. Syst. Technol.* **2010**, *18*, 1422–1429. [CrossRef]
50. Meskin, N.; Khorasani, K. Actuator Fault Detection and Isolation for a Network of Unmanned Vehicles. *IEEE Trans. Autom. Control* **2009**, *54*, 835–840. [CrossRef]
51. Negash, L.; Kim, S.H.; Choi, H.L. Distributed Unknown-Input-Observers for Cyber Attack Detection and Isolation in Formation Flying UAVs. *arXiv* **2017**, arXiv:1701.06325. [CrossRef]
52. Azizi, S.; Khorasani, K. A Decentralized Cooperative Actuator Fault Accommodation of Formation Flying Satellites in Deep Space. In Proceedings of the 2009 3rd Annual IEEE Systems Conference, Vancouver, BC, Canada, 23–26 March 2009; pp. 230–235. [CrossRef]
53. Azizi, S.; Khorasani, K. A Distributed Kalman Filter for Actuator Fault Estimation of Deep Space Formation Flying Satellites. In Proceedings of the 2009 3rd Annual IEEE Systems Conference, Vancouver, BC, Canada, 23–26 March 2009; pp. 354–359. [CrossRef]
54. Azizi, S.; Khorasani, K. A Hierarchical Architecture for Cooperative Actuator Fault Estimation and Accommodation of Formation Flying Satellites in Deep Space. *IEEE Trans. Aerosp. Electron. Syst.* **2012**, *48*, 1428–1450. [CrossRef]
55. Azizi, S.; Khorasani, K. A Hierarchical Architecture for Cooperative Fault Accommodation of Formation Flying Satellites in Deep Space. In Proceedings of the 2009 American Control Conference, St. Louis, MO, USA, 10–12 July 2009; pp. 4178–4183. [CrossRef]
56. Azizi, S.; Khorasani, K. Cooperative Actuator Fault Accommodation of Formation Flying Vehicles with Absolute Measurements. In Proceedings of the 49th IEEE Conference on Decision and Control (CDC), Atlanta, GA, USA, 15–17 December 2010; pp. 6299–6304. [CrossRef]

57. Azizi, S.; Khorasani, K. Cooperative Actuator Fault Accommodation in Formation Flight of Unmanned Vehicles Using Relative Measurements. *Int. J. Control* **2011**, *84*, 876–894. [CrossRef]
58. Shakouri, A.; Assadian, N. Fault Detection and Isolation of Satellite Gyroscopes Using Relative Positions in Formation Flying. *Aerosp. Sci. Technol.* **2018**, *78*, 403–417. [CrossRef]
59. Gao, Z.; Wang, S. Fault Estimation and Fault Tolerance Control for Spacecraft Formation Systems with Actuator Fault and Saturation. *Optim. Control. Appl. Methods* **2021**, *42*, 1591–1611. [CrossRef]
60. Barzegar, A.; Rahimi, A. Distributed Simultaneous Fault Estimation and Cluster Consensus Control of Small Satellites. *IEEE Trans. Aerosp. Electron. Syst.* **2023**, *59*, 7654–7665. [CrossRef]
61. Barzegar, A.; Rahimi, A. A Distributed Fault Detection and Estimation for Formation of Clusters of Small Satellites. In Proceedings of the 2023 IEEE International Conference on Prognostics and Health Management (ICPHM), Montreal, QC, Canada, 9–11 June 2023; pp. 1–11. [CrossRef]
62. Barzegar, A.; Rahimi, A. A dissipativity-based robust fault diagnosis approach for clusters of small satellites. *Acta Astronaut.* **2024**, *214*, 182–195. [CrossRef]
63. Valdes, A.; Khorasani, K. A Pulsed Plasma Thruster Fault Detection and Isolation Strategy for Formation Flying of Satellites. *Appl. Soft Comput.* **2010**, *10*, 746–758. [CrossRef]
64. Meskin, N.; Khorasani, K. Fault Detection and Isolation of Actuator Faults in Spacecraft Formation Flight. In Proceedings of the Proceedings of the 45th IEEE Conference on Decision and Control, San Diego, CA, USA, 13–15 December 2006; pp. 1159–1164. [CrossRef]
65. Azizi, S.; Khorasani, K. Cooperative State and Fault Estimation of Formation Flight of Satellites in Deep Space Subject To Unreliable Information. *IFAC Pap. Online* **2019**, *52*, 206–213. [CrossRef]
66. Barua, A.; Khorasani, K. Hierarchical Fault Diagnosis and Health Monitoring in Multi-Platform Space Systems. In Proceedings of the 2009 IEEE Aerospace conference, Big Sky, MT, USA, 7–14 March 2009; pp. 1–13. [CrossRef]
67. Nemati, F.; Safavi Hamami, S.; Zemouche, A. A Nonlinear Observer-Based Approach To Fault Detection, Isolation and Estimation for Satellite Formation Flight Application. *Automatica* **2019**, *107*, 474–482. [CrossRef]
68. Zaeri Amirani, M.; Bigdeli, N.; Haeri, M. Distributed Fault Detection and Isolation in Time-Varying Formation Tracking Uav Multi-Agent Systems. *Asian J. Control* **2022**, *25*, 604–622. [CrossRef]
69. Muthusamy, V.; Kumar, K. Failure Prognosis and Remaining Useful Life Prediction of Control Moment Gyroscopes Onboard Satellites. *Adv. Space Res.* **2022**, *69*, 718–726. [CrossRef]
70. Rahimi, A.; Kumar, K.; Alighanbari, H. Failure Prognosis for Satellite Reaction Wheels Using Kalman Filter and Particle Filter. *J. Guid. Control. Dyn.* **2020**, *43*, 585–588. [CrossRef]
71. Alidadi, M.; Rahimi, A. Fault Diagnosis of Lubrication Decay in Reaction Wheels Using Temperature Estimation and Forecasting Via Enhanced Adaptive Particle Filter. *Sensors* **2023**, *23*, 1474. [CrossRef]
72. Shen, F.; Li, W.; Jiang, D.; Mao, H. Autonomous Predictive Maintenance of Quadrotor Uav with Multi-Actuator Degradation. *Aeronaut. J.* **2024**, 1–25. [CrossRef]
73. Wang, B.; Liu, D.; Peng, Y.; Peng, X. Multivariate Regression-Based Fault Detection and Recovery Of Uav Flight Data. *IEEE Trans. Instrum. Meas.* **2020**, *69*, 3527–3537. [CrossRef]
74. Carvalho, L.; Toriumi, F.; Angélico, B.; Costa, O. Model-Based Fault Detection Filter for Markovian Jump Linear Systems Applied To A Control Moment Gyroscope. *Eur. J. Control* **2021**, *59*, 99–108. [CrossRef]
75. Li, Y.; Jia, Z.; Liu, Z.; Shao, H.; Zhao, W.; Liu, Z.; Wang, B. Interpretable Intelligent Fault Diagnosis Strategy for Fixed-Wing Uav Elevator Fault Diagnosis Based on Improved Cross Entropy Loss. *Meas. Sci. Technol.* **2024**, *35*, 076110. [CrossRef]
76. Azizi, S.; Khorasani, K. A hybrid and switching framework for cooperative actuator fault estimation of formation flying satellites in deep space. In Proceedings of the 2009 7th Asian Control Conference, Hong Kong, China, 27–29 March 2009; pp. 1126–1131.
77. Ghasemi, S.; Khorasani, K. Fault Detection and Isolation of the Attitude Control Subsystem of Spacecraft Formation Flying Using Extended Kalman Filters. *Int. J. Control* **2015**, *88*, 2154–2179. [CrossRef]
78. Han, W.; Wang, Z.; Shen, Y. Robust H_∞ Fault Diagnosis Observer Design for Multiple Satellites Attitude Synchronization with Disturbance. In Proceedings of the 2016 35th Chinese Control Conference (CCC), Chengdu, China, 27–29 July 2016; pp. 6575–6580. [CrossRef]
79. Barua, A.; Khorasani, K. Intelligent Model-Based Hierarchical Fault Diagnosis for Satellite Formations. In Proceedings of the 2007 IEEE International Conference on Systems, Man and Cybernetics, Montreal, QC, Canada, 7–10 October 2007; pp. 3191–3196. [CrossRef]
80. Rahimi, A.; Saadat, A. Fault Isolation Of Reaction Wheels Onboard Three-Axis Controlled In-Orbit Satellite Using Ensemble Machine Learning. *Aerosp. Syst.* **2020**, *3*, 119–126. [CrossRef]
81. Vaz Carneiro, J.; Schaub, H.; Lahijanian, M.; Lang, K.; Borozdin, K. Machine Learning For Reaction Wheel Fault Detection Using Simulated Telemetry Data. In Proceedings of the AIAA SCITECH 2022 Forum, San Diego, CA, USA & Virtual, 3–7 January 2022. [CrossRef]
82. Kenneally, P.; Piggott, S.; Schaub, H. Basilisk: A Flexible, Scalable and Modular Astrodynamics Simulation Framework. *J. Aerosp. Inf. Syst.* **2020**, *17*, 496–507. [CrossRef]
83. Suo, M.; Zhu, B.; An, R.; Sun, H.; Xu, S.; Yu, Z. Data-Driven Fault Diagnosis Of Satellite Power System Using Fuzzy Bayes Risk and Svm. *Aerosp. Sci. Technol.* **2019**, *84*, 1092–1105. [CrossRef]

84. Hedayati, M.; Barzegar, A.; Rahimi, A. Mitigating Data Scarcity for Satellite Reaction Wheel Fault Diagnosis with Wasserstein Generative Adversarial Networks. In Proceedings of the 2024 IEEE International Conference on Prognostics and Health Management (ICPHM), Washington, DC, USA, 11–13 October 2024; pp. 367–376. [CrossRef]
85. Pan, D.; Liu, D.; Zhou, J.; Zhang, G. Anomaly Detection For Satellite Power Subsystem with Associated Rules Based on Kernel Principal Component Analysis. *Microelectron. Reliab.* **2015**, *55*, 2082–2086. [CrossRef]
86. Ossenfort, J. Advanced Diagnostic and Prognostic Testbed (ADAPT) Testability Analysis Report. In *Technical Report*; Ames Research Center: Moffett Field, CA, USA, 2008.
87. Huang, G.; Wang, D.; Lan, Y. Extreme Learning Machines: A Survey. *Int. J. Mach. Learn. Cybern.* **2011**, *2*, 107–122. [CrossRef]
88. Ferrell, B.; Santuro, S. NASA Shuttle Valve Data. 2005. Available online: http://www.cs.fit.edu/~pkc/nasa/data/ (accessed on 10 October 2023).
89. Wu, Z.; Pan, S.; Chen, F.; Long, G.; Zhang, C.; Yu, P. A Comprehensive Survey on Graph Neural Networks. *IEEE Trans. Neural Netw. Learn. Syst.* **2021**, *32*, 4–24. [CrossRef]
90. Liu, W.; Yuan, L.; Wang, S.; Liu, C. Research on Fault Diagnosis Method Of Control Moment Gyroscope Based on K-Means Algorithm. In Proceedings of the 2020 39th Chinese Control Conference (CCC), Shenyang, China, 27–29 July 2020; pp. 4119–4123. [CrossRef]
91. Taimoor, M.; Lu, X.; Maqsood, H.; Sheng, C. Adaptive Rapid Neural Observer-Based Sensors Fault Diagnosis and Reconstruction Of Quadrotor Unmanned Aerial Vehicle. *Aircr. Eng. Aerosp. Technol.* **2021**, *93*, 847–861. [CrossRef]
92. Wang, Y.; Yao, Q.; Kwok, J.; Ni, L. Generalizing From A Few Examples. *Acm Comput. Surv.* **2020**, *53*, 1–34. [CrossRef]
93. Huang, Y.; Ferguson, P. Automated One-Sided Learning Fault Detection System for Reaction Wheel Bearing Friction Anomalies. *J. Aerosp. Inf. Syst.* **2024**, *21*, 380–391. [CrossRef]
94. Bialke, B. High Fidelity Mathematical Modeling of Reaction Wheel Performance. In Proceedings of the Annual AAS Rocky Mountain Guidance and Control Conference, Breckenridge, CO, USA, 4–8 Janurary 1998; pp. 483–496.
95. Rahimi, A.; Kumar, K.; Alighanbari, H. Fault Isolation of Reaction Wheels for Satellite Attitude Control. *IEEE Trans. Aerosp. Electron. Syst.* **2020**, *56*, 610–629. [CrossRef]
96. Rahimi, A. Fault Isolation and Identification of A Four-Single-Gimbal Control Moment Gyro On-Board a 3-Axis Stabilized Satellite. *Int. J. Progn. Health Manag.* **2021**, *12*. [CrossRef]
97. Jansch-Porto, J.; Hu, B.; Dullerud, G. Convergence Guarantees of Policy Optimization Methods for Markovian Jump Linear Systems. In Proceedings of the 2020 American Control Conference (ACC), Denver, CO, USA, 1–3 July 2020; pp. 2882–2887. [CrossRef]
98. Norris, J.R., Discrete-time Markov chains. In *Markov Chains*; Cambridge Series in Statistical and Probabilistic Mathematics; Cambridge University Press: Cambridge, UK, 1997; p. 1–59.
99. Wang, R.; Zhao, C.; Bai, Y.; Du, W.; Wang, J. An Actuator Fault Detection and Reconstruction Scheme for Hex-Rotor Unmanned Aerial Vehicle. *IEEE Access* **2019**, *7*, 93937–93951. [CrossRef]
100. Abbasi Nozari, H.; Sadati Rostami, S.; Castaldi, P.; Simani, S. Hybrid Robust Fault Detection and Isolation of Satellite Reaction Wheel Actuators. *J. Control Decis.* **2022**, *11*, 117–131. [CrossRef]
101. Mattsson, S.; Elmqvist, H.; Otter, M. Physical System Modeling with Modelica. *Control Eng. Pract.* **1998**, *6*, 501–510. [CrossRef]
102. Meskin, N.; Khorasani, K. Fault Detection and Isolation of Discrete-Time Markovian Jump Linear Systems with Application to a Network of Multi-Agent Systems Having Imperfect Communication Channels. *Automatica* **2009**, *45*, 2032–2040. [CrossRef]

applied sciences

MDPI

Article

Imbalanced Diagnosis Scheme for Incipient Rotor Faults in Inverter-Fed Induction Motors

Ignacio Martin-Diaz [1], Tomas Garcia-Calva [2], Óscar Duque-Perez [1] and Daniel Morinigo-Sotelo [1,*]

[1] Research Group ADIRE, Institute of Advanced Production Technologies, University of Valladolid, 47002 Valladolid, Spain; ignacio.martin.diaz@uva.es (I.M.-D.); oscar.duque@uva.es (Ó.D.-P.)

[2] Department of Electrical Engineering, University of Valladolid, 47002 Valladolid, Spain; tomasalberto.garcia@uva.es

* Correspondence: daniel.morinigo@uva.es

Abstract: Recently, fault diagnosing supervised classifiers have been widely proposed to diagnose both electric and mechanical faults in induction motors (IM). However, many of them require a large amount of data, which implies a great effort required for processing fault-related features and building the training set. Furthermore, in real-world datasets, it is required to deal with highly skewed data distributions, also known as class imbalance, which is a limiting issue and can misguide the tuning of machine learning algorithms. Resampling techniques based on a synthetic generation of minority class observations aim to address this problem. Last but not least is the fact that inverter-fed IM introduces undesired harmonics in the monitoring signal altering the diagnosis patterns. This diagnosis scheme is evaluated on experimental imbalanced data oriented to deal with the diagnosis of a rotor in situations where it is fed with an inverter. The results show how this imbalanced approach determines the actual diagnosis performance on a small amount of data. The experimental results demonstrate that balanced training sets built with class balancing techniques improve the classifier and therefore its performance for diagnosing incipient rotor faults in inverter-fed IM with studied and interpretable features recently proposed in this field of study.

Keywords: rotor faults; diagnosis features; induction motor; inverter; small data; imbalance correction

1. Introduction

Condition monitoring is a growing concern in predictive maintenance tasks to prevent unexpected machinery breakdowns and production losses. Consequently, strengthening the ability to detect and diagnose machine faults can reduce costs in modern production planning and operations. As a primary energy consumer, induction motors (IM) are a crucial element in electromechanical equipment and currently dominate in both the industrial and commercial sectors [1]. Therefore, a well-designed condition-based monitoring system can minimize maintenance costs and correctly target necessary maintenance tasks [2].

Motor current signature analysis (MCSA) has been preferred for extracting fault signatures at both transient [3] and stationary regimes [4]. This technique has several advantages over vibration analysis, such as remote sensing, low implementation cost, and online monitoring.

Furthermore, voltage source inverters are increasingly used to feed IM in variable speed applications. This, however, introduces a new set of challenges. Using inverters introduces undesired harmonics into the stator current, negatively altering the extraction of features based on spectral analysis. Consequently, many diagnosis techniques that worked for line-fed IM do not perform appropriately on inverter-fed IM. As a result, the methodologies proposed to date as well as novel ones have to take into account this challenge to avoid the loss of performance of the diagnosis systems.

Nowadays, automatic diagnosis through machine learning techniques is one of the most promising fields for this challenge. Automatic diagnosis has been conceived as a tool for improving the efficiency and reliability of a balanced or unbiased fault detection scheme. In fact, an important research community has proposed supervised classifiers to diagnose the health of the electric machine [5–7]. Basically, a classifier consists of a function (linear or nonlinear) that tries to separate the classes defined in the problem by minimizing the expected loss. In other words, it tries to identify which class corresponds to each observation with the lowest classification error. The different motor states form classes of the problem to diagnose.

Hence, it is clear that data-driven approaches oriented to an automatic diagnosis have emerged recently [8]. Authors in [9] demonstrate an interest in data-driven methods where actual field data are obtained from railway electrification networks. The data is continuously collected from the most important variables, which gives a data-driven detection approach for low-frequency oscillation events in alternating current railway catenaries. In addition, there are works where the simulation-driven machine learning training phase has also been developed by generating raw input data. This kind of approach tries to overcome limitations such as data scarcity and data imbalance [10].

Data imbalance is a crucial concern, since actual datasets are not entirely balanced [11], and it is common for actual industrial data obtained both from destructive tests of the machine and during its useful life to be imbalanced [12]. When the minority class representation is extremely low, the problem is usually denominated as an anomaly detection case [13]. Unlike the case of balanced classification, under an imbalanced scenario, the classifier directly labels new observations to the majority class because of the existing inherent bias as the imbalance proportion between classes increases. This fact is detrimental to the performance of diagnosis systems and requires much more attention. To deal with class imbalance, the data science literature has presented several techniques relating to the following approaches: embedded methods, data processing, and ensemble methods. Within data processing approaches, resampling methods have offered adequate characteristics to balance previously imbalanced training sets. Moreover, in engineering scenarios, the reduced machine information about abnormal conditions limits intelligent fault diagnosis performances since diagnosis tools with small and imbalanced data reduce the accuracy of fault identification [12,14].

This paper aims to improve the diagnosis of faults in inverter-fed induction motors, mainly focusing on rotor bar failure. It proposes using the SMOTE oversampling technique to address imbalances in datasets to enhance diagnostic performance and interpretability. The proposed method is intended to work with a small and unbalanced dataset. Additionally, the fault features will be derived from the stator current in an inverter-fed motor, which is a source of high harmonic and polluted content in this signal (as mentioned before). Hence, this study aims to create an effective diagnostic tool for the circumstances previously described. This work begins by reviewing various techniques used to generate synthetic data to balance the initial dataset, focusing on the computed feature space. The implemented techniques will be Safe-Level SMOTE, Relocating Safe-Level SMOTE, and Majority weighted minority oversampling technique (MWMOTE), which have been proposed for balancing datasets. Firstly, the ReliefF algorithm performs the feature selection. Then, the developed diagnosis methodology uses a resampling technique for balancing data. When the training set is balanced, a supervised classifier is used for the fault diagnosis of rotor asymmetries with different degrees of bar breakage. The fault signatures proposed in the bibliography are introduced in the next section. These signatures are computed from signals collected during experimentation from an induction motor fed from inverters and the line.

2. Background

An appropriate condition monitoring system can detect the machine's potential rotor breakage in its early stages, and its diagnosis can be automated based on fault-related features. The monitored signal is the stator currents, which have proven effective for stationary and non-stationary operating conditions. In this paper, the fault signatures are derived from the frequency and time domains and calculated to diagnose issues during stationary operation.

2.1. Diagnosis Signals

From the scientific literature, it seems to be a consensus that MCSA is a noninvasive diagnostic method [3,4,8,12,15] with several advantages: easy acquisition of signals; signal collection does not interfere with the operation of the motor; the obtained signal is reliable under noisy environments; and sensors with reasonable accuracy are not too expensive. Once an IM is manufactured, free of rotor faults, and it is fed from an ideal stator current signal, the temporal expression of this current contains exclusively the fundamental component of its power supply:

$$i_0(t) = I_m \cos(\omega_0 t + \phi_0), \tag{1}$$

where ω_0 denotes the power supply angular pulsation, I_m denotes the amplitude of the stator current, and ϕ_0 is the phase. However, when a rotor breakage develops, fault-related harmonics emerge in the observed signal and can be used for diagnosis purposes. Nevertheless, when the IM supply comes from an inverter or from the mains with the presence of considerable harmonic content, the amplitude quantification of these fault patterns complicates their identification [16].

This project will utilize a single stator phase as a monitoring signal, which will be obtained from the motor when it is powered by an inverter or the grid. A data acquisition system will then digitize the signal in the time domain. Following this, the digital signal will undergo processing in both the time and frequency domains to identify fault patterns.

2.2. Time-Domain Fault Signatures

Statistics computed from time domain signals have been employed as features for several fault diagnosis approaches in literature [6,7,11]. High-order statistics are sensitive to non-Gaussian distribution measurements. On the other hand, low-order statistics (e.g., first and second moments) are more robust. In Table 1, the time-domain statistical features used in this study are defined.

Table 1. Statistics computed from the time-domain signal.

Variables in Time-Domain		
Variable	**Notation**	**Formula**
1st. Moment	m_1	$\frac{1}{N}\sum x[n]$
2nd. Moment	m_2	$\frac{1}{N}\sum(x[n]-\bar{x}[n])^2$
3rd. Moment	m_3	$\frac{1}{N}\sum(x[n]-\bar{x}[n])^3$
4th. Moment	m_4	$\frac{1}{N}\sum(x[n]-\bar{x}[n])^4$
2nd. Cumulant	c_2	$m_2 - m_1^2$
3rd. Cumulant	c_3	$m_3 - 3\cdot m_1 m_2 + 2\cdot m_1^3$
4th. Cumulant	c_4	$m_4 + 3m_3m_1 - 3\cdot m_2^2 + 12\cdot m_2 m_1^2 - 6\cdot m_1^4$
Skewness	$Skew$	$\frac{m_3}{(\sqrt{m_2})^3}$
Kurtosis	$Kurt$	$\frac{m_4}{(\sqrt{m_2})^4}$
Absolute mean (ABM)	$\lvert\bar{x}\rvert$	$\frac{1}{N}\sum\bar{x}$
Peak value	PV	$\frac{1}{2}(\max(x[n]) - \min(x[n]))$
Squared root value	SRV	$\left(\frac{1}{N}\sum\sqrt{x\rvert}\right)^2$
RMS value	RMS	$\sqrt{\frac{1}{N}\sum_{n=0}^{N-1}[x[n]-\bar{x}]^2}$
Crest factor	CF	$\frac{PV}{RMS}$
Shape factor	SF	$\frac{RMS}{\lvert\bar{x}\rvert}$

2.3. Fault Signatures from Spectra

Fault signatures from the stator current can also be obtained in the frequency domain. When a rotor bar breakage occurs, a current is induced, and torque and speed pulsations originate, giving place to two sidebands around the fundamental frequency and other harmonics. These fault-related frequency components are observable in the frequency spectrum [15]. This motivates the use of these sidebands' amplitude to be considered a fault severity indicator of the rotor condition. For their frequency location, Equation (2) is used.

$$f_{BRB} = (1 \pm 2s)f, \quad (2)$$

where f is the supply frequency and s is the rotor slip. The frequency corresponding to the minus sign is known as the left sideband harmonic (LSH); the other one is known as the right sideband harmonic (RSH). Both sidebands are separated from the integer harmonics, a distance that depends on rotor slip and, thus, on the motor load level. Nonetheless, when an IM is fed from an inverter, the stator current contains time harmonics. These harmonics modify the amplitude of the present harmonics or they generate new airgap spatial harmonics. The author in [17] proposed additional faulty signatures for diagnosing rotor cage asymmetries in inverter-fed IM. Some of them are used as candidate features for the diagnosis methodology and they are defined as follows:

$$\Gamma_5 = \frac{G7\ (LSH)}{G_5} \quad (3)$$

$$\Gamma_7 = \frac{G5\ (RSH)}{G_7} \quad (4)$$

$$\Gamma_{11} = \frac{G13\ (LSH)}{G_{11}} \quad (5)$$

$$\Gamma_{13} = \frac{G11\ (RSH)}{G_{13}} \quad (6)$$

The above expressions' numerators correspond to the amplitudes of the fault-related sidebands around the fifth, seventh, eleventh, and thirteenth harmonics. The nomenclature used is LSH for the left sideband harmonic and RSH for the right sideband harmonic. The denominators are the amplitudes of the above-mentioned integer harmonics. Figures 1 and 2 show the spectrum of one stator phase current to illustrate the presence of these harmonics. Figures 1 and 2 are the spectra for a line-fed and an inverter-fed induction motor respectively. As can be seen, the signal that comes from the inverter supply introduces different harmonics, inter-, and sub-harmonics, into the stator current spectrum, affecting more negatively the extraction of the above-defined indicators and, thus, the diagnosis compared with the line-fed IM case.

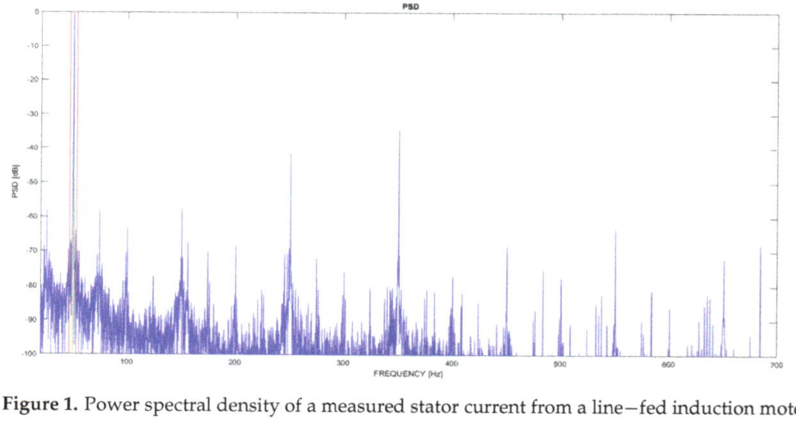

Figure 1. Power spectral density of a measured stator current from a line−fed induction motor.

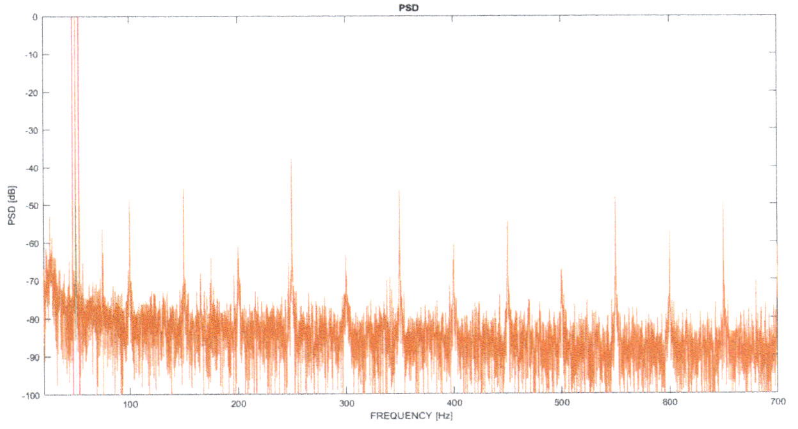

Figure 2. Power spectral density of a measured stator current from an inverter−fed induction motor.

2.4. Feature Selection through ReliefF Algorithm

The fault signatures mentioned above are obtained through a manual inspection process assisted by MATLAB© software version number R2023a. After collecting these fault signatures, a feature selection process is commonly used to eliminate noisy, irrelevant, and redundant variables from the set. This selective analysis aims to improve subsequent classification. For this purpose, the Algorithm 1 (ReliefF algorithm [18]) is applied. This technique scores the impact of changes based on how their values distinguish closely labelled observations.

ReliefF [18] is an extension of the Relief algorithm designed for binary problems. However, ReflieфF can deal with more than two classes and is more robust considering that it can deal with incomplete and noisy data [19]. Given a randomly selected observation S_o, ReliefF searches for k of its nearest neighbours in the same class. These are called the closest results R_j, hits or successes. This algorithm also searches for k nearest neighbours of each different class. They are called nearest misses $M_j(C)$. The quality estimate $W[F]$ is updated for all features F based on their values for S_o, $M_j(C)$ and R_o. The contributions of all successes and misses are averaged through the update expression. That is to say, the contribution for each class of the misses is weighted with the prior probability of that class $P(C)$ that, using the training set, is estimated. If the observations S_o and R_j have different values of the characteristic F, then the characteristic F will separate two observations belonging to the same class, which is undesirable. The estimation of quality $W[F]$ is reduced. This process is repeated m times. The choice of k successes and failures ensures higher algorithm reliability concerning noise. The difference between the values of the variable or feature F for two observations O_1 and O_2 is computed by the function $diff(F, O_1, O_2)$. For numerical variables, the function is defined as follows:

$$diff(F, O_1, O_2) = \frac{|value(F, O_1) - value(F, O_2)|}{max(F) - min(F)} \quad (7)$$

The $diff$ function calculates the distance between observations to determine the nearest neighbor's location. The total distance is calculated as the sum of the distances across all features or variables, known as the Manhattan distance.

Algorithm 1 ReliefF [18]

Input: Training observations with their respective vector of features
Output: The vector W with the quality estimation of the features
Set all feature weights to zero $W[F] = 0$
for $o \leftarrow 1, \ldots, m$ **do**
  Random selection of an observation S_o
  Locate k nearest results, named as R_j
  for each class $C \neq class(S_o)$ **do**
    From class C identify k nearest misses named as $M_j(C)$
    for $F = 1$ to f **do**
      $W[F] = W[F] - \sum_{j=1}^{k} \frac{diff(F,S_o,R_j)}{m \cdot k} + \sum_{C \neq class(S_o)} \frac{\left[\frac{P(C)}{1-P(class(S_o))} \sum_{j=1}^{k} diff(F,S_o,M_j(C))\right]}{m \cdot k}$
    end for
  end for
end for

3. Problem Statement for a Class-Imbalanced Scenario

The dataset of the IM obtained from the laboratory setup leads to a normal distribution as the data collection increases, as is the case for healthy motor instances. However, instances corresponding to faulty rotor states are fewer than healthy ones. Thus, a classifier cannot find the core of the distribution from the minority set of observations since it is too reduced to be identified. Therefore, minority-class observations must be over-sampled to perform successfully in a small data diagnosis scheme. This task even increases in complexity when features are computed from a polluted stator current signal due to the presence of undesired inverter harmonics [20]. The following sections describe the different approaches that have led to the already presented fault-related feature space treatment of the problem.

3.1. SMOTE and Following Extensions for Balancing Datasets

Synthetic Minority Over-Sampling Technique (SMOTE), proposed by Chawla et al. in [21], is a technique that oversamples the observations belonging to a minority class creating synthetic samples. This oversampling is made iteratively through a selection approach where each observation from the minority class is iterated until the total number of classes is balanced. The synthetic generation procedure consists of applying the k nearest neighbours algorithm. One of the k nearest neighbours is randomly selected, and later, it is synthesized, letting the algorithm continue until every needed synthetic sample is built. In the proposed methodology, an imbalanced dataset is characterized by its imbalanced ratio (IR), which is the quotient between the number of healthy class observations and the number of faulty cases belonging to each rotor state. For imbalanced datasets, a new observation denoted as x_{n+1} is generated from the features of the random sample x_j and the original x_i observation used to generate it. The generation of new features is created by multiplying a random number r between 0 and 1 and the difference between feature x_i and x_j as indicated in (8).

$$x_{n+1} = x_i + r(x_j - x_i), \qquad 0 \leq r \leq 1 \tag{8}$$

This random multiplication is the main disadvantage of SMOTE when used for diagnosis purposes. However, the borderline-SMOTE (BLS) method acts selectively, and it was useful as inspiration for the following density-based SMOTE. The BLS method focuses on observations close to the boundary areas prone to misclassification. This method defines a danger group for minority instances in danger of misclassification, which makes it possible to take these ones as candidates to oversample.

3.2. Safe Level-SMOTE

The Safe-Level SMOTE algorithm is an extension of the original SMOTE method. This version considers the "safe" observations found in the minority class [22]. The observations corresponding to this "safe" group are from the set of minority class observations that k-NN finds by computing the Euclidean distance. Observations that fall outside of this term coined as "safe" are considered noise. For this reason, a "safe" level ratio (slr) needs to be defined to obtain synthesized candidates. This ratio is defined as $slr = slp/sln$. The notation is as follows: slp is the "safe" level for a given observation x_i, and sln is the level for the selected neighbour of x_i, which is called x_j. In general, only candidates with minority class observations for x_i and x_j are considered. The generation of new synthetic observations goes under Equation (8). But now, r is calculated as

$$r \in \begin{cases} (0,(1/slr)), & \text{if } slr \geq 1 \\ ((1-slr),1), & \text{if } slr < 1 \end{cases} \tag{9}$$

The ratio $slr = 1$ means that x_i and x_j are equally important or "safe". Therefore, the aggregate observations can be randomly selected. On the one hand, if this ratio is greater than 1, the newly aggregated data will be weighted in the direction of the minority class, which brings the new aggregation closer to the x_i example. On the other hand, if $slr < 1$, it is weighted in the direction of the majority class, which brings the most recent sample closer to the example x_j.

3.3. Relocating Safe Level SMOTE

A limitation of the Safe Level-SMOTE method may be that it places some synthetic observations where they may be too close to the nearby majority class, leading to classifier confusion [23]. The location of each surrounding majority class observation is not considered in the safety level SMOTE, i.e., the generation of these synthetic observations may be close to some majority class observations. To prevent this situation, a later extension was proposed in [23]. This SMOTE extension is called Relocating Safe-Level SMOTE (RSLS). RSLS is conceived as a procedure to relocate synthetic observations once they are located close to minority-class observations. This approach follows the SMOTE steps on the safe side, starting with finding the k-NN of each minority class observation and computing its corresponding proportions. This approach follows the SMOTE steps on the safe side, starting with finding the k-NN of each minority class observation and computing its corresponding proportions. Once the observation O and its minority class neighbour $\hat{m}$ are determined, the "safe" level values and their safe level ratio are used to determine the possible range of observed aggregation. Once the location of the synthetic observation is randomly selected from this range, this new location is used to calculate the distance between the new synthetic observation and the observations of the surrounding majority group. These distance values are compared with the distance of these aggregate observations to their nearest point, O or $\hat{m}$. Suppose the distance between the composite observation and one of the observations of the surrounding majority group is less than the distance to its nearest point. In that case, the composite observation under consideration is moved closer to this point until it is excluded from the observations of the surrounding majority group.

3.4. Density-Based SMOTE

This method is inspired by the borderline-SMOTE. This technique applies density-based clustering by combining it with SMOTE. The synthetic observations are generated by calculating the shortest path from each instance of the minority class to the pseudo-centroid of the cluster corresponding to the minority class.

3.5. Majority Weighted Minority Oversampling Technique

The method was introduced in [24] and is also known as MWMOTE. The idea behind it is to ensure that the synthetically generated minority observations fall into the correct class cluster. To achieve this, a mechanism calculates the weight according to the distance to the nearest majority class point. Then, synthetic examples are generated from the minority class instance with weighted information using the clustering method.

Figure 3 illustrates the operation of the MWMOTE algorithm for synthetically generating data from the minority class. The figure shows two scatter plots presenting the amplitude of the LSH failure indicator as a function of motor slip (which also represents the motor load level). The green circles represent the majority class, i.e., the LSH value when the motor is healthy. The blue points represent the minority class, i.e., the LSH value when the motor experiences a broken bar. The synthetic values generated using this technique to balance the two classes are represented in red on the right side of the figure (Figure 3b).

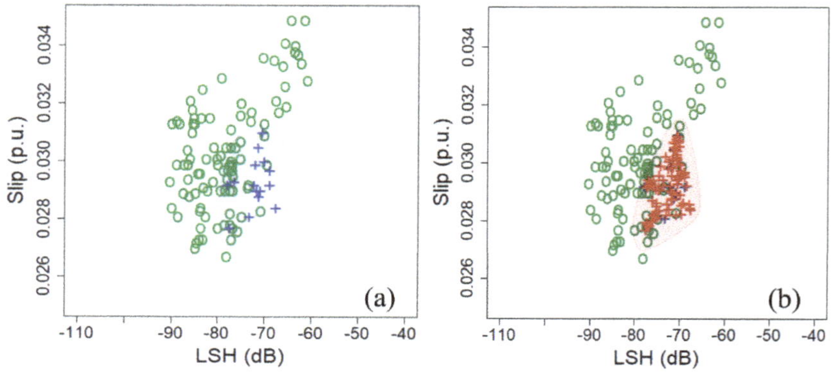

Figure 3. Demonstration of the operation of the MWMOTE algorithm. A dataset from an inverter—fed motor is used. The green markers represent the majority class (healthy motor), and the blue markers represent the minority class (motor with failure). (**a**) Case with imbalanced classes; (**b**) synthetically balanced case (synthetically created markers are shown in red).

4. Test Bench and Signal Acquisition

The test bench designed for this research is shown in Figure 4. The IM is fed from the line supply and from inverters whose specifications are defined in the Appendix A. In both cases, the voltage main frequency is 50 Hz. The motor load is a magnetic powder brake with its control unit. A Hall Effect current transducer by LEM® (Meyrin, Switzerland) is used to measure the stator current. A National Instruments® (Austin, TX, USA) NI cDAQ-9174 base platform with an NI 9215 acquisition module is used for the signal collection. The sampling frequency is 50 kHz.

The rotor fault is simulated by drilling a hole in one of the bars near the short circuit ring. Three rotor condition states are considered: (i) The rotor is healthy in the first tests. This condition is labeled as R1; (ii) After that, a 4.2 mm depth hole is drilled in one of the bars. This intermediate-severity rotor condition is labeled as R3; (iii) Finally, a more severe rotor fault is provoked by drilling the whole rotor bar. This rotor condition is known (R5). The stator current signal is collected while the motor is operating in a steady state for 10 s with a constant load torque. During these tests, the motor slip is in the range of 3–5%, according to the motor specifications mentioned in Appendix A.

Different tests, considering the type of power supply, the state of health of the rotor, and the level at which the motor is loaded, have been developed to develop an automatic steady-state diagnosis tool.

Figure 4. Test bench employed in this research. (1) Induction motor; (2) magnetic powder brake and load control unit (3) custom-made hall effect sensor board; (4) DAQ board; (5) laptop and software; (6) inverters; (7) simulated broken rotor bars.

5. Experimental Results

The diagnosis methodology begins with a feature selection stage, the training of a supervised classifier, and finally, an evaluation phase, in which the global performance of the diagnosis tool is evaluated. Table 2 shows the main characteristics of the dataset. The columns describe the voltage supply, faults severities, the imbalance ratio (IR), and the number of samples for the healthy and faulty cases, also expressed as a percentage. The last column is a label for each case.

The IR for each case is the rate between the number of healthy and faulty class instances of each rotor condition. The IRs considered in this study are those indicated in the corresponding column in Table 2. This selection of IR analyses the influence of the level of imbalance under this particular small data problem. The chosen IRs depend on the observations from the experimental trials, which are then processed to obtain the characteristic fault features or variables.

Table 2. Description of the experimental datasets

Supply	Severities	Imbalance Ratio (IR)	#Samples (Healthy–Faulty)	% (Healthy–Faulty)	Case Number
Line-fed	R1–R3	12	195–15	92.85–07.15	1
	R1–R3	7	105–15	87.50–15.50	2
	R1–R3	3	45–15	75.00–25.00	3
Line-fed	R1–R5	12	195–15	92.85–07.15	4
	R1–R5	7	105–15	87.50–15.50	5
	R1–R5	3	45–15	75.00–25.00	6
Inverter-fed	R1–R3	12	195–15	92.85–07.15	7
	R1–R3	7	105–15	87.50–15.50	8
	R1–R3	3	45–15	75.00–25.00	9
Inverter-fed	R1–R5	12	195–15	92.85–07.15	10
	R1–R5	7	105–15	87.50–15.50	11
	R1–R5	3	45–15	75.00–25.00	12

5.1. Feature Selection

After applying the ReliefF algorithm, the feature selection results for both power supplies are shown in Figure 5. In this figure, the red dashed line is the threshold for the features selected for the classification stage. Although a threshold choice is arbitrary, this threshold is set to 0.06, representing the percentile of best-scoring features to select. The analysis of this figure provides some interesting results. The algorithm highlights more relevant features for the line supply case than the inverter-fed case. It is also noticeable what happens with the m_3 and c_3 statistics: they were the worst features in the analysis with line-fed IM data but are selected for the inverter-fed case. The loss of relevance for *Skew* in the inverter-fed IM case is also appreciable. However, as was expected, *LSH* and *RSH* are the most significant features for all power supply cases. Furthermore, the Γ features proposed in [17] do not show outstanding relevance for the diagnosis.

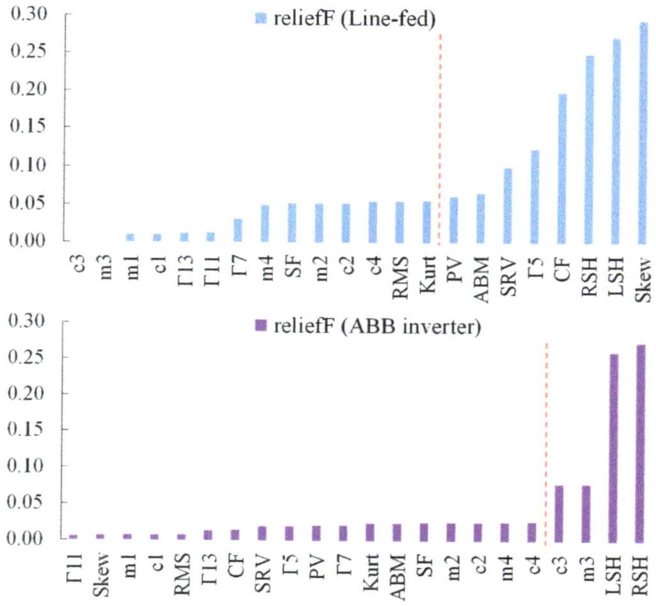

Figure 5. Results of the feature selection with the ReliefF algorithm: a line-fed case at the (**top**) and an inverter-fed case at the (**bottom**) of the figure.

5.2. Effect on the Feature Subspace

After the feature selection stage, the IR is improved by the generation of synthetic data for the minority class using the different techniques previously described: SLSMOTE, RSLSMOTE, and MWMOTE. The balanced datasets are shown in Figures 6 and 7. The first figure is for the motor line-fed case, and the second is for the inverter-fed case. The LSH fault feature vs. the motor slip is represented in all these figures for different failure severities. The different rotor conditions are depicted as follows: healthy case (O); original rotor fault data (+); synthetic rotor fault data (+).

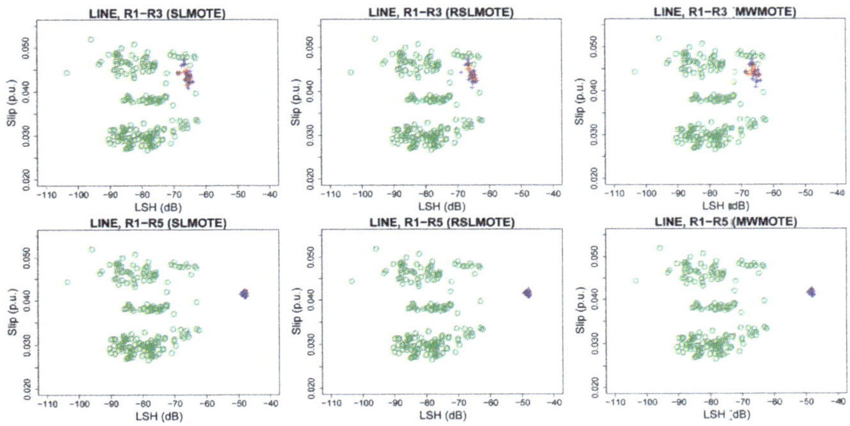

Figure 6. Bidimensional feature scatterplot (slip and LSH) for synthetically balanced data from line supply with the resampling techniques: Safe Level-SMOTE (SLMOTE), Relocating Safe Level SMOTE (RSLMOTE), and Majority weighted minority oversampling technique (MWMOTE). An imbalanced ratio of 12. Top figures: R3 rotor severity. Bottom figures: R5 rotor severity. The different rotor conditions are depicted as follows: healthy case (O); original rotor fault data (+); synthetic rotor fault data (+).

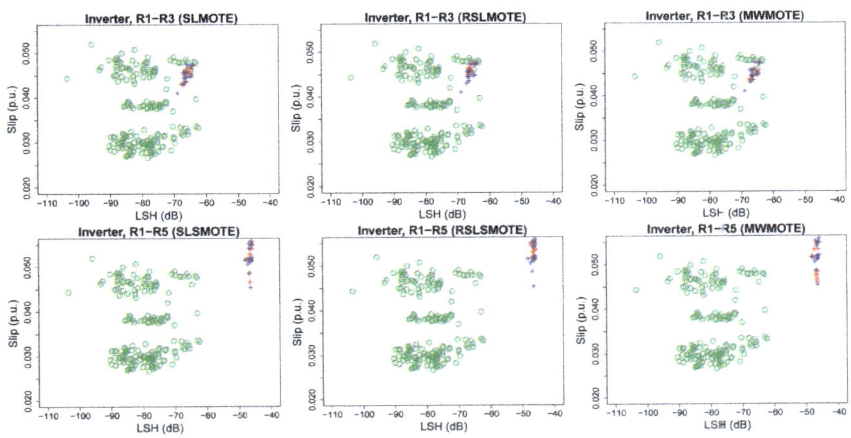

Figure 7. Bidimensional feature scatterplot (slip and LSH) for synthetically balanced data from inverter-based supply with the resampling techniques: Safe Level-SMOTE (SLMOTE), Relocating Safe Level SMOTE (RSLMOTE), and Majority weighted minority oversampling technique (MWMOTE). An imbalanced ratio of 12. Top figures: R3 rotor severity. Bottom figures: R5 rotor severity. The different rotor conditions are depicted as follows: healthy case (O); original rotor fault data (+); synthetic rotor fault data (+).

In both figures, the LSH feature is represented vs. the motor slip. In the first row, the following two-rotor conditions are represented: healthy or R1 and intermediate severity or R3. In the second row, these two conditions are plotted: healthy or R1 and fully broken bar or R5. In all cases, the healthy observations outnumber the faulty ones, producing an IR of 12.

The resampling results for each algorithm are shown with red markers for each algorithm. The synthetic data generated with the SMOTE technique for the line supply case shows dispersed decision boundaries for the two-rotor severity (R3 and R5). However, with Safe-Level SMOTE and Relocating Safe-Level SMOTE variants, this does not happen. In the inverter-fed case, the results are quite similar.

5.3. Classification Stage

After balancing the datasets, three classification techniques are trained: Support Vector Machine (SVM) with Gaussian kernel as a radial basis function, Adaptive Boosting (AdaBoost), and instance-based k-nearest Neighbors (k-NN). The tuning parameters are $C = 1$ for SVM, the number of neighbours for k-NN, and SMOTE set to $k = 5$. Previously, different combinations of tuning parameters were explored during the parameter tuning procedure. The amount of oversampling is predetermined from the beginning for the imbalanced ratios indicated in Table 2.

6. Discussion

The twelve case studies are summarized in Table 2, while Table 3 presents the metrics used to evaluate the classification results of these cases. The performance of three classification algorithms is compared with datasets obtained after variable selection and class balancing using three oversampling algorithms. The comparison metrics were obtained through a 10-fold cross-validation process and were used for the comparative study of imbalanced classes. The metrics used for comparison are Area Under the ROC Curve (AUC) and G-mean1. AUC is derived from ROC analysis, which evaluates the performance of classifiers by plotting the True Positive Rate against the False Positive Rate while varying a score threshold. It is a scalar value representing the probability that a classifier performs better at classifying a fault instance over a healthy one, providing a single score for evaluating which algorithm is better on average. G_{mean1} is used to evaluate the relative balance of classifier performance on all classes, computed as negative (Healthy) and positive (Fault):

$$G_{mean1} = \sqrt{TPR \times TNR}, \tag{10}$$

where TPR and TNR are True Positive Rate and True Negative Rate, respectively.

For study cases 8 and 9, the combination of MWMOTE for class balancing and the AdaBoost classifiers performs better than the rest of the oversampling algorithms and classifiers. These cases correspond to the motor fed by inverter and intermediate inverter feeding, intermediate fault severity (R3), and imbalance ratios IR = 7 and IR = 3, respectively. However, when the imbalance ratio increases, every classifier and oversampling tool performs quite similarly, i.e., no statistical significance is found in their results. The statistical significance is checked with the Holm test. This test is applied for a pairwise performance among each oversampling approach whose p-values are shown in Table 3. The outcome is to reject the null hypothesis when this p-value falls below a threshold. This outcome implies that the corresponding SMOTE extension is outperformed by the remaining ones. This situation is highlighted with an asterisk in the p-value and found that only happens for the MWMOTE approach. Studies 4–6 and 10–12 cover both power sources, the maximum fault severity (R5), and all class imbalance rates. The classification metrics obtained are good for the various combinations of oversampling techniques and classification algorithms.

Table 3. Classification results with feature selection and oversampling with SLSMOTE, RSLSMOTE, and MWMOTE.

Classifier	Supply	Case Number	Severity	Imbalance Ratio (IR)	AUC (%)	p-Value	Gmean1 (%)
SVM	Line-fed	1	R3	12	97 ± 2.55	0.085	98 ± 1.58
	Line-fed	2	R3	7	97 ± 2.52	0.251	97 ± 2.63
	Line-fed	3	R3	3	97 ± 1.36	0.651	97 ± 2.14
	Line-fed	4	R5	12	100 ± 0.00	0.156	100 ± 0.00
	Line-fed	5	R5	7	100 ± 0.00	0.746	100 ± 0.00
	Line-fed	6	R5	3	100 ± 0.00	0.365	100 ± 0.00
	Inverter	7	R3	12	97 ± 2.11	0.711	97 ± 2.77
	Inverter	8	R3	7	96 ± 1.23	<0.01	97 ± 2.98
	Inverter	9	R3	3	96 ± 2.58	<0.01	97 ± 2.73
	Inverter	10	R5	12	100 ± 0.00	0.411	100 ± 0.00
	Inverter	11	R5	7	100 ± 0.00	0.709	100 ± 0.00
	Inverter	12	R5	3	100 ± 0.00	0.177	100 ± 0.00
AdaBoost	Line-fed	1	R3	12	97 ± 1.23	0.785	98 ± 1.42
	Line-fed	2	R3	7	97 ± 2.58	0.150	97 ± 2.40
	Line-fed	3	R3	3	98 ± 1.32	0.255	98 ± 1.84
	Line-fed	4	R5	12	100 ± 0.00	0.456	100 ± 0.00
	Line-fed	5	R5	7	100 ± 0.00	0.606	100 ± 0.00
	Line-fed	6	R5	3	100 ± 0.00	0.865	100 ± 0.00
	Inverter	7	R3	12	97 ± 1.10	0.545	97 ± 2.66
	Inverter	8	R3	7	97 ± 1.23	<0.01 *	97 ± 2.09
	Inverter	9	R3	3	98 ± 2.58	<0.01 *	97 ± 2.20
	Inverter	10	R5	12	100 ± 0.00	0.811	100 ± 0.00
	Inverter	11	R5	7	100 ± 0.00	0.709	100 ± 0.00
	Inverter	12	R5	3	100 ± 0.00	0.387	100 ± 0.00
k-NN	Line-fed	1	R3	12	97 ± 1.02	0.113	98 ± 1.88
	Line-fed	2	R3	7	96 ± 2.52	0.852	97 ± 2.81
	Line-fed	3	R3	3	97 ± 2.02	0.673	97 ± 2.88
	Line-fed	4	R5	12	98 ± 1.56	0.555	100 ± 0.00
	Line-fed	5	R5	7	100 ± 0.00	0.143	100 ± 0.00
	Line-fed	6	R5	3	100 ± 0.00	0.861	100 ± 0.00
	Inverter	7	R3	12	97 ± 1.10	0.219	97 ± 2.97
	Inverter	8	R3	7	95 ± 3.32	<0.01 *	97 ± 2.62
	Inverter	9	R3	3	96 ± 3.88	<0.01	96 ± 3.66
	Inverter	10	R5	12	100 ± 0.00	0.815	100 ± 0.00
	Inverter	11	R5	7	100 ± 0.00	0.113	100 ± 0.00
	Inverter	12	R5	3	100 ± 0.00	0.770	100 ± 0.00

* Outcome of the test from pairwise performance where MWMOTE outperforms the rest.

7. Conclusions

This article reviews and applies several resampling techniques for diagnosing rotor faults in line-fed and inverter-fed induction motors. These resampling techniques are SMOTE extensions and produce synthetic observations belonging to the minority fault class for balancing training sets and improving the synthetic data generation according to boundary area behavior.

The evaluation of the classifier performance with the metrics AUC and G_{mean1} metrics shows that for the highest fault severity of the broken rotor bar fault (R5), all resampling techniques considered have similar behavior, which implies that in terms of performance, any of them can obtain successful diagnostic results independent of the motor power supply and the imbalanced ratio analyzed.

Therefore, by incorporating any of these oversampling techniques into the diagnosis methodology to predict the rotor's condition, it is possible to construct a more precise and robust diagnosis of the rotor faults of inverter-powered induction motors with small and unbalanced data.

In addition, from this experimental study, the authors observed that one area of exploration is the application of these oversampling methods to more incipient rotor breakages, where the feature differences regarding the healthy class are much more reduced.

Author Contributions: Conceptualization, I.M.-D. and D.M.-S.; data curation, I.M.-D.; methodology, I.M.-D. and D.M.-S.; software, T.G.-C.; validation, D.M.-S. and T.G.-C.; writing—original draft, I.M.-D.; writing—reviewing and editing, D.M.-S. and Ó.D.-P. All authors have read and agreed to the published version of the manuscript.

Funding: This research was partially funded by the University of Valladolid.

Institutional Review Board Statement: Not applicable.

Informed Consent Statement: Not applicable.

Data Availability Statement: The data presented in this study are available on request.

Acknowledgments: The authors are grateful for the institutional support received from the University of Valladolid as well as the Department of Electrical Engineering, which has made it possible to have the necessary material and human resources, who, in one way or another, contributed to the experimental design. In addition, we would like to thank the editor and reviewers for taking the time and effort necessary to review the manuscript. We sincerely appreciate all valuable comments and suggestions, which helped us to improve the quality of the manuscript.

Conflicts of Interest: The authors declare no conflicts of interest.

Abbreviations

The following abbreviations are used in this manuscript:

AUC	Area Under the ROC Curve
BRB	Broken Rotor Bar
DBSMOTE	Density-Based Synthetic Minority Oversampling Technique
FD	Fault Data
HT	Holm Test
k-NN	k-Nearest Neighbors
MCSA	Motor Current Signature Analysis
MWMOTE	Majority weighted minority oversampling technique
RSLS	Relocating Safe-Level SMOTE
SLSMOTE	Safe-Level SMOTE
SMOTE	Synthetic Minority Oversampling Technique
SVM	Support Vector Machine

Appendix A. Equipment Specifications

Three-phase asynchronous squirrel cage motor (Siemens®, Munich, Germany), star connected. Nominal characteristics: power = 0.75 kW, f = 50 Hz, voltage = 400 V, current = 1.9 A, nominal speed = 1395 rpm, and the number of pole pairs is two. Three three-phase inverters from different brands were used. The first one is an ABB® (model ACS355, Zürich, Switzerland), the second one is an Allen Bradley® (model PowerFlex 40, Milwaukee, WI, USA), and the last one is a Telemecanique® (model Altivar 66, Rueil-Malmaison, France) with range power from 0.37 to 4 kW, 0.4 to 2.2 kW and 0.75 to 250 kW, respectively. The switching frequency of all inverters is 4 kHz.

References

1. Park, Y.; Jeong, M.; Bin Lee, S.; Teska, M.; Antonino-Daviu, J.A. Influence of Blade Pass Frequency Vibrations on MCSA-Based Rotor Fault Detection of Induction Motors. *IEEE Trans. Ind. Appl.* **2017**, *53*, 2049–2058. [CrossRef]
2. Zhou, Z.J.; Hu, C.H.; Wang, W.B.; Zhang, B.C.; Xu, D.L.; Zheng, J.F. Condition-based maintenance of dynamic systems using online failure prognosis and belief rule base. *Expert Syst. Appl.* **2012**, *39*, 6140–6149. [CrossRef]
3. Pons-Llinares, J.; Antonino-Daviu, J.; Roger-Folch, J.; Morinigo-Sotelo, D.; Duque-Perez, O. Mixed eccentricity diagnosis in Inverter-Fed Induction Motors via the Adaptive Slope Transform of transient stator currents. *Mech. Syst. Signal Process.* **2014**, *48*, 423–435. [CrossRef]

4. Benbouzid, M.E.H. A review of induction motors signature analysis as a medium for faults detection. *IEEE Trans. Ind. Electron.* **2000**, *47*, 984–993. [CrossRef]
5. Cunha Palacios, R.; Da Silva, I.; Goedtel, A.; Godoy, W. A comprehensive evaluation of intelligent classifiers for fault identification in three-phase induction motors. *Electr. Power Syst. Res.* **2015**, *127*, 249–258. [CrossRef]
6. Prieto, M.D.; Cirrincione, G.; Espinosa, A.G.; Ortega, J.A.; Henao, H. Bearing fault detection by a novel condition-monitoring scheme based on statistical-time features and neural networks. *IEEE Trans. Ind. Electron.* **2013**, *60*, 3398–3407. [CrossRef]
7. Ghate, V.N.; Dudul, S.V. Cascade neural-network-based fault classifier for three-phase induction motor. *IEEE Trans. Ind. Electron.* **2011**, *58*, 1555–1563. [CrossRef]
8. Gonzalez-Jimenez, D.; del Olmo, J.; Poza, J.; Garramiola, F.; Madina, P. Data-Driven Fault Diagnosis for Electric Drives: A Review. *Sensors* **2021**, *21*, 4024. [CrossRef]
9. Gonzalez-Jimenez, D.; Del-Olmo, J.; Poza, J.; Garramiola, F.; Madina, P. Data-Driven Low-Frequency Oscillation Event Detection Strategy for Railway Electrification Networks. *Sensors* **2023**, *23*, 254. [CrossRef]
10. Gonzalez-Jimenez, D.; del Olmo, J.; Poza, J.; Garramiola, F.; Sarasola, I. Machine Learning-Based Fault Detection and Diagnosis of Faulty Power Connections of Induction Machines. *Energies* **2021**, *14*, 4886. [CrossRef]
11. Martin-Diaz, I.; Morinigo-Sotelo, D.; Duque-Perez, O.; Romero-Troncoso, R.J. Early fault detection in induction motors using AdaBoost with imbalanced small data and optimized sampling. *IEEE Trans. Ind. Appl.* **2016**, *53*, 3066–3075. [CrossRef]
12. Niu, G.; Dong, X.; Chen, Y. Motor Fault Diagnostics Based on Current Signatures: A Review. *IEEE Trans. Instrum. Meas.* **2023**, *72*, 1–19. [CrossRef]
13. Rivera, W.A. Noise Reduction A Priori Synthetic Over-Sampling for class imbalanced data sets. *Inform. Sci.* **2017**, *408*, 146–161. [CrossRef]
14. Zhang, T.; Chen, J.; Li, F.; Zhang, K.; Lv, H.; He, S.; Xu, E. Intelligent fault diagnosis of machines with small & imbalanced data: A state-of-the-art review and possible extensions. *ISA Trans.* **2022**, *119*, 152–171. [CrossRef] [PubMed]
15. Kliman, G.; Stein, J. Methods of motor current signature analysis. *Electr. Mach. Power Syst.* **1992**, *20*, 463–474. [CrossRef]
16. Martin-Diaz, I.; Morinigo-Sotelo, D.; Duque-Perez, O.; Delgado-Arredondo, P.; Camarena-Martinez, D.; Romero-Troncoso, R. Analysis of various inverters feeding induction motors with incipient rotor fault using high-resolution spectral analysis. *Electr. Power Syst. Res.* **2017**, *152*, 18–26. [CrossRef]
17. Bruzzese, C. Analysis and Application of Particular Current Signatures (Symptoms) for Cage Monitoring in Nonsinusoidally Fed Motors With High Rejection to Drive Load, Inertia, and Frequency Variations. *IEEE Trans. Ind. Electron.* **2008**, *55*, 4137–4155. [CrossRef]
18. Robnik-Sikonja, M.; Kononenko, I. Theoretical and empirical analysis of ReliefF and RReliefF. *Mach. Learn.* **2003**, *53*, 23–69. [CrossRef]
19. Kononenko, I.; Robnik-Sikonja, M.; Pompe, U. *ReliefF for Estimation and Discretization of Attributes in Classification, Regression, and ILP Problems*; IOS Press: Amsterdam, The Netherlands, 1996; Volume 35.
20. Murcia-Sepúlveda, N.; Cruz-Duarte, J.M.; Martin-Diaz, I.; Garcia-Perez, A.; Rosales-García, J.J.; Avina-Cervantes, J.G.; Correa-Cely, C.R. Fractional Calculus-Based Processing for Feature Extraction in Harmonic-Polluted Fault Monitoring Systems. *Energies* **2019**, *12*, 3736. [CrossRef]
21. Chawla, N.V.; Bowyer, K.W.; Hall, L.O.; Kegelmeyer, W.P. SMOTE: Synthetic minority over-sampling technique. *J. Artif. Intell. Res.* **2002**, *16*, 321–357. [CrossRef]
22. Bunkhumpornpat, C.; Sinapiromsaran, K.; Lursinsap, C. Safe-Level-SMOTE: Safe-Level-Synthetic Minority Over-Sampling TEchnique for Handling the Class Imbalanced Problem. In Proceedings of the Advances in Knowledge Discovery and Data Mining, Bangkok, Thailand, 27–30 April 2009; Theeramunkong, T., Kijsirikul, B., Cercone, N., Ho, T.B., Eds.; Springer: Berlin/Heidelberg, Germany, 2009; pp. 475–482.
23. Siriseriwan, W.; Sinapiromsaran, K.; Tipawanna, M.; Bunkhumpornpat, C.; Keatruangkamala, K. The effective redistribution for imbalance dataset: Relocating Safe-level SMOTE with minority outcast handling. *Chiang Mai J. Sci.* **2016**, *43*, 1288–1300.
24. Barua, S.; Islam, M.M.; Yao, X.; Murase, K. MWMOTE—Majority weighted minority oversampling technique for imbalanced data set learning. *IEEE Trans. Knowl. Data Eng.* **2014**, *26*, 405–425. [CrossRef]

applied sciences

MDPI

Article

Quaternion Signal Analysis for Detection of Broken Rotor Fault Degrees in Induction Motors

Jose Luis Contreras-Hernandez *, Dora Luz Almanza-Ojeda, Rogelio Castro-Sanchez and Mario Alberto Ibarra-Manzano

Telematics and Digital Signal Processing Research Groups (CAs), Electronics Engineering Department, DICIS, University of Guanajuato, Salamanca 36885, Mexico; dora.almanza@ugto.mx (D.L.A.-O.); castro@ugto.mx (R.C.-S.); ibarram@ugto.mx (M.A.I.-M.)

* Correspondence: jose.contreras@ugto.mx; Tel.: +52-464-647-9940

Abstract: Fault detection in induction motors is essential for maintaining the reliability of industrial operations. In practical applications, induction motors experience gradual wear on critical components, such as rotor bars, affecting their performance. This paper introduces a new methodology for modeling predictive wear functions related to rotor faults in induction motors, providing accurate forecasts and optimal performance through Quaternion Signal Analysis in the time domain. Our approach accurately detects wear in broken rotor bars and anticipates their degradation over time. The methodology involves coupling four vibration signals from the motor, representing them as quaternion coefficients, and calculating their rotational attributes to derive a statistical mean. We employ polynomial and Fourier regression techniques to construct a predictive wear function. We assess its accuracy through root mean square error (RMSE) analysis, which improves with increased sample size and regression complexity. Our findings indicate that polynomial regression, particularly at the second degree, achieves superior RMSE results compared to Fourier regression, even within limited sample windows. This approach offers a robust framework for early fault detection and wear prediction in induction motors, supporting enhanced maintenance strategies.

Keywords: bar quantification; induction motor fault; QSA; quaternions; regression

1. Introduction

Induction motors are important electric machines in industry processes and must be kept in optimal operation. Signal analysis techniques are essential for early fault detection in rotating machinery, helping to prevent unexpected downtime, economic losses, and potential major failures [1,2]. Common faults in induction motors include bearing defects [3], stator winding failures, air-gap eccentricity, unbalanced bars, and broken rotor bar (BRB) faults [4]. BRB faults are particularly significant in the literature due to their high risk of causing additional damage [5]. Numerous methodologies have been developed to detect or characterize BRB faults, focusing on spectral analysis, machine learning, and statistical approaches [6].

Spectral analysis methods involve extensive data processing and are primarily oriented toward fault detection and classification. Fourier and wavelet transforms are methodologies in spectral analysis valued for their ability to associate specific signal frequencies with BRB faults. In addition, studies in the literature often integrate statistical analysis into spectral techniques to enhance fault detection accuracy, as in [7], where a methodology combining the digital Taylor–Fourier transform (DTFT) with variance analysis effectively identified rotor bar wear across states ranging from healthy to BRB. In wavelet algorithms, the authors of [8] successfully classified multiple BRB faults by extracting statistical features directly from raw signals using advanced techniques such as Matching Pursuit (MP) and Discrete Wavelet Transform (DWT). Moreover, several studies leverage machine learning methods like k-Nearest Neighbor (KNN) and support vector machines (SVMs) for BRB classification. For instance, the approach in [9] integrates Discrete Wavelet Transform (DWT) with these machine learning algorithms to achieve robust multi-class BRB classifications.

Recently, there has been a surge of interest in utilizing machine learning and deep learning techniques for fault detection in induction motors. This trend reflects these approaches' proven accuracy and efficiency in identifying and diagnosing faults, marking significant progress in predictive maintenance [6]. Specifically, for BRB fault detection using time-domain signals, studies such as [10] have explored various machine learning algorithms—including fuzzy ARTMAP networks, SVM, KNN, and Multilayer Perceptron (MLP) networks—without requiring prior domain knowledge. This adaptability underscores the capability of machine learning methods to capture complex patterns related to BRB faults. Additionally, ref. [11] presents an innovative algorithm that classifies BRB conditions by applying statistical metrics derived from motor vibration signals, such as kurtosis, skewness, and variance. These metrics are then processed through classifiers like Naive Bayes, MLP, SVM, and Optimum-Path Forest, further demonstrating the efficacy of machine learning in accurate BRB fault classification. In [12], homogeneity and kurtosis metrics were employed to classify motor conditions, from healthy states to the presence of two BRBs, using a backpropagation artificial neural network. This approach demonstrates the power of neural networks in identifying complex data patterns and accurately diagnosing motor faults. Similarly, ref. [13] leveraged stator current signals to obtain comprehensive two-dimensional insights into BRB severity across time and frequency domains, calculating entropy at various load levels to assess fault impact. Conversely, in [14], the authors proposed a degradation model for BRB faults by applying principal component analysis (PCA). Using raw torque, stator current, and speed signals, they estimated the remaining useful life of the BRBs, highlighting PCA's utility in modeling the degradation process.

The works presented above classify BRB using methods that involve a certain level of complexity and require significant computational resources. Utilizing quaternions could simplify this process, as they provide a unique mathematical framework because they are an extension of complex numbers, consisting of a real part and a three-dimensional vector of imaginary components [15]. This distinctive configuration has spurred interest across various fields, revealing quaternions' versatility and potential in advanced applications. For example, in image processing, quaternions have been shown to facilitate image enhancement and manipulation in ways that surpass conventional methods [16], offering new data representation and transformation possibilities. In robotic motion tracking, as detailed in [17], quaternions provide a sophisticated framework for accurately monitoring and controlling robot movement, delivering the precision essential for executing complex tasks. In control systems, quaternions are equally valuable, enabling the design of robust control mechanisms with high accuracy and reliability [18]. The versatility of quaternions in various domains highlights their impact on advancing technological innovation. In spectral analysis, quaternions have been effectively combined with the Fourier Transform, as demonstrated in [19], to perform comprehensive spectral analysis of three-phase currents in induction motors, achieving a bearing fault classification accuracy of approximately 94.2%. Likewise, in [20], researchers developed the Quaternion Empirical Wavelet Transform (QEWT) based on the Quaternion Fourier Transform to diagnose outer ring, inner ring, and rolling bearing faults, illustrating quaternions' capability in high-precision fault detection.

We have demonstrated the versatility of quaternions in previous works as [21,22], where we employed Quaternion Signal Analysis (QSA) to detect multiple faults and their combinations in induction motors. This methodology involves rotating quaternion signals and performing statistical analyses, calculating mean, variance, kurtosis, skewness, standard deviation, root mean square, and shape factor. Subsequently, machine learning techniques are applied to classify various fault conditions and combinations. This approach highlights the effectiveness of QSA in capturing and classifying complex fault patterns. Similarly, the QSA was used in [23] to diagnose short-circuit damage in transformer windings demonstrating that this method is not exclusive to classifying faults in induction motors. QSA has shown high accuracy in fault classification through simple mathematical operations in the time domain, even using four statistical values from three-phase currents as in [24], where four levels of inter-turn short circuit faults per phase and healthy states were classified.

On the other hand, regression methods utilize statistical processes to model relationships between variables, providing simplicity and low computational costs, especially with polynomial regressions. Furthermore, techniques like Fourier regressions can improve the effectiveness of stepped models. Studies have introduced methodologies for modeling predictive wear functions capable of quantifying BRB faults providing accurate forecasts and optimal performance as in [25]. In this, the authors present regression-based methods to evaluate BRB degradation using curve-fitting techniques such as linear, polynomial, and least squares regression with a power law. The method utilizes fault sideband parameters from the current spectrum, machine rating, and load torques of 50%, 75%, and 100% to gauge BRB severity levels from one to six, providing a structured model for fault severity assessment.

As BRB faults in induction motors develop gradually, establishing an effective model to predict future behavior with straightforward calculations becomes crucial. This approach not only helps prevent severe damage but also ensures timely maintenance, safeguarding your equipment and investment. Our method combines QSA with statistical feature analysis and polynomial and Fourier regressions to deliver reliable, practical predictive wear functions for motor maintenance. The proposed method utilizes four vibration signals with 29,999 samples measured from an induction motor operating at steady state and full load (100%): tangential mechanical vibration speed in the housing (V_h), axial mechanical vibration speed on the driven side (V_a), radial mechanical vibration speed on the non-driven side (V_s), and radial mechanical vibration speed on the driven side (V_d). The use of four signals to adapt them to the quaternions allows us to analyze different elements of the motor in various axes. This approach allows us to cover more variables, which may reflect the characteristics of the analyzed failures. However, a disadvantage of using quaternions is that they require four signals, which means that it is necessary to use at least two or more sensors. Not all databases have access to this number of sensors, which can limit the analysis. The method is evaluated across motor conditions ranging from healthy states to one, two, and three BRB faults, demonstrating its effectiveness in fault quantification and progression tracking.

This paper is structured into four sections: Section 1 introduces BRB fault detection. Section 2 describes the background of QSA and statistical methodology. Section 3 presents and discusses the results, comparing the results of polynomial and Fourier regressions. Finally, Section 4 presents conclusions and future work.

2. Methodology

As BRB faults in induction motors develop gradually, establishing an effective model to predict future behavior with straightforward calculations becomes crucial. This approach not only helps prevent severe damage but also ensures timely maintenance, safeguarding your equipment and investment. By modeling quaternion-based statistical features, the proposed method characterizes BRB through predictive wear functions from a healthy motor state to three BRBs. Unlike traditional approaches that provide only discrete fault levels, this model enables a continuous fault severity assessment. The algorithm comprises two main stages: (1) the Quaternion Signal Analysis (QSA) method and (2) the fitting and evaluation procedures. Figure 1 illustrates the method's flowchart for detecting BRB severity, utilizing four vibration signals as input. Each stage of the method, as shown in Figure 1, is detailed in the following subsections.

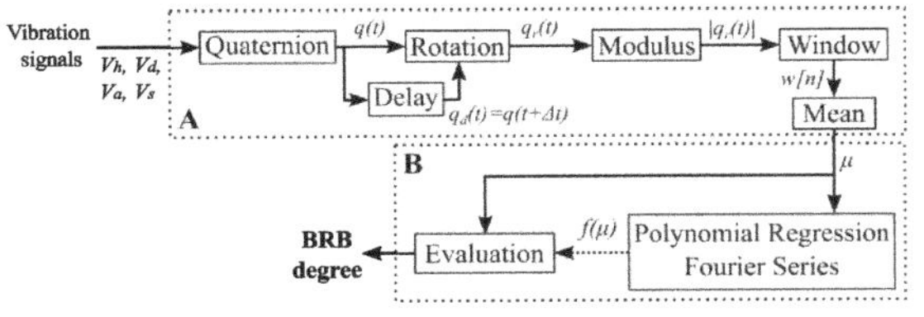

Figure 1. Proposed workflow diagram. (**A**) QSA method. (**B**) Fitting and evaluation procedures.

2.1. QSA Method

Quaternions model the three-dimensional movements of a point in space. In a complementary way, the QSA method allows modeling the signals that generate the movements, facilitating a multidimensional analysis adapted to spatial and temporal problems. This approach detects spatial signatures using quaternion arithmetic and temporal signatures by estimating statistical displacement models. Combining both elements enables the method to achieve greater precision and robustness in fault detection than the classic techniques.

This section outlines the methodology for defining quaternion signals and calculating their rotations. Subsequently, our approach computes modulus and statistical measures based on these rotations. The quaternion q is constructed as shown in (1), using the coefficients q_0, q_1, q_2, and q_3 [26].

$$q = q_0 + q_1 i + q_2 j + q_3 k \tag{1}$$

The quaternion signal $q(t)$ shown in section A of Figure 1 is constructed from the vibration signals $V_h(t)$, $V_d(t)$, $V_a(t)$, and $V_s(t)$, as formulated in (2). Signals $V_h(t)$, $V_d(t)$, and $V_a(t)$ were chosen to assess vibrations along the x-, y-, and z-axes of the induction motor, while $V_s(t)$ provides insight into rear motor vibrations. By selecting signals from different motor sections, our methodology enables a more comprehensive analysis, as Quaternion Signal Analysis (QSA) integrates the distinct characteristics of each signal. The vibration signals are represented discreetly with a total of 29,999 samples.

$$q(t) = V_h(t) + V_d(t)i + V_a(t)j + V_s(t)k \tag{2}$$

Then, the shifted quaternion signal $q_d(t)$ is obtained by the displacement of Δt samples in $q(t)$. Equation (3) presents the equality [22].

$$q_d(t) = q(t + \Delta t) \tag{3}$$

The coefficients V_{hd}, V_{dd}, V_{ad}, and V_{sd} form the quaternion $q_d(t)$ shown in Figure 1 with 29,998 samples, as defined in (4). These coefficients are the vibration coefficients V_h, V_d, V_a, and V_s in $t + \Delta t$.

$$q_d(t) = V_{hd}(t) + V_{dd}(t)i + V_{ad}(t)j + V_{sd}(t)k \tag{4}$$

A quaternion model can represent the transition from a current quaternion sample, q, to a delayed quaternion sample, q_d. This means that we can identify the quaternion model that enables q to achieve the values of q_d through quaternion rotation operations, as illustrated in Figure 2.

Rotation quaternions represent three-dimensional rotations and they are related to the axis and angle. Based on Euler's rotation theorem, any rotation can be defined using unit vectors $\hat{x}, \hat{y}$, and $\hat{z}$ and angle θ. The axis-angle components can be converted into components of a rotation quaternion q_r by Equations (5)–(8).

$$q_{r0} = \cos\left(\frac{\theta}{2}\right) \tag{5}$$

$$q_{r1} = \hat{x}\sin\left(\frac{\theta}{2}\right) \tag{6}$$

$$q_{r2} = \hat{y}\sin\left(\frac{\theta}{2}\right) \tag{7}$$

$$q_{r3} = \hat{z}\sin\left(\frac{\theta}{2}\right) \tag{8}$$

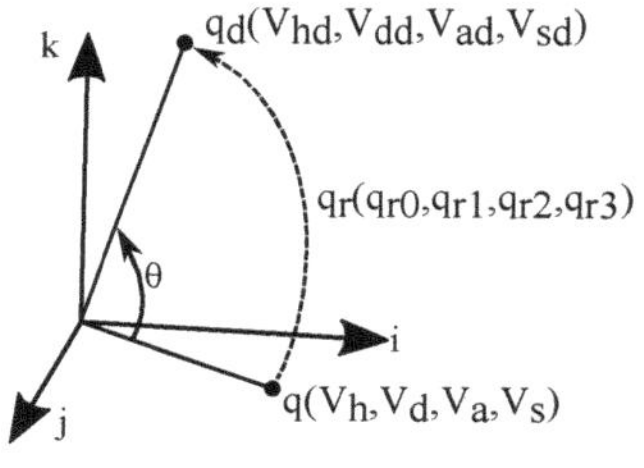

Figure 2. Rotation of $q(t)$ to obtain $q_d(t)$.

The above-mentioned process generates a rotated quaternion for each sample and its corresponding delayed value. This operation is applied to all samples of the signals to characterize the rotational behavior of the signal over time.

Taking into account the above described, q and q_d are algebraically related by (9) [21], where q_r^{-1} is the inverse of rotation quaternion.

$$q_d = q_r \otimes q \otimes q_r^{-1} \tag{9}$$

The operator $\otimes$ refers to the Hamilton product which is described as in (10).

$$\begin{aligned} q_r \otimes q = &(q_{r0}V_h - q_{r1}V_d - q_{r2}V_a - q_{r3}V_s) \\ &+(q_{r0}V_d + q_1V_h + q_{r2}V_s - q_{r3}V_a)i \\ &+(q_{r0}V_a - q_1V_s + q_{r2}V_h - q_{r3}V_d)j \\ &+(q_{r0}V_s + q_1V_a - q_{r2}V_d + q_{r3}V_h)k. \end{aligned} \tag{10}$$

By the properties of quaternions, the relation of q_r, q_d, and q can be written as (11) [23].

$$q_d(t) = q_r(t) \cdot q(t) \tag{11}$$

This means that if we know the values of q and q_d, we can obtain q_r through the product of general rotation matrix and q presented in (12) [21] using the coefficients values shown in (2) and (4). This value contains the two-sample quaternion model, which helps us obtain statistical values of multiple models to quantify BRBs later.

$$q_r = \begin{bmatrix} 1-2V_{ad}^2-2V_{sd}^2 & 2(V_{dd}V_{ad}+V_{hd}V_{sd}) & 2(V_{dd}V_{sd}-V_{hd}V_{ad}) \\ 2(V_{dd}V_{ad}-V_{hd}V_{sd}) & 1-2V_{dd}^2-2V_{sd}^2 & 2(V_{ad}V_{sd}+V_{hd}V_{dd}) \\ 2(V_{dd}V_{sd}+V_{hd}V_{ad}) & 2(V_{ad}V_{sd}-V_{hd}V_{dd}) & 1-2V_{dd}^2-2V_{ad}^2 \end{bmatrix} \begin{bmatrix} V_d \\ V_a \\ V_s \end{bmatrix} \tag{12}$$

Next, we compute and organize the rotational models for the signal based on the pure quaternion $q_r(t)$ as defined in (13) [23]. Here, q_{r0} is set to zero, reflecting that the rotation is confined to the three-dimensional space of the imaginary components.

$$q_r(t) = q_{r1}(t)i + q_{r2}(t)j + q_{r3}(t)k \tag{13}$$

We have vibration signals containing 29,999 samples, and the quaternion q_r is calculated for each sample with the next one, resulting in 29,998 pure quaternions. After this, we calculate the modulus of each quaternion using Equation (14), which generates a constant value for each one.

$$|q_r| = \sqrt{q_{r1}^2 + q_{r2}^2 + q_{r3}^2} \tag{14}$$

Using these values of consecutive models from $q_r(t)$, we can assemble a vector **w**[n] by selecting N samples from the signals as in (15) [15], where $t + N\Delta t$ refers to the sample of the rotation obtained.

$$\mathbf{w}[n] = [|q_r(t)|, |q_r(t+\Delta t)|, |q_r(t+2\Delta t)| \ldots |q_r(t+N\Delta t)|] \tag{15}$$

Statistical analysis of **w**[n] is essential for quantitatively characterizing the models and defining diagnostic value ranges for each BRB condition. Following the workflow in Figure 1, the mean (μ) was selected as the statistical measure with the best-fitting performance. The mean of **w**[n] reflects the evolution of signals over time and is calculated according to (16).

$$\mu = \frac{1}{N}\sum_{n=1}^{N} \mathbf{w}[n] \tag{16}$$

Each test file for each BRB condition contains 29,999 samples, providing 29,998 moduli of $q_r(t)$. The window **w**[n] utilizes N values from these 29,998 moduli to compute μ. By shifting the window **w**[n] across the test file, multiple values of μ can be calculated. This approach enables calculating and evaluating polynomial and Fourier models that describe BRB behavior, as shown in Section 2 of Figure 1.

2.2. Fitting and Evaluation

The values of μ are derived from test files corresponding to a healthy motor, one BRB, two BRBs, and three BRBs to model the function $f(\mu)$ using polynomial regression and Fourier series. This function quantitatively assesses the BRB damage level in the induction motor. Polynomial regressions and the Fourier series were calculated in the second to ninth order, which is the range supported by Matlab's function capabilities. Results from the second to sixth order are presented, as higher-order fits produce similar algebraic expressions and accuracy, making them unnecessary for further consideration.

The remaining files are evaluated in $f(\mu)$ by calculating μ to determine the extent of BRB damage. Figure 3 displays the fitted functions generated through polynomial and Fourier series methods, comparing these fitted functions with the actual data samples. The boxplot graphs present position results from one file for each fault state, analyzed using a fitted function. They distinctly showcase the minimum and maximum values, as well as the mean, providing clear insights into the data. As anticipated, the estimated curves closely align with the observed data, achieving the accuracy required for reliable damage assessment.

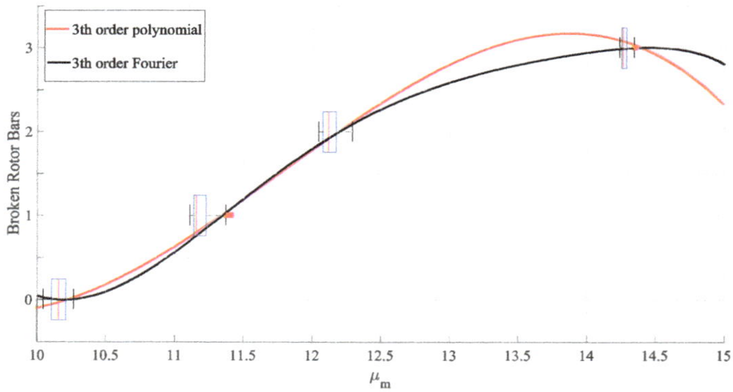

Figure 3. Polynomial and Fourier regressions.

2.3. Experimental Setup

Our work is tested with a dataset of the workbench presented in [27] to validate its effectiveness. The experimental setup, illustrated in Figure 4, includes a three-phase induction motor and a direct current machine acting as a generator to simulate load torque. They are connected by a shaft with a rotary torque wrench. Additionally, various sensors are connected at different locations within the setup.

Figure 4. Experimental setup of the induction motor with one, two, and three BRBs [27].

The workbench consists of vibration, current, and voltage measurements from which we selected four vibration measurements. These are taken from various induction motor parts to analyze its mechanical behavior. This analysis enables the adaptation of four signals into a quaternion format. In past studies [22,23], a combination of current and mechanical vibration signals was utilized due to a scarcity of signals. However, this work aims to use signals of the same type for a more consistent analysis.

These four signals are acquired by four vibrocontrol uniaxial accelerometers model PU2001, located on the housing, on the driven side, and on the non-driven side to obtain tangential, radial, and axial vibration. The sensors present a sensitivity of 10 mV/mm/s, with a frequency range from 5 to 2000 Hz. The data acquisition system utilized in the experimental setup is a National Instruments (Austin, TX, USA) data acquisition card, model NI DAQmx PCIe-6259. This card features 32 analog input channels, each with a resolution of 16 bits and an acquisition rate of 1.25 MHz. Additionally, it includes four channels of 16-bit analog output, operating at a speed of 2.8 MHz, and 48 digital I/O channels capable of functioning at a clock frequency of up to 10 MHz.

The signals V_h, V_d, V_a, and V_s are acquired from a three phase induction motor of the W22 standard line from manufacturer WEG, with 1 hp, 220 V/380 V, 3.02 A/1.75 A, four poles, 34 rotor bars in the squirrel cage rotor, a rated speed of 1715 rpm, and a nominal torque of 4.1 Nm. The database includes two additional vibration signals. These signals were tested using various combinations, but the resulting statistics were insufficient for proper characterization. Therefore, the signals that yielded the best results were those recorded from the engine chassis. Different directional measurements were selected to enhance the results. The placement of the sensors and the orientation of the signals are shown in Figure 5.

The induction motor is driven by a control panel, for which the type of drive can be selected as star or triangle, and the type of supply as direct mains voltage or via a three-phase inverter. A shaft containing a rotary torque wrench Transtec model MT-103 (Arlington, VA, USA) connects the motor with a direct-current machine, simulating the load in the motor of 100% of full load in all tests.

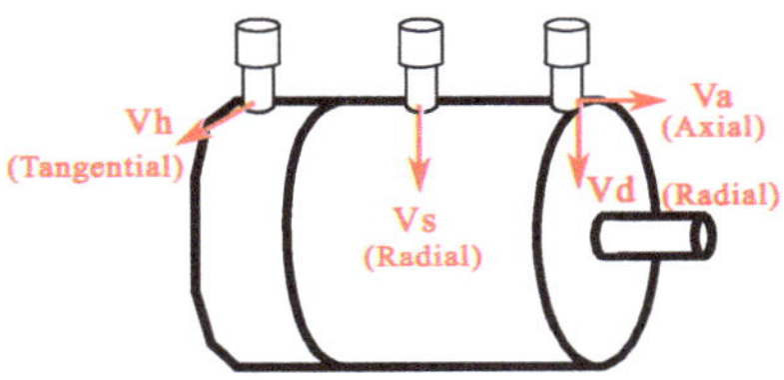

Figure 5. The position of the sensors and the orientation of the signals.

The workbench shows between one and four adjacent rotor bar failures caused by 6 mm diameter drilling in the squirrel cage rotor to ensure that the rotor bar was completely drilled alongside the healthy state by means of a bench drill mounted with the tip centered at half the longitudinal length of the rotor, as is shown in Figure 6.

Figure 6. Provoked BRB drilling [27].

We selected the sets of healthy and from one to three BRBs because they presented a gradual increase in the results, while the four BRB states provoked regressions with instability.

The dataset presents ten repetitions of each state acquired during 18 s with a sample rate of 8.326 kHz. We selected thirty thousand samples in a steady state of the four vibration signals to test.

3. Results

In the experimental tests, a random test file from each rotor state was selected to create a sub-workbench for calculating $f(\mu_w)$. Our method then adapts one sample from each of the four vibration signals to construct the current quaternion $q(t)$ and determines the rotation $q_r(t)$ based on the previous quaternion sample $q_d(t)$, as shown in (12). By calculating the modulus of $q_r(t)$ for each delayed quaternion sample, we obtain a total of 29,999 samples.

Multiple shifted windows $w[n]$—utilizing 4000, 6000, 8000, 10,000, 12,000, 14,000, and 20,000 samples—are employed to calculate the mean (μ) for each window, forming a function $f(\mu)$ that characterizes fault behavior. This model function is then evaluated on the remaining test files (excluding those used in model formation). The dispersion of data points around the fitted function can be measured by root mean square error (RMSE), allowing us to evaluate how closely the data are concentrated around the fitted function [28]. We consider this measurement to be a metric to record the joint accuracy of our method across different rotor states. The process is repeated forty times using the original dataset with polynomial regression and Fourier series. The mean RMSE values from these forty trials are analyzed across polynomial orders from second to sixth.

Figure 7 shows the average of the RMSE values obtained from the forty tests, with the different window samples tested along with multiple orders of the predictive wear functions. Analyzing the graph mentioned above, it is observed that as window size increases, the mean RMSE values remain relatively stable for each polynomial order, indicating that sample size has minimal impact on the efficiency of our method at a given order.

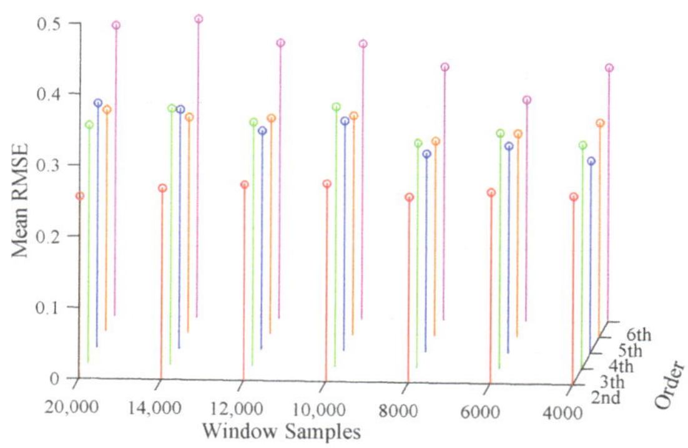

Figure 7. Mean RMSE behavior with polynomial modeling.

Table 1 presents the average of the RMSE for each window size and polynomial order, along with the minimum and maximum RMSE values across repetitions. The lowest average of the RMSE observed is 0.256, achieved with 20,000 window samples in the second order, though a comparable result is obtained with 8000 samples, yielding a mean RMSE of 0.259 in the same order. In the sixth order, mean RMSE increases in all cases due to a broader spread in minimum and maximum values. Polynomial regression demonstrates consistent behavior, largely unaffected by sample size or polynomial order.

Similarly, we calculated the mean of forty RMSE results to examine the behavior of the Fourier series as a regression method, ranging from the second to the sixth order. Figure 8 illustrates these results, showing that the method's efficiency is impacted by sample size, particularly from the fourth to the sixth order. While the Fourier series modeling displays consistent behavior at lower orders, its performance varies as the order increases.

Table 1. Mean, minimum, and maximum RMSE in each order using polynomial modeling.

Window	Order				
	2	3	4	5	6
Samples	*Mean (min, max)*	*Mean (min, max)*	*Mean (min, max)*	*Mean (min, max)*	*Mean (min, max)*
4000	0.263 (0.22, 0.39)	0.315 (0.22, 0.47)	0.270 (0.17, 0.47)	0.301 (0.20, 0.56)	0.357 (0.17, 1.29)
6000	0.267 (0.21, 0.34)	0.330 (0.21, 0.66)	0.290 (0.19, 0.56)	0.285 (0.18, 0.66)	0.310 (0.15, 0.66)
8000	0.259 (0.22, 0.32)	0.315 (0.20, 0.80)	0.278 (0.17, 0.74)	0.273 (0.17, 0.68)	0.355 (0.17, 0.88)
10,000	0.277 (0.22, 0.47)	0.364 (0.21, 0.78)	0.322 (0.18, 0.79)	0.308 (0.18, 0.65)	0.387 (0.15, 1.41)
12,000	0.274 (0.22, 0.47)	0.341 (0.21, 0.83)	0.307 (0.18, 0.85)	0.303 (0.18, 0.68)	0.387 (0.18, 1.08)
14,000	0.268 (0.21, 0.50)	0.359 (0.20, 0.71)	0.335 (0.19, 0.76)	0.302 (0.18, 0.59)	0.419 (0.14, 0.93)
20,000	0.256 (0.21, 0.43)	0.334 (0.20, 0.81)	0.343 (0.17, 0.86)	0.312 (0.18, 0.79)	0.409 (0.17, 0.96)

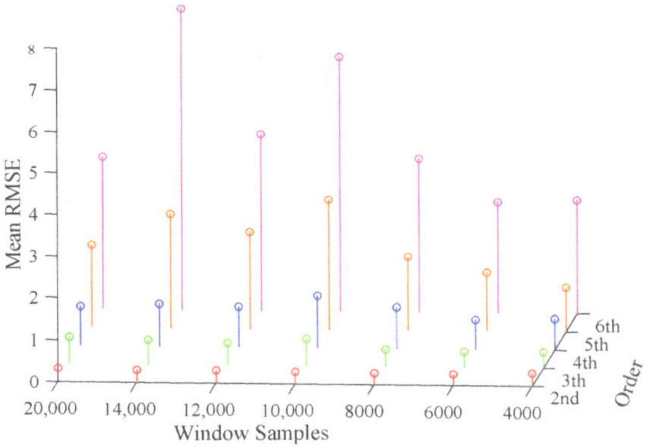

Figure 8. Mean RMSE behavior with Fourier series modeling.

Table 2 presents the mean RMSE for each window size and modeling order, along with the minimum and maximum RMSE values across forty repetitions. The lowest mean RMSE observed in the table is 0.268, achieved with 6000 window samples in the second order. It has been observed that the difference between the minimum and maximum values of orders is greater than the fourth order increases when using Fourier series modeling. Additionally, the average values of the RMSE for these orders change differently compared to those using polynomial modeling. This indicates that the functions derived from the Fourier series have less adaptation to the verification values.

Comparing the average of the RMSE values in Tables 1 and 2 reveals that increasing the number of samples minimally impacts the mean RMSE across each order. However, polynomial modeling consistently yields a lower mean RMSE, indicating that the function $f(\mu)$ achieves a closer fit to the original values and exhibits stable results as the order and window sample parameters increase. The polynomial method yields more stable values. Therefore, even when different samples are used to create the function that describes the behavior of the bars, the resulting functions adapt more effectively compared to those generated by the Fourier series. While a lower order in Fourier series with fewer samples may produce similar values in variation, the polynomial method generally demonstrates better adaptability.

Table 2. Mean, minimum, and maximum RMSE in each order from two to six using Fourier series modeling.

Window	Order				
	2	3	4	5	6
Samples	*Mean (min, max)*	*Mean (min, max)*	*Mean (min, max)*	*Mean (min, max)*	*Mean (min, max)*
4000	0.303 (0.16, 0.66)	0.391 (0.16, 1.28)	0.739 (0.18, 3.06)	1.051 (0.17, 4.48)	2.727 (0.07, 19.73)
6000	0.268 (0.19, 0.53)	0.379 (0.17, 1.34)	0.702 (0.16, 3.11)	1.396 (0.18, 6.00)	2.663 (0.12, 36.69)
8000	0.273 (0.16, 0.69)	0.414 (0.15, 2.45)	0.978 (0.23, 7.26)	1.765 (0.14, 12.63)	3.684 (0.12, 68.57)
10,000	0.293 (0.17, 0.70)	0.643 (0.16, 3.60)	1.227 (0.18, 4.69)	3.107 (0.21, 26.46)	6.125 (0.11, 81.33)
12,000	0.308 (0.18, 0.74)	0.524 (0.17, 4.67)	0.951 (0.18, 8.13)	2.321 (0.15, 26.73)	4.252 (0.10, 80.28)
14,000	0.301 (0.17, 0.64)	0.579 (0.14, 3.13)	0.998 (0.16, 4.89)	2.728 (0.17, 20.43)	7.288 (0.09, 162.18)
20,000	0.325 (0.18, 0.70)	0.634 (0.15, 3.37)	0.921 (0.15, 5.55)	1.972 (0.15, 21.71)	3.653 (0.09, 29.71)

A lower order results in fewer operations during modeling and the QSA method demonstrates that a higher order is not necessary to achieve good adaptation results with the functions. Additionally, it has been shown that, given a sampling rate of 8.326 kHz, a duration of less than one second is sufficient in a real-world scenario to obtain a diagnosis of the wear degree in rotor bars.

4. Conclusions

Few studies have developed methodologies to quantify gradual levels of faults, such as BRB faults, in induction motors. The method presented here accurately quantifies BRB failure severity using Quaternion Signal Analysis (QSA), polynomial regression, and Fourier series modeling. This approach provides incremental records of motor behavior, enabling precise estimation and timely prevention of severe faults in the time domain.

The mean RMSE results indicate consistent performance across regression orders, window sample sizes, and modeling techniques. Polynomial modeling, in particular, demonstrates that changes in function order and window size have minimal impact on method effectiveness. This robustness allows for low-order and low-sample implementations, which are computationally efficient as they rely solely on fundamental addition and multiplication operations, unlike Fourier series, which require sine and cosine harmonics.

Future work will test this methodology by expanding the number of BRB conditions to detect partial BRBs and validating it with breaks in the junction between the bar and the side rings. In addition, we are interested in applying our development to assess wear levels in various rotating machine components, such as bearings and gearboxes, to validate their performance across diverse failure sources.

The QSA method was designed to adapt to different load conditions in induction motors; however, this adaptability needs to be validated. Consequently, we will conduct an analysis to examine how the ability to detect fault signatures depends on various load levels in the induction motor.

Author Contributions: Conceptualization, J.L.C.-H. and M.A.I.-M.; methodology, J.L.C.-H. and D.L.A.-O.; software, J.L.C.-H.; validation, D.L.A.-O. and R.C.-S.; formal analysis, J.L.C.-H. and M.A.I.-M.; investigation, J.L.C.-H.; resources, M.A.I.-M.; data curation, D.L.A.-O.; writing—original draft preparation, J.L.C.-H.; writing—review and editing, D.L.A.-O. and M.A.I.-M.; visualization, J.L.C.-H. and R.C.-S.; supervision, R.C.-S.; funding acquisition, M.A.I.-M. All authors have read and agreed to the published version of the manuscript.

Funding: The APC was funded by the Electronic Engineering Department of Universidad de Guanajuato under POA 2024.

Institutional Review Board Statement: Not applicable.

Informed Consent Statement: Not applicable.

Data Availability Statement: The original contributions presented in the study are included in the article, further inquiries can be directed to the corresponding author.

Acknowledgments: The authors thank the University of Guanajuato for the facilities and support to develop this project. Special thanks to the DAIP CIIC 048/2024 project: "Implementation of an Automatic Braking System for Electric Vehicles for People with Disabilities" (valid from 1 January to 31 December 2024) for its financial and technical support.

Conflicts of Interest: The authors declare no conflicts of interest. The funders had no role in this study's design, in the collection, analyses, or interpretation of data, in the writing of the manuscript, or in the decision to publish the results.

Abbreviations

The following abbreviations are used in this manuscript:

BRB	Broken Rotor Bar
FFT	Fast Fourier Transform
MUSIC	Multiple Signal Classification
DTFT	Digital Taylor–Fourier Transform
MP	Matching Pursuit
KNN	k-Nearest Neighbor
SVM	Support Vector Machine
DWT	Discrete Wavelet Transform
QEWT	Quaternion Empirical Wavelet Transform
ART	Adaptive Resonance Theory
MAP	Mapping
MLP	Multi-Layer Perceptron
QSA	Quaternion Signal Analysis
V_h	Tangential mechanical vibration speed in the housing
V_a	Axial mechanical vibration speed on the driven side
V_s	Radial mechanical vibration speed on the non-driven side
V_d	Radial mechanical vibration speed on the driven side
$q(t)$	Quaternion signal in time domain
$q_d(t)$	Delayed quaternion in time domain
$q_r(t)$	Rotated quaternion in time domain
$\mathbf{w}[n]$	Windows vector
μ	Mean
RMSE	Root Mean Square Error

References

1. Aguayo-Tapia, S.; Avalos-Almazan, G.; Rangel-Magdaleno, J.d.J.; Ramirez-Cortes, M.J. Physical variable measurement techniques for fault detection in electric motors. *Energies* **2023**, *16*, 4780. [CrossRef]
2. Chen, J.; Wang, J.; Zhu, J.; Lee, T.H.; de Silva, C.W. Unsupervised Cross-Domain Fault Diagnosis Using Feature Representation Alignment Networks for Rotating Machinery. *IEEE ASME Trans. Mech.* **2021**, *26*, 2770–2781. [CrossRef]
3. Zheng, H.; Wang, R.; Yin, J.; Li, Y.; Lu, H.; Xu, M. A new intelligent fault identification method based on transfer locality preserving projection for actual diagnosis scenario of rotating machinery. *Mech. Syst. Signal Process.* **2020**, *135*, 106344. [CrossRef]
4. Altaf, S.; Ahmad, S.; Zaindin, M.; Huda, S.; Iqbal, S.; Soomro, M.W. Multiple industrial induction motors fault diagnosis model within powerline system based on wireless sensor network. *Sustainability* **2022**, *14*, 10079. [CrossRef]
5. Gu, B.G. Development of broken rotor bar fault diagnosis method with sum of weighted fourier series coefficients square. *Energies* **2022**, *15*, 8735. [CrossRef]
6. Gangsar, P.; Tiwari, R. Signal-based condition monitoring techniques for fault detection and diagnosis of induction motors: A state-of-the-art review. *Mech. Syst. Signal Process.* **2020**, *144*, 106908. [CrossRef]

7. Aguayo-Tapia, S.; Avalos-Almazan, G.; Rangel-Magdaleno, J.d.J.; Paternina, M.R.A. Broken bar fault detection using taylor-fourier filters and statistical analysis. *Entropy* **2023**, *25*, 44. [CrossRef] [PubMed]
8. Ibrahim, A.; Anayi, F.; Packianather, M.; Alomari, O.A. New hybrid invasive weed optimization and machine learning approach for fault detection. *Energies* **2022**, *15*, 1488. [CrossRef]
9. Yaman, O. An automated faults classification method based on binary pattern and neighborhood component analysis using induction motor. *Measurement* **2021**, *168*, 108323. [CrossRef]
10. Godoy, W.F.; da Silva, I.N.; Goedtel, A.; Palácios, R.H.C.; Lopes, T.D. Application of intelligent tools to detect and classify broken rotor bars in three-phase induction motors fed by an inverter. *IET Electr. Power Appl.* **2016**, *10*, 430–439. [CrossRef]
11. Nascimento, N.M.M.e.; Medeiros, C.M.S.; Filho, P.P.R. A comparison of machine learning methods to identify broken bar failures in induction motors using statistical moments. *Adv. Intell. Syst. Comput.* **2018**, *736*, 124–133. [CrossRef]
12. Martinez-Herrera, A.L.; Ferrucho-Alvarez, E.R.; Ledesma-Carrillo, L.M.; Mata-Chavez, R.I.; Lopez-Ramirez, M.; Cabal-Yepez, E. Multiple fault detection in induction motors through homogeneity and kurtosis computation. *Energies* **2022**, *15*, 1541. [CrossRef]
13. Singh, M.; Shaik, A.G. Entropy-based broken rotor-bar fault detection and estimation of its severity in a three-phase induction motor. In Proceedings of the 2020 IEEE International Conference on Power Electronics, Drives and Energy Systems, Jaipur, India, 16–19 December 2020; pp. 1–6. [CrossRef]
14. Bejaoui, I.; Bruneo, D.; Xibilia, M.G. Remaining useful life prediction of broken rotor bar based on data-driven and degradation model. *Appl. Sci.* **2021**, *11*, 7175. [CrossRef]
15. Kartiwa, A.; Supriatna, A.K.; Rusyaman, E.; Sulaiman, J. Review of quaternion differential equations: Historical development, applications, and future direction. *Axioms* **2023**, *12*, 483. [CrossRef]
16. Huang, C.; Li, J.; Gao, G. Review of quaternion-based color image processing methods. *Mathematics* **2023**, *11*, 2056. [CrossRef]
17. Abbassi, R.; Jerbi, H.; Kchaou, M.; Simos, T.E.; Mourtas, S.D.; Katsikis, V.N. Towards higher-order zeroing neural networks for calculating quaternion matrix inverse with application to robotic motion tracking. *Mathematics* **2023**, *11*, 2756. [CrossRef]
18. Li, J.; Chen, P.; Chang, Z.; Zhang, G.; Guo, L.; Zhao, C. Trajectory tracking control of quadrotor based on fractional-order s-plane model. *Machines* **2023**, *11*, 672. [CrossRef]
19. Rikam, L.E.; Bitjoka, L.; Nketsa, A. Quaternion Fourier transform spectral analysis of electrical currents for bearing faults detection and diagnosis. *Mech. Syst. Signal Process.* **2022**, *168*, 108656. [CrossRef]
20. Zhang, K.; Deng, Y.; Chen, P.; Ma, C.; Xu, Y. Quaternion empirical wavelet transform and its applications in rolling bearing fault diagnosis. *Measurement* **2022**, *195*, 111179. [CrossRef]
21. Contreras-Hernandez, J.L.; Almanza-Ojeda, D.L.; Ledesma-Orozco, S.; Garcia-Perez, A.; Romero-Troncoso, R.J.; Ibarra-Manzano, M.A. Quaternion Signal Analysis Algorithm for Induction Motor Fault Detection. *IEEE Trans. Ind. Electron.* **2019**, *66*, 8843–8850. [CrossRef]
22. Contreras-Hernandez, J.L.; Almanza-Ojeda, D.L.; Ledesma, S.; Garcia-Perez, A.; Castro-Sanchez, R.; Gomez-Martinez, M.A.; Ibarra-Manzano, M.A. Geometric analysis of signals for inference of multiple faults in induction motors. *Sensors* **2022**, *22*, 2622. [CrossRef] [PubMed]
23. Contreras-Hernandez, J.L.; Almanza-Ojeda, D.L.; Ibarra-Manzano, M.A.; Amezquita-Sanchez, J.P.; Valtierra-Rodriguez, M.; Camarena-Martinez, D. Short-Circuit Damage Diagnosis in Transformer Windings Using Quaternions: Severity Assessment through Current and Vibration Signals. *Appl. Sci.* **2023**, *13*, 12622. [CrossRef]
24. Cardenas-Cornejo, J.J.; Ibarra-Manzano, M.A.; González-Parada, A.; Castro-Sanchez, R.; Almanza-Ojeda, D.L. Classification of inter-turn short-circuit faults in induction motors based on quaternion analysis. *Measurement* **2023**, *222*, 113680. [CrossRef]
25. Sinha, A.K.; Das, S.; Chatterjee, T.K. Empirical relation for broken bar determination in scim. *COMPEL-Int. J. Comput. Math. Electr. Electron. Eng.* **2018**, *37*, 242–265. [CrossRef]
26. Talebi, S.P.; Werner, S.; Mandic, D.P. Quaternion-valued distributed filtering and control. *IEEE Trans. Autom. Control* **2020**, *65*, 4246–4257. [CrossRef]
27. Treml, A.E.; Flauzino, R.A.; Suetake, M.; Maciejewski, N.A.R. Experimental database for detecting and diagnosing rotor broken bar in a three-phase induction motor. *IEEE Dataport* **2020**. [CrossRef]
28. Ren, T.; Kang, X.; Sun, W.; Song, H. Study of Dynamometer Cards Identification Based on Root-Mean-Square Error Algorithm. *Int. J. Pattern Recognit. Artif. Intell.* **2018**, *32*, 1850004. [CrossRef]

applied sciences

MDPI

Article

Inter-Turn Fault Diagnosis of Induction Motors Based on Current Vector Pattern Analysis in Stationary Coordinate Frame

Inyeol Yun [1], Hyunwoo Kim [2], Ju Lee [1] and Sung-Gu Lee [3,*]

[1] Department of Electrical Engineering, Hanyang University, Seoul 04763, Republic of Korea
[2] Elmore Family School of Electrical and Computer Engineering, Purdue University, West Lafayette, IN 47907, USA; kim4852@purdue.edu
[3] Department of Electrical Engineering, Dong-A University, Busan 49315, Republic of Korea
* Correspondence: sunggu@dau.ac.kr

Abstract: In this study, a current vector pattern is analyzed for inter-turn fault (ITF) diagnosis of induction machines (IMs), and an ITF diagnosis algorithm is proposed. When an ITF occurs in IMs, a negative-sequence current is generated due to fault resistance, even though a positive-sequence voltage is applied to IMs. Based on the mathematical model of IMs with an ITF, the current vector patterns in the stationary coordinate frame are analyzed. The superposition of positive- and negative-sequence components results in an elliptical current vector trajectory, and its orientation varies depending on the fault conditions. The co-simulation using finite element analysis and circuit simulation is implemented to analyze the current vector pattern of IMs with an ITF. The ITF diagnosis is proposed based on the current vector pattern. A 12 kW, four-pole, three-phase IM and terminal box, which was used to implement an ITF, is manufactured, and an experiment setup is established to verify the ITF algorithm. The effectiveness of the proposed ITF algorithm is validated through experimental verification of the manufactured IM and terminal box.

Keywords: co-simulation; current vector analysis; induction motor; inter-turn fault diagnosis (ITF); negative-sequence current

1. Introduction

In recent years, transportation electrification has advanced at an astonishing pace, driven by increasing environmental concerns, regulatory pressures, and rapid technological innovation. Governments around the world are introducing stringent emission standards and providing financial incentives to encourage the adoption of eco-friendly mobility solutions [1,2]. At the same time, advances in battery technology, power electronics, and lightweight materials have made electric drivetrains more practical and economical than ever before. As a result, electric vehicles (EVs), hybrid electric vehicles (HEVs), and other electric mobility solutions are rapidly replacing conventional internal combustion engine vehicles in markets around the world [3,4]. While transportation electrification brings about improvements in energy efficiency and reductions in carbon emissions, it also places increasing demands on the reliability and resilience of the underlying propulsion system.

As a core component of transportation electrification, electric motors, especially those used for traction, must have reliability and the ability to detect failures to ensure safe and reliable operation [5]. Among the various types of electric motors, induction motors (IMs) have maintained their dominance in traction and industrial applications [6,7]. Their simple structure, lack of permanent magnets, and cost-effectiveness make them a favorable choice for many commercial systems [8]. However, the reliability of IMs is becoming an increasingly important concern, especially in failure-prone environments. As electric vehicles become more widespread, ensuring the robust operation of these motors under a variety of conditions is becoming increasingly important [9].

There are various types of motor faults that can occur during operation, and the appropriate diagnostic method depends on the fault type. In general, motor faults are classified into electrical, magnetic, and mechanical categories [10]. Among these, stator-related faults in IMs account for more than 30% of all failures [11]. An inter-turn fault (ITF) in IMs occurs when the insulation between adjacent stator windings deteriorates, leading to a localized short circuit. This fault induces an unbalanced magnetic field, resulting in localized overheating and magnetic flux distortion. If not detected at an early stage, the fault may evolve into more severe conditions such as phase-to-phase or phase-to-ground faults, ultimately causing complete winding burnout and system failure [12–15].

Several studies have investigated methods to diagnosis ITFs in IMs; these methods include transient-response-based methods [16–18], motor current signature analysis (MCSA) [19–25], and machine learning-based methods [24–32]. In [17–19], the transient characteristics of IMs with an ITF are analyzed, and fault diagnosis methods are proposed based on measured physical properties such as current or temperature. Fault diagnosis based on MCSA has been widely adopted due to its simplicity and cost-effectiveness, as it only relies on current sensors. MCSA can be conducted based on the time domain (e.g., current envelope analysis) [19,20], frequency domain (e.g., Fast Fourier transform) [21,22], and time–frequency domain (e.g., Hilbert transform, Discrete Wavelet transform, and Short-Time Fourier Transform) [23–25]. Machine learning-based methods, such as Support Vector Machine (SVM), have been employed for ITF detection [24–28]. These techniques utilize features derived from a mathematical model and frequency-domain characteristics of the IM under ITF. Although these methods demonstrate high diagnostic performance, their reliance on high sampling rates and considerable computational resources may limit their feasibility for implementation in embedded systems. Recently, fault monitoring of inverter-fed IMs has been proposed using differential-mode (DM) and common-mode (CM) impedance analysis [5,29–32]. In [29,30], DM and CM impedance modeling and measurement methods are presented to facilitate high-frequency impedance analysis for stator health monitoring. The influence of stator faults on DM impedance was investigated, and variations in DM noise were accurately predicted [31]. In contrast, CM impedance measured at specific frequencies has proven to be effective for detecting ITFs. This approach is particularly advantageous because it does not require direct electrical connections to the motor terminals [32]. Instead, it relies on externally injected radio-frequency sinusoidal signals, offering high reliability and robustness, as it remains unaffected by the supply voltage quality. Although impedance analyzers can measure up to 120 MHz [33], additional equipment, such as inductive couplers and impedance analyzers, is necessary for implementation.

In this paper, a fault diagnosis algorithm based on the current vector pattern is proposed for detecting ITFs in IMs. When an ITF occurs in IM, the fault resistance occurs in the parallel connection of the faulted phase in the IM. Based on a mathematical model of an IM with an ITF, the current in the steady state is analyzed. Due to the faulted phase, a negative-sequence current is generated, even though a positive-sequence voltage is applied to the IM. Since both positive- and negative-sequence currents are generated, the current vector pattern in the stationary coordinate frame has a specific shape, which can be obtained from the mathematical model. Furthermore, the co-simulation with a finite element analysis (FEA) and circuit simulation is implemented to analyze the current vector pattern, and an ITF diagnosis algorithm is proposed based on the current vector analysis. An experimental setup is implemented using a 12 kW, four-pole, three-phase IM, inverter, fault resistance, and terminal box for winding connection. The effectiveness of the proposed ITF diagnosis algorithm is validated through experiments conducted under the fault conditions.

The rest of this paper is organized as follows. The current vector analysis of an IM with an ITF based on a mathematical model is discussed in Section 2. Based on the current vector analysis, an ITF diagnosis algorithm is proposed, and the co-analysis is described in Section 3. Section 4 presents experimental verification using the 12 kW, four-pole, three-phase IM and terminal box. Section 5 provides the conclusion of this study.

2. Current Vector Analysis of IMs with ITF

2.1. Mathematical Model for ITF

Figure 1 shows the mathematical model for an induction motor with an ITF in Phase A. Since the fault resistance is connected to the fault phase parallelly, the voltage equation is expressed as follows [34]:

$$\begin{aligned} &[v_s] = [R_s][i_s] + \tfrac{d}{dt}[\lambda_s] - [\mu R_s\ 0\ 0]^T\ i_f \\ &R_f i_f = \mu R_s\left(i_{as} - i_f\right) + \tfrac{d}{dt}\lambda_{as2} \\ &0 = [R_r][i_r] + \tfrac{d}{dt}[\lambda_r] \end{aligned} \quad (1)$$

where

$$\begin{aligned} [v_s] &= [v_{as}\ v_{bs}\ v_{cs}]^T,\ [i_s] = [i_{as}\ \ i_{bs}\ \ i_{cs}]^T \\ [\lambda_s] &= [L_{ss}][i_s] + [L_{sr}][i_r] + \mu[-(L_{ls} + L_{ms})\ L_{ms}/2\ L_{ms}/2]^T i_f, \\ [\lambda_r] &= [L_{rs}][i_s] + [L_{rr}][i_r] - \mu L_{ms}[\cos\theta_e\ \ \cos(\theta_e + 2\pi/3)\ \ \cos(\theta_e - 2\pi/3)]^T i_f, \\ \lambda_{as2} &= -\mu[-(L_{ls} + L_{ms})\ \ L_{ms}/2\ \ L_{ms}/2][i_s] \\ &\quad + \mu L_{ms}[\cos\theta_r\ \ \cos(\theta_r - 2\pi/3)\ \ \cos(\theta_r + 2\pi/3)][i_r] - \mu(L_{ls} + \mu L_{ms})i_f, \end{aligned}$$

v_{as}, v_{bs}, and v_{cs} are the phase voltages; i_{as}, i_{bs}, and i_{cs} are the phase currents; i_r is the rotor current; i_f is the fault current; λ_s and λ_r are the flux linkages of the stator and rotor, respectively; L_{ss} and L_{rr} are the self-inductances of the stator and rotor, respectively; L_{sr} and L_{rs} are the mutual inductances between the stator and rotor; R_s and R_r are the resistances of the stator and rotor, respectively; L_{ls} and L_{ms} are the leakage and magnetize inductance, respectively; $\mu = N_{fault}/N_{ph}$ is the ratio of the number of turns in the fault winding; and θ_e is the electrical angle.

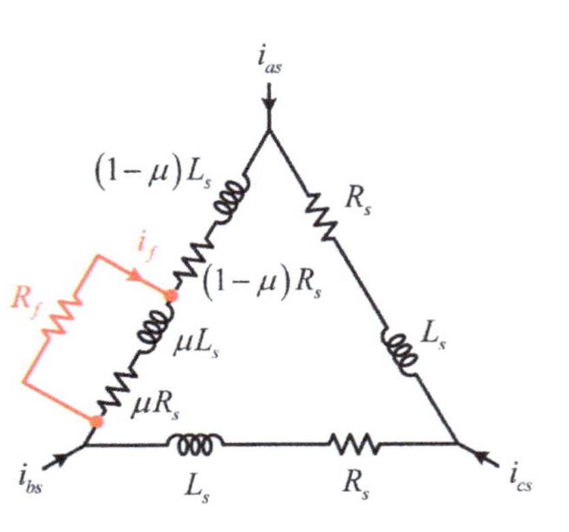

Figure 1. Configuration of ITF in IM with Δ connection.

To analyze the current vector pattern in the stationary coordinate frame (the $\alpha\beta$-axis plane), the coordinate transformation is considered as illustrated in Figure 2. From Figure 2, the coordinate transformation can be expressed as follows:

$$\left[C_{abc\to\alpha\beta}\right]=\frac{2}{3}\begin{bmatrix}1 & -\frac{1}{2} & -\frac{1}{2}\\ 0 & \frac{\sqrt{3}}{2} & -\frac{\sqrt{3}}{2}\end{bmatrix},\ \left[C_{\alpha\beta\to abc}\right]=\begin{bmatrix}1 & -\frac{1}{2} & -\frac{1}{2}\\ 0 & \frac{\sqrt{3}}{2} & -\frac{\sqrt{3}}{2}\end{bmatrix}^{T} \tag{2}$$

where $[C_{abc\to\alpha\beta}]$ is the coordinate transformation from abc to $\alpha\beta$, and $[C_{\alpha\beta\to abc}]$ is the coordinate transformation from $\alpha\beta$ to abc. From Equation (2), three-phase components are transformed into stationary coordinate frame as shown in Equation (3).

$$[f_{\alpha\beta}]=\left[C_{abc\to\alpha\beta}\right][f_{abc}],\ [f_{abc}]=\left[C_{\alpha\beta\to abc}\right][f_{\alpha\beta}] \tag{3}$$

where $[f_{abc}]$ is the physical value, such as the voltage and current, in *abc*, and $[f_{\alpha\beta}]$ is the physical value, such as the voltage and current, in $\alpha\beta$. By subtracting Equation (3) from Equation (1), the voltage equations in the stationary coordinate frame can be obtained using the stationary coordination transformations:

$$\begin{array}{l}[v_{\alpha\beta s}]=[R_s][i_{\alpha\beta s}]+\frac{d}{dt}[\lambda_{\alpha\beta s}]-\frac{2}{3}\mu R_s i_f\\ 0=[R_r][i_{\alpha\beta r}]+\frac{d}{dt}[\lambda_{\alpha\beta r}]-j\omega_e[\lambda_{\alpha\beta r}]\end{array} \tag{4}$$

where

$$[\lambda_{\alpha\beta s}]=[L_s][i_{\alpha\beta s}]+[L_m][i_{\alpha\beta r}]-\frac{2}{3}\mu\begin{bmatrix}L_s\\0\end{bmatrix}i_f,\ [\lambda_{\alpha\beta r}]=[L_m][i_{\alpha\beta s}]+[L_r][i_{\alpha\beta r}]-\frac{2}{3}\mu\begin{bmatrix}L_m\\0\end{bmatrix}i_f,$$

$[v_{\alpha\beta s}]$ is the voltage in the stationary coordinate frame; $[i_{\alpha\beta s}]$ and $[i_{\alpha\beta r}]$ are the current of the stator and rotor in the stationary coordinate frame, respectively; $[\lambda_{\alpha\beta s}]$ and $[\lambda_{\alpha\beta r}]$ are the flux linkage of the stator and rotor in the stationary coordinate frame, respectively; and ω_e is the electrical speed.

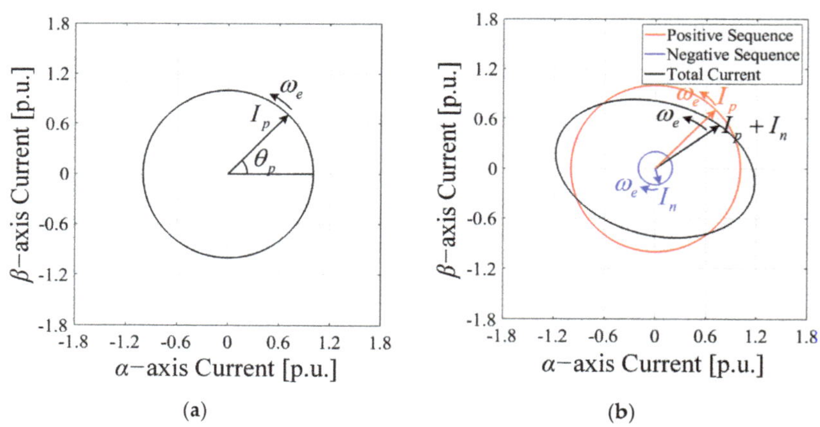

Figure 2. Current vector pattern: (**a**) health condition; (**b**) fault condition.

In a steady-state IM, the voltage and current can be expressed as a complex form:

$$\mathbf{V}_s=V_{ps}e^{j\omega_e t}+V_{ns}e^{-j\omega_e t},\ \ \mathbf{I}_s=I_{ps}e^{j\omega_e t}+I_{ns}e^{-j\omega_e t},\ \ \mathbf{I}_r=I_{pr}e^{js\omega_e t}+I_{nr}e^{-js\omega_e t} \tag{5}$$

where vs is the phase form of the voltage; $\mathbf{I}_s$ is the phase form of the stator current; $\mathbf{I}_r$ is the phase form of the rotor current; V_{ps} and V_{ns} are the positive and negative sequences of the voltage, respectively; I_{ps} and I_{ns} are the positive and negative sequences of the stator current, respectively; and I_{pr} and I_{nr} are the positive and negative sequences of the rotor current, respectively.

By subtracting Equation (5) from Equation (4), the phase equation in terms of the positive and negative sequences can be expressed as follows:

$$\begin{aligned} V_{ps} &= (R_s + j\omega_e L_s)\left(I_{ps} - \tfrac{2}{3}\mu I_f\right) + j\omega_e L_m I_{pr} \\ V_{ns} &= (R_s + j\omega_e L_s)\left(I_{ns} - \tfrac{2}{3}\mu I_f\right) + j\omega_e L_m I_{nr} \\ 0 &= \left(\tfrac{R_r}{s} + j\omega_e L_r\right) I_{pr} + j\omega_e L_m \left(I_{ps} - \tfrac{2}{3}\mu I_f\right) \\ 0 &= \left(\tfrac{R_r}{2-s} + j\omega_e L_r\right) I_{nr} + j\omega_e L_m \left(I_{ns} - \tfrac{2}{3}\mu I_f\right) \\ \mu(V_{sp} + V_{sn}) &= \mu\left(1 - \tfrac{2}{3}\mu\right)(R_s + j\omega_e L_{ls}) I_f + R_f I_f \end{aligned} \quad (6)$$

From Equation (6), the stator current can be expressed as follows:

$$\begin{bmatrix} I_{ps} \\ I_{ns} \end{bmatrix} = \begin{bmatrix} Y_{pp} & Y_{pn} \\ Y_{np} & Y_{nn} \end{bmatrix} \begin{bmatrix} V_{ps} \\ V_{ns} \end{bmatrix} \quad (7)$$

where

$$\begin{aligned} Y_{pp} &= \frac{1}{R_s + j\omega_e L_s - \frac{\omega_e^2 L_m^2}{R_r/s + j\omega_e L_r}} + Y_{pn}, \\ Y_{nn} &= \frac{1}{R_s + j\omega_e L_s - \frac{\omega_e^2 L_m^2}{R_r/(2-s) + j\omega_e L_r}} + Y_{np}, \\ Y_{pn} = Y_{np} &= \frac{2}{3}\frac{\mu^2}{R_f + \mu\left(1 - \frac{2\mu}{3}\right)(R_s + j\omega_e L_{ls})}, \end{aligned} \quad (8)$$

Y_{pp} and Y_{nn} are the admittances of the positive and negative sequences, respectively, and Y_{pn} and Y_{np} are the admittances between the positive and negative sequences.

If the positive-sequence voltage is applied to the IM, the positive and negative currents can be expressed as shown in Equation (7) because the negative-sequence voltage is zero.

$$I_{ps} = Y_{pp} V_{ps},\ I_{ns} = Y_{np} V_{ps} \quad (9)$$

When a fault occurs in Phase A, the admittance between the positive and negative sequences becomes non-zero due to the presence of μ in Equation (8). As a result, the negative-sequence current is induced by the applied positive-sequence voltage and is influenced by fault parameters such as the fault resistance and the ratio of shorted turns. The presence of a negative-sequence component under fault conditions changes the current vector pattern differently from that of a healthy condition.

2.2. Current Vector Pattern Analysis

Under ITF conditions in the IM, both positive- and negative-sequence currents are generated, even when only a positive-sequence voltage is applied. These sequence currents can be represented as phasors in the complex plane.

$$I_{ps} = I_p e^{j\theta_p},\ I_{ns} = I_n e^{-j\theta_n},\ I_p = Y_{pp} V_{ps},\ I_n = Y_{np} V_{ps} \quad (10)$$

where I_p and I_n are the magnitudes of the positive- and negative-sequence currents, respectively, and θ_p and θ_n are the phase angles of the positive- and negative-sequence currents.

The total current, comprising both positive- and negative-sequence components, can be expressed as follows:

$$\begin{aligned} I_t &= I_{ps}e^{j\omega_e t} + I_{ns}e^{-j\omega_e t} \\ I_\alpha + jI_\beta &= I_p e^{j(\omega_e t+\theta_p)} + I_n e^{-j(\omega_e t+\theta_n)} \\ I_\alpha &= I_p\cos(\omega_e t+\theta_p) + I_n\cos(\omega_e t+\theta_n) \\ I_\beta &= I_p\sin(\omega_e t+\theta_p) - I_n\sin(\omega_e t+\theta_n) \end{aligned} \quad (11)$$

From Equation (9), the relationship between the $\alpha\beta$-axis current can be expressed as follows:

$$AI_\alpha^2 + BI_\alpha I_\beta + CI_\beta^2 = \left(I_p^2 - I_n^2\right)^2 \quad (12)$$

where

$$\begin{aligned} A &= I_p^2 + I_n^2 - 2I_pI_n\cos(\theta_p - \theta_n) \\ B &= -4I_pI_n\sin(\theta_p - \theta_n) \\ C &= I_p^2 + I_n^2 + 2I_pI_n\cos(\theta_p - \theta_n). \end{aligned} \quad (13)$$

Equation (12) represents a quadratic form, where the relationship between the α- and β-axis currents is governed by the coefficients defined in Equation (13). These coefficients in Equation (12) satisfy the following relationships:

$$B^2 - 4AC = -4\left(I_p^2 - I_n^2\right)^2 < 0. \quad (14)$$

In the quadratic form presented in Equation (12), the α- and β-axis currents exhibit an elliptical relationship if the condition given in Equation (14) is satisfied. Under healthy operating conditions, Equation (12) simplifies to a circular form, as the negative-sequence current becomes zero according to Equations (8) and (9).

$$I_\alpha^2 + I_\beta^2 = I_p^2 \quad (15)$$

Therefore, the current vector pattern can be represented as a circle under healthy conditions and as an ellipse under ITF conditions, as illustrated in Figure 2. Moreover, both the magnitude and phase of the current vector are influenced by the fault conditions. Although the ITF conditions are analyzed using Equations (8), (9) and (12), accurately calculating the admittance between the positive and negative sequences is challenging due to the coupling between the stator and rotor currents. To address this, the next section presents a fault diagnosis approach based on current vector pattern analysis using co-simulation with FEA and circuit simulation under various fault conditions.

3. ITF Diagnosis Based on Current Vector Pattern

As shown in Section 2, the current vector pattern is analyzed when an ITF occurs in Phase A. Due to the negative-sequence current, the current vector pattern has an ellipse shape. In this section, the current vector pattern is analyzed based on the co-analysis with FEA and circuit simulation using ANSYS Maxwell 2024R1 and Simplorer 2024R1.

3.1. FEA Model

To construct the FEA model, the 12 kW, four-pole, three-phase IM used in the RENAULT TWIZY is considered, as summarized in Table 1. Figure 3a presents the FEA model, in which an ITF is introduced in Phase A by incorporating a fault resistance connected in parallel with the faulted winding. To implement the ITF condition, co-simulation is conducted using ANSYS Simplorer 2024R1, as shown in Figure 3b. The stator winding configuration is a Δ-connection, and the fault resistance is connected in parallel with the winding of the faulted phase. The co-simulation model includes a three-phase inverter, an FEA-based IM block, a stationary coordinate transformation block, and a V/f control block. The V/f control applies a positive-sequence voltage to the IM through the three-phase inverter. Based on the measured stator currents, the $\alpha\beta$-axis currents are calculated using the stationary coordinate transformation block.

Table 1. Specification of IM for Twizy.

Content	Value	Unit
Rated power	12	kW
Rated speed	2010	rpm
Maximum speed	7250	rpm
Rated frequency	70	Hz
DC voltage	52	V
Rated voltage	20	V_{rms}
Rated current	248	A_{rms}
DC link voltage	48	V
Switching frequency	10	kHz
Modulation method	SVPWM	-
Stator outer/inner diameter	200/125	mm
Rotor outer/inner diameter	124/36	mm
Airgap length	0.5	mm
Stack length	120	mm
Number of poles/slots	4/36	-
Total series turns	13	
Number of parallel branches	4	-
Core material	35PN230	-

3.2. FEA Results

As discussed in Section 2, the current vector exhibits an elliptical pattern under ITF conditions in an IM, primarily due to the presence of negative-sequence currents. These negative-sequence currents are influenced by fault parameters such as the fault resistance, the number of shorted turns, and the specific faulted phase. To investigate the current vector pattern under various fault conditions, co-simulations are conducted. Figure 4 illustrates the current vector patterns for each faulted phase. Under healthy conditions, the current vector forms a circle. However, in the presence of an ITF, the pattern transitions into an ellipse due to the unbalanced impedance introduced by the fault resistance. The main axis of the ellipse varies depending on the faulted phase. Specifically, when Phase A is faulted, the ellipse's main axis appears between 0° and 60°; if Phase B is faulted, it appears between −60° and 0°; and if Phase C is faulted, it appears between −120° and −60°. This main axis of the ellipse corresponds to a 30° angular displacement between the ellipse main axis and the magnetic motive force center of the faulted phase, as depicted in Figure 4. In the case of a Y-connected winding instead of a Δ-connected one, the current phase lags by 30°, which shifts the main axis. In this case, the main axis of the ellipse for Phase A lies between −30° and 30°; for Phase B, it lies between 30° and 90°; and for Phase C, it lies between −90° and −30°. A summary of the main axis angle for each faulted phase is provided in Table 2. The magnitude of the negative-sequence current is influenced by the fault parameters, such as the number of shorted turns and the fault resistance, as illustrated in Figure 5a,b. As the number of shorted turns increases or the fault resistance decreases, the elliptical distortion becomes more pronounced, indicating a higher degree of current imbalance. Furthermore, the positive- and negative-sequence currents are influenced by the speed and slip, as shown in Equation (8). The current vector pattern is analyzed for speed and slip, as illustrated in Figure 5c,d. Although the magnitude of the current vector depends on the speed and slip, the main axis of the ellipse appears between 0° and 60°. These trends are consistently observed across all simulation scenarios, confirming that the proposed method not only detects the occurrence of faults but also provides insights into their severity.

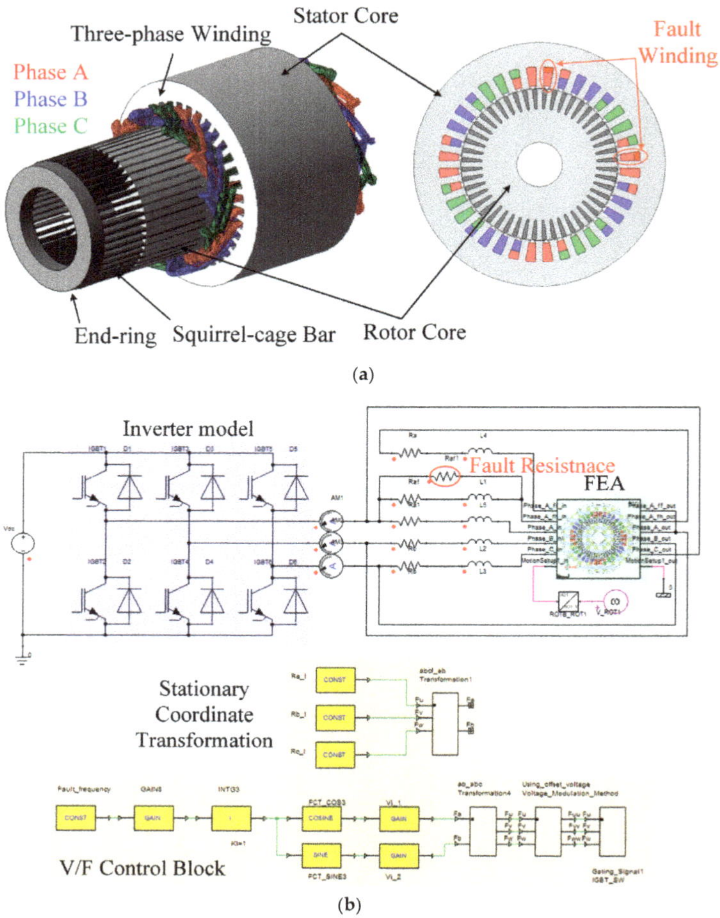

Figure 3. Co-analysis used in analysis of the current vector pattern in an ITF in an IM: (**a**) IM model (right: 3D view; left: 2D view); (**b**) circuit model for implementing the ITF.

Table 2. The angle of the maximum current vector for the fault phase and winding connection.

Winding Connection	Fault Phase	The Angle of the Maximum Current Vector
	A	0–60° & 180–240°
Δ	B	300–360° & 120–180°
	C	240–300° & 60–120°
	A	0–30° & 150–210° & 330–360°
Y	B	270–300° & 90–150°
	C	30–90° & 210–270°

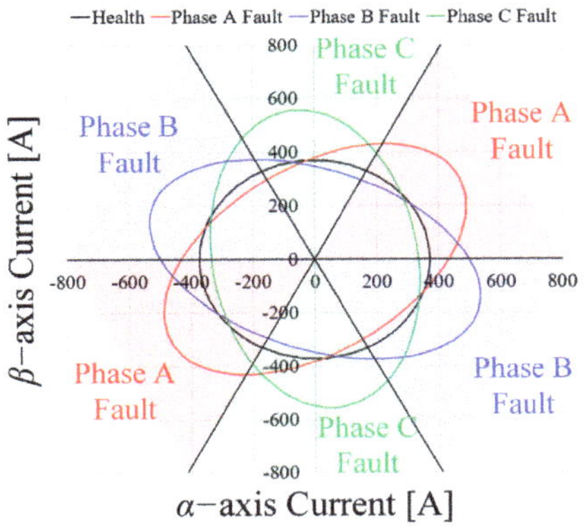

Figure 4. Current vector pattern for healthy and faulty conditions.

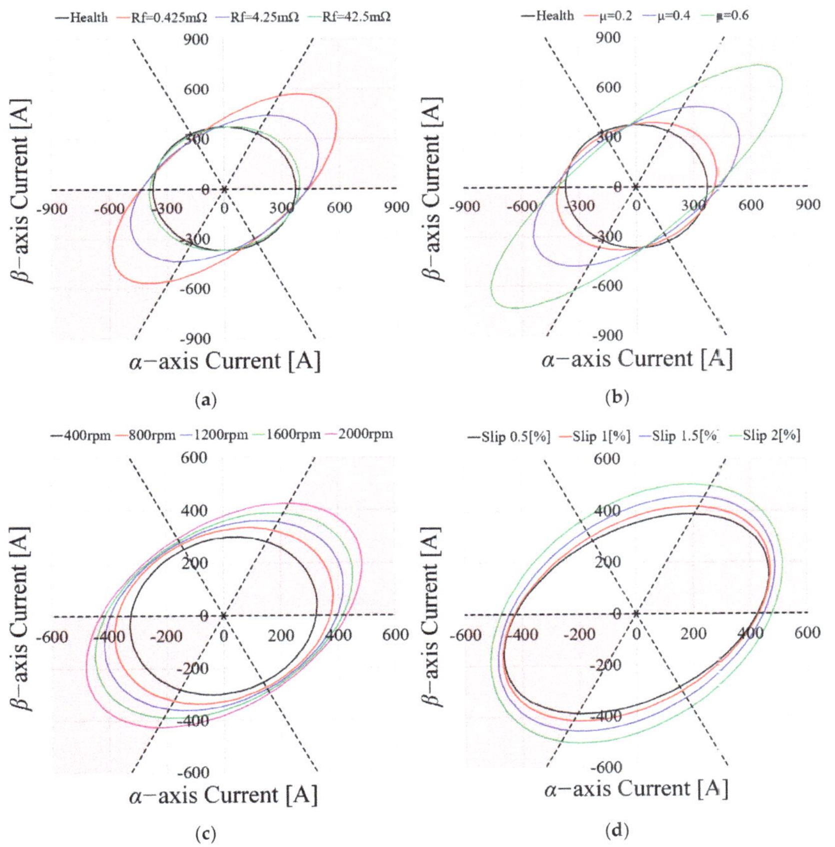

Figure 5. Current vector pattern for the fault resistance and number of fault turns when the ITF occurs in Phase A: (**a**) fault resistance; (**b**) the ratio of shorted turns; (**c**) speed; (**d**) slip.

3.3. Fault Diagnosis Algorithm

Through the co-simulation of the FEA model and circuit analysis, the current vector pattern under ITF conditions is analyzed, and a fault diagnosis algorithm is proposed.

Based on the co-simulation results, an ITF diagnosis algorithm tailored for Δ-connected windings is developed, as illustrated in Figure 6. In this approach, a positive-sequence voltage is applied to the IM, and the stator current in the stationary coordinate frame is obtained using the measured three-phase currents. Since the stationary coordinate transformation is employed, the use of an encoder is not required, which simplifies the implementation of the diagnostic method. The angle of the maximum current vector is calculated from the stationary-frame current, and fault detection is performed based on this angle. As summarized in Table 2 and illustrated in Figure 6, the proposed algorithm enables effective identification of the faulted phase without the need for complex signal processing or rotor position information.

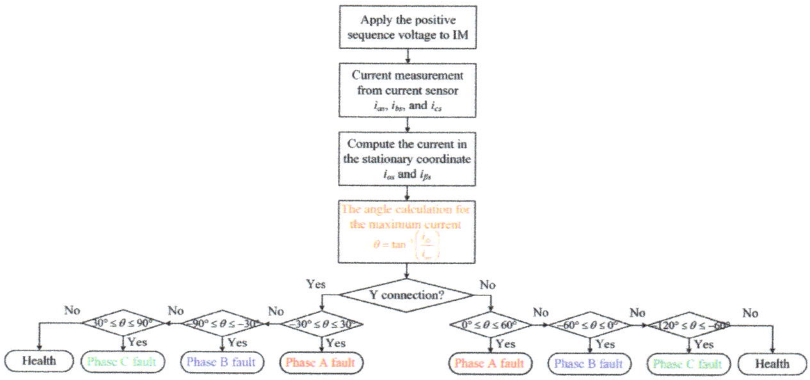

Figure 6. ITF diagnosis based on the current vector pattern.

4. Experimental Verification

To validate the proposed ITF diagnosis method, experiments were conducted using a 12 kW, four-pole, three-phase IM with both Y and Δ winding configurations. Figure 7a shows the manufactured machine. The stator windings were separated to allow for fault injection in each phase as shown in Figure 7b, and the winding connections were configured via a terminal box as shown in Figure 7a. A TELEDYNE oscilloscope (Teledyne Technologies, Thousand Oaks, CA, USA) was used to capture the current waveforms, while a control desk was employed to manage the inverter and IM operation. The inverter utilized in the experiments was equipped with an SPS600GB10S6 dual-pack MOSFET module from SEMI POWEREX (Seattle, WA, USA). A digital signal processor (DSP), specifically the TMS320F28069, was employed to compute the α- and β-axis current based on the measured three-phase currents, as shown in Figure 8a. The three-phase currents were digitized using an analog-to-digital converter (ADC), and the resulting $\alpha\beta$-axis currents were obtained through coordinate transformation on the DSP. The α- and β-axis currents were converted to analog signals through a digital-to-analog converter (DAC) and monitored using the oscilloscope. The manufactured IM was coupled with a dynamometer to conduct the fault experiment during motor operation, as shown in Figure 8b. ITFs were implemented at the terminal box by connecting a fault resistance (TGHGCR0050FE from OHMITE, Warrenville, IL, USA) in parallel with the fault phase of the IM. Although the nominal value of the fault resistor is 0.5 mΩ, the total resistance increased to approximately 3.5 mΩ due to additional lead wire resistance. To activate the fault during motor operation, an electrical relay was connected in series with the fault resistance and the faulted phase winding, as illustrated in Figure 8c. Figure 8d presents the measured current waveforms before and after the fault through the electrical relay. The experimental parameters, including operating speed, switching frequency, and motor specifications, are shown in Table 3. The experiments were conducted under low-speed conditions, approximately 200 rpm, to guarantee safety. For each fault phase, the stator currents were measured using the current sensor and observed from the oscilloscope. The current vector patterns during the steady-state operation are illustrated in Figures 9 and 10 for Y and Δ connections, respectively. Under healthy conditions, the current vector formed a nearly circle pattern as shown in Figures 9a and 10a. Upon implementing the faulty condition, the current vector pattern distorted into an ellipse, consistent with the FEA and simulation results, as illustrated in Figures 9b–d and 10b–d. The main axis of the ellipse varied with the faulted phase, demonstrating the algorithm's capability to distinguish not only between healthy and faulty conditions, but also to accurately identify the specific faulted phase. These results validate that the current vector pattern-based fault diagnosis method is effective across different winding configurations.

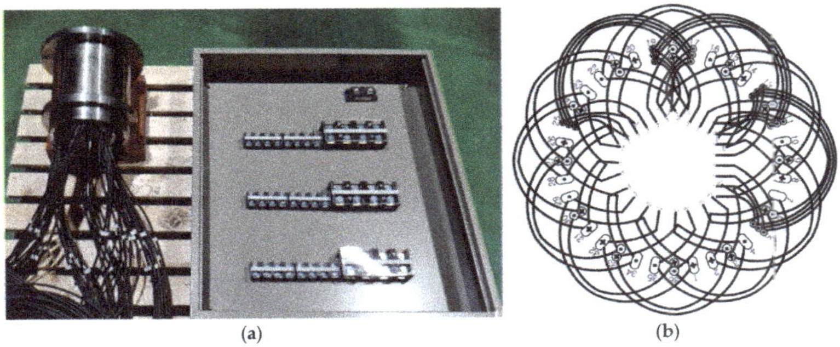

Figure 7. Manufactured machine: (**a**) IM and terminal box; (**b**) winding configuration.

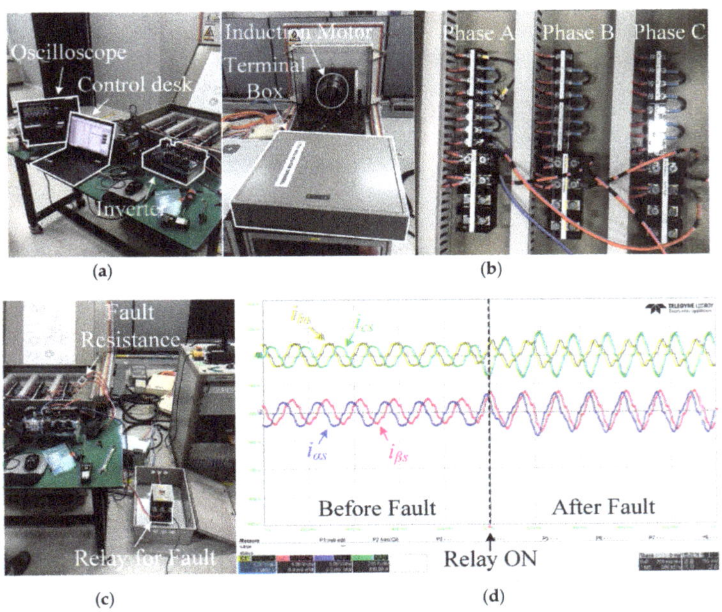

Figure 8. Experiment environment: (**a**) inverter, control desk, and oscilloscope; (**b**) IM, terminal box, and winding; (**c**) fault resistance and relay; (**d**) current waveform.

Table 3. Experimental parameters.

Content	Value	Unit
Operating speed	200	rpm
Switching frequency	500	Hz
Phase resistance	11.5	mΩ
Phase inductance	0.6	mH
Fault resistance wo/w lead wire	0.5/3.5	mΩ

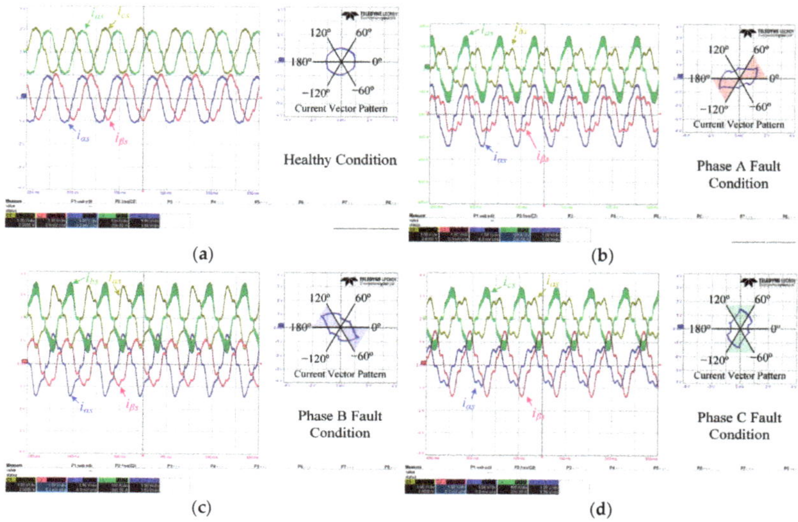

Figure 9. Experiment results of IM with Δ connection: (**a**) health condition; (**b**) Phase A fault condition; (**c**) Phase B fault condition; (**d**) Phase C fault condition.

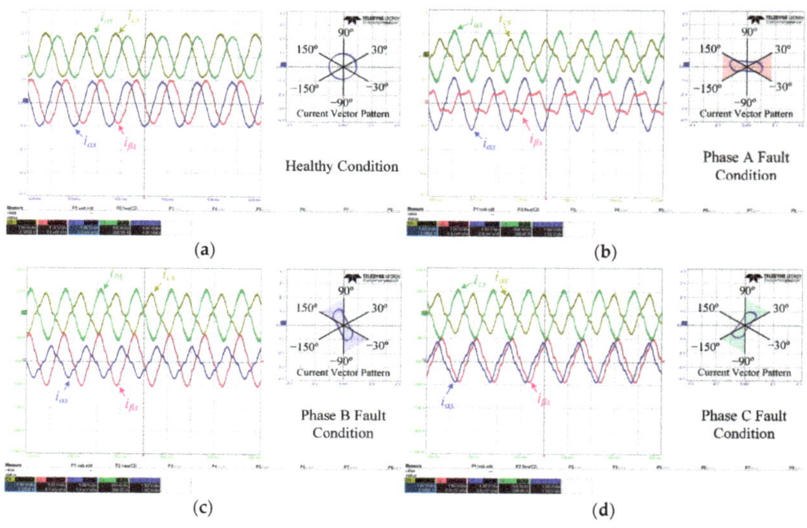

Figure 10. Experiment results of IM with Y connection: (**a**) health condition; (**b**) Phase A fault condition; (**c**) Phase B fault condition; (**d**) Phase C fault condition.

5. Conclusions

In this study, the current vector patterns of IM with the ITF were analyzed based on a mathematical model, co-simulation, and experiment. When an ITF occurs in an IM, a negative-sequence current is generated, even though only a positive-sequence voltage is applied. The superposition of positive- and negative-sequence components results in an elliptical current vector pattern, which is characteristic of ITFs. The co-simulation with the FEA and circuit simulation was established to analyze the current vector pattern for an ITF in an IM. The angle of the maximum current vector depends on the fault phase and parameters. Based on this relationship, a fault diagnosis algorithm was proposed, which identifies the faulted phase by detecting the angle of the major axis in the stationary coordinate frame. An experimental setup was established using a 12 kW, four-pole, three-phase IM from a RENAULT Twizy and a terminal box for implementing the winding connections to verify the ITF diagnosis algorithm. The current in the stationary coordinate frame was computed in the DSP based on the measured current from current sensor. The current vector pattern was observed using the DAC and ACD of DSP and oscilloscope. The ITF conditions were implemented through the terminal box, and the current vector pattern was analyzed for the fault phase. As a result, the fault diagnosis algorithm was verified by the co-simulation and experimental results. This approach provides an ITF diagnosis during the operation of the IM, and the position sensor, such as encoder and resolver, is not needed because the stationary coordinate transformation is used. Although current vector pattern analysis has been employed for the diagnosis of ITFs in IMs, the detection of other types of faults also requires further investigation. Therefore, future research will focus on extending current pattern analysis methodologies to encompass various fault conditions in IMs, with consideration of the high-frequency IM model and constraints of motor drive systems. In addition, machine learning-based methods will be explored to improve fault classification accuracy, particularly for distinguishing between ITFs, broken rotor bars, and turn-to-turn faults, by utilizing features extracted from current vector patterns.

Author Contributions: Conceptualization, I.Y. and H.K.; methodology, H.K.; software, I.Y., S.-G.L. and H.K.; validation, I.Y., S.-G.L. and H.K.; formal analysis, S.-G.L. and H.K.; investigation, I.Y.; resources, I.Y. and J.L.; data curation, I.Y. and H.K.; writing—original draft preparation, I.Y. and H.K.; writing—review and editing, S.-G.L., J.L. and H.K.; visualization, I.Y.; supervision, J.L. and H.K.; project administration, J.L. and H.K.; funding acquisition, J.L. All authors have read and agreed to the published version of the manuscript.

Funding: This work was supported by the Korea Institute of Energy Technology Evaluation and Planning (KETEP) grant funded by the Korea government (MOTIE) (No. RS-2023-00237024: Development and Demonstration of Medium-Sized Variable Speed High-Power Motor).

Institutional Review Board Statement: Not applicable.

Informed Consent Statement: Not applicable.

Data Availability Statement: The original contributions presented in this study are included in the article. Further inquiries can be directed to the corresponding author.

Conflicts of Interest: The authors declare no conflicts of interest.

References

1. Xue, C.; Zhou, H.; Wu, Q.; Wu, X.; Xu, X. Impact of Incentive Policies and Other Socio-Economic Factors on Electric Vehicle Market Share: A Panel Data Analysis from the 20 Countries. *Sustainability* **2021**, *13*, 2928. [CrossRef]
2. Zaino, R.; Ahmed, V.; Alhammadi, A.M.; Alghoush, M. Electric Vehicle Adoption: A Comprehensive Systematic Review of Technological, Environmental, Organizational and Policy Impacts. *World Electr. Veh. J.* **2024**, *15*, 375. [CrossRef]

3. Teske, S.; Bratzel, S.; Tellermann, R.; Stephan, B.; Vargas, M. Net Zero: The Remaining Global Market Volume for Internal Combustion Engines in Light-Duty Vehicles under a 1.5 °C Carbon Budget Trajectory. *Energies* **2022**, *15*, 8037. [CrossRef]
4. Memon, M.; Rossi, C. A Review of EV Adoption, Charging Standards, and Charging Infrastructure Growth in Europe and Italy. *Batteries* **2025**, *11*, 229. [CrossRef]
5. Issa, R.; Clerc, G.; Hologne-Carpentier, M.; Michaud, R.; Lorca, E.; Magnette, C.; Messadi, A. Review of Fault Diagnosis Methods for Induction Machines in Railway Traction Applications. *Energies* **2024**, *17*, 2728. [CrossRef]
6. Joksimović, G.; Zajec, P.; Tessarolo, A.; Ambrožič, V.; Rihar, A. Stator Current Spectral Content of Inverter-Fed Cage Rotor Induction Motor. *IEEE Access* **2022**, *10*, 23112–23121. [CrossRef]
7. Asad, B.; Vaimann, T.; Belahcen, A.; Kallaste, A.; Rassõlkin, A.; Ghafarokhi, P.S.; Kudelina, K. Transient Modeling and Recovery of Non-Stationary Fault Signature for Condition Monitoring of Induction Motors. *Appl. Sci.* **2021**, *11*, 2806. [CrossRef]
8. Park, D.-H.; Song, C.-H.; Won, Y.-J.; Park, J.-C.; Kim, H.-S.; Park, H.-R.; Yoon, Y.-D.; Lim, M.-S. Magnetizing Inductance Estimation Method of Induction Motor for EV Traction Considering Magnetic Saturation Changes According to Current and Slip Frequency. *IEEE Trans. Magn.* **2024**, *60*, 1–5. [CrossRef]
9. Yatsugi, K.; Pandarakone, S.E.; Mizuno, Y.; Nakamura, H. Common Diagnosis Approach to Three-Class Induction Motor Faults Using Stator Current Feature and Support Vector Machine. *IEEE Access* **2023**, *11*, 24945–24952. [CrossRef]
10. Jin, L.; Wang, X.; Mao, Y.; Lu, L.; Wang, Z. Online Attribute Matching Based Few-Sample Data-Driven Diagnosis of Electrical Faults in PMSM Drive. *IEEE Trans. Power Electron.* **2024**, *39*, 2620–2631. [CrossRef]
11. Rengifo, J.; Moreira, J.; Vaca-Urbano, F.; Alvarez-Alvarado, M.S. Detection of Inter-Turn Short Circuits in Induction Motors Using the Current Space Vector and Machine Learning Classifiers. *Energies* **2024**, *17*, 2241. [CrossRef]
12. Allal, A.; Lammouchi, Z.; Khechekhouche, A.; Mohamed, N.; Alqunun, K.; Alghadhban, A.; Marouani, I.; Guesmi, T. Advanced Residual Optimal Mapping Approach for Precise Detection of Stator Faults in Induction Motors. *IEEE Access* **2024**, *12*, 143515–143530. [CrossRef]
13. Aswad, R.A.K.; Jassim, B.M.H.; Zahawi, B.; Gadoue, S. Hybrid Model-Based Fuzzy Logic Diagnostic System for Stator Faults in Three-Phase Cage Induction Motors. *IEEE Access* **2023**, *11*, 75707–75714. [CrossRef]
14. Dybkowski, M.; Bednarz, S. Modified Rotor Flux Estimators for Stator-Fault-Tolerant Vector Controlled Induction Motor Drives. *Energies* **2019**, *12*, 3232. [CrossRef]
15. Goh, Y.J.; Kim, O. Linear Method for Diagnosis of Inter-Turn Short Circuits in 3-Phase Induction Motors. *Appl. Sci.* **2019**, *9*, 4822. [CrossRef]
16. Singh, G.; Anil Kumar, T.C.; Naikan, V.N.A. Induction motor inter turn fault detection using infrared thermographic analysis. *Infrared Phys. Technol.* **2016**, *77*, 277–282. [CrossRef]
17. Chen, L.; Shen, J.; Xu, G.; Chi, C.; Feng, Q.; Zhou, Y.; Deng, Y.; Wen, H. Induction Motor Stator Winding Inter-Tern Short Circuit Fault Detection Based on Start-Up Current Envelope Energy. *Sensors* **2023**, *23*, 8581. [CrossRef]
18. Zaparoli, I.O.; Rabelo Baccarini, L.M.; Lamim Filho, P.C.M.; Batista, F.B. Transient envelope current analysis for inter-turn short-circuit detection in induction motor stator. *J. Braz. Soc. Mech. Sci. Eng.* **2020**, *42*, 97. [CrossRef]
19. Niu, G.; Dong, X.; Chen, Y. Motor Fault Diagnostics Based on Current Signatures: A Review. *IEEE Trans. Instrum. Meas.* **2023**, *72*, 1–19. [CrossRef]
20. da Silva, A.M.; Povinelli, R.J.; Demerdash, N.A.O. Induction Machine Broken Bar and Stator Short-Circuit Fault Diagnostics Based on Three-Phase Stator Current Envelopes. *IEEE Trans. Ind. Electron.* **2008**, *55*, 1310–1318. [CrossRef]
21. Ghanbari, T.; Mehraban, A.; Farjah, E. Inter-Turn Fault Detection of Induction Motors Using a Method Based on Spectrogram of Motor Currents. *Measurement* **2022**, *205*, 112180. [CrossRef]
22. Surya, G.N.; Khan, Z.J.; Ballal, M.S.; Suryawanshi, H.M. A simplified frequency-domain detection of stator turn fault in squirrel-cage induction motors using an observer coil technique. *IEEE Trans. Ind. Electron.* **2017**, *64*, 1495–1506. [CrossRef]
23. Rajan Babu, W.; Senthil Kumar, R.; Satheesh Kumar, R. Rigorous investigation of stator current envelope of an induction motor using Hilbert spectrum analysis. *Mater. Today Proc.* **2021**, *45*, 2474–2478. [CrossRef]
24. Almounajjed, A.; Sahoo, A.K.; Kumar, M.K. Diagnosis of stator fault severity in induction motor based on discrete wavelet analysis. *Measurement* **2021**, *182*, 109780. [CrossRef]
25. Jung, J.-H.; Lee, J.-J.; Kwon, B.-H. Online diagnosis of induction motors using MCSA. *IEEE Trans. Ind. Electron.* **2006**, *53*, 1842–1852. [CrossRef]
26. Saucedo-Dorantes, J.J.; Jaen-Cuellar, A.Y.; Perez-Cruz, A.; Elvira-Ortiz, D.A. Detection of Inter-Turn Short Circuits in Induction Motors under the Start-Up Transient by Means of an Empirical Wavelet Transform and Self-Organizing Map. *Machines* **2023**, *11*, 958. [CrossRef]
27. Morales-Perez, C.J.; Camarena-Martinez, D.; Amezquita-Sanchez, J.P.; Rangel-Magdaleno, J.D.J.; Ramírez, E.E.S.; Valtierra-Rodriguez, M. Early Detection of Inter-Turn Short Circuits in Induction Motors Using the Derivative of Stator Current and a Lightweight 1D-ResNet. *Computation* **2025**, *13*, 140. [CrossRef]

28. Cardenas-Cornejo, J.J.; Almanza-Ojeda, D.L.; Gonzáalez-Parada, A.; Hernandez-Ramirez, V.; Ibarra-Manzano, M.A. Complex Signal Analysis for Inter-Turn Short-Circuits Faults on Induction Motors. *IEEE Sens. J.* **2025**, *25*. 13433–13440. [CrossRef]
29. Creux, J.; Haje Obeid, N.; Boileau, T.; Meibody-Tabar, F. Online High Frequency Impedance Identification Method of Inverter-Fed Electrical Machines for Stator Health Monitoring. *Appl. Sci.* **2024**, *14*, 10911. [CrossRef]
30. Vinayak, B.A.; Alam, M.F.; Jagadanand, G.; Anand, K.A. Stator Inter Turn Fault Diagnosis by High-Frequency Modeling of Inverter Fed Induction Motor. In Proceedings of the 2020 IEEE International Conference on Power Electronics, Smart Grid and Renewable Energy (PESGRE2020), Cochin, India, 2–4 January 2020; pp. 1–5.
31. Zhao, Z.; Fan, F.; Wang, W.; Liu, Y.; See, Y.K. Detection of Stator Interturn Short-Circuit Faults in Inverter-Fed Induction Motors by Online Common-Mode Impedance Monitoring. *IEEE Trans. Instrum. Meas.* **2021**, *70*, 1–10. [CrossRef]
32. Fan, F.; Zhao, Z.; Jie, H.; Sun, Q.; Tu, P.; Shu, Z.; Wang, W.; See, K.Y. Impact of Motor Stator Winding Faults on Motor Differential-Mode Impedance and Mode Transformation. *Chin. J. Electr. Eng.* **2022**, *8*, 12–21. [CrossRef]
33. Jie, H.; See, K.Y.; Chang, Y.; Gao, R.X.-K.; Fan, F.; Zhao, Z. Enhancing motor impedance measurements: Broadening the spectrum from low to high frequencies. *Meas. Sci. Technol.* **2024**, *35*, 86008. [CrossRef]
34. Tallam, R.M.; Habetler, T.G.; Harley, R.G. Transient Model for Induction Machines with Stator Winding Turn Faults. *IEEE Trans. Ind. Appl.* **2002**, *38*, 632–637. [CrossRef]

applied sciences

MDPI

Article

Impedance-Based Inter-Turn Fault Diagnosis in Integrated Induction Motor and Drive Systems Using Space Voltage Vectors

Jungho Ahn [1], Ju Lee [1] and Hyunwoo Kim [2,*]

1 Department of Electrical Engineering, Hanyang University, Seoul 04763, Republic of Korea; ahnleo@hanyang.ac.kr (J.A.); julee@hanyang.ac.kr (J.L.)
2 Elmore Family School of Electrical and Computer Engineering, Purdue University, West Lafayette, IN 47907, USA
* Correspondence: kim4852@purdue.edu

Abstract: In this paper, a winding inter-turn fault (ITF) diagnosis algorithm for electrical machines in integrated motor and drive (IMD) systems is proposed based on impedance analysis using space voltage vectors. An ITF alters the stator resistance, causing an imbalance in the motor's impedance depending on the phase connection. This impedance asymmetry can be effectively utilized for fault diagnosis. However, in IMD systems, direct impedance measurement through mechanical terminal access is difficult due to the integrated structure. To address this, the impedance of the induction motor is analyzed electrically, without the need for physical disconnection, allowing practical implementation within integrated systems. The specific phase angle of the space voltage vector in the three-phase inverter is analyzed to replicate the electrical conditions of mechanical terminal configurations. Based on this approach, a fault diagnosis algorithm was developed by analyzing the variation in stator current and impedance with respect to space voltage vector angles. The effectiveness of the proposed method was verified through experimental validation using a 12kW three-phase induction motor and terminal box.

Keywords: fault diagnosis; integrated motor and drive (IMD); intern-turn fault (ITF); three-phase inverter; space voltage vector

1. Introduction

With the recent rapid growth of the electric vehicle (EV) industry, integrated motor and drive (IMD) systems, which simultaneously provide high power density and efficiency, have attracted significant attention across various industrial sectors [1,2]. IMD systems offer several advantages, including reduced installation space, lower wiring losses, suppression of electromagnetic interference, and improved overall system efficiency [3,4]. Furthermore, an IMD system enables the elimination of components such as transformers and filters and offers the benefits of an optimized cooling structure and modular design, making it particularly well suited for EV applications that demand high power density [5,6]. In line with this trend, extensive research has been conducted on compact, high-frequency IMD systems that employ PCB-based stators and GaN-based switching devices. Integrated designs incorporating GaN devices capable of operating at frequencies above 1 MHz have emerged as promising solutions for both reducing system size and suppressing current ripple [7,8].

Electrical faults such as inter-turn faults (ITFs), bearing failures, and insulation breakdowns in motors—critical components in EV systems—can lead to system shutdowns, fires, and serious safety incidents during operation. As a result, diagnostic technologies, including fault-tolerant design approaches, have become increasingly important [9]. In particular, autonomous driving systems require immediate fault response and sustained control system stability due to the limited possibility of driver intervention [10,11]. Recent studies have applied hybrid optimization algorithms for fault diagnosis and parameter estimation in power systems and electrical drives. These approaches offer improved robustness and accuracy in noisy or nonlinear environments [11,12].

Motor faults that can occur during operation are diverse, and diagnostic approaches and countermeasures vary depending on the fault type. In general, motor faults are categorized into electrical, magnetic, and mechanical faults [13]. Electrical faults include inter-turn short circuits (ITSCs), phase-to-phase short circuits, and phase-to-ground faults [14–17]. Magnetic faults consist of demagnetization and magnetic asymmetry [18,19], while mechanical faults include bearing failures, rotor eccentricity, broken rotor bars, and air-gap asymmetry. Other possible faults include sensor failures and inverter faults [20–22]. Among these, ITSC faults are considered the most severe, as they can lead to total winding failure within a short period and generate significant torque ripple [23,24]. This occurs because a localized short between turns produces extremely high circulating currents, resulting in severe overheating, insulation breakdown, and ultimately catastrophic motor failure or even fire [25,26]. In particular, ITFs can rapidly destroy the entire winding and, in the case of permanent magnet (PM) motors, may even cause magnet demagnetization, posing a serious threat to overall system reliability [27,28].

Various methods have been developed to diagnose ITFs due to their severity. These methods are considered to be among the most critical aspects of motor fault diagnosis [28–33]. In [28–30], the widely adopted approach of motor current signature analysis (MCSA) is introduced. MCSA techniques are generally categorized into FFT-based spectrum analysis, time–frequency analysis (e.g., wavelet transform and Hilbert–Huang transform), and other methods. However, spectrum-analysis-based methods are highly sensitive to high-frequency noise and load variations, which can degrade diagnostic reliability. In [27], artificial intelligence (AI)-based diagnostic techniques, including neural networks and support vector machines (SVMs), have also been actively explored. However, AI-based approaches require large amounts of labeled fault data for effective training, and acquiring sufficient and diverse fault data under various operating conditions remains a major challenge in real-world industrial environments. Furthermore, conventional model-based methods require complex mathematical modeling and are susceptible to uncertainties in motor parameters [31,32].

ITF diagnosis based on impedance analysis in induction motors (IMs) was introduced in a previous study [33]. The impedance characteristics of six winding connections were analyzed, and the phase of impedance was calculated to diagnose the ITF. The six winding connections are required to calculate the phase of impedance for ITF diagnosis. Although the introduced ITF diagnosis method is simplified compared to other methods, it presents practical limitations when applied to real industrial systems such as electric vehicles (EVs), which commonly utilize IMD systems. In these systems, the physical disconnection of the inverter from the motor is required to perform offline measurements, which significantly limits the applicability of the approach in IMD systems.

This paper proposes an offline ITF diagnosis algorithm for IMD systems based on impedance analysis and the space voltage vector control of a three-phase inverter. The proposed method enables fault detection without mechanical separation between the motor and the inverter, making it highly suitable for IMD-based applications. Assuming an ITF occurs in phase A, the terminal impedances are analyzed and compared to identify the maximum and minimum impedance values. Although ITFs can be diagnosed through terminal resistance measurements, the terminal configuration in IMD systems differs due to the integrated structure of the motor and inverter. Therefore, the space voltage vector of the three-phase inverter is analyzed for each switching state. A fault diagnosis algorithm is proposed based on both the impedance characteristics and the space voltage vector analysis. To validate the proposed algorithm, an experimental setup was constructed using a 12 kW three-phase induction motor (IM), an inverter, fault resistance, and a terminal box for winding connections. The algorithm was verified for both Y and Δ connections without requiring access to mechanical terminal connections.

Here, we provide a summary of the main contributions of our work:

(1) To diagnose the ITF, the impedance of electrical machine is analyzed for the terminal connection. The terminal connection is implemented electrically through the space vector control of the three-phase inverter, eliminating the need for inverter disconnection or access to terminal connections when the ITF of the electrical machine is diagnosed.

(2) The proposed method does not require the tracking of time-varying operational parameters, thereby simplifying implementation compared to online diagnostic approaches. Furthermore, a low-complexity algorithm based on phase impedance comparison across six space voltage vectors is employed, enhancing compatibility with embedded systems and real-time control platforms.

(3) A 12 kW three-phase IM was manufactured, and an experimental environment was implemented to verify the proposed algorithm for ITF diagnosis. The proposed ITF diagnosis algorithm was verified through the measurement of impedance for the space voltage vector for Y and Δ connections.

This paper is organized as follows: Section 2 describes the analysis of the impedance characteristics under ITF conditions in phase A for both Y and Δ connections. Section 3 presents the space voltage vector analysis corresponding to inverter switching states and proposes the ITF diagnosis algorithm based on impedance variation. Section 4 describes the experimental validation using the fabricated IM and terminal box. Section 5 concludes the paper.

2. Impedance Analysis of Inter-Turn Fault of Induction Motor

2.1. Mathematical Model of Induction Motor with Inter-Turn Fault

Figure 1 shows the configuration of an induction motor under ITF conditions with Y and delta connections. In the event of an ITF in phase A, fault resistance appears in parallel with the winding of the faulty phase. This additional parallel path results in reduced impedance in the fault phase, leading to an imbalance in the per-phase impedance of the stator. The resulting asymmetry in impedance can be effectively utilized as a diagnostic feature for identifying ITF in induction motors. The impedance of healthy and fault phases can be expressed as follows:

$$Z_h = R_a + j\omega L_a \tag{1}$$

$$Z_f = \frac{R_f(R_{a1} + j\omega L_{a1})}{R_{a1} + j\omega L_{a1} + R_f} + R_{a2} + j\omega L_{a2} \tag{2}$$

where

$$R_a = R_{a1} + R_{a2}, \ L_a = L_{a1} + L_{a2}, \tag{3}$$

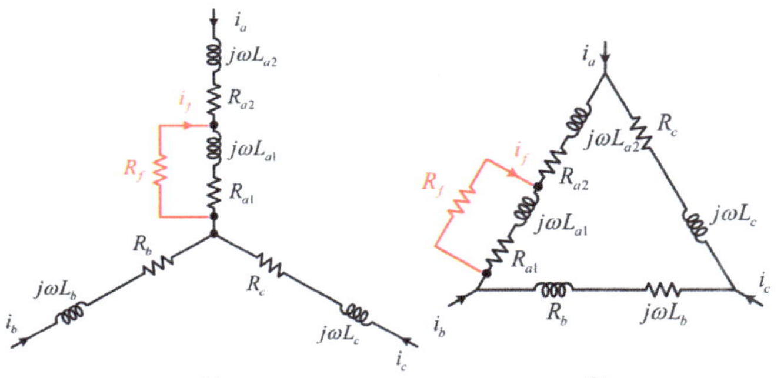

Figure 1. Configuration of inter-turn fault in IM: (**a**) Y connection; (**b**) Δ connection.

Z_h and Z_f denote the impedances of healthy and fault phases, respectively. R_a and L_a represent the phase resistance and inductance of the healthy phase, respectively. R_{a1} and L_{a1} are the resistance and inductance of the fault phase with the fault resistance connected in parallel. R_{a2} and L_{a2} denote the resistance and inductance of the fault phase without the fault resistance. R_f is the fault resistance, and ω is the electrical angular speed. By subtracting Equation (3) from Equation (2), the impedance of the fault phase is expressed as follows:

$$\begin{aligned} Z_f &= \frac{R_f(R_{a1}+j\omega L_{a1})}{R_{a1}+j\omega L_{a1}+R_f} + R_a - R_{a1} + j\omega(L_a - L_{a1}) \\ &= Z_h - \frac{(R_{a1}+j\omega L_{a1})(R_{a1}+j\omega L_{a1}-R_f)}{R_{a1}+j\omega L_{a1}+R_f}. \end{aligned} \quad (4)$$

Since the fault resistance is connected in parallel with the phase impedance, the fault phase impedance becomes lower than that of the healthy phase, as shown in Equation (4). This indicates that impedance imbalance occurs under ITF conditions, and the impedance of the fault phase is lower than that of the healthy phase. In the next subsection, the impedance analysis of the motor connection is provided to propose the fault diagnosis.

2.2. Impedance Analysis with Y Connection

When an ITF occurs in phase A, the terminal connection is considered to analyze the impedance for the Y connection, as shown in Figure 2. The impedance analysis is conducted for the terminal connection configuration. When the two phases are connected as shown in Figure 2a–c, the terminal impedance can be expressed as follows:

$$Z_{a-b} = Z_{c-a} = Z_f + Z_h, \quad Z_{b-c} = 2Z_h \quad (5)$$

where $Z_{a\text{-}b}$, $Z_{c\text{-}a}$, and $Z_{b\text{-}c}$ denote the terminal impedances corresponding to the connections between phases A and B, C and A, and B and C, respectively. Since the fault impedance is lower than the healthy impedance, as shown in Equation (4), the terminal impedance comparison is as follows.

$$Z_{a-b} - Z_{b-c} = Z_f - Z_h < 0 \ \rightarrow \ Z_{a-b} = Z_{c-a} < Z_{b-c} \quad (6)$$

From Equation (6), the specific terminal impedance is greater than the other terminal impedances. If an ITF occurs in a different phase, the specific terminal impedance increases when the two corresponding phases are connected.

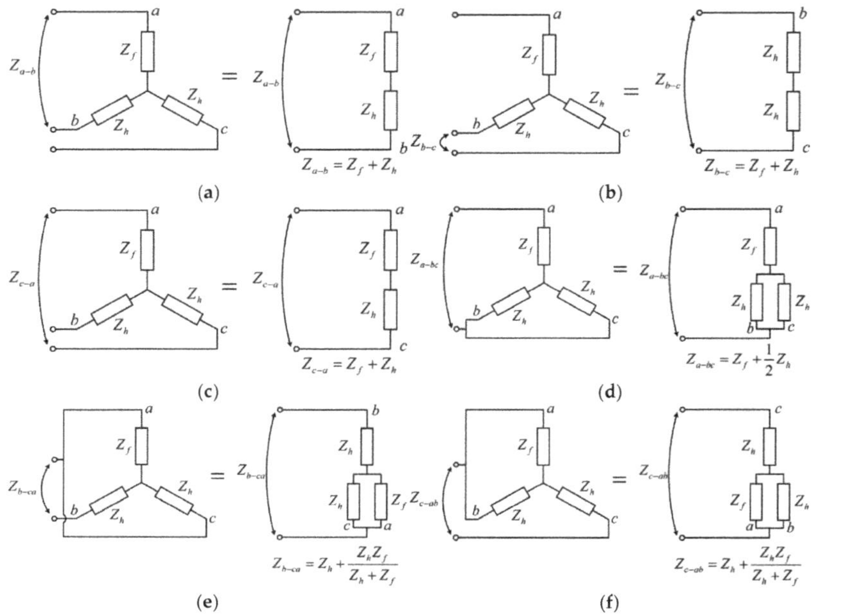

Figure 2. Impedance analysis for a Y connection with a phase A fault: (**a**) *a-b*; (**b**) *b-c*; (**c**) *c-a*; (**d**) *a-bc*; (**e**) *b-ca*; (**f**) *c-ab*.

The impedance for the terminal connection between phase A and the midpoint of phases B and C, denoted as $Z_{a\text{-}bc}$ and shown in Figure 2f, is calculated as in Equation (7).

$$Z_{a-bc} = Z_f + Z_h \,||\, Z_h = Z_f + \frac{1}{2} Z_h \quad (7)$$

Similarly, the impedances for connections *b-ca* and *c-ab*, represented in Figure 2d,e, are given by Equation (8).

$$Z_{b-ca} = Z_{c-ab} = Z_h + Z_f \,||\, Z_h = Z_h + \frac{Z_f Z_h}{Z_f + Z_h} \quad (8)$$

A comparison between (7) and (8) reveals that the impedance of the *a-bc* configuration is greater than those of the *b-ca* and *c-ab*, as summarized in Equation (9).

$$\begin{aligned} Z_{a-bc} - Z_{b-ca} &= \frac{(Z_f - Z_h)(2Z_f + Z_h)}{2(Z_f + Z_h)} < 0 \\ Z_{a-bc} &< Z_{b-ca} = Z_{c-ab} \end{aligned} \quad (9)$$

From Equation (9), the specific impedance of a given terminal connection configuration is lower than that of the other configurations. If a fault occurs in phase B or C, the impedance becomes either maximum or minimum, depending on the specific terminal connection. Table 1 presents the maximum and minimum impedance values for each fault phase and terminal connection when an ITF occurs in phase A with the Y connection.

Table 1. Impedance analysis results for a Y connection with a phase A fault.

Fault Phase	Maximum Impedance among $Z_{a\text{-}b}$, $Z_{b\text{-}c}$, and $Z_{c\text{-}a}$	Minimum Impedance among $Z_{a\text{-}bc}$, $Z_{b\text{-}ca}$, and $Z_{c\text{-}ab}$
A	$Z_{b\text{-}c}$	$Z_{a\text{-}bc}$
B	$Z_{c\text{-}a}$	$Z_{b\text{-}ca}$
C	$Z_{a\text{-}b}$	$Z_{c\text{-}ab}$

2.3. Impedance Analysis with Δ Connection

Figure 3 illustrates the terminal configuration of an IM with a Δ connection when an ITF occurs in phase A. As shown in Figure 3, three terminals are available, enabling various inter-terminal connection configurations. Depending on the selected terminal pair, the measured impedance varies due to the presence of fault resistance in the affected phase. The impedance measured between phases A and B, as shown in Figure 3a, is expressed in Equation (10).

$$Z_{a-b} = Z_f \,||\, 2Z_h = \frac{2Z_f Z_h}{Z_f + 2Z_h} \tag{10}$$

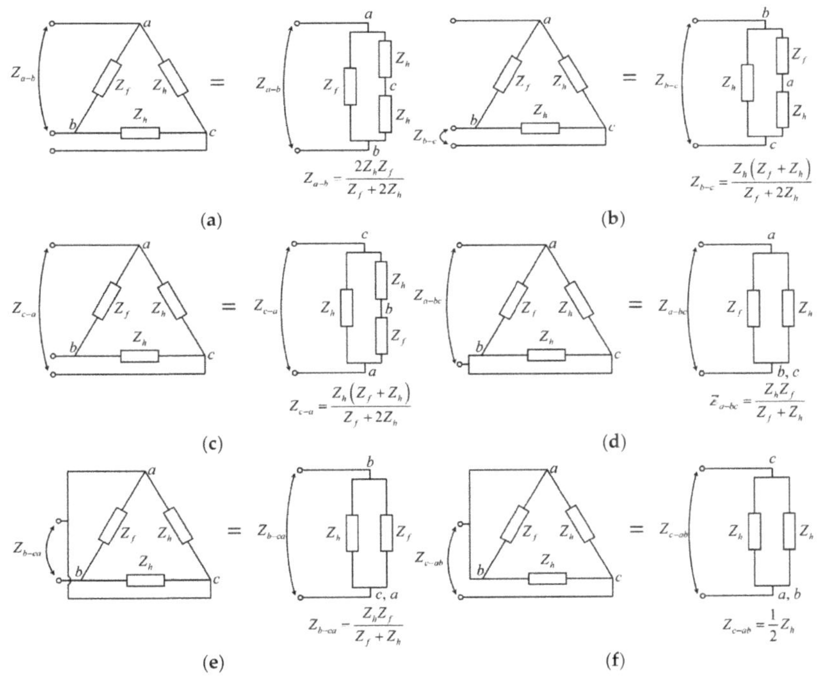

Figure 3. Impedance analysis for a Δ connection with a phase A fault: (**a**) *a-b*; (**b**) *b-c*; (**c**) *c-a*; (**d**) *a-bc*; (**e**) *b-ca*; (**f**) *c-ab*.

In the same way, the impedance values for terminal pairs *b-c* and *c-a*, corresponding to Figure 3b,c, are described in Equation (11).

$$Z_{b-c} = Z_{c-a} = Z_h \,||\, (Z_f + Z_h) = \frac{Z_h(Z_f + Z_h)}{Z_f + 2Z_h} \tag{11}$$

Under ITF conditions, the impedance of the fault phase decreases due to the fault resistance being placed in parallel with the stator winding. As a result, the impedance of the *a-b* connection becomes lower than those of the *b-c* and *c-a* connections, as demonstrated in Equation (12).

$$\begin{aligned} Z_{a-b} - Z_{b-c} &= \frac{Z_h(Z_f - Z_h)}{Z_f + 2Z_h} < 0 \\ Z_{a-b} &< Z_{b-c} = Z_{c-a} \end{aligned} \tag{12}$$

The impedance for the terminal connection between c and the midpoint of *a-b*, denoted as *c-ab* and shown in Figure 3f, is calculated as in Equation (13).

$$Z_{c-ab} = Z_h \,||\, Z_h = \frac{1}{2} Z_h \tag{13}$$

Likewise, the impedances for connections *a-bc* and *b-ca*, represented in Figure 3d,e, are given by Equation (14).

$$Z_{a-bc} = Z_{b-ca} = Z_f \,||\, Z_h = \frac{Z_f Z_h}{Z_f + Z_h} \tag{14}$$

A comparison of Equations (13) and (14) reveals that the impedance of the *c-ab* configuration is greater than those of the *a-bc* and *b-ca* connections, as summarized in Equation (15).

$$\begin{aligned} Z_{c-ab} - Z_{a-bc} &= \frac{Z_h}{2} \frac{Z_h - Z_f}{Z_f + Z_h} > 0 \\ Z_{c-ab} &> Z_{a-bc} = Z_{b-ca} \end{aligned} \tag{15}$$

When a fault occurs in phase B or C, the impedance becomes either a maximum or a minimum, depending on the specific terminal connection. Table 2 presents the maximum and minimum impedance values for each fault phase and terminal configuration. This information enables effective ITF diagnosis based on impedance characteristics.

Table 2. Impedance analysis results for Δ connection with phase A fault.

Fault Phase	Minimum Impedance among $Z_{a\text{-}b}$, $Z_{b\text{-}c}$, and $Z_{c\text{-}a}$	Maximum Impedance among $Z_{a\text{-}bc}$, $Z_{b\text{-}ca}$, and $Z_{c\text{-}ab}$
A	$Z_{a\text{-}b}$	$Z_{c\text{-}ab}$
B	$Z_{b\text{-}c}$	$Z_{a\text{-}bc}$
C	$Z_{c\text{-}a}$	$Z_{b\text{-}ca}$

However, in IMD systems, such as those used in electric vehicles, mechanical access to motor terminals is often limited or impractical. A diagnostic approach based on space voltage vector control is proposed to address this limitation by emulating terminal connection conditions electrically, without physical rewiring.

3. Inter-Turn Fault Diagnosis of Integrated Motor and Drive Systems

As shown in Section 2, impedance is analyzed for the terminal configuration when an ITF occurs in phase A with both Y and Δ connections. Physical terminal access is required to measure impedances. However, such access is often limited or impractical in IMD systems. In this section, the space voltage vector method is employed to analyze the impedance.

3.1. Space Voltage Vector in Three-Phase Inverter

There are six switching components in a three-phase inverter, as shown in Figure 4a. Each phase has two switching states (ON and OFF), resulting in three possible configurations: (1) upper switch ON, lower switch OFF: '1'; (2) upper switch OFF, lower switch

ON: '0'; (3) both switches OFF: '-'. In a three-phase inverter, the possible voltage states are expressed as shown in Figure 4b. The space voltage vector is defined as follows:

$$\mathbf{V} = V_m e^{j\theta_m} \tag{16}$$

where **V** is the space voltage vector, V_m is the magnitude of the space voltage vector, and θ_m is the phase angle of the space voltage vector. The space voltage vector is determined by the switch states for each phase. Table 3 shows the phase angles of the space voltage vector corresponding to each switching state.

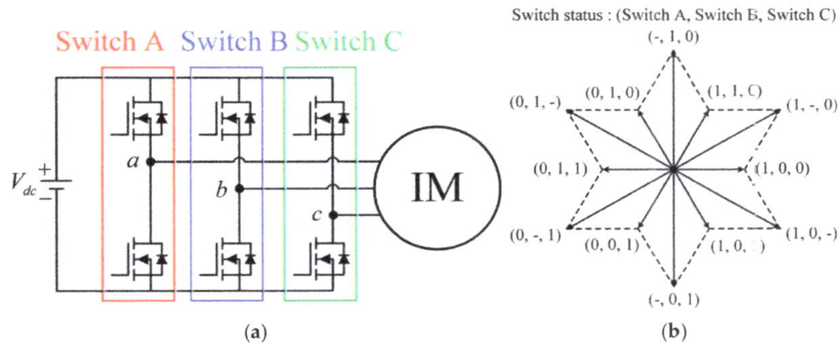

Figure 4. Three-phase inverter configuration and space voltage vectors for switching conditions: (**a**) integrated motor and drive system; (**b**) space voltage vectors based on switching conditions for each phase.

Table 3. Phase of space voltage vector for switching conditions.

Phase A Switch	Phase B Switch	Phase C Switch	Phase of Voltage Vector [°]
1	0	0	0
1	-	0	30
1	1	0	60
-	1	0	90
0	1	0	120
0	1	-	150
0	1	1	180
0	-	1	210
0	0	1	240
-	0	1	270
1	0	1	300
1	0	-	330

3.2. Terminal Connection for Space Voltage Vector

Since the space voltage vector is determined by the switching states for each phase, as shown in Figure 4 and Table 3, the terminal connection is implemented for the space voltage vector as shown in Figure 5. In the phase with a space voltage vector of 0° the resulting terminal connection corresponds to *a-bc*, as illustrated in Figure 5a. At 30°, the configuration results in a terminal connection equivalent to *c-a*, as shown in Figure 5b. In the same way, each space voltage vector position shown in Figure 5 yields an electrical connection that corresponds directly to a specific mechanical terminal connection described in Figures 2 and 3. Table 4 summarizes the relationship between the phase of the space voltage vector and the equivalent terminal connections. This correspondence demonstrates that terminal connections can be realized based on the space voltage vector in the three-phase inverter. The alternating voltage is applied to the motor for the space voltage vector, and the calculated impedance from the measured phase current can be used in the fault diagnosis, as shown in Section 2. As a result, ITF diagnosis can be achieved electrically without the need for the physical reconnection of motor terminals in IMD systems.

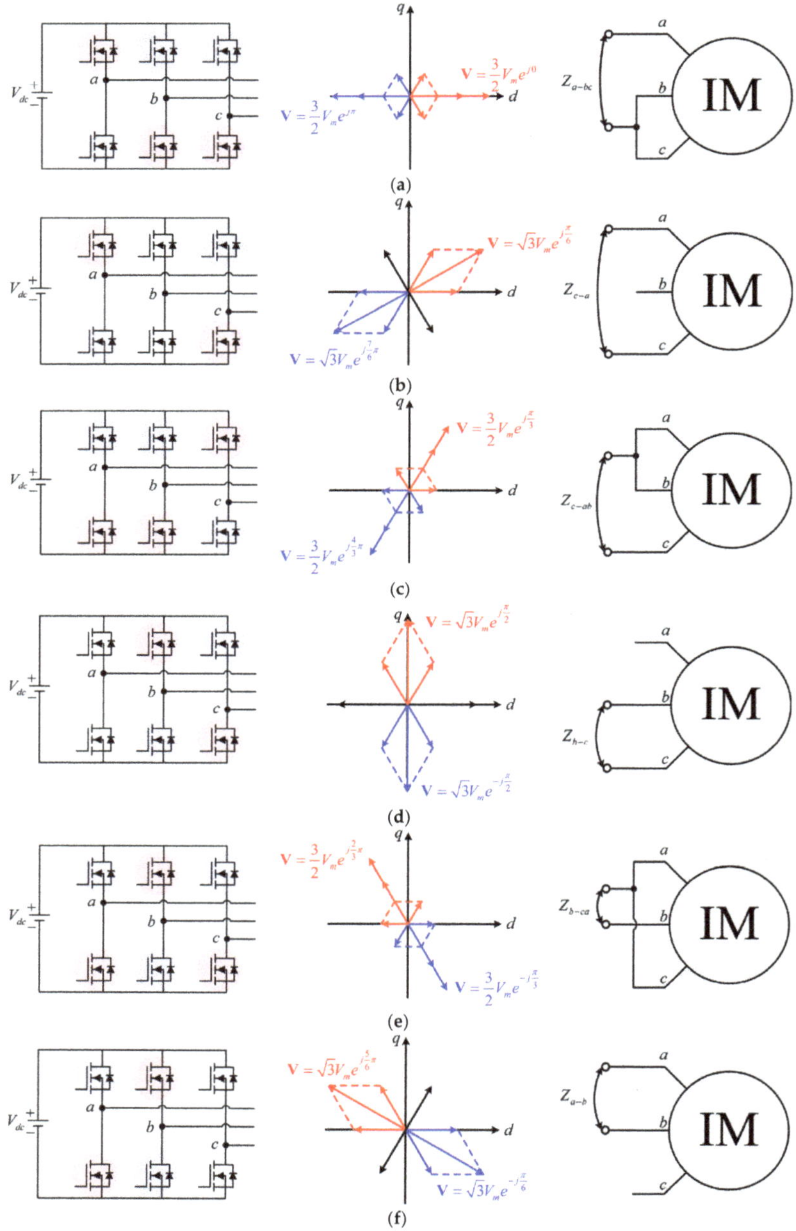

Figure 5. Switching status, space voltage vector, and terminal connection in IMD with three-phase inverter: (**a**) 0°; (**b**) 30°; (**c**) 60°; (**d**) 90°; (**e**) 120°; (**f**) 150°.

Table 4. Terminal connections for phases of space voltage vector.

Phase of Space Voltage Vector [°]	Terminal Connection
0 and 180	*a-bc*
30 and 210	*c-a*
60 and 240	*c-ab*
90 and 270	*b-c*
120 and 300	*b-ca*
150 and 330	*a-b*

3.3. Fault Diagnosis Algorithm

The terminal connections are implemented based on the space voltage vector by the three-phase inverter. Based on the impedance and space voltage vector, the ITF algorithm is proposed as shown in Figure 6. The alternating voltage is applied to the motor according to the phase of the space voltage vector through the three-phase inverter. The impedances corresponding to each phase of the space voltage vector are calculated from the input voltage and measured phase currents. The fault diagnosis is performed by identifying the maximum and minimum impedances associated with each phase of the space voltage vector. For example, when an ITF occurs in phase A with the Δ connection, the phases of the space voltage vector corresponding to the maximum and minimum impedance values are 60° and 150°, respectively, as shown in Figure 6b. In the same way, when the ITF occurs in phase B or C, the fault diagnosis can be performed as shown in Figure 6.

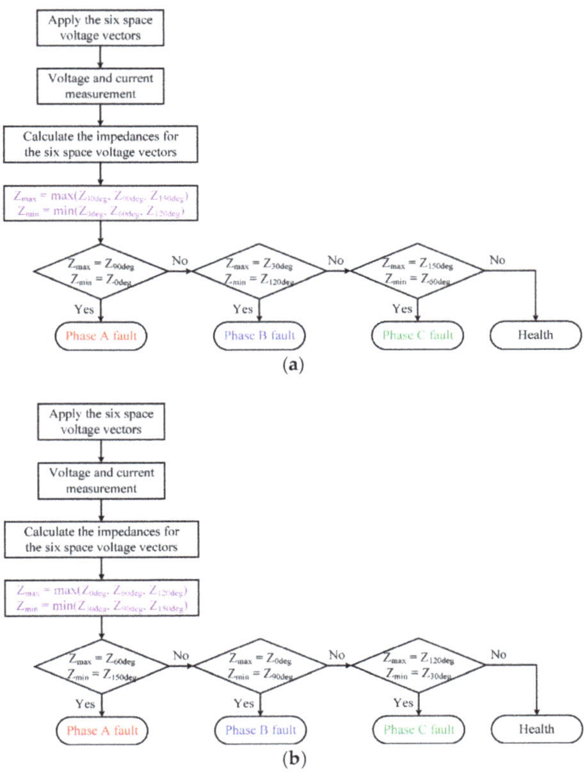

Figure 6. Inter-turn fault diagnosis algorithm: (**a**) Y connection; (**b**) Δ connection.

4. Experimental Verification

This section presents the experimental verification of the proposed fault diagnosis algorithm. A 12 kW three-phase IM from RENAULT TWIZY, used for EV applications, is described in Table 5. The motor parameters and fault resistance were measured using the impedance analyzer. The phase resistance of the IM was 11.5 mΩ, and a fault resistance of 3.6 mΩ was considered. Figure 7 shows the manufactured IM and terminal box used to implement the winding connections and simulate fault conditions. The windings of the IM were separated, as shown in Figure 8, and connected through the terminal box. The experimental setup for ITF diagnosis is shown in Figure 9. The test motor was coupled to a dynamometer system, and both Y and Δ connections were implemented through the terminal box. A three-phase inverter was used to apply the space voltage vector to the IM. Fault diagnosis was conducted with both Y and Δ connections, as illustrated in Figure 10. The fault resistance was connected in parallel with the faulty phase. The voltage and current were measured using voltage and current sensors, such as the HAH1DR series from LEM, and the impedance was calculated for each space voltage vector. The measured voltage and current were converted to digital values using an analog-to-digital converter in a digital signal processor (DSP), specifically the TMS320F28069, to enable impedance calculation. Moreover, the calculated impedance, the phase of the space voltage vector, and the diagnostic status were converted back to analog signals via the DSP's digital-to-analog converter and displayed on an oscilloscope. The experimental results for the Y and Δ connections are shown in Figures 11 and 12, respectively. As shown in Figure 11, when an ITF occurs in phase A, the impedance reaches its maximum at a phase angle of 90° and its minimum at 0°, corresponding to the specific phase angles of the space voltage vector in the three-phase inverter. When the fault occurs in phase B, the maximum impedance is observed at 30° and the minimum at 120°. Similarly, for a fault in phase C, the impedance is maximized at 150° and minimized at 60°. As demonstrated in Figure 12, the measured impedance values for each fault condition were analyzed based on measured currents. By comparing the impedance at each space voltage vector phase, the presence and location of the fault are determined based on the fault diagnosis algorithm shown in Figure 6. As shown in Figures 11 and 12, the fault diagnosis algorithm was verified and conducted within 3.5 s.

Table 5. Specification of test motor.

Contents	Value	Unit
DC voltage	52	V
Power	12	kW
Current	248	Arms
Phase resistance	11.5	mΩ
Phase inductance	0.6	mH
Fault resistance	3.6	mΩ
Frequency of space voltage vector	500	Hz

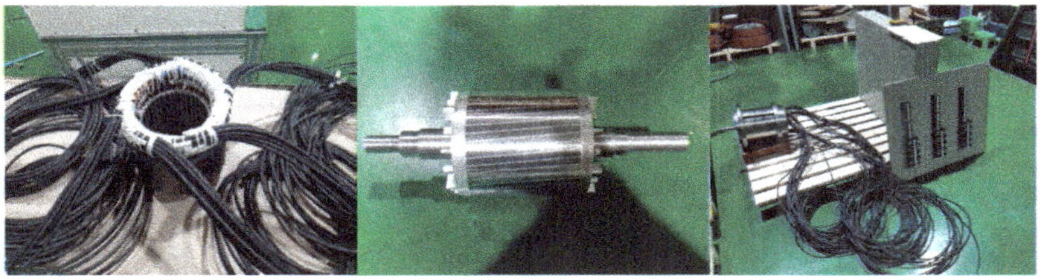

(**a**) (**b**) (**c**)

Figure 7. Manufacture and experimental environment: (**a**) stator; (**b**) rotor; (**c**) terminal box.

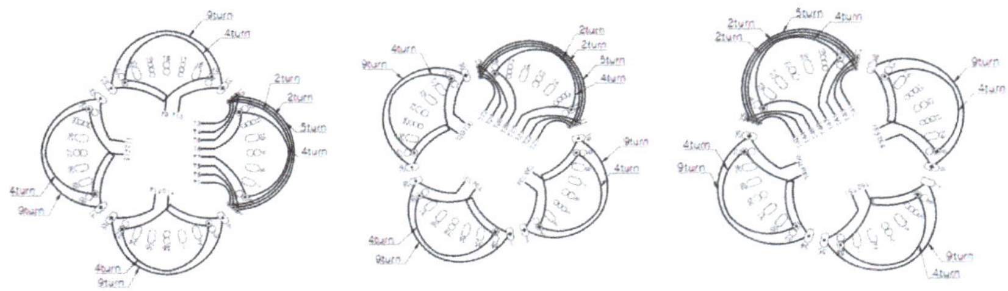

(**a**) (**b**) (**c**)

Figure 8. Winding configuration for validation of inter-turn fault algorithm: (**a**) phase A; (**b**) phase B; (**c**) phase C.

(**a**) (**b**) (**c**)

Figure 9. Experimental environment for inter-turn fault diagnosis: (**a**) motor and terminal box; (**b**) fault situation; (**c**) inverter for fault diagnosis.

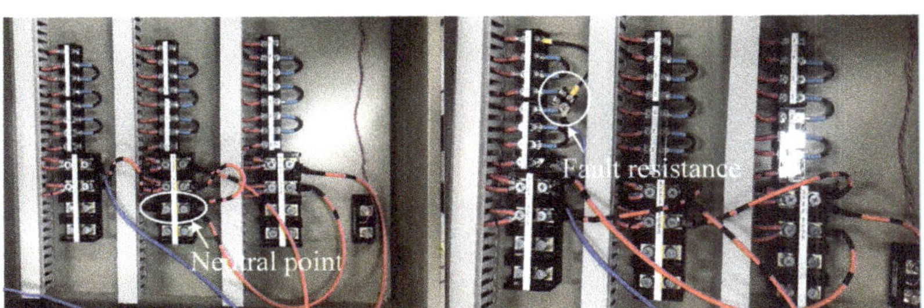

(**a**) (**b**)

Figure 10. Connections for experiment: (**a**) Y connection; (**b**) Δ connection.

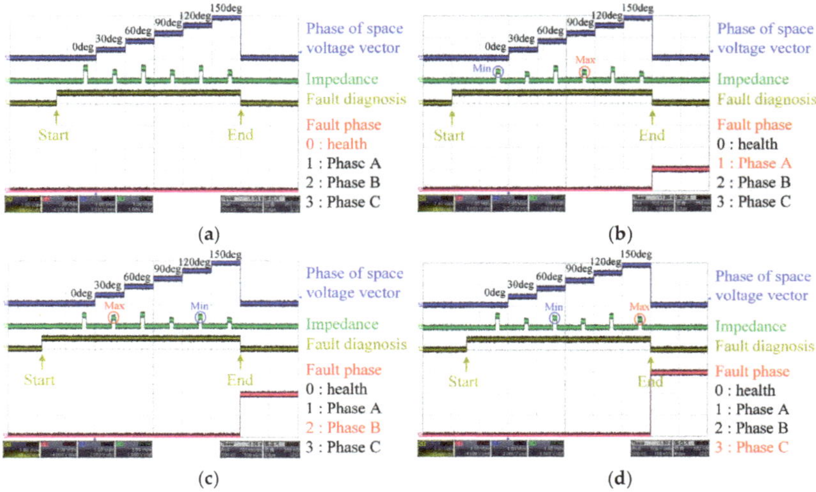

Figure 11. Experiment results for Y connection: (**a**) healthy; (**b**) fault phase A; (**c**) fault phase B; (**d**) fault phase C.

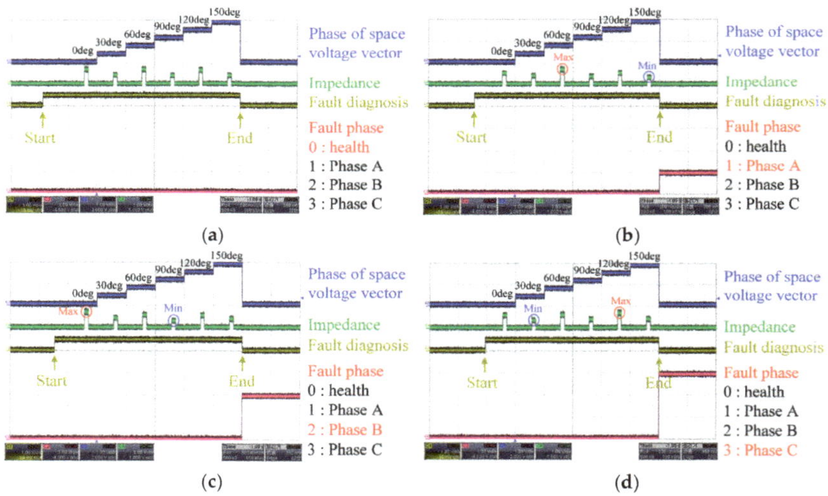

Figure 12. Experiment results for Δ connection: (**a**) healthy; (**b**) fault phase A; (**c**) fault phase B; (**d**) fault phase C.

5. Conclusions

In this paper, an inter-turn fault diagnosis algorithm is proposed based on impedance analysis and space voltage vector control for IMD systems. The terminal impedances when an inter-turn fault occurs in phase A are analyzed, revealing specific maximum and minimum impedance patterns with both Y and Δ connections. The space voltage vectors of a three-phase inverter are also analyzed according to the switching states of each phase. A fault diagnosis algorithm is proposed by associating the terminal connections with the phases of the space voltage vector. An experimental setup, including a 12 kW three-phase induction motor from RENAULT TWIZY and a terminal box for winding connections, was implemented to validate the proposed algorithm. The impedances for each space voltage vector were calculated using a DSP (TMS320F28069) based on measured current and voltage and they were displayed on the oscilloscope using the processor's DAC. Inter-turn faults and various winding connections were implemented through the terminal box to emulate different fault conditions. The proposed diagnosis algorithm was verified through the experimental results for faults in each phase for both Y and Δ connections, and the fault diagnosis was conducted within 3.5 s. This approach provides a practical and reliable solution for fault detection in highly integrated EV drive systems where mechanical access to motor terminals is restricted. Furthermore, the proposed ITF diagnosis does not require prior information about the electrical machine, such as healthy impedance, as it relies on the maximum and minimum impedance values associated with six space voltage vectors. Future research will focus on analyzing the frequency dependence of measured impedance to determine the optimal frequency range for reliable fault detection. To enhance diagnostic robustness, meta-heuristic techniques, such as genetic algorithms for optimization and support vector machines for classification, will also be investigated.

Author Contributions: Conceptualization, J.A. and H.K.; methodology, H.K.; software, J.A.; validation, J.A. and H.K.; formal analysis, H.K.; investigation, J.A.; resources, J.L.; data curation, J.A.; writing—original draft preparation, J.A. and H.K.; writing—review and editing, J.L. and H.K.; visualization, J.A.; supervision, J.L. and H.K.; project administration, H.K.; funding acquisition, J.L. All authors have read and agreed to the published version of the manuscript.

Funding: This work was supported by the Korea Institute of Energy Technology Evaluation and Planning (KETEP) grant funded by the Korea government (MOTIE) (RS-2023-00237024, Development and demonstration of medium-sized variable speed high-power motor).

Institutional Review Board Statement: Not applicable.

Informed Consent Statement: Not applicable.

Data Availability Statement: Data are contained within the article.

Conflicts of Interest: The authors declare no conflicts of interest.

References

1. Azab, M. A Review of Recent Trends in High-Efficiency Induction Motor Drives. *Vehicles* **2025**, *7*, 15. [CrossRef]
2. Zhang, B.; Song, Z.; Liu, S.; Huang, R.; Liu, C. Overview of Integrated Electric Motor Drives Opportunities and Challenges. *Energies* **2022**, *15*, 8299. [CrossRef]
3. Wu, X.; Geng, Y.; Li, M.; Wang, W.; Tu, M. Inter-Turn Short Circuit Diagnosis of Permanent Magnet Synchronous Motor Based on Siamese Convolutional Neural Network Under Small Fault Samples. *IEEE Sensors J.* **2024**, *24*, 26982–26993. [CrossRef]
4. Bahrami-Fard, M.; Chen, T.; Winchell, E.; Fahimi, B. Integrated Induction Motor Drive for Variable Speed Industrial Applications. *IEEE Trans. Power Electron.* **2025**, *40*, 7176–7188. [CrossRef]
5. Gulsuna, O.; Tokgoz, F.; Karakaya, F.; Keysan, O. Design and Analysis of a GaN-Based Megahertz Integrated Motor Drive for a PCB Motor. *IEEE Trans. Ind. Electron.* **2024**, *71*, 15435–15444. [CrossRef]
6. Zhang, J.; Zhang, X.; Gu, C.; Li, J.; Zhang, H.; Zhao, H. Characterizations of PM Synchronous Motor with an Integrated Common-Mode Voltage Filter. *IEEE Trans. Ind. Appl.* **2024**, *60*, 353–366.
7. Nam, D.W.; Hong, M.K.; Jo, N.R.; Jung, D.H.; Kim, W.H. Design of Coil Patterns for an Axial Flux Permanent Magnet Synchronous Motor with PCB Stator. *IEEE Access* **2025**, *13*, 12936–12944. [CrossRef]
8. Naradhipa, A.M.; Kim, S.; Yang, D.; Choi, S.; Yeo, I.; Lee, Y. Power Density Optimization of 700 kHz GaN-Based Auxiliary Power Module for Electric Vehicles. *IEEE Trans. Power Electron.* **2021**, *36*, 5610–5621. [CrossRef]
9. Hwang, K.Y.; Kim, S.I.; Song, B.K. Single Winding Type Determination of Dual Winding Three-Phase Motor Considering Overheat Problem in Integrated Electric Braking System of Autonomous Vehicles. *IEEE Trans. Transp. Electrific.* **2023**, *9*, 556–666. [CrossRef]
10. Sun, X.; Song, C.; Zhang, Y.; Sha, X.; Diao, N. An Open-Circuit Fault Diagnosis Algorithm Based on Signal Normalization Preprocessing for Motor Drive Inverter. *IEEE Trans. Instrum. Meas.* **2023**, *72*, 3513712. [CrossRef]
11. Orlowska-Kowalska, T.; Wolkiewicz, M.; Pietrzak, P.; Skowron, M.; Ewert, P.; Tarchala, G.; Krzysztofiak, M.; Kowalski, C.T. Fault Diagnosis and Fault-Tolerant Control of PMSM Drives–State of the Art and Future Challenges. *IEEE Access* **2022**, *10*, 59979–60024. [CrossRef]

12. Gupta, V.; Sultana, M.; Khanna, A. Hybrid optimization-based approach for transmission line parameter identification in the presence of noise. *Data Anal. J.* **2023**, *9*, 100182.
13. Jin, L.; Wang, X.; Mao, Y.; Lu, L.; Wang, Z. Online Attribute Matching Based Few-Sample Data-Driven Diagnosis of Electrical Faults in PMSM Drive. *IEEE Trans. Power Electron.* **2024**, *39*, 2620–2631. [CrossRef]
14. Tokgoz, F.; Gülsuna, Ö.; Karakaya, F.; Cakal, G.; Keysan, O. Mechanical and Thermal Design of an Optimized PCB Motor for an Integrated Motor Drive System with GaNFETs. *IEEE Trans. Energy Convers.* **2023**, *38*, 653–661. [CrossRef]
15. Wang, S.; Sun, Y.; Huang, Z.; Mu, S. Analysis of Stator Internal Phase-to-Phase Short Circuit in the 12-Phase Synchronous Generator with Rectifier-Load System. *IEEE Trans. Energy Convers.* **2018**, *33*, 299–311. [CrossRef]
16. Yuan, W.; Li, Y.; Xu, L.; Li, T.; Chen, X. A Fast Faulty Phase Selection Method Considering Fault Tolerance for Single Phase to Ground Fault in Distribution Networks. *IEEE Trans. Instrum. Meas.* **2023**, *72*, 3533212. [CrossRef]
17. Yuan, J.; Feng, C.; Ji, Y.; Liu, X.; Dong, X.; Liu, J. Faulty Feeder Detection for Successive Single Phase-to-Ground Faults Based on Line Model Recognition and Correlation Comparison. *IEEE Trans. Power Del.* **2025**, *40*, 750–763. [CrossRef]
18. Gao, C.; Gao, B.; Xu, X.; Si, J.; Hu, Y. Automatic Demagnetization Fault Location of Direct-Drive Permanent Magnet Synchronous Motor Using Knowledge Graph. *IEEE Trans. Instrum. Meas.* **2024**, *73*, 3502312. [CrossRef]
19. Bonet-Jara, J.; Pons-Llinares, J.; Gyftakis, K.N. Comprehensive Analysis of Principal Slot Harmonics as Reliable Indicators for Early Detection of Interturn Faults in Induction Motors of Deep-Well Submersible Pumps. *IEEE Trans. Ind. Electron.* **2023**, *70*, 11692–11702. [CrossRef]
20. Choi, J.; Kim, S.H.; Lee, W.K.; Lee, J.S. Rapid Control Prototyping and Analysis of a Fault Diagnosis and Amplifying Algorithm for Broken Rotor Bar Induction Motor Drives. *IEEE Access* **2025**, *13*, 19064–19074. [CrossRef]
21. Kang, Y.; Yao, L. Fault Diagnosis for Permanent Magnet Synchronous Motor with Demagnetization Fault and Sensor Fault. *IEEE Trans. Instrum. Meas.* **2024**, *73*, 3533311. [CrossRef]
22. Ma, J.; Bai, X.; Ma, F.; Zhuo, S.; Sun, B.; Li, C. Convolutional Neural Network Design Based on Weak Magnetic Signals and Its Application in Aircraft Bearing Fault Diagnosis. *IEEE Sens. J.* **2024**, *24*, 36031–36043. [CrossRef]
23. Niu, G.; Dong, X.; Chen, Y. Motor Fault Diagnostics Based on Current Signatures: A Review. *IEEE Trans. Instrum. Meas.* **2023**, *72*, 3520919. [CrossRef]
24. Lang, W.; Hu, Y.; Gong, C.; Zhang, X.; Xu, H.; Deng, J. Artificial Intelligence-Based Technique for Fault Detection and Diagnosis of EV Motors: A Review. *IEEE Trans. Transp. Electrific.* **2022**, *8*, 384–406. [CrossRef]
25. Chen, C.S.; Lin, C.J.; Yang, F.J.; Lin, F.C. Model Design of Inter-Turn Short Circuits in Internal Permanent Magnet Synchronous Motors and Application of Wavelet Transform for Fault Diagnosis. *Appl. Sci.* **2024**, *14*, 9570. [CrossRef]
26. Qin, Y.; Li, G.J.; Jia, C.; McKeever, P. Investigation of Inter-Turn Short-Circuit Fault of PM Machines Using PWM Voltage-Based Modeling. *IEEE Trans. Transp. Electrific.* **2024**, *10*, 1324–1334. [CrossRef]
27. Cui, R.; Fan, Y.; Li, C. On-Line Inter-Turn Short-Circuit Fault Diagnosis and Torque Ripple Minimization Control Strategy Based on OW Five-Phase BFTHE-IPM. *IEEE Trans. Energy Convers.* **2018**, *33*, 2200–2209. [CrossRef]
28. Chen, Q.; Han, X.; Liu, G.; Zhao, W.; Shi, H. Inter-Turn Fault Diagnosis and Control for Five-Phase PMSMs by Disturbance Observer. *IEEE Trans. Ind. Electron.* **2024**, *71*, 13901–13909. [CrossRef]
29. Dongare, U.V.; Umre, B.S.; Ballal, M.S.; Dongare, V.P. Online Inter-Turn Fault Detection in Wound Rotor Induction Motors Based on VI Loci Pattern. *IEEE Trans. Ind. Appl.* **2024**, *60*, 411–425. [CrossRef]
30. Asad, B.; Vaimann, T.; Belahcen, A.; Kallaste, A.; Rassõlkin, A.; Ghafarokhi, P.S.; Kudelina, K. Transient Modeling and Recovery of Non-Stationary Fault Signature for Condition Monitoring of Induction Motors. *Appl. Sci.* **2021**, *11*, 2806. [CrossRef]
31. Zhang, J.; Zhan, W.; Ehsani, M. Fault-Tolerant Control of PMSM with Inter-Turn Short-Circuit Fault. *IEEE Trans. Energy Convers.* **2019**, *34*, 2267–2275. [CrossRef]
32. Asad, B.; Vaimann, T.; Belahcen, A.; Kallaste, A.; Rassõlkin, A.; Iqbal, M.N. The Cluster Computation-Based Hybrid FEM-Analytical Model of Induction Motor for Fault Diagnostics. *Appl. Sci.* **2020**, *10*, 7572. [CrossRef]
33. Gu, B.-G. Offline Interturn Fault Diagnosis Method for Induction Motors by Impedance Analysis. *IEEE Trans. Ind. Electron.* **2018**, *65*, 5913–5920. [CrossRef]

applied sciences

MDPI

Article

Online High Frequency Impedance Identification Method of Inverter-Fed Electrical Machines for Stator Health Monitoring

Jérémy Creux [1,2], Najla Haje Obeid [1], Thierry Boileau [2,*] and Farid Meibody-Tabar [2]

1 IFP Energies Nouvelles, Institut Carnot IFPEN Transports Energie, 1 et 4 Avenue de Bois-Préau, 92852 Rueil-Malmaison, France; jeremy.creux@ifpen.fr (J.C.); najla.haje-obeid@ifpen.fr (N.H.O.)
2 Université de Lorraine, CNRS, LEMTA, F-54000 Nancy, France; farid.meibody-tabar@univ-lorraine.fr
* Correspondence: thierry.boileau@univ-lorraine.fr

Abstract: In electric powertrain traction applications, the adopted trend to improve the performance and efficiency of electromechanical power conversion systems is to increase supply voltages and inverter switching frequencies. As a result, electrical machine conductors are subjected to ever-increasing electrical stresses, leading to premature insulation degradation and eventual short-circuits. Winding condition monitoring is crucial to prevent such critical failures. Based on the scientific literature, several methods can be used for early identification of aging. A first solution is to monitor partial discharges. This method requires the use of a specific measurement device and an undisturbed test environment. A second solution is to monitor the inter-turn winding capacitance, which is directly related to the condition of the insulation and can cause a change in the stator impedance behavior. Several approaches can be used to estimate or characterize this impedance behavior. They must be performed on a machine at standstill, which limits their application. In this paper, a new characterization method is proposed to monitor the high-frequency stator impedance evolution of voltage source inverter-fed machines. This method can be applied at any time without removing the machine from its operating environment. The range and accuracy of the proposed frequency characterization depend in particular on the supply voltage level and the bandwidth of the measurement probes. The effects of parameters such as temperature, switching frequency, and DC voltage amplitude on the impedance characteristic were also studied and will be presented. Tests carried out on an automotive traction machine have shown that the first two series and parallel resonances of the high-frequency impedance can be accurately identified using the proposed technique. Therefore, by monitoring these resonances, it is possible to predict the aging rate of the conductor

Keywords: winding health monitoring; frequency response analysis; online characterization; impedance; high-frequency behavior; inverter-fed electrical machines

1. Introduction

The health monitoring of stator windings in electrical machines is one of the most important challenges in the automotive traction industry. Increasing supply voltages and inverter switching frequencies lead to accelerated aging of these windings [1]. Such aging is characterized by the degradation of insulation layers between conductors. Factors such as the duration and level of exposure to electrical stress influence the quality of these insulators [2]. In order to reduce the risks of premature insulation degradation and to increase winding robustness, several techniques can be considered: improving the conductor impregnation or using enamels that are particularly resistant to partial discharge phenomena [3]; promoting the design of hairpin winding machines that help reduce vibration phenomena in the winding heads [4]. However, it remains essential to monitor the health of these windings to identify their aging as early as possible. Therefore, if a given signature were to exceed a certain threshold, preventive maintenance could be considered to protect the system and reduce maintenance and repair costs and times [5,6].

There are several approaches to monitoring the health of an electrical machine winding. First, partial discharge measurement can be highlighted [7–9]. Some strategies use insulation resistance measurement and polarization index calculation to perform the diagnosis [10,11]. Estimation of equivalent phase-to-ground or interturn capacitance [12,13] may also be conducted. The main drawback of these methods is that they are usually performed on a machine at standstill, disconnected from the power supply. In some applications, particularly industrial applications, these test conditions cannot be met, making aging monitoring impossible. Furthermore, resistance tests can give an indication of windings state of health [14], but since these tests are destructive, they cannot be applied industrially. Statistical studies are used in [15] to estimate the remaining useful life (RUL) of a system. Finally, in recent literature, many aging monitoring methods are based on the analysis of systems frequency response to determine the winding condition. The various approaches used are usually based on the variation of the frequency signatures of the machine impedance characteristic [16–19]. This characterization over a wide frequency band takes into account leakage and magnetization inductances, iron and eddy current losses, as well as interturn and phase-to-ground capacitive couplings. As a result, the different resonance frequencies reflect these different phenomena.

It has been shown in [12] that the decrease of the winding breakdown voltage correlates with an increase of the machine equivalent capacitance. Furthermore, ref. [16] explains that this equivalent capacitance is particularly impacted by the evolution of the capacitance between windings. It can therefore be concluded that its monitoring will help to detect the aging of a machine inter-turn insulator at an early stage. Finally, the model developed in [20] shows that this inter-turn capacitance has a strong influence on the impedance frequency behavior of the machine. As a result, this characteristic will evolve as the machine age. Ref. [16] emphasizes that the first resonance frequency of the impedance characteristic decreases with time, indicating degradation of the winding insulation. It has also been observed in [18] when the windings are subjected to adverse environmental stresses, such as an exposure to a corrosive substance on their insulators. The resonance frequency decreases with prolonged exposure to the corrosive agent. The amplitude of this resonance also decreases as the conductors age. All of these results demonstrate the benefits of monitoring the evolution of frequency impedance over time to assess the aging of electrical machine windings.

Based on the above considerations, we decided to develop an online characterization and diagnostic method for monitoring the health of stator windings of electric traction machines. The advantage of an online approach is that it provides a continuous flow of information within the functional environment of the drive and without interrupting its operation. In particular, this approach can be used to estimate the "remaining useful life" (RUL) of the winding [21–23]. Therefore, an accurate and efficient condition-based maintenance strategy can be established to ensure the availability of the system.

This paper is organized as follows: Section 2 presents a review of the different types of impedance measurements in electrical machines and the associated characterization methods. Section 3 focuses on the broadband frequency behavior of these machines and reviews the different approaches found in the literature to model their high-frequency (HF) behavior. Section 4 explains the online strategy we have developed to characterize this frequency response without removing the machine from its operating environment. Section 5 examines the impact of various factors, such as operating temperature, supply voltage level, and the presence of the rotor in the machine, on the impedance frequency response of the machine. Finally, Section 6 draws conclusions from the paper and discusses future works.

2. Impedance Measurement in Electrical Drive Systems

2.1. Electrical Machine Impedance Definition

Several types of impedance measurements can be performed on a permanent magnet synchronous machine (PMSM). The most common types in the literature are common-mode (CM) and differential-mode (DM) measurements. On an electrical machine with a star-connected winding, these CM and DM measurements are carried out by connecting the different phases of the machine, as shown in Figure 1a,b. For condition monitoring purposes and to locate the fault in the windings, it is possible to successively analyze the behavior of two of the machine phases, as shown in Figure 1c. Finally, with access to the neutral point, it is possible to characterize the behavior of each phase individually, as described in Figure 1d.

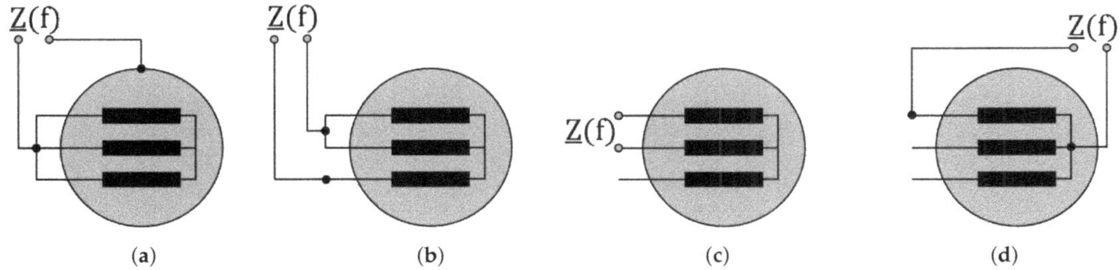

Figure 1. Connection and measurement circuit for CM (**a**), DM (**b**), phase-to-phase (**c**), and phase-to-neutral (**d**) impedance characterization in a star-connected three-phase machine.

Similarly, for machines with delta-connected windings, the CM and DM impedance behaviors are measured by connecting the machine phases as shown in Figure 2a,b. Figure 2c describes the connection scheme for characterizing the impedance behavior of each of the machine phases.

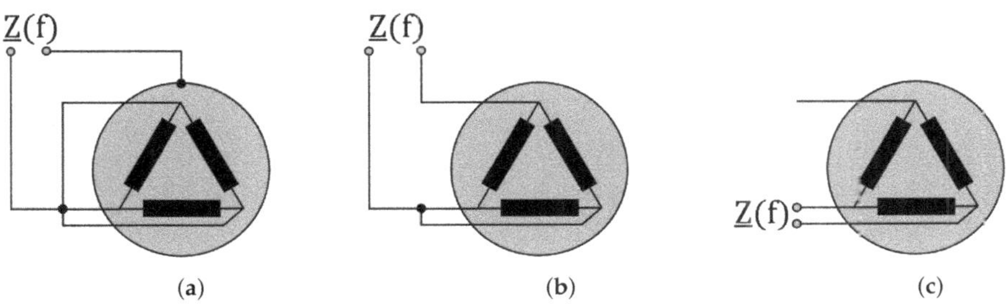

Figure 2. Connection and measurement circuit for CM (**a**), DM (**b**), and phase-to-phase (**c**) impedance characterization in a delta-connected three-phase machine.

Finally, in the case of an open-end winding, the behavior of each machine phase can be straightforwardly characterized according to the measurement scheme of Figure 3.

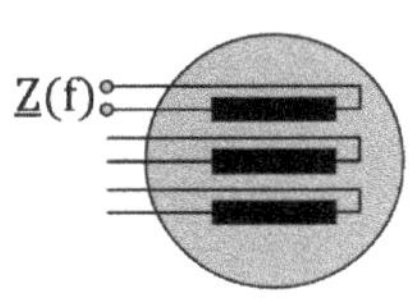

Figure 3. Connection and measurement circuit for phase-to-phase impedance characterization in an open-end winding three-phase machine.

2.2. Impedance Characterization Methods

Impedance measurement is traditionally performed on a stationary machine disconnected from any power source using an impedance analyzer [24–27], i.e., a measurement device dedicated to characterizing the winding impedance. Multiple instruments are usually required to cover a wide frequency range. The reason for this is that different calculation approaches must be implemented for certain frequency ranges in order to guarantee the best possible accuracy. It should be noted that the calibration process is a very important step in obtaining consistent and accurate results. The vector network analyzer we used in our study performs a so-called "Gain/Phase" calculation over a frequency range [5 Hz–30 MHz] to determine the impedance characteristic of a machine [28]. The approach consists of the application of a sine wave excitation and the measurement of voltages at two different points of a system to evaluate the impedance of a circuit at a given frequency. The instrument then scans multiple excitation frequencies while storing the various input and output signals obtained in order to construct a Bode diagram of the impedance over the selected frequency range.

Another approach to offline characterization of stator windings requires an oscilloscope to measure the amplitudes of the voltage across a winding and the current flowing through it, as well as the temporal phase shift between these two signals. Then, by applying Ohm law, we can calculate the amplitude and the phase of the winding impedance at a given frequency [17]. The winding must be connected to a low-frequency voltage generator (LFVG), which provides a sweep of frequencies from 1 kHz to 30 MHz to extract the Bode diagram of this impedance. The connection and measurement scheme of this method are shown in Figure 4. This approach can therefore be implemented using fairly common electrical engineering equipment. However, it is a time-consuming process that is difficult to automate and requires numerous measurement points, especially for frequencies above megahertz, to ensure the same accuracy as the results obtained with the impedance analyzer. Moreover, the accuracy of the characterization around certain resonances may be limited by the current amplitude that the LFVG can deliver.

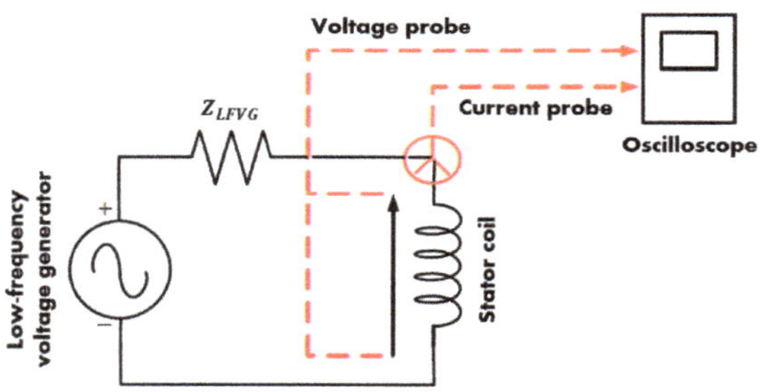

Figure 4. Connection and measurement scheme for impedance characterization, using a LFVG and an oscilloscope—redraw from [17].

The main disadvantage of these characterization methods is that they must be performed on a machine that is disconnected from the power supply. This can be particularly challenging in an industrial application. However, online approaches have the advantage of providing continuous information measurement without interrupting the drive's nominal operation. When monitoring the health of windings in an electrical machine, behavioral deviations synonymous with aging can therefore be identified as early as possible. The online approach to impedance characterization most widely used in the literature involves superimposing an HF voltage signal on the machine's main power supply. For this purpose, active [29,30], passive [31,32], or inductive [33] coupling systems are used. An equivalent circuit of the injection and measurement principle taken from [29] is shown in Figure 5.

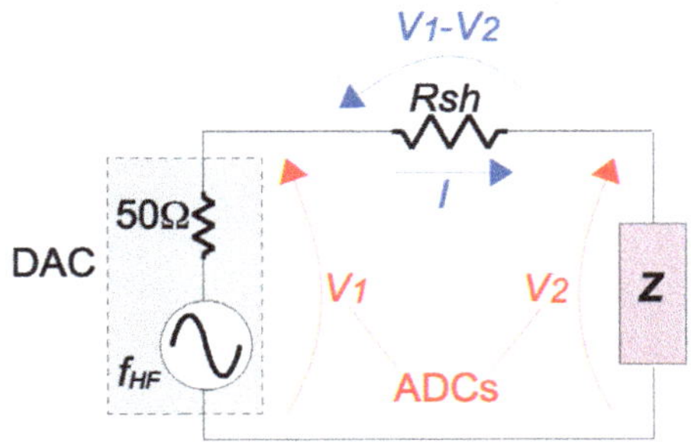

Figure 5. Equivalent circuit for online impedance measurement and characterization, using HF signal injection—taken from [29].

Based on Kirchhoff laws, the impedance for each excitation frequency can be calculated. The derived expression is given by (1).

$$Z = \frac{V_2}{I} = R_{sh}.\frac{V_2}{V_1 - V_2} \quad (1)$$

where Z is the impedance amplitude of the system to be characterized, V_1 and V_2 are the voltage amplitudes measured at the HF generator and machine terminals respectively and R_{sh} is the shunt resistor used to determined the amplitude of the current I.

One of the disadvantages of this approach is its invasiveness, as external HF injection has a significant impact on the nominal operation of the machine and can even accelerate the aging of its windings.

3. Broadband Stator Model for Data Storage

For health monitoring, comparison of a large number of data sets is needed. This requires the ability to store them in the form of models with a limited number of parameters. This would allow to reconstruct the impedance characteristic and keep only the essential information for aging assessment. In this section, the impedance frequency response will be first displayed to highlight the information that needs to be stored. Then, the various models suitable for condition monitoring that meet the aforementioned conditions are detailed.

3.1. Impedance Frequency Response

As mentioned in the previous section, common-mode and differential-mode impedance behaviors are generally characterized and modeled. These works address electromagnetic interference and compatibility issues [34–36], particularly during the interconnection of the various subsystems that form an electromechanical powertrain. The impedance characteristics outlined are intended to evaluate the CM and DM currents that will circulate and then to size the components that will limit them. However, there are also approaches to characterize the impedant or admittant behavior of the machine, carried out from a health monitoring perspective [17,37].

The impedant behavior is generally displayed as a Bode diagram over a wide frequency range (from hundreds of hertz to several tens of megahertz), as shown in the example in Figure 6 taken from [25]. In this example, the low-frequency behavior of CM is purely capacitive, while that of DM is mostly inductive. Moreover, at higher frequencies, there are numerous resonant frequencies that represent the wealth of the information provided by this characteristic.

Ref. [24] explains that the impedance characteristics of mechatronic systems are largely impacted by parasitic effects such as inductive loops, capacitive couplings, or dissipative phenomena. Consequently, it is difficult to accurately predict and model this behavior during the design phase of these systems because there are too many manufacturing uncertainties. Therefore, most studies are based on actual measurements of the impedance behavior and attempt to capture and represent it in a simplified manner.

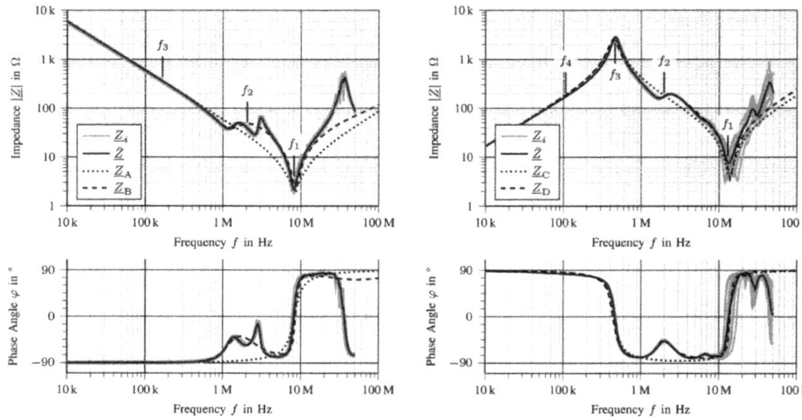

(a) (b)

Figure 6. Example of Bode diagram depicting CM (**a**) and DM (**b**) impedance behaviors—taken from [25].

3.2. High-Frequency Modeling of PMSM

Classical electrical equivalent circuit (EEC) models of electrical machines, either sound [38,39] or tailored to the study of interturn short-circuit faults [40], are external machine models. They can also be described as "low-frequency", since high-frequency phenomena, which are of less importance for the purpose of the modeling, are generally neglected. As mentioned in the previous section and recalled in [41], the HF behavior is particularly sensitive to "parasitic" effects linked to the quality of the conductor insulation, which means that it must be properly modeled to monitor the evolution of the winding state of health. There are several approaches in the literature to model and/or characterize the HF behavior of a PMSM. These have interesting properties for identifying behavioral deviations associated with insulator aging. They can be divided into two categories: those based on a physical model of the machine phases and those based on an experimental frequency response of the machine behavior. The most popular are HF EEC models [19,25,34,42–45]. There are many such methods in the literature, mainly because they can be used to evaluate either overvoltages at the terminals of inverter-fed machines [46], CM currents circulating through bearings [47], or electromagnetic interference phenomena within power conversion chains [35]. The accuracy of these methods depends on their complexity, i.e., the number of parameters to be evaluated. Depending on the purpose of the model and the accuracy required, these parameters are calculated using hybrid methods that combine analytical and numerical calculations with experimental characterization of the machine behavior [48].

There are also transfer function approaches to the impedance frequency response of PMSM, as presented in [17,35]. Finally, there is an HF approach based on the physical design of the machine phases, which takes into account capacitive coupling phenomena between coils and with the machine ground [13].

The next parts will describe these different HF modeling approaches in detail.

3.2.1. PMSM HF Electrical Equivalent Circuit Model

This HF approach of EEC models is similar to conventional low-frequency circuits, but the proposed EEC models are much more complex and take into account a number of phenomena that are considered parasitic and which are neglected in the standard approach. Thus, the complex association of electrical components (resistance, inductance, and capacitance) connected in series or in parallel means that the following phenomena can be taken into account:

- Interturn capacitive couplings;
- Phase-to-ground capacitive couplings;
- Leakage inductances in the first turns of the windings;
- Magnetizing inductances;
- Iron and eddy current losses.

The model complexity is determined by the purpose of the modeling and the accuracy required. The first approaches identified in the literature are based on π modeling of each machine phase, similar to what is used to describe the behavior of transmission lines [49]. For example, we can identify the HF EEC models developed in [34,43,50] and presented in Figure 7.

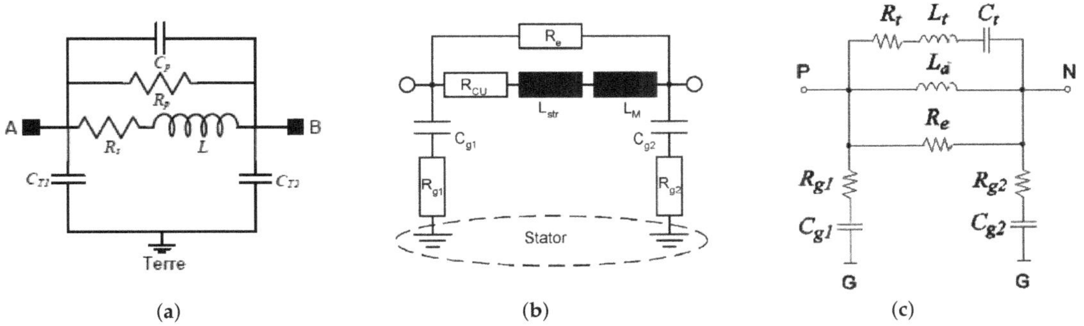

Figure 7. Examples of π-based HF EEC models of machine phases—taken from [50] (**a**), [34] (**b**), and [43] (**c**).

Figure 8a compares the impedance frequency response estimated using the model presented in Figure 7a with its experimental profile for the three parallelized phases of the 750 W induction machine studied in [50]. Figure 8b compares the experimental and estimated frequency response characteristics, using the model described in Figure 7b of the CM (curve with the lowest mean impedance over the considered frequency range) and DM (curve with the highest mean impedance over the considered frequency range) impedances, of the 15 kW induction machine tested in [34]. Finally, Figure 8c compares the experimental and estimated CM (left curve) and DM (right curve) impedance profiles calculated using the model shown in Figure 7c for the 2.2 kW electric machine studied in [43]. It should be noted that the accuracy of the various models presented remains rather limited and does not perfectly characterize the impedant frequency behavior of these machines over the entire frequency range considered. Building on this postulate, more recent work has sought to construct more complex HF EEC models capable of taking into account more HF phenomena. This is particularly the case in [44,51], whose models have been improved in [45].

Figure 9 shows the complete HF EEC model developed in [45] to characterize the impedant frequency behavior of an 8 kW, star-connected PMSM. Note the following in this model:

- R_g is the resistance of each phase.
- C_{g_i} represents the parasitic capacitances between the windings and the stator frame.
- "EMF" blocks describe the electromotive forces generated in each of the machine phases during operation.
- "Part 0" blocks characterize the CM and the DM behavior around resonance and anti-resonance frequencies at very high frequencies (between one and a few tens of megahertz).
- "Part 1" blocks characterize the CM and the DM behavior around resonance and anti-resonance frequencies at high frequency (between 10 kHz and 1 MHz).

- "Part 2" blocks model magnetizing inductances and eddy-current losses in the ferromagnetic core.
- "Part 3" blocks characterize the DM impedant behavior at high frequencies (between 10 kHz and 1 MHz).
- "Part 4" blocks model the effects of leakage inductances in the first turns of coils.

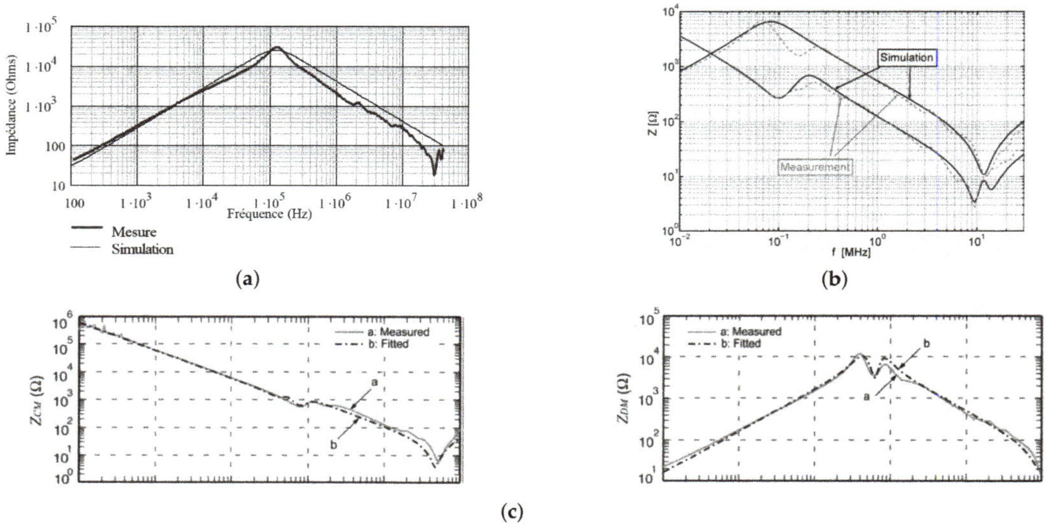

Figure 8. Experimental and modeled impedance frequency response of three parallel phases in [50] (**a**), of CM and DM in [34] (**b**), and of CM and DM in [43] (**c**).

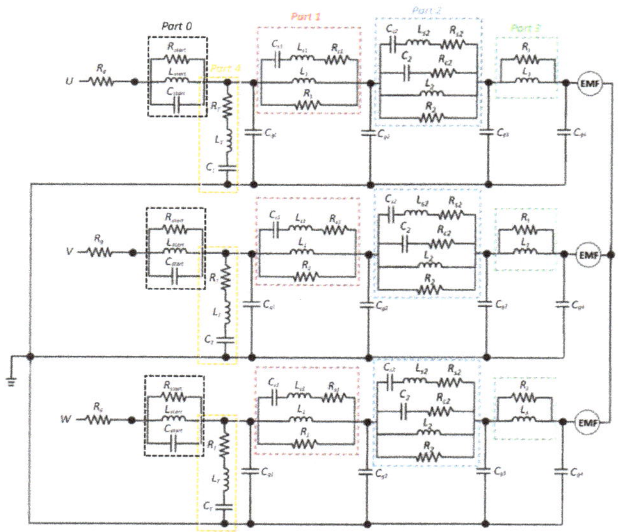

Figure 9. Example of a complete HF EEC model of a star-connected PMSM—taken from [45].

Figure 10 compares the experimental and estimated frequency response of the CM (a) and DM (b) impedances of the machine studied in [45], using the model shown in Figure 9.

Comparing the accuracy of these plots with those of Figure 8, a much better correlation of the impedance behavior is observed over the entire frequency range considered, especially at very high frequencies (beyond the megahertz). The various resonances are perfectly described, both in terms of frequency and amplitude.

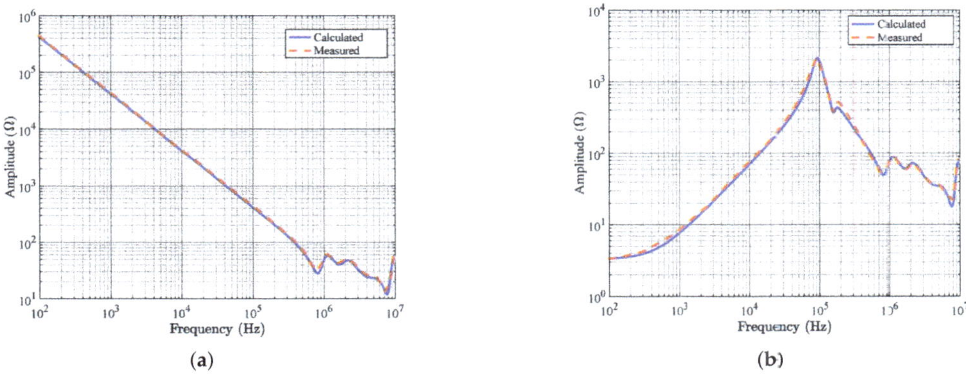

Figure 10. Comparison of experimental and modeled CM (**a**) and DM (**b**) impedance frequency behavior—taken from [45].

To evaluate the different parameters of the models presented in Figures 7 and 9, the usual approach is to experimentally characterize the impedance frequency response of the machine under test, then to simplify certain frequency ranges according to the assumptions made, and finally to estimate the various equivalent parameters, as explained by M. Schinkel in [34]. Such a model cannot be considered generic and must be developed for each machine topology. Finally, when the model becomes more complex, it may be necessary to couple this approach with analytical and/or numerical calculations in order to evaluate all the different parameters of the HF EEC model designed [45,43].

3.2.2. Transfer Function Approach to PMSM Behavior

In the interest of early diagnosis of interturn short-circuit faults in PMSM, N. Haje Obeid proposes in [52] to develop a transfer function multi-model to accurately model the behavior of a machine stator winding over a wide frequency range from a few kilohertz to over ten megahertz. The idea is to determine the transfer function that matches the experimental behavior of the winding. The method is based on a graphical model that uses the experimental admittance plot (inverse of the impedance) to determine the linear, time-invariant transfer function characteristic of the machine behavior. The transfer function of such a system is described by the Equation (2) below:

$$H(s) = \frac{a_n s^n + a_{n-1} s^{n-1} + \cdots + a_0}{b_m s^m + b_{m-1} s^{m-1} + \cdots + b_0} \tag{2}$$

where n and m are the order of the numerator and denominator polynomials, respectively, while the coefficients a_i and b_j that characterize the frequency response are calculated by solving an optimization problem.

In the example developed in [52], the optimization problem is solved using a genetic algorithm. This type of algorithm uses the principle of natural selection to ensure that the initial population of the problem evolves towards the best possible solution after a given number of generations N_g. To find an optimal solution, the order of the transfer function is increased if the transfer function of the best individual from the previous generation does not satisfy the system frequency response. The solution is adopted when this condition is satisfied. It should be noted that if the optimization problem is solved by a relatively high-order transfer function, it is possible to split the study into two different frequency ranges in order to obtain simpler solutions that describe the desired behavior as accurately. The resulting multi-model transfer function can then be stored and reused in a diagnostic process.

Another approach is to use vector fitting techniques to extract a generic transfer function for characterizing the CM impedance of an electrical machine [53]. This transfer function is defined by the relation (3).

$$H(s) = A + B.s + \sum_{k=1}^{N} \frac{r_k}{s - p_k} \quad (3)$$

where A denotes a resistive behavior, B refers to an inductance type of behavior, N is a natural integer that characterizes the order of the polynomial, and r and p are the various residues and poles of the transfer function, respectively.

To go further, ref. [35] proposes to use this transfer function as a basis for an equivalent circuit model, where the number of parameters to be estimated depends on the order of the resulting transfer function. Two approaches are then considered to evaluate these parameters:

- A constrained optimization, to minimize the error between the model and the experimental measurement of the machine CM impedance [35];
- An analytical calculation of the various parameters, taking into account the type and design of the winding as well as the material and method of impregnation of the conductors [24].

The author argues that for the machine studied, a 4th-order transfer function satisfies the conditions to accurately model the CM impedance behavior. The resulting model is compared in Figure 11 with the impedance frequency response measured on the machine.

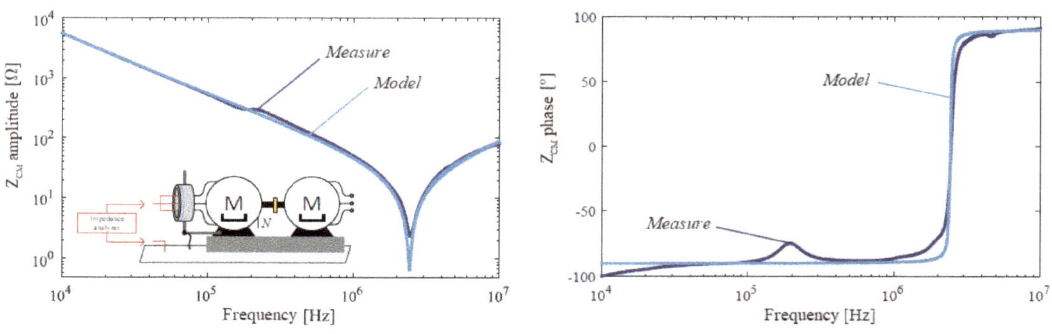

Figure 11. Comparison of the experimental frequency response and the proposed transfer function model of CM impedance—taken from [35].

3.2.3. HF Approach Based on PMSM Phase Design

The approach based on the machine phase design developed in [13] takes into account capacitive coupling phenomena between the coils and with the machine ground, which are neglected when building a traditional EEC model. The consideration of these couplings increases the accuracy of the model. Furthermore, the formulation in state equations makes the numerical resolution of the problem fast and straightforward. Finally, because the model is based on a portion of the winding, its symmetry properties provide a simple and effective tool for representing the behavior of a single coil or of the entire stator winding by placing state models in series or in parallel.

W. Liu explains in [13] the problem formulation for three healthy turns of a winding. The drawing of the studied structure helps to consider the different phenomena encountered (capacitive phase-to-ground couplings, couplings between two adjacent turns, and inductive couplings between adjacent half-turns). This structure is described in Figure 12.

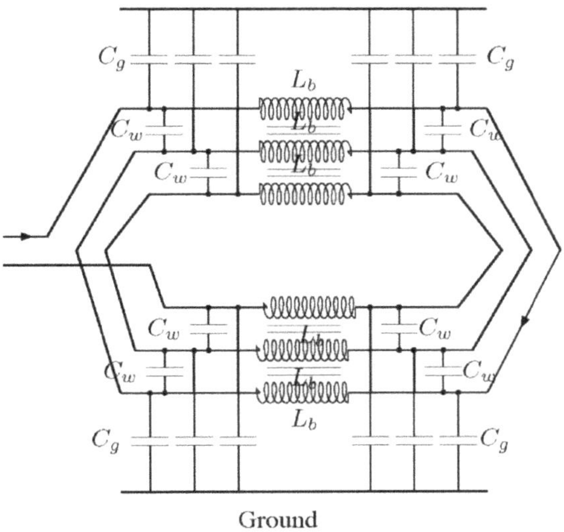

Figure 12. Structure of a three-turn coil model with inductive and capacitive couplings—taken from [13].

To simplify the formulation, an equivalent circuit of the three machine turns is defined. Then, the state variables of the model are the voltages across the equivalent capacitors and the currents flowing in each of the equivalent inductances. For a three-turn coil, as in [13], there are 13 such variables. The state vector aggregates these different variables, allowing us to formulate the problem as (4). Finally, after linear transformations, we obtain a state model (5) capable of describing the behavior of 3 inductively and capacitively coupled PMSM turns.

$$\dot{X} = G.\dot{X} + H.X + K.U \tag{4}$$

$$\begin{cases} \dot{X} = A.X + B.U \\ Y = C.X + D.U \end{cases} \tag{5}$$

where X is the state vector, Y is the output vector, and U is the input vector.

Therefore, thanks to this state–space formulation of the behavior of the coupled turns, the precise behavior of both the low-frequency and high-frequency behavior of the complete MSAP can be evaluated with fast numerical resolution. The intermediate step is the estimation of the various model parameters. This estimation is done in [13] using an output error optimization algorithm. The advantage of this formulation, which is based on the machine design, is that it takes into account the winding topology and capacitive couplings, which are generally neglected. As explained earlier, these couplings are representative of the quality of the insulation between the conductors and consequently characteristic of the winding aging and the development of interturn short circuits.

4. Proposed Online Method for Impedance Characterization of Electrical Machines

The impedance characterization method developed in this section has been designed to monitor the aging of windings in electrical machines powered by voltage inverters. The aim of this online approach is to observe, at each moment of the machine operation, the evolution of the different resonance frequencies, which are typical of insulation degradation [16,18]. To do this, we assumed that a supply through a voltage source inverter would intrinsically excite the stator windings of electric machines over a very wide frequency range. Actually, as W. Chen points out in [54], every multiple of the switching frequency is found in the voltage signal applied to the machine terminals. Therefore, unlike the online methods presented in Section 2.2, the proposed method does not require any dedicated HF injection device. However, for accurate broadband impedance characterization of the windings, it relies on wideband gap probes and a dedicated data acquisition system.

In the following, we will present the system under study and the different measurements required for the characterization. Then, we will describe the proposed characterization process, step by step, to obtain the information needed to monitor the winding state of health.

4.1. Studied System and Measurements

The system under investigation is described in Figure 13. The electrical machine is powered by a voltage source inverter (VSI), made of SiC power components, and driven by pulse–width modulation (PWM). The only variables required for frequency impedance characterization are the voltages at machine terminals and the currents circulating in the windings. As mentioned above, these signals are evaluated using wideband gap probes and processed to calculate the frequency response of the windings. Finally, the resistance value of each machine phase is also required to implement the proposed method.

The experimental test bench used in this study is shown in Figure 14. The prototype in our experimental setup is star-connected. Also, we have made the neutral point accessible in order to measure the voltage at this point of the winding. Therefore, two types of characterization presented in Section 2.1 can be undertaken with our method: phase-to-phase impedance and phase-to-neutral impedance.

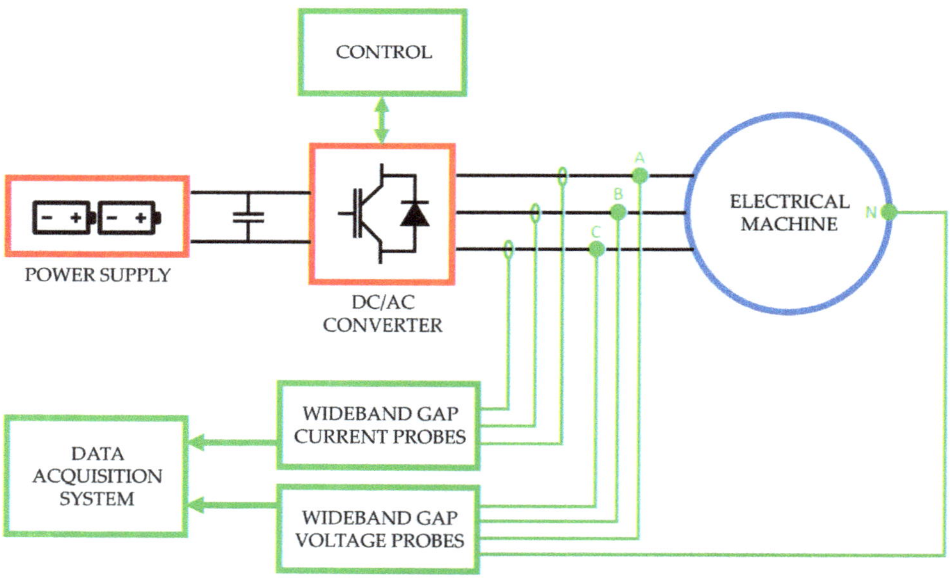

Figure 13. System description and measured signals.

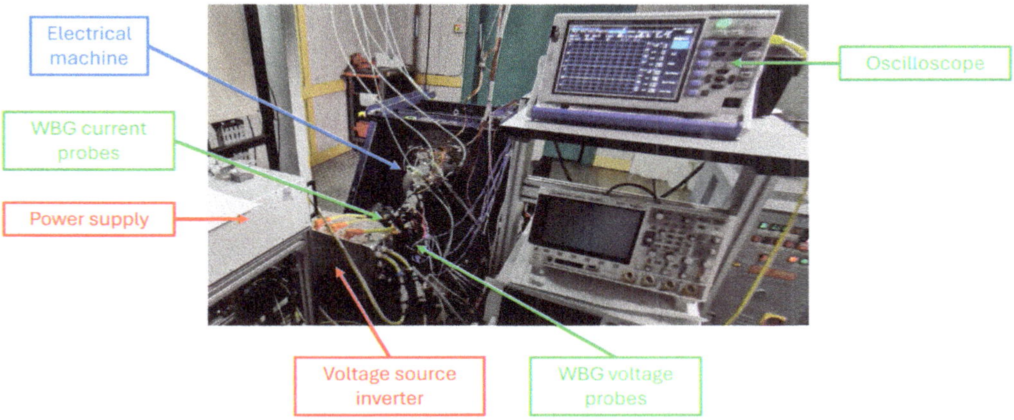

Figure 14. Experimental test bench.

4.2. Flow Chart of the Proposed Characterization Method

The online impedance characterization method developed in this work is based on the flowchart presented in Figure 15.

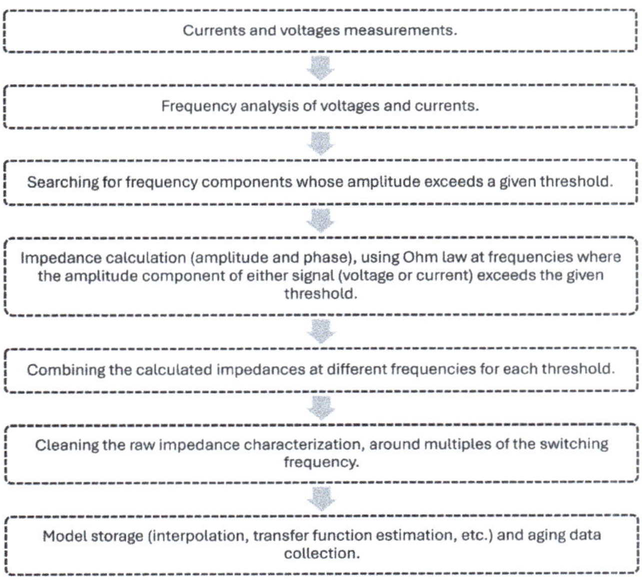

Figure 15. Flowchart of the proposed characterization method.

The various steps in this process are detailed in the following paragraphs.

Step 1—Measurements

Voltages are measured at points A, B, and C (and N, if the neutral point is accessible) of the system described in Figure 13. The currents flowing in each phase of the machine are also measured. It should be noted that the frequency range of the characterization depends directly on the bandwidth (BW) of these sensors. In our study, their BW is [12 Hz, 50 MHZ]. To measure these data, the acquisition is performed using an oscilloscope with a sampling frequency of 62.5 MHz. Therefore, the Nyquist theorem states that only frequency components up to 31.25 MHz (f_{max}) are properly displayed in the discrete signals obtained [55].

The machine under test is fed by a VSI with a switching frequency between 10 kHz and 35 kHz. The electrical fundamental frequency (f_e) of this traction machine is in the frequency range [100 Hz, 500 Hz]. Also, in these applications, the DC bus voltage ranges from 300 V to 750 V. When the neutral point is not accessible, we can consider a simple voltage system by applying the relation 6.

$$V_{abc} = \frac{U_{abc}}{\sqrt{3}} \quad (6)$$

where U_{abc} are phase-to-phase voltages and V_{abc} are simple voltages.

Note that the proposed method can be performed with simple voltages or phase-to-phase voltages. For the remainder of this study, we will only consider simple voltages. Figure 16 depicts the waveforms of the simple voltages (a) and the currents circulating in the stator windings (b), respectively. Per unit has been used for confidentiality reasons. For time variables, the signal amplitude is divided by the amplitude of the fundamental component. However, for frequential spectrums, the signal amplitude is divided by the maximum amplitude over the considered frequency range.

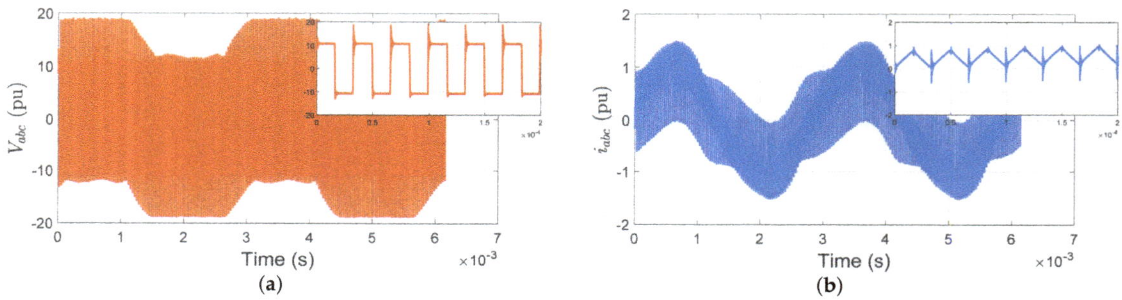

Figure 16. Waveforms of the simple voltages (**a**) and measured currents (**b**).

Step 2—Frequency analysis

To perform the frequency analysis of voltages and currents, we considered these signals to be stationary, i.e., measured on a machine in steady-state operation. Therefore, we decided to use the Fast Fourier Transform (FFT) to extract their frequency spectrum. However, under transient conditions, time–frequency transformations should be used to extract the frequency spectrum of these signals. These include Short-Time Fourier Transform (STFT); Wavelet Transform (WT); Wigner-Ville Distribution (WVD); Choi-Williams Distribution (CWD); Hilbert-Huang Transform (HHT); etc.

This transformation is expressed mathematically by the relation 7 [56].

$$\left[\quad f[n] = \frac{2n}{N\Delta t} \quad , \quad y[n] = \sum_{k=0}^{\frac{N}{2}-1} x[k].e^{-j\frac{4\pi}{N}kn} \quad \right] \quad (7)$$

where $f[n]$ is the frequency associated with the nth point of the transform, $y[n]$ is the result at the nth point of the transform, N is the maximum number of points of the transform to satisfy the Nyquist criterion, and Δt is the sampling period of the measured time signal.

The frequency content of the various signals analyzed is shown in Figure 17. The logarithmic scale on the x-axis and the transformation of the processed signal into decibel amplitude have been used for visual clarity.

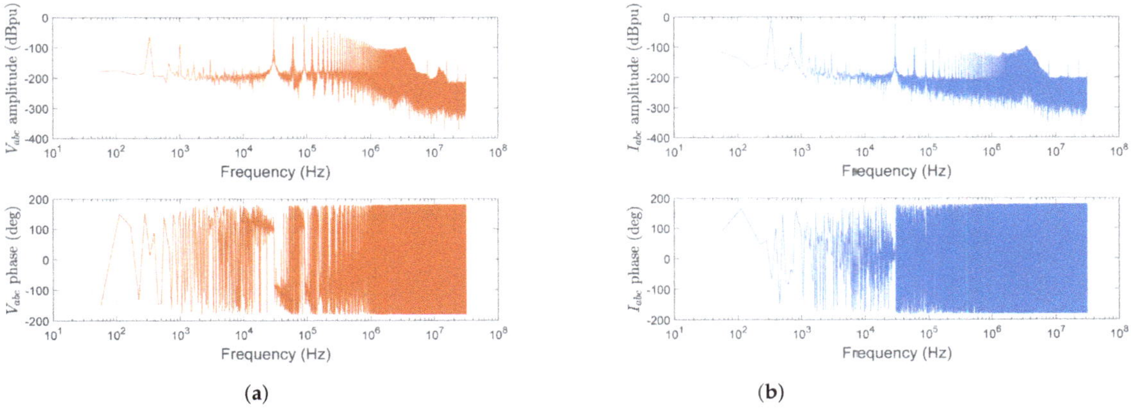

Figure 17. Frequential analysis of the simple voltages (**a**) and measured currents (**b**).

Step 3—Search for dominant frequency components

Then, the idea is to extract the frequency components whose amplitude is predominant depending on the frequency range considered. For this purpose, we have defined adaptive thresholds for the frequency spectrum of voltages and currents. These thresholds are characterized by relations (8) and (9), respectively. The graphical representation of this search for the dominant frequency components is explained for the simple voltages in Figure 18. Note that these thresholds have been defined using the DC bus voltage (U_{DC}, which for this excitation corresponds to the amplitude at the switching frequency) and the maximum sampling frequency (to satisfy the Nyquist criterion).

$$S_V(f) = \frac{U_{DC}}{\sqrt{3}}.\frac{f_{max}}{f_{PWM}}.\frac{1}{f} \quad (8)$$

$$S_I(f) = S_V(f).R_{ph} \quad (9)$$

where S_V and S_I are the thresholds for simple voltages and currents, respectively, f is the given frequency, U_{DC} is the DC bus voltage, f_{max} is the maximum sampling frequency, f_{PWM} is the switching frequency, and R_{ph} is the electrical resistance of the machine phases.

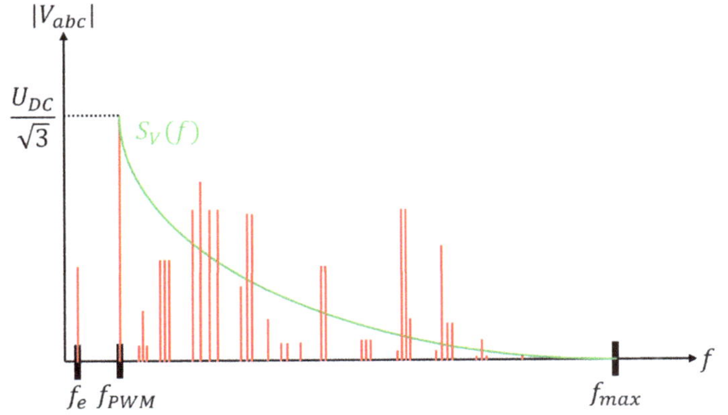

Figure 18. Graphical representation of the adaptive threshold defined to extract the predominant frequency components within single voltages.

Finally, for each of the processed signals (simple voltages and currents), once the amplitude of the given component exceeds the threshold set at a given frequency (f_{prep}), the following quintet is stored:

$$Q = [\ f_{prep}\ ,\ |V_{abc}(f_{prep})|\ ,\ |I_{abc}(f_{prep})|\ ,\ \angle V_{abc}(f_{prep})\ ,\ \angle I_{abc}(f_{prep})\ ]$$

Step 4—Impedance calculation

To characterize the impedance behavior as a function of the excitation frequency, we must determine the corresponding impedance for each frequency f_{prep} contained in the vectors Q_V and Q_I. The Ohm law is used to calculate the amplitude and the phase of this impedance, as described in Equations (10) and (11).

$$|Z(f_{prep})| = \frac{|V_{abc}(f_{prep})|}{|I_{abc}(f_{prep})|} \tag{10}$$

$$\angle Z(f_{prep}) = \angle V_{abc}(f_{prep}) - \angle I_{abc}(f_{prep}) \tag{11}$$

The raw impedance characteristics obtained independently from the two thresholding methods are shown in Figure 19. In particular, it can be seen that the point densities are not evenly distributed depending on the threshold used. On the one hand, around the so-called "parallel" resonances, i.e., when the inductive reactance is equal to the capacitive reactance, the current amplitude is minimal and its signal is in phase with the applied voltage. This results in a local impedance maximum and more points with the threshold at simple voltages. On the other hand, around the so-called "series" resonances, the applied AC voltage generates the maximum current amplitude. These two signals are then in phase. This results in a local impedance minimum and more points with the threshold at the currents.

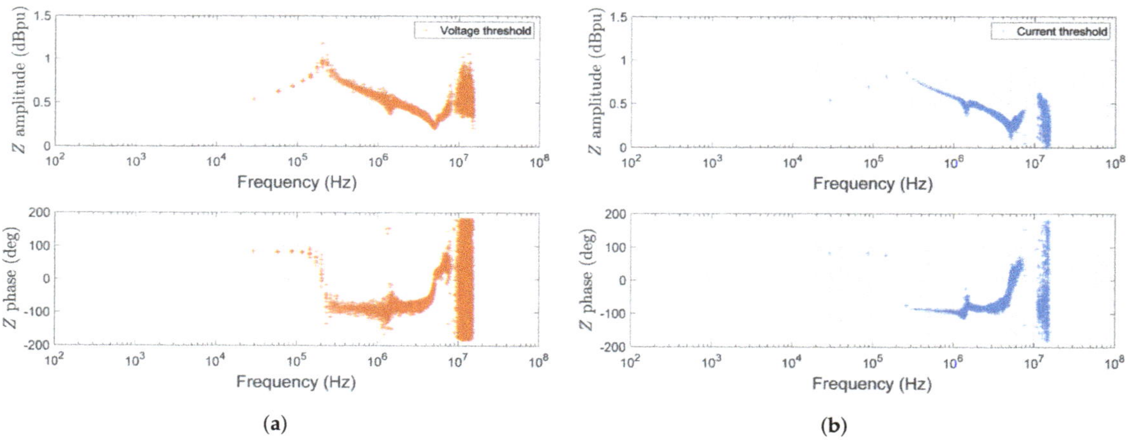

Figure 19. Raw frequency impedance characterization obtained by thresholding simple voltages (**a**) and currents (**b**).

Also, as expected, the maximum frequency of these characteristics is equal to $\frac{f_{max}}{2}$, i.e., around 15.6 MHz. Finally, these two profiles are particularly noisy at very high frequencies (over a dozen megahertz).

Step 5—Combining impedances

To take advantage of the precision of both plots around the different resonance frequencies, we will concatenate these two characterizations. The superposition of the two curves is shown in Figure 20. Despite the relative clarity of the full impedance characterization, we can see that it remains fairly noisy, which is why it is necessary to perform a cleanup step on the raw data.

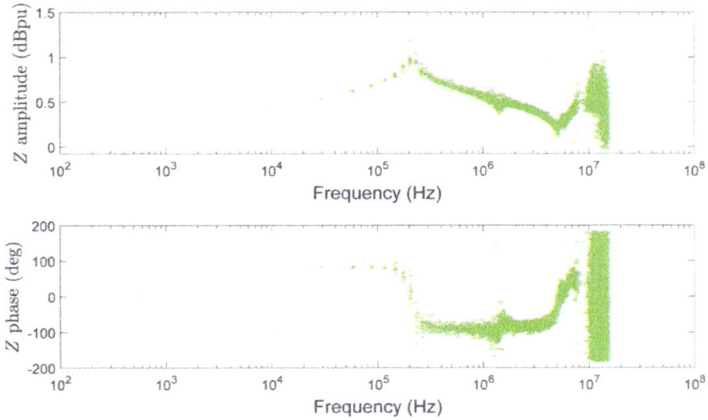

Figure 20. Complete raw frequency impedance characterization.

Step 6—Cleaning the raw impedance characterization

To extract a more precise impedance characteristic from the raw pattern generated in the previous step, we need to perform a "cleaning" step. The strategy adopted is to cluster the various points gravitating around multiples of the switching frequency. In this study, an interval of 5% on either side of these multiples was considered. Thus, the impedance amplitude and phase for each of these multiples is defined as the average of the values within this interval. If no point is close to one of the switching frequency multiples, then no value is assigned to it.

By applying this strategy to the raw impedance characterization obtained in Figure 20, we obtain a much cleaner frequency spectrum, especially at very high frequencies. To complete the resulting spectrum, the value of the impedance calculated at the fundamental excitation frequency (f_e) is added. This new characterization is shown in Figure 21.

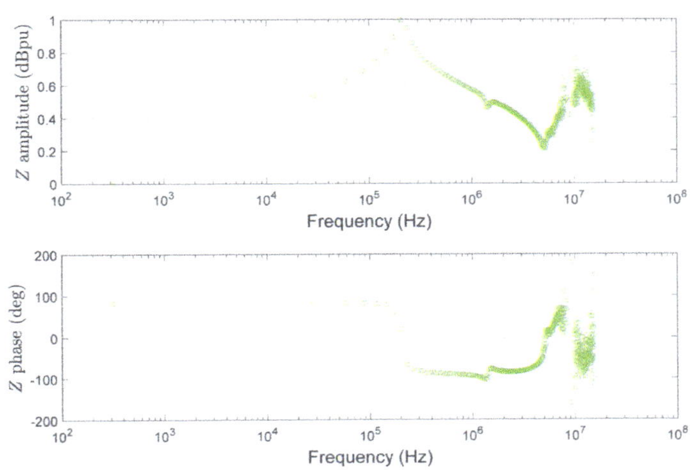

Figure 21. Improved frequency impedance characterization.

Step 7—Model storage and aging data collection

Finally, for aging monitoring purposes, it is essential to accurately evaluate the various parallel and series resonances of this improved impedance characterization. Since each point represents a multiple of the switching frequency, it is not possible to precisely assess these resonance values. There are two ways to recover these informations. The first is to interpolate these profiles and extract local extremes in order to analyze their evolution. The second is to model the impedance frequency behavior of the machine phases using the various approaches presented in Section 3.2 and look for behavioral deviations synonymous with aging.

Figure 22 shows the impedance behavior obtained by interpolating the improved characterization calculated in step 6. The type of interpolation used is the "Akima spline" [57]. The first two parallel and series resonances can be properly characterized, and data can be stored to monitor the health of the winding. Finally, the overall behavior at very high frequencies is more erratic and fails to accurately identify the other resonances.

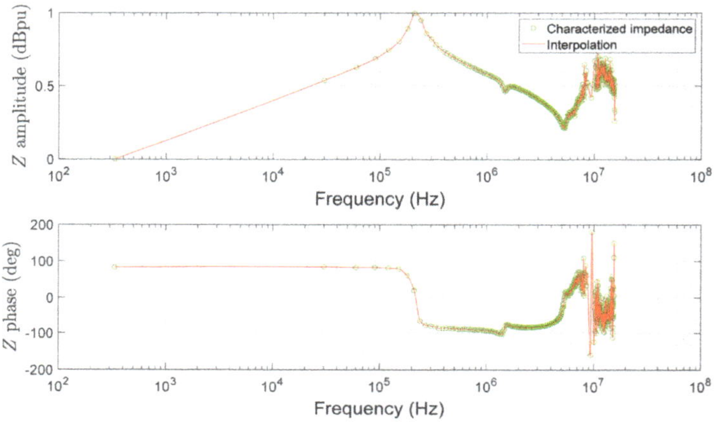

Figure 22. Interpolation of the impedance frequency response.

4.3. Validation and Interest of the Proposed Method

The objective was to verify the reliability and consistency of our strategy. For this purpose, we performed about 15 measurements on the same test machine at different times during its operation (at different operating points and on different days). Interpolation was used to determine the amplitude and the frequency of the various resonances found.

Figure 23 shows the different characteristics acquired during these tests. It can be seen that the different profiles are superimposed regardless of the machine operating point. The first two resonances can be identified. Based on the different results, we highlight that the uncertainty to characterize the first parallel resonance is less than or equal to 12.33 kHz and 4.4 dBΩ (<3%). The frequency uncertainty to characterize the first series resonance is identical to that observed for the first parallel resonance. However, in terms of amplitude, it is less than or equal to 233 kHz.

Furthermore, in order to validate the proposed online characterization approach, we compared the resulting plot with that measured offline using a vector network analyzer. The results are shown in Figure 24.

There is a strong correlation between the two frequency behaviors. The relative difference between the two characteristic frequencies of the first parallel resonance is less than 10%. In terms of amplitude, the two approaches give the same result. On the other hand, there is a deviation between the behavior observed with our online method and with the impedance analyzer around the first series resonance. This is intrinsically related to the use of fixture adapters and a measurement cable between the impedance analyzer and the test machine, as explained in [58] and confirmed by Figure 25, which compares the impedance response of a traction machine as a function of the cable used. Based on these results, it is essential to emphasize the importance of using the same measurement conditions if we want to properly exploit the results of these frequency analyses and draw the correct conclusions about the health of the windings.

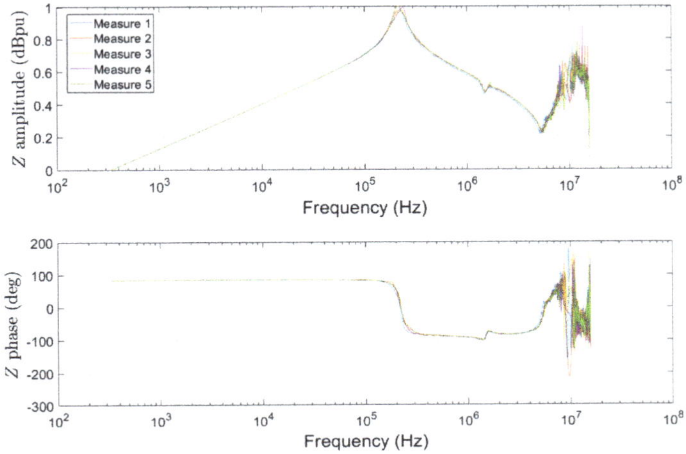

Figure 23. Comparison of impedance interpolations obtained at different times and operating points.

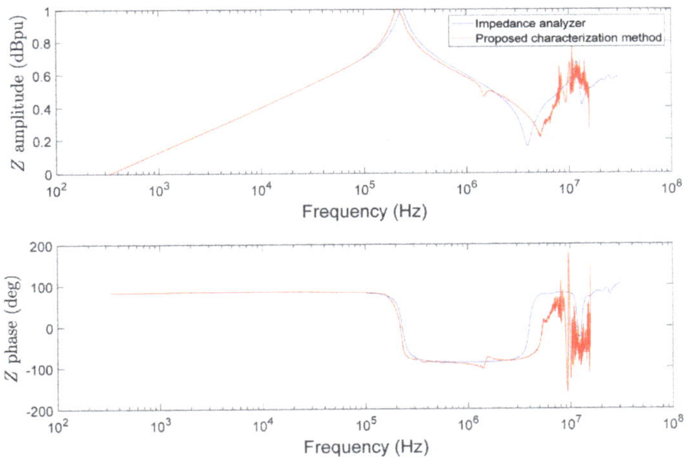

Figure 24. Comparison of impedance characterizations obtained with the proposed online method and with an impedance analyzer.

Finally, the benefits of this online impedance characterization approach can be summarized in three points:

- The HF excitation is naturally provided by the voltage source inverter. The method is therefore non-invasive and has no effect on actuator operation or aging.
- No measurement cables or fixture adapters are required, eliminating the need for a calibration phase and making the characterization process highly reliable.
- Finally, health monitoring is performed at all times without removing the machine from its operating environment, allowing early detection of insulation degradation in the windings.

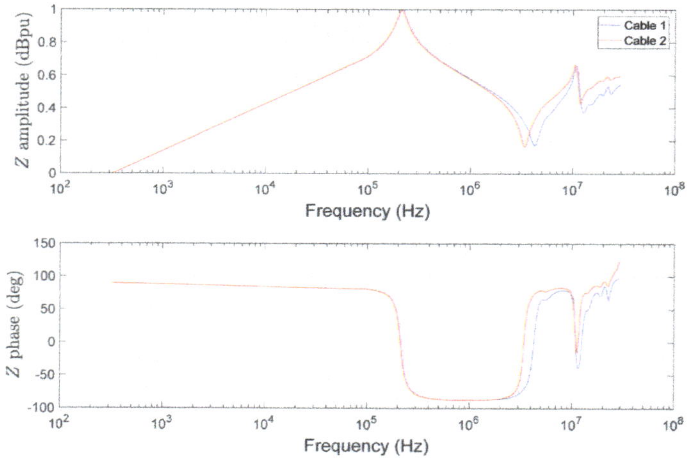

Figure 25. Measurement cable effect on the frequency response measured by the impedance analyzer.

5. Parameters Influencing Impedance Frequency Response

To extend our study, we wanted to understand the influence of some parameters on the impedance characteristics obtained using the proposed strategy. The three parameters investigated are the machine operating temperature, the inverter switching frequency, and the DC bus voltage amplitude. Findings are discussed below.

5.1. Temperature

To investigate the effect of temperature on the HF impedance frequency response of the machine, we carried out two characterizations at different temperatures. The first at room temperature ($T = 25\ °C$), after a few moments of operation, and the second after a few tens of minutes of circulating a current in the windings, so that the thermal regime is established. The second test was performed at a temperature of 80 °C. The results are shown in Figure 26. From the characterization plot after cleaning, as well as its interpolation, we can see that the HF impedance behavior remains identical, regardless of the operating temperature.

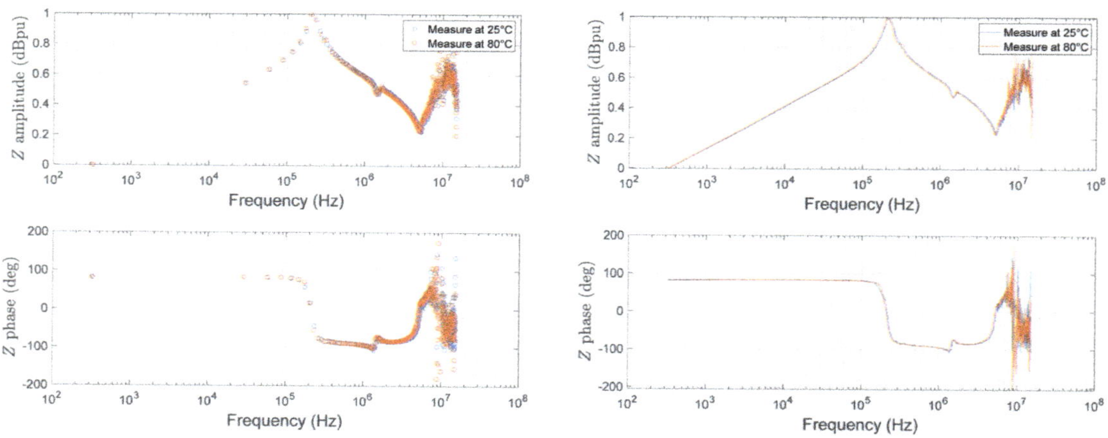

Figure 26. Comparison of the impact of the operating temperature on the cleaned impedance characteristic (**a**) and its interpolation (**b**).

5.2. Switching Frequency

We have also studied the influence of the inverter switching frequency on the accuracy of the proposed impedance identification method. Since it is the various multiples of the switching frequency that are intrinsically excited, this means that for a fixed measurement frequency range, the characteristic obtained is enriched or depleted, depending on the switching frequency used. This assertion is confirmed by comparing the cleaned impedance characteristics shown in Figure 27a, obtained with $f_{PWM} = 15$ kHz and $f_{PWM} = 30$ kHz. The number of blue points is twice as high when the switching frequency is set to 15 kHz.

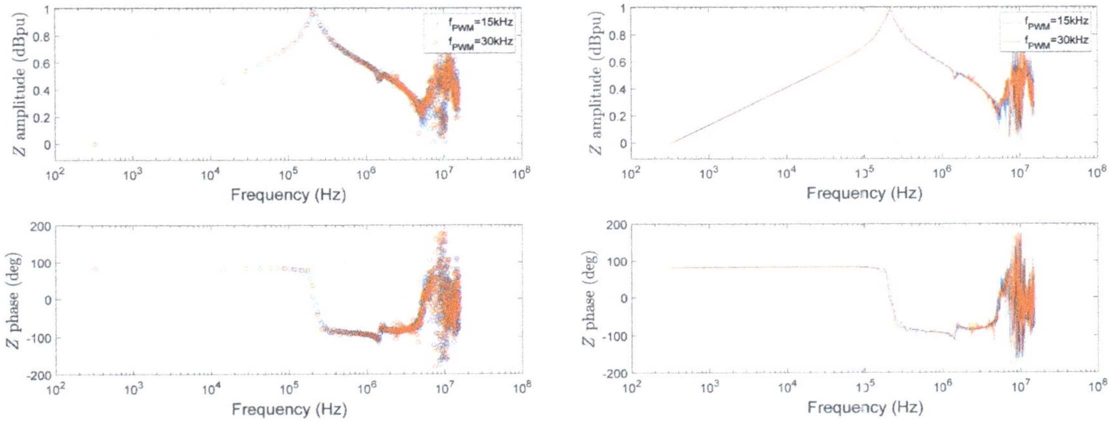

(a) (b)

Figure 27. Comparison of the impact of the switching frequency on the cleaned impedance characteristic (**a**) and its interpolation (**b**).

This increased point density provides a better approximation of the different series and parallel resonance frequencies. As a result, their estimation is all the more accurate and reliable. However, the interpolations presented in Figure 27b show that the difference in accuracy obtained with these two switching frequencies is not significant enough to bias the characterization, as the deviation is less than 3% (the uncertainty range calculated in Section 4.3 during reliability tests of the method).

5.3. DC Voltage

Finally, we have investigated the impact of the DC bus voltage on the impedance frequency response of the machine. For this purpose, we performed two characterizations with $U_{DC} = 300$ V and $U_{DC} = 600$ V. The results presented in Figure 28 show that the amplitude of the DC bus voltage has little effect on the accuracy of the impedance frequency response method. The reason for this lies in the very definition of the proposed method, since the adaptive threshold used to identify the various dominant frequency components that are used to derive the frequency characterization is defined using this parameter. The approach is therefore inherently independent of the DC bus voltage amplitude.

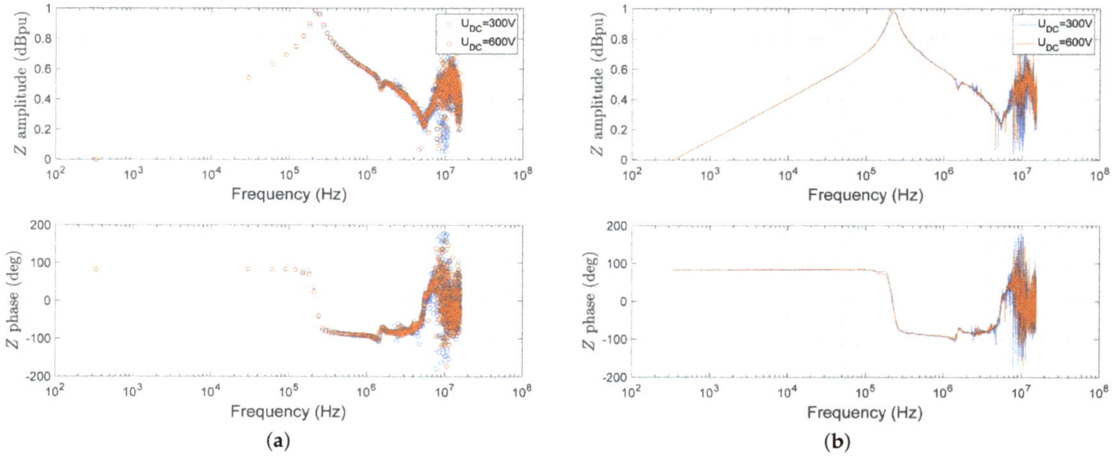

Figure 28. Comparison of the impact of the DC bus voltage on the cleaned impedance characteristic (**a**) and its interpolation (**b**).

6. Conclusions

In this paper, after reviewing the state of the art in frequency characterization and modeling of stator winding impedance behavior in electrical machines, we have presented the online characterization principle we have developed for monitoring conductor aging. This characterization method is applicable to any machine fed by a voltage inverter. When the inverter is controlled by PWM, it naturally excites all the multiples of its switching frequency up to very high frequencies (i.e., above tens of megahertz), and thus we are able to extract the impedance frequency response of the machine phases using time measurements of voltages and currents.

The proposed method consists of several steps. First, a frequency analysis of the signals is performed. Then we look for frequency components whose amplitude is above a defined threshold. Then, using these data, we apply Ohm law to calculate the impedance amplitude and phase for each highlighted frequency. To reduce noise and uncertainties around multiples of the switching frequency, we must refine the raw characterization. Finally, modeling or interpolation is performed to store this impedance behavior and thus gather useful information for aging monitoring (frequencies and amplitudes of the first parallel and series resonances).

We have demonstrated the reliability and accuracy of our characterization method compared with other offline approaches available in the literature. We have also studied the influence of various parameters, such as operating temperature, inverter switching frequency, or DC bus voltage, on the impedance frequency response of the machine windings.

Finally, the advantage of the proposed broadband impedance characterization approach designed for aging monitoring lies in the fact that unlike other online methods in the literature, it does not require any additional injection system to excite the windings. The invasiveness of the method is therefore limited since it does not affect the operation or aging of the machine. Calibration is the most critical step to ensure good data analysis when using an impedance analyzer. Our approach eliminates this step, making it all the more reliable. Finally, aging monitoring can be done at any time without removing the machine from its operating environment, allowing early detection of insulation degradation in the windings.

Author Contributions: Conceptualization, J.C. and N.H.O.; methodology, J.C., N.H.O., T.B. and F.M.-T.; software, J.C.; validation, J.C., N.H.O., T.B. and F.M.-T.; formal analysis, J.C. and N.H.O.; investigation, J.C. and N.H.O.; resources, N.H.O.; data curation, J.C.; writing—original draft preparation, J.C.; writing—review and editing, N.H.O., T.B. and F.M.-T.; visualization, J.C.; supervision, N.H.O., T.B. and F.M.-T.; project administration, N.H.O., T.B. and F.M.-T.; funding acquisition, N.H.O., T.B. and F.M.-T. All authors have read and agreed to the published version of the manuscript.

Funding: This research received no external funding.

Institutional Review Board Statement: Not applicable.

Informed Consent Statement: Not applicable.

Data Availability Statement: The data presented in this study are available on request from the corresponding author. The data are not publicly available due to privacy concerns for the participants involved in the study.

Conflicts of Interest: The authors declare no conflicts of interest.

References

1. Bose, B. *Power Electronics and Motor Drives*, 2nd ed.; Academic Press: Cambridge, MA, USA, 2020; ISBN 978-0-12-821360-5.
2. Manns, D.; Galioto, S.; Weeber, K.; Yagielski, J.High frequency life testing of stator coil insulation. In Proceedings of the 2008 IEEE International Symposium on Electrical Insulation, Vancouver, BC, Canada, 9–12 June 2008.
3. Guastavino, F.; Ratto, A.; Torello, E.; Biondi, G. Aging tests on nanostructured enamels for winding wire insulation. *IEEE Trans. Ind. Electr.* **2014**, *61*, 5550–5557. [CrossRef]
4. Gerlach, M.E.; Weber, S.; Ponick, B. Influence of Hairpin winding and insulation on the vibration behavior of electric machines. *IEEE Trans. Ind. Appl.* **2024**, *60*, 206–214.
5. Besnard, F.; Bertling, L. An approach for condition-based maintenance optimization applied to wind turbine blades. *IEEE Trans. Sustain. Energy* **2010**, *1*, 77–83. [CrossRef]
6. Rafiee, K.; Feng, Q.; Coit, D.W. Condition-based maintenance for repairable deteriorating systems subject to a generalized mixed shock model. *IEEE Trans. Reliab.* **2015**, *64*, 1164–1174. [CrossRef]
7. Kaufhold, M.; Borner, G.; Eberhardt, M.; Speck, J. Failure mechanism of the interturn insulation of low voltage electric machines fed by pulse-controlled inverters. *IEEE Electr. Ins. Mag.* **1996**, *12*, 9–16. [CrossRef]
8. Hudon, C.; Amyot, N.; Lebey, T.; Castelan, P.; Kandev, N. Testing of low-voltage motor turn insulation intended for pulse-width modulated applications. *IEEE Trans. Dielectr. Electr. Insul.* **2000**, *7*, 783–789.
9. Stone, G.C. Partial discharge diagnostics and electrical equipment insulation condition assessment. *IEEE Trans. Dielectr. Electr. Insul.* **2005**, *12*, 891–903. [CrossRef]
10. Boileau, T. *3rd Year Engineering Cycle*; Lecture on diagnosis and fault detection in PMSM; ENSEM: Vandoeuvre-lès-Nancy, France, 2019.
11. IEEE Standard 43-2013. Recommended Practice for Testing Insulation Resistance of Rotating Machinery. 2014. Available online: https://ieeexplore.ieee.org/document/836297 (accessed on 4 February 2023).
12. Werynski, P. Vieillissement des Diélectriques et Surveillance in situ des Machines Electriques. Ph.D. Thesis, Université d'Artois (LSEE), Arras, France, 2006. Available online: https://www.researchgate.net/publication/272857976 (accessed on 28 January 2021).
13. Liu, W.; Schaeffer, E.; Loron, L.; Chanemouga, P. High Frequency modelling of stator windings dedicated to machine insulation diagnosis by parametric identification. In Proceedings of the 2007 IEEE International Symposium on Diagnostics for Electric Machines Power Electronics and Drives, Cracow, Poland, 6–8 September 2007.
14. Yang, J.; Kang, T.-J.; Kim, B.; Lee, S.B.; Yoon, Y.-W.; Kang, D.; Cho, J.; Kim, H. Experimental evaluation of using the surge PD test as a predictive maintenance tool for monitoring turn insulation quality in random wound AC motor stator windings. *IEEE Trans. Dielectr. Electr. Insul.* **2012**, *19*, 53–60. [CrossRef]
15. Lahoud, N.; Nguyen, M.Q.; Maussion, P.; Malec, D.; Mary, D. Lifetime model of the inverter-fed motors secondary insulation by using a design of experiments. *IEEE Trans. Dielectr. Electr. Insul.* **2015**, *22*, 3170–3176. [CrossRef]
16. Perisse, F.; Mercier, D.; Lefevre, E.; Roger, D. Robust diagnostics of stator insulation based on high frequency resonances measurements. *IEEE Trans. Dielectr. Electr. Insul.* **2009**, *16*, 1496–1502. [CrossRef]
17. Haje Obeid, N. Contribution à la Détection des Défauts Statoriques des Actionneurs à Aimants Permanents—Application à la Détection d'un Défaut Inter-Spires Intermittent et au suivi de Vieillissement. Ph.D. Thesis, Université de Lorraine (GREEN), Nancy, France, 2016. Available online: https://tel.archives-ouvertes.fr/tel-01530619/ (accessed on 1 December 2020).
18. Jameson, N.J.; Azarian, M.H.; Pecht, M. Impedance-based condition monitoring for insulation systems used in low-voltage electromagnetic coils. *IEEE Trans. Ind. Electr.* **2017**, *64*, 3748–3757. [CrossRef]
19. Zou, Z.; Liu, S.; Kang, J. Degradation mechanism and online electrical monitoring techniques of stator winding insulation in inverter-fed machines. *World Electr. Veh. J.* **2024**, *15*, 444–466.[CrossRef]
20. Perisse, F.; Werynski, P.; Roger, D. A new method for AC machine turn insulation diagnostic based on high frequency resonances. *Trans. Dielectr. Electr. Insul.* **2007**, *14*, 1308–1315. [CrossRef]

21. Ahmadzadeh, F.; Lundberg, J. Remaining useful life estimation: review. *Int. J. Syst. Assur. Eng. Manag.* **2014**, *5*, 461–474. [CrossRef]
22. Nguyen, V.; Seshadrinath, J.; Wang, D.; Nadarajan, S.; Vaiyapuri, V. Model-based diagnosis and RUL estimation of induction machines under interturn fault. *IEEE Trans. Ind. Appl.* **2017**, *53*, 2690–2701. [CrossRef]
23. Venkatesan, S.; Manickavasagam, K.; Tengenkai, N.; Vijayalakshmi, N. Health monitoring and prognosis of electric vehicle motor using intelligent-digital twin. *IET Electr. Power Appl.* **2019**, *13*, 1328–1335. [CrossRef]
24. Ramos Chavez, J.I. Design of High Frequency Operating Mechatronic Systems: Tools and Methods of Characterization of Electromagnetic Couplings Between Electromechanic Converters and Power Electronics Converters. Ph.D. Thesis, Institut National Polytechnique de Toulouse (ENIT-LGP), Toulouse, France, 2016. Available online: https://www.theses.fr/2016INPT0075 (accessed on 25 January 2023).
25. Hoffmann, A.; Ponick, B. Statistical deviation of high-frequency lumped model parameters for stator windings in three-phase electrical machines. In Proceedings of the 2020 International Symposium on Power Electronics, Electrical Drives, Automation and Motion, (SPEEDAM), Sorrento, Italy, 24–26 June 2020.
26. Al-Ameri, S.M.; Abdul-Malek, Z.; Salem, A.A.; Noorden, Z.A.; Alawady, A.A.; Yousof, M.F.M.; Mosaad, M.I.; Abu-Siada, A.; Thabit, H.A. Frequency response analysis for three-phase star and delta induction motors—Pattern recognition and fault analysis using statistical indicators. *Machines* **2023**, *11*, 106–123. [CrossRef]
27. Ranzinger, L.; Uhrig, S.; Tenbohlen, S. Analysis and modeling the frequency response of rotating machines regarding fault diagnosis using SFRA. *IEEE Trans. Energy Convers.* **2024**, *39*, 1–10. [CrossRef]
28. Keysight Technologies Incorporation. E5061B ENA Vector Network Analyzer—Data Sheet. 2021. Available online: https://www.keysight.com/us/en/assets/7018-02242/data-sheets/5990-4392.pdf (accessed on 4 July 2024).
29. Sant'Ana, W.C.; Lambert-Torres, G.; Bonaldi, E.L.; Gama, B.R.; Zacarias, T.G.; Areias, I.A.; Arantes, D.; Assuncao, F.O.; Campos, M.M.; Steiner, F.M. Online frequency response analysis of electric machinery through an active coupling system based on power electronics. *Sensors* **2021**, *21*, 8057. [CrossRef]
30. Gama, B.R.; Sant'ana, W.C.; Lambert-Torres, G.; Salomon, C.P.; Bonaldi, E.L.; Borges-Da-Silva, L.E.; Carvalho, R.B.B.; Steiner, F.M. FPGA prototyping using the STEMlab board with application on frequency response analysis of electric machinery. *IEEE Access* **2021**, *9*, 26822–26838. [CrossRef]
31. Sant'Ana, W.C.; Salomon, C.P.; Lambert-Torres, G.; Da Borges Silva, L.E.; Bonaldi, E.L.; Lacerda Oliveira, L.E.; Da Borges Silva, J.G. Early detection of insulation failures on electric generators through online Frequency Response Analysis. *Electr. Power Syst. Res.* **2016**, *140*, 337–343. [CrossRef]
32. Bucci, G.; Ciancetta, F.; Fioravanti, A.; Fiorucci, E.; Mari, S.; Silvestri, A. Online SFRA for reliability of power systems. *Sensors* **2023**, *23*, 2583. [CrossRef] [PubMed]
33. Rathnayaka, S.B.; See, K.Y. Early detection of induction motor's defects using an inductively coupled impedance extraction method. In Proceedings of the 2017 IEEE International Electric Machines and Drives Conference (IEMDC), Miami, FL, USA, 21–24 May 2017.
34. Schinkel, M.; Weber, S.-P.; Guttowski, S.; John, W.; Reichl, H. Efficient HF modeling and model parameterization of induction machines for time and frequency domain simulations. In Proceedings of the 21st Annual IEEE Applied Power Electronics Conference and Exposition, (APEC), Dallas, TX, USA, 19–23 March 2006.
35. Dos Santos, V. Modélisation des émissions Conduites de mode Commun D'une Chaîne électromécanique—Optimisation Paramétrique de l'Ensemble Convertisseur Filtres sous Contraintes CEM. Ph.D. Thesis, Institut National Polytechnique de Toulouse (LAPLACE), Toulouse, France, 2019. Available online: https://www.theses.fr/2019INPT0021 (accessed on 5 December 2022).
36. Chen, H.; Yan, Y.; Zhao, H. Extraction of Common-Mode impedance of an inverter-fed induction motor. *IEEE Trans. Electr. Comp.* **2016**, *58*, 599–-606. [CrossRef]
37. Zheng, D.; Lu, G.; Zhang, P. A noninvasive interturn insulation condition monitoring method based on the Common-Mode impedance spectrum of inverter-fed machines. *IEEE Trans. Ind. Appl.* **2021**, *57*, 4786–4795. [CrossRef]
38. Grenier, D.; Labrique, F.; Buyse, H.; Matagne, E. *Electromécanique—Convertisseurs D'énergie et Actionneurs*, 1st ed.; Dunod: Paris, France, 2001; ISBN 978-2-10-005325-4.
39. Fudeh, H.R.; Ong, C.-M. Modeling and analysis of induction machines containing space harmonics - Part I: Modeling and Transformation. *IEEE Trans. Power App. Syst.* **1983**, *PAS-102*, 2608–2615. [CrossRef]
40. Boileau, T. Contribution à la Continuité de Service des Actionneurs Synchrones à Aimants Permanents. Tolérance au Défaut de Capteur Mécanique: Détection de Défauts électriques. Ph.D. Thesis, Institut National Polytechnique de Lorraine (GREEN), Nancy, France, 2010. Available online: https://hal.univ-lorraine.fr/tel-01749443 (accessed on 1 December 2020).
41. Ruiz-Sarrio, J.E.; Antonino-Daviu, J.A.; Navarro-Navarro, A.; Biot-Monterde, V. A review of broadband frequency techniques for insulation monitoring and diagnosis in rotating electrical machines. *IEEE Trans. Ind. Appl.* **2024**, *60*, 6092–6102. [CrossRef]
42. Weber, S.-P.; Hoene, E.; Guttowski, S.; John, W.; Reichl, H. Modeling induction machines for EMC-Analysis. In Proceedings of the IEEE 35th Annual Power Electronics Specialists Conference, Aachen, Germany, 20–25 June 2004.
43. Wang, L.; Ngai-Man Ho, C.; Canales, F.; Jatskevich, J. High-Frequency modeling of the long-cable-fed induction motor drive system using TLM approach for predicting overvoltage transients. *IEEE Trans. Power Electr.* **2010**, *25*, 2653–2664. [CrossRef]

44. Wu, Y.; Bi, C.; Jia, K.; Jin, D.; Li, H.; Yao, W.; Liu, G. High-frequency modelling of permanent magnet synchronous motor with star connection. *IET Electric Power Appl.* **2018**, *12.4*, 539–546. [CrossRef]
45. Rahimi, A.; Kanzi, K. Improved High-Frequency modeling of PMSM using 3-D finite element analysis. In Proceedings of the 2019 International Power System Conference (PSC), Teheran, Iran, 9–11 December 2019.
46. Ferreira, R.S.; Ferreira, A.C. Transient model to study voltage distribution in electrical machine windings considering the rotor. *Electric Power Syst. Res.* **2021**, *195*, 107155–107162. [CrossRef]
47. Vidmar, G.; Miljavec, D. A universal high-frequency three-phase electric-motor model suitable for the delta- and star-winding connections. *IEEE Trans. Power Electr.* **2015**, *30*, 4365–4376. [CrossRef]
48. Moreno, Y.; Almandoz, G.; Egea, A.; Arribas, B.; Urdangarin, A. Analysis of permanent magnet motors in High Frequency—A review. *Appl. Sci.* **2021**, *11*, 6334. [CrossRef]
49. Yang, Z. PI-based simulation modeling for performance testing of the power transmission line. *Energy* **2024**, *301*, 131753. [CrossRef]
50. Revol, B. Modélisation et Optimisation des Performances CEM D'une Association Variateur de Vitesse—Machine Asynchrone. Ph.D. Thesis, Université Joseph Fourier (Laboratoire d'électrotechnique de Grenoble), Grenoble, France, 2003. Available online: https://theses.hal.science/tel-00006396v1 (accessed on 17 March 2024).
51. Toulabi, M.S.; Wang, L.; Bieber, L.; Filizadeh, S.; Jatskevich, J. A universal High-Frequency induction machine model and characterization method for arbitrary stator winding connections. *IEEE Trans. Energy Conv.* **2019**, *34*, 1164–1177. [CrossRef]
52. Haje Obeid, N.; Boileau, T.; Nahid-Mobarakeh, B. Comparison of high frequency winding modeling for stator health monitoring. In Proceedings of the 14th International Symposium on Diagnostics for Electrical Machines, Power Electronics and Drives (SDEMPED), Chania, Greece, 28–31 August 2023.
53. Gustavsen, B.; Semlyen, A. Rational approximation of frequency domain responses by vector fitting. *IEEE Trans. Power Del.* **1999**, *14*, 1052–1061. [CrossRef]
54. Chen, W.; Sun, H.; Gu, X.; Xia, C. Synchronized space-vector PWM for three-level VSI with lower harmonic distortion and switching frequency. *IEEE Trans. Power Electr.* **2016**, *31*, 6428–6441. [CrossRef]
55. Mishali, M.; Eldar, Y. Sub-Nyquist Sampling. *IEEE Signal Proc. Mag.* **2011**, *28*, 98–124. [CrossRef]
56. Cottet, F. *Traitement du Signal*, 1st ed.; Dunod: Paris, France, 2005; ISBN 978-2-10-049690-5.
57. Akima, H. A new method of interpolation and smooth curve fitting based on local procedures. *J. ACM* **1970**, *17*, 589–602. [CrossRef]
58. Jie, H.; See, K.Y.; Chang, Y.; Gao, R.X.-K.; Fan, F.; Zhao, Z. Enhancing motor impedance measurements: broadening the spectrum from low to high frequencies. *Meas. Sci. Technol.* **2024**, *35*, 86008. [CrossRef]

applied sciences

MDPI

Article

Demagnetization Fault Location of Direct-Drive Permanent Magnet Synchronous Motor Based on Search Coil Data-Driven

Caixia Gao [1], Zhen Jiang [1], Xiaozhuo Xu [1,*] and Jikai Si [2]

[1] School of Electrical Engineering and Automation, Henan Polytechnic University, Jiaozuo 454150, China; gcx@hpu.edu.cn (C.G.); 13137176411@163.com (Z.J.)
[2] College of Electrical Engineering, Zhengzhou University, Zhengzhou 450000, China; sijikai@zzu.edu.cn
* Correspondence: xxz@hpu.edu.cn

Abstract

Demagnetization faults in direct-drive permanent magnet synchronous motors (DDPMSM) can cause significant performance degradation and unplanned downtime. Traditional fault location methods, which rely on manual feature extraction, exhibit limited accuracy and efficiency in complex and variable operating conditions. To address these limitations, this study presents a demagnetization fault location method based on a search coil employing a data-driven one-dimensional convolutional neural network (1D-CNN). Firstly, the arrangement of search coils was determined, and a partitioned mathematical model was established, using the residual back electromotive force (back-EMF) of the search coil over a single electrical cycle as the fundamental unit. Secondly, the residual back-EMF in the search coil is analyzed under various demagnetization parameters and operating conditions to assess the robustness of the proposed method. Furthermore, a 1D-CNN-based fault location model was developed using residual back-EMF signals as the input and targeting the identification of demagnetized permanent magnet types. Simulation and experimental results demonstrate that the proposed method can effectively detect and locate demagnetization faults across different operating conditions. When the demagnetization degree is not less than 10%, the fault location accuracy reaches 99.58%, and the minimum detectable demagnetization degree is 8%. The approach demonstrates excellent robustness and generalization, offering an intelligent solution for demagnetization fault location in PMSMs.

Keywords: direct-drive permanent magnet synchronous motor; convolutional neural network; demagnetization fault location; search coil; residual electromotive force

1. Introduction

With its direct load connection eliminating mechanical transmission, the direct-drive permanent magnet synchronous motor (DDPMSM) offers key advantages, including direct drive, high power density, and a compact structure. These merits have led to its broad adoption in aerospace, rail transport, industrial automation, and other fields [1,2]. However, motors are subject to the combined effects of electrical, thermal, and mechanical stresses and other factors during operation, posing a risk of fault. Common fault types include demagnetization, inter-turn short circuits, eccentricity, etc. As the core excitation component of a PMSM, permanent magnets can lead to a decline in motor output torque when they undergo an irreversible demagnetization fault. Under the constant load torque constraints, to maintain stable torque output from the motor, the armature current needs to increase simultaneously to compensate for the weakening of the excitation magnetic field. At this point, the motor temperature will gradually rise, which not only exacerbates the demagnetization fault but also damages the winding insulation and induces winding short-circuit faults. If the motor operates in this state for a long time, it will result in motor burnout, cause unscheduled downtime, and even cause catastrophic accidents. Therefore, achieving timely detection and accurate location of demagnetization faults not only facilitates early warning but also provides an important basis for post-fault operation and maintenance strategies formulation and accident prevention, thereby effectively ensuring the stable operation of the motor system. Furthermore, for DDPMSM, with a larger number of magnetic poles, the precise location of faulty magnetic poles becomes more challenging. Therefore, conducting in-depth research on demagnetization fault detection and location methods holds significant theoretical and practical engineering value.

Currently, online demagnetization detection methods have primarily relied on spectral analysis of signals, such as current [3,4], voltage [5–7], torque [8], and noise [9,10]. However, the characteristic harmonics in these signals that indicate demagnetization are mostly suitable for local demagnetization and are less sensitive to uniform demagnetization. Moreover, most methods can only achieve fault detection but struggle with precise location. This limitation stems primarily from their reliance on characteristic harmonics for analysis, which are susceptible to interference from factors such as motor winding configuration, structural design, and fault type. Consequently, the reliability of these methods is often compromised in complex and variable real-world industrial environments for diagnosis.

To address the limitations of spectral analysis, some researchers have proposed detection strategies based on magnetic signals. Demagnetization faults lead to alterations in the air gap magnetic flux density distribution, and such variations are measurable through detection coils installed in the motor. Search coils offer advantages such as low cost and simple structure (made solely of fine copper wire winding) and have been widely used for demagnetization fault diagnosis via induced voltage. By winding search coils around stator teeth, localized and uniform demagnetization can be detected based on the induced voltage; however, fault location remains unachievable [11]. To improve location capabilities, researchers have attempted to install independent coils on each stator tooth, extracting tooth flux to diagnose the quantity of demagnetized permanent magnets and decouple the induced electromotive force to determine their positions [12,13]. However, these methods currently only study the fault mode of a single demagnetized PM, and the installation of multiple coils results in complex system structures and high invasiveness. To balance location accuracy, implementation invasiveness, and early fault detection capability, our previous research proposed a detection solution based on toroidal yoke search coils [14]. This method involves placing three coils in stator slots to generate three positioning signals and determining the demagnetization location by matching with predefined waveforms. However, the multi-coil structure and reliance on multiple signals also lead to high algorithm complexity.

With the aim of coil reduction and diagnostic system simplification, a novel demagnetization fault diagnosis strategy is proposed in this paper, utilizing only two series-connected toroidal yoke search coils, with their residual voltage serving as the fault characteristic [15]. By matching the measured waveform with predefined waveforms in a fault dictionary, precise identification of the demagnetization type and location is achieved. Although the method effectively reduces the number of coils and simplifies the process, its performance still heavily relies on the accurate setting of predefined waveform templates and manual thresholds. Therefore, an intelligent approach capable of autonomously adapting to varying operating conditions and automatically extracting features is required.

https://doi.org/10.3390/app16020870

In recent years, artificial intelligence algorithms have been widely applied in fault diagnosis, primarily attributed to their powerful capabilities in feature learning and pattern recognition. In traditional machine learning, the Support Vector Machine (SVM) is suitable for small-sample classification tasks but has limitations such as weak interpretability and reliance on manual feature extraction [16]. On the other hand, shallow neural networks can realize automatic feature learning. The Probabilistic Neural Network (PNN) is appropriate for pattern classification and offers fast training speed, but its classification accuracy is highly affected by the choice of the smoothing factor [17,18]. Backpropagation Neural Network (BPNN) has excellent function approximation capabilities but exhibits instability in classification tasks and tends to fall into local optima [19]. Shallow neural networks generally lack sufficient representation capability, whereas deep learning models like Recurrent Neural Network (RNN), Long Short-Term Memory (LSTM) network, and CNN significantly improve feature representation by increasing network depth and structural complexity. RNNs excel in long-time-series prediction but face challenges in spatiotemporal feature extraction and are prone to gradient vanishing or explosion, making it hard to establish effective long-term dependencies [20]. LSTM networks address these issues through gating mechanisms and strengthen the ability of long-term dependency modeling; however, they have high computational resources and longer training times [21]. CNNs, leveraging their structural advantages, reduce sequence dimensionality while preserving key information, leading to improved training efficiency [22]. Notably, as an effective feature extractor, CNN is capable of extracting high-level abstract features directly from raw signals, providing a more accurate characterization of fault features compared to raw data [23–27].

To address the issues of traditional methods relying on precise threshold setting and the high computational complexity and poor real-time performance of intelligent algorithms, this paper proposes a fault location method based on 1-D CNN. The main contributions and innovations of this study are as follows:

(1) A precise method for demagnetization fault localization in DDPMSM based on the back-EMF signal of the search coil is proposed. This method utilizes the sensitivity of the residual back-EMF signal to changes in the internal magnetic field of the motor, significantly enhancing the sensitivity of fault location.

(2) This study adopts a CNN to construct an end-to-end fault diagnosis model. This approach avoids the reliance on manual feature extraction and threshold setting in traditional methods, thereby improving the development efficiency of the diagnostic system.

(3) In response to the raw one-dimensional time series signals as input and the need for real-time diagnosis, this paper proposes a lightweight 1D-CNN structure. The model directly processes the original one-dimensional time series signals, avoiding complex feature engineering and two-dimensional image conversion steps, thereby reducing computational complexity and enhancing the real-time processing performance of the location system.

This article is organized as follows: Section 2 establishes a mathematical model for the search coil residual back-EMF and analyzes the influence patterns of different fault types, operating conditions, and severity levels affect signal characteristics based on the model. Section 3 details the construction process of building a comprehensive fault sample database. and elaborates on the architecture and operational principles of the proposed one-dimensional CNN diagnostic network. Section 4 presents experimental validation and a comparative analysis of the proposed method's effectiveness, conducted on a developed experimental platform. Finally, Section 5 concludes the article.

https://doi.org/10.3390/app16020870

2. Analysis of Search Coil Residual Back-EMF Under Demagnetization Fault

2.1. Search Coil Arrangement

The demagnetization fault in PMSM weakens the magnetic field of the demagnetized magnet, a change detectable via stator-mounted coils. The waveform of the coil's back-EMF, being directly dependent on the magnetic state of its coupled permanent magnet, carries substantial demagnetization fault information and is therefore ideal for demagnetization diagnosis. To accurately locate the demagnetized magnet, a search coil configuration is employed, consisting of two identical toroidal yoke coils. These coils are installed within the stator slots of the PMSM and connected in forward series to constitute a search coil, as shown in Figure 1. Each of the toroidal yoke coils is arranged in the inner and outer stator slots of the PMSM. The two yoke coils should be spaced approximately one pole pitch apart to enhance the signal amplitude, which improves the Signal-to-Noise Ratio (SNR). The operating mechanism of the novel searching coil will be analyzed in the following section.

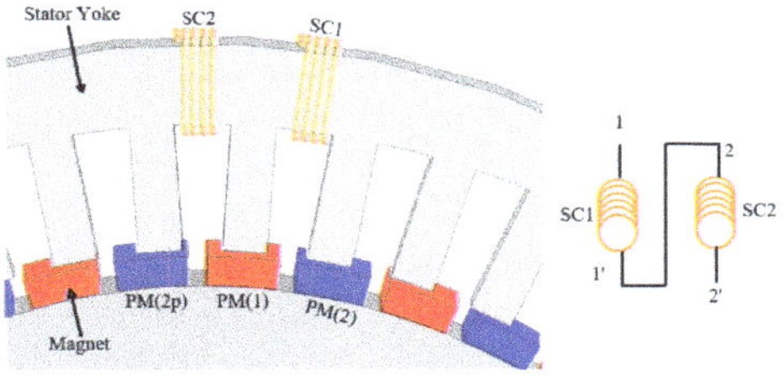

Figure 1. The arrangement of the search coil.

2.2. Residual Back-EMF Model of SC

The search coil's back-EMF arises from the time-varying magnetic flux linked with the coil. Under no-load conditions, the spatial distribution of this magnetic field is primarily determined by the arrangement of PMs. For analytical convenience, the PMs are sequentially numbered from 1 to 2p, as illustrated in Figure 1. To achieve synchronous sampling over the entire mechanical cycle, the initial data acquisition point is defined when PM1 and PM2p are positioned such that their geometric centers are alignment with the geometric center of coil SC1. The back-EMFs of SC1 and SC2 for a full motor rotation cycle are collected, and they are subtracted from those under healthy conditions to obtain the residuals of these EMFs. The residual back-EMFs induced by PM_i in SC1 and SC2 are defined as:

$$e_{1-\mathrm{PM}i-\mathrm{rs}} = N_{\mathrm{s}} l_{\mathrm{st}} v k_{\mathrm{demi}} B_{\mathrm{g-i-h}} \quad (1)$$

$$e_{2-\mathrm{PM}i-\mathrm{rs}} = N_{\mathrm{s}} l_{\mathrm{st}} v k_{\mathrm{demi}} B_{\mathrm{g-i-h}} e^{j\theta_{\mathrm{s}}} \quad (2)$$

where N_s denotes toroidal yoke coil turns, l_{st} represents the motor stack length, and v is the motor rotational speed. k_{demi} is the demagnetization severity coefficient, and $k_{demi} \in [0, 1]$, which is 0 when the PMi is healthy and is 1 when the PMi is totally demagnetized. As the severity of demagnetization increases, k_{demi} approaches 1. $B_{\text{g-i-h}}$ is the airgap flux density produced by the healthy permanent magnet PMi. The residual back-EMF induced by PM_i in SC2 is calculated by phase shifting (1). θ_s is the electrical angle between SC1 and SC2, and this angle equals or nearly approximates one pole pitch.

https://doi.org/10.3390/app16020870

The SC is obtained by summing the residual EMFs from SC1 and SC2. The SC for PMi demagnetization (eSC-PMi) is obtained by subtracting (2) from (1) as:

$$\begin{cases} e_{SC-\mathrm{PM}i} = N_s l_{st} v k_{\mathrm{demi}} B_{\mathrm{g-i-h}}(1 + e^{j\theta_s}) \\ i\pi \leq \alpha \leq i\pi + \theta_s + \theta_{\mathrm{pm}} \end{cases} \quad (3)$$

where α denotes the electrical angle relative to the coordinate origin. θ_{pm} is the electrical angle corresponding to the surface angle of PM, and it is close to π.

For simplicity, the residual EMF is assumed to be a sinusoidal wave. Based on Fourier series decomposition, $e_{SC\text{-PM}i}$ may be simplified as:

$$\begin{cases} e_{SC-\mathrm{PM}i} = (-1)^{i-1} E_1 \mathrm{Sin}(\alpha - (i-1)\pi) \\ i\pi \leq \alpha \leq (i+2)\pi \end{cases} \quad (4)$$

where E_1 denotes the amplitude of $e_{sc\text{-PM}i}$'s fundamental component.

Reliable fault location features must effectively distinguish fault locations under various demagnetization modes. Therefore, aiming to acquire more reliable location information, this paper investigates the influence of various demagnetization modes on SC residual back-EMF. Considering the alternating arrangement of N and S poles in permanent magnet motors, the same fault pattern exhibits similar characteristics across different electrical cycles. For this reason, this paper adopts one electrical cycle of the SC as the fundamental unit for analysis and segments the signal accordingly. The kth interval of the SC (SC_k) corresponds to the [(2k − 2) π, 2$k\pi$]. According to (4), the waveform of SC_k is determined by the demagnetization state of PM(2k), PM(2k − 1), and PM(2k − 2). Specifically, if k satisfies k = 1, then the PM(2k − 2) corresponds to PM2p. Based on the superposition principle, the model for the kth electrical period (SC$_k$) is expressed as:

$$\begin{cases} SC_k = e_{SC-\mathrm{PM}(2k)} + e_{SC-\mathrm{PM}(2k-1)} + e_{SC-\mathrm{PM}(2k-2)} \\ (2k-2)\pi \leq \alpha \leq 2k\pi \end{cases} \quad (5)$$

where k indicates the number of electrical cycles.

This model reveals the core mapping correlation between the SC signal and the condition of the PMs. This relationship establishes a theoretical foundation for analyzing the impact of different demagnetization patterns (including the quantity, position, and severity of demagnetized PMs) on the SC signals, thereby providing a theoretical basis for precisely locating demagnetized PMs.

2.3. Analysis of Demagnetization Faults on the Residual Back-EMF of SC

For exploring the influence of the quantity, location, and severity of demagnetized PMs on the residual back-EMF of search coils, a prototype equipped with search coils is established via Infolytica Magnet (R2017) under preconfigured healthy states and demagnetization fault scenarios. The key parameters of the prototype for testing are listed in Table 1.

https://doi.org/10.3390/app16020870

Table 1. Main structure parameters of the DDPMSM.

Items	Values	Unit
Number of stator slots	48	
Number of pole pairs	22	
Stator outer-diameter	270	mm
Rotor outer-diameter	201	mm
Stator inter-diameter	230	mm
Rotor inter-diameter	176	mm
Axial length	100	mm
Thickness of PM	4.5	mm
Rated frequency	66	Hz
Rated speed	180	rpm
Rated power	1.5	kW
Number of coils	24	/
Coil turns	70	/
Magnet type	N45M	/

Based on the analytical model of SC in Section 2.2, the residual back-EMF waveform of the search coil over one electrical cycle is affected by three consecutive PMs. According to the combination of the three PMs' conditions, there are one healthy and seven demagnetized types, as presented in Table 2.

Table 2. Combination of demagnetization fault types for three PMs coupled with residual back-EMF of SC.

	Health Condition and Demagnetization Fault Modes	PM Numbers and Their Corresponding Condition		
		$2k$	$2k-1$	$2k-2$
The kth Electrical Period	Health	0	0	0
	Type 1	0	1	1
	Type 2	1	0	0
	Type 3	1	1	0
	Type 4	0	0	1
	Type 5	0	1	0
	Type 6	1	0	1
	Type 7	1	1	1

As shown in Table 2, 0 represents a healthy condition of the PM, and 1 represents the demagnetized condition. For instance, Type 6 means PM($2k$) and PM($2k-2$) exhibit demagnetization, while PM($2k-1$) remains healthy.

In this study, the 8th electrical period of SC is a case for analysis, which is subject to the health and fault conditions of PM14, PM15, and PM16. Demagnetization faults are simulated through uniform reduction in the PMs' coercivity, and the demagnetization degree is set to 10% for each faulty PM. The simulation of search coil residual back-EMF under healthy conditions and different fault types is illustrated in Figure 2.

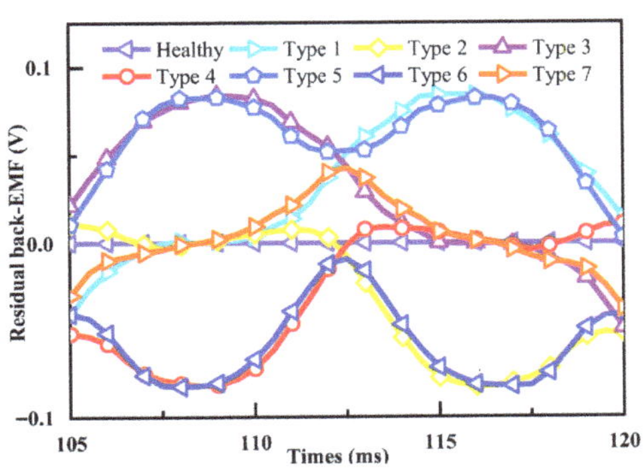

Figure 2. Residual back-EMF of SC in healthy condition and seven demagnetization fault types.

https://doi.org/10.3390/app16020870

As illustrated in Figure 2, the partitioned residual back-EMF provides distinct signatures that enable reliable discrimination among the healthy condition and all seven demagnetization fault types.

2.4. Robustness Analysis of Demagnetization Fault Characteristic Signal

(1) Effect of fault severity on the characteristic signal of demagnetization fault.

For exploring the influence of fault severity on the residual back-EMF of SC, five demagnetization fault severities are considered, which include 10%, 30%, 50%, 70%, and 100% demagnetization. Taking Type 1 as an example, the residual back-EMF of SC with different levels of severity is illustrated in Figure 3.

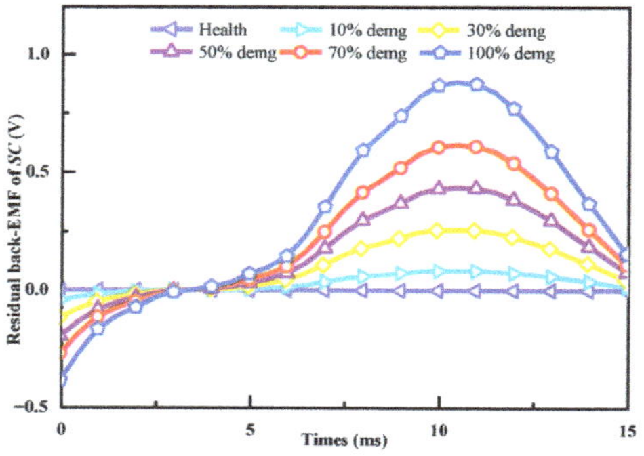

Figure 3. Residual back-EMF of SC under Type 1 with different demagnetization degrees.

As illustrated in Figure 3, the residual back-EMF of SC increases proportionally with the severity of the demagnetization fault. In Type 1, the search coil back-EMF at different fault severities has identical waveform morphology, with only amplitude variations. Therefore, the fault severity does not affect the ability to distinguish between demagnetization fault types.

(2) Effect of load on the characteristic signal of demagnetization fault.

For exploring the influence of rated load on SC residual back-EMF, six loads of demagnetization fault are considered under constant speed, which include 0%, 25%, 50%, 75%, 100%, and 125% load. Taking Type 3 as an example, the residual back-EMF of SC with different rated loads is illustrated in Figure 4.

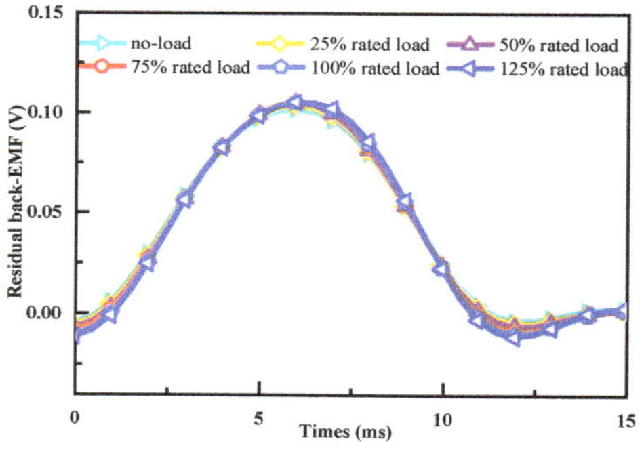

Figure 4. Residual back-EMF of SC under Type 3 with different loads.

As illustrated in Figure 4, the back-EMF in the SC exhibits minimal variation under different load conditions for demagnetization faults of identical severity. In Type 3, both

https://doi.org/10.3390/app16020870

the waveform morphology and amplitude remain largely consistent across the tested load range. Therefore, the load condition has a negligible impact on the detection of demagnetization faults.

(3) Effect of speed on the characteristic signal of demagnetization fault.

For exploring the influence of speed on the residual back-EMF of the SC, four speed conditions are considered under rated conditions, which include 90 rpm, 144 rpm, 180 rpm, and 216 rpm. The back-EMF of the SC1 in a healthy motor under different rotational speeds is illustrated in Figure 5.

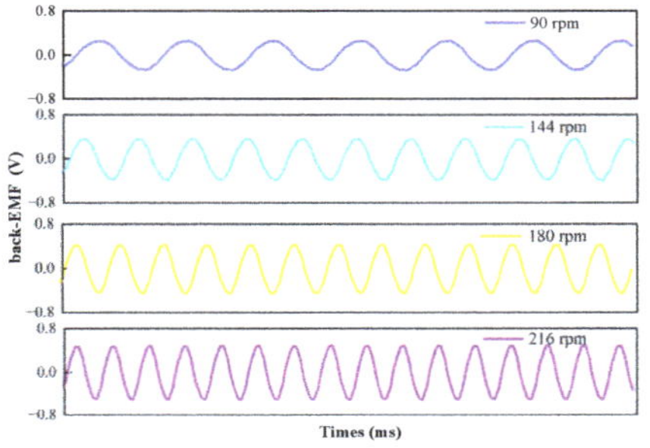

Figure 5. Back-EMF waveforms of SC1 at different speeds.

As illustrated in Figure 5, the back-EMF waveforms of a healthy SC1 at different rotational speeds exhibit significant variations in both amplitude and period with speed: The amplitude difference stems from the proportionality between back-EMF and speed (as given in Equation (1)), while the period variation results from the use of a fixed sampling frequency under varying speeds. These discrepancies complicate the selection of a unified reference signal for differential-based analysis, thereby limiting the diagnostic model's generalization capability.

To overcome this limitation, equiangular resampling and amplitude normalization were applied to the SC1 back-EMF signal. Following resampling, as shown in Figure 6, the angular-domain waveforms exhibit consistent periods across speeds, eliminating period-related discrepancies. For amplitude normalization, the measured signal was multiplied by a normalization coefficient k_v, given by

$$k_v = V / V_n \tag{6}$$

where V_n is the rated speed, and V is the actual motor speed.

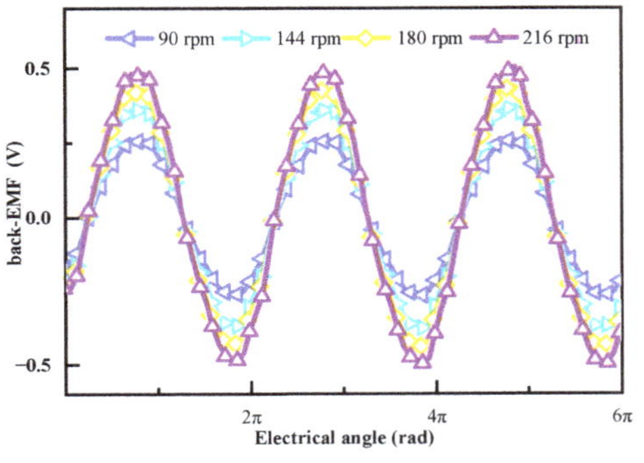

Figure 6. The back-EMF waveform of SC1 after equiangular resampling.

https://doi.org/10.3390/app16020870

After these processing steps, both the amplitude and waveform of the SC1 back-EMF become independent of speed, endowing the proposed diagnostic method with robustness against speed fluctuations.

3. Permanent Magnet Fault Diagnosis Based on CNN

Extracting fault features from characteristic signals is a crucial step in fault diagnosis. However, traditional fault diagnostic methods rely on manually extracting features based on signal processing techniques, with diagnostic performance constrained by expert expertise and exhibit limited generalization capability. In contrast, deep learning techniques possess a remarkable ability to extract discriminative features adaptively and have been widely adopted in various equipment fault diagnosis applications. Among these, the CNN, as one of the most representative architectures in deep learning, has achieved outstanding performance in motor fault diagnosis scenarios. Therefore, this study proposes an intelligent diagnosis method based on CNN. The workflow can be generally split into data processing and CNN model design, as illustrated in Figure 7. Specifically, it consists of three main steps: (1) constructing an enhanced fault sample library by injecting noise; (2) designing a CNN architecture suitable for residual sequences of search coils back-EMF; and (3) conducting model training and performance evaluation. Each step is detailed in the subsequent subsections.

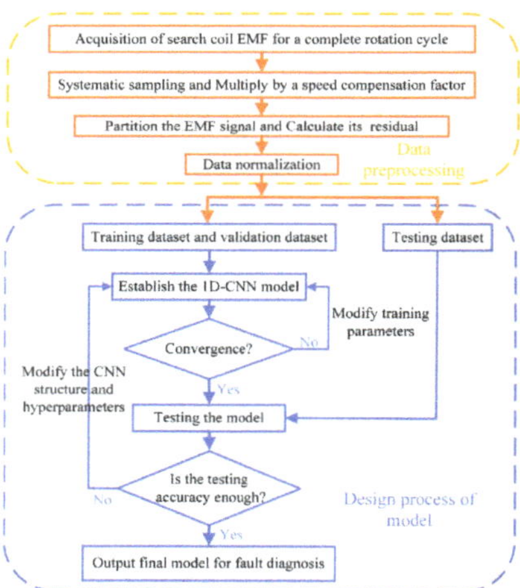

Figure 7. Flowchart of the 1-D CNN model design process.

3.1. Fault Sample Database Construction with Noise Injection

The limited availability of experimental samples (only one healthy prototype and one with a Type 5 demagnetization fault) restricted the comprehensive coverage of fault types, severity levels, and locations. To overcome this constraint, we employed finite element simulations to generate 720 samples for establishing a comprehensive initial sample library.

This sample library encompasses eight operational states (one healthy and seven fault Types, Type1-Type7), six load levels (0%, 25%, 50%, 100%, 125% of rated load), four speed conditions (40%, 60%, 80%, 100% rated speed), and five fault severity levels (10%, 30%, 50%, 70%, 100% demagnetization). A full-factorial design was used to evaluate all combinations of these states and conditions. Both healthy and faulty samples cover identical ranges of load and speed, with data in each group being mutually independent and non-overlapping.

Since motors are susceptible to external environmental noise during actual operation, a factor not considered in finite element model simulations. To validate the model's

 https://doi.org/10.3390/app16020870

robustness, Gaussian white noise is injected into the simulated data. This type of noise is widely adopted for CNN robustness testing due to its flat power spectral density (simulates random interference) and computational efficiency. To better emulate real-world noise conditions, SNR levels of 5 dB and 15 dB are applied, representing strong electromagnetic interference (5 dB) and moderate background noise (15 dB), respectively. The noise intensity is quantified by the SNR, defined as:

$$SNR = 10 \times \log_{10}(\frac{P_s}{P_n}) \tag{7}$$

where P_s and P_n denote the average power of the signal component and noise component, respectively, within a specified time interval.

The resampled signal x(m) constitutes a discrete sequence of length L_m. The average power P_x of this discrete signal x(m) is computed as follows:

$$P_x = \frac{1}{L_m} \times \sum_0^{L_m-1} x(m)^2 \tag{8}$$

The discrete SC fault signal s(m) were substituted into Equation (8) to obtain the original signal power (P_s). Substituting the predefined SNR threshold and Ps into Equation (7) yields the target noise power (P_n). Gaussian white noise with a power of P_n is generated via Equation (9).

$$N_{oise} = \sqrt{P_{noise}} \times randn(1, L_m) \tag{9}$$

where the *randn* () function generates random samples drawn from a standard normal distribution, and noise represents the synthesized noise data.

The generated noise was superimposed onto all raw signals to produce noise-contaminated signals at the target SNRs. Taking Type 2 as an example, the residual back-EMF of the search coil under different SNRs is illustrated in Figure 8.

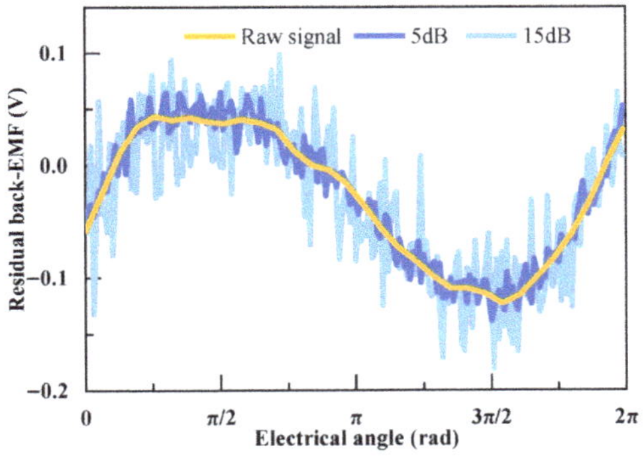

Figure 8. Residual back-EMF of the SC at different SNRs.

As shown in Figure 8, the raw signal is the residual back-EMF waveform without noise interference, exhibiting a smooth and continuous trend with stable periodic characteristics. After adding noise, the smoothness of the residual back-EMF waveform significantly decreases and waveform amplitude fluctuation intensifies. As the SNR decreases, the signal quality deteriorates significantly, the smoothness of the waveform further decreases, and the amplitude fluctuation becomes more severe, and the greater the deviation from the raw signal trend. Under strong noise interference, if the fault feature is weak, it will be difficult to effectively extract, which is prone to misdiagnosis and seriously affects the accuracy of diagnosis.

https://doi.org/10.3390/app16020870

To facilitate analysis of these reliable fault features, a fault sample library was developed using experimental data from healthy and faulty prototypes as well as finite element simulation data, as detailed in Table 3.

Table 3. Fault sample.

Category Label	Type	Details	SNR (dB)	Sample Quantity
1	Healthy	Speed(r/min): 90/144/180/216 Load(A): 0/1/2/3/4/5 Demagnetization Degree: -	-	240
2	Type 1	Speed(r/min): 90/144/180/216 Load(A): 0/1/2/3/4/5 Demagnetization Degree: 10%/30%/50%/70%/100%	5/15	240
3	Type 2		5/15	240
4	Type 3		5/15	240
5	Type 4		5/15	240
6	Type 5		-	240
7	Type 6		5/15	240
8	Type 7		5/15	240

3.2. Proposed 1D-CNN Model Design

Based on the established dataset, this study focuses on the core component of the intelligent diagnostic model: a 1-D CNN architecture for end-to-end demagnetization fault diagnosis. When designing the CNN, the following concepts were adhered to: uniform filter size, the number of filters should increase layer by layer, and each convolutional layer must be followed by an activation function and a pooling layer. Therefore, we propose a CNN that adjusts parameters based on the above principles, relying on experience and previous experimental tuning. After experimental tuning, the finalized CNN architecture generally comprises one input layer, two convolutional layers, two max-pooling layers, one fully connected layer, and one output layer, as illustrated in Figure 9.

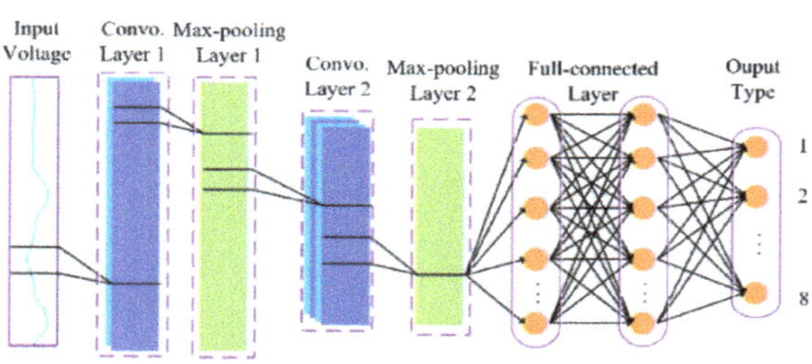

Figure 9. The basic structure of 1-D CNN.

(1) The input layer is configured to receive a 200 × 1 residual signal from the search coil. This dimension was selected because, firstly, the features of demagnetization faults are most distinctly exhibited over one full electrical cycle; Secondly, integer-period sampling was employed to eliminate spectral leakage. This requires the number of samples N to satisfy N $= f_s/f_0$. Consequently, with a fundamental frequency f_0 of 66 Hz and a sampling frequency fs of 13.2 kHz, N is determined to be 200. This approach can retain the complete information of the phase and amplitude, thereby providing input data for the convolutional layers to learn the essential discriminative features.

(2) The convolution layer utilizes the convolution operation to automatically extract features from the output of the preceding layer or the original signal. The convolution operation is given by

$$x_j^l = f(\sum_{x \in M_j} x_i^{l-1} * k_j^l + b_j^l) \quad (10)$$

 https://doi.org/10.3390/app16020870

where x_i^{l-1} indicates the input data for the lth layer and $*$ represents the convolution operation, x_j^l is the output of layer l, M_j denotes the set of input features, k_j^l and b_j^l represent the convolutional kernel weight and bias corresponding to the jth kernel in the lth convolutional layer, respectively. where $f(\cdot)$ refers to a nonlinear activation function, applied to enhance the network's nonlinear expression capability.

The ReLU activation function is commonly used to accelerate the convergence of the model during its training process, as given by

$$\mathrm{ReLU}(x) = \max(0, x) \tag{11}$$

After determining the basic network composition, the effectiveness of feature extraction primarily depends on two key architectural parameters: convolutional depth and kernel size. These parameters collectively determine the receptive field size and the granularity of feature extraction, which are essential to accurately extracting the local patterns exhibited by demagnetization faults in residual back-EMF sequences. Therefore, this study adopts a stepwise systematic optimization strategy: first determining the optimal network depth, and then optimizing the kernel size based on this foundation.

First, network depth optimization was conducted. Considering the limited length of the input sequence (200 × 1) and the need to precisely capture the localized pulse signatures characteristic manifested in the residual back-EMF under demagnetization faults, this study employed a small initial convolutional kernel (kernel size = 3, stride = 1) to preserve the temporal details of the input signal. With this configuration, this study evaluated architectures with 1 to 4 convolutional layers. The different convolutional layers' diagnostic accuracy is presented in Table 4. The results indicated that the two-layer network achieved significantly higher diagnostic accuracy compared to the single-layer model. However, increasing the depth to 3 or 4 layers yielded no substantial improvement in accuracy. This suggests that the two-layer structure already provides a sufficient receptive field to capture fault characteristics, while deeper structures increase the risks of vanishing/exploding gradients. Consequently, a two-layer architecture was identified as optimal, employing 16 and 32 filters in its first and second layers, respectively.

Table 4. Diagnostic accuracy with different convolutional layers.

Layer	1	2	3	4
Accuracy	98.20%	99.50%	99.60%	99.62%

Subsequently, within the two-layer framework, this study compared four kernel sizes (2 × 1, 3 × 1, 5 × 1, 7 × 1). The results indicate that CNN models with different kernel sizes all achieve classification accuracy of approximately 99%. The kernel size has minimal impact on the classification accuracy of the CNN model. Balancing computational efficiency and result accuracy, this paper selects a 3 × 1 kernel size for the model design.

(3) The pooling layer is placed after the convolutional layer to reduce the dimension of the extracted features. The pooling operation is treated as a down-sampling operation. This reduces the computational and parametric overhead, thereby mitigating overfitting and improving model generalization capability. The most widely used in CNN is maximum pooling. The mathematical expression is

$$P_i^{l+1} = \underset{(j-1)W+1<t<jW}{max} \{q_i^l(t)\} \tag{12}$$

https://doi.org/10.3390/app16020870

where P_i^{l+1} represents the output value of the $(l + 1)$th layer, W refers to the width of the pooling window. $q_i^l(t)$ represents the output of the tth neurons in the ith channel of the l layer.

Following the convolutional operation, a max pooling layer (kernel size = 2, stride = 2) is employed to downsample the feature maps. This configuration is specifically optimized for 1-D residual back-EMF time-series signals, with its key advantages being: First, its compact receptive field preserves fine-grained temporal features more effectively than larger kernels, maintaining essential structural details during down-sampling. Second, a stride of 2 halves the length of the feature maps, reducing computational complexity and memory consumption in subsequent layers by approximately 50%. Finally, for an input sequence length of 200, this stride ensures non-overlapping and complete coverage of the entire sequence, preventing information loss at the boundaries. In summary, this empirically validated configuration achieves an optimal balance between computational efficiency and feature preservation.

(4) The purpose of a fully connected layer is to project the feature space to the sample label space; the features are converted into 1-D vectors. Each neuron in this layer is fully connected to all neurons in the preceding layer. The output vector of the l layer is

$$x_j^l = f(\sum_{i=1}^{M} x_i^{l-1} \times w_{ij}^l + b_j^l), j = 1,\ 2,\ \ldots,\ N \tag{13}$$

where M is the input feature dimensionality, and N is the output vector dimensionality. w_{ij}^l denotes the weight of the connection between the ith neuron of the $(l-1)$ th layer and the jth neuron of the lth layer, and b_j^l indicates the bias term of the jth neuron of the lth layer.

The SoftMax is connected after the fully connected layer and acts as a classifier. The SoftMax function calculates the probability that the input data belongs to each class. The output expression of SoftMax regression is

$$p(x_j^l) = \frac{e^{x_j^{l-1}}}{\sum_{k=1}^{N} e^{x_k^{l-1}}} \tag{14}$$

where N is the number of classes, x_j^{l-1} is the output of the upper layer, and $p\left(x_j^l\right)$ is the output probability.

The classification module, comprising fully connected layers and an output layer, utilizes the Softmax activation function, maps the high-dimensional features to a probability distribution over fault classes, enabling detection and location of demagnetization faults. The model's detailed architecture is specified in Table 5.

Table 5. Network architecture and parameters.

Layer	Parameter	Output Dim
Input Layer	\	200 × 1
Convolution Layer(C1)	16 filters, kernel = 3 × 1	200 × 1 × 16
Activation Function	ReLU	\
Maximum Pooling	Kernel = 2 Stride = 2	100 × 1 × 16
Convolution Layer(C2)	32 filters, kernel = 3 × 1	100 × 1 × 32
Activation Function	ReLU	\
Maximum Pooling	Kernel = 2 Stride = 2	50 × 1 × 32
Fully connected Layer	\	1 × 1 × 8
SoftMax Layer	\	1 × 1 × 8
Output Layer	\	1 × 1 × 8

 https://doi.org/10.3390/app16020870

3.3. Model Training and Performance Evaluation

To evaluate the performance and generalization capability of the designed 1D-CNN model, we employed the comprehensive fault database detailed in Table 3. The dataset was partitioned randomly into training, validation, and test sets at a 2:1:1 proportion to ensure an unbiased model evaluation process, then fed into a CNN model constructed with the MATLAB 2024a Deep Learning Toolbox for training. Training of the 1D-CNN model was using the Adaptive Moment Estimation (Adam) optimizer. Key hyperparameters are set as follows: 0.01 for the learning rate, 40 for the maximum epochs, and 64 for the batch size. The training and validation progress are illustrated in Figure 10, and the confusion matrix corresponding to the 1D-CNN test results is illustrated in Figure 11.

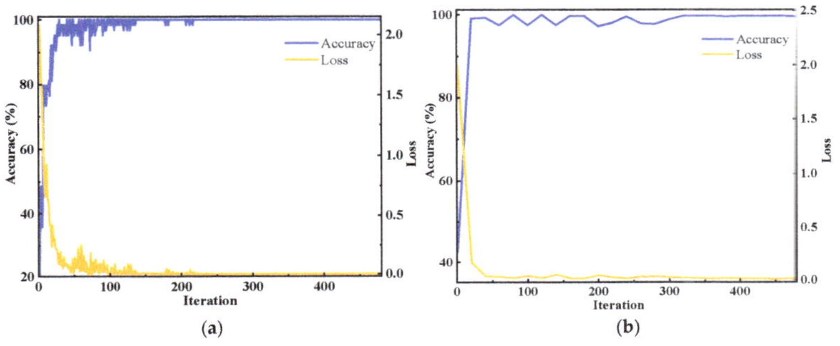

Figure 10. Accuracy and loss curves during training and validation. (**a**) Training process. (**b**) Validation process.

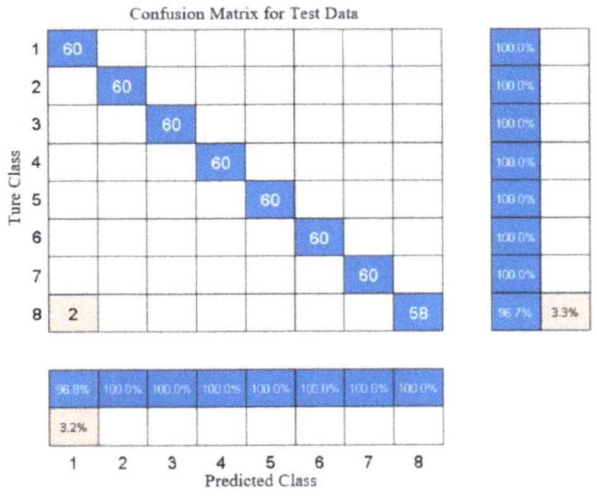

Figure 11. Confusion matrix of the 1D-CNN test results.

As shown in Figure 10, the model approached convergence after approximately 300 iterations. Both training and validation accuracy exceeded 99%. Furthermore, the training and validation losses decreased consistently and stabilized around 0.001. These results suggest that the model exhibits strong convergence and generalization performance.

As illustrated in Figure 11, the model demonstrated excellent performance on the test set, achieving an overall accuracy of 99.5% and a classification accuracy exceeding 96.5% for all eight fault states. However, confusion matrix analysis revealed one misclassification: a small number of samples from Type 7 (uniform demagnetization of three permanent magnets within the same electrical cycle) were incorrectly identified as normal. This misclassification can be attributed to the preservation of magnetic field symmetry in Type 7, which only causes variations in signal amplitude without waveform distortion.

https://doi.org/10.3390/app16020870

In contrast, demagnetization of single or dual permanent magnets disrupts magnetic symmetry, resulting in more distinguishable features.

Notably, all misclassifications of Type 7 occurred at the 10% demagnetization level. At this early stage, the fault feature closely resembles normal conditions, making differentiation difficult. From an engineering safety perspective, a 10% demagnetization degree generally does not immediately cause system failure or severe performance degradation, thus providing enough maintenance time before performance deteriorates.

4. Experimental Result

4.1. Experimental Setup

The experimental platform is illustrated in Figure 12. It primarily comprises two comparative prototypes (healthy and defective, with key parameters summarized in Table 1). During the experiment, the driver powers the test prototype while an independent motor provides the load. Both prototypes were mechanically coupled to a torque transducer for simultaneous speed and torque measurement. Search coils (diameter: 0.5 mm, turns per coil: 5) were installed in the test prototype to sense the back-EMF. The back-EMF data, recorded by the acquisition card, is routed to the computer via USB. An oscilloscope, connected to the search coils via probes in parallel, provided visual monitoring of the back-EMF. To simulate demagnetization faults, permanent magnets PM9 and PM31 were detached from the test prototype. These magnets were symmetrically positioned relative to the center line, thereby avoiding potential eccentricity effects.

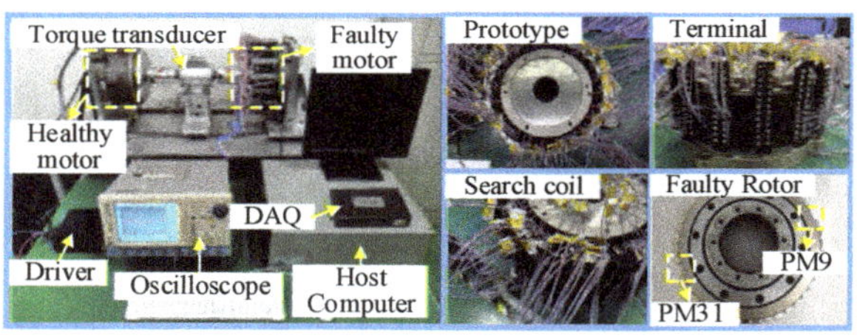

Figure 12. Experiment platform.

4.2. Experimental Results and Analysis

To verify the proposed location method, the experimental back-EMF signals from the healthy prototype under rated load conditions were compared with the corresponding simulation results. A low-pass filter realized in MATLAB was applied to process the measured back-EMF signal. The back-EMF signals of SC1, SC2, and SC for the healthy prototype under rated load conditions are illustrated in Figure 13.

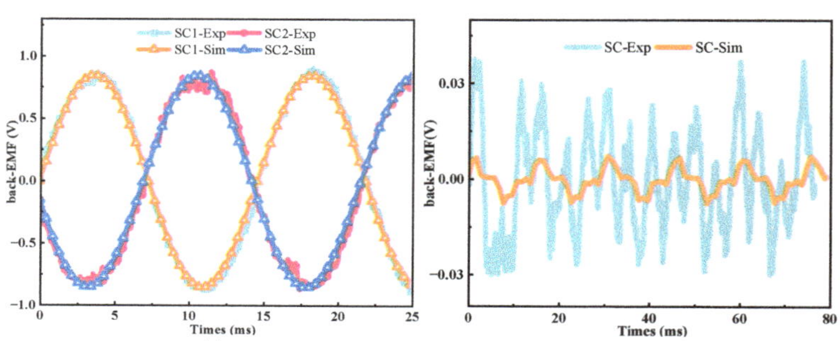

Figure 13. Experimental results of search coil back-EMF (**a**) SC1 and SC2 back-EMF, (**b**) SC back-EMF.

https://doi.org/10.3390/app16020870

As illustrated in Figure 13a, the experimental back-EMF results for search coils agree well with the theoretical predictions, with a maximum relative error of 4.4% between the actual and theoretical peak values. As illustrated in Figure 13b, the maximum discrepancy between simulated and experimental results of SC was 0.04 V. This discrepancy stems from environmental noise, PMSM magnetic field non-uniformity, and stator winding imbalances, inherent phenomena present in both healthy and faulty states. However, these minor discrepancies have a negligible impact on diagnostic outcomes, as the 1-D CNN model is not dependent on precise peak-value measurements. Instead, it utilizes an end-to-end learning approach to extract discriminative features directly from the raw signal data.

After verifying the reliability of the healthy baseline measurements under rated load, the influence of load variations on diagnostic signals was further analyzed. The SC waveforms of the healthy motor at rated speed under four rated load conditions are illustrated in Figure 14.

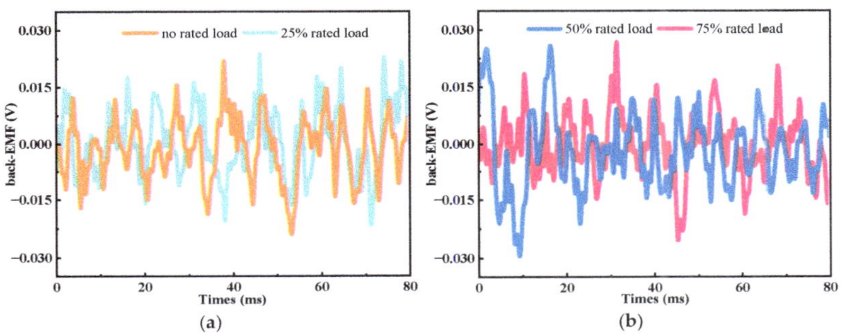

Figure 14. Experimental results of SC residual back-EMF for the different loads. (**a**) no rated load and 25% rated load (**b**) 50% rated load and 75% rated load.

As shown in Figure 14, the SC signal of the healthy motor is largely independent of load variations, and these minor differences do not affect the diagnostic outcome.

To demonstrate its objective of fault diagnosis, the method was subsequently applied to fault location experiments. In the faulty prototype, permanent magnets PM9 and PM31 were completely removed. The fault location signal exhibits distinct features at the 5th and 16th electrical cycles, which correspond precisely to the magnetic pole partitions of the removed PMs. The waveforms of the fault location signal measured under rated operating conditions are illustrated in Figure 15.

As illustrated in Figure 15a,b, the simulated and experimental residual back-EMF waveforms obtained from the search coil are in strong agreement. Specifically, both SC1 and SC2 exhibit pronounced signals during the intervals of [60 ms, 75 ms] and [225 ms, 245 ms], which correspond to the 5th and 16th electrical periods, respectively. As illustrated in Figure 15c,d, the SC waveform within these critical intervals displays the same characteristics previously defined for Type 5 and matches the fault signature (0, 1, 0) listed in Table 2. Substituting $k = 5$ and $k = 16$ into the expression $(2k - 2, 2k - 1, 2k)$, the demagnetized permanent magnets are identified as PM9 and PM31.

https://doi.org/10.3390/app16020870

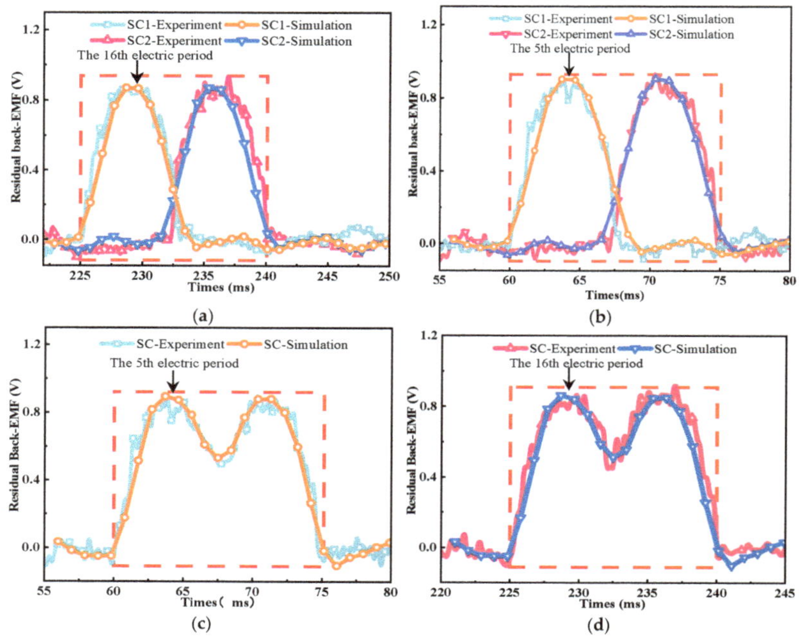

Figure 15. Experimental results of search coils' residual back-EMFs for defective prototype. (**a**) Residual back-EMF in the 5th electrical cycle (**b**) Residual EMF of SC1 and SC2 in the 16th electrical cycle (c) Residual EMF of SC in the 5th electrical cycle (**d**) Residual EMF of SC in the 16th electrical cycle.

4.3. Comparison with Other Methods

To validate the performance advantages of the 1-D CNN model, LSTM, BPNN, PNN, and SVM are used as comparative models. Meanwhile, to avoid random errors caused by a single test, each model underwent ten independent experiments, with the average classification accuracy serving as the performance evaluation metric. The per-class accuracy of these models is illustrated in Figure 16, and overall accuracy and computational time are summarized in Table 6.

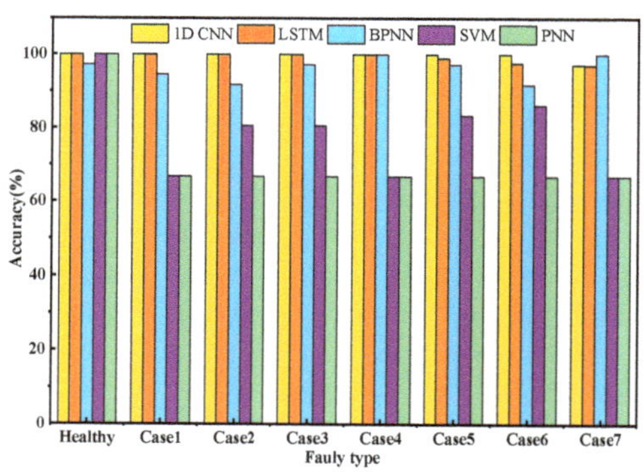

Figure 16. Comparison of diagnostic accuracy among different algorithms.

https://doi.org/10.3390/app16020870

Table 6. The result of different methods.

Methods	Testing Accuracy	Time
1-D CNN	99.58%	3 s
LSTM	98.93%	9 s
BPNN	96.18%	12 s
PNN	78.84%	0.06 s
SVM	70.86%	0.2 s

As illustrated in Figure 16, the proposed 1D-CNN model achieved superior performance in classification, with all fault-type accuracies exceeding 96.5%. Although the BPNN marginally outperformed our model in Type 7, the 1D-CNN maintained the highest accuracy across all other fault types.

As summarized in Table 6, the proposed 1D-CNN achieved the highest testing accuracy (99.58%) with a computational time of only 3 s, outperforming other methods. In contrast, the LSTM achieves 98.93% accuracy but requires 9 s for computation, while the BPNN reaches 96.18% accuracy but requires 12 s for computation. Although the PNN and SVM computed faster (0.06 s and 0.2 s), their accuracy was more than 21% lower than the proposed model. This demonstrates that the CNN model holds significant advantages when applied to demagnetization fault detection and location in DDPMSMs.

5. Conclusions

This study proposes a 1-D CNN location method based on the residual back-EMF of SC to locate demagnetization faults in DDPMSM. Experiments and simulations corroborated the efficacy and reliability of the proposed method.

(1) SC1 and SC2, two toroidal coils, are installed in corresponding positions within adjacent pole-pairs in the stator slots, connected in series to constitute the SC assembly. This configuration enables fault location through residual back-EMF analysis. This approach significantly reduces intrusiveness and implementation cost compared to conventional methods that involve installing search coils and applies to permanent magnet synchronous motors with any pole-slot combination.

(2) The rotor is partitioned into several sectors based on the pole pairs of the permanent magnets, dividing the full mechanical cycle accordingly. Previous analysis indicates that the residual back-EMF waveform is related to the state of three adjacent permanent magnets. Therefore, by analyzing the location signal waveform within each sector, the magnetic state of the permanent magnets in that sector can be effectively identified, thereby enabling accurate location of the demagnetized permanent magnets.

(3) This paper proposes a precise location method based on a 1D-CNN, which is capable of achieving both detection and location of demagnetization faults. Compared with the other classification approaches, the experimental result shows that the proposed method's fault type location accuracy reaches 99.58%, significantly outperforming methods such as LSTM, BPNN, PNN, and SVM. Although the computational efficiency of this method is slightly lower than that of PNN and SVM, its diagnostic accuracy shows an improvement of over 20%, and both its computational efficiency and comprehensive diagnostic performance are superior to those of BPNN and LSTM.

The proposed method demonstrates strong capabilities in detecting and locating demagnetization faults. While the method is invasive, it applies to safety-sensitive scenarios, including rail transit systems, aerospace power systems, and industrial automation systems.

Future work will primarily focus on the refining and enhancing the existing location model to improve its performance and investigating fault severity assessment methodolo-

https://doi.org/10.3390/app16020870

gies to achieve a comprehensive diagnostic procedure from fault location to severity evaluation.

Author Contributions: C.G. supervised the study, reviewed the manuscript, and proposed the research direction of the paper. Z.J. wrote the main manuscript text, prepared the figures and tables, performed coding work, and conducted data processing. X.X. and J.S. supervised the study, reviewed the manuscript. All authors have read and agreed to the published version of the manuscript.

Funding: This work was supported by National Natural Science Foundation of China (52177039) and, Natural Science Foundation of Henan Province (252300421328). Fundamental Research Funds for the Universities of Henan Province (NSFRF240306).

Institutional Review Board Statement: Not applicable.

Informed Consent Statement: Not applicable.

Data Availability Statement: The data that support the findings are available from the corresponding author upon reasonable request.

Conflicts of Interest: The authors have no conflicts to disclose.

References

1. Mohamed, Y.A.-R.I. Direct Instantaneous torque control in direct drive permanent magnet synchronous motors—A new approach. *IEEE Trans. Energy Convers.* **2007**, *22*, 829–838. [CrossRef]
2. Mohamed, Y.A.-R.I. A Hybrid-Type variable-structure instantaneous torque control with a robust adaptive torque observer for a high performance direct-drive PMSM. *IEEE Trans. Ind. Electron.* **2007**, *55*, 2491–2499. [CrossRef]
3. Goktas, T.; Zafarani, M.; Akin, B. Discernment of broken magnet and static eccentricity faults in permanent magnet synchronous motors. *IEEE Trans. Energy Convers.* **2016**, *31*, 578–587. [CrossRef]
4. Kao, I.H.; Wang, W.J.; Lai, Y.H.; Perng, J.W. Analysis of permanent magnet synchronous motor fault diagnosis based on learning. *IEEE Trans. Instrum. Meas.* **2018**, *68*, 310–324. [CrossRef]
5. Urresty, J.-C.; Riba, J.-R.; Romeral, L. A back-EMF based method to detect magnet failures in PMSMs. *IEEE Trans. Magn.* **2013**, *49*, 591–598. [CrossRef]
6. Bossio, J.; Ruschett, C.; Bossio, G.; Verucchi, C.; De Angelo, C. Rotor fault diagnosis in permanent magnet synchronous machine using the midpoint voltage of windings. *IET Electr. Power Appl.* **2020**, *14*, 256–261. [CrossRef]
7. Usman, A.; Rajpurohit, B.S. Modeling and classification of stator inter-turn fault and demagnetization effects in BLDC motor using rotor back-EMF and radial magnetic flux analysis. *IEEE Access* **2020**, *8*, 118030–118049. [CrossRef]
8. Ebrahimi, M.; Faiz, J. Demagnetization fault diagnosis in surface mounted permanent magnet synchronous motors. *IEEE Trans. Magn.* **2013**, *49*, 1185–1192. [CrossRef]
9. Zhu, M.; Yang, B.; Hu, W.; Feng, G.; Kar, N.C. Vold–Kalman Filtering Order Tracking Based Rotor Demagnetization Detection in PMSM. *IEEE Trans. Ind. Appl.* **2019**, *55*, 5768–5778. [CrossRef]
10. Barmpatza, A.C.; Kappatou, J.C. Study of the total demagnetization fault of an AFPM wind generator. *IEEE Trans. Energy Convers.* **2021**, *36*, 916–928. [CrossRef]
11. Rafaq, M.S.; Lee, H.; Park, Y.; Lee, S.B.; Zapico, M.O.; Fernandez, D.; Diaz-Reigosa, D.; Briz, F. Airgap search coil based identification of PM synchronous motor defects. *IEEE Trans. Ind. Electron.* **2022**, *69*, 6551–6560. [CrossRef]
12. Zeng, C.; Huang, S.; Lei, J.; Wan, Z.; Yang, Y. Online rotor fault diagnosis of permanent magnet synchronous motors based on stator tooth flux. *IEEE Trans. Ind. Appl.* **2021**, *57*, 2366–2377. [CrossRef]
13. Da, Y.; Shi, X.; Krishnamurthy, M. A New approach to fault diagnostics for permanent magnet synchronous machines using electromagnetic signature analysis. *IEEE Trans. Power Electron.* **2013**, *28*, 4104–4112. [CrossRef]
14. Chen, H.; Gao, C.; Si, J.; Nie, Y.; Hu, Y. A novel method for diagnosing demagnetization fault in PMSM using toroidal-yoke-type search coil. *IEEE Trans. Instrum. Meas.* **2021**, *71*, 750101. [CrossRef]
15. Gao, C.; Li, B.; Chen, H.; Xu, Y.; Xu, X.; Si, J. A less-invasive method for accurately diagnosing demagnetization fault in PMSM using rotor partition. *IEEE Trans. Transp. Electrif.* **2021**, *9*, 2356–2366. [CrossRef]
16. Shi, Q.; Zhang, H. Fault diagnosis of an autonomous vehicle with an improved SVM algorithm subject to unbalanced datasets. *IEEE Trans. Ind. Electron.* **2021**, *68*, 6248–6256. [CrossRef]
17. Guo, F.; Fu, W.; Wang, Y.; Chen, J. Research on fault diagnosis method for photovoltaic array based on model fusion. *Electr. Eng.* **2025**, *107*, 8189–8199. [CrossRef]

 https://doi.org/10.3390/app16020870

18. Yang, X.; Chen, W.; Li, A.; Yang, C. A Hybrid machine-learning method for oil-immersed power transformer fault diagnosis. *IEEJ Trans. Electr. Electron. Eng.* **2020**, *15*, 501–507. [CrossRef]
19. Liu, F.; Wu, R.; Teng, F.; Liu, Y.; Lu, S.; Ju, B.; Cao, Z. A Two-stage learning model for track-side acoustic bearing fault diagnosis. *IEEE Trans. Instrum. Meas.* **2021**, *70*, 3516512. [CrossRef]
20. Ma, L.; Dong, J.; Peng, K. A practical root cause diagnosis framework for quality-related faults in manufacturing processes with irregular sampling measurements. *IEEE Trans. Instrum. Meas.* **2022**, *71*, 3511509. [CrossRef]
21. Dao, F.; Zeng, Y.; Qian, J. Fault diagnosis of hydro-turbine via the incorporation of Bayesian algorithm optimized CNN-LSTM neural network. *Energy* **2024**, *290*, 130326. [CrossRef]
22. Skowron, M.; Orlowska-Kowalska, T.; Kowalski, C.T. Detection of permanent magnet damage of PMSM drive based on direct analysis of the Stator phase currents using Convolutional Neural Network. *IEEE Trans. Ind. Electron.* **2022**, *69*, 13665–13675. [CrossRef]
23. Dai, Y.; Zhang, L.; Xu, D.; Chen, Q.; Li, J. Development of deep learning-based cooperative fault diagnosis method for multi-PMSM drive system in 4WID-EVs. *IEEE Trans. Instrum. Meas.* **2024**, *73*, 3506513. [CrossRef]
24. Lv, K.; Wang, D.; Huang, W.; Liu, H.; Jiang, Y.; Hu, J. Multiple fault diagnosis of PMSM based on stator tooth flux and parallel residual convolutional neural network. *IEEE Trans. Transp. Electrif.* **2024**, *10*, 9727–9748. [CrossRef]
25. Eker, M.; Gündogan, B. Demagnetization fault detection of permanent magnet synchronous motor with convolutional neural network. *Electr. Eng.* **2023**, *105*, 1695–1708. [CrossRef]
26. Ni, S.; Wang, H.; Liu, S.; Wang, Y.; Yu, J.; Wang, L. Efficient multimodal motor demagnetization diagnosis framework with half-wave fundamental wave division transformation. *IEEE Trans. Ind. Electron.* **2025**, *74*, 2533116. [CrossRef]
27. Ince, T.; Kiranyaz, S.; Eren, L.; Askar, M.; Gabbouj, M. Real-Time motor fault detection by 1-D convolutional neural network. *IEEE Trans. Ind. Electron.* **2016**, *63*, 7067–7075. [CrossRef]

https://doi.org/10.3390/app16020870

applied sciences

MDPI

Article

Influence of Temperature on Brushless Synchronous Machine Field Winding Interturn Fault Severity Estimation

Rubén Pascual [1], Eduardo Rivero [2], José M. Guerrero [3], Kumar Mahtani [1] and Carlos A. Platero [1,*]

[1] Department of Automation, Electronic and Electrical Engineering and Industrial Informatics, Escuela Técnica Superior de Ingenieros Industriales, Universidad Politécnica de Madrid, 28006 Madrid, Spain; ruben.pascualj@alumnos.upm.es (R.P.); kumar.mahtani@upm.es (K.M.)

[2] ESELEC Ingenieros, 28008 Madrid, Spain; eduribar@hotmail.com

[3] Department of Electrical Engineering, Escuela de Ingeniería de Bilbao, Universidad del País Vasco, 48013 Bilbao, Spain; josemanuel.guerrerog@ehu.eus

* Correspondence: carlosantonio.platero@upm.es

Abstract: There are numerous methods for detecting interturn faults (ITFs) in the field winding of synchronous machines (SMs). One effective approach is based on comparing theoretical and measured excitation currents. This method is unaffected by rotor temperature in static excitation SMs. However, this paper investigates the influence of rotor temperature in brushless synchronous machines (BSMs), where rotor temperature significantly impacts the exciter excitation current. Extensive experimental tests were conducted on a special BSM with measurable rotor temperature. Given the challenges of measuring rotor temperature in industrial machines, this paper explores the feasibility of using stator temperature in the exciter field current estimation model. The theoretical exciter field current is calculated using a deep neural network (DNN), which incorporates electrical brushless synchronous generator output values and stator temperature, and it is subsequently compared with the measured exciter field current. This method achieves an error rate below 0.5% under healthy conditions, demonstrating its potential for simple implementation in industrial BSMs for ITF detection.

Keywords: AC machines; brushless synchronous machines; condition monitoring; deep learning; fault detection; generators; deep neural networks; rotor temperature; stator temperature; power generation

1. Introduction

In power generation applications, brushless synchronous machines (BSMs) usually consist of a main machine and an exciter, mechanically coupled on the same shaft. The exciter's function is to supply the necessary field power to the main machine for its operation. In the conventional topology, the exciter's field winding is rolled around the stator poles, and its armature is an AC winding located on the rotor. The AC output of the exciter armature is converted into the DC excitation input for the main machine by a rotating diode rectifier bridge. The field winding of the main machine is located on the rotor and its armature on the stator.

Currently, several excitation systems for BSMs are identified by the field power source supplied to the exciter [1–3]. The primary advantages of removing brushes and slip rings in the excitation system [4] are reduced maintenance costs and increased operational safety, as no sparks are produced. In some facilities, such as oil refineries and chemical plants, the use of BSMs is mandatory.

The market for BSMs is extensive. High-power BSMs are widely used in nuclear power plants, while low-power units are employed in emergency diesel generators and hydro power plants [5,6]. Furthermore, BSMs can address issues arising from the high penetration of renewable generation plants in the grid, such as the reduction of the short-circuit ratio (SCR). BSMs used as synchronous condensers can help maintain a minimum SCR level in global grids [7,8], especially those equipped with a PMG as a pilot exciter. Additionally, BSMs can help renewable power plants comply with various grid code requirements, which are largely based on specific standards [9–11].

In all these applications, BSMs must meet industry reliability standards, including protection against faults and abnormal operating conditions [12,13]. BSMs are susceptible to various faults due to their operating conditions. This manuscript focuses on the influence of temperature on estimating the theoretical field exciter excitation current, with the ultimate goal of detecting interturn faults (ITFs) in the field winding of BSMs. The number of ITFs is directly proportional to the severity of the fault [14,15].

The main difficulty in detecting ITFs is that they do not usually result in large overcurrent in the machine, so the machine can operate normally if the interturn fault severity level is lower than a tolerable limit. However, the existence of a magnetic unbalance causes additional mechanical stress on the machine, which can lead to large vibrations that can damage the generator.

The main origin of ITFs is the weakness of the insulation [16,17]. A short-circuit between turns is usually caused by deterioration of the insulation, which is usually caused by the high stresses applied to the insulation. The ITFs cause a drop in the electrical resistance between the poorly insulated turns, which leads to an increase in the excitation current of the generator.

Differences in electrical potential, mechanical stresses due to vibrations and frictions, and thermal stresses due to a temperature gradient are the causes of high stresses in the insulation, which can lead to damage to it. In addition, the presence of fine particles or liquids such as water or acids can increase the likelihood of ITFs.

There are some previous model-based approaches for rotor ITF detection and severity estimation in BSMs [14,15]. Some of these methods are based on the comparison of two variables: the theoretical model estimation obtained from the machine electrical output values and the measured exciter field current. This comparison is very useful for attesting deviations from the healthy condition and for calculating the fault severity level employing the relative difference between both values. So, accurate assessment of the excitation current in BSMs is essential for proper machine characterization. This significance arises from the inherent nature of brushless excited SMs, where the field winding becomes physically inaccessible during operation, making direct measurement of the field current extremely challenging.

Another previous article [18] investigates the idea of implementing deep neural network (DNN) algorithms to detect ITFs in BSMs. The combination of large numbers of experimental data and the use of deep learning methods achieves excellent results for detecting ITFs in BSMs. However, preceding algorithms fail to account for the impact rotor winding temperature may have on field current estimation. During machine operation, a notable increase in rotor winding temperature occurs, gradually causing a decline in the field current. It is interesting to highlight that, while extensive studies have explored field current estimation, limited research investigates temperature-induced alterations in these values. In [19], field current and field winding temperature are estimated using a machine model, and then the estimates are compared with the actual currents measured in the stator and adjusted through a correction mechanism using a Kalman filter, demonstrating efficacy in simulations. In [20], field current and temperature estimation is achieved through the DC link current, initially computed via an algorithm and afterward fed back to the controller and compared with the measured value. The resulting error facilitates temperature estimation refinement, thereby improving both field and DC link current estimates.

Furthermore, due to the difficulty of accurate modeling posed by SMs combined with recent advancements in computational learning, contemporary studies explore methodologies such as neural networks for identifying faults within the field winding. Works such as [18,21] collect healthy machine data alongside post-fault data to use these data for early fault detection. Nonetheless, these methods do not work in the absence of real-time consideration of BSM operating temperatures, as situations may arise where a fault between turns in a cold machine exhibits field current values like those of a healthy machine operating at high temperatures.

Taking into account this difficulty, the main objective of this work is to demonstrate and verify the influence of temperature on the exciter excitation current estimation algorithm. This will substantially improve the results studied in previous works [18], where a neural network model was presented that allowed the estimation of the excitation correlation of a brushless machine but did not consider the rotor temperature. Considering how the rotor temperature affects the excitation current of the machine, the results obtained by taking into account the machine temperature as input to the model will be much closer to the real machine measurements. The second objective is to validate the use of stator temperature instead of rotor temperature in the estimation algorithm. This is a very important validation because the stator temperature can be measured during operation in industrial BSMs, where it is not possible to measure the rotor temperature.

For this purpose, this paper proposes using DNN architecture algorithms. It was decided to use DNNs to estimate the excitation current because the introduction of artificial intelligence in the field of diagnosis of synchronous machines is on the rise, as can be seen in several recent research works [22–24].

The developed DNN uses voltage, active and reactive power, and temperature (U, P, Q, and θ) as inputs of the model to estimate the field excitation current of the BSM. The proposed method was developed using a special laboratory setup, firstly to collect all the necessary data to train the DNN under healthy conditions.

In Section 2, the influence of temperature is analyzed in different excitation systems of synchronous machines (SMs). It also highlights the importance of considering temperature as an input in developing the theoretical excitation current estimation model of a BSM. The proposed method and the steps to develop the estimation method are discussed in Section 3. Then, Section 4 describes the experimental setup and the experimental tests performed on it under healthy conditions. Finally, Section 5 summarizes the paper's conclusions.

2. Influence of Rotor Temperature on Different Excitation Systems of SMs

This section presents the need to consider the rotor temperature in the model to estimate the exciter field current for detecting ITFs in BSMs.

In an SM with static excitation, as shown in Figure 1, the excitation current (I_f) is controlled directly by a controlled rectifier which modifies the field winding voltage (U_f). As the temperature increases, the excitation field winding resistance increases, and consequently, the excitation voltage must be raised to maintain a constant excitation current.

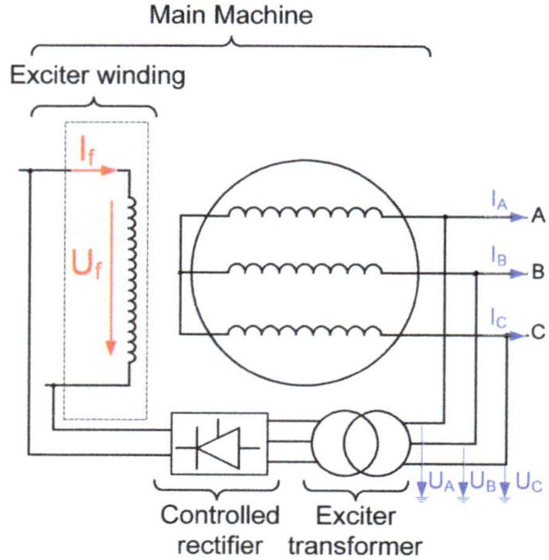

Figure 1. Simplified schema of an SM with static excitation under a self-excited topology.

However, in a BSM, as can be seen in Figure 2, the main machine field winding current is regulated through the exciter. When the temperature increases, the resistance of the field winding of the main machine increases, following Equation (1) [25]:

$$R_\theta = R_{\theta_0} \cdot [1 + \alpha \cdot (\theta - \theta_0)] \tag{1}$$

where R_θ is the resistance at the operating temperature θ, R_{θ_0} is the resistance at the reference temperature θ_0, and α is the temperature coefficient of resistance for the conductor material.

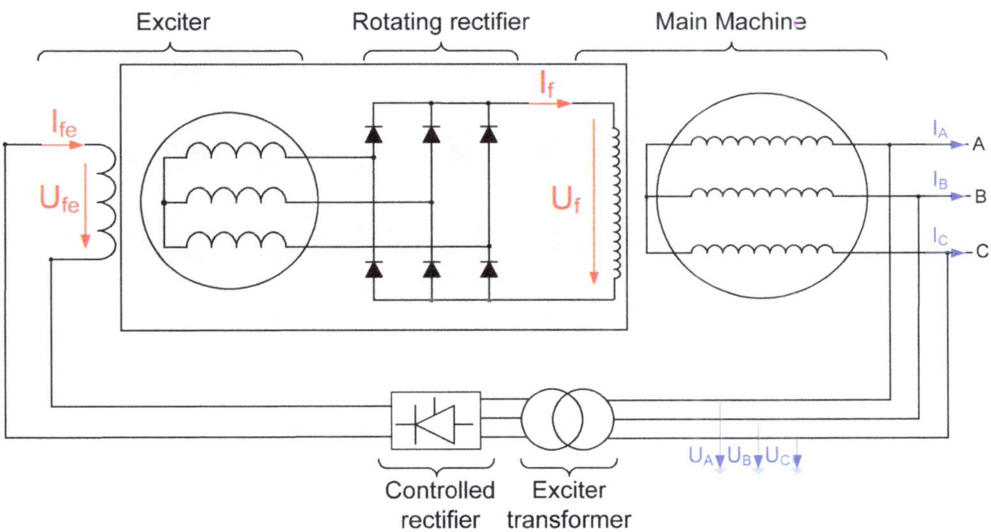

Figure 2. Simplified schema of a BSM under a self-excited topology.

This increase in the field winding resistance leads to an increase in the field winding voltage (U_f), assuming the excitation current (I_f) remains constant. Therefore, to maintain the same operating point (I_f constant), it is necessary to increase the exciter excitation current (I_{fe}) to raise the induced voltage in the armature winding of the exciter and consequently the field winding voltage (U_f). Additionally, because, as said, the resistance of the exciter field winding increases with the exciter's temperature, this also affects the exciter field voltage (U_{fe}).

3. Inclusion of Temperature in the DNN Model-Based Method

3.1. Foundations

This section pursues two main objectives. The first is to utilize DNN algorithm architectures to predict the value of the theoretical exciter field current in a BSM, considering the effect of temperature on the machine, with temperature as an additional input. This approach aims to improve the results obtained in other research [18]. The results shown in this previous article are good because of the short-term tests in which the machine did not have time to warm up and the effect of temperature on the estimation of fault severity was not too high. However, in the current work it has been verified that if continuous monitoring of the BSM is to be carried out, the effect of the rotor temperature must be considered, as shown in Section 5.

The second objective is to validate the feasibility of using the stator temperature in the model instead of the rotor temperature, since measuring rotor temperature in industrial BSMs is not possible.

Based on the principles described in [18], the theoretical exciter field current ($I_{fe,cal}$) is estimated using a DNN model under healthy condition calculations. The inputs to this model are a combination of machine output voltage (U) and current (I) measurements that fully define a [U, P, Q] operating point. In addition to these inputs, the machine temperature measurement (θ) will be included as an input in the present work, following the reasoning presented in Section 3. The comparison of $I_{fe,cal}$ with the actual measured exciter field current ($I_{fe,mea}$) enables ITF severity (α) estimation, as the ratio of faulty turns is directly related to the field current increase imposed by the automatic voltage regulator (AVR) with respect to the estimated current. Equation (2) shows the explicit calculation of the severity of the interturn fault in a BSM:

$$\alpha\ [\%] = \frac{N_{Faulty}}{N_{total}} \cdot 100 = \left(\frac{I_{fe,\ cal}}{I_{fe,mea}} - 1 \right) \cdot 100 \tag{2}$$

where N_{Faulty} is the number of equivalent faulty turns with $R_f = 0\ \Omega$, and N_{total} is the total number of turns.

Figure 3 illustrates the simplified layout of the field winding ITF detection and severity estimation technique, with inclusion of the influence of temperature.

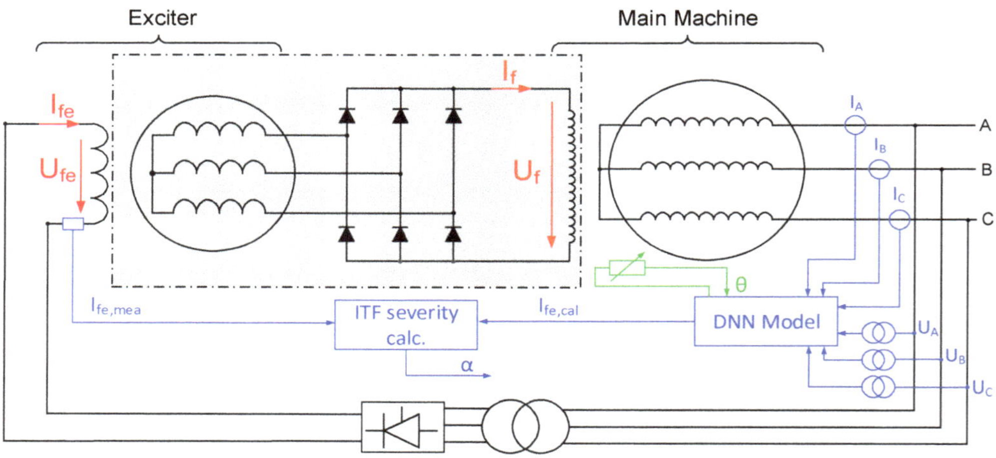

Figure 3. Simplified layout of the field winding ITF detection and severity estimation method, with inclusion of the influence of temperature.

The purposely developed DNN allows one to adaptively model the complex relationships among the mentioned variables through approximation of non-linear functions, with a reduced error. In accordance with the complexity of the problem, the DNN model is composed of four neurons in the input layer, one for each input variable (U, P, Q, and θ), and one output neuron which employs the sigmoid as the activation function. The developed DNN has two hidden layers, with their neurons using the rectified linear unit (ReLU) activation function, and was optimized for computational efficiency.

To determine the optimal number of neurons in each hidden layer, the model was trained with all combinations of neurons ranging from 3 to 20 per layer, utilizing the backpropagation method. The backpropagation learning rate was set at 0.001, and the number of epochs was fixed at 40.000. The optimum configuration was selected by evaluating the trained DNNs based on the minimization of the mean square error (MSE) and the minimization of points with errors greater than 5%. The general form of the involved DNN architecture is presented in Figure 4.

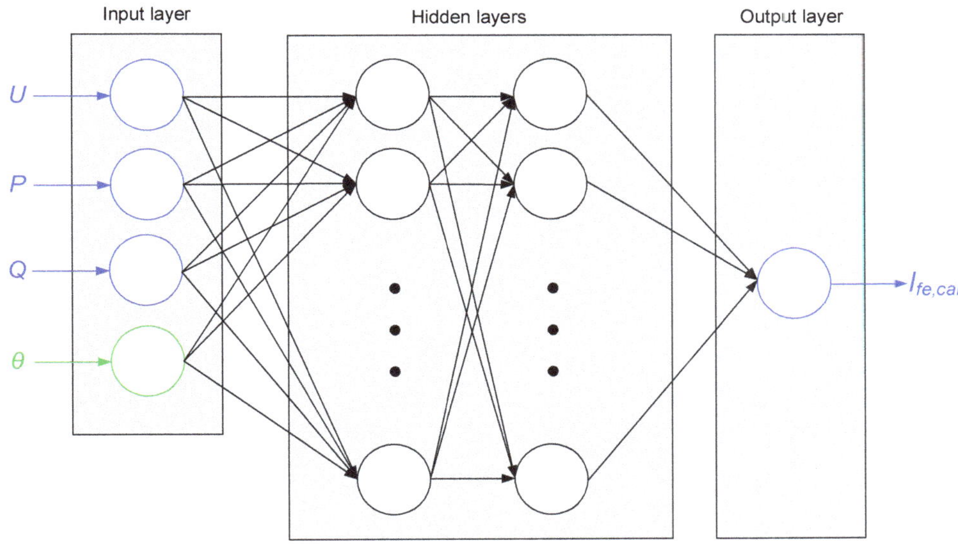

Figure 4. Generalized form of the proposed DNN architecture.

The main features of the development stages (data collection and processing, training with validation, and testing) are described in the following subsections.

3.2. Data Collection and Processing

The data were obtained through experimental testing on a special laboratory test bench, as described in Section 4, under healthy conditions for BSMs. This dataset was collected by conducting several series of long-term tests, with different grid voltage ($380 \leq U \leq 420$ V), real power ($0 \leq P \leq 3000$ W), and reactive power ($-2000 \leq Q \leq 3000$ var) values.

V, P, and Q were obtained using a Camile Bauer SINEAX converter M563, and temperatures were measured using $Pt-100$ commercial sensors with converters. The converters of the temperature sensors and the transducer give an output signal proportional to the measured variables in the range of [4–20] mA. To adapt these signals to the measurement range of the Arduino © microcontroller, a series resistor of 249 ohms was connected to each current signal, so the signals are finally in the range [1–5] V, which is a suitable range for the Arduino © microcontroller. Finally, $I_{fe,mea}$ was measured using a Hall effect sensor (INA219) directly compatible with the Arduino © microcontroller.

After acquisition, the data were filtered to remove anomalous measurements. In total, approximately 400,000 refined records were analyzed under healthy conditions. The removed data include transients, anomalous measures, and communication failures.

This data pool was randomly divided into three subsets for training, validation, and testing, as shown in Figure 5.

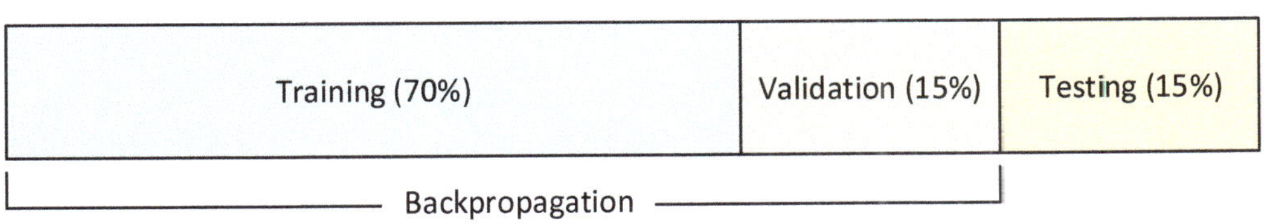

Figure 5. Distribution of the refined data pool.

3.3. Training with Validation

The feed-forward DNN is trained using the popular backpropagation method [25–27], in order to align the outputs ($I_{fe,cal}$) as closely as possible with the measured values ($I_{fe,mea}$). Backpropagation is an iterative gradient descent optimization technique applied to the network's weight space, enabling the weights to be updated in each iteration and to progressively minimize the cost function (CF).

At the first stage of each epoch, forward propagation is performed, where the inputs (U, P, Q, and θ) are fed to the DNN and the output value is calculated based on those inputs and with the initialization values of the weight, as stated in Equation (3).

$$I_{fe,cal} = F\cdot\left[\sum_{k=1}^{K} w_{F,\,2k} h_{2k}\left(\sum_{j=1}^{J} w_{2k,\,1} h_{1j}\left(w_{1j,U}\cdot U + w_{1j,P}\cdot P + w_{1j,Q}\cdot Q + w_{1j,\theta}\cdot\theta + w_{j,0}\right) + w_{k,0}\right) + w_{F,0}\right] \quad (3)$$

where $wF,2k$, $w2k,1$, and $w1j,U$, $w1j,P$, $w1j,Q$, $w1j,\theta$ are the synaptic weights from the neurons of the second hidden layer to the output neuron, from the neurons of the first hidden layer to the second hidden layer, and from the neurons of the input layer to the ones of the first hidden layer, respectively.

To summarize Equation (3), it is better to write the weights vector as in Equation (4):

$$W_t = \left[w_{F,\,2k}\; w_{2k,\,1j}\; w_{1j,\,U}\; w_{1j,\,P}\; w_{1j,\,Q}\; w_{1j,\,\theta}\right] \quad (4)$$

with j from 1 to J, and k from 1 to K.

At the second stage, the algorithm calculates the value adopted by the CF and subsequently evaluates its gradient by propagating the errors from the output layer back to the input layer through the chain rule. Through stochastic gradient descent, accordingly, moving towards the negative direction of the gradient, the iteration is concluded by updating the weights and biases.

The progressive adjustment of the network is performed by minimizing the mean square error (MSE) between the predicted and the measured exciter field currents, serving as the CF. The cost function calculation is shown in (5).

$$CF = \frac{1}{N}\sum_{n=1}^{N}\left(I_{fe,\,cal} - I_{fe,\,mea}\right)^2 \quad (5)$$

The adaptive moment estimation (ADAM) algorithm, which employs the exponential moving average of both the gradient or first momentum (∇CF) and the squared gradient or second momentum (∇^2CF), is used for gradient descent optimization. The diagram of the employed backpropagation method is provided in Figure 6.

Also, the following equations allow the weights vector at epoch t to be determined from the vectors at the previous epoch ($t-1$):

$$W_t = W_{t-1} - l\cdot\frac{\hat{m}_t}{\sqrt{\hat{v}_t + \varepsilon}} \quad (6)$$

$$\hat{m}_t = m_t\cdot\frac{1}{1-\beta_m} \quad (7)$$

$$m_t = \beta_m\cdot m_{t-1} + (1-\beta_m)\cdot\nabla CF \quad (8)$$

$$\hat{v}_t = v_t\cdot\frac{1}{1-\beta_v} \quad (9)$$

$$v_t = \beta_v\cdot v_{t-1} + (1-\beta_v)\cdot(\nabla CF)^2 \quad (10)$$

where l represents the learning rate that allows speeding up of the convergence of the CF to its minimum, v_t and m_t are variables updated in each epoch, and ε, β_m, and β_v are constant.

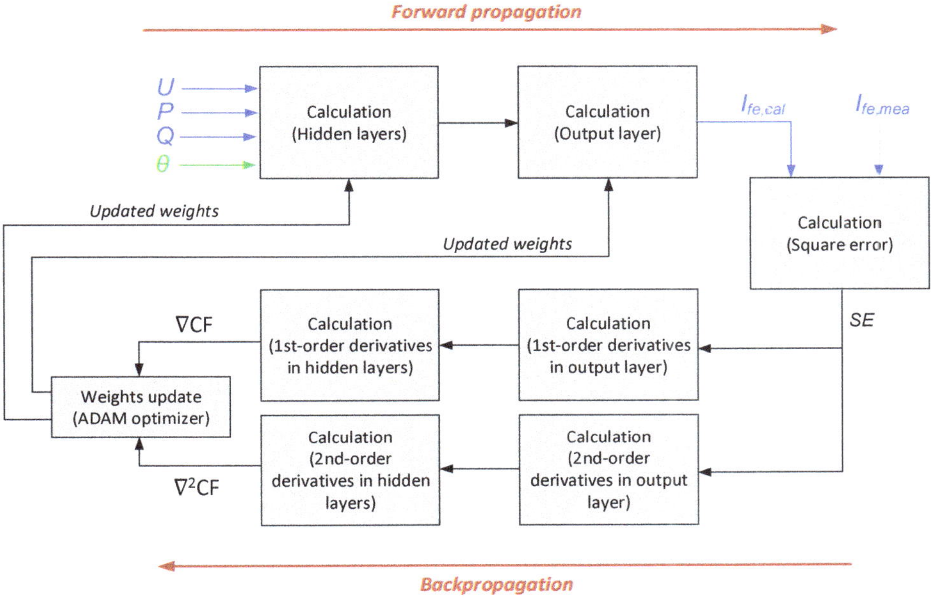

Figure 6. DNN training process through backpropagation.

After training, hyperparameter tuning is performed to achieve the best-fitting network. During this validation phase, various neural networks were evaluated to determine the optimal number of hidden layers and nodes in each layer. The network that reduced the maximum absolute error (MAE) to 0.1% was selected.

3.4. Tests

After the training and validation process, an unbiased evaluation of the model performance was carried out. The absolute maximum error is approximately 0 5%, so if this method was used to detect ITFs, faults with a severity greater than 0.5% could be detected. In other words, because 0.5% is the smallest severe detectable fault, the tripping limit should never be fixed beneath this value.

Table 1 shows the evaluation results of the DNN network for the machine in healthy conditions, including the value of the mathematical metrics used to evaluate its accuracy: the mean average error (MAE) and the root mean square error (RMSE).

Table 1. Results obtained from the DNN predictions under healthy conditions.

Set of Data	MAE [%]	RMSE [mA]
Tests under healthy conditions	0.1	0.04

The errors obtained through DNN prediction are significantly beneath the errors obtained in previous papers where the temperature was not considered as an input in the model [16]. An MAE of 0.1% shows that the trained DNN has a fairly high overall estimation accuracy, in terms of overall closeness to null error values, while an RMSE of 0.06 implies a fair mean value of prediction dispersion with respect to the actual values, as this metric penalizes outliers or large deviations.

4. Experimental Tests

4.1. Experimental Setup

Numerous long-duration tests under constant operation points (P, Q, and U) were carried out on a special 4-salient pole, 5-kVA, 400-V BSM. A simplified diagram of the machine is shown in Figure 7.

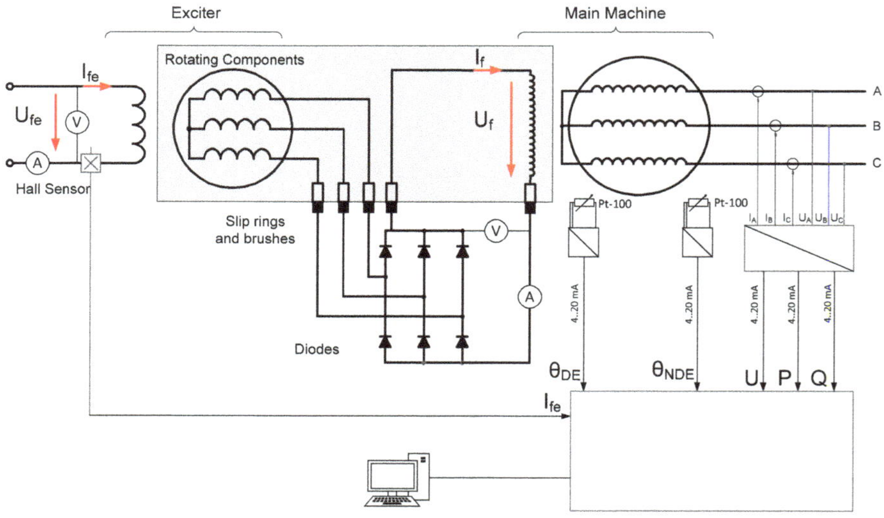

Figure 7. Simplified diagram of the experimental setup.

This BSM was modified by adding 5 slip rings and brushes to permit access to the field winding of the main machine and to the "rotating" diodes (which were actually made static in this setup and placed outside of the machine). This modification allows for the measurement of the voltage (U_f) and the current (I_f) of the field winding of the main machine and consequently the temperature by observing variations in its resistance. Additionally, two Pt−100 type temperature sensors were installed at both ends of the armature windings of the main synchronous machine, and these temperatures were also recorded.

From the armature currents and voltages, the values of stator voltage (U), active power (P), and reactive power (Q) are measured by a commercial 3-channel programmable transducer, SINEAX M563. Finally, the exciter field current is measured using a Hall effect sensor. All these measurements are recorded on a computer through an Arduino© microcontroller, communicating via the COM port of the computer.

During the long-duration test, the temperature of the rotor and the stator of the main synchronous machine increased. To maintain the machine at the same operating point, the exciter excitation current (I_{fe}) needs to be increased as the temperature rises.

The ratings of the main machine and the exciter are detailed in Tables 2 and 3, respectively.

There are five slip rings and brushes (4): three are connected to the armature winding of the exciter, and the other two are connected to the main machine's field winding. The voltage and current of the main machine's field winding are measured (5). The temperature of the main machine's armature windings is monitored by two Pt−100 sensors (6). These sensors are placed in the end winding of the machine at both sides of the generator, the non-drive end NDE and the drive end DE. In element 6 of Figure 8, there is a zoom of this area. This is the best non-invasive location of the temperature sensors that allows estimating an average temperature with which to compare the estimations of the DNN.

This modified synchronous machine is actioned by an induction motor (7), which is controlled with a frequency variable converter. A DC source provides the excitation power to the machine, and the machine is connected to the grid.

The experimental laboratory setup is shown in Figure 8. The BSM (1) and the exciter (2) are mechanically coupled with a common shaft. The diodes (3) are located outside the machine in this layout to be accessible.

Table 2. Main machine data.

Alternator Type	Synchronous 3-Phase	
Rated power	5	kVA
Rated speed	1500	rpm
Rated voltage	400	V
Rated current	7.2	A
Pole pairs	2	
Rated frequency	50	Hz
IP	21	
Insulation class	F	
Rated excitation voltage	33	V
Rated excitation current	4.1	A

Table 3. Exciter data.

Alternator Type	Synchronous 3-Phase	
Rated power	277	VA
Rated speed	1500	rpm
Rated voltage	40	V
Rated current	4	A
Pole pairs	4	
Rated frequency	100	Hz
IP	21	
Insulation class	F	
Rated excitation voltage	33	V
Rated excitation current	0.61	A

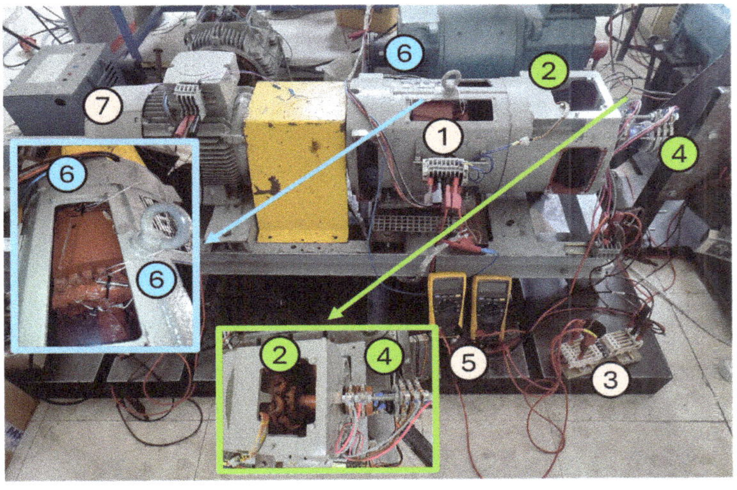

Figure 8. Experimental setup: (1) main machine, (2) exciter, (3) diodes, (4) slip rings and brushes, (5) main field winding voltage and current measurement, (6) stator temperature measurement, and (7) driver (induction motor).

4.2. Experimental Results

Numerous tests have been conducted to verify that the measured exciter field current (I_{fe}) must be significantly increased in BSMs under warm conditions to maintain a constant reactive power delivered by the machine. The tests involve initially setting an operating point for the BSM in terms of active and reactive power, and then allowing the machine to operate as it warms up, while continuously keeping the reactive power constant. To maintain constant Q as the temperature rises, it is necessary to increase the exciter field current (I_{fe}).

All the aforementioned electrical variables (P, Q, U, U_{fe}, I_{fe}, I_f, and U_f) and the rotor and stator temperatures are continuously monitored. Figure 9 illustrates the increase in the exciter field winding current due to the rising rotor temperature during a test.

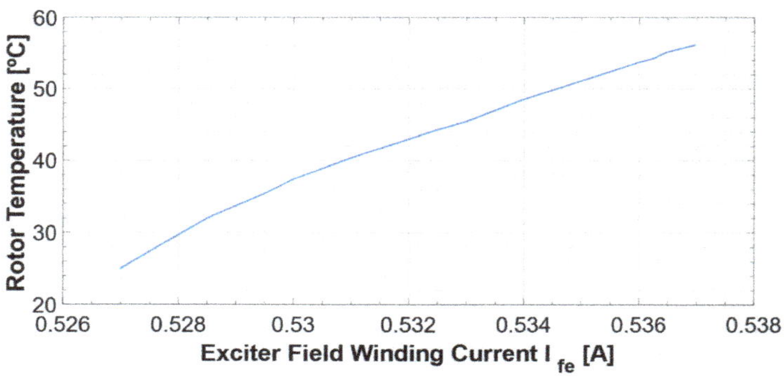

Figure 9. Exciter field current (I_{fe}) with respect to rotor temperature during a 2 h operation test (P = 2.9 kW, Q = 2.3 kvar, U = 400 V).

Since it is impossible to access the voltage and current of the field winding of the main machine in industrial BSMs and consequently to calculate the rotor temperature, the stator temperature was measured to assess whether it could be used instead of the rotor temperature to estimate the excitation current (I_{fe}).

Figures 10 and 11 show the exciter field winding current variation with the increment of the stator and rotor temperatures, respectively. It can be seen that, in this BSM, the exciter field current (I_{fe}) consumption increases by 0.75% every 10 K degrees of temperature rise. Thus, to develop a model capable of estimating the excitation current consumption at an electrical operating point of the machine, in the case of BSMs it is necessary to add the machine temperature variable in the estimation model, as said above.

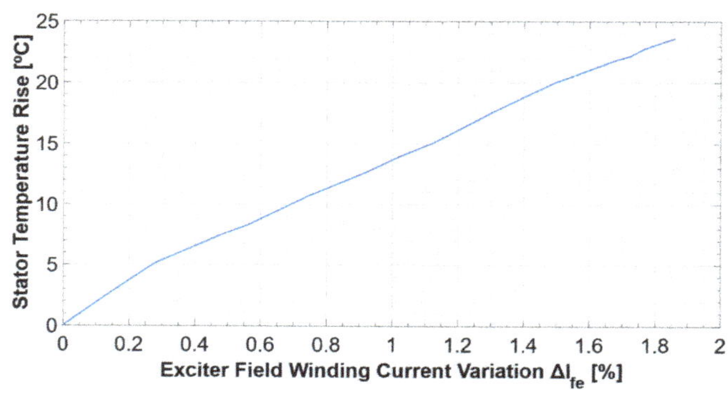

Figure 10. Exciter field current rise (ΔI_{fe}) with respect to stator temperature rise during a 2 h operation test (P = 2.9 kW, Q = 2.3 kvar, U = 400 V).

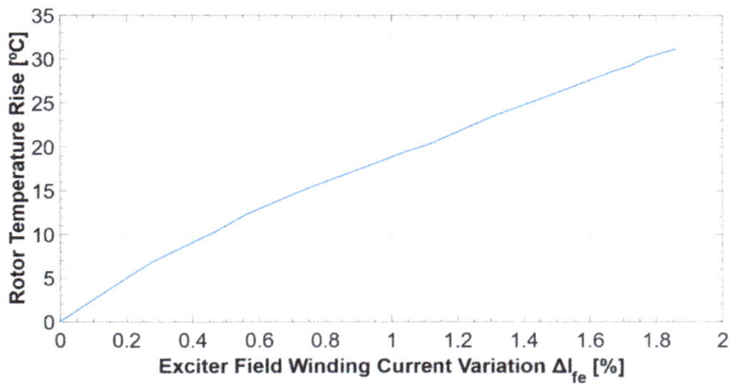

Figure 11. Exciter field current rise (ΔI_{fe}) with respect to rotor temperature rise during a 2 h operation test (P = 2.9 kW, Q = 2.3 kvar, U = 400 V).

Comparing Figures 10 and 11, it is verified that the variation in excitation current consumption with increases in stator and rotor temperatures is very similar. Importantly, the slope remains constant with both temperatures, indicating that it is feasible to use the stator temperature in the estimation model.

Additionally, Figure 12 displays the rotor and stator temperatures during a constant reactive power test. The difference between them is also represented. This difference is not constant and varies depending on the machine temperature.

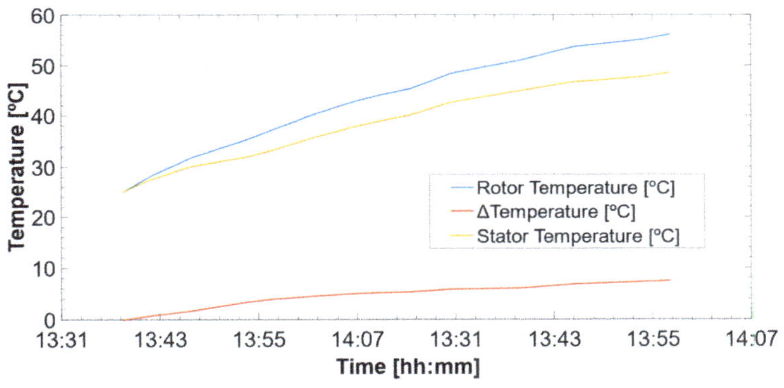

Figure 12. Temperature variations during a 2 h operation test (P = 2.9 kW, Q = 2.3 kvar, U = 400 V).

As described in Section 4, to validate that this method is significantly more efficient when using temperature as an input to the model, a DNN was trained with real operating points measured using the experimental system described in the previous subsection.

Figure 13 shows the measured excitation current, $I_{fe,mea}$ (in blue), and the calculated excitation current, $I_{fe,cal}$ (in orange), for all test operational points. The calculated excitation current value is derived from the DNN previously trained with the machine operational data under healthy conditions.

As shown in Figure 13, the DNN excitation current prediction is highly accurate, with a maximum error of 3 mA, which is an error of 1.5%. The MSE obtained is 0.5%, which means that the minimum detectable fault severity is 0.5%, considering the fault severity of Equation (2).

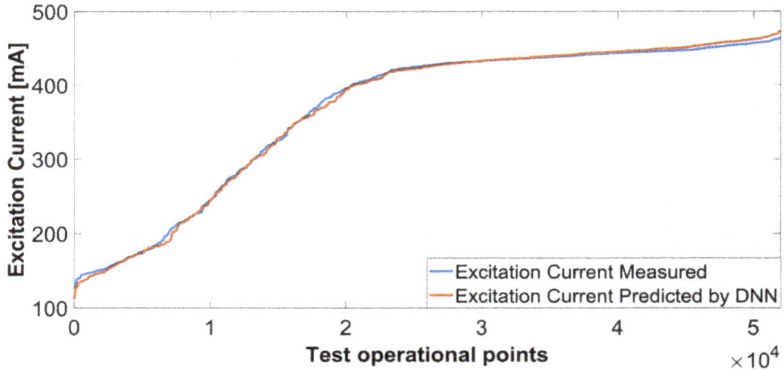

Figure 13. Prediction of the excitation current using a DNN.

5. Conclusions

In BSMs, rotor temperature is a crucial variable for detecting ITFs based on exciter field current estimation. As the machine heats up, the rotor's resistance increases, necessitating a rise in the rotor field voltage to maintain a constant field current and reactive power. Consequently, the exciter field current must also be increased.

Previous research did not consider the rotor temperature as a variable in the development of the online model-based fault severity estimation algorithm for BSMs. This omission is significant, as the presented results indicate that the exciter excitation current varies by 0.75% for every 10 °C increase in temperature. This could result in errors exceeding 5% in an industrial machine, where the normal operating temperature range can vary from 20 °C in cold conditions to 100 °C in warm conditions. This fact implies that the algorithm could only detect faults from 5% fault severity or higher if unwanted trips are to be avoided.

To enhance the accuracy of the detection method, an alternative online model has been developed for detecting ITFs in BSMs, incorporating the machine's temperature. Given the difficulty of measuring rotor temperature, the use of stator temperature was evaluated as an alternative, yielding satisfactory results.

To validate the equivalence of using rotor versus stator temperature, the variation in the excitation current with different temperatures was compared in several laboratory tests, using 400,000 operational points for training the neural network.

The results obtained in this paper indicate that the DNN using stator temperature as an input can estimate the exciter field current with errors below 0.5%. Consequently, ITFs with a severity of approximately 0.5% or higher can also be detected.

However, this study leaves the gates open to future research, which should focus on carrying out tests with known interturn faults in the brushless generator. Thus, the DNN developed for this article could be used to verify the accuracy of the model as a function of the severity of the fault.

Author Contributions: Conceptualization, R.P. and C.A.P.; methodology, R.P., K.M. and C.A.P.; software, R.P. and E.R.; validation, R.P., K.M. and E.R.; formal analysis, J.M.G. and C.A.P.; investigation, R.P. and K.M.; resources, C.A.P.; data curation, R.P.; writing—original draft preparation, R.P. and J.M.G.; writing—review and editing, J.M.G., R.P., K.M., E.R. and C.A.P.; visualization, K.M.; supervision, C.A.P.; project administration, C.A.P.; funding acquisition, C.A.P. All authors have read and agreed to the published version of the manuscript.

Funding: This research received no external funding.

Institutional Review Board Statement: Not applicable.

Informed Consent Statement: Not applicable.

Data Availability Statement: The original contributions presented in this study are included in the article; further inquiries can be directed to the corresponding author.

Conflicts of Interest: Author Eduardo Rivero was employed by the company ESELEC Ingenieros. The remaining authors declare that the research was conducted in the absence of any commercial or financial relationships that could be construed as a potential conflict of interest.

References

1. Nøland, J.K.; Nuzzo, S.; Tessarolo, A.; Alves, E.F. Excitation System Technologies for Wound-Field Synchronous Machines: Survey of Solutions and Evolving Trends. *IEEE Access* **2019**, *7*, 109699–109718. [CrossRef]
2. Nuzzo, S.; Bolognesi, P.; Decuzzi, G.; Giangrande, P.; Galea, M. A Consequent-Pole Hybrid Exciter for Synchronous Generators. *IEEE Trans. Energy Convers.* **2021**, *36*, 368–379. [CrossRef]
3. Chakraborty, C.; Basak, S.; Rao, Y.T. Synchronous Generator with Embedded Brushless Synchronous Exciter. *IEEE Trans. Energy Convers.* **2019**, *34*, 1242–1254. [CrossRef]
4. Chapman, S.J. *Electric Machinery Fundamentals*, 5th ed.; McGraw-Hill: New York, NY, USA, 2012; ISBN 9780073529547.
5. *IEEE Std C50.12-2005*; IEEE Standard for Salient-Pole 50 Hz and 60 Hz Synchronous Generators and Generator/Motors for Hydraulic Turbine Applications Rated 5 MVA and Above. IEEE: New York, NY, USA, 2006; pp. 1–45.
6. *IEEE Std C50.13-2014*; IEEE Standard for Cylindrical-Rotor 50 Hz and 60 Hz Synchronous Generators Rated 10 MVA and Above. IEEE: New York, NY, USA, 2014; pp. 1–63.
7. Igbinovia, F.O.; Fandi, G.; Müller, Z.; Švec, J.; Tlustý, J. Optimal location of the synchronous condenser in electric-power system networks. In Proceedings of the IEEE 17th International Scientific Conference on Electric Power Engineering (EPE), Prague, Czech Republic, 16–18 May 2016; pp. 1–6.
8. Marrazi, E.; Yang, G.; Weinreich-Jensen, P. Allocation of synchronous condensers for restoration of system short-circuit power. *J. Mod. Power Syst. Clean Energy* **2018**, *6*, 17–26. [CrossRef]
9. Jiménez, R.P.; Platero, C.A.; Sánchez, M.E.; Barneto, E.R. Synchronous Condenser for European Grid Code Compliance: A Case Study of a PV Power Plant in Spain. *IEEE Access* **2023**, *11*, 54713–54723. [CrossRef]
10. Commission Regulation (EU) 2016/631—Of 14 April 2016—Establishing a Network Code on Requirements for Grid Connection of Generators. Available online: https://op.europa.eu/en/publication-detail/-/publication/1267e3d1-0c3f-11e6-ba9a-01aa75ed71a1/language-en (accessed on 15 May 2024).
11. *IEEE Std 2800-2022*; IEEE Standard for Interconnection and Interoperability of Inverter-Based Resources (IBRs) Interconnecting with Associated Transmission Electric Power Systems. IEEE: New York, NY, USA, 2022; pp. 1–180.
12. Stone, G.; Boulter, E.A.; Culbert, I.; Dhirani, H. *Electrical Insulation for Rotating Machines: Design, Evaluation, Aging, Testing, and Repair*, 4th ed.; Wiley-IEEE Press: Piscataway, NJ, USA, 2004; pp. 1–41.
13. *IEEE Std 56-2016*; IEEE Guide for Insulation Maintenance of Electric Machines. IEEE: New York, NY, USA, 2016; pp. 1–86.
14. Mahtani, K.; Guerrero, J.M.; Beites, L.F.; Platero, C.A. Model-Based On-Line Protection Method for Brushless Synchronous Generators. In Proceedings of the 2021 IEEE 13th International Symposium on Diagnostics for Electrical Machines, Power Electronics and Drives (SDEMPED), Dallas, TX, USA, 22–25 August 2021; pp. 126–132.
15. Mahtani, K.; Guerrero, J.M.; Beites, L.F.; Platero, C.A. Model-Based Field Winding Interturn Fault Detection Method for Brushless Synchronous Machines. *Machines* **2022**, *10*, 1227. [CrossRef]
16. *IEEE Std 1129-2014*; IEEE Guide for Online Monitoring of Large Synchronous Generators (10 MVA and Above). IEEE: New York, NY, USA, 2014; pp. 1–62.
17. Sottile, J.; Trutt, F.C.; Leedy, A.W. Condition Monitoring of Brushless Three-Phase Synchronous Generators with Stator Winding or Rotor Circuit Deterioration. *IEEE Trans. Ind. Appl.* **2006**, *42*, 1209–1215. [CrossRef]
18. Pascual, R.; Mahtani, K.; Rivero, E.; Platero, C.A. Brushless Synchronous Machine Field Winding Interturn Fault Severity Estimation through Deep Neural Networks. *IEEE Trans. Energy Convers.* **2023**, *39*, 1341–1351. [CrossRef]
19. Tang, J.; Liu, Y.; Lundberg, S. Estimation Algorithm for Current and Temperature of Field Winding in Electrically Excited Synchronous Machines with High-Frequency Brushless Exciters. *IEEE Trans. Power Electron.* **2021**, *36*, 3512–3523. [CrossRef]
20. Tang, J.; Jiang, B.; Boscaglia, L.; Chen, H.; Liu, Y. Observations of Field Current and Field Winding Temperature in Electrically Excited Synchronous Machines with Brushless Excitation. In Proceedings of the 2022 International Conference on Electrical Machines (ICEM), Valencia, Spain, 5–8 September 2022; pp. 841–847.
21. Tun, P.P.; Kumar, P.S.; Pratama, R.A.; Shuyong, L. Brushless Synchronous Generator Turn-to-Turn Short Circuit Fault Detection Using Multilayer Neural Network. In Proceedings of the 2018 Asian Conference on Energy, Power and Transportation Electrification (ACEPT), Singapore, 30 October–2 November 2018; pp. 1–8.
22. Ferreira, J.G.; Warzecha, A. An application of machine learning approach to fault detection of a synchronous machine. In Proceedings of the International Symposium on Electrical Machines (SME), Naleczow, Poland, 18–21 June 2017; pp. 1–6.
23. Evangeline, S.I.; Darwin, S.; Raj, E.F.I. A deep residual neural network model for synchronous motor fault diagnostics. *Appl. Soft Comput.* **2024**, *160*, 111683. [CrossRef]
24. Yin, W.; Xia, H.; Huang, X.; Zhang, J.; Miyombo, M.E. A fault diagnosis method for nuclear power plant rotating machinery based on adaptive deep feature extraction and multiple support vector machines. *Prog. Nucl. Energy* **2023**, *164*, 104862. [CrossRef]

25. Ansermet, J.P.; Brechet, S.D. Principles of Thermodynamics. Cambridge University Press: Cambridge, UK, 2019; ISBN 9781108620932.
26. Zhang, X.; Xu, N.; Dai, W.; Zhu, G.; Wen, J. Turbofan Engine Health Prediction Model Based on ESO-BP Neural Network. *Appl. Sci.* **2024**, *14*, 1996. [CrossRef]
27. Taïbi, Z.M.; Hasni, M.; Hamdani, S.; Rahmani, O.; Touhami, O.; Ibtiouen, R. Optimization of the feedforward neural network for rotor cage fault diagnosis in three-phase induction motors. In Proceedings of the 2011 IEEE International Electric Machines & Drives Conference (IEMDC), Niagara Falls, ON, Canada, 15–18 May 2011; pp. 194–199.

applied sciences

MDPI

Article

Detection, Isolation, and Identification of Multiplicative Faults in a DC Motor and Amplifier Using Parameter Estimation Techniques

Sanja Antić *, Marko Rosić, Branko Koprivica, Alenka Milovanović and Milentije Luković

Faculty of Technical Sciences Čačak, University of Kragujevac, 32000 Čačak, Serbia;
marko.rosic@ftn.kg.ac.rs (M.R.); branko.koprivica@ftn.kg.ac.rs (B.K.); alenka.milovanovic@ftn.kg.ac.rs (A.M.);
milentije.lukovic@ftn.kg.ac.rs (M.L.)
* Correspondence: sanja.antic@ftn.kg.ac.rs

Featured Application: The developed laboratory platform and associated algorithms can be applied in educational and research settings for real-time simulation, detection, and isolation of multiplicative faults in electromechanical systems. Its remote-access capability allows engineering students and researchers to conduct fault diagnosis experiments on DC motor systems without physical presence, supporting distance learning and development of robust FDII methods for safety-critical applications.

Abstract: The increasing complexity of modern control systems highlights the need for reliable and robust fault detection, isolation, and identification (FDII) methods, particularly in safety-critical and industrial applications. The study focuses on the FDII of multiplicative faults in a DC motor and its electronic amplifier. To simulate such scenarios, a complete laboratory platform was developed for real-time FDII, using relay-based switching and custom LabVIEW software 2009. This platform enables real-time experimentation and represents an important component of the study. Two estimation-based fault detection (FD) algorithms were implemented: the Sliding Window Algorithm (SWA) for discrete-time models and a modified Sliding Integral Algorithm (SIA) for continuous-time models. The modification introduced to the SIA limits the data length used in least squares estimation, thereby reducing the impact of transient effects on parameter accuracy. Both algorithms achieved high model output-to-measured signal agreement, up to 98.6% under nominal conditions and above 95% during almost all fault scenarios. Moreover, the proposed fault isolation and identification methods, including a decision algorithm and an indirect estimation approach, successfully isolated and identified faults in key components such as amplifier resistors (R_1, R_9, R_{12}), capacitor (C_8), and motor parameters, including armature resistance (R_a), inertia (J), and friction coefficient (B). The decision algorithm, based on continuous-time model coefficients, demonstrated reliable fault isolation and identification, while the reduced Jacobian-based approach in the discrete model enhanced fault magnitude estimation, with deviations typically below 10%. Additionally, the platform supports remote experimentation, offering a valuable resource for advancing model-based FDII research and engineering education

Keywords: fault detection, isolation, and identification; multiplicative faults; DC motor; electronic amplifier; Sliding Window Algorithm; Sliding Integral Algorithm; remote laboratory

1. Introduction

As control systems continue to grow in complexity, ensuring their safety and robustness becomes increasingly critical across a wide range of industrial processes. The reliability, robustness, availability, and environmental safety of such systems are often evaluated based on their sensitivity to faults, which may occur randomly during operation. This growing need has resulted in a distinct area within engineering theory and practice–fault detection and isolation (FDI) [1–5].

From a practical engineering standpoint, faults are commonly categorized into additive and multiplicative types. Additive faults are unexpected deterministic inputs (e.g., sudden load change or sensor biases) that affect the system output without altering its internal structure. These faults, which include actuator jamming or sensor drift, are typically absent under normal conditions but produce clear effects on outputs once present [6–12]. Multiplicative faults, by contrast, involve changes in system parameters, such as resistance, inductance, inertia, or friction coefficients. These faults may appear gradually (due to aging or wear) or abruptly (as in structural damage) and are more challenging to detect due to their parameter-driven, indirect, and often masked effects on system output [13–17].

In DC motors, multiplicative faults include misalignment of brushes, wear of the commutator, armature winding faults, variations in armature resistance, and changes in mechanical friction. They may also occur in the accompanying amplifier—making accurate real-time diagnosis a challenging but essential task [18–21].

Various FDI methods have been developed to address these issues, which can be classified into four broad categories:

1. Signal-based techniques rely on analyzing vibration or current signals, often using spectral, statistical, or wavelet-based methods [19,22–25]. While model-free and simple to implement, such methods are generally limited to offline analysis, and their performance deteriorates in low-power or noisy systems. They are also less suitable for detecting multiplicative faults, which often do not generate prominent signal features.
2. Model-based techniques exploit the system's dynamic model and include:
    - Parity equation methods, which compare measured and predicted outputs to generate residuals sensitive to specific faults [3,10,11,26]. These methods are simple to implement and effective for additive fault isolation, but they perform poorly for multiplicative faults due to their reliance on accurate models and sensitivity to noise and parameter uncertainties. Moreover, they typically lack fault identification capability, as they do not quantify fault magnitude.
    - Observer and filter-based methods (e.g., Luenberger observers, Kalman filters, sliding-mode observers) that estimate system states and detect deviations due to faults [7,14,27,28]. Such approaches provide good dynamic tracking and are suitable for real-time use, but they often require precise model tuning and can become unstable or inaccurate in the presence of model uncertainty or high noise.
    - Parameter estimation methods (e.g., Least Squares (LS), Recursive Least Squares (RLS), Extended Kalman Filters (EKF), Unscented Kalman Filters (UKF), Algebraic-Geometric Technique (AGT) [18,26,29–38].
3. Knowledge and data-driven approaches such as fuzzy logic, expert systems, and neural networks can learn or encode relationships between faults and observable features [6,12,18,39,40]. While effective for nonlinear and complex systems, these methods require large training datasets and may lack interpretability.

4. Optimization-based techniques such as Genetic Algorithms (GA), Particle Swarm Optimization (PSO), and Whale Optimization Algorithm (WOA)) that track deviations in estimated parameters, particularly relevant for multiplicative faults [41,42].

Among model-based approaches, parameter estimation techniques offer significant potential for detecting multiplicative faults, as they enable direct monitoring of physical changes within the system. Compared to traditional parameter estimation-based fault detection methods, the advantages of the proposed Sliding Window Algorithm (SWA) [43] and Sliding Integral Algorithm (SIA) [44] become more apparent.

Classical parameter estimation techniques such as LS [18,32,38] and RLS [30,31] are widely used due to their simplicity and low computational requirements. LS is primarily an offline method, as it processes a fixed batch of data and assumes time-invariant parameters during estimation. In contrast, RLS is designed for real-time application, updating estimates recursively as new data become available and using a forgetting factor to adapt to changing system dynamics. However, despite its real-time capability, RLS often struggles with transient disturbances and abrupt parameter changes caused by faults, especially when the forgetting factor is not optimally tuned. These limitations can lead to delayed or inaccurate fault detection in systems subject to abrupt changes [26,30].

Advanced methods such as EKF [29,34,35] and UKF [36,37] offer improved performance in noisy and nonlinear environments. However, their reliance on accurate system models and careful tuning of noise covariances increases implementation complexity and computational burden, limiting their suitability for embedded or real-time applications.

Similarly, algebraic techniques such as the AGT offer analytical insight and enable fault localization. However, they are generally unsuitable for real-time implementation due to their dependence on signal derivatives, symbolic computations, and high sensitivity to measurement noise [33].

Optimization-based techniques such as GA, PSO, and WOA offer global optimization capabilities for parameter estimation, but their high computational complexity makes them impractical for real-time applications [41,42].

In contrast, SWA and SIA offer robust and low-complexity alternatives that preserve interpretability and can be scaled to more complex systems with relative ease. SWA, applied to discrete-time models with short data windows, enhances responsiveness to abrupt faults, while SIA, using continuous-time integral formulations, reduces noise and transient sensitivity—making it suitable for gradual parameter changes. Both methods are non-intrusive, require minimal tuning, and are computationally efficient, making them ideal for embedded real-time applications. Furthermore, by linking estimated model parameters (e.g., a_1, b_0, b_1) to physical quantities (e.g., resistance, inertia), they enable not only fault detection and isolation but also identification of fault severity.

This paper presents a real-time, parameter-estimation-based framework for detection, isolation, and identification (FDII) of multiplicative faults in a DC motor and its amplifier. The system is implemented on a hardware-in-the-loop experimental platform using LabVIEW 2009 and NI CompactRIO CRIO-9075 (National Instruments, Austin, TX, USA), enabling remote access, real-time fault injection, and online testing.

The key contributions include:

1. Development of an interactive laboratory platform for real-time simulation and testing of FDII algorithms, specifically designed to support both practical implementation and engineering education. This is achieved through a hardware-in-the-loop, remotely accessible setup built using LabVIEW and NI CompactRIO, enabling real-time fault injection, detection, isolation, and identification in both the DC motor and its electronic amplifier—representing a significant advancement in applied FDI research and training environments.

2. Comprehensive Evaluation and Comparative Analysis of two complementary parameter estimation methods under identical experimental conditions: a modified Sliding Integral Algorithm (SIA) for continuous-time system modeling, which limits the data window to reduce the influence of transients and improves robustness of parameter estimation; and a Sliding Window Algorithm (SWA) for discrete-time modeling, which enhances responsiveness to abrupt parameter changes. This enables a comparative analysis of their performance and limitations, which is rarely addressed in the existing literature [19–21,33].
3. A modification of the SIA by limiting the data window, which reduces the influence of transients and improves the robustness of parameter estimation.
4. Demonstration of two complementary fault identification approaches that enable quantification of the change in affected physical parameters—a capability rarely addressed in the existing literature [13,14,16,18,19]. In addition to detecting and isolating faults, the proposed methods also estimate fault magnitude, which significantly enhances diagnostic resolution. This is achieved through:
    - A novel decision algorithm, based on the model obtained using SIA, which maps deviations in estimated model parameters to specific physical components;
    - An indirect estimation method, applied in the SWA framework, supported by a selective Jacobian matrix approach that improves the accuracy of physical parameter identification and strengthens the overall diagnostic capability of the system, especially when the relation between model and physical parameters is nonlinear or non-negligible. This overcomes limitations of prior indirect approaches that typically rely on the full Jacobian matrix [3,7].

In summary, this work presents a practical and robust methodology for detecting and identifying multiplicative faults using simple yet effective parameter estimation techniques. Its real-time implementation, experimental validation, and educational applicability make it a valuable contribution to both academic research and industrial practice.

2. Materials and Methods

Figure 1 presents the full experimental setup of the measurement equipment, comprising a motor, electronic amplifier, relays, and the CRIO-9075 controller.

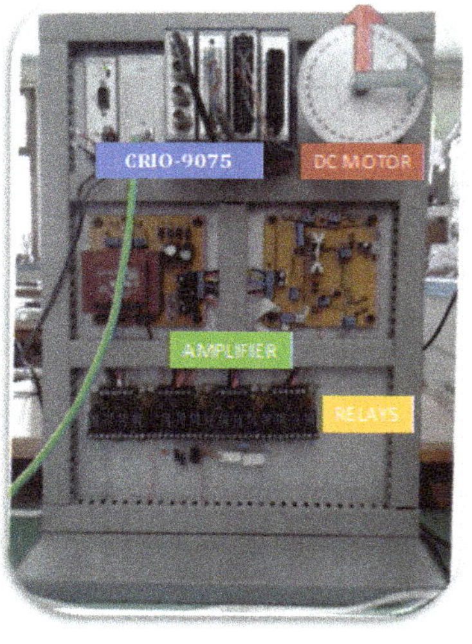

Figure 1. Experimental setup of the measuring equipment.

2.1. Laboratory Setup

2.1.1. DC Motor

For the planned experiments, a DC motor with permanent magnets A-max 26 A Ø26 mm Precious Metal Brushes CLL (Maxon Motor AG, Sachseln, Switzerland), 4.5 W was used. Figure 2a shows a DC micro-motor (1) equipped with an MR-128-1000 encoder (2), offering a resolution of 1000 impulses per revolution, and a Gear GP-26-B_144032 gearbox (3) with a gear ratio of 3.8:1.

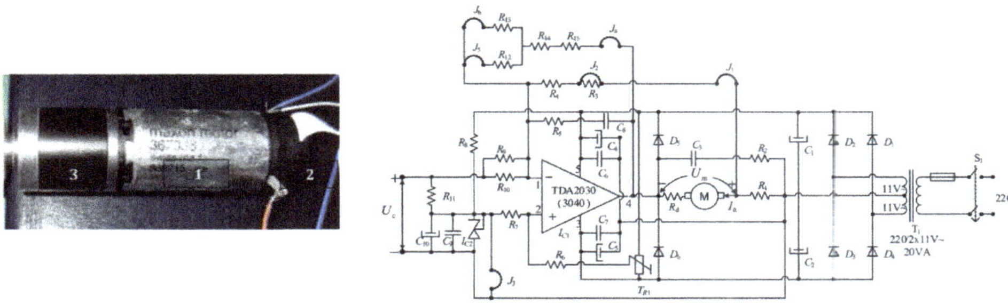

(a) (b)

Figure 2. Main components of the laboratory setup: (**a**) DC micro-motor (1), with encoder (2), and gearbox (3); (**b**) Electronic amplifier-schematic representation.

2.1.2. Electronic Amplifier

The developed amplifier is in the form of a linear electronic converter. It represents a linear electronic circuit whose main component is the operational amplifier TDA2030(A) (STMicroelectronics, Geneva, Switzerland), powered by a DC supply via a single-phase transformer with two secondary windings and a diode rectifier. This amplifier can be used either as a voltage-to-current or voltage-to-voltage converter, for operation in two-quadrant or four-quadrant mode.

Figure 2b shows the electronic schematic of the amplifier used to supply the DC motor. The input voltage of the amplifier represents the control voltage U_c. The output of the amplifier is, depending on the mode, either the motor armature current I_a or the reference armature voltage U_m.

The gain values of the amplifier are: in current modes K_{UcIa} = 1/5 A/V, and in voltage mode K_{UcUm} = 12/5 V/V. The accuracy of the amplifier operation is 1% of the range. In the experiments, the electronics were set to voltage-to-voltage, two-quadrant mode. Table 1 presents the data specifications of the DC motor and the amplifier.

Table 1. Specifications of the A-max 26 A DC motor (Ø26 mm, 4.5 W, with CLL precious metal brushes) and the amplifier.

Nominal armature voltage	$U_{an} = 12$ V
Nominal armature current	$I_{an} = 0.629$ A
Nominal motor speed	$n_n = 2750\ \mathrm{min}^{-1}$
Motor moment of inertia	$J_m = 1.3\cdot10^{-6}$ kgm^2 *
Gear reducer moment of inertia	$J_{gr} = 0.6\cdot10^{-7}$ kgm^2 *
Encoder moment of inertia	$J_e = 0.7\cdot10^{-7}$ kgm^2 *
Viscous friction coefficient	$B = 1.8\cdot10^{-6}$ Nm/rad/s **
Coulomb friction	$M_c = 0.48\cdot10^{-3}$ Nm **
Motor torque constant	$K_T = 25.5\cdot10^{-3}$ Nm/A
Back electromotive force (EMF) constant	$K_e = 25.5\cdot10^{-3}$ V/rad/s
Armature resistance	$R_a = 7.41\ \Omega$

Table 1. *Cont.*

Armature inductance	$L_a = 0.77 \cdot 10^{-3}$ H
Gear ratio	$N = 3.8$
Amplifier resistor R_1	$R_1 = 1\ \Omega$
Amplifier resistor R_5	$R_5 = 1\ k\Omega$
Amplifier resistors R_9, R_{10}, and R_{12}	$R_9 = R_{10} = R_{12} = 10\ k\Omega$
Amplifier resistors R_{14} and R_{15}	$R_{14} = R_{15} = 1\ k\Omega$
Equivalent resistor R_{e1}	$R_{e1} = R_9 \cdot R_{10}/(R_9 + R_{10}) = 5\ k\Omega$
Equivalent resistor R_{e2}	$R_{e2} = R_{12} + R_{14} + R_{15} = 12\ k\Omega$
Amplifier capacitor C_8	$C_8 = 48$ pF

* The total moment of inertia of the DC motor is expressed as the sum $J = J_m + J_{gr} + J_e = 1.43 \cdot 10^{-6}$ kgm^2.
** This values of B and M_c are not specified in the datasheet for the particular motor but were determined experimentally [45].

2.1.3. CompactRIO 9075

The CompactRIO 9075 integrated system consists of a real-time processor and a reconfigurable FPGA (Field-Programmable Gate Array) on the same chassis, enabling control and monitoring capabilities, Figure 1. The CRIO-9075 integrates a 400 MHz industrial real-time processor with an LX25 FPGA and includes four slots for NI C Series I/O modules.

The modules used in the experimental setup are:

- NI 9402 (LVTTL, bidirectional, 4-channel, 55 ns digital input-output module)—for measuring motor speed after the gearbox, based on the frequency of the pulse train from the optical encoder;
- NI 9403 (5 V/TTL, bidirectional, 32-channel, 7 μs digital input-output module)—for controlling the activation of the corresponding relays;
- NI 9205 (32-channel, ±200 mV to ±10 V, 16-bit, 250 kS/s voltage input module)—for measuring the armature voltage of the motor and the armature current (by measuring the voltage across an additional resistor);
- NI 9263 (4-channel, ±10 V, 16-bit voltage output module)—for supplying the control voltage to the amplifier.

2.1.4. Relays

In order to develop a real-time FDII system, relays were designed to introduce multiplicative faults into the system by abruptly changing one of the amplifier parameters. Figure 3a shows the electronic schematic of the main part of the relay board used for connecting a resistance R_i (i = 1, 9, 12) into the amplifier circuit.

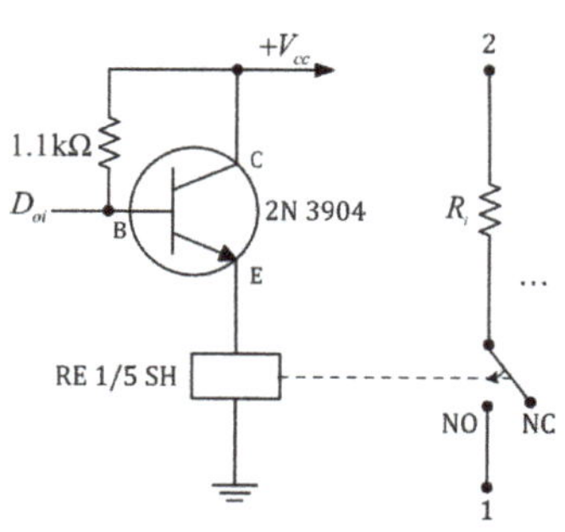

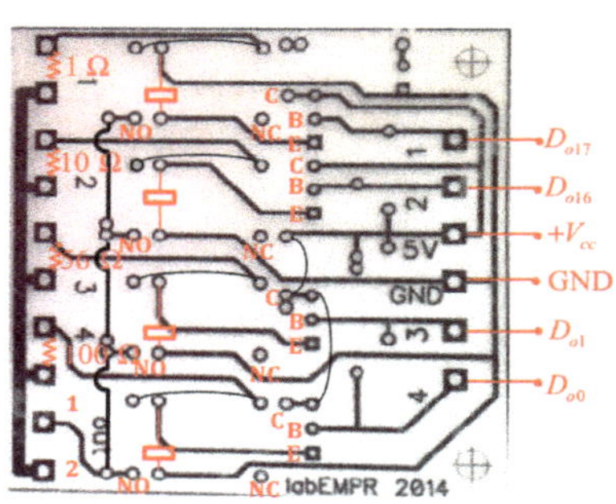

Figure 3. Relay for switching resistance: (**a**) Electronic schematic; (**b**) Physical board.

Figure 3b represents an example of a printed circuit board used to switch on or off resistance R_1 in the amplifier circuit by assigning a binary value 0/1 to the corresponding digital outputs.

When the value of the i-th digital output (D_{oi})of the NI 9403 module is 0, the 2N3904 transistor is in the conducting state, and the relay's electromagnet shifts the contact from the NC (normally closed) position to the NO (normally open) position, thereby connecting the desired resistance R_i (i = 1, 9, 12) (or capacitance C_8) into the amplifier circuit.

When the value of the digital output is 1, the bipolar transistor does not conduct, and no current flows through the relay coil. In this case, the NO contacts remain open, while the NC contacts are closed.

There are four printed circuit boards (PCBs), and each can switch between four different resistance values R_i (i = 1, 9, 12), or four capacitance values C_8 of the electronic amplifier, as specified in Table 2.

Table 2. Amplifier circuit parameter values that can be adjusted using relays.

D_{oi}	R_1	D_{oi}	R_9
D_{o17}	1 Ω	D_{o19}	10 kΩ
D_{o16}	10 Ω	D_{o18}	33 kΩ
D_{o1}	56 Ω	D_{o3}	56 kΩ
D_{o0}	100 Ω	D_{o2}	100 kΩ
D_{oi}	C_8	D_{oi}	R_{12}
D_{o21}	47 pF	D_{o6}	10 kΩ
D_{o20}	1 μF	D_{o7}	5.6 kΩ
D_{o5}	10 μF	D_{o22}	3.3 kΩ
D_{o4}	47 μF	D_{o23}	2 kΩ

2.2. System Modeling in the Presence of Multiplicative Faults

The electronic amplifier model, which uses an operational amplifier as its core component (Figure 2b), is influenced by the characteristics of the motor connected to its output. The total motor friction is described by a nonlinear function that combines static, dynamic, and viscous components, as defined by the Karnopp–Reynolds model (1) and illustrated in Figure 4a [46]. Moreover, since viscous friction is temperature-dependent (the value listed in Table 1 corresponds to a measurement at 23 °C), the combined motor–amplifier system exhibits nonlinear behavior [45].

$$M_f(\omega) = \begin{cases} M_s \mathrm{sgn}\omega; \text{ for } |\omega| < \Delta\omega \\ (M_c + (M_s - M_c)e^{-(\omega/\omega_k)^\delta})\mathrm{sgn}\omega + B\omega; \text{ for } |\omega| > \Delta\omega \end{cases} \tag{1}$$

where M_f is total friction in the motor, M_s is static friction, M_c is Coulomb friction, B is viscous friction coefficient, ω is the angular velocity of the motor, and $\Delta\omega$ is area around ω = 0. Parameter δ is estimated empirically. Some forms of this model are: Tustin's (δ = 1) and Gauss's (δ = 2).

After assigning constant values for B and M_c, provided in Table 1, and after linearizing the friction curve (Figure 4b), a linearized model of the amplifier can be represented in the form:

$$\begin{aligned} U_m(s) = & \frac{R_{e2}(C_8R_5s+1)[(R_a+sL_a)(B+Js)+K_eK_T]}{R_{e1}[(R_{e2}+R_5)C_8s+1][(R_a+R_1+sL_a)(B+Js)+K_eK_T]}U_c(s) \\ & -\frac{R_1K_e}{[(R_a+R_1+sL_a)(B+Js)+K_eK_T]}M_c(s). \end{aligned} \tag{2}$$

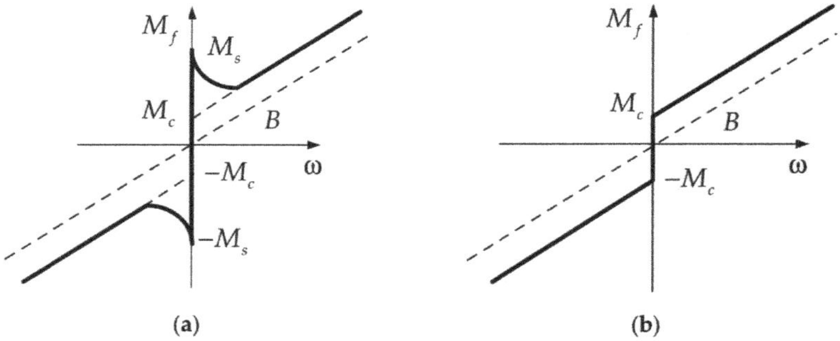

Figure 4. Dependence of total motor friction torque on motor rotational speed: (**a**) actual nonlinear model; (**b**) linear approximation.

By initially neglecting the inductance L_a of the motor inductor circuit and subsequently reducing the resulting second-order system to a first-order one, a simplified model is obtained:

$$U_m(s) = \frac{b_{11}s + b_{01}}{a_{11}s + 1}U_c(s) - \frac{b_{02}}{a_{12}s + 1}M_c(s), \tag{3}$$

where

$$\begin{aligned} b_{01} &= \tfrac{R_{2e}}{R_{1e}} \tfrac{BR_a + K_e K_T}{B(R_a + R_1) + K_e K_T}, \\ b_{11} &= \tfrac{R_{2e}}{R_{1e}} \tfrac{[JR_a + C_8 R_5 (BR_a + K_e K_T)]}{B(R_a + R_1) + K_e K_T}, \\ a_{11} &= \tfrac{J(R_a + R_1) + C_8(R_5 + R_{2e})[B(R_a + R_1) + K_e K_T]}{B(R_a + R_1) + K_e K_T}, \end{aligned} \tag{4}$$

$$\begin{aligned} b_{02} &= \tfrac{R_1 K_e}{B(R_a + R_1) + K_e K_T}, \\ a_{12} &= \tfrac{J(R_a + R_1)}{B(R_a + R_1) + K_e K_T}. \end{aligned} \tag{5}$$

By omitting Coulomb friction, simplified continuous model suitable for analyzing multiplicative faults can be derived:

$$U_m(s) = \frac{b_1 s + b_0}{a_1 s + 1}U_c(s), \tag{6}$$

$$b_0 = b_{01}, b_1 = b_{11}, a_1 = a_{11}. \tag{7}$$

Discretizing the continuous amplifier model with a sampling period T = 10 ms then produces:

$$U_m(z) = \frac{b_{0d} + b_{1d}z^{-1}}{1 + a_{1d}z^{-1}}U_c(z), \tag{8}$$

where the coefficients in (8) are defined by the discretization process:

$$\begin{aligned} b_{0d} &= b_1/a_1, \\ b_{1d} &= b_0(1 - e^{-T/a_1}) - b_1/a_1, \\ a_{1d} &= e^{-T/a_1}. \end{aligned} \tag{9}$$

The resulting discrete form (8) is directly applicable for implementing discrete multiplicative-fault detection algorithms.

2.3. Sliding Window Algorithm

The SWA is a method for estimating the parameters of a discrete model [47,48]. This algorithm is a variant of the least squares algorithm [38], where the data length K is relatively small.

Consider a system model with a single input and a single output, expressed in the form:

$$y(k) = \frac{b_{0d} + b_{1d}z^{-1} + b_{2d}z^{-2} + \ldots + b_{vd}z^{-v}}{1 + a_{1d}z^{-1} + a_{2d}z^{-2} + \ldots + a_{vd}z^{-v}} u(k), \tag{10}$$

where:
$u(k) = U_c(k)$ is the system input,
$y(k) = U_m(k)$ is the system output,
v is the system order.

Defining the regression vector

$$\boldsymbol{\varphi}'(k) = [u(k)u(k-1)\ldots u(k-v); -y(k-1) - y(k-2)\ldots - y(k-v)], \tag{11}$$

and the parameter vector as

$$\boldsymbol{\pi} = [b_{0d}\ b_{1d}\ b_{2d}\ \cdots\ b_{vd};\ a_{1d}\ a_{2d}\ \cdots\ a_{vd}]', \tag{12}$$

the output can be defined as follows:

$$y(k) = \boldsymbol{\varphi}'(k)\boldsymbol{\pi}. \tag{13}$$

The parameter estimation using the least squares algorithm is provided in the form:

$$\hat{\boldsymbol{\pi}}(k, K) = \left[\sum_{j=0}^{K-1} \boldsymbol{\varphi}(k-j)\boldsymbol{\varphi}'(k-j)\right]^{-1} \left[\sum_{j=0}^{K-1} \boldsymbol{\varphi}(k-j)y(k-j)\right], \tag{14}$$

$$\hat{\boldsymbol{\pi}}(k, K) = \left[\boldsymbol{\Phi}'(k, K)\boldsymbol{\Phi}(k, K)\right]^{-1}\boldsymbol{\Phi}'(k, K)\mathbf{Y}(k, K), \tag{15}$$

$$\boldsymbol{\Phi} = \begin{bmatrix} \boldsymbol{\varphi}'(k) \\ \vdots \\ \boldsymbol{\varphi}'(k-K+1) \end{bmatrix}, \mathbf{Y} = \begin{bmatrix} y(k) \\ \vdots \\ y(k-K+1) \end{bmatrix}, \tag{16}$$

where: k denotes current time step, and K represents the length of the data used in the algorithm.

When estimating parameters to detect and analyze multiplicative faults, a transient response occurs following a parameter jump, lasting for $K-1$ samples. As a result, the detection of such faults is delayed by at least $K-1$ samples. To reduce this delay, the window length K in the least squares algorithm must be limited, prompting a shift to the SWA. A shorter window reduces detection delay but increases sensitivity to noise and variance in parameter estimates. Additionally, successful parameter estimation requires the input signal to be sufficiently informative, satisfying the persistent excitation condition [49].

2.4. Sliding Integral Algorithm

The SIA [3,44,50,51] is a control or signal processing method that computes the integral (i.e., the accumulated sum) of a signal over a moving (sliding) window of time or samples, Figure 5. It can identify the parameters of a continuous-time model directly from sampled data, which is often more complex than discrete-time identification but yields physically meaningful parameters. To reduce sensitivity to noise from derivative approximations, an equivalent integral form of the differential equation is used, typically over a sliding window, to avoid the need for initial conditions, as proposed in [52].

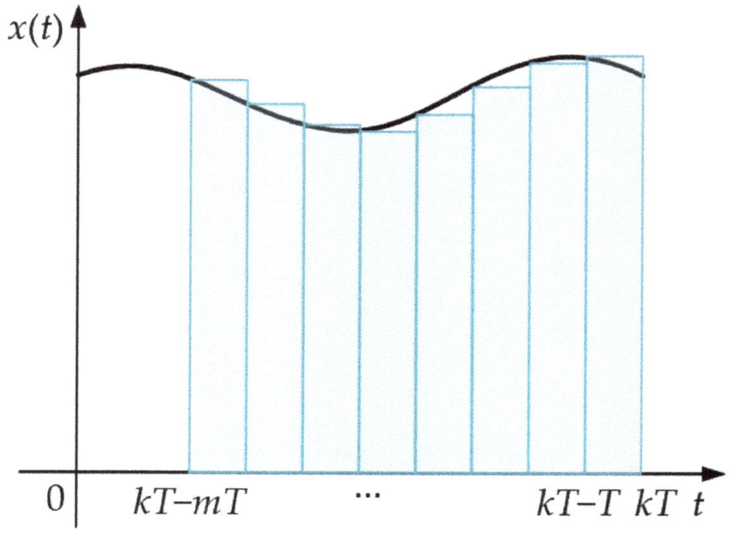

Figure 5. Rectangular approximation of the sliding integral.

The SIA is explained in detail in [3,51], and a more concise version is presented here. The sliding integral of a continuous-time variable $x(t)$ is:

$$I_1[x,k] = \int_{t=kT-mT}^{kT} x(t)dt, \tag{17}$$

where T is sampling interval, k is current time step, and m is window length and sample counts representing the number of samples within the integration interval.

With rectangular approximation, this can be written as

$$I_1[x,k] = T\sum_{i=1}^{m} x(kT-iT) = TQ(z)x(k), \tag{18}$$

where

$$Q(z) = (z^{-1} + z^{-2} + \ldots + z^{-m}).$$

In general, the j-th sliding integral is expressed as:

$$I_j[x,k] = T^j Q^j(z)x(k). \tag{19}$$

The following expression also holds [3,51]

$$I_j[x^{(j)},k] = (1-z^{-m})^j x(k), \tag{20}$$

where $x^{(j)} = d^j x(t)/dt^j$.

By combining Equations (19)–(21), this equation can be obtained:

$$I_n[x^{(j)},k] = I_{n-j}[I_j(x^{(j)},k\prime),k] = \rho_j(z)x(k), \tag{21}$$

where

$$\rho_j(z) = T^{n-j}Q^{n-j}(1-z^{-m})^j, j = 0, 1, \ldots, n. \tag{22}$$

For a single-input single-output (SISO) linear time-invariant (LTI) system, the system dynamics can be described by the following differential equation:

$$y(t) + a_1\frac{dy(t)}{dt} + \ldots + a_n\frac{d^n y(t)}{dt^n} = b_0 u(t) + b_1\frac{du(t)}{dt} + \ldots + b_n\frac{d^n u(t)}{dt^n}, \tag{23}$$

where $u(t)$ is input, $y(t)$ is output of the system.

By applying the sliding integral to the differential Equation (23) n times, it is obtained

$$I_n[y,k] + a_1 I_n[y^{(1)},k] + \ldots + a_n I_n[y^{(n)},k] = b_0 I_n[u,k] + b_1 I_n[u^{(1)},k] + \ldots + b_n I_n[u^{(n)},k]. \tag{24}$$

Using (21), (24) can be rewritten as follows:

$$\rho_0(z)y(k) + a_1\rho_1(z)y(k) + \ldots + a_n\rho_n(z)y(k) = b_0\rho_0(z)u(k) + b_1\rho_1(z)u(k) + \ldots + b_n\rho_n(z)u(k). \tag{25}$$

This is alternatively represented as

$$v(k) = \boldsymbol{\varphi}'(k)\boldsymbol{\theta}, \tag{26}$$

where

$$v(k) = \rho_0(z)y(k), \tag{27}$$

$$\boldsymbol{\varphi}'(k) = [\rho_0(z)u(k)\ldots\rho_n(z)u(k); -\rho_1(z)y(k)\ldots-\rho_n(z)y(k)], \tag{28}$$

$$\boldsymbol{\theta} = [b_0\cdots b_n; a_1\cdots a_n]'. \tag{29}$$

Accordingly, the least squares algorithm can be formulated as:

$$\boldsymbol{\theta} = (\boldsymbol{\Phi}'\boldsymbol{\Phi})^{-1}\boldsymbol{\Phi}'V, \tag{30}$$

where

$$\boldsymbol{\Phi} = \begin{bmatrix} \boldsymbol{\varphi}'(k) \\ \vdots \\ \boldsymbol{\varphi}'(k-K+1) \end{bmatrix}, V = \begin{bmatrix} v(k) \\ \vdots \\ v(k-K+1) \end{bmatrix}. \tag{31}$$

Although the identification procedure still relies on sampled data, the SIA is more complex and computationally intensive than parameter identification in the discrete-time case. Due to the inherent trade-off between computational complexity and noise attenuation, a window length m of up to 20 is recommended [3]. Furthermore, the literature does not impose any specific constraints on the length of the identification dataset K.

2.5. The Main Steps of the Proposed FDII Methodology

A flowchart representing the main steps of the proposed FDII methodology is presented in Figure 6, providing a clear visual overview of the detection, isolation, and identification process.

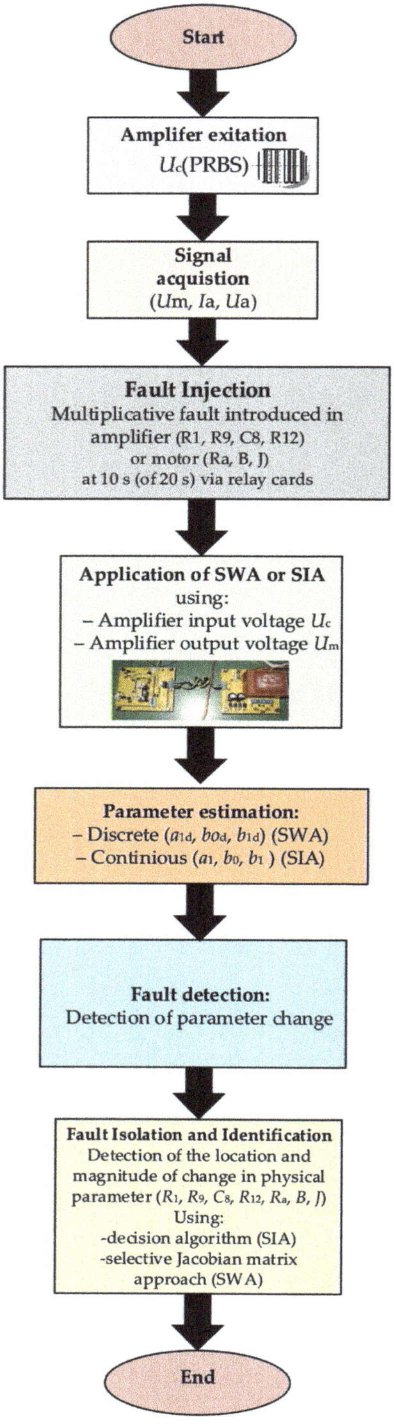

Figure 6. Flowchart of the main steps of the FDII.

3. Results

To demonstrate the effectiveness of detecting and isolating individual multiplicative amplifier faults, the amplifier parameters were deliberately and abruptly altered (one at a time) at the midpoint of the recording interval using relay cards (Figure 3). In this experimental setup, two parameter estimation approaches were developed.

The first approach can be described as offline. Signal data were collected in the LabVIEW environment and stored as text files. The estimation process was then carried out in batch mode using MATLAB 2020b, where the estimation algorithm was implemented through dedicated m-files.

Figure 7 presents a portion of the LabVIEW block diagram for generating and recording signals under multiplicative fault conditions. It illustrates the generation of the amplifier's input signal U_c, using a PRBS (Pseudo-Random Binary Sequence) and the measurement of the amplifier's output signal U_m. The system also enables monitoring of the armature voltage U_a, the armature current I_a (by monitoring the voltage on the series resistor R_d, Figures 2b and 7), and the motor speed n.

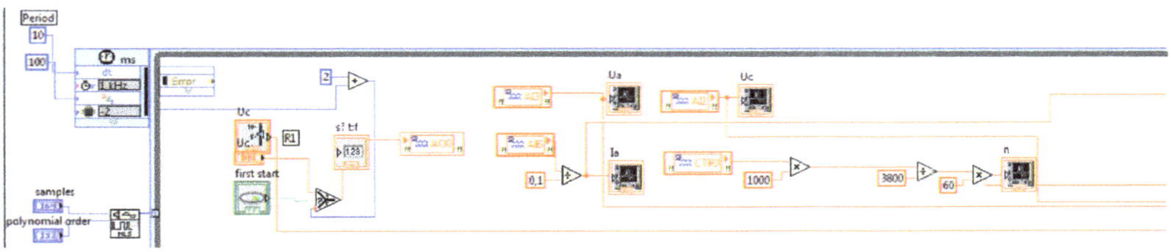

Figure 7. A segment of the LabVIEW Block Diagram developed for generating and recording signals under multiplicative fault conditions.

Figure 8 presents a portion of the LabVIEW Block Diagram, which activates specific amplifier parameters by switching the corresponding relays, thereby inducing individual multiplicative faults.

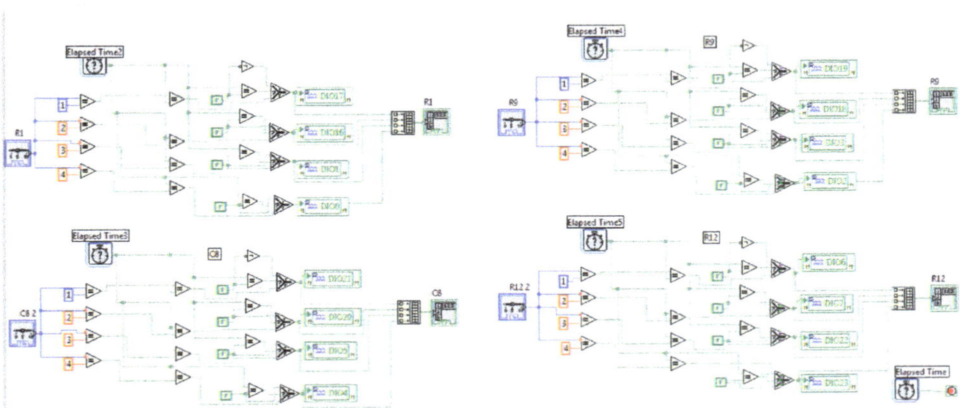

Figure 8. A segment of the LabVIEW Block Diagram developed to inject multiplicative faults into the system.

Figure 9 shows the Front Panel of the LabVIEW program used for monitoring input-output signals, featuring controls for the abrupt changing of one of the amplifier's parameter values in the event of a multiplicative fault. The software is configured so that all parameters are set to their nominal values (slider position 1) prior to the occurrence of a fault. At the fault moment, one of the amplifier parameters is changed. In the example shown, the resistance R_{12} is in 10 s reduced from its nominal value of 10 kΩ to a lower value of 5.6 kΩ (slider position 3) by automatically activating the corresponding relay.

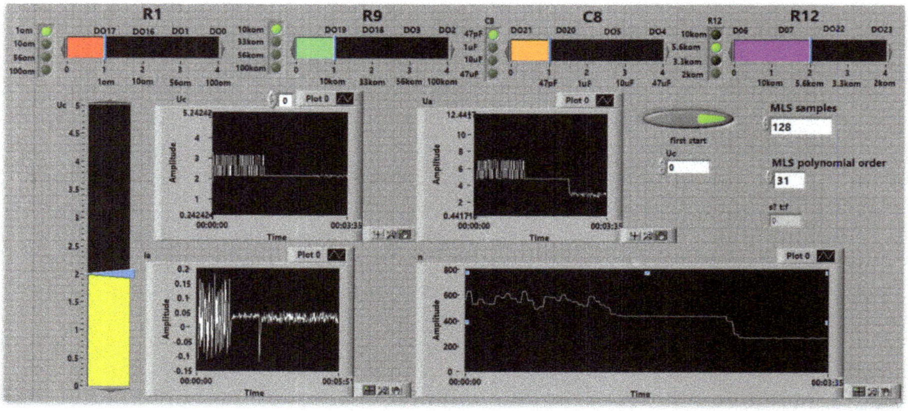

Figure 9. LabVIEW Front Panel for acquisition of input–output signals under multiplicative fault conditions.

The second approach to detecting multiplicative faults in the amplifier and DC motor was developed entirely within the LabVIEW environment and can be classified as a real-time method. After acquiring the input–output signals (Figures 7 and 9) during the injection of specific multiplicative faults using relays (Figures 8 and 9), the signals are passed to the next frame of the Flat Sequence structure. Within this structure, the desired estimation algorithm, either discrete (SWA), based on (14)–(16), or continuous (SIA), based on (30) and (31), is implemented using a MathScript Node within the LabVIEW software (Figure 10). The estimated parameters of the amplifier b_{0d}, b_{1d}, and a_{1d} from the discrete model of the amplifier (8) (SWA), or b_0, b_1, and a_1 from the continuous model of the amplifier (6) (SIA), are then displayed in the Chart block of the LabVIEW program. Figure 11 displays section of the LabVIEW Front Panel displaying the estimation of continuous parameters b_0, b_1, and a_1 using the SIA.

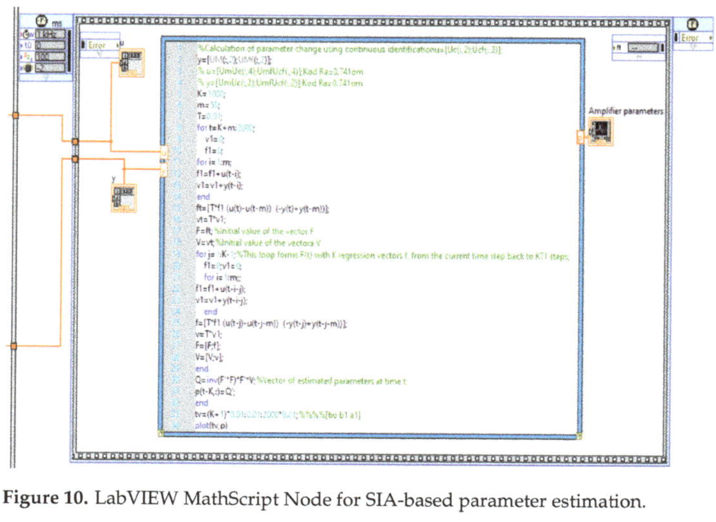

Figure 10. LabVIEW MathScript Node for SIA-based parameter estimation.

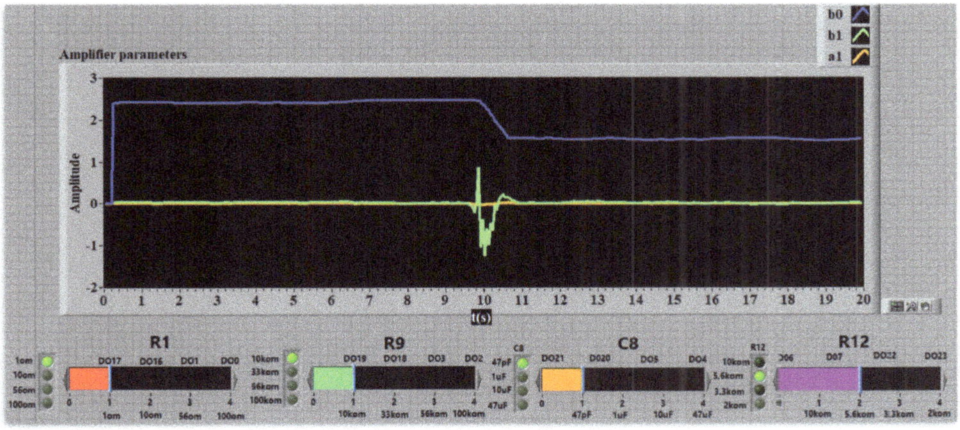

Figure 11. Section of the LabVIEW Front Panel displaying the estimation of parameters b_0, b_1, and a_1 using the SIA.

The procedure for generating multiplicative motor faults, specifically changes in inductance (L), armature resistance (R_a), moment of inertia (J), and viscous friction coefficient (B), was implemented at the simulation level using the Multisim environment. This setup utilized the amplifier model shown in Figure 12 and the motor model illustrated in Figure 13. In this configuration, the input signal for parameter estimation was a Pseudo-Random Binary Sequence (PRBS) control voltage (U_c), while the output voltage of the amplifier (U_m) was recorded using the Parameter Sweep faults. This approach leverages the mapping of motor parameters onto the electronic amplifier model.

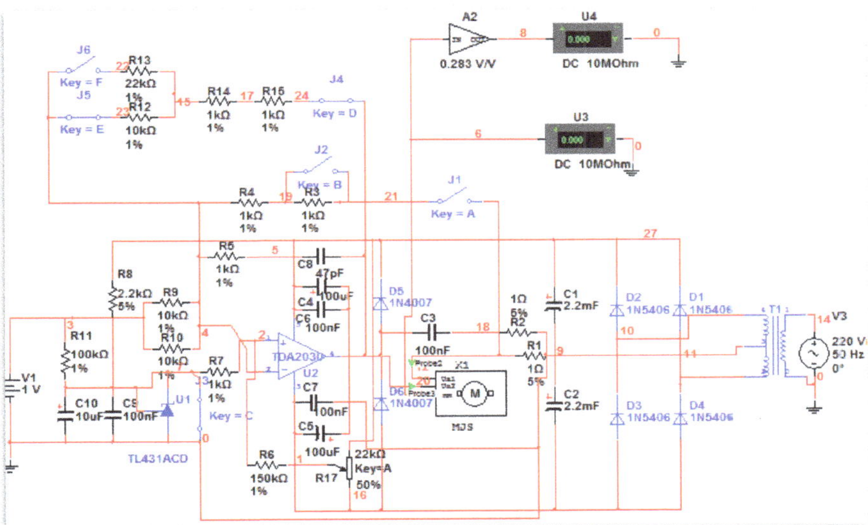

Figure 12. Multisim model of the electronic amplifier.

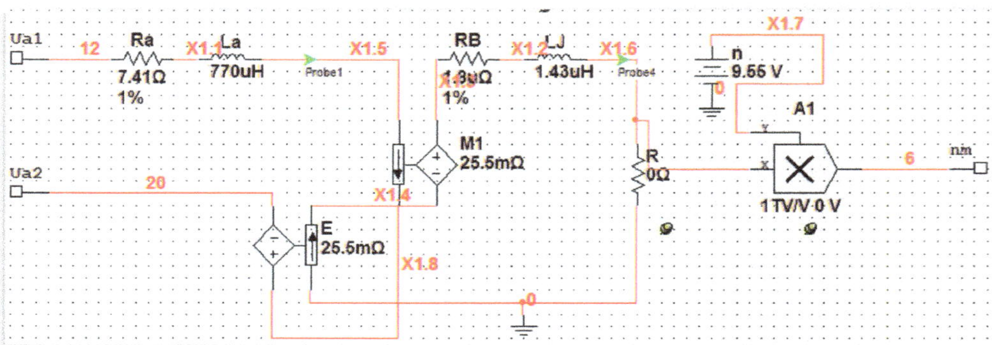

Figure 13. Multisim model of the DC motor.

3.1. Multiplicative Fault Detection

3.1.1. Input–Output Signals Under Multiplicative Faults

To ensure accurate estimation of sudden multiplicative faults, the input signal must be persistently exciting, i.e., rich enough to guarantee a unique and identifiable set of parameter estimates [49]. The Pseudo-Random Binary Sequence (PRBS), which is readily available in most software libraries, is, therefore, widely adopted for system identification.

A PRBS signal with values between 2 V and 3 V, as shown in Figure 14, was generated using LabVIEW and applied as the input to the electronic amplifier. Figures 15 and 16 illustrate the amplifier and motor output voltages over a 20 s period, during which a sudden multiplicative fault is introduced at the 10th s. This fault corresponds to an abrupt change in a selected amplifier or motor parameter from its nominal value to a predefined fault value. For the amplifier, faults were introduced in the following components: R_1 (from the nominal value of 1 Ω to fault values 10 Ω, 56 Ω, and 100 Ω), Figure 15a; R_9 (from the nominal value of 10 kΩ to the faults values 33 kΩ, 56 kΩ and 100 kΩ), Figure 15b, C_8 (from the nominal value of 47 pF to the fault values 1 μF, 10 μF, 47 μF), Figure 15c; R_{12} (from the nominal value of 10 kΩ to the fault values 2 kΩ, 3.3 kΩ and 5.6 kΩ), Figure 15d. For the motor, faults were introduced in the parameters: armature resistance R_a (from the nominal value 7.41 Ω to the fault values 0.741 Ω and 74.1 Ω), Figure 16a,b, viscous friction coefficient B (from the nominal value $1.8 \cdot 10^{-6}$ Nm/rad/s to the fault values $1.8 \cdot 10^{-4}$ Nm/rad/s and $1.8 \cdot 10^{-8}$ Nm/rad/s), Figure 16c, moment of inertia J (from the nominal value $1.43 \cdot 10^{-6}$ kgm^2 to the fault values $1.43 \cdot 10^{-4}$ kgm^2 and $1.43 \cdot 10^{-8}$ kgm^2), Figure 16d.

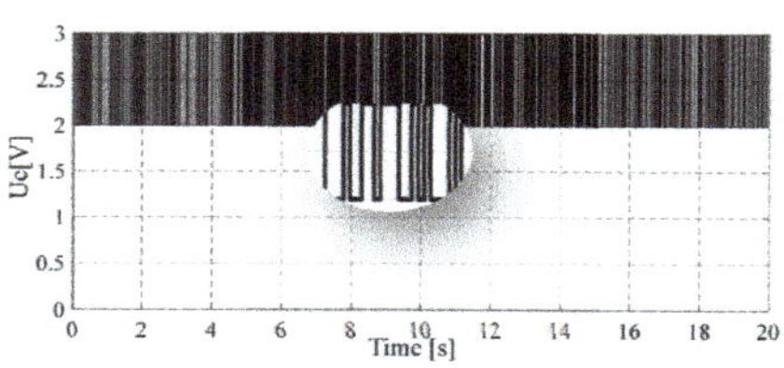

Figure 14. Amplifier input voltage $U_c(t)$—PRBS excitation signal with a magnified view of the waveform.

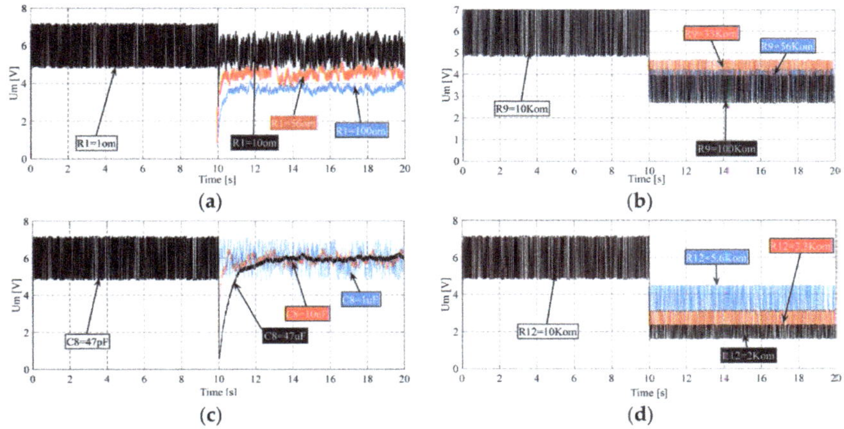

Figure 15. Amplifier output voltage $U_m(t)$ during multiplicative faults on: (**a**) R_1; (**b**) R_9; (**c**) C_8; (**d**) R_{12}.

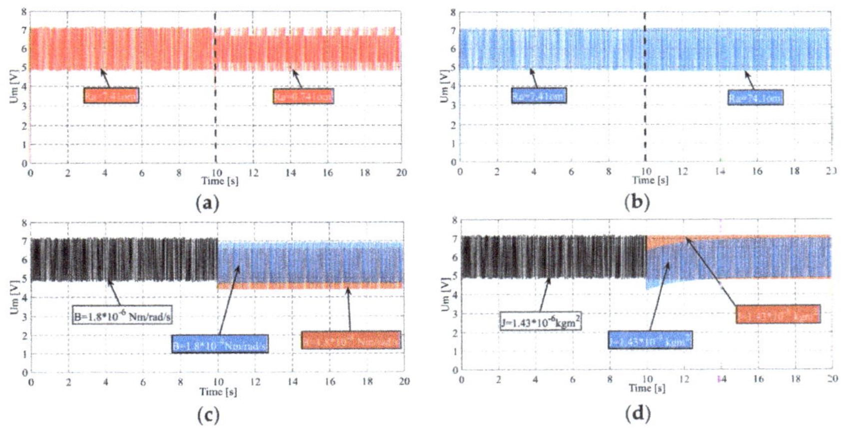

Figure 16. Amplifier output voltage U_m(t) during multiplicative faults on: (**a**) Ra; (**b**) Ra; (**c**) B; (**d**) J. The symbol "*" denotes a multiplication sign.

3.1.2. Multiplicative Fault Detection Using Sliding Window Algorithm

The following diagrams present selected results from the application of the SWA for estimating amplifier parameters using various window lengths (K). The result of applying the SWA is the set of discrete amplifier parameters, a_{1d}, b_{0d}, and b_{1d}, described in model (8). Figures 17 and 18 show SWA-based parameter estimation in response to a multiplicative fault in the amplifier resistance R_1 and the motor's viscous friction coefficient B, respectively.

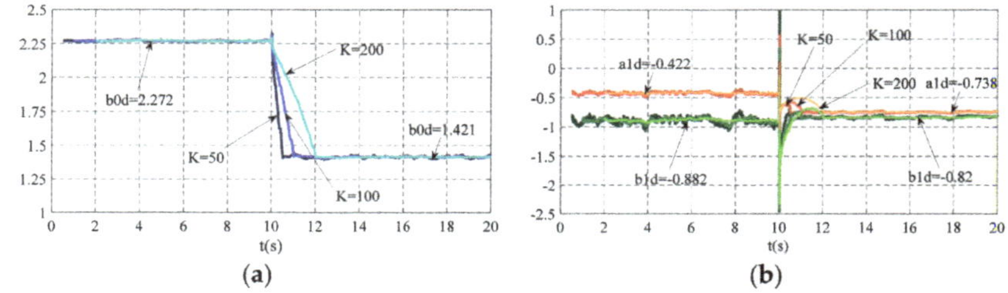

Figure 17. SWA amplifier parameters estimation: (**a**) b_{0d} (**b**) a_{1d}, b_{1d} during the change of R_1 from 1 Ω to 10 Ω.

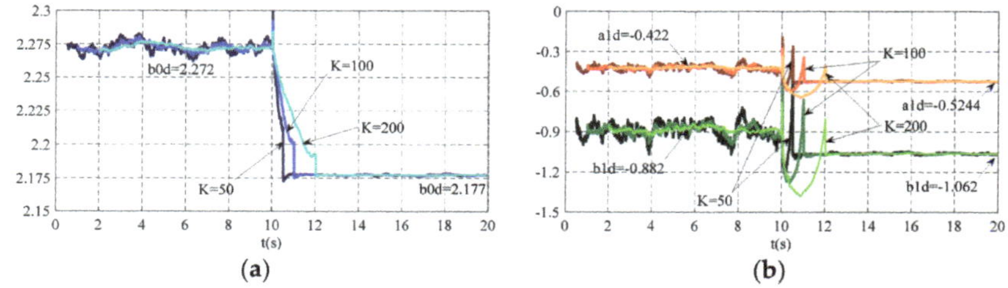

Figure 18. SWA amplifier parameter estimation: (**a**) b_{0d}, (**b**) a_{1d}, b_{1d}, during the change of B from $1.8{\cdot}10^{-6}$ kgm^2 to $1.8{\cdot}10^{-5}$ kgm^2.

In Figures 17 and 18, the time interval from 0 to 10 s represents the estimated parameters before the fault occurs, while the interval from 10 to 20 s corresponds to the post-fault parameter estimation. A transient interval is noticeable within this post-fault period, lasting approximately $(K-1){\cdot}T$ s, where $T = 0.01$ s denotes the sampling period.

The graphical estimation results indicated that a sudden change in the amplifier or motor parameters triggers a transient response in the identification algorithm during the measurement period. Consequently, selecting a shorter window length is more advantageous in terms of reducing fault detection delay. However, it is also evident that as the window length decreases, the variance of the estimates increases and the sensitivity to noise becomes more pronounced. In the conducted experiments, $K = 50$; $K = 100$; and $K = 200$, different window lengths were tested. Since the total measurement period was 20 s, the chosen window lengths corresponded to transient durations of approximately 0.49 s, 0.99 s, and 1.99 s. In this particular case, the parameter choice of $K = 100$ proved to be the most effective.

Tables 3–6 summarize the complete results of applying the SWA for various multiplicative faults in the four amplifier parameters R_1, R_9, R_{12}, and C_8, and the three motor parameters R_a, J, and B. The obtained results pertain to the discrete amplifier parameters b_{0d}, b_{1d}, and a_{1d} from model (8).

Table 3. Amplifier parameter estimation using the SWA for multiplicative faults in R_1 and R_9.

$R_1 = 1\ \Omega$				$R_9 = 10$ kΩ			
Fault	b_{0d}	b_{1d}	a_{1d}	**Fault**	b_{0d}	b_{1d}	a_{1d}
10 Ω	1.421	−0.82	−0.738	33 kΩ	1.488	−0.5	−0.373
56 Ω	0.472	−0.279	−0.896	56 kΩ	1.337	−0.412	−0.35
100 Ω	0.29	−0.18	−0.925	100 kΩ	1.258	−0.359	−0.316

Table 4. Amplifier parameter estimation using the SWA for multiplicative faults in R_{12} and C_8.

R_{12} = 10 kΩ				C_8 = 47 pF			
Fault	b_{0d}	b_{1d}	a_{1d}	**Fault**	b_{0d}	b_{1d}	a_{1d}
2 kΩ	0.757	−0.368	−0.514	1 µF	1.148	−0.013	−0.526
3.3 kΩ	1	−0.458	−0.492	10 µF	0.312	−0.142	−0.929
5.6 kΩ	1.434	−0.784	−0.572	47 µF	0.21	−0.168	−0.982

Table 5. Estimation of motor parameters using the SWA for multiplicative faults in R_a and J.

R_a = 7.41 Ω				$J = 1.43 \cdot 10^{-6}$ kgm^2			
Fault	b_{0d}	b_{1d}	a_{1d}	**Fault**	b_{0d}	b_{1d}	a_{1d}
0.741 Ω	1.9	0.356	−0.06	$1.43 \cdot 10^{-4}$ kgm^2	2.12	−2.13	−1
74.1 Ω	2.369	−2.227	−0.94	$1.43 \cdot 10^{-8}$ kgm^2	2.34	−0.192	−0.083

Table 6. Estimation of motor parameters using the SWA for multiplicative fault in B.

$B = 1.8 \cdot 10^{-6}$ Nm/rad/s			
Fault	b_{0d}	b_{1d}	a_{1d}
$1.8 \cdot 10^{-4}$ Nm/rad/s	2.162	−0.34	−0.172
$1.8 \cdot 10^{-5}$ Nm/rad/s	2.177	−1.062	−0.524

3.1.3. Multiplicative Fault Detection Using Sliding Integral Algorithm

The following figures show selected results from applying the SIA to estimate amplifier parameters under different window lengths (m) and data lengths (K). In each case, multiplicative faults were introduced abruptly at 10 s. In Figures 19 and 20, the time interval from 0 to 10 s corresponds to the parameter estimates prior to the fault occurrence, while the interval from 10 to 20 s reflects the parameter estimation after the fault has been introduced. During this post-fault period, a transient response can be observed, lasting approximately $(K + m - 1) \cdot T$ seconds, where T = 0.01 s is the sampling period.

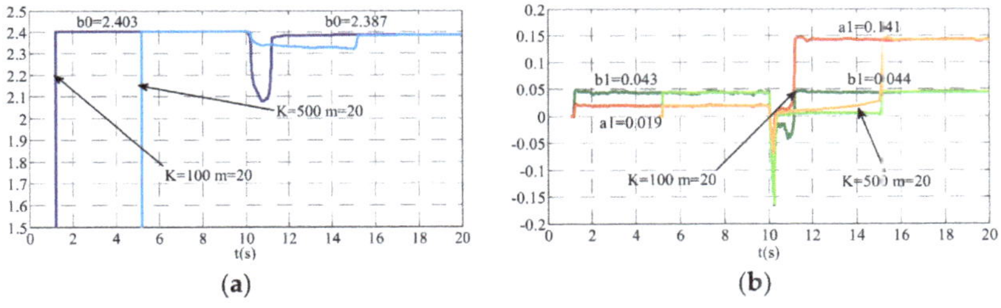

Figure 19. SIA amplifier parameters estimation: (**a**) b_0, (**b**) a_1, b_1 during the change of C_8 from 47 pF to 10 µF.

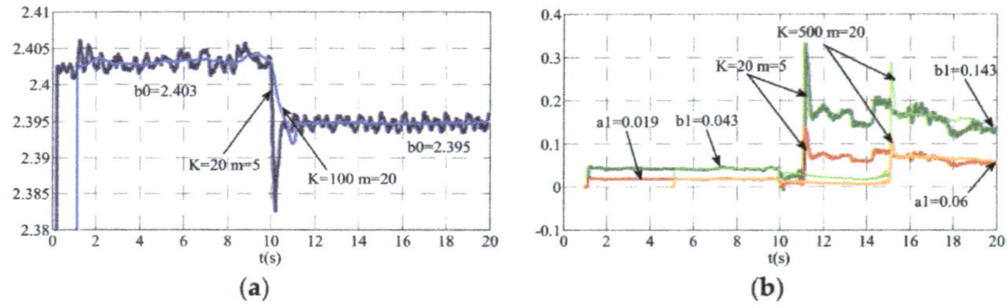

Figure 20. SIA amplifier parameters estimation: (**a**) b_0, (**b**) a_1, b_1 during the change of R_a from 7.41 Ω to 74.1 Ω.

Figure 19 shows amplifier parameter estimation using the SIA in response to a multiplicative fault in the amplifier capacitor C_8, while Figure 20 corresponds to a multiplicative fault in the motor armature resistance R_a.

When applying the estimation algorithm for the detection of abrupt parameter changes, it has been observed that the length of the identification dataset K should be selected to be sufficiently small (yet greater than m, $K > m$), as the estimation process effectively operates as a "window of length m" nested within a broader window of length K. Such changes induce a transient in the identification process lasting $K + m - 1$ samples. The used window configurations were: $K = 20$, $m = 5$; $K = 100$, $m = 20$, and $K = 500$, $m = 20$. Since the sampling period $T = 10$ ms, these configurations correspond to transient durations of approximately 0.24 s, 1.19 s, and 5.19 s, respectively.

Tables 7–10 summarize the complete results of applying the SIA for various multiplicative faults in the four amplifier parameters R_1, R_9, R_{12}, and C_8, and the three motor parameters R_a, J, and B. The obtained results pertain to the continuous amplifier parameters b_0, b_1, and a_1 from model (6).

Table 7. Amplifier parameter estimation using the SIA for multiplicative faults in R_1 and R_9.

$R_1 = 1$ Ω				$R_9 = 10$ kΩ			
Fault	b_0	b_1	a_1	**Fault**	b_0	b_1	a_1
10 Ω	2.306	0.0546	0.038	33 kΩ	1.331	0.0245	0.0199
56 Ω	1.85	0.047	0.1	56 kΩ	1.425	0.03	0.0221
100 Ω	1.495	0.0352	0.126	100 kΩ	1.575	0.032	0.022

Table 8. Amplifier parameter estimation using the SIA for multiplicative faults in R_{12} and C_8.

$R_{12} = 10$ kΩ				$C_8 = 47$ pF			
Fault	b_0	b_1	a_1	**Fault**	b_0	b_1	a_1
2 kΩ	0.8	0.0175	0.023	1 μF	2.39	0.024	0.0209
3.3 kΩ	1.059	0.0228	0.0228	10 μF	2.387	0.044	0.141
5.6 kΩ	1.521	0.032	0.022	47 μF	2.386	0.1142	0.56

Table 9. Estimation of motor parameters using the SIA for multiplicative faults in R_a and J.

$R_a = 7.41$ Ω				$J = 1.43{\cdot}10^{-6}$ kgm^2			
Fault	b_0	b_1	a_1	**Fault**	b_0	b_1	a_1
0.741 Ω	2.393	0.02	0.011	$1.43{\cdot}10^{-4}$ kgm^2	2.392	3.794	1.824
74.1 Ω	2.395	0.143	0.06	$1.43{\cdot}10^{-8}$ kgm^2	2.394	0.018	0.008

Table 10. Estimation of motor parameters using the SIA for multiplicative faults in B.

B = 1.8·10^{-6} Nm/rad/s			
Fault	b_0	b_1	a_1
1.8·10^{-4} Nm/rad/s	2.2	0.026	0.012
1.8·10^{-5} Nm/rad/s	2.346	0.0462	0.0213

3.2. Multiplicative Fault Isolation and Identification

3.2.1. Decision Algorithm

The applied techniques for detecting multiplicative faults provide information about changes in the system model parameters but do not reveal the values of the specific physical parameters that caused those changes. Hence, they are capable of fault detection, i.e., they confirm that a fault has occurred in the system.

In order to isolate faults (i.e., identify the fault location), a decision algorithm based on the amplifier model, (6), with parameters provided with (4) and (7), has been developed as shown in Figure 21.

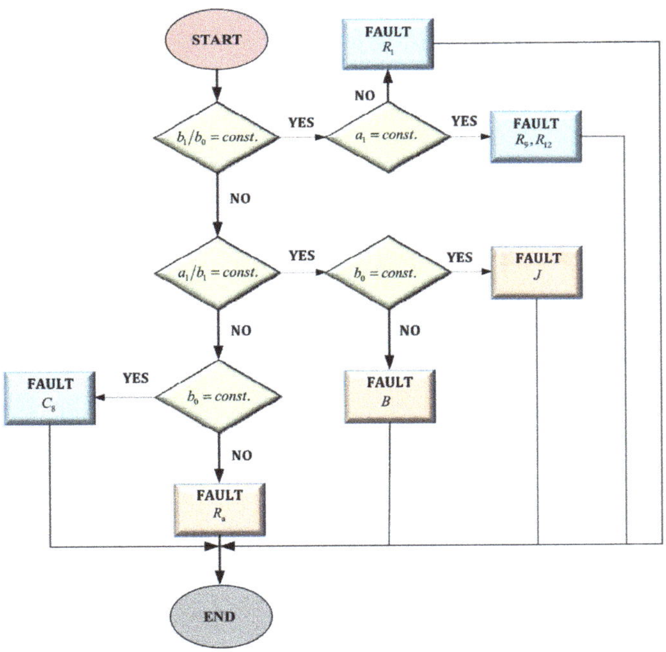

Figure 21. Multiplicative fault isolation using a decision algorithm with the SIA.

This algorithm can be performed in the case of application of continuous fault detection procedure, i.e., SIA. Fault identification, in terms of determining the magnitude of the change in the corresponding physical parameter, is straightforward in this case. The actual faulty physical parameter can be directly determined from the relations provided in (4) and (7), thereby completing the identification.

3.2.2. Indirect Estimation Procedure

An alternative solution to the fault identification problem, which can also be applied in the case of discrete parameter estimation using the SWA, the actual physical parameters can be determined through one of the indirect estimation procedures [3]. This approach is fully applicable when the relationship between the estimated and physical parameters is linear. However, if the relationship is nonlinear (which is more common), the procedure remains applicable only when the change (i.e., the fault) in the physical parameter is relatively small compared to its nominal value.

The starting point is the relationship between the physical parameters $\boldsymbol{\theta}$, whose number is L, and the model parameters $\boldsymbol{\pi}$, whose number is N, expressed through the relation:

$$\boldsymbol{\pi} = f(\boldsymbol{\theta}). \tag{32}$$

When $L = N$, the actual physical parameters can be indirectly estimated according to:

$$\Delta\hat{\boldsymbol{\theta}} = \boldsymbol{R}^{-1}\Delta\hat{\boldsymbol{\pi}}, \tag{33}$$

$$\boldsymbol{R} = \left[\frac{\partial\boldsymbol{\pi}}{\partial\boldsymbol{\theta}}\right]_{\Delta\boldsymbol{\theta}=0} = \begin{bmatrix} \frac{\partial\pi_1}{\partial\theta_1} & \cdots & \frac{\partial\pi_1}{\partial\theta_L} \\ \vdots & & \vdots \\ \frac{\partial\pi_N}{\partial\theta_1} & \cdots & \frac{\partial\pi_N}{\partial\theta_L} \end{bmatrix}_{\Delta\boldsymbol{\theta}=0}, \tag{34}$$

where:

$\Delta\hat{\boldsymbol{\pi}} = \hat{\boldsymbol{\pi}} - \boldsymbol{\pi}_{\mathrm{n}}$ is estimation in model parameters variation,
$\boldsymbol{\pi}_{\mathrm{n}}$ is the nominal model parameters,
$\hat{\boldsymbol{\pi}}$ is estimating model parameters,
$\Delta\hat{\boldsymbol{\theta}}$ is estimation of actual physical parameters.

The partial derivatives in (34) are calculated using the nominal values of the parameters. If $L < N$, a slightly modified form of (33) defined in [3] can be used:

$$\Delta\hat{\boldsymbol{\theta}} = (\boldsymbol{R}'\boldsymbol{R})^{-1}\boldsymbol{R}'\Delta\hat{\boldsymbol{\pi}}. \tag{35}$$

In the specific example, the vector of actual physical parameters is defined as:

$$\boldsymbol{\theta} = \begin{bmatrix} R_1 & R_9 & R_{12} & C_8 & R_{\mathrm{a}} & J & B \end{bmatrix}', \tag{36}$$

so, the columns of the Jacobian matrix (matrix of partial derivatives) are:

$$\boldsymbol{R}_1 = \begin{bmatrix} \frac{db_0}{dR_1} \\ \frac{db_1}{dR_1} \\ \frac{da_1}{dR_1} \end{bmatrix} = \begin{bmatrix} -\frac{R_{\mathrm{e2}}B}{R_{\mathrm{e1}}} \frac{R_{\mathrm{a}}B+K_{\mathrm{T}}K_{\mathrm{e}}}{((R_1+R_{\mathrm{a}})B+K_{\mathrm{T}}K_{\mathrm{e}})^2} \\ \frac{R_{\mathrm{e2}}}{R_{\mathrm{e1}}} \frac{R_{\mathrm{a}}JB}{((R_1+R_{\mathrm{a}})B+K_{\mathrm{T}}K_{\mathrm{e}})^2} \\ \frac{JK_{\mathrm{T}}K_{\mathrm{e}}(R_1+R_{\mathrm{a}})}{((R_1+R_{\mathrm{a}})B+K_{\mathrm{T}}K_{\mathrm{e}})^2} \end{bmatrix}; \boldsymbol{R}_2 = \begin{bmatrix} \frac{db_0}{dR_9} \\ \frac{db_1}{dR_9} \\ \frac{da_1}{dR_9} \end{bmatrix} = \begin{bmatrix} -\frac{R_{\mathrm{e2}}}{R_9^2} \frac{R_{\mathrm{a}}B+K_{\mathrm{T}}K_{\mathrm{e}}}{(R_1+R_{\mathrm{a}})B+K_{\mathrm{T}}K_{\mathrm{e}}} \\ -\frac{R_{\mathrm{e2}}}{R_9^2} \frac{JR_{\mathrm{a}}}{(R_1+R_{\mathrm{a}})B+K_{\mathrm{T}}K_{\mathrm{e}}} \\ 0 \end{bmatrix} \tag{37}$$

$$\boldsymbol{R}_3 = \begin{bmatrix} \frac{db_0}{dR_{12}} \\ \frac{db_1}{dR_{12}} \\ \frac{da_1}{dR_{12}} \end{bmatrix} = \begin{bmatrix} \frac{1}{R_{\mathrm{e1}}} \frac{R_{\mathrm{a}}B+K_{\mathrm{T}}K_{\mathrm{e}}}{((R_1+R_{\mathrm{a}})B+K_{\mathrm{T}}K_{\mathrm{e}})^2} \\ \frac{1}{R_{\mathrm{e1}}} \frac{R_{\mathrm{a}}J}{((R_1+R_{\mathrm{a}})B+K_{\mathrm{T}}K_{\mathrm{e}})^2} \\ 0 \end{bmatrix}; \boldsymbol{R}_4 = \begin{bmatrix} \frac{db_0}{dC_8} \\ \frac{db_1}{dC_8} \\ \frac{da_1}{dC_8} \end{bmatrix} = \begin{bmatrix} \frac{R_{\mathrm{e2}}}{R_{\mathrm{e1}}} \frac{R_5(R_{\mathrm{a}}B+K_{\mathrm{T}}K_{\mathrm{e}})}{(R_1+R_{\mathrm{a}})B+K_{\mathrm{T}}K_{\mathrm{e}}} \\ 0 \\ R_5 + R_{2\mathrm{e}} \end{bmatrix}; \tag{38}$$

$$\boldsymbol{R}_5 = \begin{bmatrix} \frac{db_0}{dR_{\mathrm{a}}} \\ \frac{db_1}{dR_{\mathrm{a}}} \\ \frac{da_1}{dR_{\mathrm{a}}} \end{bmatrix} = \begin{bmatrix} \frac{R_{\mathrm{e2}}}{R_{\mathrm{e1}}} \frac{B^2R_1}{((R_1+R_{\mathrm{a}})B+K_{\mathrm{T}}K_{\mathrm{e}})^2} \\ \frac{R_{\mathrm{e2}}}{R_{\mathrm{e1}}} \frac{J(R_{\mathrm{a}}B+K_{\mathrm{T}}K_{\mathrm{e}})}{((R_1+R_{\mathrm{a}})B+K_{\mathrm{T}}K_{\mathrm{e}})^2} \\ \frac{JK_{\mathrm{T}}K_{\mathrm{e}}(R_1+R_a)}{((R_1+R_{\mathrm{a}})B+K_{\mathrm{T}}K_{\mathrm{e}})^2} \end{bmatrix}; \boldsymbol{R}_6 = \begin{bmatrix} \frac{db_0}{dJ} \\ \frac{db_1}{dJ} \\ \frac{da_1}{dJ} \end{bmatrix} = \begin{bmatrix} 0 \\ \frac{R_{\mathrm{e2}}}{R_{\mathrm{e1}}} \frac{R_{\mathrm{a}}}{(R_1+R_{\mathrm{a}})B+K_{\mathrm{T}}K_{\mathrm{e}}} \\ \frac{(R_1+R_{\mathrm{a}})}{(R_1+R_{\mathrm{a}})B+K_{\mathrm{T}}K_{\mathrm{e}}} \end{bmatrix}; \tag{39}$$

$$R_7 = \begin{bmatrix} \frac{db_0}{dB} \\ \frac{db_1}{dB} \\ \frac{da_1}{dB} \end{bmatrix} = \begin{bmatrix} -\frac{R_{e2}}{R_{e1}} \frac{K_T K_e R_1}{((R_1+R_a)B+K_T K_e)^2} \\ -\frac{R_{e2}}{R_{e1}} \frac{J R_a (R_1+R_a)}{((R_1+R_a)B+K_T K_e)^2} \\ -\frac{J(R_1+R_a)}{((R_1+R_a)B+K_T K_e)^2} \end{bmatrix}. \quad (40)$$

After substituting the nominal values of the amplifier and motor model parameters (provided in Table 1), the Jacobian matrix with respect to the actual physical parameters was obtained as:

$$R = \begin{bmatrix} -6.5 \cdot 10^{-3} & -1.197 \cdot 10^{-4} & 1.995 \cdot 10^{-4} & 2.39 \cdot 10^{3} & 1.76 \cdot 10^{-5} & 0 & -3.52 \cdot 10^{3} \\ -1 \cdot 10^{-4} & -1.9 \cdot 10^{-6} & 3.2 \cdot 10^{-6} & 0 & 5.1 \cdot 10^{-3} & 2.67 \cdot 10^{4} & -4.83 \cdot 10^{2} \\ 2.1 \cdot 10^{-4} & 0 & 0 & 1.3 \cdot 10^{3} & 2.1 \cdot 10^{-3} & 1.26 \cdot 10^{4} & -2.28 \cdot 10^{2} \end{bmatrix}. \quad (41)$$

It has been concluded that modifying (35) into the form:

$$\Delta\hat{\theta}_i = (R'_i R_i)^{-1} R'_i \Delta\hat{\pi}_i, \quad (42)$$

leads to significantly more accurate parameter estimation, resulting in improved precision and reliability of the results compared to the original approach, where:

$\Delta\hat{\theta}_i$ is the estimate of the change in the i-th physical parameter,
$\mathbf{R}_i$ is the i-th column of the Jacobian matrix (matrix of partial derivatives)
$\Delta\hat{\pi}_i = \hat{\pi}_i - \pi_n$ is the estimate of the change in model parameters due to the i-th fault,
π_n is nominal model parameters,
$\hat{\pi}_i$ is estimated model parameters in the presence of a fault in the i-th physical parameter.

Tables 11 and 12 present the estimates of the actual physical parameters of the amplifier and DC motor in the presence of individual multiplicative faults, using the described indirect method provided in (42). Thereby, the parameter estimates were calculated using:

$$\hat{\theta}_i = \Delta\hat{\theta}_i + \theta_n, \quad (43)$$

where θ_n represent nominal values of the actual physical parameters of the system.

Table 11. Estimation of amplifier physical parameters under multiplicative faults.

Parameter	**R_1 = 1 Ω**			**R_9 = 10 kΩ**			**R_{12} = 10 kΩ**			**C_8 = 47 pF**		
Fault	10 Ω	56 Ω	100 Ω	33 kΩ	56 kΩ	100 kΩ	2 kΩ	3.3 kΩ	5.6 kΩ	1 μF	10 μF	47 μF
Estimated fault	9.8 Ω	49 Ω	79 Ω	17 kΩ	18.2 kΩ	19 kΩ	2 kΩ	3.3 kΩ	5.6 kΩ	0.97 μF	9.7 μF	46 μF

Table 12. Estimation of motor physical parameters under multiplicative faults.

Parameter	**R_a = 7.41 Ω**		**J = 1.43·10^{-6} kgm^2**		**B = 1.8·10^{-6} Nm/rad/s**	
Fault	0.741 Ω	74.1 Ω	1.43·10^{-4} kgm^2	1.43·10^{-8} kgm^2	1.8·10^{-4} Nm/rad/s	1.8·10^{-5} Nm/rad/s
Estimated fault	0.62 Ω	64.4 Ω	1.43·10^{-4} kgm^2	1.45·10^{-8} kgm^2	0.57·10^{-4} Nm/rad/s	1.53·10^{-5} Nm/rad/s

Based on the values presented in Tables 11 and 12, it can be concluded that the indirect estimation of physical parameters R_{12}, C_8, and J is most accurate in the presence of multiplicative faults. This accuracy arises from the linear relationship with parameters b_0, b_1, and a_1 of the continuous amplifier model. During the estimation of R_1, R_9, R_a, and B, due to the nonlinear relationship with the amplifier parameters and the large deviation of these parameters from their nominal values, the estimation results were less accurate. Nevertheless, the deviations obtained using the proposed solution (42) are significantly smaller than the completely unacceptable results produced by applying the approach available in the literature (35).

4. Discussion

4.1. Validation of Fault Detection Measurement Results

To evaluate the effectiveness of different methods for detecting multiplicative faults in the motor and amplifier, the outputs of the amplifier model (3), obtained by substituting system parameters into Equations (4) and (5) for three different values of Coulomb friction [45] (M_c = 0, M_c = 0.48 mNm and M_c = 0.33 mNm), with the outputs of model (6), which incorporates parameters estimated using the SIA, and model (8), based on the SWA results, have been compared. These outputs were compared with the measured output voltage U_m, and the percentage agreement was calculated using MATLAB's System Identification Toolbox, as shown in Tables 13–20.

Table 13. Percentage match of amplifier output before fault occurrence.

SWA	SIA	$M_c = 0$	M_c = 0.48 mNm (M_c = 0.33 mNm)
98.6%	98.59%	92.27%	91.92% (92%)

Table 14. Percentage match of system output during parametric fault in R_1.

$R_1 = 1\ \Omega$	SWA	SIA	$M_c = 0$	M_c = 0.48 mNm (M_c = 0.33 mNm)
10 Ω	95.71%	96.42%	70.78%	69% (72.11%)
56 Ω	81.65%	86.5%	−30.61%	35.13% (76.74%)
100 Ω	74.44%	76.75%	−150.2%	−17.67% (79%)

Table 15. Percentage match of system output during parametric fault in R_9.

R_9 = 10 kΩ	SWA	SIA	$M_c = 0$	M_c = 0.48 mNm (M_c = 0.33 mNm)
33 kΩ	97.7%	97.14%	90.54%	89.01% (89.52%)
56 kΩ	97.28%	96.72%	91.07%	89.26% (89.87%)
100 kΩ	93.09%	97.57%	90.25%	88.27% (88.93%)

Table 16. Percentage match of system output during parametric fault in R_{12}.

R_{12} = 10 kΩ	SWA	SIA	$M_c = 0$	M_c = 0.48 mNm (M_c = 0.33 mNm)
2 kΩ	98.32%	98.57%	92.14%	90% (90.89%)
3.3 kΩ	95.89%	98.59%	92%	91.4% (91.6%)
5.6 kΩ	98.77%	97.96%	92.48%	91.52% (91.89)

Table 17. Percentage match of system output during parametric fault in C_8.

C_8 = 47 pF	SWA	SIA	$M_c = 0$	M_c = 0.48 mNm (M_c = 0.33 mNm)
1 μF	97.85%	90.6%	75.41%	75.4% (75.43%)
10 μF	95.07%	97.57%	88.53%	89.04% (88.98%)
47 μF	87.96%	96.72%	96.3%	96.69% (96.7%9

Table 18. Percentage match of system output during parametric fault in R_a.

R_a = 7.41 Ω	SWA	SIA	M_c = 0	M_c = 0.48 mNm (M_c = 0.33 mNm)
0.741 Ω	99.68%	89.53%	95.43%	95.01% (95.25%)
74.1 Ω	95.84%	99.31%	99.75%	98.53% (98.93%)

Table 19. Percentage match of system output during parametric fault in J.

$J = 1.43 \cdot 10^{-6}$ kgm^2	SWA	SIA	M_c = 0	M_c = 0.48 mNm (M_c = 0.33 mNm)
$1.43 \cdot 10^{-4}$ kgm^2	85.84%	96.63%	99.59%	98.72% (99.12%)
$1.43 \cdot 10^{-8}$ kgm^2	82.76%	82.69%	75.73%	75.71% (75.73%)

Table 20. Percentage match of system output during parametric fault in B.

B = $1.8 \cdot 10^{-6}$ Nm/rad/s	SWA	SIA	M_c = 0	M_c = 0.48 mNm (M_c = 0.33 mNm)
$1.8 \cdot 10^{-4}$ Nm/rad/s	99.54%	99.21%	96.88%	96.84% (96.86%)
$1.8 \cdot 10^{-5}$ Nm/rad/s	99.45%	99.1%	96.5%	96.2% (96.35%)

An example of the Simulink models and the comparison interface for model outputs with the measured output voltage, in the case of a fault R_1 = 100 Ω, is shown in Figure 22. The corresponding output signals are presented in Figure 23.

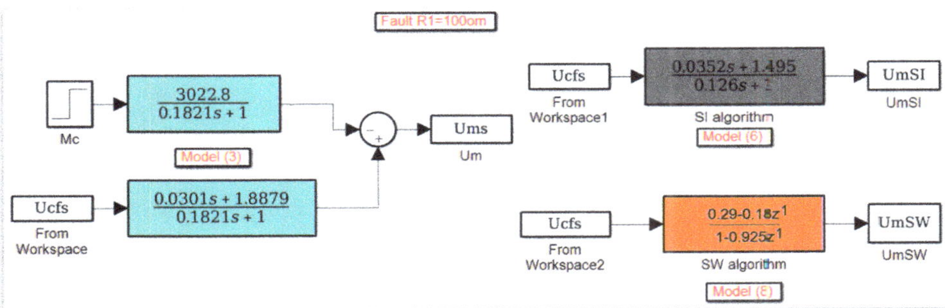

Figure 22. Simulation models of the amplifier under a fault R_1 = 100 Ω.

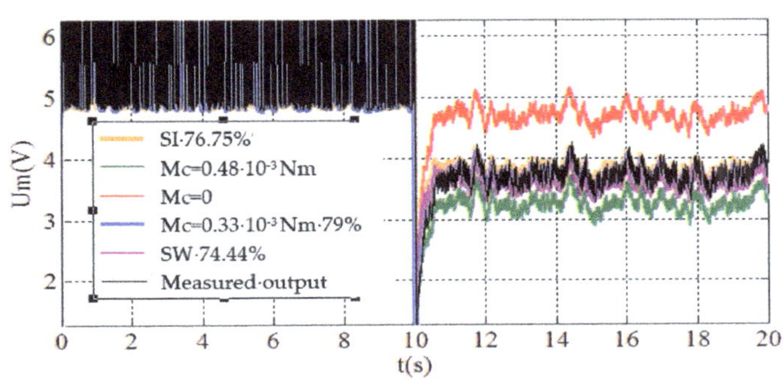

Figure 23. Measured and simulated outputs of the amplifier during the fault R_1 = 100 Ω.

Based on the results, it is clear that a multiplicative fault in resistor R_1 = 100 Ω (Table 14) significantly amplifies the effect of Coulomb friction (Figure 23). In this scenario, the model incorporating M_c = 0.33 mNm achieves the closest match to the measured output (79%), compared with the SIA-based model (76.8%) and the SWA-based model (74.4%). By contrast, for faults in C_8, R_9, or R_{12}, including Coulomb friction makes virtually no difference: the percentage match to the measured output is effectively the same whether friction is modeled or not.

A comparison of the fault-detection algorithms and the linearized models (with and without Coulomb friction) yields the following insights:

- Although the SIA is computationally more complex and more sensitive to noise than the discrete-time approach, it proved almost as effective at detecting multiplicative faults. As it has been observed, it is important to ensure that the length of the identification dataset K is chosen to be sufficiently small yet greater than m (i.e., $K > m$).
- The SWA is computationally simpler and more robust to measurement noise, making it better suited for real-time applications. Its parameter estimates also showed strong agreement with the actual system response.
- Obtained results confirm that, at the nominal resistance R_1 = 1 Ω, Coulomb friction is negligible. When R_1 increases abruptly, however, Coulomb friction can no longer be ignored (which is consistent with Equation (2)). Hence, simplified SIA and SWA models based on (6) and (8) that omit Coulomb friction lose accuracy. Also, the Coulomb friction value M_c = 0.33 mNm, which yields the best fit, appears to be more accurate than the value reported in [45]. However, the agreement between the measured system output and the output of the linearized model (2), with M_c = 0.33 mNm, does not exceed 80% for any fault magnitude in R_1, indicating that the system can only be linearized approximately.

4.2. Discussion of Multiplicative Fault Isolation and Identification

4.2.1. Discussion of the Decision Algorithm

After the fault isolation procedure is carried out using the decision algorithm, fault identification can be performed (when using one of the continuous estimation procedures, such as the SIA) based on a known, generally relatively simple relationship with the model parameters.

4.2.2. Discussion of the Indirect Estimation Procedure

When fault detection is performed through parameter estimation of a discrete model, such as the SWA, faults can only be identified using indirect approaches. However, due to the higher complexity and nonlinearity of the relationship with the model parameters in this case, the accuracy of the method is directly affected by how much the fault magnitude deviates from nominal values.

Here, too, the advantage of the continuous estimation method over the discrete one becomes evident, particularly in terms of the complexity of diagnosing multiplicative faults.

Also, more accurate parameter estimation in the indirect identification procedure has been achieved by extracting only selected columns from the Jacobian matrix (matrix of partial derivatives), instead of applying the full matrix as suggested in the literature. This is supported by the analyzed case involving large multiplicative faults that significantly deviate from the nominal operating point, where the literature-proposed method proved inapplicable. In such cases of large faults, when the relationship between model parameters and actual physical parameters is linear, the proposed method provides highly accurate identification. In the case of a nonlinear relationship, the result remains approximate but is still more acceptable than the entirely inaccurate solution obtained using the literature-based approach.

4.3. Comparison of the Results

A summary of key parameter estimation, knowledge-based, and optimization methods used in fault detection, isolation, and identification (FDII) is presented in Table 21, highlighting their real-time applicability, computational complexity, robustness to noise and transients, and overall FDII capability.

As shown in Table 21, each method presents a trade-off between real-time capability, complexity, and fault coverage. The proposed SWA and SIA approaches demonstrate a favorable balance, offering full diagnostic functionality with low-to-moderate complexity and strong robustness to noise.

Table 21. Comparison of Proposed FDII Framework with Existing Literature.

Method	Real-Time Capability	Effectiveness	Computational Complexity	Characteristics	Fault Detection/Isolation/ Identification
Least Squares (LS) [18,32,38]	Offline	Effective for static faults	Low	Not suitable for real time; sensitive to noise; poor performance under transients	Detection only
Recursive Least Squares (RLS) RLS [30,31]	Real time	Effective for gradual parameter changes	Moderate	Sensitive to noise and abrupt faults; requires forgetting factor tuning	Detection + Identification
Extended Kalman Filter (EKF) [29,34,35]	Real time (with tuning)	Suitable for nonlinear systems	High	Requires accurate model; sensitive to noise covariance; tuning required	Full (Detection, Isolation, Identification)
Unscented Kalman Filter (UKF) UKF [36,37]	Real time (with tuning)	High estimation accuracy for nonlinear systems	Very High	Computationally intensive; complex tuning process	Full (Detection, Isolation, Identification)
Algebraic-Geometric Technique (AGT) [33]	Offline	High analytical accuracy	High	Requires symbolic derivation; not robust to noise	Full (Detection, Isolation, Identification)
Genetic Algorithm (GA) [42].	Offline (training/optimisation)	Effective in feature selection, parameter tuning	High	Not suitable for real time; used primarily for offline training/optimisation	Detection, Optimisation (offline)
Particle Swarm Optimisation (PSO) [41].	Offline (training/optimisation)	Effective for weight tuning in ML models	Moderate–High	Not real time; applied offline for tuning; performance depends on swarm settings	Detection, Optimisation (offline)
Neural Networks (NN) [6,18,41]	Possible (inference only)	High accuracy with nonlinear and complex patterns	High (training), Low–Mod (inference)	Training is offline; real time only feasible if model is simple and pretrained	Full (if combined with logic/rules)
Proposed method: Sliding Window Algorithm (SWA)+ selected Jacobian	Real-time	Robust to noise; suited for abrupt faults;	Low	Performance depends on window size;	Full (Detection, Isolation, Identification)
Proposed method: Sliding Integral Algorithm (SIA)+ Decision algorithm	Real-time	Robust to noise; suited for abrupt faults; clear model-parameter link	Moderate	Performance depends on window lengths and data lengths	Full (Detection, Isolation, Identification)

5. Conclusions

The paper discusses diagnosis of multiplicative faults caused by sudden changes in one of the parameters of the converter or motor, which significantly deviate from the nominal operating point. Four amplifier faults were considered (changes in resistances R_1, R_9, R_{12}, and capacitance C_8), along with three motor faults (changes in moment of inertia J, viscous friction coefficient B, and armature resistance R_a). The method leverages the advantageous fact that the electromechanical parameters of the motor are reflected in the parameters of the amplifier model, which is realized as a linear electronic circuit with an operational amplifier. Therefore, the fault detection algorithms could be based solely on parameter estimation of the discrete or continuous model of the amplifier, which is a unique approach to the problem.

Diagnosing faults caused by abrupt parameter changes is particularly demanding. Classical parameter estimation techniques yield unsatisfactory results because a sudden change in one system parameter initiates a transient process during which the effect of the change spreads to other parameters as well. To address this, more advanced estimation techniques were employed for detecting multiplicative faults induced by sudden parameter changes in the amplifier or motor. These included: the SWA for estimating parameters of the discrete model and the SIA for estimating parameters of the continuous model.

An additional requirement for the application of these techniques is the necessity of a sufficiently informative input signal (i.e., persistent excitation) of adequate order to enable parameter identification. To satisfy this, a PRBS (Pseudo-Random Binary Sequence) signal was used in the experiments.

This paper implements a modified version of the SIA, which introduces a constraint on the data length in the least squares algorithm, within which the sliding integral is applied. This modification aims to reduce the impact of transients on parameter estimation.

To compare the two approaches to multiplicative fault detection and to assess estimation quality, as well as the influence of Coulomb friction disturbances on system behavior under fault conditions, the responses of the respective models were compared with the measured system response. It was concluded that although the continuous parameter estimation method is more demanding to implement (particularly in real-time applications), it proved equally effective in detecting multiplicative faults in the specific case studied. Furthermore, the influence of Coulomb friction as a disturbance was most pronounced in cases involving a fault in resistance R_1, where both the SWA and SIA algorithms yielded the best results.

To support fault isolation in case of SIA, a new decision algorithm was developed to enable accurate fault localization. In the case of continuous parameter estimation, fault identification—defined as the determination of both the faulty physical parameter and the magnitude of its change—is straightforward. The faulty parameter can be directly identified using the analytical relationships that define the amplifier model.

An alternative approach, applicable when fault detection is performed using the SWA method, involves indirect estimation of changes in physical parameters. This technique is based on evaluating the partial derivatives (i.e., sensitivity coefficients) of the model parameters with respect to the physical parameters at the nominal operating point. The method yields accurate results when the relationship between estimated model parameters and physical parameters is linear. However, in the more common case of nonlinear dependencies, the estimation becomes an approximation, valid primarily for small deviations around the nominal point. Consequently, its accuracy diminishes as the magnitude of the physical fault increases.

It was observed that better estimation results are obtained by selecting specific columns of the Jacobian matrix (i.e., the matrix of partial derivatives), rather than using the full matrix, as suggested in the literature. This approach proved especially effective in cases of nonlinear dependencies, even when the multiplicative faults are large and significantly deviate from the nominal operating point—representing one of the key contributions of this paper.

For the considered multiplicative faults, LabVIEW codes with partial real-time implementation were developed, using the Flat Sequence Frame structure. After signal acquisition, they are processed within LabVIEW MathScript Nodes for implementing the required parameter estimation algorithms. This results in practical solutions suitable for remote laboratories and training in the FDII.

Based on the above, it can be concluded that FDII techniques originally intended for LTI models, with appropriate modifications, can be successfully applied to certain classes of nonlinear dynamic systems, such as the analyzed permanent magnet DC motor with the corresponding amplifier.

Supported by the experimental results, which include precise fault detection, reliable fault isolation, and identification, it can be concluded that the presented findings are fully supported by the obtained data and the observed system behavior.

Feature work will be based on development of a fault-tolerant control (FTC) system that is robust against faults by incorporating a compensation mechanism between the faulty plant and the nominal controller. This compensation mechanism should allow the fault to be accommodated by the controller, ensuring that the faulty plant, under nominal control, remains globally stable and delivers satisfactory performance in both the transient and steady-state regimes.

Author Contributions: Conceptualization, S.A., M.R. and A.M.; methodology, S.A., M.R. and A.M.; software, S.A., B.K. and M.L.; validation, S.A. and M.R.; formal analysis, B.K., A.M. and M.L.; investigation, S.A. and M.R.; resources, B.K. and M.L.; data curation, S.A. and M.L.; writing—original draft preparation, S.A. and A.M.; writing—review and editing, M.R., B.K. and M.L.; visualization, S.A., M.R., B.K. and M.L.; supervision, A.M. All authors have read and agreed to the published version of the manuscript.

Funding: This study was supported by the Ministry of Science, Technological Development, and Innovation of the Republic of Serbia, and these results are part of the Grant No. 451-03-136/2025-03/200132, with University of Kragujevac—Faculty of Technical Sciences Čačak.

Institutional Review Board Statement: Not applicable.

Informed Consent Statement: Not applicable.

Data Availability Statement: The original contributions presented in this study are included in the article. Further inquiries can be directed to the corresponding author.

Conflicts of Interest: The authors declare no conflict of interest.

References

1. Park, Y.J.; Fan, S.K.S.; Hsu, C.Y. A review on fault detection and process diagnostics in industrial processes. *Processes* **2020**, *8*, 1123. [CrossRef]
2. Du, X. Fault detection using bispectral features and one-class classifiers. *J. Process Control* **2019**, *83*, 1–10. [CrossRef]
3. Gertler, J.J. *Fault Detection and Diagnosis in Engineering Systems*; CRC Press: Boca Raton, FL, USA, 2017. [CrossRef]
4. Mariano-Hernández, D.; Hernández-Callejo, L.; Zorita-Lamadrid, A.; Duque-Pérez, O.; Santos García, F. A review of strategies for building energy management system: Model predictive control, demand side management, optimization, and fault detect & diagnosis. *J. Build. Eng.* **2021**, *33*, 101692. [CrossRef]
5. Mercorelli, P. Recent advances in intelligent algorithms for fault detection and diagnosis. *Sensors* **2024**, *24*, 2656. [CrossRef]
6. Adouni, A.; Ben Hamed, M.; Flah, A.; Sbita, L. Sensor and actuator fault detection and isolation based on artificial neural networks and fuzzy logic applicated on induction motor. In Proceedings of the 2013 International Conference on Control, Decision and Information Technologies (CoDIT), Hammamet, Tunisia, 6–8 May 2013; pp. 917–922. [CrossRef]

7. Guzmán-Rabasa, J.A.; López-Estrada, F.R.; González-Contreras, B.M.; Valencia-Palomo, G.; Chadli, M.; Pérez-Patricio, M. Actuator fault detection and isolation on a quadrotor unmanned aerial vehicle modeled as a linear parameter-varying system. *Meas. Control* **2019**, *52*, 1228–1239. [CrossRef]
8. Debele, Y.; Shi, H.Y.; Wondosen, A.; Ku, T.W.; Kang, B.S. Deep learning-based robust actuator fault detection and isolation scheme for highly redundant multirotor UAVs. *Drones* **2023**, *7*, 437. [CrossRef]
9. Tirandaz, H.; Keliris, C.; Polycarpou, M. Actuator fault detection and isolation in a class of nonlinear interconnected systems. *Int. J. Control* **2024**, *97*, 2914–2934. [CrossRef]
10. Antic, S.; Djurovic, Z.; Kvascev, G. Application of structured and directional residuals for fault detection and isolation on permanent-magnet DC motor with amplifier. *Qual. Reliab. Eng. Int.* **2016**, *32*, 2601–2621. [CrossRef]
11. Antic, S.; Rosic, M.; Djurovic, Z.; Bozic, M. Comparison of structured residuals design techniques for actuator and sensor fault detection and isolation in a permanent magnet DC motor. *Electr. Eng.* **2024**, *106*, 2087–2105. [CrossRef]
12. Antić, S.; Luković, V.; Đurović, Ž. Expert system for FDI of DC motor faults using structured residuals design technique. *Serb. J. Electr. Eng.* **2023**, *20*, 93–105. [CrossRef]
13. Boulkroune, B.; Gálvez-Carrillo, M.; Kinnaert, M. Additive and multiplicative fault diagnosis for a doubly-fed induction generator. In Proceedings of the IEEE International Conference on Control Applications (CCA), Denver, CO, USA, 28–30 September 2011; pp. 1302–1308. [CrossRef]
14. Ding, S.X.; Frank, P.M.; Ding, E.L. An approach to the detection of multiplicative faults in uncertain dynamic systems. In Proceedings of the IEEE Conference on Decision and Control, Las Vegas, NV, USA, 10–13 December 2002; Volume 4, pp. 4371–4376. [CrossRef]
15. Huang, H.P.; Li, C.C.; Jeng, J.C. Multiple multiplicative fault diagnosis for dynamic processes via parameter similarity measures. *Ind. Eng. Chem. Res.* **2007**, *46*, 4517–4530. [CrossRef]
16. Zhu, C.C.; Li, L.; Ding, S.X. Multiplicative fault detection and isolation in dynamic systems using data-driven K-gap metric based KNN algorithm. *IFAC-Pap.* **2022**, *55*, 169–174. [CrossRef]
17. Ben Brahim, A.; Dhahri, S.; Ben Hmida, F.; Sellami, A. Multiplicative fault estimation-based adaptive sliding mode fault-tolerant control design for nonlinear systems. *Complexity* **2018**, *2018*, 1462594. [CrossRef]
18. Liu, X.Q.; Zhang, H.Y.; Liu, J.; Yang, J. Fault detection and diagnosis of permanent-magnet DC motor based on parameter estimation and neural network. *IEEE Trans. Ind. Electron.* **2000**, *47*, 1021–1030. [CrossRef]
19. Lu, L.; Wang, W. Fault diagnosis of permanent magnet DC motors based on multi-segment feature extraction. *Sensors* **2021**, *21*, 7505. [CrossRef]
20. Dobra, P.; Trușcă, M.; Sita, I.V.; Munteanu, R.A.; Munteanu, M. Model based approach in fault detection and diagnosis for DC motor. In Proceedings of the 8th IEEE Symposium on Diagnostics for Electrical Machines, Power Electronics and Drives (SDEMPED), Bologna, Italy, 5–8 September 2011; pp. 229–233. [CrossRef]
21. Liu, M.; Yang, X.; Cui, S. Online fault detection and diagnosis of permanent-magnetic DC motor. In Proceedings of the 2008 IEEE Vehicle Power and Propulsion Conference (VPPC), Harbin, China, 3–5 September 2008. [CrossRef]
22. Nandi, S.; Toliyat, H.A.; Li, X. Condition monitoring and fault diagnosis of electrical motors—A review. *IEEE Trans. Energy Convers.* **2005**, *20*, 719–729. [CrossRef]
23. Akin, B.; Choi, S.; Orguner, U.; Toliyat, H.A. A simple real-time fault signature monitoring tool for motor-drive-embedded fault diagnosis systems. *IEEE Trans. Ind. Electron.* **2011**, *58*, 1990–2001. [CrossRef]
24. Park, J.-K.; Seo, I.-M.; Hur, J. Fault type detection using frequency pattern of stator current in IPM-type BLDC motor under stator inter-turn, dynamic eccentricity, and coupled faults. In Proceedings of the 2013 IEEE Energy Conversion Congress and Exposition (ECCE), Denver, CO, USA, 15–19 September 2013; pp. 2516–2521. [CrossRef]
25. Saadaoui, W.; Jelassi, K. Induction motor bearing damage detection using stator current analysis. In Proceedings of the 2011 International Conference on Power Engineering, Energy and Electrical Drives (POWERENG), Málaga, Spain, 11–13 May 2011; pp. 1–6. [CrossRef]
26. Chan, C.W.; Hua, S.; Hong-Yue, Z. Application of fully decoupled parity equation in fault detection and identification of DC motors. *IEEE Trans. Ind. Electron.* **2006**, *53*, 1277–1284. [CrossRef]
27. Lee, J.H.L.J.; Lyou, J.L. Fault Diagnosis and Fault Tolerant Control of DC Motor Driving System. In Proceedings of the 2001 IEEE International Symposium on Industrial Electronics (ISIE 2001), Pusan, Republic of Korea, 8–11 July 2001; Volume 3, pp. 1719–1723. [CrossRef]
28. Najafabadi, T.A.; Salmasi, F.R.; Jabehdar-Maralani, P. Detection and Isolation of Speed-, DC-Link Voltage-, and Current-Sensor Faults Based on an Adaptive Observer in Induction-Motor Drives. *IEEE Trans. Ind. Electron.* **2011**, *58*, 1662–1672. [CrossRef]
29. Kargar, S.M.; Salahshoor, K.; Yazdanpanah, M.J. Integrated Nonlinear Model Predictive Fault-Tolerant Control and Multiple-Model-Based Fault Detection and Diagnosis. *Chem. Eng. Res. Des.* **2014**, *92*, 340–349. [CrossRef]
30. Jiang, Z.; Huang, X.; Cao, W. RLS-Based Algorithm for Detecting Partial Demagnetization under Both Stationary and Nonstationary Conditions. *Energies* **2022**, *15*, 3509. [CrossRef]

31. Kuric, M.; Lacevic, B.; Osmic, N.; Tahirovic, A. RLS-Based Fault-Tolerant Tracking Control of Multirotor Aerial Vehicles. In Proceedings of the IEEE/ASME International Conference on Advanced Intelligent Mechatronics (AIM), Munich, Germany, 24 August 2017; pp. 1148–1153. [CrossRef]
32. Leite, D.F.; Hell, M.B.; Diez, P.H.; Ganglio, B.S.L.; Nascimento, L.O.; Costa, P. Real-time model-based fault detection and diagnosis for alternators and induction motors. In Proceedings of the IEEE International Electric Machines and Drives Conference (IEMDC), Miami, FL, USA, 3–5 May 2007; Volume 1, pp. 202–207. [CrossRef]
33. Menini, L.; Possieri, C.; Tornambè, A. Application of algebraic geometry techniques in permanent magnet DC motor fault detection and identification. *Eur. J. Control* **2015**, *25*, 39–50. [CrossRef]
34. Fernández Gómez, A.J.; Jaramillo, V.H.; Ottewill, J.R. Fault Detection in Electric Motors by Means of the Extended Kalman Filter as Disturbance Estimator. In Proceedings of the 2014 UKACC International Conference on Control (CONTROL 2014), Loughborough, UK, 9–11 July 2014; pp. 432–437. [CrossRef]
35. Miniach, M.; Orlowska-Kowalska, T. Innovative Extended Kalman Filter in a Coherent System for Detection and Compensation of Various Current Sensor Faults in the Induction Motor Drive. *IEEE Trans. Ind. Electron.* **2025**, *72*, 7772–7784. [CrossRef]
36. Xiong, K.; Chan, C.W.; Zhang, H.Y. Unscented Kalman Filter for Fault Detection. *IFAC Proc. Vol.* **2005**, *38*, 113–118. [CrossRef]
37. Karami, F.; Poshtan, J.; Poshtan, M. Model-Based Fault Detection in Induction Motors. In Proceedings of the IEEE International Conference on Control Applications (CCA), Yokohama, Japan, 8–10 September 2010; pp. 1957–1962. [CrossRef]
38. Rao, C.R.; Toutenburg, H. *Linear Models: Least Squares and Alternatives*; Springer: New York, NY, USA, 1999. [CrossRef]
39. Isermann, R. *Fault-Diagnosis Applications: Model-Based Condition Monitoring; Actuators, Drives, Machinery, Plants, Sensors, and Fault-Tolerant Systems*; Springer: Berlin/Heidelberg, Germany, 2011. [CrossRef]
40. de Miguel, L.J.; Blázquez, L.F. Fuzzy Logic-Based Decision-Making for Fault Diagnosis in a DC Motor. *Eng. Appl. Artif. Intell.* **2005**, *18*, 423–450. [CrossRef]
41. Lee, C.Y.; Maceren, E.D.C. Induction motor bearing fault classification using deep neural network with particle swarm optimization-extreme gradient boosting. *IET Electr. Power Appl.* **2024**, *18*, 297–311. [CrossRef]
42. Toma, R.N.; Prosvirin, A.E.; Kim, J.M. Bearing fault diagnosis of induction motors using a genetic algorithm and machine learning classifiers. *Sensors* **2020**, *20*, 1884. [CrossRef]
43. Patton, R.J.; Frank, P.M.; Clark, R.N. (Eds.) *Issues of Fault Diagnosis for Dynamic Systems*; Springer: London, UK, 2000.
44. Isermann, R. *Fault-Diagnosis Systems: An Introduction from Fault Detection to Fault Tolerance*; Springer: Berlin/Heidelberg, Germany, 2006.
45. Bjekic, M.; Antic, S.; Milovanovic, A. Permanent Magnet DC Motor Friction Measurement and Analysis of Friction's Impact. *Int. Rev. Electr. Eng. (IREE)* **2011**, *6*, 2261–2269.
46. Karnopp, D. Computer Simulation of Stick-Slip Friction in Mechanical Dynamic Systems. *J. Dyn. Syst. Meas. Control* **1985**, *107*, 100–103. [CrossRef]
47. Ljung, L. *System Identification: Theory for the User*, 2nd ed.; Prentice Hall: Upper Saddle River, NJ, USA, 1999.
48. Gholami, A.; Haddadnia, J.; Ghasemi, J.B. A Sliding Window Approach for Signal Processing in Fault Detection. *Signal Process.* **2013**, *93*, 2760–2771. [CrossRef]
49. Zhao, J.; Wang, X.; Gao, G.; Na, J.; Liu, H.; Luan, F. Online Adaptive Parameter Estimation for Quadrotors. *Algorithms* **2018**, *11*, 167. [CrossRef]
50. Yang, B.; Wang, J.; Sun, Y. Fault Detection and Diagnosis Based on Sliding Mode Integral Algorithm for Nonlinear Systems. *Appl. Sci.* **2020**, *10*, 1234. [CrossRef]
51. Antić, S.; Đurović, Ž. Application of Sliding Integral Algorithm for Parametric Fault Detection of a DC Motor Amplifier. In Proceedings of the 2nd International Conference on Electrical, Electronic and Computing Engineering (IcETRAN 2015), Silver Lake, Serbia, 8–11 June 2015; pp. AUI1.5.1–AUI1.5.6, ISBN 978-86-80509-71-6.
52. Sagara, S.; Zhao, Z.Y. Numerical Integration Approach to On-Line Identification of Continuous-Time Systems. *Automatica* **1990**, *26*, 63–74. [CrossRef]

MDPI AG
Grosspeteranlage 5
4052 Basel
Switzerland
Tel.: +41 61 683 77 34

Applied Sciences Editorial Office
E-mail: applsci@mdpi.com
www.mdpi.com/journal/applsci

www.ingramcontent.com/pod-product-compliance
Lightning Source LLC
Chambersburg PA
CBHW061508210926
51850CB00008B/1855
* 9 7 8 3 7 2 5 8 8 0 0 1 0 *

Strategies to Improve the Security and Nutritional Quality of Crop Species

Strategies to Improve the Security and Nutritional Quality of Crop Species

Guest Editors

Grażyna Podolska
Anna Szafrańska

MDPI

Basel • Beijing • Wuhan • Barcelona • Belgrade • Novi Sad • Cluj • Manchester

Guest Editors
Grażyna Podolska
Institute of Soil Science and Plant Cultivation—State Research Institute
Puławy
Poland

Anna Szafrańska
Department of Grain Processing and Bakery
Prof. Wacław Dąbrowski Institute of Agricultural and Food Biotechnology—State Research Institute
Warsaw
Poland

Editorial Office
MDPI AG
Grosspeteranlage 5
4052 Basel, Switzerland

This is a reprint of the Special Issue, published open access by the journal *Agriculture* (ISSN 2077-0472), freely accessible at: https://www.mdpi.com/journal/agriculture/special_issues/04DW5K15VU.

For citation purposes, cite each article independently as indicated on the article page online and as indicated below:

Lastname, A.A.; Lastname, B.B. Article Title. *Journal Name* **Year**, *Volume Number*, Page Range.

ISBN 978-3-7258-5831-6 (Hbk)
ISBN 978-3-7258-5832-3 (PDF)
https://doi.org/10.3390/books978-3-7258-5832-3

Contents

agriculture

MDPI

ticle

'he Influence of *Fusarium culmorum* on the Technological Value f Winter Wheat Cultivars

lyta Aleksandrowicz [1], Krzysztof Dziedzic [2], Anna Szafrańska [3,*] and Grażyna Podolska [1,*]

1 Institute of Soil Science and Plant Cultivation—State Research Institute, Czartoryskich 8, 24-100 Puławy, Poland
2 Department of Food Technology of Plant Origin, Poznań University of Life Sciences, Wojska Polskiego 28, 60-637 Poznań, Poland; krzysztof.dziedzic@up.poznan.pl
3 Prof. Wacław Dąbrowski Institute of Agricultural and Food Biotechnology—State Research Institute, 36 Rakowiecka St., 02-532 Warsaw, Poland
* Correspondence: anna.szafranska@ibprs.pl (A.S.); grazyna.podolska@iung.pulawy.pl (G.P.)

Abstract: The research hypothesis assumes that *Fusarium culmorum* infection affects the baking value of wheat. The aim of the research was to determine the effect of the cultivar on the rheological properties of wheat dough in response to *Fusarium culmorum* infection of wheat. A two-factor experiment conducted during the 2018–2020 growing seasons in Osiny, Poland, was set up using the completely randomized block design with three replications. The first factor was winter wheat cultivars (six cultivars), while the second factor was inoculation (two levels—*Fusarium culmorum* and distilled water—control). The immunoenzymatic ELISA method was used to determine the content of deoxynivalenol (DON) in grain. The DON content in the grain varied between cultivars. *Fusarium culmorum* inoculation resulted in an increase in protein, ash content, and flour water absorption, changes in dough rheological properties, and a decrease in the sedimentation index. Inoculation also caused negative changes in starch properties. The observed interaction between *Fusarium culmorum* inoculation and cultivars in shaping the qualitative parameters and rheological properties of the dough indicates that there are wheat cultivars less susceptible to *Fusarium* infection, which do not show any significant changes as a result of infection.

Keywords: artificial inoculation; food safety; *Fusarium culmorum*; grain quality; deoxynivalenol; rheological properties; wheat

1. Introduction

Wheat is one of the most important crops in the world, ranking as the third largest in terms of production, after sugar cane and maize. In 2022, global wheat production reached 808 million tons, cultivated on an area of 219.2 million hectares [1]. The data indicate that over 50% of wheat harvests are used by food enterprises for production of flour, bread, breakfast cereals, pasta, and others [2]. Wheat intended for bread production must meet specific quality indicators that determine its suitability for flour production and baking. For this purpose, the grain undergoes various analyses, which provide information on the quantity and quality of gluten, the quality of the protein complex (sedimentation index), the activity of amylolytic enzymes (falling number), and the rheological properties of the dough [3].

The raw material delivered to flour mills must not only meet the requirements for baking quality but also be completely safe for human health. The quality of the raw material, collected from the field, is influenced by the presence of physical, chemical, or biological contaminants. Chemical contaminations, such as pesticide residues, heavy metals, and nitrites, are not acceptable in final food products and feed. Grain infection by *Fusarium* spp. causing *Fusarium* head blight (FHB) may be indicated by high levels of mycotoxins.

Mycotoxins are a group of harmful compounds, which are secondary metabolites of filamentous fungi. These substances can have strong toxic, mutagenic, or teratogenic effects. Secondary metabolites are organic compounds produced by organisms, such as plants, fungi, and bacteria, that are not directly involved in their normal growth, development, or reproduction. The presence of mycotoxins in food poses a potential threat to human health. It is estimated that about 25% of global cereal production is contaminated with these metabolites, leading to a deterioration in the health quality of grain and its elimination from consumption and feed [4]. To ensure consumer safety, the European Commission has established unified standards and legal regulations regarding the maximum levels of some of the most important mycotoxins in cereal grains and products. The maximum level of DON is set at 1000 $\mu g\ kg^{-1}$ for unprocessed cereals, except for durum wheat with a limit of 1500 $\mu g\ kg^{-1}$ and oats with a limit up to 1750 $\mu g\ kg^{-1}$ [5]. For zearalenone (ZEN), the maximum level for these products is 100 $\mu g\ kg^{-1}$. The maximum levels of mycotoxins in cereal-based products are lower. The DON level in flour, pasta, and bran must not exceed 600 $\mu g \cdot kg^{-1}$, in bread 400 $\mu g \cdot kg^{-1}$, and in products for children 150 $\mu g \cdot kg^{-1}$. Similarly, the ZEN level in flour and bran must not exceed 75 $\mu g \cdot kg^{-1}$, and in bread it must not exceed 50 $\mu g \cdot kg^{-1}$ [6].

Wheat cultivation is primarily aimed at producing grain for consumption, especially for flour and bread production. Wheat grain is rich in various components that differ in molecular structure, physicochemical, and functionality. Its dry mass consists of 60–70% starch, 10–17% proteins, 3–6% sugars, 2–2.5% lipids, vitamin E, B group vitamins (riboflavin, niacin, thiamine), dietary fiber, 1.4–2.3% minerals (Cu, Mg, Zn, P, and Fe), and other phytochemical compounds [7,8]. The chemical composition of wheat grain is influenced by environmental conditions, agricultural practices, and genotype [7].

Fusarium fungal infection of grain can lead to biochemical changes and structural alteration in the grain [9]. The extent of these changes depends on the level of fungal invasion [10]. Infected grain may have fungal spores on its surface, and its internal structure may contain mycotoxins. Infested grain shows a reduction in the thickness of the aleurone layer, altering the proportion of seed coat to endosperm, which results in a decrease in thousand kernel weight and an increase in ash content in whole-grain flour [11–13]. A positive correlation between the degree of *Fusarium culmorum* infection and the ash content, which includes potassium and calcium salts in wheat grain, was previously reported [14,15].

Increased activity of amylase, protease, and other enzymes responsible for the degradation of non-starch polysaccharides in infected samples was also observed [16–19]. Fungal proteases produced by *Fusarium culmorum* are active in a wide range of temperatures (10–100 °C) and pH values (4.5–8.5), potentially impairing dough functionality and bread quality throughout processing. Kreuzberger et al. [17], in an in vitro study, observed a clear effect of *Fusarium graminearum* proteases on glutenin and gliadin fractions, with a higher preference for glutenins, which are responsible for dough elasticity and viscosity. Fusarium-infected ears, during kernel growth, may show little visible damage or macromolecules inside the kernel but still contain high levels of DON [20]. The fungal infection may destroy the cellular structures of the starchy endosperm, damaging starch granules. These compositional may reduce flour quality, resulting in a low falling number [21]. The presence of *Fusarium* leads to starch degradation, which correlates with the activity of α-amylase in the grain and a decrease in the falling number value [9,22,23].

Starch damage caused by *Fusarium* may contribute to increased water absorption in the samples [18]. The quality of wheat flour primarily depends on gluten proteins, which play an important role in carbon dioxide retention, dough development, and baking quality [24]. *Fusarium* fungi and their toxins can affect both protein quantity and disturb the mutual proportions of individual protein fractions, thereby impacting the quality of the raw material and the final product [25]. In some cultivars (Greina, Lona, and Brusino), the sedimentation index decreases after *Fusarium* infection, indicating an effect on the protein quantity and quality, while no such negative changes were observed in others. Studies indicate that *Fusarium* decreases gluten quantity and quality, as evidenced by the reduction in the Zeleny sedimentation index in *Fusarium*-infected grain compared to healthy seeds [14,21]. The reduction in the quality of gluten proteins due to *Fusarium* inoculation also affected the rheological properties of wheat flour dough. The increase in protease activity found in wheat flour infected by *Fusarium* had a significant effect on the increase in dough softening and decrease in dough stability of flour [13,14,18]. However, the *Fusarium* infection did not influence the water absorption of flour [14,18]. The deterioration of qualitative characteristics and rheological properties of the dough made by infected flour was also noted previously. *Fusarium* infection also negatively impacted dough energy, maximum tensile strength, and the strength to extensibility ratio, reducing these parameters significantly. Additionally, some authors observed a 20-fold increase in the gliadin to glutenin ratio in flour samples with a DON content of 3.98 mg·kg^{-1}, resulting in reduced bread volume [13,26]. Contrary to the abovementioned results, other authors noted an increase of loaf volume under *Fusarium* infection regarding the dose of *Fusarium* infection and different bread production technologies [18,27]. Limited studies using Mixolab, which allows for the assessment of the protein–starch complex in a single test, showed that *Fusarium* spp. infection of the grain led to a reduction in the rheological properties of dough. Other research showed that an increasing intensity of *Fusarium* spp. contamination had negative effects on protein and mainly on the starch part of grain which was noted on the Mixolab curves [13]. The authors of [13,28] suggested that the cause may be the degree of starch damage granules in the inoculated samples, which affects the dough characteristics, making it stickier and potentially resulting in poor baking quality.

The research hypothesis posits that *Fusarium* infection and mycotoxins in grain and flour affect wheat's baking quality, causing changes in both starch and protein structure. These effects are cultivar-dependent. Therefore, the aim of this research is to determine the impact of cultivar on the rheological quality of wheat dough in response to stress caused by *Fusarium* infection of wheat.

2. Materials and Methods

A two-factor experiment was established using the completely randomized block method in three replications. The factors were winter wheat cultivars and spray. In the experiment, six winter wheat cultivars were used, with each cultivar being treated with either *Fusarium culmorum* inoculation or distilled water (as a control sample).

Plant material. The research work concerned winter wheat cultivars representing two technological groups approved by the Polish National List of Agricultural Plant Varieties issued yearly by COBORU (the Research Centre for Variety Testing). The individual cultivars differed in morphological structure, utility features, genotype and place of cultivation (Table 1). Group A comprises quality bread cultivars, namely: 'Legenda', 'Pokusa', 'Tonacja', and 'Sailor'. Group B comprises the following bread cultivars: 'Muszelka' and 'KWS Ozon'.

The cultivars differed in earliness and the earing stage; the fastest were cv. 'Muszelka', 'Pokusa', 'Sailor', and the latest was cv. 'Tonacja'. The maturity stage was reached the fastest by cv. 'Muszelka' and 'Pokusa' (202 days from 1 January), and the latest by: cv.

'KWS Ozon' and 'Sailor' (203 days from 1 January). The cultivars differed in resistance to *Fusarium* head blight (FHB). Resistance is estimated by the Central Research Centre for Cultivar Testing on a nine-point scale, where 1 means susceptible and 9 completely resistant. The cv. 'Legenda' was characterized by the highest resistance (7.9) and the cv. 'Muszelka' (6.5) showed the highest susceptibility. Moreover, the cultivars differed in height. The height difference was 35 cm. The tallest cultivar was cv. 'Legenda' (115 cm) and the shortest was cv. 'Muszelka' (80 cm) (Table 1).

Table 1. Characteristics of winter wheat cultivars.

Cultivar	Breeding Place	Resistance to FHB *	Height (cm)	Wheat Quality Group	Flowering, Maturing (Days from 1st January)
KWS Ozon	KWS Lochow GmbH	7.1	83	B	151 203
Legenda	Poznańska Plant Breeding	7.9	115	A	151 203
Muszelka	DANKO Plant Breeding	6.5	80	B	150 202
Pokusa	Strzelce Plant Breeding	7.9	97	A	150 202
Tonacja	Strzelce Plant Breeding	7.8	104	A	152 204
Sailor	DANKO Plant Breeding	7.7	101	A	150 203

* FHB—Fusarium head blight.

Field trials. The field experiment was conducted in 2018/2019 and 2019/2020 growing seasons at the Experimental Station in Osiny (51°35′, 21°55′), belonging to the Institute of Soil Science and Plant Cultivation—State Research Institute (IUNG-PIB) Puławy, Poland. The experiment was established on pseudopodzolic soil, class IIIb, agricultural soil suitability complex—very good rye complex, with a pH of 6.1. The mineral content in the soil was: extractable phosphorus (P: 12.6 mg kg^{-1}), exchangeable potassium (K: 15.3 mg kg^{-1}), magnesium (Mg 4.9 mg kg^{-1}). The forecrop was winter rape. The plot size was 24 m^2, 11.25 m^2 was designed for harvest. In autumn, after harvesting the forecrop, pre-sowing of crops was carried out—cultivation with a disc cultivator, ploughing and cultivation with a seed drill. Wheat sowing was performed on optimal dates in subsequent years of the experiment: 2 October 2018 and 7 October 2019, respectively. Sowing density was 450 grains per m^2. The plots were fertilized with: N (160 kg ha^{-1}), P (70 kg ha^{-1}), and K (105 kg ha^{-1}). Nitrogen fertilizer was applied three times in individual vegetation stages: beginning of vegetation (BBCH 23)—70 kg N ha^{-1}, stem shooting (BBCH 32)—60 kg N ha^{-1}, full earing (BBCH 53)—30 kg N ha^{-1}. Plants were protected against weeds, diseases, and pests using Bizon 400 L (active substance: tralkoxydim) 1.0 l·ha^{-1}; Mustang Forte 195 SE (active substance: florasulam, aminopyralid, 2.4D) 0.5 l ha^{-1}; Unix 75WG (active substance: cyprodinil) 0.6 kg ha^{-1}; Fury 100 EC (active substance: zeta-cypermethrin) 0.1 L ha^{-1}. Harvesting was carried out at full maturity on 20 July 2019 and 1 August 2020.

Weather conditions. The weather conditions during the wheat growing period, including the heading (May), flowering (June), and maturity (July) stages, varied across different growing seasons (Table 2). In 2019, air temperature was higher in May and June by 1.8 °C and 3.3 °C, respectively compared to the 2020 harvest year. Precipitation in the mentioned months of 2019 was lower by 26.6 mm, 150.8 mm, and 15.9 mm, respectively, compared to 2020. The high air temperature in June and July, together with very low precipitation, led to the occurrence of drought. These were unfavorable conditions for the growth and development of *Fusarium culmorum*.

Table 2. Weather conditions in the growing seasons.

Month	Temperature (°C)		Precipitation (mm)	
	Growing Season			
	2019	**2020**	**2019**	**2020**
March	5.5	4.5	22.7	25.3
April	9.6	8.5	35.5	11.9
May	12.9	11.1	86.1	112.7
June	21.7	18.4	38.7	189.5
July	18.6	18.6	33.9	49.8

Preparation of Fusarium culmorum inoculum and inoculation in the field. The inoculation suspension was obtained from grain infected with *Fusarium culmorum* spores, prepared at the Plant Breeding and Acclimatization Institute (IHAR)—National Research Institute in Radzików, Poland. To produce the inoculum, a *Fusarium culmorum* isolate forming deoxynivalenol was prepared according to the methodology followed in [29]. The isolate was incubated on autoclaved wheat grain in glass flasks for approximately 4 weeks and then irradiated with continuous UV light for 4 to 7 days at a temperature of about 18 °C. Then the grain infected with *Fusarium culmorum* was dried and stored at 4 °C until the inoculates were prepared for spraying. On the day of inoculation, grain with mycelium and conidial spores of *Fusarium culmorum* was soaked in sterile tap water for two hours and then filtered through gauze to obtain a spore suspension without mycelium fragments. After thoroughly mixing the spore suspension, the number of colony-forming units (CFU) was determined using a Thom counting chamber. The field experiment used artificial inoculation by spraying with *Fusarium culmorum*. It was performed during the flowering phase of the wheat cultivars (BBCH 65). The plants were sprayed manually with a suspension of spores prepared in the laboratory at a concentration of 5×10^5 CFU·mL^{-1} at 100 mL per 1 m^2. The inoculation method used imitated the natural conditions of ear infection by *Fusarium* and allowed for determining type I resistance (resistance to infections) and type II resistance (resistance to the spread of the pathogen in tissues). Inoculations were performed in the late afternoon hours, when the temperature decreased and the relative air humidity increased. The experiment included a control, on which spraying with distilled water was used during the flowering period in order to maintain similar humidity conditions to the inoculated field.

Determination of deoxynivalenol in wheat grain. The immunoenzymatic ELISA method was used to determine the content of DON in wheat grains, using the AgraQuant test from RomerLabs® (Getzersdorf, Austria). The optical density of the samples was measured using a Stat Fax 303 Plus microplate reader (photometer) (Awareness Technology, Inc., Palm City, FL, USA) at a wavelength of 650 nm. According to the recommendation, the results were read within 20 min of adding the Red Stop solution. The standard curve was prepared by using a standard (optical density of mycotoxin concentration). For each analysis the correlation coefficient R (dependence of absorbance on A/c concentration) was obtained. According to recommendations, the results are considered reliable when the value of the R coefficient is not less than 0.996. Representative samples of grain were ground in a Laboratory Mill 3100 by Perten and stored at 2–8 °C until extraction. After adding the extractant (distilled water), the samples were shaken vigorously for 3 min. The extracts were filtered by passing at least 5 mL through a Whatman 1 filter. The estimation was carried out in accordance with AgraQuant standards in duplicate. The contents of DON in inoculated and uninoculated samples (2019) and the control samples (2020) were below the detection limit (LOD < 200 µg kg^{-1}), therefore, these values are not presented in the manuscript.

Technological quality analysis. The quality analysis was determined after harvest. The physical and chemical properties were tested in the samples obtained only from the 2020 harvest year. In 2019 *Fusarium* infection was not observed at a level that would affect the technological properties of the grain, therefore the date from this year was abandoned. The thousand kernel weight was determined according to ISO 520:2010 [30]. A grinder (FN 3100, Perten Instruments AB, Hägersten, Sweden) was used for preparation of whole-grain flours, and next the following quality parameters were marked: moisture content (determined according to ISO 712:2009) [31], protein content (ISO 20483:2013) [32], gluten quantity (ISO 21415-2:2015) [33], Zeleny sedimentation index (ISO 5529:2007) [34], and ash content (ISO 2171:2023) [35] were determined in order to evaluate the technological quality of tested wheat cultivars. Wheat flour samples for the Zeleny sedimentation test were prepared by using a Sedimat laboratory mill (Brabender GmbH & Co. KG, Duisburg, Germany).

Rheological properties of a dough. Mixolab ChopinWheat+ Protocol according to ISO 17718:2013 [36] (KPM Analytics, Villeneuve-la-Garenne, France) was used to analyze the rheological properties of a dough. Grain was ground by an FN 3100 grinder (Perten InstrumentsAB, Hägersten, Sweden) to obtain whole-grain flour to perform the Mixolab test. The whole-grain flour samples (50 g) based on 14% moisture content were used to determine the water absorption of flour at the consistency of 1.1 ± 0.05 Nm. The mixing speed was 80 rpm, and bowl temperature was set at 30 °C. Protein properties related to water absorption, dough development time, stability, and weakening (C2, C1–C2, slope α) were determined. The starch properties correspond to starch gelatinization during the increase in the temperature from 30 to 90 °C (C3, C3–C2, slope β) and enzymatic activities (C4, C3–C4, slope γ) and retrogradation (C5, C5–C4) were also evaluated. The following Mixolab Profiler indexes were evaluated to better characterize the baking quality potential of tested cultivars: water absorption, mixing, gluten+, viscosity, amylase, and retrogradation.

Statistical analyses. The results were statistically analyzed by the one-way analysis of variance (ANOVA) with a subsequent Tukey's HSD test with the significance level of $p < 0.05$ to compare the means of control treatments and inoculations with *Fusarium culmorum*. The second factor was the wheat cultivar used. Principal component analysis (PCA) was performed to reduce the dimensionality of the data and represent the samples in a new coordinate system. All results were analyzed using Statistica software (v. 13.3, StatSoft, Tulsa, OK, USA).

3. Results

Wheat grain samples from the 2020 harvest year showed significant infection by *Fusarium* spp. and were also characterized by relatively high DON content. Variation in DON levels and basic quality parameters were observed. These characteristics are essential for assessing their suitability for grain processing for consumption. The results of the analysis of variance (Table 3) indicated that both the cultivars and *Fusarium* inoculation significantly affected the measured technological parameters of wheat grain. However, *Fusarium* inoculation had no significant effect on the gluten content. Table 3 does not include the results of wheat cultivars harvested in 2019 because no ear infection was observed, which was confirmed by DON determinations below the LOD. Additionally, no difference was found in the amount of gluten and the Zeleny sedimentation index between inoculated and control samples.

Table 3. Significance of effect in ANOVA for wheat samples from the 2020 harvest.

Parameter/Factor	A (Cultivar)	B (Inoculation)	A × B	V (%)
Thousand kernel weight	***	***	*	6.28
Protein content	***	**	***	3.29
Gluten content	***	n.s.	***	4.43
Zeleny sedimentation index	***	***	***	4.24
Ash content	***	***	***	7.55
Water absorption	***	***	***	3.49
Time T1	***	***	***	4.45
Stability	***	***	***	5.36
Slope α	***	***	***	3.85
Protein weakening, C2	**	***	**	5.63
C1–C2	***	***	***	4.40
Slope β	***	***	***	4.02
Slope γ	***	***	***	7.92
C3	***	***	***	3.44
C4	***	***	***	2.54
C5	***	***	***	4.83
C3–C2	***	***	***	3.47
C3–C4	***	***	***	2.54
C5–C4	***	***	***	3.12

Abbreviation: *—significant at $p < 0.05$; **—significant at $p < 0.01$; ***—significant at $p < 0.001$; n.s.—non-significant; V—coefficient of variation.

3.1. Deoxynivalenol Content in Wheat Grain

The deoxynivalenol (DON) content of individual grain wheat cultivars varied across different years. Grain samples harvested in 2019 were characterized by a DON level below the limit of detection (LOD) of 200 $\mu g \cdot kg^{-1}$. In 2020, the grain of the tested wheat cultivars contained higher levels of DON compared to 2019. The lowest amounts of DON were accumulated in the cv. 'Sailor' (6847 $\mu g \cdot kg^{-1}$), cv. 'Pokusa' (7477 $\mu g \cdot kg^{-1}$), and cv. 'Legenda' (8790 $\mu g \cdot kg^{-1}$), while the highest in cv. 'Muszelka' (11,021 $\mu g \cdot kg^{-1}$), cv. 'KWS Ozon' (11,033 $\mu g \cdot kg^{-1}$), and cv. 'Tonacja' (11,513 $\mu g \cdot kg^{-1}$) (Figure 1). In 2020, in uninoculated samples (control), DON content was below limit of detection (200 $\mu g \cdot kg^{-1}$).

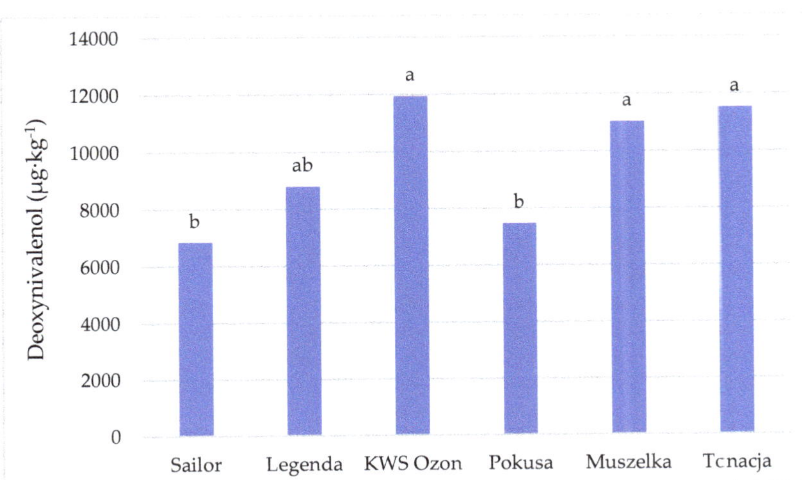

Figure 1. Deoxynivalenol (DON) content in wheat grain inoculated with *Fusarium culmorum* in 2020. The same letters indicate no statistically significant differences between the data, $p < 0.05$.

3.2. Quality Characteristics of Wheat Cultivars

The effect of *Fusarium culmorum* inoculation on technological features of grain is presented in Table 4. Thousand kernel weight (TKW) is related to grain yield, milling efficiency, and seedling vigor. TKW differed significantly between control samples of the cultivars, ranging from 38.93 g (cv. 'Legenda') to 47.51 g (cv. 'KWS Ozon'). *Fusarium* inoculation caused a significant reduction in TKW in all cultivars, except for cv. 'KWS Ozon' (Table 4).

Table 4. The quality characteristics of winter wheat cultivars control and *F. culmorum*-infected samples.

Parameter	Inoculation	Cultivar						Mean
		Sailor	Legenda	KWS Ozon	Pokusa	Muszelka	Tonacja	
Thousand kernel weight (g)	Control	42.15 ^{a}	38.93 ^{a}	47.51 ^{a}	46.92 ^{a}	46.84 ^{a}	41.67 ^{a}	44.00 ^{B}
	Fusarium	38.52 ^{b}	33.24 ^{b}	43.52 ^{a}	43.73 ^{b}	42.15 ^{b}	36.17 ^{b}	39.56 ^{A}
	Average	40.3 ^{AB}	36.08 ^{B}	45.51 ^{A}	45.32 ^{A}	44.49 ^{A}	38.92 ^{B}	
Protein content (%)	Control	13.9 ^{b}	12.9 ^{b}	13.1 ^{a}	13.4 ^{b}	13.3 ^{b}	11.9 ^{a}	13.1 ^{B}
	Fusarium	14.8 ^{a}	16.4 ^{a}	13.2 ^{a}	14.3 ^{a}	14.1 ^{a}	12.4 ^{a}	14.2 ^{A}
	Average	14.4 ^{A}	14.6 ^{A}	13.1 ^{D}	13.9 ^{B}	13.6 ^{C}	12.1 ^{E}	
Gluten content (%)	Control	29.4 ^{b}	35.6 ^{a}	28.6 ^{a}	29.6 ^{b}	28.5 ^{a}	28.8 ^{a}	30.08 ^{A}
	Fusarium	30.5 ^{a}	34.2 ^{b}	27.4 ^{a}	31.8 ^{a}	28.7 ^{a}	30.0 ^{a}	30.43 ^{A}
	Average	30.0 ^{A}	34.9 ^{A}	28.0 ^{B}	30.7 ^{A}	28.6 ^{B}	29.4 ^{A}	
Zeleny sedimentation index (cm^3)	Control	48 ^{a}	45 ^{a}	52 ^{a}	55 ^{a}	47 ^{a}	46	48.8 ^{A}
	Fusarium	45 ^{a}	51 ^{a}	34 ^{b}	46 ^{b}	42 ^{a}	44 ^{a}	43.6 ^{B}
	Average	47.0 ^{A}	48.0 ^{A}	43.0 ^{A}	50.7 ^{A}	44.5 ^{A}	45.5 ^{A}	
Ash content (%)	Control	1.42 ^{b}	1.48 ^{a}	1.51 ^{b}	1.27 ^{b}	1.41 ^{b}	1.45 ^{b}	1.42 ^{B}
	Fusarium	1.56 ^{a}	1.59 ^{a}	1.91 ^{a}	1.71 ^{a}	1.71 ^{a}	1.58 ^{a}	1.68 ^{A}
	Average	1.49 ^{B}	1.53 ^{B}	1.71 ^{A}	1.49 ^{B}	1.56 ^{B}	1.52 ^{B}	

$^{a,b,A-E}$—homogeneous groups, statistically significant differences at $p = 0.05$.

The protein content significantly depended on the cultivar. An interaction between *Fusarium culmorum* inoculation and cultivar was observed for protein content. Wheat samples inoculated with *Fusarium* had protein content in the range of 12.4% (cv. 'Tonacja') to 16.4% (cv. 'Legenda'). The protein content in non-inoculated samples ranged from 11.9 (cv. 'Tonacja') to 13.9 (cv. 'Sailor'). *Fusarium* inoculation caused a significant increase in protein content in grain, except for cv. 'KWS Ozon', where no effect was observed. The greatest increase in protein content was found in cv. 'Legenda', where the difference between inoculated and control samples was 3.5 percentage points. The difference for cv. 'Sailor', 'Pokusa', and 'Muszelka' ranged from 0.8 to 0.9 percentage points.

The gluten content in control wheat samples ranged from 28.6% (cv. 'Muszelka') to 35.6% (cv. 'Legenda') (Table 3). *Fusarium* inoculation caused a significant increase in gluten content in cv. 'Sailor' and cv. 'Pokusa' by 1.1, and 2.2 percentage points, respectively. In cv. 'Muszelka', 'Tonacja', and 'KWS Ozon', no significant differences were observed compared to the control samples.

The Zeleny sedimentation index values for flour from wheat grain samples ranged from 45 cm^3 (cv. 'Legenda') to 55 cm^3 (cv. 'Pokusa') (Table 3). *Fusarium* inoculation reduced the sedimentation index value on average by 5.2 cm^3 across the cultivars. The highest reduction was observed in cv. 'KWS Ozon' (18 cm^3) and cv. 'Pokusa' (9 cm^3), while in cv. 'Legenda', there was an increase of 6 cm^3.

Ash content in grain was influenced by both the cultivar and *Fusarium* inoculation (Table 3). Wheat samples were characterized by ash content in the range of 1.27% (cv. 'Pokusa') to 1.51% (cv. 'KWS Ozon'). Infected grain showed significantly higher ash content compared to the control samples, with differences ranging from 0.11 percentage points (cv. 'Legenda') to 0.44 percentage points (cv. 'Pokusa').

3.3. Rheological Properties of the Protein Complex of Tested Wheat Cultivars

The rheological properties of whole-grain flour dough were studied using the Chopin-Wheat+ protocol of Mixolab (Chopin Technologies), which allows for the determination of both protein and starch properties of wheat flour in a single test.

Determination of flour water absorption is an important criterion for wheat flour, as it provides information about baking absorption, which is related to bread yield and quality of bakery products. In the presented studies, the water absorption of the flour samples from the cultivars ranged from 60.5% (cv. 'Sailor') to 62.9% (cv. 'Tonacja'). A significant increase in this indicator was observed in all inoculated cultivars compared to the control samples. Flour obtained from grain infected with *Fusarium* showed higher water absorption, on average by 3.6 percentage points, compared to control grain samples (Table 5). The highest increases in this parameter were observed for cv. 'Pokusa' and cv. 'Muszelka' (4.6 and 5 percentage points, respectively).

Table 5. The rheological properties of dough obtained from winter wheat cultivar control and *F. culmorum*-infected samples.

Parameter	Inoculation	Cultivar						Mean
		Sailor	Legenda	KWS Ozon	Pokusa	Muszelka	Tonacja	
Water absorption (%)	Control	60.5 ^{b}	61.4 ^{b}	62.6 ^{b}	62.1 ^{a}	61.8 ^{b}	62.9 ^{b}	61.9 ^{B}
	Fusarium	64.6 ^{a}	63.4 ^{a}	66.4 ^{a}	66.7 ^{b}	66.8 ^{a}	64.9 ^{a}	65.5 ^{A}
	Average	62.6 ^{A}	62.4 ^{A}	64.5 ^{A}	64.4 ^{A}	64.3 ^{A}	63.9 ^{A}	
Time T1 (min)	Control	3.37 ^{b}	4.4 ^{a}	5.0 ^{a}	4.1 ^{a}	5.4 ^{a}	8.2 ^{a}	5.0 ^{A}
	Fusarium	4.0 ^{a}	2.6 ^{b}	1.9 ^{b}	3.9 ^{a}	1.2 ^{b}	3.2 ^{b}	2.8 ^{B}
	Average	3.66 ^{B}	3.50 ^{B}	3.47 ^{C}	4.00 ^{B}	3.25 ^{D}	5.65 ^{A}	
Stability (min)	Control	11.0 ^{a}	9.0 ^{a}	10.3 ^{a}	8.4 ^{a}	9.8 ^{a}	11.8 ^{a}	10.1 ^{A}
	Fusarium	7.7 ^{b}	8.2 ^{b}	7.0 ^{b}	6.2 ^{b}	5.3 ^{b}	5.2 ^{b}	6.6 ^{B}
	Average	9.3 ^{A}	8.6 ^{A}	8.6 ^{A}	7.3 ^{B}	7.6 ^{B}	8.5 ^{A}	
Slope α (Nm min^{-1})	Control	−0.102 ^{b}	−0.088 ^{a}	−0.101 ^{b}	−0.083 ^{b}	−0.090 ^{a}	−0.091 ^{b}	−0.092 ^{B}
	Fusarium	−0.082 ^{a}	−0.091 ^{a}	−0.091 ^{a}	−0.070 ^{a}	−0.090 ^{a}	−0.060 ^{a}	−0.080 ^{A}
	Average	−0.092 ^{B}	−0.090 ^{B}	−0.096 ^{B}	−0.076 ^{A}	−0.090 ^{B}	−0.075 ^{A}	
Protein weakening, C2 (Nm)	Control	0.53 ^{a}	0.42 ^{a}	0.46 ^{a}	0.40 ^{a}	0.47 ^{a}	0.55 ^{a}	0.47 ^{A}
	Fusarium	0.21 ^{b}	0.22 ^{b}	0.17 ^{b}	0.18 ^{b}	0.15 ^{b}	0.13 ^{b}	0.18 ^{B}
	Average	0.36 ^{A}	0.32 ^{B}	0.31 ^{B}	0.29 ^{C}	0.31 ^{B}	0.33 ^{B}	
C1–C2 (Nm)	Control	0.58 ^{b}	0.66 ^{b}	0.63 ^{b}	0.67 ^{b}	0.62 ^{b}	0.55 ^{b}	0.62 ^{B}
	Fusarium	0.91 ^{a}	0.89 ^{a}	0.94 ^{a}	0.95 ^{a}	0.97 ^{a}	0.97 ^{a}	0.94 ^{A}
	Average	0.74 ^{B}	0.77 ^{B}	0.78 ^{A}	0.81 ^{A}	0.79 ^{A}	0.76 ^{AB}	

$^{a,b,A–D}$—homogeneous groups, statistically significant differences at $p = 0.05$.

The protein characteristics of whole-meal flour samples were determined based on time T1 (dough development time), stability, and protein weakening (C2, C1–C2, slope α). *Fusarium* inoculation significantly influenced the protein characteristics of the tested winter wheat cultivars.

Dough development time (T1), which is primarily related to protein content, varied between cultivars and was significantly affected by *Fusarium* inoculation (Figures 2–4). Time T1 for the cultivar samples ranged from 3.3 min (cv. 'Sailor') to 8.2 min (cv. 'Tonacja') (Table 5). However, the influence of *Fusarium* inoculation varied between cultivars. A reduction in time T1 was observed for four of the six tested cultivars. The highest difference was stated for cv. 'Tonacja' (on average 5 min), while for cv. 'Pokusa', there was no change in this parameter, and for cv. 'Sailor', T1 increased by 0.7 min.

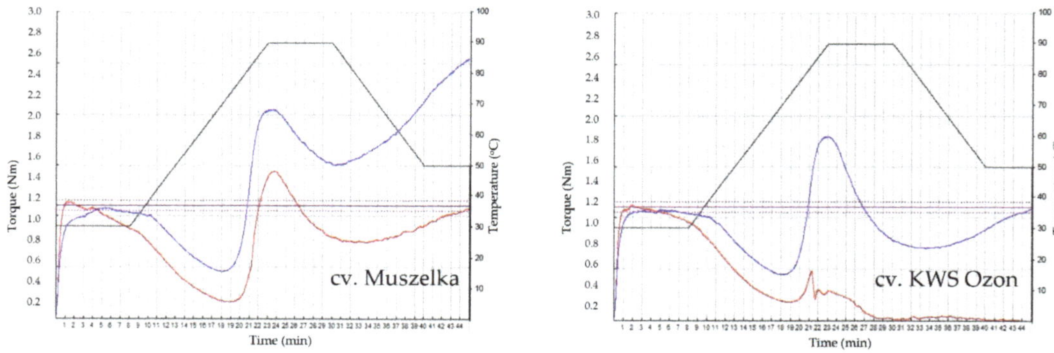

(**a**) (**b**)

Figure 2. Comparison of Mixolab curves between control (blue lines) and *Fusarium culmorum*-inoculated (red lines) cultivars: (**a**) 'Muszelka'; (**b**) 'KWS Ozon'.

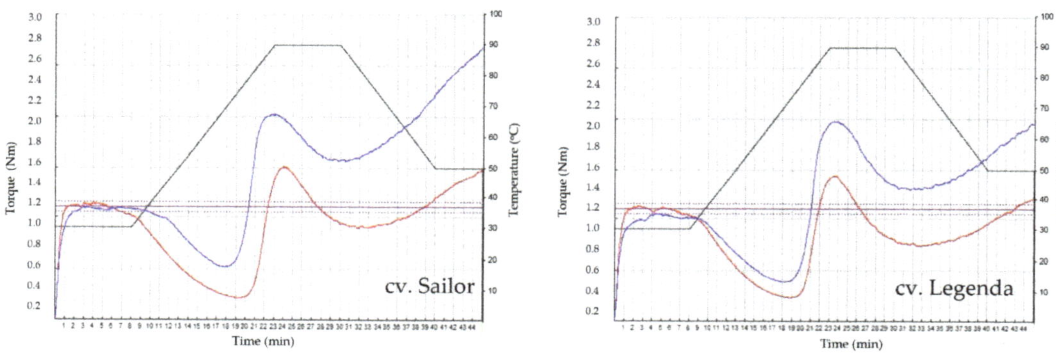

(**a**) (**b**)

Figure 3. Comparison of Mixolab curves between the control (blue line) and *Fusarium culmorum*-inoculated (red line) cultivars: (**a**) 'Sailor'; (**b**) 'Legenda'.

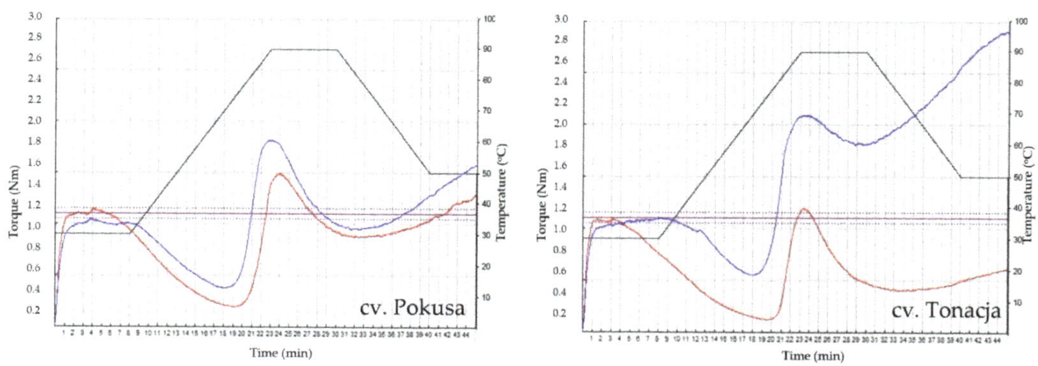

(**a**) (**b**)

Figure 4. Comparison of Mixolab curves between control (blue line) and *Fusarium culmorum*-inoculated (red line) cultivars: (**a**) 'Pokusa'; (**b**) 'Tonacja'.

Dough stability, measured as the time until the upper frame of the Mixolab curve decreases by 11% from the point C1, corresponds to the dough strength. The value of this parameter significantly depended on the experimental factors (Table 4). Among the control samples of wheat cultivars, the highest stability during dough mixing was found for the cv. 'Tonacja' (11.8 min), while the lowest value was observed for cv. 'Pokusa' (8.4 min). Inoculation with *Fusarium* fungi caused a significant decrease in dough stability for all tested cultivars, by an average reduction of 3.5 min. The greatest decrease was observed for cv. 'Tonacja' (6.6 min), while the lowest for cv. 'Legenda' (0.8 min), which was characterized by one of the lowest stability values among the control cultivars.

Protein weakening, as measured by Mixolab, was determined based on the change in dough consistency during temperature increases (C2, C1–C2, and slope α). The C2 parameter for the tested wheat flours obtained from control samples was within an appropriate level for baking, varying between cultivars. The lowest protein weakening (C2) was observed in cv. 'Pokusa' (0.40 Nm), while the highest was found in cv. 'Tonacja' (0.55 Nm). *Fusarium* inoculation caused a significant decrease in the C2 value (on average 0.30 Nm). The largest decrease was found for cv. 'Tonacja' (0.42 N m) and the lowest for cv. 'Legenda' (0.20 Nm). However, the C2 parameter for all tested samples ranged from 0.13 to 0.22 Nm, which corresponds to weak dough and potential issues during mechanical processing in dough preparation.

To better characterize the changes in dough properties, and referring to the studies [13,37,38], the differences in dough resistance measured at points C1 and C2, as well as C3 and C2, C3 and C4, and C5 and C4, were determined. The low quality of gluten proteins in samples inoculated with *Fusarium culmorum* was confirmed by increased differences in torque values between points C1–C2 (0.62 Nm for control samples vs. 0.94 for inoculated samples) (Table 4). The difference in dough resistance at points C1 and C2 (C1–C2) indicates changes in the structure of gluten proteins due to heating [13]. According to Koksel et al. [38], this corresponds to dough resistance to mixing. The greatest increase in the C1–C2 parameter was observed in cv. 'Tonacja' (0.42 Nm), which had the lowest C2 value. The smallest increase was stated for cv. 'Legenda' and 'Pokusa' (0.23 and 0.28 Nm, respectively). Winter wheat cultivars inoculated with *Fusarium culmorum* also exhibited a higher rate of protein weakening under heat, as measured by slope α (-0.080 Nm min^{-1} for inoculated grain samples vs. -0.092 Nm min^{-1} for control samples).

3.4. Rheological Properties of the Starch Complex of Tested Wheat Cultivars

The properties of starch and the activity of amylolytic enzymes are crucial for the dough fermentation process and for obtaining high-quality wheat bread. The following Mixolab parameters were used to determine the starch complex of wheat flour: starch gelatinization (C3), amylase activity (C4), and retrogradation (C5). Starch gelatinization speed and enzyme degradation speed are determined as slope β and slope γ, respectively.

The β index, which characterizes the increase in dough resistance due to the swelling of starch granules as the temperature rises from 30 to 90 °C, ranged from 0.587 Nm min^{-1} (cv. 'KWS Ozon') to 0.782 Nm min^{-1} (cv. 'Muszelka'). Inoculation of ears with the fungus caused a significant decrease in this parameter in all analyzed wheat cultivars (Table 6). The greatest decrease in slope β was observed in cv. 'KWS Ozon' which also had the lowest value (0.087 Nm min^{-1}) among the cultivars.

Table 6. The effect of *Fusarium culmorum* contamination of grain on the rheological properties of dough from winter wheat cultivars determined using Mixolab, which characterizes the properties of starch.

Parameter	Inoculation	Cultivar						Mean
		Sailor	Legenda	KWS Ozon	Pokusa	Muszelka	Tonacja	
Slope β (Nm/min)	Control	0.658 ^{a}	0.602 ^{a}	0.587 ^{a}	0.673 ^{a}	0.782 ^{a}	0.699 ^{a}	0.667 ^{A}
	Fusarium	0.583 ^{b}	0.563 ^{b}	0.087 ^{b}	0.419 ^{b}	0.518 ^{b}	0.602 ^{b}	0.462 ^{B}
	Average	0.621 ^{B}	0.583 ^{C}	0.337 ^{E}	0.546 ^{D}	0.650 ^{A}	0.651 ^{A}	
Slope γ (Nm/min)	Control	0.129 ^{b}	−0.138 ^{b}	−0.111 ^{b}	−0.111 ^{b}	−0.089 ^{a}	−0.071 ^{a}	−0.108 ^{B}
	Fusarium	−0.089 ^{a}	−0.111 ^{a}	−0.499 ^{a}	−0.069 ^{a}	−0.052 ^{a}	−0.089 ^{a}	−0.152 ^{A}
	Average	−0.109 ^{B}	−0.125 ^{B}	−0.305 ^{A}	−0.090 ^{C}	−0.071 ^{C}	−0.080 ^{C}	
C3 (Nm)	Control	2.02 ^{a}	1.99 ^{a}	1.81 ^{a}	1.84 ^{a}	2.06 ^{a}	2.09 ^{a}	1.97 ^{A}
	Fusarium	1.50 ^{b}	1.43 ^{b}	0.47 ^{b}	1.47 ^{b}	1.43 ^{b}	1.20 ^{b}	1.25 ^{B}
	Average	1.76 ^{A}	1.71 ^{A}	1.14 ^{C}	1.66 ^{B}	1.74 ^{A}	1.64 ^{B}	
C4 (Nm)	Control	1.56 ^{a}	1.320 ^{a}	0.72 ^{a}	0.97 ^{a}	1.50 ^{a}	1.79 ^{a}	1.31 ^{A}
	Fusarium	0.88 ^{b}	0.72 ^{b}	0.20 ^{b}	0.86 ^{a}	0.73 ^{b}	0.41 ^{b}	0.63 ^{B}
	Average	1.22 ^{A}	1.02 ^{D}	0.46 ^{F}	0.92 ^{E}	1.12 ^{B}	1.10 ^{C}	
C5 (Nm)	Control	2.67 ^{a}	1.97 ^{a}	1.10 ^{a}	1.61 ^{a}	2.56 ^{a}	2.91 ^{a}	2.14 ^{A}
	Fusarium	1.47 ^{b}	1.20 ^{b}	0.01 ^{b}	1.28 ^{b}	1.05 ^{b}	0.62 ^{b}	0.94 ^{B}
	Average	2.07 ^{A}	1.59 ^{C}	0.55 ^{E}	1.44 ^{D}	1.81 ^{B}	1.77 ^{B}	
C3–C2 (Nm)	Control	1.50 ^{a}	1.58 ^{a}	1.35 ^{a}	1.43 ^{a}	1.59 ^{a}	1.54 ^{a}	1.50 ^{A}
	Fusarium	1.29 ^{b}	1.20 ^{b}	0.30 ^{a}	1.29 ^{a}	1.28 ^{b}	1.07 ^{b}	1.07 ^{B}
	Average	1.39 ^{A}	1.39 ^{A}	0.82 ^{B}	1.36 ^{A}	1.43 ^{A}	1.30 ^{A}	
C3–C4 (Nm)	Control	0.46 ^{b}	0.68 ^{a}	1.09 ^{a}	0.87 ^{a}	0.56 ^{b}	0.29 ^{b}	0.66 ^{A}
	Fusarium	0.62 ^{a}	0.71 ^{a}	0.27 ^{b}	0.61 ^{b}	0.70 ^{a}	0.79 ^{a}	0.62 ^{B}
	Average	0.54 ^{D}	0.69 ^{B}	0.68 ^{B}	0.74 ^{A}	0.63 ^{C}	0.54 ^{D}	
C5–C4 (Nm)	Control	1.11 ^{a}	0.65 ^{a}	0.39 ^{a}	0.63 ^{a}	1.06 ^{a}	1.11 ^{a}	0.83 ^{A}
	Fusarium	0.59 ^{b}	0.48 ^{a}	−0.20 ^{b}	0.42 ^{a}	0.32 ^{b}	0.21 ^{b}	0.30 ^{B}
	Average	0.851 ^{A}	0.569 ^{C}	0.097 ^{D}	0.525 ^{C}	0.690 ^{B}	0.663 ^{B}	

$^{a,b,A-F}$—homogeneous groups, statistically significant differences at $p = 0.05$.

The values of slope γ, which characterize the rate of enzymatic starch decomposition, ranged from −0.111 Nm min^{-1} (cv. 'KWS Ozon' and 'Pokusa') to −0.071 Nm min^{-1} (cv. 'Tonacja'). Inoculation with *Fusarium culmorum* significantly affected the values of slope γ in four cultivars, causing an increase in this parameter. However, for the cv. 'KWS Ozon' and cv. 'Tonacja', the value of slope γ decreased by −0.388 and −0.018 Nm min^{-1}, respectively (Table 6).

Regarding starch gelatinization (C3), control wheat flours exhibited medium to low α-amylase activity, with values ranging from 1.81 Nm (cv. 'KWS Ozon') to 2.09 Nm (cv. 'Tonacja'). *Fusarium culmorum* inoculation caused significant reductions in C3 values (from 1.92 Nm for control samples to 1.25 Nm for inoculated samples). The greatest reduction was stated for cv. 'KWS Ozon' (−1.34 Nm to 0.47 Nm) (Figure 2) compared to other cultivars, where the reduction was on average −0.6 Nm. The C3 value for cv. 'KWS Ozon', with the lowest C3 in inoculated samples, was characterized by flour from sprouted grain with very high α-amylase activity.

The control winter wheat samples were more variable in terms of amylase activity (C4) and retrogradation (C5) than starch gelatinization (C3). C4 values for control samples ranged from 0.72 Nm (cv. 'KWS Ozon') to 1.79 Nm (cv. 'Tonacja') (Table 6). *Fusarium* inoculation caused a significant decrease in amylase activity (C4) (from 1.31 Nm in control samples to 0.63 Nm in inoculated samples). The highest drop was found for cv. 'Tonacja' (1.79 Nm to 0.41 Nm), and the lowest for cv. 'Pokusa' (0.97 Nm to 0.86 Nm). The lowest value of C4 was stated for cv. 'KWS Ozon' (0.20 Nm), confirming the stickiness of the dough.

Starch retrogradation in the cooling stage of Mixolab analysis varied between control winter wheat samples. The highest C5 value was recorded for cv. 'Tonacja' (2.91 Nm), compared to the very low value for cv. 'KWS Ozon' (1.10 Nm). Similar to other starch complex characteristics, starch retrogradation was strongly influenced by *Fusarium culmorum* inoculation (on average 2.14 vs. 0.94 Nm, respectively). The largest difference was found for cv. 'Tonacja' (a drop of 2.29 Nm) and the smallest for cv. 'Pokusa' (a drop of 0.33 Nm).

In partial agreement with the previous evaluation of the Mixolab curve, the differences between C3–C2, C3–C4, and C5–C4 were also analyzed. Dough properties related to starch gelatinization are determined at point C3 of the graph, and the difference in dough resistance at points C3 and C2 (C3–C2) are determined [39]. The value of the C3–C2 parameter was influenced by the cultivar and *Fusarium* inoculation. Inoculation caused a significant decrease in the value of this parameter. Flour from control cv. 'Muszelka' exhibited the highest C3–C2 value (1.59 Nm), while cv. 'KWS Ozon' showed the lowest (1.35 Nm). The greatest decrease in C3–C2 was observed in cv. 'KWS Ozon' (a drop of 1.05 Nm), whereas the reduction for inoculated cv. 'Pokusa' was only 0.14 Nm (Table 6).

Control samples of cv. 'KWS Ozon' were characterized by the highest differences between C3–C4 (1.09 Nm), indicating the highest α-amylase activity or less resistance to shear thinning. In contrast, cv. 'Tonacja' had the lowest C3–C4 difference (0.29 Nm), corresponding to low α-amylase activity. *Fusarium* inoculation caused an increase in C3–C4 difference in four of the six cultivars. In the case of cv. 'KWS Ozon', the dough became sticky at points C3 and C4, making it difficult to analyze, which explains the opposite change in this parameter.

The difference in dough resistance between points C5 and C4 (C5–C4) reflects the susceptibility of bread to staling and may indicate its shelf life, as described by Papoušková et al. [13]. The smallest difference was observed in the flour of cv. 'Pokusa' and the largest in cv. 'Tonacja' and cv. 'Sailor'. *Fusarium* inoculation significantly affected the values of this parameter (Table 6). Among the cultivars after inoculation, cv. 'KWS Ozon' was assessed as the most favorable, as it exhibited the lowest C5–C4 value, suggesting the potential for the longest shelf life of the obtained bread. On the other hand, cv. 'Sailor' and cv. 'Tonacja' were considered the least favorable. For cv. 'KWS Ozon', negative values of the C5–C4 parameter in the flour after grain inoculation indicated very high amylolytic activity and technological problems in bread production.

3.5. Mixolab Wheat Flour Profiles

The Mixolab profiler allows the evaluation of dough consistency over time and determines, in the same assay, both the mixing and pasting characteristics of flour during a steady increase in temperature [40]. The Mixolab profiler provides users with the ability to classify the quality of tested flours based on six quality criteria: water absorption, mixing, gluten+, viscosity, amylase, and retrogradation, with values ranging from 0 to 9. This quality control tool, based on both protein and starch characteristics, helps to better screen and detect differences between tested samples and identify underperforming flours.

The Mixolab profilers obtained in our research significantly differed between the control cultivar samples and those inoculated with *Fusarium culmorum* (Figures 5–7). Differences were also noted between the tested cultivars.

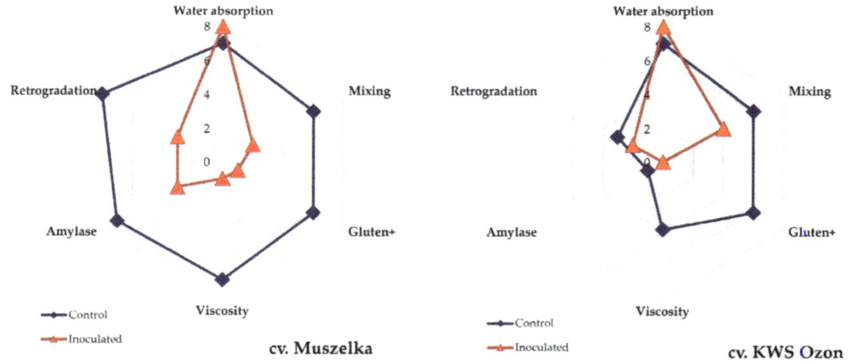

(**a**) (**b**)

Figure 5. Mixolab profiler values of whole-grain wheat flours obtained from: (**a**) cv. 'Muszelka'; (**b**) cv. 'KWS Ozon' from both control samples and samples inoculated by *Fusarium culmorum*.

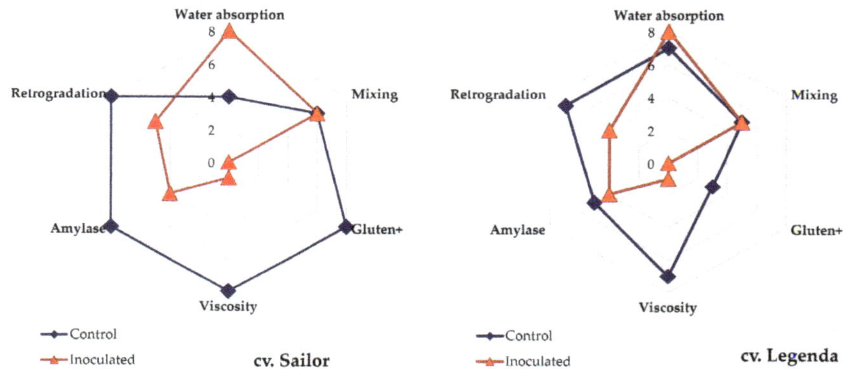

(**a**) (**b**)

Figure 6. Mixolab profiler values of whole-grain wheat flours obtained from: (**a**) cv. 'Sailor'; (**b**) cv. 'Legenda' from both control samples and samples inoculated by *Fusarium culmorum*.

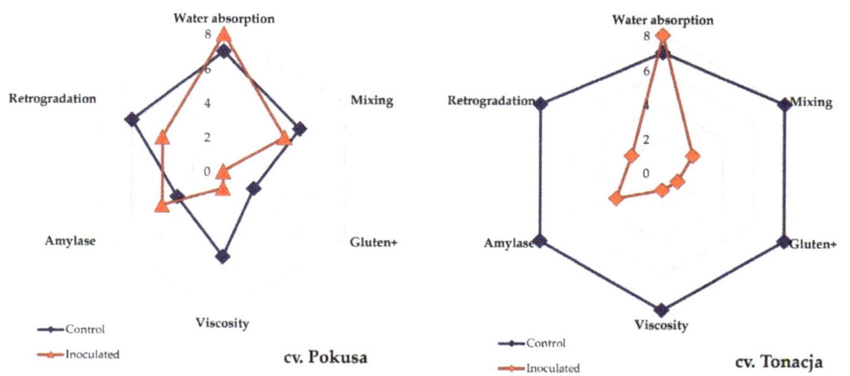

(**a**) (**b**)

Figure 7. Mixolab profiler values of whole-grain wheat flours obtained from: (**a**) cv. 'Pokusa'; (**b**) cv. 'Tonacja' from both control samples and samples inoculated by *Fusarium culmorum*.

The water absorption index, which corresponds to the ability of flour to absorb water, was in the range of 4 (cv. 'Sailor') to 7 (other cultivars) among the control samples. *Fusarium* inoculation significantly increased the water absorption index to 8 in each cultivar.

In contrast, other parameters showed the opposite effect. The mixing index, which corresponds to the stability of flour dough during kneading, is higher for stronger dough. Control samples of wheat cultivars had a mixing index ranging from 5 (medium quality—cv. 'Legenda', cv. 'Pokusa') to 8 (strong gluten structure—cv. 'Tonacja'). *Fusarium* inoculation weakened the gluten structure, lowering the mixing index (an average reduction of 2 units). However, different effects were observed across cultivars. The greatest reduction in this parameter was seen in cv. 'Tonacja' (Figure 7), while cv. 'Sailor' and cv. 'Legenda' showed no effect (Figure 6).

The gluten+ index describes the resistance of gluten to heat and the changes in the quality of gluten proteins under the influence of an increase in temperature from 30 °C to 60 °C. In non-inoculated samples, this index ranged from 2 to 8, with the highest value observed in the control samples of cv. 'Sailor' and cv. 'Tonacja' and the lowest in cv. 'Pokusa'. *Fusarium* inoculation significantly reduced the heat-induced weakening of gluten proteins, lowering the gluten+ index by an average of 5 units to 0 (cv. 'Sailor', 'Legenda', 'KWS Ozon', 'Pokusa') and 1 (cv. 'Muszelka' and cv. 'Tonacja'), indicating a reduction in stability to heat.

The next index with significant differences observed was viscosity, which corresponds to changes in dough characteristics as the temperature increases from 60 °C to 80 °C. This index is influenced by both amylase activity and starch quality, ranging from 4 (cv. 'KWS Ozon') to 8 (cv. 'Sailor' and cv. 'Tonacja'). *Fusarium* inoculation significantly lowered the viscosity index to 0 (cv. 'KWS Ozon') and 1 (the rest of the cultivars). The average drop of 6 units was the highest among the other Mixolab Profiler indexes.

The amylase index, reflecting changes in dough properties due to the α-amylase enzyme, ranged from 1 (cv. 'KWS Ozon') to 8 (cv. 'Sailor' and cv. 'Tonacja'). *Fusarium* inoculation caused a significant decrease in this index in cv. 'Muszelka' (Figure 5), cv. 'Sailor' (Figure 6), and cv. 'Tonacja' (Figure 7) compared to the control samples. However, an increase in the viscocity index was observed in cultivars with the lowest amylase index in the control samples: cv. 'KWS Ozon' (from 1 to 4) and cv. 'Pokusa' (from 3 to 4).

The retrogradation index, which depends on the characteristics of the starch and its hydrolysis during the test, ranged from 3 (cv. 'KWS Ozon') to 8 (cv. 'Sailor', cv. 'Muszelka', cv. 'Tonacja') in non-inoculated cultivars. In flour samples from grain contaminated with *Fusarium culmorum*, a significant reduction in this index was observed (on average a drop of 3 units).

3.6. Principal Component Analyses (PCAs)

Almost all investigated factors were presented in a two-dimensional coordinate system (Figure 8). Based on the results, we identified four distinct clusters of samples. The first group consisted of KWSOI (blue color), the second group included TC, MC, and SC (red color), the third group comprised PI, MI, SI, TI, and LI (green color), and the fourth group included LC, KWSOC, and PC (black color). Notably, opposite properties were observed between the first and fourth groups of samples. A similar trend was noted between the second and third clusters. The KWSOI sample was characterized by a high ash content and water absorption. In contrast, the LC, KWSOC, and PC samples exhibited higher rheological parameters (C3 and C3–C2), sedimentation index, and gluten content. The samples PI, MI, SI, TI, and LI stood out due to their total protein content and specific rheological parameters (C1–C2 and C3–C4). On the other hand, the TC, MC, and SC samples were distinguished by TKW, sedimentation time, and other rheological parameters (T1, C2, C4, C5, and C5–C4).

It is worth noting that all inoculated samples were positioned on the left side of the score plot, while the control samples were located on the right side. This distribution indicates a negative interdependence between the different plant treatments.

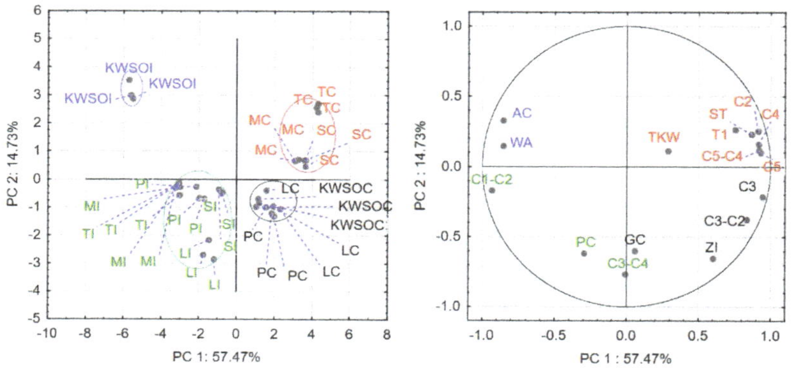

Figure 8. Principal component analyses represented by two-dimensional plot: score plot (left side) and loading plot (right side). Explanation for cases: KWSOI—KWS Ozon Inoculated; KWSOC—KWS Ozon Control; TI—Tonacja Inoculated; TC—Tonacja Control; MI—Muszelka Inoculated; MC—Muszelka Control; SI—Sailor Inoculated; SC—Sailor Control; PI—Pokusa Inoculated; PC—Pokusa Control; LI—Legenda Inoculated; LC—Legenda Control. Explanation for variables: TKW—Thousand kernel weight; PC—Protein content; GC—Gluten content; ZI—Zeleny sedimentation index; AC—Ash content; T1—Time T1 (rheological parameter); ST—Stabilization time (rheological parameter); WA—Water absorption; C2, C3, C4, C5, C1–C2, C3–C2, C3–C4, C5–C4—Rheological parameters of dough determined by Mixolab equipment.

4. Discussion

The literature review and the results of our studies indicate that *Fusarium culmorum* is pathogenic to winter wheat and produces the mycotoxin deoxynivalenol (DON) in the grain [21,41,42]. The results indicate differences in the response of wheat cultivars to stress caused by *Fusarium culmorum* inoculation. This is confirmed by changes in the properties of the starch and protein complex induced by *Fusarium culmorum* inoculation. The research hypothesis was tested through conducting a field experiment and laboratory tests. They included the inoculation of *Fusarium culmorum*, which, according to the literature, is the species that most commonly infects wheat in Polish weather conditions and produces the largest amounts of mycotoxins [29,43].

4.1. Mycotoxin Content in Grain of Winter Wheat Cultivars

The presented studies focused on the inoculation of *Fusarium culmorum* as it is a predominant species in Polish conditions and produces the most mycotoxins [29,44,45]. Bryła et al. [44] examined the levels of 26 mycotoxins in cereal grains in Poland and found the presence of DON in all tested wheat samples. The studies described different levels of DON depending on the crop year. For the infection of wheat by *Fusarium ssp.*, the most crucial factors are the weather conditions during the flowering period and the initial ripeness of wheat (May and June). As shown in Table 2, the weather conditions in 2019 were not favorable for the development of *Fusarium* spp. The air temperature was higher in May and June, by 1.8 and 3.3 °C, respectively, compared to 2020. There were very large differences in the amount of precipitation—2019 was a very dry year. The amount of rainfall in 2020 in May and June was 26.6 mm and 150.8 mm higher, respectively, than in 2019. During the flowering period, it was too dry for the fungi to develop in the wheat grains even after artificial inoculation. Weather conditions caused the DON level to be very low in 2019, therefore only seeds from 2020 were used in our experiment.

The studies found different levels of DON depending on the cultivars. The lowest content of this mycotoxin was found in cv. 'Sailor', while the highest was in cv. 'Muszelka' and cv. 'Legenda'. Selection of genotypes resistant to Fusarium head blight (FHB) is one of the methods used in wheat cultivation to reduce the occurrence of mycotoxins in wheat grain. Resistance to FHB is controlled by major and minor genes located on all wheat chromosomes, except for chromosome 7D [46]. The presented research results confirm the literature data indicating differences in the amounts of *Fusarium* mycotoxins in individual wheat cultivars grown under the same weather conditions [29,42,44]. The differences in the amount of mycotoxins in wheat grain may result from different types of resistance [47–49]. Therefore, there is not always a highly significant relationship between resistance to the *Fusarium* disease complex and the accumulation of mycotoxins. This was confirmed in our studies, as it was shown that cultivars with high resistance to FHB (cv. 'Legenda') had higher levels of mycotoxins than cultivars characterized by lower resistance (cv. 'Sailor'). Most likely, these cultivars are characterized by resistance to the accumulation of *Fusarium* toxins in the grain. Studies by Siou et al. [50] report that the cropping of cultivars with a short vegetation period or early flowering allows for the avoidance or significant reduction of infection with fungi of the genus *Fusarium* and the accumulation of toxins. This relationship was not confirmed in the presented studies. It may have resulted from the fact that the difference in the earliness of cultivars was only 5 days. Another factor that is most likely the reason for the varying amounts of toxin accumulation is the structure and composition of grain. Walter et al. show that cell walls and the aleurone layer are significant physical barriers to the penetration of the fungus [51]. Literature data also indicate that other morphological features such as grain color, hulls, wax coating, and ferulic acid concentration affect the levels of mycotoxins in wheat grain [42].

The wheat microbiome plays a crucial role in limiting infections caused by *Fusarium* spp., which are responsible for diseases such as Fusarium head blight and root rot. The plant microbiota can influence plant health through various mechanisms, including competition with pathogens, induction of plant immunity, and the production of antimicrobial substances. Microorganisms in the wheat microbiome, particularly bacteria and fungi, can produce a range of antimicrobial compounds that inhibit pathogen growth. Some *Bacillus* strains produce lipopeptides that act as antimicrobial agents, inhibiting the development of *Fusarium* [52]. Other bacteria produce enzymes, such as chitinases, which can degrade the cell walls of *Fusarium* fungi, limiting their growth and increasing biodiversity within the microbiome. The biodiversity of the wheat microbiome may also play a key role in protecting against *Fusarium* infections. Studies have shown that plant microbiomes with higher biodiversity can improve plant health and enhance resistance to diseases caused by *Fusarium* [53,54]. The biodiversity of the wheat microbiome may also play a key role in protecting against *Fusarium* infections. Diverse microbial communities in the rhizosphere can collaborate to more effectively suppress pathogens.

4.2. *Influence of Fusarium culmorum Infection on the Grain Quality and Rheological Properties of Dough*

The technological value of wheat depends on the basic quality indicators, including, among others, the total protein content and its quality, the quantity and quality of gluten, the Zeleny sedimentation index, and the degree of starch damage [7,55]. To assess wheat flour quality more precisely, tests can be extended to include the rheological properties of the dough, which are performed using devices such as the farinograph, extensograph, and alveograph. Based on the data obtained from these tests, changes in the dough's consistency and extensibility can be assessed. The viscosity of the flour suspension in water during heating is determined using an amylograph [56]. A comprehensive assessment of rheological properties requires the use of several devices. In the studies discussed, Mixolab was used to assess the quality of whole-grain wheat flour. This device allows for the simultaneous evaluation of both protein and starch properties [40,56].

The impact of wheat infection by fungi of the *Fusarium* genus and grain contamination with mycotoxins on technological value has been analyzed in only a few studies. Typically, the effect of *Fusarium* on individual quality features, such as nutrient content, protein content, gluten quantity and quality, Zeleny sedimentation index, or dough rheological properties, has been taken into account [21,57,58]. According to the literature, *Fusarium* infection of grains causes changes in the proportion of seed coat to endosperm, which results in an increase in ash content [11–13]. These relationships were confirmed in the present study and were visualized in PCA charts. In all cultivars, inoculation with *Fusarium culmorum* led to an increase in ash content. The highest increase was recorded in the cv. 'Pokusa' (0.44%), while the lowest was in the cv. 'Tonacja' (0.13%) and cv. 'Legenda' (0.11%).

Another aspect of the study was to identify wheat genotypes less susceptible to the reduction of thousand kernel weight (TKW) due to *Fusarium* infection. This is of significant importance for both yield size and quality, as TKW affects the milling and baking value of flour [15]. Cultivars with low TKW are characterized by low flour extraction due to disturbed proportions of endosperm to fruit-seed coat. While they have higher protein content, these proteins are not accumulated in the endosperm but in the fruit-seed coat, which is irrelevant for flour production. In the present study, the greatest reduction in TKW due to *Fusarium* inoculation was observed in the cv. 'Legenda', 'Muszelka', and 'Tonacja', which was proved in PCA charts. The results of these studies are consistent with those of Packa et al. [59], who found a reduction in TKW under *Fusarium* infection, with an average of 13% over several years.

Proteins stored in wheat grain are an important source of nutrients for humans and play a major role in determining bread baking quality [7]. The present research showed an increase in protein content after *Fusarium culmorum* inoculation, with an average increase of 1.1% percentage point across the cultivars (PCA chart). The highest increase, 3.5% compared to the control, was observed in the cv. 'Legenda', while the lowest increases were seen in the 'Pokusa' and 'Muszelka' cultivars. In the cv. 'KWS Ozon', the difference was not significant. The results suggest a significant change in one of the most important quality indicators [7]. These findings align with the literature, which indicates an increase in protein content due to the *Fusarium* contamination, as well as showing varietal differences [14,25]. The increase in protein content in the grain may be a response to stress, with plant cells synthesizing specific proteins and potentially altering their fractional and amino acid composition [60,61]. These proteins help mitigate the effects of stress by neutralizing its impact and protecting essential cellular structures and metabolic processes [62].

From the perspective of flour suitability for baking, both the quantity and quality of protein are important. The distinguishing factor determining both values is the sedimentation index. In the present studies, the sedimentation index was found to decrease by an average of 10.7% in cultivars subjected to *Fusarium* inoculation (PCA chart). Similar results were obtained by El Chami et al. [21]. Research by Gärtner et al. indicates varietal differences in this index, a finding confirmed in this work [14]. Kreuzberger et al. and Wang et al. obtained different results, suggesting no effect of *Fusarium* mycotoxins on the sedimentation index [17,18]. This discrepancy may be attributed to the selection of cultivars.

Based on the results, cultivars were classified into those where the sedimentation index was significantly reduced (cv. 'KW Ozon', cv. 'Pokusa'), those with an increase (cv. 'Legenda'), and those that showed no effect of *Fusarium* on the sedimentation index (cv. 'Muszelka' and cv. 'Tonacja'). Changes in the sedimentation index reflect the impact of *Fusarium* on the balance between gliadins and glutenins that form gluten. Therefore, it can be concluded that, among the cultivars compared, 'Muszelka' and 'Tonacja' were more resistant to changes in protein quality due to *Fusarium* infection, as evidenced by the negligible change in gluten content in these cultivars.

In addition to determining the effect of *Fusarium culmorum* inoculation on the qualitative characteristics of grain, the effect on the quality of starch–protein complex was also examined. Laboratory tests were performed using Mixolab, which allowed for a comprehensive assessment of the experimental factors' impact on the baking value and dough quality. It is worth noting that the quality was assessed based on the quality parameters of the starch and protein complex.

According to Gärtner et al., flour obtained from wheat cultivars containing mycotoxins, compared to flours from uncontaminated cultivars, is characterized by reduced baking quality parameters such as water absorption and dough softening [14]. Similar results were obtained by Horvat et al., who compared the rheological properties of dough from contaminated and uncontaminated flours [26]. They found a reduction in water absorption by 1.6%, dough development time by 17%, and quality number by 19.9%. The authors also showed that *Fusarium* contamination reduced dough energy, maximum tensile strength, and the strength-to-extensibility ratio by 57.7%. Additionally, they observed a negative effect on the specific volume of bread. Similarly, the presented studies demonstrated a negative effect of *Fusarium* on flour and dough properties. Water absorption of flours from inoculated cultivars increased by an average of 3.6% compared to uninfected grain, and dough development time was shortened by an average of 44.5%. A decrease in dough stability was also observed, with an average reduction of 34.4%, alongside changes in dough resistance to kneading depending on the cultivars. These changes were likely caused by proteolytic enzymes of *Fusarium*, which remain inactive in the grain but can be activated during dough kneading [18,19]. In the research by Peršić et al. [63], FHB infection did not affect the water absorption of wheat flour from either location. However, FHB infection caused the most significant increase in dough softening, by over 400% compared to control samples in one of the locations, suggesting significant genotypic variability.

The results also indicate a negative effect of *Fusarium* on starch properties, particularly the swelling of starch granules and the rate of enzymatic degradation. Changes in starch properties under the influence of *Fusarium* have been confirmed in the literature [20–22]. The presented study found that the effect of *Fusarium* on starch degradation varied depending on the genotype. The β index, which characterizes the increase in dough resistance due to starch granule swelling, decreased significantly in all analyzed wheat cultivars. In contrast, the γ index, indicating the high amylolytic enzyme activity, varied between the cultivars. The greatest decrease in the γ index, indicating high amylolytic enzyme activity, was found in cv. 'KWS Ozon', the cultivar with the highest cumulative level of DON. As a result, the β index, indicating a reduction in dough resistance, was also reduced. Similar conclusions were drawn by Gärtner et al. and Wang et al. [14,18].

5. Conclusions

The effect of *Fusarium culmorum* inoculation on the grain quality, starch properties, and protein complex was investigated, resulting in changes in flour properties and dough rheological characteristics. *Fusarium culmorum* inoculation led to an increase in protein, gluten, and ash content, a decrease in the sedimentation index, an increase in flour water absorption, as well as changes in dough rheological properties: a shorter dough development time, reduced dough stability time, decreased dough resistance to mixing, and reduced dough consistency. Inoculation caused changes in starch properties, as indicated by the determination of the β slope parameter in Mixolab (the rate of enzymatic starch decomposition) and the γ slope (starch gelatinization). It also led to a weakening of dough consistency. The observed interaction between *Fusarium culmorum* inoculation and the wheat cultivars in shaping the qualitative parameters and rheological properties of the dough indicates that some wheat cultivars include less susceptible genotypes, which do not exhibit any changes as a result of *Fusarium* infection.

Analyses of the technological value of grain showed that there were no significant differences between cv. 'KWS Ozon', cv. 'Tonacja', and cv. 'Muszelka' in terms of protein content, gluten content, and Zeleny sedimentation index. It was shown that, in the case of 'Pokusa' cv., the effect of *Fusarium* inoculation was not significant in contrast to 'KWS Ozon' cv., where a significant effect on starch–protein complex was observed.

The studies have shown that, under the same environmental and cultivation conditions, different cultivars accumulate various amounts of mycotoxins. From a scientific perspective, it is important to further investigation the underlying causes of this variability. The 'Pokusa' and 'Sailor' cultivars may serve as a valuable source of resistance to mycotoxin accumulation in grain, as they exhibited the lowest contamination with DON among all tested cultivars.

A review of the literature suggests that the cultivar's microbiome may play a key role in this process, therefore, future research should consider this aspect.

Author Contributions: Conceptualization, G.P.; resources, G.P., E.A. and A.S.; writing—original draft preparation, G.P. and E.A.; writing—review and editing, G.P., K.D. and A.S.; supervision, G.P. and A.S.; project administration, G.P. and A.S.; funding acquisition, G.P. and A.S. All authors have read and agreed to the published version of the manuscript.

Funding: This research received no external funding.

Institutional Review Board Statement: Not applicable.

Data Availability Statement: The original contributions presented in this study are included in the article. Further inquiries can be directed to the corresponding authors.

Conflicts of Interest: The authors declare no conflicts of interest.

References

1. FAOSTAT. Crops and Livestock Products. *FAOSTAT 2023*. Available online: https://www.fao.org/faostat/en/#data/qc (accessed on 15 May 2024).
2. Borneo, R.; León, A.E. Whole Grain Cereals: Functional Components and Health Benefits. *Food Funct.* **2012**, *3*, 110–119. [CrossRef] [PubMed]
3. Carson, G.R.; Edwards, N.M. Criteria of Wheat and Flour Quality. In *Wheat*, 4th ed.; Khan, K., Shewry, P.R., Eds.; AACC International Press: St. Paul, MN, USA, 2009; pp. 97–118.
4. Eskola, M.; Kos, G.; Elliott, C.T.; Hajšlová, J.; Mayar, S.; Krska, R. Worldwide Contamination of Food-Crops with Mycotoxins: Validity of the Widely Cited 'FAO Estimate' of 25%. *Crit. Rev. Food Sci. Nutr.* **2020**, *60*, 2773–2789. [CrossRef] [PubMed]
5. European Union. Commission Regulation (EU) 2024/1022 of 8 April 2024 Amending Regulation (EU) 2023/915 as Regards Maximum Levels of Deoxynivalenol in Food. *Off. J. Eur. Union* **2024**, *2024/1022*, 1–4.
6. European Union. Commission Regulation (EU) 2023/915 of 25 April 2023 on Maximum Levels for Certain Contaminants in Food and Repealing Regulation (EC) No 1881/2006 (Text with EEA Relevance). *OJ L* **2023**, *119*, 103–157.
7. Khalid, A.; Hameed, A.; Tahir, M.F. Wheat Quality: A Review on Chemical Composition, Nutritional Attributes, Grain Anatomy, Types, Classification, and Function of Seed Storage Proteins in Bread Making Quality. *Front. Nutr.* **2023**, *10*, 1053196. [CrossRef]
8. Shewry, P.R.; Hawkesford, M.J.; Piironen, V.; Lampi, A.-M.; Gebruers, K.; Boros, D.; Andersson, A.A.M.; Åman, P.; Rakszegi, M.; Bedo, Z.; et al. Natural Variation in Grain Composition of Wheat and Related Cereals. *J. Agric. Food Chem.* **2013**, *61*, 8295–8303. [CrossRef]
9. Alconada, T.M.; Moure, M.C.; Ortega, L.M. Fusarium Infection in Wheat, Aggressiveness and Changes in Grain Quality: A Review. *Vegetos* **2019**, *32*, 441–449. [CrossRef]

Jackowiak, H.; Packa, D.; Wiwart, M.; Perkowski, J. Scanning Electron Microscopy of Fusarium Damaged Kernels of Spring Wheat. *Int. J. Food Microbiol.* **2005**, *98*, 113–123. [CrossRef]

McMullen, M.; Jones, R.; Gallenberg, D. Scab of Wheat and Barley: A Re-Emerging Disease of Devastating Impact. *Plant Dis.* **1997**, *81*, 1340–1348. [CrossRef]

McMullen, M.; Bergstrom, G.; De Wolf, E.; Dill-Macky, R.; Hershman, D.; Shaner, G.; Van Sanford, D. A Unified Effort to Fight an Enemy of Wheat and Barley: Fusarium Head Blight. *Plant Dis.* **2012**, *96*, 1712–1728. [CrossRef]

Papoušková, L.; Capouchová, I.; Kostelanská, M.; Škeříková, A.; Prokinová, E.; Hajšlová, J.; Salava, J.; Faměra, O. Changes in Baking Quality of Winter Wheat with Different Intensity of *Fusarium* spp. Contamination Detected by Means of New Rheological System. *Czech J. Food Sci.* **2011**, *29*, 420–429. [CrossRef]

Gärtner, B.H.; Munich, M.; Kleijer, G.; Mascher, F. Characterisation of Kernel Resistance against Fusarium Infection in Spring Wheat by Baking Quality and Mycotoxin Assessments. *Eur. J. Plant Pathol.* **2007**, *120*, 61–68. [CrossRef]

Matthäus, K.; Dänicke, S.; Vahjen, W.; Simon, O.; Wang, J.; Valenta, H.; Meyer, K.; Strumpf, A.; Ziesenib, H.; Flachowsky, G. Progression of Mycotoxin and Nutrient Concentrations in Wheat after Inoculation with *Fusarium culmorum*. *Arch. Anim. Nutr.* **2004**, *58*, 19–35. [CrossRef] [PubMed]

Eggert, K.; Rawel, H.M.; Pawelzik, E. In Vitro Degradation of Wheat Gluten Fractions by Fusarium Graminearum Proteases. *Eur. Food Res. Technol.* **2011**, *233*, 697–705. [CrossRef]

Kreuzberger, M.; Limsuwan, S.; Eggert, K.; Karlovsky, P.; Pawelzik, E. Impact of Fusarium Spp. Infection of Bread Wheat (*Triticum aestivum* L.) on Composition and Quality of Flour in Association with EU Maximum Level for Deoxynivalenol. *J. Appl. Bot. Food Qual.* **2015**, *88*, 177185. [CrossRef]

Wang, J.; Wieser, H.; Pawelzik, E.; Weinert, J.; Keutgen, A.J.; Wolf, G.A. Impact of the Fungal Protease Produced by Fusarium Culmorum on the Protein Quality and Breadmaking Properties of Winter Wheat. *Eur. Food Res. Technol.* **2005**, *220*, 552–559. [CrossRef]

Koga, S.; Rieder, A.; Ballance, S.; Uhlen, A.K.; Veiseth-Kent, E. Gluten-Degrading Proteases in Wheat Infected by *Fusarium Graminearum*—Protease Identification and Effects on Gluten and Dough Properties. *J. Agric. Food Chem.* **2019**, *67*, 11025–11034. [CrossRef]

Wegulo, S.N. Factors Influencing Deoxynivalenol Accumulation in Small Grain Cereals. *Toxins* **2012**, *4*, 1157–1180. [CrossRef]

El Chami, J.; El Chami, E.; Tarnawa, Á.; Kassai, K.M.; Kende, Z.; Jolánkai, M. Effect of Fusarium Infection on Wheat Quality Parameters. *Cereal Res. Commun.* **2023**, *51*, 179–187. [CrossRef]

Hareland, G.A. Effects of Pearling on Falling Number and α-Amylase Activity of Preharvest Sprouted Spring Wheat. *Cereal Chem.* **2003**, *80*, 232–237. [CrossRef]

Kochiieru, Y.; Mankevičienė, A.; Cesevičienė, J.; Semaškienė, R.; Ramanauskienė, J.; Gorash, A.; Janavičienė, S.; Venslovas, E. The Impact of Harvesting Time on Fusarium Mycotoxins in Spring Wheat Grain and Their Interaction with Grain Quality. *Agronomy* **2021**, *11*, 642. [CrossRef]

Žilić, S. Wheat Gluten: Composition and Health Effects. In *Gluten*; Later, D.B., Ed.; Nova Science Publishers, Inc.: Hauppauge, NY, USA, 2013; Chapter IV; pp. 73–86. ISBN 978-62618-343-8.

Boyacioğlu, D.; Hettiarachchy, N.S. Changes in Some Biochemical Components of Wheat Grain That Was Infected with Fusarium Graminearum. *J. Cereal Sci.* **1995**, *21*, 57–62. [CrossRef]

Horvat, D.; Spanic, V.; Dvojkovic, K.; Simic, G.; Magdic, D.; Nevistic, A. The Influence of *Fusarium* Infection on Wheat (*Triticum aestivum* L.) Proteins Distribution and Baking Quality. *Cereal Res. Commun.* **2015**, *43*, 61–71. [CrossRef]

Prange, A.; Birzele, B.; Krämer, J.; Meier, A.; Modrow, H.; Köhler, P. Fusarium-Inoculated Wheat: Deoxynivalenol Contents and Baking Quality in Relation to Infection Time. *Food Control* **2005**, *16*, 739–745. [CrossRef]

Capouchová, I.; Papoušková, L.; Konvalina, P.; Vepříková, Z.; Dvořáček, V.; Zrcková, M.; Janovská, D.; Škeříková. A.; Pazderů, K. Effect of *Fusarium* spp. Contamination on Baking Quality of Wheat. In *Wheat Improvement, Management and Utilization*; Wanyera, R., Owuoche, J., Eds.; InTech: London, UK, 2017; ISBN 978-953-51-3151-9.

Góral, T.; Wiśniewska, H.; Ochodzki, P.; Walentyn-Góral, D. Higher Fusarium Toxin Accumulation in Grain of Winter Triticale Lines Inoculated with Fusarium Culmorum as Compared with Wheat. *Toxins* **2016**, *8*, 301. [CrossRef]

ISO 520; Cereals and Pulses—Determination of the Mass of 1 000 Grains. International Organization for Standardization: Geneva, Switzerland, 2010.

ISO 712; Cereals and Cereal Products—Determination of Moisture Content—Reference Method. International Organization for Standardization: Geneva, Switzerland, 2009.

ISO 20483; Cereals and Pulses—Determination of the Nitrogen Content and Calculation of the Crude Protein Content—Kjeldahl Method. International Organization for Standardization: Geneva, Switzerland, 2013.

ISO 21415-2:2015; Wheat and Wheat Flour—Gluten Content Part 2: Determination of Wet Gluten and Gluten Index by Mechanical Means. International Organization for Standardization: Geneva, Switzerland, 2015.

34. *ISO 5529*; Wheat—Determination of the Sedimentation Index—Zeleny Test. International Organization for Standardizati Geneva, Switzerland, 2007.
35. *ISO 2171*; Cereals, Pulses and by-Products—Determination of Ash Yield by Incineration. International Organization for Stand ization: Geneva, Switzerland, 2023.
36. *ISO 17718*; Wholemeal and Flour from Wheat (Triticum aestivum L.)—Determination of Rheological Behaviour as a Functio Mixing and Temperature Increase. International Organization for Standardization: Geneva, Switzerland, 2013.
37. Ozturk, S.; Kahraman, K.; Tiftik, B.; Koksel, H. Predicting the Cookie Quality of Flours by Using Mixolab®. *Eur. Food Res. Tech* **2008**, *227*, 1549–1554. [CrossRef]
38. Koksel, H.; Kahraman, K.; Sanal, T.; Ozay, D.S.; Dubat, A. Potential Utilization of Mixolab for Quality Evaluation of Bread Wh Genotypes. *Cereal Chem.* **2009**, *86*, 522–526. [CrossRef]
39. Codină, G.G.; Mironeasa, S.; Bordei, D.; Leahu, A. Mixolab versus Alveograph and Falling Number. *Czech J. Food Sci.* **2** *28*, 185–191. [CrossRef]
40. Dubat, A. A New AACC International Approved Method to Measure Rheological Properties of a Dough Sample. *CFW* **2** *55*, 150–153. [CrossRef]
41. Bottalico, A.; Perrone, G. Toxigenic Fusarium Species and Mycotoxins Associated with Head Blight in Small-Grain Cere in Europe. In *Mycotoxins in Plant Disease*; Logrieco, A., Bailey, J.A., Corazza, L., Cooke, B.M., Eds.; Springer: Dordrecht, T Netherlands, 2002; pp. 611–624. ISBN 978-94-010-3939-0.
42. Czaban, J.; Wróblewska, B.; Sułek, A.; Mikos, M.; Boguszewska, E.; Podolska, G.; Nieróbca, A. Colonisation of Winter Wh Grain by *Fusarium* spp. and Mycotoxin Content as Dependent on a Wheat Variety, Crop Rotation, a Crop Management Syst and Weather Conditions. *Food Addit. Contam. Part A* **2015**, *32*, 874–910. [CrossRef]
43. Marzec-Schmidt, K.; Börjesson, T.; Suproniene, S.; Jędryczka, M.; Janavičienė, S.; Góral, T.; Karlsson, I.; Kochiieru, Y.; Ochodzki, Mankevičienė, A.; et al. Modelling the Effects of Weather Conditions on Cereal Grain Contamination with Deoxynivalenol in Baltic Sea Region. *Toxins* **2021**, *13*, 737. [CrossRef] [PubMed]
44. Bryła, M.; Waśkiewicz, A.; Podolska, G.; Szymczyk, K.; Jędrzejczak, R.; Damaziak, K.; Sułek, A. Occurrence of 26 Mycotoxins the Grain of Cereals Cultivated in Poland. *Toxins* **2016**, *8*, 160. [CrossRef] [PubMed]
45. Stępień, Ł.; Chełkowski, J. Fusarium Head Blight of Wheat: Pathogenic Species and Their Mycotoxins. *WMJ* **2010**, *3*, 107–1 [CrossRef]
46. Buerstmayr, H.; Ban, T.; Anderson, J.A. QTL Mapping and Marker-assisted Selection for *Fusarium* Head Blight Resistance Wheat: A Review. *Plant Breed.* **2009**, *128*, 1–26. [CrossRef]
47. Mesterházy, Á. Role of deoxynivalenol in aggressiveness of *Fusarium graminearum* and *F. culmorum* and in resistance to Fusari head blight. *Eur. J. Plant Pathol.* **2002**, *108*, 675–684. [CrossRef]
48. Boutigny, A.-L.; Richard-Forget, F.; Barreau, C. Natural Mechanisms for Cereal Resistance to the Accumulation of Fusari Trichothecenes. *Eur. J. Plant Pathol.* **2008**, *121*, 411–423. [CrossRef]
49. Foroud, N.A.; Eudes, F. Trichothecenes in Cereal Grains. *Int. J. Mol. Sci.* **2009**, *10*, 147–173. [CrossRef]
50. Siou, D.; Gélisse, S.; Laval, V.; Repinçay, C.; Canalès, R.; Suffert, F.; Lannou, C. Effect of Wheat Spike Infection Timing on Fusari Head Blight Development and Mycotoxin Accumulation. *Plant Pathol.* **2014**, *63*, 390–399. [CrossRef]
51. Walter, S.; Nicholson, P.; Doohan, F.M. Action and Reaction of Host and Pathogen during *Fusarium* Head Blight Dise *New Phytol.* **2010**, *185*, 54–66. [CrossRef]
52. Bulgarelli, D.; Schlaeppi, K.; Spaepen, S.; van Themaat, E.V.L.; Schulze-Lefert, P. Structure and Functions of the Bacter Microbiota of Plants. *Annu. Rev. Plant Biol.* **2013**, *64*, 807–838. [CrossRef]
53. Trivedi, P.; Leach, J.E.; Tringe, S.G.; Sa, T.; Singh, B.K. Plant–Microbiome Interactions: From Community Assembly to Pla Health. *Nat. Rev. Microbiol.* **2020**, *18*, 607–621. [CrossRef] [PubMed]
54. Baard, V.; Bakare, O.O.; Daniel, A.I.; Nkomo, M.; Gokul, A.; Keyster, M.; Klein, A. Biocontrol Potential of Bacillus Subtilis a Bacillus Tequilensis against Four Fusarium Species. *Pathogens* **2023**, *12*, 254. [CrossRef] [PubMed]
55. DuPont, F.M.; Vensel, W.H.; Chan, R.; Kasarda, D.D. Characterization of the 1B-Type ω-Gliadins from *Triticum aestivum* Cultiv Butte. *Cereal Chem.* **2000**, *77*, 607–614. [CrossRef]
56. Szafrańska, A. Comparison of Alpha-Amylase Activity of Wheat Flour Estimated by Traditional and Modern Techniqu *Acta Agrophysica* **2014**, *4*, 493–505.
57. Martínez, M.; Ramírez Albuquerque, L.; Arata, A.F.; Biganzoli, F.; Fernández Pinto, V.; Stenglein, S.A. Effects of *Fusari Graminearum* and *Fusarium Poae* on Disease Parameters, Grain Quality and Mycotoxins Contamination in Bread Wheat (Part *J. Sci. Food Agric.* **2020**, *100*, 863–873. [CrossRef]
58. Polišenská, I.; Vaculová, K.; Jirsa, O.; Sedláčková, I.; Frydrych, J. The Effect of Fusarium Culmorum on Yield and Gra Characteristics of Winter Wheat Cultivars. *Zemdirb. Agric.* **2020**, *107*, 113–122. [CrossRef]
59. Packa, D.; Załuski, D.; Graban, Ł.; Lajszner, W.; Hośnik, M. Reakcja Diploidalnych, Tetraploidalnych i Heksaploidalnych Pszer Na Inokulację Fusarium Culmorum (W.G.Smith) Sacc. *Pol. J. Agron.* **2013**, *12*, 38–48. [CrossRef]

0. Daniel, C.; Triboi, E. Effects of Temperature and Nitrogen Nutrition on the Grain Composition of Winter Wheat: Effects on Gliadin Content and Composition. *J. Cereal Sci.* **2000**, *32*, 45–56. [CrossRef]
1. Daniel, C.; Triboï, E. Changes in Wheat Protein Aggregation during Grain Development: Effects of Temperatures and Water Stress. *Eur. J. Agron.* **2002**, *16*, 1–12. [CrossRef]
2. Ponts, N. Mycotoxins Are a Component of Fusarium Graminearum Stress-Response System. *Front. Microbiol.* **2015**, *6*, 1234. [CrossRef]
3. Peršić, V.; Božinović, I.; Varnica, I.; Babić, J.; Španić, V. Impact of Fusarium Head Blight on Wheat Flour Quality: Examination of Protease Activity, Technological Quality and Rheological Properties. *Agronomy* **2023**, *13*, 662. [CrossRef]

agriculture

MDPI

Article

Foliar Spraying with Endophytic Trichoderma Biostimulant Increases Drought Resilience of Maize and Sunflower

András Csótó [1], György Tóth [1,†], Péter Riczu [2], Andrea Zabiák [3], Vera Tarjányi [4], Erzsébet Fekete [5], Levente Karaffa [5] and Erzsébet Sándor [3,*]

[1] Institute of Plant Protection, Faculty of Agricultural and Food Science and Environmental Management, University of Debrecen, H-4032 Debrecen, Hungary; csoto.andras@agr.unideb.hu (A.C.); gyuribali12@gmail.com (G.T.)

[2] KITE Zrt., H-4181 Nádudvar, Hungary; riczupeter@kite.hu

[3] Institute of Food Science, Faculty of Agricultural and Food Science and Environmental Management, University of Debrecen, H-4032 Debrecen, Hungary; zabiak.andrea@agr.unideb.hu

[4] Department of Pharmacology and Pharmacotherapy, Faculty of Medicine, University of Debrecen, H-4032 Debrecen, Hungary; tarjanyi.vera@med.unideb.hu

[5] Department of Biochemical Engineering, Faculty of Science and Technology, University of Debrecen, Egyetem tér 1, H-4032 Debrecen, Hungary; kicsizsoka@yahoo.com (E.F.); levente.karaffa@science.unideb.hu (L.K.)

* Correspondence: karaffa@agr.unideb.hu

† Present address: Agropoint Kft., Hunyadi János utca 10, H-4026 Debrecen, Hungary.

Abstract: Microbial biostimulants that promote plant growth and abiotic stress tolerance are promising alternatives to chemical fertilizers and pesticides. Although *Trichoderma* fungi are known biocontrol agents, their biostimulatory potential has been scarcely studied in field conditions. Here, the mixture of two endophytic *Trichoderma* strains (*Trichoderma afroharzianum* TR04 and *Trichoderma simmonsii* TR05) was tested as biostimulant in the form of foliar spray on young (BBCH 15-16) maize (5.7 ha) and sunflower (5.7 and 11.3 ha) fields in Hungary. The stimulatory effect was characterized by changes in plant height, the number of viable leaves, and the chlorophyll content, combined with yield sensor collected harvest data. In all trials, the foliar treatment with *Trichoderma* spores increased photosynthetic potential: the number of viable leaves increased by up to 6.7% and the SPAD index by up to 19.1% relative to the control. In extreme drought conditions, maize yield was doubled (from 0.587 to 1.62 t/ha, $p < 0.001$). The moisture content of the harvested seeds, as well as sunflower height, consistently increased post-treatment. We concluded that foliar spraying of young plants with well-selected endophytic *Trichoderma* strains can stimulate growth, photosynthesis, and drought tolerance in both monocot maize and dicots sunflower crops in field conditions.

Keywords: *Trichoderma afroharzianum*; *Trichoderma simmonsii*; drought tolerance; chlorophyll content; maize yield; fungal biostimulant; foliar spray

1. Introduction

The overuse of synthetic chemical fertilizers and pesticides poses significant environmental challenges, contributing to biodiversity loss and raising concerns about soil health, plant integrity, and food safety [1–4]. Additionally, climate change—characterized by severe droughts, storms, and declining biodiversity—further exacerbates these issues by reducing crop yields, diminishing soil fertility, and increasing the prevalence of pests and disease outbreaks [5–8]. In response to these challenges, sustainable crop production must strike a delicate balance by minimizing reliance on chemical fertilizers and pesticides while ensuring stable yields under increasingly adverse environmental conditions [7].

Beneficial microorganisms offer a promising solution by promoting plant growth and alleviating environmental stresses through diverse mechanisms. These microorganisms enhance nutrient uptake and improve the solubilization of phosphorus, potassium, iron, and zinc by producing organic acids and siderophores. Additionally, they help mitigate biotic and abiotic stresses by synthesizing phytohormones, inducing systemic resistance, boosting the activity of defense-related enzymes, and increasing the accumulation of osmotically active substances in plants [9–13]. Furthermore, these fungi can mitigate abiotic stress by activating endogenous plant defense responses and altering plant metabolism [14–16].

Among these microorganisms, *Trichoderma* spp. stand out as versatile fungi capable of functioning as biocontrol agents, plant growth stimulators, and biofertilizers [17–19]. These fungi enhance nutrient availability by producing organic acids and siderophores, which foster robust root development [11]. Furthermore, *Trichoderma* spp. directly promote plant growth through phytohormone production [10,20] and mitigate abiotic stresses by activating endogenous plant defenses and modifying plant metabolism [14–16].

The biocontrol and biostimulant properties of *Trichoderma* spp. have been investigated across various crops, including maize and sunflower, through soil or seed treatments [21–25]. While most research has been conducted in controlled environments, further field studies are necessary to validate these findings and assess the long-term impacts of *Trichoderma* applications. This aspect is particularly significant, as *Trichoderma* spp. predominantly colonize root tissues; however, endophytic strains have also been identified, and their biocontrol potential in annual crops remains to be explored [26,27].

In this study, we evaluated the effects of a commercially formulated endophytic *Trichoderma* biostimulant, applied as a foliar spray, on maize and sunflower under field conditions in Hungary. The product, containing two endophytic strains—*Trichoderma afroharzianum* TR04 and *Trichoderma simmonsii* TR05—was derived from woody grapevine tissues [26]. Our results demonstrate the biostimulant efficacy of this formulation in enhancing crop performance in these key monocot and dicot crops in practice under field conditions.

2. Materials and Methods

2.1. Experimental Sites and Treatments

These experiments were conducted with maize in 2022 at Experimental Site I, with sunflower in 2022 at Experimental Site II, and in 2023, at Experimental Site I. Both experimental sites are located on the outskirts of Hajdúszovát, Hungary, in the South-Hajdúság meso-region of the Great Hungarian Plain.

The soil at both sites is loamy or clay–loam meadow chernozem with a humus content of 3.2–3.5% and a pH of 6.7–7.15 (determined by accredited laboratory). The average altitude of the sites is 92 m above sea level. The GPS coordinates for the center of Experimental Site I (Figure 1a) are 47°23′36″ N 21°24′50″ E, and for Experimental Site II (Figure 1b), they are 47°22′16″ N 21°26′12″ E. In all cases, the field was planted with maize the previous year.

The control and the *Trichoderma*-treated plots were set up in four alternating strips (Figure 1). Sunflower NK Neoma (early) hybrid was seeded on the 21st and 22nd April 2022 and 2023. Maize Dekalb DKC4897 (FAO 390–400) hybrid was also seeded on 21 April 2022. Monosem NG Plus 6 (Monosem Inc., Edwardsville, KS, USA) sowing machine was used in all cases. Plant density was 60,500 plants per hectare for sunflower and 74,000 plants per hectare for maize. All experiments were treated one week after applying post-emergent weed control at the BBCH 15-16 phenological stages of the plants [28]. This phase provides sufficient plant surface for effective colonization by the endophytic strains and allows for combination with herbicide treatment. The application was carried out using a John Deere 6120M tractor equipped with AutoTrac and a StarFire 6000 antenna (John Deere Inc., Mannheim, Germany). The navigation system was utilized, with the machine's working width set and tracks recorded for accurate operation. Control plots were sprayed with water in amounts equivalent to the water content of the Trichoderma formulation. *Trichoderma*-treated plots were sprayed with a freshly prepared solution of Tricho Immun (Danuba, Szentendre, Hungary), which contained a mixture of *Trichoderma afroharzianum* TR04 and *Trichoderma simmonsii* TR05 conidia (2×10^8 CFU/g) on a substrate consisting of glucose and perlite powder. The viability of the conidia was previously tested in the laboratory. The application rate was 1 kg/ha in 300 L of water per hectare, following the manufacturer's protocol [29]. The plot width was 18 m, corresponding to the working width of the Kertitox Revolution (Farmgép, Debrecen, Hungary) trailed field sprayer, which is three times the 6 m working width of the combine harvester (Figure 1c).

(a) (b)

<table>
<tr><td rowspan="4">Border (12 m)</td><td colspan="8">Border (12 m)</td><td rowspan="4">Border (>12 m)</td></tr>
<tr><td>C1</td><td>T1</td><td>C2</td><td>T2</td><td>C3</td><td>T3</td><td>C4</td><td>T4</td></tr>
<tr><td colspan="8">Border (12 m)</td></tr>
<tr><td>18.2 m</td><td>18.2 m</td><td>18.2 m</td><td>18.2 m</td><td>18.2 m</td><td>18.2 m</td><td>18.2 m</td><td>18.2 m</td></tr>
</table>

(c)

Figure 1. Experimental design to study the effect of the endophytic *Trichoderma* formulation. Aerial view of Experimental Site 1 (**a**) and Experimental Site 2 (**b**). Blue lines indicate the borders of the Experimental Sites, red lines indicate the borders of the plots. (**c**) Experimental designs C1–4 indicate control plots; T1–4 indicate plots treated with *Trichoderma*.

Meteorological data were collected with a Campbell Scientific (Campbell Scientific Ltd., Logan, UT, USA) meteorological station with Campbell HygroVUE10 temperature and humidity sensor, Kipp&Zonen CMP-11 pyranometer, and PG-200 weighing rain gauge. It is operated by the Centre for Precision Farming R&D Services, FAFSEM, University of Debrecen. The weather station is located in the Agrometeorological Observatory, Debrecen-Kismacs, approximately 20 km from the experimental sites. The annual meteorological diagram was created using monthly resolution data, while the differences during the vegetation period are based on weekly resolution data.

2.2. Evaluation of Experimental Site Homogeneity Based on NDVI

Normalized Differential Vegetation Index (*NDVI*) was calculated using data from Sentinel-2 satellites (launched by the Global Environment and Security Monitoring Program of the European Space Agency). The calculation was based on red and infrared spectral values from 10 m spatial grids, as described by Rouse et al. (1974) [30]:

$$NDVI = \frac{\lambda_{NIR} - \lambda_{RED}}{\lambda_{NIR} + \lambda_{RED}}$$

where λ_{NIR} is the reflectance value of near-infrared (NIR), and λ_{RED} is the reflectance value of red.

To evaluate the homogeneity of the experimental sites, firstly, NDVI images were created, and then, cloud-covered and cloud-shadowed images were ignored by the Fmask algorithm [31]. Research and practical experience have shown that a stronger correlation existed between NDVI values and yield at certain phenological phases [32]. Consequently, only NDVI time points where the field-level NDVI values exceeded 0.7 were used for homogeneity analysis. Annual average NDVI values were calculated using 10 × 10 m Sentinel-2 raster data for the years from 2017 to 2021. Each year, the analysis included more than five images. The raster data were polygonized using ArcGIS Pro before analysis. To accurately assess NDVI within experimental plots, Sentinel-2 imagery (10 m resolution) was spatially intersected with plot polygons. Grids with less than 80% overlap were excluded to minimize mixed-pixel effects. Subsequently, average NDVI values were calculated for each plot and categorized into five classes (0.2 NDVI intervals) to highlight NDVI differences in the plot. While Experimental Site I exhibited a range of NDVI values, with 5.14% of pixels in the 0.6–0.8 category, all pixels at Experimental Site II were classified as high NDVI (0.8–1.0) (Figure 2).

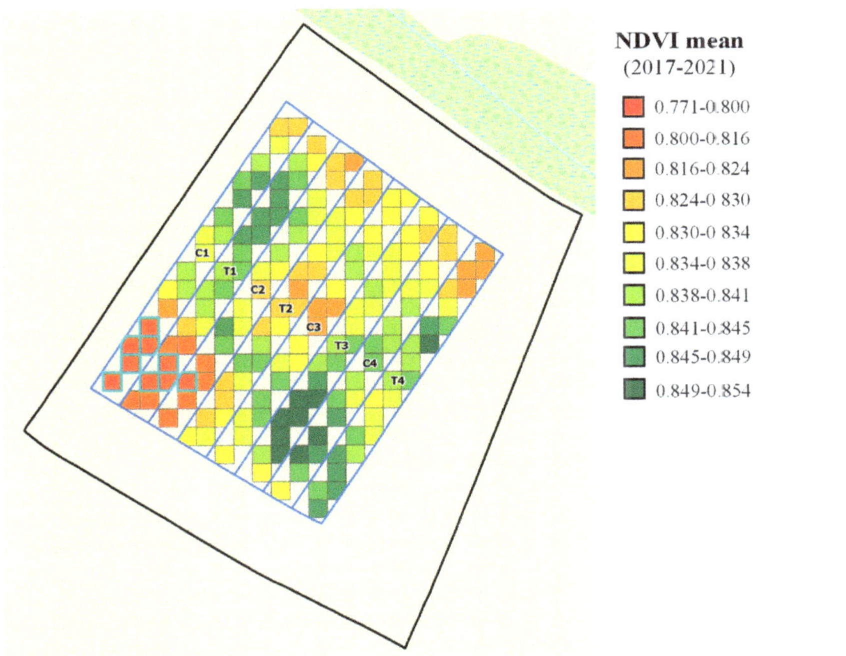

Figure 2. Average NDVI mean values with natural breaks coloring method, using 80% overlapping (Experimental Site I). Light blue polygon outline color shows the lower average NDVI values (below 0.8). C1–4 indicate control plots; T1–4 indicate Trichoderma-treated plots.

2.3. *Measurements of Photosynthetic Potential and Plant Height*

To assess photosynthetic potential, the number of viable leaves was counted, and chlorophyll content was measured weekly from treatment until harvest using a SPAD 502Plus chlorophyll meter (Konica Minolta Sensing Inc., Osaka, Japan). At harvest, plant height for sunflowers was manually measured using a Bosch GR 500 (Robert Bosch Tools GmbH, Leinfelden-Echterdingen, Germany) leveling staff. Due to severe drought damage in 2022, maize plant height could not be assessed. For each plot, SPAD measurements were taken on 20 randomly selected plants. Four points on four upper leaves of each plant were taken per plant and averaged. Viable leaf numbers were also counted on these same 20 plants. Plant height was measured similarly on 20 plants per plot right before harvest. Data from 80 control and 80 Trichoderma-treated plants were used to calculate the average and standard error for each data point.

2.4. Obtaining Harvest Data

The 2022 maize experiment harvest was gathered on 10 October 2022, while the 2022 and 2023 sunflower experiment harvests were gathered on 5 September 2022 and 13 September 2023, respectively. A John Deere S770i combine harvester equipped with an intelligent yield sensor, AutoTrac, and a StarFire 6000 RTK antenna (John Deere Inc., Mannheim, Germany) was used for all harvests. Yield and moisture content data were collected at one-second intervals and filtered using box plot analysis to eliminate outliers. Grid cells were excluded from analysis if they had less than 80% overlap with their assigned treatment area (Figure 2). Additionally, grid cells from Experimental Site I with predominantly low NDVI values (below a threshold of 0.8) were removed, as these areas were considered to have low productivity (Figure 2). Harvester data were spatially aligned with Sentinel-2 NDVI grids, and average yield and moisture values were calculated for each grid cell using the "Summarize Within" function in ArcGIS Pro.

2.5. Statistical Analysis

Data processing and visualization were performed using MS Excel 2016, while descriptive statistics and hypothesis testing were conducted in IBM SPSS 29. The normality assumption for parametric tests was assessed using Q–Q plots and the Shapiro–Wilk test. Homogeneity of variances (homoscedasticity) was evaluated using Levene's test. When both assumptions were satisfied, the parametric *t*-test was employed. The non-parametric Mann–Whitney test was used for comparisons when the data did not meet the normality or homogeneity assumptions.

3. Results

The 2022 growing season was characterized by severe drought and elevated temperatures, particularly during the critical period from May to August (Figure 3).

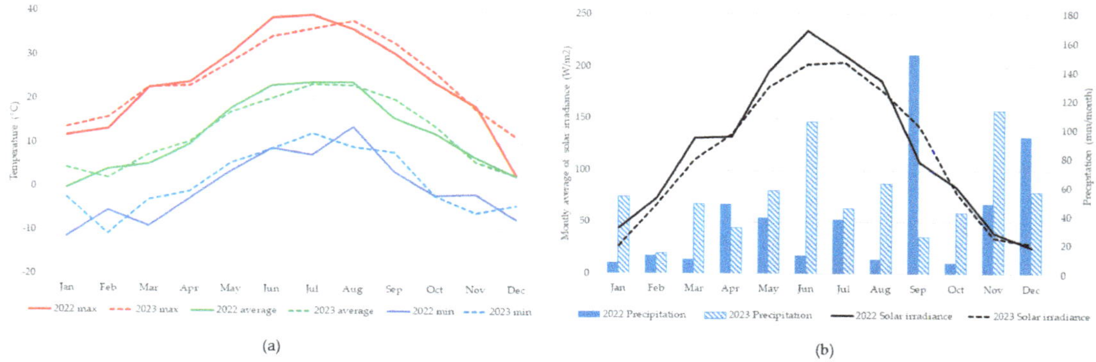

Figure 3. Monthly weather data of the experimental years 2022 and 2023. (**a**) Average, minimal and maximal temperatures. (**b**) Monthly average of solar radiation and precipitation.

The average annual precipitation is 600 mm in this region [33], but it was only 484.6 mm in 2022. A more abundant 662.2 mm precipitation was measured in 2023. Furthermore, the pre-sowing period (October 2021 to March 2022) was drier in 2022, with a precipitation deficit of 156 mm compared to the following year. The growing season of 2022 also had higher average temperatures (+1.5 °C) and increased solar irradiance compared to 2023 (Figure 3).

3.1. Homogeneity of the Experimental Sites

A five-year pre-experimental analysis revealed that all plots at Experimental Site II had highly stable and consistent NDVI values, ranging from 0.813 to 0.845 within the 0.8–0.9 category (Figure 2). However, at Experimental Site I, a small but significant proportion (5.5%) of plots showed lower NDVI values, falling within the 0.7–0.8 range. These plots were excluded from further analyses to ensure homogeneity within the dataset (Figure 2).

3.2. Plant Growth

Given the significant drought stress in 2022, which led to the premature death of corn plants, measuring plant height was not feasible. As the plants did not reach maturity, measuring height would not have provided meaningful data on growth vigor but rather on the timing of desiccation. The sunflower plants were shorter by 50 cm in 2022 compared to 2023. The application of *Trichoderma* species containing commercial products mitigated these drought effects, resulting in a 10 cm increase in sunflower height in 2022 and a 4 cm increase in 2023 (Table 1).

Table 1. Effects of *Trichoderma* TR04 and TR05 strains on the plant height of sunflower at harvest.

	Treatment	Sunflower (2022) (Mean ± SE)	Sunflower (2023) (Mean ± SE)
Plant height (cm)	*Trichoderma*	120.4 ± 2.14	174.9 ± 0.89
	Control	110.5 ± 3.37	171.1 ± 0.93
t-test *p*		0.014	0.003

3.3. Photosynthetic Potential

To evaluate the impact of treatment on photosynthetic potential, the number of viable (green) leaves was determined, and SPAD values were measured. Maize plants treated with *Trichoderma* spp. exhibited significantly increased leaf viability, with the effect becoming apparent as early as two weeks after the foliar application and persisting throughout the growing season (Figure 4a). The application of *Trichoderma* spp. to sunflower plants significantly increased viable leaf numbers, thereby stimulating plant development and prolonging leaf longevity. A statistically significant increase in leaf number was detected three weeks after treatment in 2023 and five weeks after treatment in 2022. This positive effect of *Trichoderma* spp. on leaf development and viability was maintained until the end of the vegetation period (Figure 4b).

Trichoderma treatment significantly increased SPAD values, indicating enhanced chlorophyll content in both maize and sunflower plants. Maize plants showed a rapid response to Trichoderma treatment, with increased SPAD values observed shortly after application. This positive effect persisted throughout the entire growing season (Figure 4c). Sunflower plants exhibited a delayed response, with significant increases in SPAD values observed five weeks post-treatment in 2022 and two weeks post-treatment in 2023 (Figure 4d).

While the foliar application of *Trichoderma* spp. had a positive impact on plant health and photosynthetic parameters, it did not significantly affect NDVI values. The difference in the NDVI between the treated and the control plots was less than 1%, following the treatment on either crop.

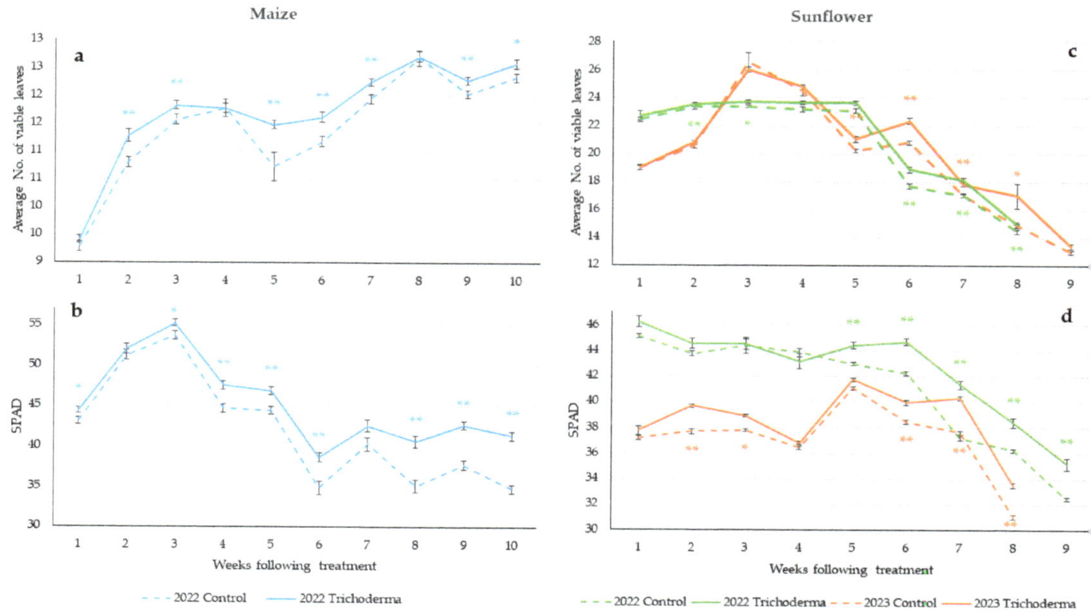

Figure 4. Changes in average number of viable leaves (**a**,**c**) and chlorophyll content (**b**,**d**, measured as SPAD values) of maize (**a**,**b**) and sunflower (**c**,**d**) plants following treatment with *Trichoderma* spp. compared to untreated control plants. Data points represent mean values, and error bars indicate standard error. *: t-test $p < 0.01$, **: t-test $p < 0.001$.

3.4. Yields and Moisture Content

The severe drought in 2022 significantly reduced maize yield to an extremely low 0.587 t/ha. However, Trichoderma treatment mitigated the negative effects of drought, significantly increasing maize yield to 1.62 t/ha (Figure 5a).

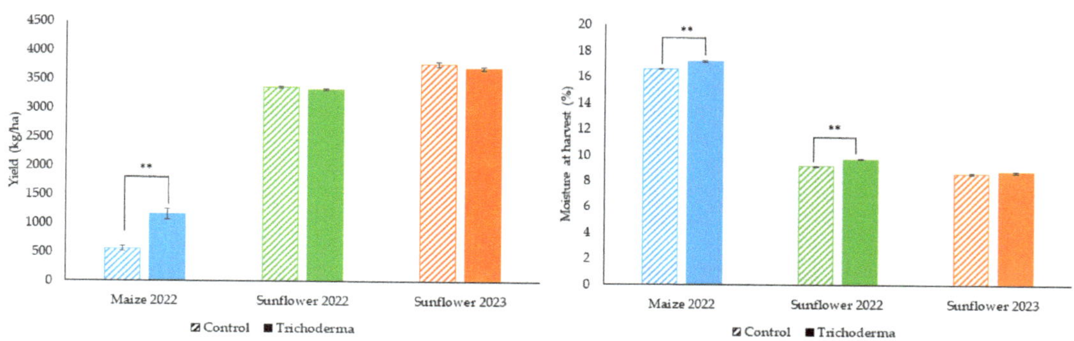

(**a**) (**b**)

Figure 5. Average yield (**a**) and average moisture content of harvested seeds (**b**) of maize and sunflower plants treated with *Trichoderma* spp. compared to untreated controls. Error bars represent standard error. **: t-test $p < 0.001$.

The moisture content of the maize increased significantly from 16.6% to 17.19%, representing only a 0.59% difference between the treated and untreated maize. However, this increase did not affect the quality of the harvested seeds. Similarly, in the drought year (2022), the moisture content of the sunflower increased slightly to 9.7% compared to the 9.16% measured in the seeds from the control plots, a difference of only 0.54%. In 2023, when normal precipitation was detected, neither the yield nor the moisture content of the sunflower showed any significant changes (Figure 5b).

4. Discussion

Climate change, with its increasing frequency and intensity of extreme weather events such as heatwaves and droughts, poses a critical threat to global food security by reducing crop yields and exacerbating pest and disease outbreaks [34,35]. To address these challenges, sustainable agricultural solutions, including the use of biostimulants, are emerging as viable alternatives to traditional chemicals [36,37]. While traditional agricultural practices often depend on chemical inputs, there is a growing demand for more sustainable and environmentally friendly approaches to secure food production in a changing climate [36,37].

Microbial biostimulants, particularly those based on *Trichoderma* species, have shown promise in mitigating the adverse effects of climate change on crop production [38]. These fungi form beneficial associations with plant tissues by producing plant hormones such as indole-3-acetic acid, gibberellic acid, abscisic acid, ethylene, jasmonic acid, and salicylic acid, as well as biostimulant metabolites like lactones and hydrophobins, which regulate growth and stress responses [18,39,40]. These interactions enhance root growth, improve water and nutrient absorption [41], and boost photosynthesis and carbohydrate metabolism, optimizing energy utilization for development [42–44]. *Trichoderma* spp. also produces siderophores, increasing iron availability crucial for metabolic functions, including photosynthesis [45]. Additionally, endophytic *Trichoderma* species have also been shown to induce systemic resistance, thereby enhancing their defense against pathogens and environmental stresses [27,40,46]. While the mechanisms of *Trichoderma* spp. in laboratory settings are well understood, its field application in crops like maize and sunflower remains underutilized. This is partly attributed to the difficulty in achieving field homogeneity, a critical factor for accurate evaluation of biostimulant efficacy in large-scale experiments evaluations [47]. Remote sensing technology employing vegetation indices, such as NDVI, can effectively assess field variability and inform experimental design, ultimately improving the reliability of field trials [48,49].

Maize and sunflower field crops were evaluated under homogeneous field conditions (5.7 and 11.3 ha, respectively). Field homogeneity was assessed using NDVI data from the previous five years. A preceding dry pre-sowing period (January–March 2022; 30.8 mm precipitation) compared to 2023 (117 mm), coupled with increased temperatures (+1.5 °C) and solar radiation during the 2022 growing season, contributed significantly to the severity of the drought. Trichoderma treatment significantly improved the plant's growth and photosynthetic capacity in sunflower plants. The increased number of viable leaves and higher SPAD index, indicative of enhanced chlorophyll content, were observed two weeks post-application and persisted throughout the growing season. These findings suggest that foliar Trichoderma treatment with endophytic strains can promote plant vigor and optimize photosynthetic efficiency. These differences highlight the sunflower's inherent drought tolerance due to its efficient water use and resilience mechanisms [50,51].

For field crops such as maize and sunflower, microbial biostimulants, including Trichoderma products, are typically applied through seed treatment or direct soil application of the root colonizing strains [25,52–55]. While these fungi are typically considered to colonize only the roots of plants [56], their presence has been detected in other plant tissues above the soil [57,58], suggesting that foliar spraying could also be an effective treatment method. While seed and soil treatments are common, foliar applications offer greater flexibility for targeted interventions during specific growth stages or in response to stress or control diseases. Disease symptom expression was not detected in the field of the studied crops. In vitro confrontation tests indicated high biocontrol indexes of both applied *Trichoderma* strains toward several plant pathogen fungi [26]. Field tests necessary for the registration of the product with plant protection effect are underway. Preliminary results indicate the disease Trichoderma formulation tested in this study can repress the Fusarium head blight and sunflower foliar disease. The biocontrol activity of the product may further increase the yield and quality parameters of the harvested grains. Foliar spraying provides the potential for reducing environmental impact and operational costs by enabling combined applications with herbicides. However, ensuring compatibility between *Trichoderma* strains and herbicides is crucial for maintaining efficacy stressors [59].

In summary, microbial biostimulants, particularly Trichoderma-based products, represent a sustainable and flexible solution for enhancing crop resilience and productivity under drought stress, with further research needed to optimize their field applications and integration into agricultural practices.

5. Conclusions

Foliar spraying with the endophytic *Trichoderma* TR04 and TR06 strains stimulates abiotic stress tolerance of the monocot maize and the dicot sunflower under field conditions and would likely have the same effect with other annual crops as well. Pest control that includes spraying with low-cost microbial agents in combination with compatible pesticides (e.g., with post-emergent weed control) may lead to reduced costs and a smaller carbon footprint.

Author Contributions: Conceptualization, A.C., P.R. and E.S.; data curation, A.C., P.R., L.K. and E.S.; formal analysis, E.F., L.K. and E.S.; investigation, G.T., P.R. and A.Z.; methodology, A.C., G.T., P.R. and E.S.; resources, A.C., P.R., L.K. and E.S.; software, A.C., P.R. and V.T.; supervision, A.C. and E.S.; validation, A.C., L.K. and E.S.; visualization, A.C. and P.R.; writing—original draft, A.C., G.T., P.R., A.Z., V.T., E.F., L.K. and E.S.; writing—review and editing, A.C., P.R., L.K. and E.S. All authors have read and agreed to the published version of the manuscript.

Funding: This research was supported by the Hungarian National Research, Development, and Innovation Fund, grants numbers K 146406 to L.K. and K 138489 to E.F.

Institutional Review Board Statement: Not applicable.

Data Availability Statement: The data presented in this study are available upon request from the corresponding author due to the ongoing nature of our research. Additional field data are currently restricted to our research team.

Acknowledgments: The authors are grateful to the Centre for Precision Farming R&D Services, FAFSEM, University of Debrecen, for providing weather data.

Conflicts of Interest: Author Péter Riczu was employed by the company KITE Zrt. E.S. receives a royalty for the Trichoderma product "TrichoImmun" tested in this study. The rest of the authors declare no conflicts of interest.

References

1. Alengebawy, A.; Abdelkhalek, S.T.; Qureshi, S.R.; Wang, M.-Q. Heavy Metals and Pesticides Toxicity in Agricultural Soil and Plants: Ecological Risks and Human Health Implications. *Toxics* **2021**, *9*, 42. [CrossRef]
2. Rani, L.; Thapa, K.; Kanojia, N.; Sharma, N.; Singh, S.; Grewal, A.S.; Srivastav, A.L.; Kaushal, J. An extensive review on the consequences of chemical pesticides on human health and environment. *J. Clean. Prod.* **2021**, *283*, 124657. [CrossRef]
3. Penuelas, J.; Coello, F.; Sardans, J. A better use of fertilizers is needed for global food security and environmental sustainability. *Agric. Food Secur.* **2023**, *12*, 5. [CrossRef]
4. Eliasson, K.; West, C.D.; Croft, S.A.; Green, J.M.H. A spatially explicit approach to assessing commodity-driven fertilizer use and its impact on biodiversity. *J. Clean. Prod.* **2023**, *382*, 135195. [CrossRef]
5. Malhi, G.S.; Kaur, M.; Kaushik, P. Impact of Climate Change on Agriculture and Its Mitigation Strategies: A Review. *Sustainability* **2021**, *13*, 1318. [CrossRef]
6. Eftekhari, M.S. Impacts of Climate Change on Agriculture and Horticulture. In *Climate Change: The Social and Scientific Construct*; Bandh, S.A., Ed.; Springer International Publishing: Cham, Switzerland, 2022; pp. 117–131.
7. Bibi, F.; Rahman, A. An Overview of Climate Change Impacts on Agriculture and Their Mitigation Strategies. *Agriculture* **2023**, *13*, 1508. [CrossRef]

Yuan, X.; Li, S.; Chen, J.; Yu, H.; Yang, T.; Wang, C.; Huang, S.; Chen, H.; Ao, X. Impacts of Global Climate Change on Agricultural Production: A Comprehensive Review. *Agronomy* **2024**, *14*, 1360. [CrossRef]

Bhupenchandra, I.; Chongtham, S.K.; Devi, E.L.; Choudhary, A.K.; Salam, M.D.; Sahoo, M.R.; Bhutia, T.L.; Devi, S.H.; Thounaojam, A.S.; Behera, C.; et al. Role of biostimulants in mitigating the effects of climate change on crop performance. *Front. Plant Sci.* **2022**, *13*, 967665. [CrossRef]

Abdullah, N.S.; Doni, F.; Mispan, M.S.; Saiman, M.Z.; Yusuf, Y.M.; Oke, M.A.; Suhaimi, N.S. Harnessing Trichoderma in Agriculture for Productivity and Sustainability. *Agronomy* **2021**, *11*, 2559. [CrossRef]

Adedayo, A.A.; Babalola, O.O. Fungi That Promote Plant Growth in the Rhizosphere Boost Crop Growth. *J. Fungi* **2023**, *9*, 239. [CrossRef] [PubMed]

Shah, A.; Nazari, M.; Antar, M.; Msimbira, L.; Naamala, J.; Lyu, D.; Rabileh, M.; Zajonc, J.; Smith, D. Corrigendum: PGPR in agriculture: A sustainable approach to increasing climate change resilience. *Front. Sustain. Food Syst.* **2024**, *8*, 1438520. [CrossRef]

Thepbandit, W.; Athinuwat, D. Rhizosphere Microorganisms Supply Availability of Soil Nutrients and Induce Plant Defense. *Microorganisms* **2024**, *12*, 558. [CrossRef] [PubMed]

Chen, M.; Liu, Q.; Gao, S.-S.; Young, A.E.; Jacobsen, S.E.; Tang, Y. Genome mining and biosynthesis of a polyketide from a biofertilizer fungus that can facilitate reductive iron assimilation in plant. *Proc. Natl. Acad. Sci. USA* **2019**, *116*, 5499–5504. [CrossRef]

Martínez-Medina, A.; Van Wees, S.C.M.; Pieterse, C.M.J. Airborne signals from Trichoderma fungi stimulate iron uptake responses in roots resulting in priming of jasmonic acid-dependent defences in shoots of Arabidopsis thaliana and Solanum lycopersicum. *Plant Cell Environ.* **2017**, *40*, 2691–2705. [CrossRef] [PubMed]

Macías-Rodríguez, L.; Contreras-Cornejo, H.A.; Adame-Garnica, S.G.; Del-Val, E.; Larsen, J. The interactions of Trichoderma at multiple trophic levels: Inter-kingdom communication. *Microbiol. Res.* **2020**, *240*, 126552. [CrossRef] [PubMed]

Druzhinina, I.S.; Seidl-Seiboth, V.; Herrera-Estrella, A.; Horwitz, B.A.; Kenerley, C.M.; Monte, E.; Mukherjee, P.K.; Zeilinger, S.; Grigoriev, I.V.; Kubicek, C.P. Trichoderma: The genomics of opportunistic success. *Nat. Rev. Microbiol.* **2011**, *9*, 749–759. [CrossRef] [PubMed]

Sood, M.; Kapoor, D.; Kumar, V.; Sheteiwy, M.S.; Ramakrishnan, M.; Landi, M.; Araniti, F.; Sharma, A. Trichoderma: The "Secrets" of a Multitalented Biocontrol Agent. *Plants* **2020**, *9*, 762. [CrossRef] [PubMed]

Andrzejak, R.; Janowska, B. Trichoderma spp. Improves Flowering, Quality, and Nutritional Status of Ornamental Plants. *Int. J. Mol. Sci.* **2022**, *23*, 15662. [CrossRef]

Garnica-Vergara, A.; Barrera-Ortiz, S.; Muñoz-Parra, E.; Raya-González, J.; Méndez-Bravo, A.; Macías-Rodríguez, L.; Ruiz-Herrera, L.F.; López-Bucio, J. The volatile 6-pentyl-2H-pyran-2-one from Trichoderma atroviride regulates Arabidopsis thaliana root morphogenesis via auxin signaling and ETHYLENE INSENSITIVE 2 functioning. *New Phytol.* **2015**, *209*, 1496–1512. [CrossRef]

Webber, H.; Ewert, F.; Olesen, J.E.; Müller, C.; Fronzek, S.; Ruane, A.C.; Bourgault, M.; Martre, P.; Ababaei, B.; Bindi, M.; et al. Diverging importance of drought stress for maize and winter wheat in Europe. *Nat. Commun.* **2018**, *9*, 4249. [CrossRef] [PubMed]

Sah, R.P.; Chakraborty, M.; Prasad, K.; Pandit, M.; Tudu, V.K.; Chakravarty, M.K.; Narayan, S.C.; Rana, M.; Moharana, D. Impact of water deficit stress in maize: Phenology and yield components. *Sci. Rep.* **2020**, *10*, 2944. [CrossRef] [PubMed]

Kim, K.A.-O.; Lee, B.M. Effects of Climate Change and Drought Tolerance on Maize Growth. *Plants* **2023**, *12*, 3548. [CrossRef]

Pilorgé, É. Sunflower in the global vegetable oil system: Situation, specificities and perspectives. *OCL* **2020**, *27*, 34. [CrossRef]

Gaikwad, N.; Verma, S. Effect of Trichoderma harzianum on Growth of Corn under Water Stress Condition. *Int. J. Plant Soil. Sci.* **2024**, *36*, 447–454. [CrossRef]

Kovács, C.; Csótó, A.; Pál, K.; Nagy, A.; Fekete, E.; Karaffa, L.; Kubicek, C.P.; Sándor, E. The Biocontrol Potential of Endophytic Trichoderma Fungi Isolated from Hungarian Grapevines. Part I. Isolation, Identification and In Vitro Studies. *Pathogens* **2021**, *10*, 1612. [CrossRef]

Harman, G.E. Integrated Benefits to Agriculture with Trichoderma and Other Endophytic or Root-Associated Microbes. *Microorganisms* **2024**, *12*, 1409. [CrossRef] [PubMed]

Meyer, R.B.D.; O'Brien, D.; Darling, R. *High Plains Sunflower Production Handbook*; MF-2384; Kansas State University Agricultural Experiment Station and Cooperative Extension Service: Manhattan, Kansas, 2009.

NÉBIH. Tricho Immun Felhasználási és Forgalomba Hozatali Okirat. 6700/1484-2/2023 and 6300/1283-1/2022. Available online: https://webadmin.danuba.hu/storage/uploads/939483a7-ab54-46b3-a81a-30cffc24f7dd/Tricho-Immun-mo%CC%81dosi%CC%81tott-NE%CC%81BIH-engede%CC%81ly-oki%CC%81rat.pdf (accessed on 14 December 2024).

Rouse, J.W., Jr.; Haas, R.H.; Schell, J.A.; Deering, D.W. Monitoring Vegetation Systems in the Great Plains with Erts. In *NASA Special Publication*; Freden, S.C., Mercanti, E.P., Becker, M.A., Eds.; NASA Special Publications: Greenbelt, MD, USA, 1974; Volume 351, p. 309.

Zhu, Z.; Wang, S.; Woodcock, C. Improvement and expansion of the Fmask algorithm: Cloud, cloud shadow, and snow detection for Landsats 4-7, 8, and Sentinel 2 images. *Remote Sens. Environ.* **2015**, *159*, 269–277. [CrossRef]

Bolton, D.K.; Friedl, M.A. Forecasting crop yield using remotely sensed vegetation indices and crop phenology metrics. *Agric. For. Meteorol.* **2013**, *173*, 74–84. [CrossRef]

HungaroMet. Precipitation Conditions of Hungary. Available online: https://www.met.hu/en/eghajlat/magyarorszag_eghajlata/altalanos_eghajlati_jellemzes/csapadek/ (accessed on 14 December 2024).

34. European Comission. Consequences of Climate Change. Available online: https://climate.ec.europa.eu/climate-chan consequences-climate-change_en (accessed on 10 November 2024).
35. FAO. Climate Related Transboundary Pests and Diseases. In Climate Change, Energy and Food. Available onl https://openknowledge.fao.org/server/api/core/bitstreams/9f31b468-7bfe-4fdf-8d2d-d61013176ef1/content (accessed or November 2024).
36. Reynolds, M.P.; Quilligan, E.; Aggarwal, P.K.; Bansal, K.C.; Cavalieri, A.J.; Chapman, S.C.; Chapotin, S.M.; Datta, S.K.; Duvei E.; Gill, K.S.; et al. An integrated approach to maintaining cereal productivity under climate change. *Glob. Food Secur.* **201** 9–18. [CrossRef]
37. Lahlali, R.; Taoussi, M.; Laasli, S.-E.; Gachara, G.; Ezzouggari, R.; Belabess, Z.; Aberkani, K.; Assouguem, A.; Meddich, A. Jarroudi, M.; et al. Effects of climate change on plant pathogens and host-pathogen interactions. *Crop Environ.* **2024**, *3*, 159– [CrossRef]
38. Woo, S.L.; Hermosa, R.; Lorito, M.; Monte, E. Trichoderma: A multipurpose, plant-beneficial microorganism for eco-sustaina agriculture. *Nat. Rev. Microbiol.* **2023**, *21*, 312–326. [CrossRef] [PubMed]
39. Harman, G.E.; Howell, C.R.; Viterbo, A.; Chet, I.; Lorito, M. Trichoderma species—Opportunistic, avirulent plant symbionts. *N Rev. Microbiol.* **2004**, *2*, 43–56. [CrossRef]
40. Nephali, L.; Moodley, V.; Piater, L.; Steenkamp, P.; Buthelezi, N.; Dubery, I.; Burgess, K.; Huyser, J.; Tugizimana, F. A Metabolo Landscape of Maize Plants Treated With a Microbial Biostimulant Under Well-Watered and Drought Conditions. *Front. Plant* **2021**, *12*, 676632. [CrossRef]
41. Fiorentino, N.; Ventorino, V.; Woo, S.L.; Pepe, O.; De Rosa, A.; Gioia, L.; Romano, I.; Lombardi, N.; Napolitano, M.; Colla, G.; e Trichoderma-Based Biostimulants Modulate Rhizosphere Microbial Populations and Improve N Uptake Efficiency, Yield, a Nutritional Quality of Leafy Vegetables. *Front. Plant Sci.* **2018**, *9*, 743. [CrossRef]
42. Shoresh, M.; Harman, G.E.; Mastouri, F. Induced systemic resistance and plant responses to fungal biocontrol agents. *Annu. R Phytopathol.* **2010**, *48*, 21–43. [CrossRef]
43. Sridharan, A.P.; Sugitha, T.; Karthikeyan, G.; Nakkeeran, S.; Sivakumar, U. Metabolites of Trichoderma longibrachiatum E inhibits soil borne pathogen, Macrophomina phaseolina by triggering amino sugar metabolism. *Microb. Pathog.* **2021**, *150*, 104 [CrossRef]
44. Shoresh, M.; Harman, G.E. The molecular basis of shoot responses of maize seedlings to Trichoderma harzianum T22 inoculati of the root: A proteomic approach. *Plant Physiol.* **2008**, *147*, 2147–2163. [CrossRef] [PubMed]
45. Srivastava, S.N.; Singh, V.; Awasthi, S.K. Trichoderma induced improvement in growth, yield and quality of sugarcane. *Su Tech.* **2006**, *8*, 166–169. [CrossRef]
46. Pieterse, C.M.; Zamioudis, C.; Berendsen, R.L.; Weller, D.M.; Van Wees, S.C.; Bakker, P.A. Induced systemic resistance by benefi microbes. *Annu. Rev. Phytopathol.* **2014**, *52*, 347–375. [CrossRef]
47. Neuhoff, D.; Neumann, G.; Weinmann, M. Testing plant growth promoting microorganisms in the field—A proposal for standa *Front. Plant Sci.* **2023**, *14*, 1324665. [CrossRef] [PubMed]
48. Omia, E.; Bae, H.; Park, E.; Kim, M.S.; Baek, I.; Kabenge, I.; Cho, B.-K. Remote Sensing in Field Crop Monitoring: A Comprehens Review of Sensor Systems, Data Analyses and Recent Advances. *Remote Sens.* **2023**, *15*, 354. [CrossRef]
49. Mohr, J.; Tewes, A.; Ahrends, H.; Gaiser, T. Assessing the Within-Field Heterogeneity Using Rapid-Eye NDVI Time Series D *Agriculture* **2023**, *13*, 1029. [CrossRef]
50. Killi, D.; Bussotti, F.; Raschi, A.; Haworth, M. Adaptation to high temperature mitigates the impact of water deficit duri combined heat and drought stress in C3 sunflower and C4 maize varieties with contrasting drought tolerance. *Physiol. Plant* **20** *159*, 130–147. [CrossRef] [PubMed]
51. He, R.; Tong, C.; Wang, J.; Zheng, H. Comparison of Water Utilization Patterns of Sunflowers and Maize at Different Fertil Stages along the Yellow River. *Water* **2024**, *16*, 198. [CrossRef]
52. Degani, O.; Dor, S. Trichoderma Biological Control to Protect Sensitive Maize Hybrids against Late Wilt Disease in the Field *Fungi* **2021**, *7*, 315. [CrossRef] [PubMed]
53. Estévez-Geffriaud, V.; Vicente, R.; Vergara-Díaz, O.; Narváez Reinaldo, J.J.; Trillas, M.I. Application of *Trichoderma asperellum* T on maize (*Zea mays*) seeds protects against drought stress. *Planta* **2020**, *252*, 8. [CrossRef] [PubMed]
54. Güçlü, T.; Özer, N. Trichoderma harzianum antagonistic activity and competition for seed colonization against seedbor pathogenic fungi of sunflower. *Lett. Appl. Microbiol.* **2022**, *74*, 1027–1035. [CrossRef]
55. Lian, H.; Chen, Y.-r.; Li, M.; Li, R.-z.; Zhang, T.; Ma, G.-s. Effects of Trichoderma on physiological characteristics of sunflow seedlings and control efficacy against Sclerotinia sclerotiorum. *Agric. Res. Arid. Areas* **2023**, *41*, 169–177. [CrossRef]
56. Alonso-Ramírez, A.; Poveda J Fau—Martín, I.; Martín I Fau—Hermosa, R.; Hermosa R Fau—Monte, E.; Monte E Fau—Nicol C.; Nicolás, C. Salicylic acid prevents Trichoderma harzianum from entering the vascular system of roots. *Mol. Plant Pathol.* **20** *15*, 823–831. [CrossRef] [PubMed]
57. Csótó, A.; Kovács, C.; Pál, K.; Nagy, A.; Peles, F.; Fekete, E.; Karaffa, L.; Kubicek, C.P.; Sándor, E. The Biocontrol Potential Endophytic Trichoderma Fungi Isolated from Hungarian Grapevines, Part II, Grapevine Stimulation. *Pathogens* **2023**, *12*, [CrossRef] [PubMed]

-. Carro-Huerga, G.; Compant, S.; Gorfer, M.; Cardoza, R.E.; Schmoll, M.; Gutiérrez, S.; Casquero, P.A. Colonization of Vitis vinifera L. by the Endophyte Trichoderma sp. Strain T154: Biocontrol Activity Against Phaeoacremonium minimum. *Front. Plant Sci.* **2020**, *11*, 1170. [CrossRef] [PubMed]
-. Preininger, C.; Sauer, U.; Bejarano, A.; Berninger, T. Concepts and applications of foliar spray for microbial inoculants. *Appl. Microbiol. Biotechnol.* **2018**, *102*, 7265–7282. [CrossRef] [PubMed]

agriculture

MDPI

Article

Growth Parameters, Yield and Grain Quality of Different Winter Wheat Cultivars Using Strip Tillage in Relation to the Intensity of Post-Harvest Soil Cultivation

Marcin Różewicz *, Jerzy Grabiński and Marta Wyzińska

Department of Crops and Yield Quality, Institute of Soil Science and Plant Cultivation—State Research Institute, 8 Czartoryskich Street, 24-100 Pulawy, Poland

* Correspondence: mrozewicz@iung.pulawy.pl

Abstract: The research has been undertaken to determine whether it is worthwhile to do a post-tillage on stubble before applying strip-till or whether tillage operations such as tillage and stubble ploughing should be performed. Therefore, ploughed tillage + strip tillage (PT), stubble discing + strip tillage (SD) and strip tillage (ST) operations were evaluated on three genetically distant winter wheat cultivars, including Formacja, Metronom and Desamo. A three-year field experiment was conducted from 2018 to 2021 at the Agricultural Experimental Station Kepa-Osiny in Pulawy, Poland. The experiment design was a split-block design with four repetitions of every treatment. The results showed that the cultivars differed in dry matter growth. However, no differences were found between the cultivar and post-harvest tillage method in terms of dry matter, plant height, and flag leaf area. Grain yield per ear was the main factor of yield variation across the cultivar and tillage systems. The extent of tillage only in the case of previously performed ploughing had an effect on the thousand grain weight. On the other hand, the omission of post-harvest tillage (ST) had a positive effect on the sedimentation index value. In terms of wheat grain yield, plough tillage (PT) proved to be the most advantageous, while reducing the intensity of tillage caused a systematic decrease in yield by 6% in the SD treatment and 9% in the ST treatment, respectively. Other quality parameters (gluten quantity, gluten index, falling number) did not depend on the applied tillage range. The response of cultivars to the applied cultivation methods was generally similar. Due to the beneficial effect of reducing the scope of cultivation on the environment, a small reduction in yield and no negative impact on the quality characteristics of grain, it is recommended to use strip-till cultivation without prior post-harvest cultivation. The results provide new insights into the growth of different winter wheat cultivars and the postharvest tillage applied, and they can be used in the future to validate existing wheat growth models.

Keywords: crop residue management; cultivation systems; dry plant mass; plant growth; yield structure; plant development

1. Introduction

Wheat is one of the most important cereal crops grown worldwide [1]. The global wheat sown area is around 214.3 million hectares. The European Union (EU) accounts for 22.9 million hectares, and Poland ranks third in terms of wheat production volume in the EU [2–4]. Due to the high popularity of the species, research is often conducted on the appropriate agrotechnology for this cereal. In recent years, research topics have focused on minimising the negative impact of agricultural production on the natural environment and on issues related to the potential of increasing carbon sequestration to limit the rate of climate change. This is the reason why research has concentrated in recent years towards reducing the use of plough tillage, whose role in carbon dioxide emission is very large [5].

Tillage also plays a decisive role in shaping plant growth conditions, which also directly determines the productive and economic effects obtained by the farmer. It can be implemented in the traditional manner with full ploughing, but it can also be limited to shallow cover crops or eliminated altogether (direct drilling). Simplified tillage reduces negative environmental effects and increases the potential for carbon sequestration in the arable layer, but it also carries possible negative consequences in terms of lower yield [6–10]. Conservative cultivation systems, including strip tillage with stubble leaving, also influence the retention of nutrients in the soil and improve their balance, especially sulphur, which has a positive effect on the sulphur balance and the sustainability of agricultural production in soil management [11]. Ploughless cultivation, along with leaving as much crop residue as possible, can have a positive effect on soil quality, including the content of organic matter, and thus can increase the yield in wheat cultivation [12]. A particular cultivation method with the possibility of simultaneous sowing that combines the advantages of ploughing and no-till is strip-till [13]. There are reasonable views that the use of strip-till produces the same higher yields as plough tillage or zero tillage [14]. Advantages of this method include: aeration of the soil in the rhizosphere, faster heating of the soil in the strip-till, prevention of wind erosion, and a reduction in the loss of soil clay and silt particles—responsible for the soil sorption complex. Increased carbon and nitrogen content is also a beneficial effect of strip-till [15,16]. The wider row spacing used in this cultivation method may favour an increased grain yield as a result of a more efficient photosynthetic process in the plants [17].

The scientific literature on minimising plough tillage and replacing it with simplification or direct drilling is relatively rich, but most of the work published so far has dealt with the effects of different tillage methods on crop yields and quality. However, there is a lack of works that fully describe the process of shaping the final yield by assessing plant growth during the growing season in relation to the different cultivation methods. Only some works describe the growth of the root system and the aboveground weight of winter wheat in relation to tillage [18–20]. In addition, Fu et al. [21] noted that monitoring plant growth and development in a given environment is an important way to understand phenomena occurring in the soil environment. In the case of wheat, cultivation intensity and cultivar choice have been found to have some influence [22,23]. The rate of biomass growth has a direct impact on wheat grain yield and the carbon cycle [24–28]. Thus, it can be said that the determination of temporal and spatial variability of aboveground plant weight in wheat provides basic information on plant growth, but furthermore allows estimation of potential yields in a given growing season [29–31]. Conventional methods for assessing plant growth involving sampling green plants per unit area and drying them and later assessing the dry matter yield are extremely time-and labour-intensive [26,27,32], and hence, there are not many examples of work using such methods in the world literature. Therefore, this study aimed to determine the growth parameters of winter wheat grown using the strip-till method in relation to the extent of post-harvest tillage and cultivar. The research hypothesis was that both the extent of post-harvest tillage applied and the cultivar would significantly affect the plant growth rate, which would ultimately determine the level of grain yield obtained.

2. Materials and Methods

2.1. Field Conditions and Setup of the Experiment

The research hypothesis was verified by field experiments carried out in the three growing seasons: 2018/2019, 2019/2020, 2020/2021. The experiment was located at the Agricultural Experimental Station Kępa-Osiny (51°27′ N; 22°2′ E) belonging to the Institute of Soil Science and Plant Cultivation-State Research Institute in Pulawy-Poland. The soil was classified as a Gleyic Phaeozems (according to the World Reference Base for Soil Resources). Winter wheat was used as a forecrop in each year.

The field experiment was set up in the split-block method in four replications for each treatment. Three different tillage methods were applied according to Figure 1. The depth of plough cultivation was 20 cm, of the cultivator 8–10 cm, and in strip-till cultivation in the cultivation strips 18–20 cm. The distance between two rows of plants in the cultivated strip was 12 cm and the distance between the planting strips was 36 cm. The second experimental factor featured three wheat cultivars of considerably varying origin (from different breeders). They were selected taking into account all available varietal traits, mainly the resistance to biotic and abiotic stresses. The characteristics of the selected cultivars are presented in Figure 2.

CHARACTERISTICS OF THE POST-HARVEST CULTIVATION IN THE VARIANTS USED IN THE EXPERIMENT

1.- Plouged tillage + strip tillage (PT)

plow post-harvest medium-deep ploughing, followed by strip tillage combined with wheat sowing

2. - Stubble discing + strip tillage (SD)

post-harvest stubble loosening with a disc harrow, followed by strip tillage combined with wheat sowing

3. Strip tillage (ST)

without any post-harvest cultivation, only strip-till cultivation was carried out together with wheat sowing

Image from an experimental plot Plouged tillage + strip tillage (PT).
Crop residues have been ploughed in and are no longer on the soil surface.

Image from an experimental plot Stubble discing + strip tillage (SD).
The stubble has been disturbed, but crop residues are still on the soil surface.

Image from an experimental plot Strip tillage (ST).
The stubble remains were not disturbed in the post-harvest cultivation, there are much more of them than in the SD treatment

Figure 1. Characteristics of treatments using different post-harvest cultivation.

Characteristics of winter wheat varieties used in the study

FORMACJA

BREEDING COMPANY: Poznańska Hodowla Roślin

- quality variety, group A
- very high yield potential
- good resistance to fungal diseases
- good tolerance to soil acidification
- medium plants with good resistance to lodging
- good frost resistance
- exceptional tillering strength
- 1000 grain mass and high uniformity

good resistance to sprouting in the spike

METRONOM

BREEDING COMPANY: RAGT Saaten

- quality variety, group A
- good and very stable yield,
- does well in all soils and climate conditions,
- high tolerance to soil acidification allows cultivation in all locations,
- good plant health,
- good winter hardiness.

DESAMO

BREEDING COMPANY:DANKO

- quality variety, group A
- high yield potential and very high profitability of cultivation;
- useful for intensive cultivation technology on medium-quality and good soils at different sowing dates;
- good disease resistance;
- earing date – medium early;
- winter hardiness: – very good.

Figure 2. Characteristics of the cultivars used in the research.

In each year of the research, the same setup was used with a combination of different types of cultivation and cultivar. There were four replicates of every treatment. Each plot measured 9 m width and 25 m length, which corresponded to 225 m^2. The detailed setup is shown in Figure 3.

The content of nutrients and pH of soil are presented in Figure 4. The detailed agri-technical calendar is shown in Figure 5.

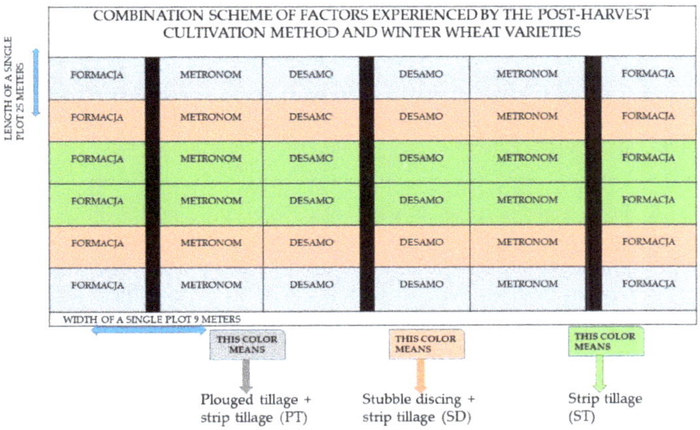

Figure 3. Plan of the applied experiment and combination of factors.

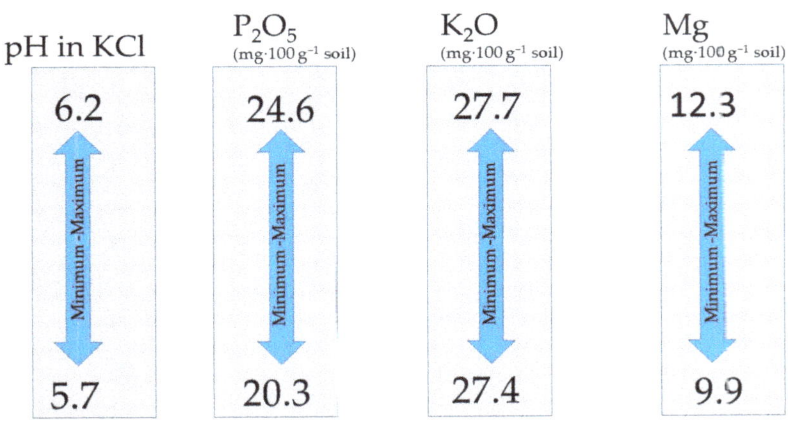

Figure 4. Characteristics of the physicochemical properties of soil.

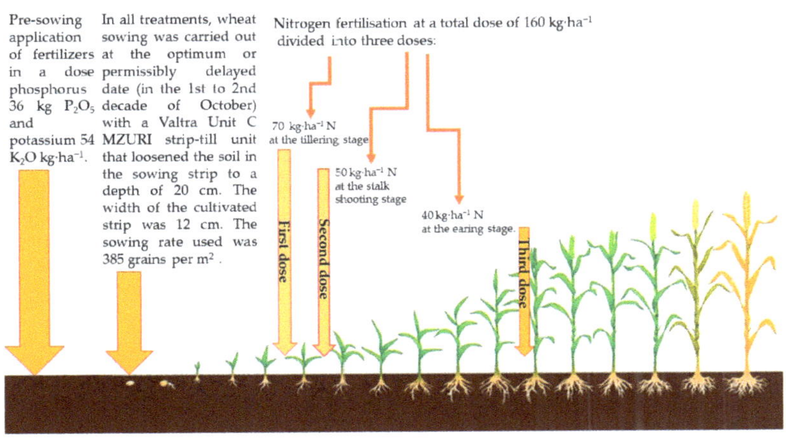

Figure 5. Agricultural technology and fertilisation used in the conducted research.

2.2. Dry Matter Yields at Selected Growth Stages

To assess adhesion parameters, plants were sampled from an area of 1 m^2, from each plot in duplicate, at the following stages

- Bush stage BBCH 29
- Stalking BBCH 32
- Flowering BBCH 59

Prior to placing the samples in the dryer, 10 plants were randomly selected from each sampled plot for measurement of height and flag leaf area, using an Area Meter AM 101 automatic leaf scanner from BioScientific LTD, Hoddesdon, UK. All green matter sampled from each plot was placed in an air-circulating dryer for 72 h, at 55 °C, and then weighed.

Meteorological conditions during the vegetation period of the plants were characterised by mean daily temperature (°C) and precipitation (mm), as well as the perennial averages of these parameters (Table 1).

Table 1. Meteorological conditions during the growing seasons in which the study was conducted.

	Temperature °C				Precipitation (mm)			
	Growing Season			Multi-Year Average	Growing Season			Multi-Year Average
Month	2018/2019	2019/2020	2020/2021	1981–2010	2018/2019	2019/2020	2020/2021	1981–2010
September	15.5	14.4	14.9	13.3	48.0	57.8	102.0	55.0
October	10.0	10.8	10.4	8.0	40.5	33.5	90.0	44.0
December	4.2	6.4	5.1	2.7	8.9	31.4	14.0	39.0
November	0.9	3.1	1.7	−1.4	61.0	47.9	19.0	37.0
January	−2.4	1.7	−1.4	−3.3	62.0	27.1	51.0	31.0
February	2.9	3.4	−2.7	−2.3	15.2	56.5	38.0	30.0
March	5.7	4.7	2.8	1.6	20.9	16.7	12.0	30.0
April	10.0	8.9	6.9	8.7	39.0	14.4	50.0	39.0
May	13.9	11.9	12.9	14.5	69.0	93.9	61.0	58.0
June	22.7	19.1	20.0	17.2	37.0	159.0	53.0	65.0
July	19.4	19.3	22.2	19.5	71.0	31.9	110.0	80.0
August	20.4	20.3	17.1	17.8	94.3	95.5	219.0	87.0

During the study, weather conditions were varied between years. Autumn and winter periods saw large differences in temperature, while meteorological conditions in the spring months were similar, not differing largely from the multi-year average. Each season saw periods with greater or lesser precipitation deficiency, but in general, precipitation totals in each season were relatively high. This was particularly the case in the third season 2020/2021, when precipitation in the preharvest was well above the multi-year average.

2.3. Yield Structure

At full maturity, plants were taken from an area of 1 m^2, two samples from each plot; thus, each treatment was represented by 8 samples in order to determine the yield components. The number of plants and the number of ears were determined in the samples. The number of grains per ear and grain weight per ear were determined in each sample on 10 randomly selected plants. The harvest index was calculated by the division of the grain yield by the sum of the grain yield and straw weight and was expressed in percentage.

Harvest was performed with a combine in the first decade of August, at the stage of full maturity. Following the harvest, grain moisture was determined at 15% moisture content.

2.4. Grain Quality

In order to determine the quality of the grain, representative samples of grain were taken after harvesting, in which the following were determined: the thousand grain weight and the bulk density of the grain (using a densimeter equipped with a 1000 mL cylinder—according to the PN-EN ISO 7971-3 standard [32]), and the quality characteristics of the flour were determined, such as: the wet gluten amount (according to standard PN-A-74042) [33], the gluten index (GI), the falling number and the Zeleny sedimentation index (according to standard PN-EN ISO 5529) [34].

2.5. Statistical Analysis of the Results

The obtained results were developed using analysis of variance (ANOVA) and the Tukey test at a significance level of $p \leq 0.05$ with STATISTICA ver. 13.1 software (StatSoft, Inc., Tulsa, OK, USA).

3. Results and Discussion

3.1. Effects of Tillage Types on Wheat Growth Parameters

Plant growth and final yield are the product of a number of processes, the most important of which is the intensity of photosynthesis, which determines the rate of growth of vegetative mass, which is essential for an adequate supply of essential assimilates to the plant during the generative phase [35]. The accumulated higher vegetative green matter of wheat increases the efficiency of sunlight utilisation, which contributes to a better use of the photosynthetic process in yield [36]. The lack of adequate green matter at any given time increases the danger of severe stress conditioning, e.g., shoot reduction and reduced ear number in wheat [37]. A fast rate of green matter growth is also important due to the higher coverage and increased competitiveness of wheat against weeds, which is also conditioned by the varietal factor [38]. It should be added that the rate of green matter accumulation is also dependent on the ability to take up nutrients from the soil with water. In the present study, the rate of green matter accumulation varied according to the cultivation method. The lowest value of this trait at the tillering stage was found in the PT plot with ploughing, while the highest was in the ST plot without post-tillage. The differences between the above-mentioned treatments exceeded 6% and were, therefore, quite large, although statistically insignificant. This indicates that the use of the plough had a negative effect on plant growth, but only in the initial period of plant development because in subsequent stages, significantly higher weight per unit area was found in the treatments with ploughing. The SD and ST treatments did not differ significantly throughout the growing season with respect to the trait in question (Table 2), so it can be concluded that the conditions for plant growth in these treatments were similar. Research by Lipiec and Nosalewicz [39] showed that the building of aboveground green matter in winter wheat depends on soil compactness and water availability. These researchers concluded that higher soil compaction has a positive effect on dry matter building in winter wheat. The results of our own research did not confirm this because it was on ploughed soil that by far the largest amount of green matter was obtained. Sha et al. [40] showed that colder and more compacted inter-row soil in strip-till cultivation was unfavourable for early root growth, resulting in limited shoot and green matter growth in maize. The fact that wheat seed sowing with a strip-till unit in ST under pre-applied ploughing conditions was slightly deeper than in the treatments with the other cultivation treatments may also have had an influence on the growth conditions associated with better water availability. Such a relationship was shown by Ali et al. [41], who found that wheat grown in furrows, where the rows into which plants were sown, gathered water and accumulated green matter better than plants grown without furrows, on a flat surface. Although, in the study described, this positive effect appeared with a delay—only at stages associated with intensive weight gain.

Table 2. Green dry matter of wheat plants (g·m^2) at different developmental stages according to cropping system, cultivar, and years of study.

Specification	Development Phase		
	Tillering	Stem Formation	Beginning of Earing
Cultivation system			
Ploughed tillage + strip tillage (PT)	97 ^{a}	434 ^{a}	1207 ^{a}
Stubble discing + strip tillage (SD)	101 ^{a}	413 $^{a\,b}$	1151 ^{b}
Strip tillage (ST)	103 ^{a}	405 ^{b}	1157 ^{b}
Cultivar			
Formacja	90 ^{b}	413 ^{b}	1181 ^{a}
Metronom	116 ^{a}	431 ^{a}	1222 ^{a}
Desamo	105 ^{b}	408 ^{b}	1112 ^{b}
Years			
2019	85 ^{b}	358 ^{c}	1329 ^{a}
2020	88 ^{b}	501 ^{a}	987 ^{c}
2021	128 ^{a}	394 ^{b}	1196 ^{b}
Factor interaction			
T	ns	*	*
C	***	ns	***
Y	***	***	***
T × C	ns	ns	**
T × Y	***	*	**
C × Y	**	***	**
T × C × Y	**	*	***

Different letters ($^{a–c}$) mean the significant difference ($p \leq 0.05$) according to ANOVA and Tukey's test. Significant interaction on level p value * ≤ 0.05, ** ≤ 0.01, *** < 0.001, and ns—not significant difference, T—tillage, C—cultivar, Y—year.

As expected, the effect of the cultivar on plant growth rate, as measured by the amount of dry matter per unit area, was also significant. At each growth stage, the highest value for this trait was recorded in the Metronom cultivar. The cultivars Desamo and Formacja had significantly less green matter. However, there was no interaction between the cultivar and the cultivation method, indicating that despite the large variation between cultivars, their response to the cultivation method was the same. Saini et al. [42] identified cultivars that were more efficient in tolerating reduced to zero tillage among the wheat and rice cultivars tested. The reasons for the variation in the response of cultivars to cultivation conditions may be very complex. It appears that the pool of free carbohydrates may play an important role in this regard, allowing plants with a larger supply of carbohydrates to survive the stress better. The size of the root system may also play an important role here. This was pointed out by Tazhibayeva et al. [43], who showed that the development of the root system varies between cultivars and may determine more efficient water uptake under drought conditions and thus mass accumulation. This was also confirmed by results obtained by other authors [17,44–46].

Kumar and Sachan [2], as well as Wilczewski et al. [47], pointed out the large role of mulch in no-tillage in shaping soil moisture. In the studies of these authors, no-tillage and direct seeding into the mulched soil surface had a more beneficial effect on wheat yield than ploughing and irrigation. In the present study, mulch on the soil surface existed in large amounts only in the inter-rows. The strongest surface coverage of mulch (cut straw) was in the inter-rows in the ST, where the degree of soil cover was about 50%. At the SD treatment, the amount of mulch remaining on the surface was already considerably less (around 15%). It seems that the positive effect of mulch on wheat growth and yield was limited due to the relatively high rainfall during the years of the study, which meant that the reduction in water loss from the mulched surface did not have a clearly positive effect in shaping plant growth parameters. Sha et al. [48] found that strip-till plants had the ability to rapidly adapt and recover from abiotic stresses due to which green matter was comparable in strip-till and plough tillage. Our research showed that the system of applied post-harvest cultivation together with strip tillage had a significant interaction in combination with the accumulated green mass, but only in two phases—the stem formation and the beginning of earing. In the case of cultivars, a highly significant interaction (p value < 0.001) was found in the tillering and the beginning of earing phases. Years also had a significant iteration in each phase studied (p value < 0.001). Interactions between individual experimental factors were also significant. We found a non-significant interaction only between T × C in two developmental phases: tillering and stem formation. Rieger et al. [49] also showed that the cultivation system had no significant interaction with the green mass of plants, which we also noted in our own studies. Plaza-Bonilla et al. [50] also demonstrated the lack of interactions, indicating that green matter and root matter develop the same in different tillage systems.

The flag leaf is an organ that plays a very important role in shaping the assimilation process and, consequently, the yield of wheat plants. Its larger surface area promotes a higher intensity of photosynthesis-related processes [51,52]. In our study, neither the crop used, nor the cultivar had a significant effect on this trait of wheat plants (Table 3). However, a strong effect of years related to the occurrence of different weather conditions was found. As indicated by Yang et al. [53], the area of the flag leaf depends precisely on weather conditions and is smaller under stress conditions. In particular, drought stress associated with water scarcity negatively affects flag leaf area in wheat [54,55]. The lack of differences in the trait in question according to the experimental factors suggests that the magnitude of drought stress was similar in all treatments. We found a significant interaction in flag leaf area and other factors only between years (Y), and also between factors C × Y and T × C × Y.

Table 3. Flag leaf area (cm^2) according to cultivation method, cultivar and years of study.

Specification	Flag Leaf Area (cm^2)
Cultivation system	
Ploughed tillage + strip tillage (PT)	22.5 [a]
Stubble discing + strip tillage (SD)	22.7 [a]
Strip tillage (ST)	23.2 [a]
Cultivar	
Formacja	25.5 [a]
Metronom	25.3 [a]
Desamo	24.9 [a]
Years	
2019	20.1 [c]
2020	23.2 [a]
2021	21.6 [b]

Table 3. *Cont.*

Specification	Flag Leaf Area (cm^2)
Factor interaction	
T	ns
C	ns
Y	***
T × C	ns
T × Y	ns
C × Y	*
T × C × Y	*

Different letters ($^{a–c}$) mean the significant difference ($p \leq 0.05$) according to ANOVA and Tukey's test. Significant interaction on level p value * ≤ 0.05, *** < 0.001, and ns—not significant difference, T—tillage, C—cultivar, Y—year.

The height of the plants at the different phases depended significantly on the cultivation method used. Wheat plants were highest at the stem formation and earing stages in the ST treatment (Table 4). Plant height, on the other hand, did not depend significantly on the cultivar, although it is worth noting the slightly higher height of the Metronom cultivar at the tillering and stem formation stages. This trend was reversed at the beginning of earing when plants of the Metronom cultivar were characterised by the lowest height compared to the other cultivars. It is a well-known fact that plants vary greatly depending on the weather conditions specific to the vegetation period. Künze et al. [56] also showed a significant effect of years on plant height than the cultivar itself. In our own research, the colder April in 2021 could have slowed down plant growth and accelerated the course of further phenological phases associated with the photoperiod, which could have resulted in a lower plant height this year compared to 2019 and 2020. The research by Wilhelm et al. [57] confirms such a relationship, which was found in our own research. All of the experimental factors, T, C and Y, showed highly significant interactions (p value < 0.001) with plant height in individual phases. We also found highly significant interactions between factors C × Y; C × Y and T × C × Y (p value < 0.001). We did not find a significant interaction only between T × C in the stem formation phase.

Table 4. Winter wheat plant height (cm) at different developmental stages according to cropping system, cultivar and years of study.

Specification	Development Phase		
	Tillering	Stem Formation	Beginning of Earing
Cultivation system			
Ploughed tillage + strip tillage (PT)	27.3 ^{a}	51.2 ^{a}	86.9 ^{b}
Stubble discing + strip tillage (SD)	26.9 ^{a}	50.6 ^{a}	86.3 ^{b}
Strip tillage (ST)	27.7 ^{a}	53.4 ^{a}	90.1 ^{a}
Cultivar			
Formacja	26.2 ^{a}	50.9 ^{a}	89.3 ^{a}
Metronom	28.6 ^{a}	52.2 ^{a}	85.4 ^{a}
Desamo	26.6 ^{a}	50.5 ^{a}	93.2 ^{a}

Table 4. *Cont.*

Specification	Development Phase		
	Tillering	Stem Formation	Beginning of Earing
Years			
2019	27.0 [a]	50.6 [b]	91.2 [b]
2020	25.9 [a]	64.7 [a]	97.1 [a]
2021	19.0 [b]	41.1 [c]	61.2 [c]
Factor interaction			
T	***	***	***
C	***	***	***
Y	***	***	***
T × C	*	ns	***
T × Y	***	***	***
C × Y	***	***	***
T × C × Y	***	***	***

Different letters ([a–c]) mean the significant difference ($p \leq 0.05$) according to ANOVA and Tukey's test. Significant interaction on level p value * ≤ 0.05, *** < 0.001, and ns—not significant difference, T—tillage, C—cultivar, Y—year.

3.2. *Effects of Tillage Types on Wheat Yield*

Grain weight was significantly higher for the PT treatment compared to the SD and ST treatments (Table 5). The other two treatments showed no significant differences between them. Straw weight was also significantly higher in the PT treatment than in the SD and ST treatments, by 9.5 and 10.1%, respectively.

The varietal factor significantly differentiated wheat grain yields. Formacja showed a significantly higher grain yield than cultivar Desamo. In contrast, the cultivars showed no significant differences between each other in terms of straw weight. The harvest index value is a trait that determines the ratio between accumulated vegetative green matter and the main yield, which in the case of wheat is grain [58]. White and Wilson [59] indicate that the limiting harvest index (HI) value for wheat to guarantee the highest possible yield is 0.55. In their study, this value was slightly lower and did not depend on the cropping system and cultivar. In some studies [60], the role of cultivation system and cultivar in shaping the HI was greater than in our study. Grain mass g·m^{-2} showed a highly significant interaction between T and Y (p value < 0.001), and a significant interaction between C (p value $\leq$ 0.05). Straw mass had a highly significant interaction with Y (p value < 0.001) and a significant interaction with T (p value $\leq$ 0.01). In addition, there was a highly significant interaction between the factors T × Y (p value < 0.001), and a significant interaction between C × Y (p value $\leq$ 0.05). In the Harvest index, we also found a highly significant interaction between C and Y (p value < 0.001). We also found a high interaction between T × Y and C × Y (p value < 0.001).

The role of the tillage method in shaping cereal yields may result from its influence on plant density [61]. In the presented study, however, the range of tillage applied did not significantly affect the size of this wheat canopy trait (Table 6), although it is undoubtedly noteworthy that there was a tendency for the number of plants per unit area to decrease in the treatment with the least intensive tillage (ST). Similarly, Wesołowski and Cierpiała [62] showed a lower wheat grain yield when post-harvest tillage was reduced in winter wheat cultivation. Also, the genetic factor did not have a significant effect on this wheat canopy trait, although the tendency towards a lower value of this trait in the Formacja cultivar was quite pronounced. However, the main role in shaping grain yield, according to cultivar, was played by yield per ear, which varied a lot over the years—a generally known relationship resulting from the influence of weather conditions on this trait [63–65].

Table 5. Grain weight, straw and harvest index value according to crop, cultivar and years of study.

	Grain Weight g·m^{-2}	Straw Weight g·m^{-2}	Harvest Index (%)
	Cultivation system		
Ploughed tillage + strip tillage (PT)	970.7 [a]	953.2 [a]	0.51 [a]
Stubble discing + strip tillage (SD)	886.3 [b]	891.9 [a b]	0.50 [a]
Strip tillage (ST)	881.0 [b]	861.8 [b]	0.51 [a]
	Cultivar		
Formacja	955.0 [a]	887.4 [a]	0.52 [a]
Metronom	916.4 [a b]	890.7 [a]	0.51 [a]
Desamo	866.7 [b]	928.8 [a]	0.49 [a]
	Years		
2019	1027.0 [a]	981.4 [b]	0.52 [a]
2020	925.6 [b]	1066.4 [a]	0.47 [b]
2021	785.5 [c]	659.0 [c]	0.55 [a]
	Factor interaction		
T	***	**	ns
C	*	ns	***
Y	***	***	***
T × C	ns	ns	ns
T × Y	ns	***	***
C × Y	ns	*	***
T × C × Y	ns	ns	ns

Different letters ([a–c]) mean the significant difference ($p \leq 0.05$) according to ANOVA and Tukey's test. Significant interaction on level p value * ≤ 0.05, ** ≤ 0.01, *** < 0.001, and ns—not significant difference, T—tillage, C—cultivar, Y—year.

In a study by Parylak and Pytlarz [66], limiting wheat pre-sowing cultivation to the use of a cultivating unit resulted in high yield decreases, compared to plough tillage, but it should be noted that the mentioned researchers used sowing by the traditional method, i.e., with a standard seed drill. Jaskulska et al. [61] achieved similar results to our study in terms of winter wheat yield depending on the tillage method. These authors also used a Mzuri unit for setting up the experiment and compared this treatment to conventional ploughing and simplified (no-plough) tillage, in which sowing was conducted with a seed drill. Similar results were also obtained by Mohammadi et al. [67]. It should be emphasised that in the strip till method, the soil in the sowing strip is cultivated to a depth of 20 cm, i.e., to a depth appropriate for ploughing and even deeper, which lowers the negative effects associated with too much soil compaction that occurs in direct sowing (sowing without prior tillage). Therefore, on the basis of the present research, it may be concluded that the strip-till method in the conditions of the clay soil, in which our research was conducted, does not create the same conditions for plant growth as with properly conducted ploughing, but the negative effects associated with its non-application are the same.

Table 6. Yield and its structure according to the crop used, cultivar and years of study.

Specification	Tillering Index	Plant Density (pcs·m^{-2})	Number of Ears (pcs·m^{-2})	Weight of Kernels per Ear (g)	Yield (t·ha^{-1})
Cultivation system					
Ploughed tillage + strip tillage (PT)	2.2 ^{a}	284 ^{a}	477 ^{a}	1.87 ^{a}	7.88 ^{a}
Stubble discing + strip tillage (SD)	1.9 ^{a}	279 ^{a}	478 ^{a}	1.63 ^{b}	7.41 ^{b}
Strip tillage (ST)	2.1 ^{a}	267 ^{a}	490 ^{a}	1.62 ^{b}	7.16 ^{b}
Cultivar					
Formacja	2.1 ^{a}	273 ^{a}	469 ^{a}	1.75 ^{a}	7.68 ^{a}
Metronom	2.1 ^{a}	278 ^{a}	488 ^{a}	1.78 ^{a}	7.53 ^{a}
Desamo	2.2 ^{a}	282 ^{a}	488 ^{a}	1.59 ^{b}	7.24 ^{b}
Years					
2019	1.6 ^{c}	310 ^{a}	505 ^{a}	1.82 ^{a}	8.10 ^{a}
2020	2.3 ^{b}	260 ^{b}	482 ^{b}	1.64 ^{b}	7.73 ^{b}
2021	2.5 ^{a}	262 ^{b}	458 ^{b}	1.66 ^{b}	6.62 ^{c}
Factor interaction					
T	ns	ns	ns	***	ns
C	ns	ns	ns	***	ns
Y	***	***	***	***	ns
T × C	ns	ns	ns	**	ns
T × Y	*	ns	*	ns	ns
C × Y	*	ns	***	ns	ns
T × C × Y	*	ns	ns	*	ns

Abbreviation 'pcs' means 'pieces'. Different letters ($^{a-c}$) mean the significant difference ($p \leq 0.05$) according to ANOVA and Tukey's test. Significant interaction on level p value * ≤ 0.05, ** ≤ 0.01, *** < 0.001, and ns—not significant difference, T—tillage, C—cultivar, Y—year.

Among the studied yield structure traits, only for the weight of kernels per ear (g) was a highly significant interaction (p value < 0.001) term was found between T and C. We found a highly significant interaction between Y (p value < 0.001) and all yield structure traits except the grain yield itself. C × Y; C × Y; T × C × Y showed significant interaction ($p \leq 0.05$) with each other in terms of tilering index. The number of ears was influenced by the interaction between T × Y ($p \leq 0.05$) and C × Y (<0.001). The weight of kernels per ear (g) showed a significant influence between experience factors T × C (p value ≤ 0.01) and T × C × Y (p value ≤ 0.05). Roohi et al. [68] indicate that the interaction between the cultivation system and the cultivar in the yield concerns the grain head from the spike, which was confirmed in our own research. Herrera et al. [69] compared the available scientific studies and showed that the percentage of reported significant T × C interactions was higher for spring wheat (71%) than for winter wheat (40%).

3.3. Effects of Tillage Types on Wheat Grain Quality

In our own research, it was found that the cultivation method as well as the cultivar had a significant effect on the thousand grain weight (TGW). The highest value of this grain trait was found under post-harvest plough tillage (PT) and the lowest was when strip tillage was combined with sowing made in no-till (ST) (Table 7). The higher grain weight and yield in the plough system relative to the no-till system were also found by other authors [70,71], but it should be noted that they did not use a strip-till unit for sowing. The effect of the cultivar on thousand grain weight was also significant. The highest thousand grain weight was found in the Metronom cultivar and the lowest in the Desamo cultivar. An important indicator of grain quality, which determines its maturity, is its bulk density. This trait, which determines its milling value, did not significantly depend on the post-harvest tillage applied, although it should be emphasised that the tendency for a higher value of this trait in the PT treatment was clear. Among the winter wheat cultivars tested, we found a significant effect on the density of grain at the bulk state, as each cultivar differed significantly in this respect. The Formacja cultivar had the highest value for this trait, which was 4% higher than that of Metronom and 9% higher than that of Desamo. Some scientific studies have shown that the value of winter wheat grain density is higher when plough tillage is applied than under reduced tillage conditions [72]. In contrast, Jaskulska et al. [8] and Taner et al. [73] showed no effect of using the tillage system (ploughing, reduced tillage and no-till) on this grain trait. Large differences between cultivars in grain density were pointed out by many authors [74–76]. Bobryk-Mamczarz et al. [77] pointed out the influence of weather conditions varying over the years on this trait, which was also confirmed by their own research, as the grain density values obtained in each year of the study differed significantly. The tests carried out showed that the method of post-harvest cultivation had no statistically significant effect on the amount of total gluten. Only a statistically insignificant tendency towards a slightly higher amount of gluten in the grain from the ST treatment was found. The results obtained in our study are in line with what was obtained in their studies by Šíp et al. [78] and Woźniak and Rachoń [79]. In studies by other authors, the effect of applied soil tillage intensity in shaping the amount of gluten varied. Amato et al. [80] showed a higher amount of gluten in grain from treatments with more intensive (plough) cultivation, while Konavko and Ruža [81] showed the opposite relation, i.e., a higher amount of gluten in wheat grain from treatments with less intensive cultivation. Our own research showed that the amount of gluten was dependent on the winter wheat cultivar. The cultivar Desamo had the statistically significant highest amount of gluten. On average, the Metronom cultivar contained 5% less gluten and the Formacja cultivar 14% less. There is a consensus among various authors dealing with the issue of wheat grain quality that the amount of gluten is a varietal trait [82–84]. A very extensive study in this field on 762 cultivars, also confirming the mentioned relationship, was conducted by Pengpeng et al. [85]. No less important than the cultivar in shaping the amount of gluten is the variable weather conditions over the years. This is confirmed by our own research. Irrespective of cultivation method and cultivar, on average, the highest amount of gluten was found in 2020, and in the other two years, 2019 and 2021, 6.3% and 3.0% less, respectively (Table 7). The important role of years in shaping the amount of gluten was also emphasised by other authors [86,87]. This is related to the weather-dependent efficiency of nitrogen use for protein synthesis in grain [88–90]. The intensity of the post-harvest tillage applied did not significantly affect the gluten index value (Table 7). In contrast, a study by Gawęda and Haliniarz [70] showed that tillage intensity can significantly modify this trait. The authors cited found higher gluten index values under no-tillage compared to plough tillage. A different relationship—a higher gluten index under reduced tillage than under no-tillage—was found by Buczek et al. [91]. In our study, a strong influence of the varietal factor on the gluten index was shown. The Formacja cultivar had by far the highest value (81% on average) of the trait in question. In the other two cultivars, the gluten index was significantly lower. The gluten index is cultivar-related in winter wheat [92], which was also confirmed in our own research. Our own research also showed, the effect of years on the gluten index value. The lowest value of this trait was found in 2020, while the other years did not differ significantly in the value of this trait (Table 7). In general, there is some consensus in the literature on the large effect of weather condition variance over the years, with this being explained by drought stress, the occurrence of which reduces the value of this trait [93]. The sedimentation index relates to both the quality and quantity of protein in the grain, and therefore has an impact on the quality of the bread obtained and, in particular, its structure [94]. The highest possible value is desirable. A high sedimentation rate should be combined with a high content of gluten proteins, especially gluten, which is the most important for baking [95]. Our study showed a significant effect of the extent of post-harvest cultivation on the Zeleny sedimentation coefficient. Its highest value was found in the ST treatment, in which strip-tillage with seed sowing was carried out directly into the stubble. Significantly lower values for this parameter were found in the reduced tillage (SD) and plough tillage (PT) treatments, by 8 and 13%, respectively. Similar trends in the effect of tillage intensity on sedimentation rate were also found by Bilalis et al. [96] and Wozniak and Rachoń [79]. In the study of Šíp et al. [78], on the other hand, the sedimentation rate was higher with a plough tillage system than with no-tillage.

Table 7. The value of the analysed grain quality traits depending on the post-harvest cultivation used, cultivar and year of research.

Specification	Thousand Grain Weight (g)	Bulk Density of Grain ($kg{\cdot}hl^{-1}$)	Amount of Gluten (%)	Gluten Index (%)	Sedimentation Index Zeleny (cm^3)	Falling Number (s)
			Cultivation system			
oughed tillage + rip tillage (PT)	39.2 [a]	74.3 [a]	33.7 [a]	66 [a]	46 [c]	358 [a]
ubble discing + rip tillage (SD)	37.9 [b]	72.1 [a]	33.6 [a]	66 [a]	49 [b]	361 [a]
rip tillage (ST)	36.1 [c]	71.9 [a]	34.0 [a]	63 [a]	53 [a]	374 [a]
			Cultivar			
Formacja	37.9 [b]	76.0 [a]	30.9 [c]	81 [a]	47 [b]	365 [b]
Metronom	41.5 [a]	72.7 [b]	34.2 [b]	59 [b]	59 [a]	321 [c]
Desamo	33.8 [c]	69.5 [c]	35.9 [a]	55 [c]	42 [c]	404 [a]
			Years			
2019	37.9 [a]	72.3 [b]	30.5 [c]	68 [a]	39 [c]	371 [a]
2020	37.8 [a]	69.5 [c]	36.8 [a]	62 [b]	60 [a]	363 [a]
2021	38.0 [a]	77.6 [a]	33.7 [b]	66 [a]	49 [b]	358 [a]
			Factor interaction			
T	***	**	ns	*	***	ns
C	***	***	***	***	***	***
Y	ns	***	***	*	***	ns
T × C	**	ns	ns	***	ns	ns
T × Y	ns	ns	ns	ns	**	ns
C × Y	***	***	***	ns	***	***
T × C × Y	ns	ns	ns	*	*	ns

Different letters ([a–c]) mean the significant difference ($p \leq 0.05$) according to ANOVA and Tukey's test. Significant interaction on level p value * ≤ 0.05, ** ≤ 0.01, *** < 0.001, and ns—not significant difference, T—tillage, C—cultivar, Y—year.

In our study, there was a significant effect of cultivar on the Zeleny sedimentation coefficient. The highest value was found in the Metronom cultivar, which had a higher index than the Formacja cultivar by 20% than on the Desamo cultivar by 29%. The strong influence of the cultivar on the value of the sedimentation index was indicated by many authors [79,95,97–99]. In addition, our study found a large variation in sedimentation rate values between years (Table 7), the highest being recorded in 2020, with significantly lower values in the other years. Tatar et al. [100] consider that the magnitude of the sedimentation index is conditioned by the occurrence of drought at the grain pouring stage. In the cited study, the sedimentation rate was significantly lower under rainfall deficit conditions. The falling number is considered to be an important discriminator for the technological value of the grain, which determines alpha-amylase activity. The minimum value of the falling number according to wheat standards (EN ISO 3093) is 250 [101]. In our study, the tested grain from each research treatment met this requirement. However, there was no statistically significant effect of the extent of post-harvest tillage on the falling number value. Weber [86] and Buczek et al. [91] found that plough tillage favoured a higher fall number compared to reduced tillage. On average, the highest falling number was characterised by the cultivar Desamo (independent of the treatment, more than 400), but in the case of the other two cultivars, the values of this trait were also high and exceeded 300. Similarly, a large role of the cultivar in shaping the falling number was shown by Knapowski et al. [102] and Amiri et al. [83]. In our own study, there was no significant effect of the study years on the value of falling number. However, the literature indicates that weather conditions have the greatest influence on the value of this trait. The lack of heavy rainfall in the pre-harvest period favours a high value of the fall number [103]. The tillage system showed significant interaction (p value < 0.001) between the thousand grain weight and Zeleny's sedimentation index, the bulk density of grain (p value < 0.001) and gluten index ($p \leq 0.05$). The cultivar showed a high significant interaction (p value < 0.001) with each of the tested grain quality traits. Years of research had a high significant interaction (p value < 0.001) with the bulk density of grain, amount of gluten and Zeleny's sedimentation index, and also significant interactions with the gluten index ($p \leq 0.05$). A highly significant interaction (p value < 0.001) was also confirmed between the C $\times$ Y experience factors on traits such as thousand grain weight, bulk density of grain, amount of gluten, Zeleny's sedimentation index and falling number. T $\times$ C had a significant interaction for gluten index (p value < 0.001) and thousand grain weight (p value $\leq$ 0.01). T $\times$ Y had a significant interaction for Zeleny's sedimentation index (p value $\leq$ 0.01). Interactions between T $\times$ C $\times$ Y had a significant impact (p value $\leq$ 0.05) on gluten index and Zeleny's sedimentation index. Buczek et al. [91] study of the interaction between T, C and Y showed a highly significant interaction of Zeleny's sedimentation index and falling number and gluten index only for T and C, which was also confirmed by our own study.

4. Conclusions

The lowest value of this trait at the tillering stage was found in the ploughed tillage + strip tillage treatment. In the subsequent growth phases, this was the treatment with the highest aboveground weight, indicating that the plant growth rate was clearly higher in this treatment. The genetic factor had a significant influence on plant growth rate. In each growth phase, the highest value for this trait was recorded in the Metronom cultivar. The cultivars Desamo and Formacja had a significantly lower green matter. The results of the present study can be used to validate existing wheat growth models. The extent of harvest tillage preceding the strip-till sowing of wheat had a significant effect on the thousand grain weight. The higher value of this trait was characterised by grain obtained from the treatment in which strip-tillage of wheat was applied after ploughing. The grain quality parameters (gluten content, gluten index, falling number) did not depend on the applied post-harvest tillage regime, except for the sedimentation index. The beneficial effect of strip-till cultivation on the environment and the slight decrease in yield and no effect on quality characteristics mean that we recommend the use of strip-till cultivation without prior post-harvest cultivation, but it is important to select the appropriate winter wheat cultivar.

Author Contributions: Conceptualization, M.R. and J.G.; methodology, M.R. and J.G.; validation, M.R. and J.G.; formal analysis, M.R. and J.G.; investigation, M.R.; data curation, M.R.; writing—original draft preparation, M.R.; writing—review and editing, M.R.; visualisation, J.G.; supervision M.W.; project administration, M.R.; funding acquisition, J.G. All authors have read and agreed to the published version of the manuscript.

Funding: This research received no external funding.

Institutional Review Board Statement: Not applicable.

Data Availability Statement: Dataset available on request from the authors.

Conflicts of Interest: The authors declare no conflicts of interest.

References

Leghari, S.J.; Han, W.; Soomro, A.A.; Shoukat, M.R.; Zain, M.; Wei, Y.; Xu, Q.; Buriro, M.; Bhutto, T.M.; Soothar, R.K.; et al. Navigating water and nitrogen practices for sustainable wheat production by model-based optimization management systems: A case study of China and Pakistan. *Agric. Water Manag.* **2024**, *300*, 108917. [CrossRef]

Kumar, A.; Sachan, S. An Empirical Analysis of Profitability of Wheat Cultivation. *Think India J.* **2019**, *22*, 263–277.

Hyles, J.; Bloomfield, M.T.; Hunt, J.R.; Trethowan, R.M.; Trevaskis, B. Phenology and related traits for wheat adaptation. *Heredity* **2020**, *125*, 417–430. [CrossRef]

Eurostat. 2022. Available online: https://ec.europa.eu (accessed on 27 July 2024).

Langridge, P.; Alaux, M.; Almeida, N.F.; Ammar, K.; Baum, M.; Bekkaoui, F.; Bentley, A.R.; Beres, B.L.; Berger, B.; Braun, H.-J.; et al. Meeting the Challenges Facing Wheat Production: The Strategic Research Agenda of the Global Wheat Initiative. *Agronomy* **2022**, *12*, 2767. [CrossRef]

Van den Putte, A.; Govers, G.; Diels, J.; Gillijns, K.; Demuzere, M. Assessing the effect of soil tillage on crop growth: A meta-regression analysis on European crop yields under conservation agriculture. *Eur. J. Agron.* **2010**, *33*, 231–241. [CrossRef]

Jaskulski, D.; Kotwica, K.; Jaskulska, I.; Piekarczyk, M.; Osiński, G.; Pochylski, B. Elementy współczesnych systemów uprawy roli i roślin–skutki produkcyjne oraz środowiskowe. *Fragm. Agron.* **2012**, *29*, 61–70.

Jaskulska, I.; Jaskulski, D.; Kotwica, K.; Wasilewski, P.; Gałężewski, L. Effect of tillage simplifications on yield and grain quality of winter wheat after different previous crops. *Acta Sci. Polonorum. Agric.* **2013**, *12*, 37–44.

Kotwica, K.; Jaskulska, I.; Galezewski, L.; Jaskulski, D.; Lamparski, R. The effect of tillage and management of post-harvest residues and biostymulant application on the yield of winter wheat in increasing monoculture. *Acta Sci. Pol. Agric.* **2014**, *13*, 65–76.

10. Kotwica, K.; Gałęzewski, L.; Wilczewski, E.; Kubiak, W. Reduced Tillage, Application of Straw and Effective Microorganisms as Factors of Sustainable Agrotechnology in Winter Wheat Monoculture. *Agronomy* **2024**, *14*, 738. [CrossRef]

11. Kumar, U.; Cheng, M.; Islam, M.J.; Maniruzzaman, M.; Nasreen, S.S.; Haque, M.E.; Hossain, M.B.; Jahiruddin, M.; Bell, R.W.; Jahangir, M.M.R. Retention of Crop Residue Increases Crop Productivity and Maintains Positive Sulfur Balance in Intensive Rice-Based Cropping Systems. *J. Soil Sci. Plant Nutr.* **2024**, 1–12. [CrossRef]

12. Yan, Y.; Li, H.; Zhang, M.; Liu, X.; Zhang, L.; Wang, Y.; Yang, M.; Cai, R. Straw Return or No Tillage? Comprehensive Meta-Analysis Based on Soil Organic Carbon Contents, Carbon Emissions, and Crop Yields in China. *Agronomy* **2024**, *14*, 2263. [CrossRef]

13. Wade, T.; Claassen, R. No-Till or Strip-Till Use Varies by Region. Amber Waves 2016, C1. Available online: https://www.proquest.com/openview/43a856d4dfb759ff72e472df9a4646e8/1.pdf?pq-origsite=gscholar&cbl=42620 (accessed on 27 July 2024).

14. Dou, S.; Wang, Z.; Tong, J.; Shang, Z.; Deng, A.; Song, Z.; Zhang, W. Strip tillage promotes crop yield in comparison with no tillage based on a meta-analysis. *Soil Tillage Res.* **2024**, *240*, 106085. [CrossRef]

15. Dębska, B.; Jaskulska, I.; Jaskulski, D. Method of Tillage with the Factor Determining the Quality of Organic Matter. *Agronomy* **2020**, *10*, 1250. [CrossRef]

16. Wojewódzki, P.P.; Kondratowicz-Maciejewska, K.; Dębska, B.; Jaskulska, I.; Jaskulski, D.; Pakuła, J. Luvisol soil macroaggregates under the influence of conventional, strip-till, and reduced tillage practice. *Int. Agrophys.* **2024**, *38*, 311–324. [CrossRef] [PubMed]

17. Sun, Y.; Yang, C.; Liang, H.; Yang, Y.; Bu, K.; Dong, Y.; Hai, J. The Border Effects of Dry Matter, Photosynthetic Characteristics, and Yield Components of Wheat under Hole Sowing Condition. *Agronomy* **2023**, *13*, 766. [CrossRef]

18. Wang, Y.; Hu, W.; Zhang, X.; Li, L.; Kang, G.; Feng, W.; Guo, T. Effects of cultivation patterns on winter wheat root growth parameters and grain yield. *Field Crops Res.* **2014**, *156*, 208–218. [CrossRef]

19. Fang, Y.; Du, Y.; Wang, J.; Wu, A.; Qiao, S.; Xu, B.; Zhang, S.; Siddique, K.H.M.; Chen, Y. Moderate drought stress affected root growth and grain yield in old, modern and newly released cultivars of winter wheat. *Front. Plant Sci.* **2017**, *8*, 672. [CrossRef] [PubMed]

20. Slafer, G.A.; García, G.A.; Serrago, R.A.; Miralles, D.J. Physiological drivers of responses of grains per m2 to environmental and genetic factors in wheat. *Field Crops Res.* **2022**, *285*, 108593. [CrossRef]

21. Fu, Y.Y.; Yang, G.J.; Wang, J.H.; Song, X.Y.; Feng, H.K. Winter wheat biomass estimation based on spectral indices, band depth analysis and partial least squares regression using hyperspectral measurements. *Comput. Electr. Agric.* **2014**, *100*, 51–59. [CrossRef]

22. Trethowan, R.M.; Mahmood, T.; Ali, Z.; Oldach, K.; Garcia, A.G. Breeding wheat cultivars better adapted to conservation agriculture. *Field Crops Res.* **2012**, *132*, 76–83. [CrossRef]

23. Iqbal, M.M.; Khan, I.; Chattha, M.U.; Hassan, M.U.; Iqbal, M.; Farooq, M. Performance of Wheat Cultivars Under Different Tillage and Crop Establishment Methods. *Int. J. Plant Prod.* **2021**, *16*, 287–297. [CrossRef]

24. Eitel, J.U.; Magney, T.S.; Vierling, L.A.; Brown, T.T.; Huggins, D.R. Lidar based biomass and crop nitrogen estimates for rapid, non-destructive assessment of wheat nitrogen status. *Field Crops Res.* **2014**, *159*, 21–32. [CrossRef]

25. Schirrmann, M.; Hamdorf, A.; Garz, A.; Ustyuzhanin, A.; Dammer, K.-H. Estimating wheat biomass by combining ima clustering with crop height. *Comput. Electron. Agric.* **2016**, *121*, 374–384. [CrossRef]
26. Walter, J.; Edwards, J.; McDonald, G.; Kuchel, H. Photogrammetry for the estimation of wheat biomass and harvest index. *Fie Crops Res.* **2018**, *216*, 165–174. [CrossRef]
27. Zhang, C.; Liu, J.; Shang, J.; Dong, T.; Tang, M.; Feng, S.; Cai, H. Improving winter wheat biomass and evapotranspirati simulation by assimilating leaf area index from spectral information into a crop growth model. *Agric. Water Manag.* **2021**, *2* 107057. [CrossRef]
28. Li, H.; Korohou, T.; Liu, Z.; Geng, J.; Ding, Q. Analysis of Multiangle Wheat Density Effects Based on Drill Single-Seed Seedir *Agriculture* **2024**, *14*, 176. [CrossRef]
29. Huang, H.; Huang, J.; Li, X.; Zhuo, W.; Wu, Y.; Niu, Q.; Su, W.; Yuan, W. A Dataset of Winter Wheat Aboveground Biomass China during 2007–2015 Based on Data Assimilation. *Sci. Data* **2022**, *9*, 200. [CrossRef] [PubMed]
30. Wang, F.; Yang, M.; Ma, L.; Zhang, T.; Qin, W.; Li, W.; Zhang, Y.; Sun, Z.; Wang, Z.; Li, F. Estimation of Above-Ground Biomass Winter Wheat Based on Consumer-Grade Multi-Spectral UAV. *Remote Sens.* **2022**, *14*, 1251. [CrossRef]
31. Łopatka, A.; Koza, P.; Suszek-Łopatka, B.; Siebielec, G.; Jadczyszyn, J. Assessment of soil impact on pre-and post-harvest ND extrema by machine learning. *Soil Sci. Ann.* **2024**, *75*, 189540. [CrossRef]
32. Casadesús, J.; Villegas, D. Conventional digital cameras as a tool for assessing leaf area index and biomass for cereal breedir *J. Integr. Plant Biol.* **2014**, *56*, 7–14. [CrossRef]
33. *ISO 21415-2:2015*; Wheat and Wheat Flour—Gluten Content Part 2: Determination of Wet Gluten and Gluten Index by Me-chanic Means. International Organization for Standardization: Geneva, Switzerland, 2015.
34. *ISO 5529*; Wheat—Determination of the Sedimentation Index—Zeleny Test. International Organization for Standardizatio Geneva, Switzerland, 2007.
35. Donmez, E.; Sears, R.G.; Shroyer, J.P.; Paulsen, G.M. Genetic gain in yield attributes of winter wheat in the Great Plains. *Crop S* **2001**, *41*, 1412–1419. [CrossRef]
36. Jimenez-Berni, J.A.; Deery, D.M.; Pablo, R.L.; Condon, A.G.; Rebetzke, G.J.; James, R.A.; Bovill, W.D.; Furbank, R.T.; Sirault, X.R. High throughput determination of plant height, ground cover, and above-ground biomass in wheat with LiDAR. *Front. Plant Sc* **2018**, *9*, 237. [CrossRef]
37. Almeida, M.L.D.; Sangoi, L.; Merotto Jr, A.; Alves, A.C.; Nava, I.C.; Knopp, A.C. Tiller emission and dry mass accumulation wheat cultivars under stress. *Sci. Agric.* **2004**, *61*, 266–270. [CrossRef]
38. Hendriks, P.-W.; Gurusinghe, S.; Ryan, P.R.; Rebetzke, G.J.; Weston, L.A. Competitiveness of Early Vigour Wheat (*Triticu aestivum* L.) Genotypes Is Established at Early Growth Stages. *Agronomy* **2022**, *12*, 377. [CrossRef]
39. Lipiec, J.; Nosalewicz, A. Wzrost pędów pszenicy w zależności od miejscowego zagęszczenia gleby. *Acta Agrophysica* **2002**, *7* 151–157.
40. Sha, Y.; Liu, Z.; Hao, Z.H.; Huang, Y.W.; Shao, H.; Feng, G.Z.; Chen, F.J.; Mi, G.H. Root growth, root senescence and root syster architecture in maize under conservative strip tillage system. *Plant Soil* **2023**, *495*, 253–269. [CrossRef]
41. Ali, S.; Xu, Y.; Jia, Q.; Ahmad, I.; Wei, T.; Ren, X.; Zhang, P.; Din, R.; Cai, T.; Jia, Z. Cultivation techniques combined wit deficit irrigation improves winter wheat photosynthetic characteristics, dry matter translocation and water use efficiency unde simulated rainfall conditions. *Agric. Water Manag.* **2018**, *201*, 207–218. [CrossRef]
42. Saini, A.; Manuja, S.; Kumar, S.; Hafeez, A.; Ali, B.; Poczai, P. Impact of Cultivation Practices and Varieties on Productivit Profitability, and Nutrient Uptake of Rice (*Oryza sativa* L.) and Wheat (*Triticum aestivum* L.) Cropping System in India. *Agricultu* **2022**, *12*, 1678. [CrossRef]
43. Tazhibayeva, T.; Abugalieva, A.; Morgunov, A.; Kozhakhmetov, K. Introgressive forms-approach for biotechnology advance c winter wheat on environmental adaptability. *Int. Multidiscip. Sci. GeoConference* **2016**, *1*, 607–614.
44. Mullan, D.J.; Reynolds, M.P. Quantifying genetic effects of ground cover on soil water evaporation using digital imaging. *Func Plant Biol.* **2010**, *37*, 703–712. [CrossRef]
45. Yang, Y.; Wan, H.; Yang, F.; Xiao, C.; Li, J.; Ye, M.; Chen, C.; Deng, G.; Wang, Q.; Li, A.; et al. Mapping QTLs for enhancing earl biomass derived from Aegilops tauschii in synthetic hexaploid wheat. *PLoS ONE* **2020**, *15*, e0234882. [CrossRef] [PubMed]
46. Chen, Y.; Palta, J.; Prasad, P.V.V.; Siddique, K.H.M. Phenotypic variability in bread wheat root systems at the early vegetativ stage. *BMC Plant Biol.* **2020**, *20*, 185. [CrossRef] [PubMed]
47. Wilczewski, E.; Jug, I.; Lipiec, J.; Gałęzewski, L.; Đurđević, B.; Kocira, A.; Brozović, B.; Marković, M.; Jug, D. Tillage systen regulates the soil moisture tension, penetration resistance and temperature responses to the temporal variability of precipitatio during the growing season. *Int. Agrophys.* **2023**, *37*, 391–399. [CrossRef]
48. Sha, Y.; Hao, Z.; Liu, Z.; Huang, Y.; Feng, G.; Chen, F.; Mi, G. Regulation of maize growth, nutrient accumulation an remobilization in relation to yield formation under strip-till system. *Arch. Agron. Soil Sci.* **2023**, *69*, 2615–2630. [CrossRef]
49. Rieger, S.; Richner, W.; Streit, B.; Frossard, E.; Liedgens, M. Growth, yield, and yield components of winter wheat and the effect of tillage intensity, preceding crops, and N fertilisation. *Eur. J. Agron.* **2008**, *28*, 405–411. [CrossRef]
50. Plaza-Bonilla, D.; Álvaro-Fuentes, J.; Hansen, N.C.; Lampurlanés, J.; Cantero-Martínez, C. Winter cereal root growth an aboveground–belowground biomass ratios as affected by site and tillage system in dryland Mediterranean conditions. *Plant Soi* **2014**, *374*, 925–939. [CrossRef]

Sun, Y.; Zhang, S.; Yan, J. Contribution of green organs to grain weight in dryland wheat from the 1940s to the 2010s in Shaanxi Province, China. *Sci. Rep.* **2021**, *11*, 3377. [CrossRef] [PubMed]
Ding, Y.; Zhang, X.; Ma, Q.; Li, F.; Tao, R.; Zhu, M.; Li, C.; Zhu, X.; Guo, W.; Ding, J. Tiller Fertility Is Critical for Improving Grain Yield, Photosynthesis, and Nitrogen Efficiency in Wheat. *J. Integr. Agric.* **2023**, *22*, 2054–2066. [CrossRef]
Yang, D.; Liu, Y.; Cheng, H.; Chang, L.; Chen, J.; Chai, S.; Li, M. Genetic dissection of flag leaf morphology in wheat (*Triticum aestivum* L.) under diverse water regimes. *BMC Genet.* **2016**, *17*, 94. [CrossRef]
Subhani, G.M.; Chowdhry, M.A. Correlation and path coefficient analysis in bread wheat under drought stress and normal conditions. *Pak. J. Biol. Sci.* **2000**, *3*, 72–77. [CrossRef]
Khanna-Chopra, R.; Singh, K.; Shukla, S.; Kadam, S.; Singh, N.K. QTLs for cell membrane stability and flag leaf area under drought stress in a wheat RIL population. *J. Plant Biochem. Biotechnol.* **2019**, *29*, 276–286. [CrossRef]
Künzel, A.; Münzel, S.; Böttcher, F.; Spengler, D. Analysis of Weather-Related Growth Differences in Winter Wheat in a Three-Year Field Trial in North-East Germany. *Agronomy* **2021**, *11*, 1854. [CrossRef]
Wilhelm, E.P.; Boulton, M.I.; Al-Kaff, N.; Balfourier, F.; Bordes, J.; Greenland, A.J.; Powell, W.; Mackay, I.J. Rht-1 and Ppd-D1 associations with height, GA sensitivity, and days to heading in a worldwide bread wheat collection. *Appl. Genet.* **2013**, *126*, 2233–2243. [CrossRef]
Jensen, S.M.; Svensgaard, J.; Ritz, C. Estimation of the harvest index and the relative water content—Two examples of composite variables in agronomy. *Eur. J. Agron.* **2020**, *112*, 125962. [CrossRef]
White, E.M.; Wilson, F.E.A. Responses of grain yield, biomass and harvest index and their rates of genetic progress to nitrogen availability in ten winter wheat varieties. *Ir. J. Agric. Food Res.* **2006**, *45*, 85–101.
Porker, K.; Straight, M.; Hunt, J.R. Evaluation of G × E × M interactions to increase harvest index and yield of early sown wheat. *Front. Plant Sci.* **2020**, *11*, 994. [CrossRef] [PubMed]
Jaskulska, I.; Jaskulski, D.; Różniak, M.; Radziemska, M.; Gałęzewski, L. Zonal Tillage as Innovative Element of the Technology of Growing Winter Wheat: A Field Experiment under Low Rainfall Conditions. *Agriculture* **2020**, *10*, 105. [CrossRef]
Wesołowski, M.; Cierpiała, R. Plonowanie pszenicy ozimej w zależności od sposobu wykonania uprawy przedsiewnej. *Fragm. Agron.* **2011**, *28*, 106–118.
Evers, T.; Millar, S. Cereal grain structure and development: Some implications for quality. *J. Cereal Sci.* **2002**, *36*, 261–284. [CrossRef]
Philipp, N.; Weichert, H.; Bohra, U.; Weschke, W.; Schulthess, A.W.; Weber, H. Grain number and grain yield distribution along the spike remain stable despite breeding for high yield in winter wheat. *PLoS ONE* **2018**, *13*, e0205452. [CrossRef] [PubMed]
Rachoń, L.; Bobryk-Mamczarz, A.; Kiełtyka-Dadasiewicz, A.; Woźniak, A.; Stojek, Z.; Zajdel-Stępień, P. Plonowanie i jakość wybranych gatunków i odmian pszenicy makaronowej. Cz. I. Plonowanie. *Ann. UMCS Sect. E Agric.* **2022**, *77*, 55–63. [CrossRef]
Parylak, D.; Pytlarz, E. Skutki produkcyjne monokultury pszenicy ozimej w warunkach upraszczania uprawy roli. *Fragm. Agron* **2013**, *30*, 114–121.
Mohammadi, R.; Rajabi, R.; Haghparast, R. On-farm assessment of agronomic performance of rainfed wheat cultivars under different tillage systems. *Soil Tillage Res.* **2024**, *235*, 105902. [CrossRef]
Roohi, E.; Mohammadi, R.; Niane, A.A.; Vafabakhsh, J.; Roustaee, M.; Jalal Kamali, M.R.; Sohrabi, S.; Fatehi, S.; Tarimoradi, H. Genotype × tillage interaction and the performance of winter bread wheat genotypes in temperate and cold dryland conditions. *J. Integr. Agric.* **2022**, *21*, 3199–3215. [CrossRef]
Herrera, J.M.; Verhulst, N.; Trethowan, R.M.; Stamp, P.; Govaerts, B. Insights into Genotype × Tillage Interaction Effects on the Grain Yield of Wheat and Maize. *Crop Sci.* **2013**, *53*, 1845–1859. [CrossRef]
Gawęda, D.; Haliniarz, M. Grain Yield and Quality of Winter Wheat Depending on Previous Crop and Tillage System. *Agriculture* **2021**, *11*, 133. [CrossRef]
Yousefian, M.; Shahbazi, F.; Hamidian, K. Crop Yield and Physicochemical Properties of Wheat Grains as Affected by Tillage Systems. *Sustainability* **2021**, *13*, 4781. [CrossRef]
Woźniak, A.; Gos, M. Yield and quality of spring wheat and soil properties as affected by tillage system. *Plant Soil Environ.* **2014**, *60*, 141–145. [CrossRef]
Taner, A.; Zafer, R.; Kaya, Y.; Gültekin, İ.; Partigöç, F. The effects of various tillage systems on grain yield, quality parameters and energy indices in winter wheat production under the rainfed conditions. *Fresenius Environ. Bull.* **2015**, *24*, 1463–1473.
Stępniewska, S.; Abramczyk, D. The correlation between quality parameters of selected winter wheat grain. *Postępy Nauk. I Technol. Przemysłu Rolno-Spożywczego* **2013**, *1*, 65–78.
Stępniewska, S.; Słowik, E. Ocena wartości technologicznej wybranych odmian pszenicy ozimej i jarej. *Acta Agrophys.* **2016**, *23*, 275–286.
Noworól, M. Reakcja Odmian Pszenicy Ozimej na Poziom Intensywności Technologii Produkcji. Ph.D. Thesis, UR Rzeszów, Rzeszów, Poland, 2018.
Bobryk-Mamczarz, A.; Rachoń, L.; Kiełtyka-Dadasiewicz, A.; Szydłowska-Tutaj, M.; Lewko, P.; Woźniak, A. Plonowanie i jakość wybranych gatunków i odmian pszenicy makaronowej. Cz. II. Wartość technologiczna ziarna. *Agron. Sci.* **2022**, *77*, 65–78. [CrossRef]
Šíp, V.; Vavera, R.; Chrpová, J.; Kusá, H.; Růžek, P. Winter wheat yield and quality related to tillage practice, input level and environmental conditions. *Soil Tillage Res.* **2013**, *132*, 77–85. [CrossRef]

79. Woźniak, A.; Rachoń, L. Effect of Tillage Systems on the Yield and Quality of Winter Wheat Grain and Soil Properties. *Agricul* **2020**, *10*, 405. [CrossRef]
80. Amato, G.; Di Miceli, G.; Frenda, A.S.; Giambalvo, D.; Stringi, L. Wheat yield and grain quality as affected by tillage, sowing t and nitrogen fertilization under rainfed Mediterranean conditions. *Options Méditerr* **2004**, *60*, 151–155.
81. Konavko, A.; Ruža, A. Influence of tillage and crop rotation on winter wheat grain quality. In Proceedings of the Scientific Practical Conference "Harmonious Agriculture", Jelgava, Latvia, 23 February 2017; pp. 50–54.
82. Simic, G.; Horvat, D.; Jurkovic, Z.; Drezner, G.; Novoselovic, D.; Dvojkovic, K. The genotype effect on the ratio of wet glu content to total wheat grain protein. *J. Cent. Eur. Agric.* **2006**, *7*, 13–18.
83. Amiri, R.; Sasani, S.; Jalali-Honarmand, S.; Rasaei, A.; Seifolahpour, B.; Bahraminejad, S. Genetic diversity of bread wh genotypes in Iran for some nutritional value and baking quality traits. *Physiol. Mol. Biol. Plants* **2018**, *24*, 147–157. [Cross [PubMed]
84. Desheva, G.; Deshev, M. Correlation and Regression Relationships between Main Grain Quality Characteristics in Bread Wi Wheat. *Agric. Conspec. Sci.* **2022**, *87*, 135–143.
85. Pengpeng, L.; Wei, S.; Hongjun, X.U.; Fengjuan, C.; Xinnian, H.; Yingbin, N.; Dezhen, K.; Bo, Z.; Peiyuan, M.U. Effects of Genot and Environment on Protein Qualities of Winter Wheat in Xinjiang. *Xinjiang Agric. Sci.* **2022**, *59*, 45–54. [CrossRef]
86. Weber, R. Influence of tillage system and stubble height on grain properties of selected winter wheat cultivars. *Nauka Pr Technol.* **2013**, *7*, 18–23.
87. Buczek, J. Quality and productivity of hybrid wheat depending on the tillage practices. *Plant Soil Environ.* **2020**, *66*, 415– [CrossRef]
88. Johansson, E.; Prieto-Linde, M.L.; Svensson, G. Influence of Nitrogen Application Rate and Timing on Grain Protein Composit and Gluten Strength in Swedish Wheat Cultivars. *J. Plant Nutr. Soil. Sci.* **2004**, *167*, 345–350. [CrossRef]
89. Liu, D.; Shi, Y. Effects of different nitrogen fertilizer on quality and yield in winter wheat. *Adv. J. Food Sci. Technol.* **2013**, *5*, 646– [CrossRef]
90. Gagliardi, A.; Carucci, F.; Masci, S.; Flagella, Z.; Gatta, G.; Giuliani, M.M. Effects of Genotype, Growing Season and Nitro Level on Gluten Protein Assembly of Durum Wheat Grown under Mediterranean Conditions. *Agronomy* **2020**, *10*, 755. [Cross
91. Buczek, J.; Migut, D.; Jańczak-Pieniążek, M. Effect of Soil Tillage Practice on Photosynthesis, Grain Yield and Quality of Hyb Winter Wheat. *Agriculture* **2021**, *11*, 479. [CrossRef]
92. Šekularac, A.; Torbica, A.; Živančev, D.; Tomić, J.; Knežević, D. The influence of wheat genotype and environmental factors gluten index and the possibility of its use as bread quality predictor. *Genetika* **2018**, *50*, 85–93. [CrossRef]
93. Mahdavi, S.; Arzani, A.; Mirmohammady Maibody, S.A.M.; Kadivar, M. Grain and flour quality of wheat genotypes grown un heat stress. *Saudi J. Biol. Sci.* **2022**, *29*, 103417. [CrossRef] [PubMed]
94. Knapowski, T.; Ralcewicz, M.; Spychaj-Fabisiak, E.; Ložek, O. Ocena jakości ziarna pszenicy ozimej uprawianej w warunka zróżnicowanego nawożenia azotem. *Fragm. Agron.* **2010**, *27*, 73–80.
95. Murawska, B.; Spychaj-Fabisiak, E.; Keutgen, A.; Wszelaczyńska, E.; Pobereżny, J. Cechy technologiczne badanych odmian zia pszenicy ozimej uprawianych w warunkach Polski i Wielkiej Brytanii. *Inż. Ap. Chem.* **2014**, *53*, 96–98.
96. Bilalis, D.; Karkanis, A.; Patsiali, S.; Agriogianni, M.; Konstantas, A.; Triantafyllidis, V. Performance of wheat varieties (*Tritic aestivum*, L.) under conservation tillage practices in organic agriculture. *Not. Bot. Horti Agrobot. Cluj-Napoca* **2011**, *39*, [CrossRef]
97. Li, Y.; Huang, C.; Sui, X.; Fan, Q.; Li, G.; Chu, X. Genetic variation of wheat glutenin subunits between landraces and variet and their contributions to wheat quality improvement in China. *Euphytica* **2009**, *169*, 159–168. [CrossRef]
98. Zecevic, V.; Knezevic, D.; Boskovic, J.; Madic, M. Effect of Genotype and Environment on Wheat Quality. *Genetika* **2009**, 247–253. [CrossRef]
99. Zecevic, V.; Boskovic, J.; Knezevic, D.; Micanovic, D.; Milenkovic, S. Influence of cultivar and growing season on quality propert of winter wheat (*Triticum aestivum* L.). *Afr. J. Agric. Res.* **2013**, *8*, 2545–2550.
100. Tatar, O.; Cakalogullari, U.; Aykut Tonk, F.; Istipliler, D.; Karakoc, R. Effect of Drought Stress on Yield and Quality Traits Common Wheat During Grain Filling Stage. *Turk. J. Field Crops* **2020**, *25*, 236–244. [CrossRef]
101. *ISO 3093:2009*; Wheat, Rye and Their Flours, Durum Wheat and Durum Wheat Semolina—Determination of the Falling Numb According to Hagberg-Perten. International Organization for Standardization: Geneva, Switzerland, 2009.
102. Knapowski, T.; Kozera, W.; Murawska, B.; Wszelaczyńska, E.; Pobereżny, J.; Mozolewski, W.; Keutgen, A.J. Ocena parametr technologicznych wybranych odmian pszenicy ozimej pod względem wypiekowym. *Inżynieria I Apar. Chem.* **2015**, *5*, 255–25
103. Dojczew, D. Wpływ czasu porastania ziarniaków na wartość technologiczną mąki pszennej. *Postępy Tech. Przetwórstwa Spożywcze* **2010**, *1*, 31–35.

agriculture

MDPI

ticle

mpacts of Weed Resistance to Glyphosate on Herbicide Commercialization in Brazil

rgio de Oliveira Procópio [1], Robson Rolland Monticelli Barizon [1,*], Ricardo Antônio Almeida Pazianotto [1], arcelo Augusto Boechat Morandi [2] and Guilherme Braga Pereira Braz [3]

1 Research Department, EMBRAPA Meio Ambiente, Jaguariúna 13918-110, Brazil; sergio.procopio@embrapa.br (S.d.O.P.); ricardo.pazianotto@embrapa.br (R.A.A.P.)

2 Resarch Department, EMBRAPA Assessoria de Relações Internacionais, Brasília 70770-901, Brazil; marcelo.morandi@embrapa.br

3 Postgraduate Program in Plant Production, Universidade de Rio Verde, Rio Verde 75901-970, Brazil; guilhermebrag@gmail.com

* Correspondence: robson.barizon@embrapa.br; Tel.: +55-19-3311-2700

Abstract: Herbicides are essential tools for the phytosanitary security of agricultural areas, but their excessive use can cause problems in agricultural production systems and have negative impacts on human health and the environment. The objective of this study was to present and discuss the main causes behind the increase in herbicide commercialization in Brazil between 2010 and 2020. Data from the Brazilian pesticide database, provided by *the Instituto Brasileiro do Meio Ambiente e dos Recursos Naturais Renováveis (IBAMA)*, were used. In 2010 and 2020, Brazil sold 157,512 and 329,697 tons of herbicide active ingredients, respectively, representing a 128.1% increase in commercialization over 11 years. Some herbicides, such as clethodim, haloxyfop-methyl, triclopyr, glufosinate, 2,4-D, diclosulam, and flumioxazin, showed increases in sales volumes between 2010 and 2020 of 2672.8%, 896.9%, 953.5%, 290.2%, 233.8%, 561.3%, and 531.6%, respectively, percentages far exceeding the expansion of Brazil's agricultural area. The primary reason for this sharp increase in herbicide sales was the worsening cases of weeds resistant and tolerant to glyphosate, with species such as *Conyza* spp., *Amaranthus* spp., *Digitaria insularis*, and *Eleusine indica* standing out. This situation created the necessity of the use of additional herbicides to achieve effective chemical control of these weed species.

Keywords: chemical control; genetically modified crops; integrated management; pesticide market

1. Introduction

Herbicides are a class of pesticides used to control weeds, serving as a crucial tool in integrated management [1,2], particularly in large-scale agricultural operations. Their importance lies in the fact that, without weed control, agricultural activities suffer severe damage, leading to significant yield losses that can even make crop harvests unfeasible [3,4]. On a global scale, inefficient weed control causes yield losses of approximately 32% across various crops [5]. In this context, herbicide application aims to protect crops from interference by weed communities [6,7], which tend to occur at higher densities in agricultural areas and exhibit superior adaptation to the competitive process for essential natural resources [8]. Weeds can also serve as alternative hosts for insect pests and plant disease-causing microorganisms [9,10] and exhibit allelopathic activity [11,12].

Due to their efficacy and ease of application, allowing high operational efficiency [13], herbicide use has rapidly intensified in Brazil's agricultural, forestry, and pasture areas. The widespread adoption of chemical control also results from the labor shortage in rural areas caused by the ongoing migration to urban centers, driving a preference for herbicide use over manual weed control [14]. In recent decades, glyphosate has become prominent among herbicides registered in Brazil, primarily due to its favorable agronomic characteristics, such as broad-spectrum activity [15,16], effectiveness in controlling weeds at advanced developmental stages [17], efficiency in managing perennial species that propagate vegetatively [18,19], and the absence of soil active residues [20,21]. Two transformative events in Brazilian agriculture have driven the significant increase in glyphosate use. The first was the introduction and rapid adoption of the no-tillage system in grain production areas, where pre-sowing burndown became a routine operation, initially with glyphosate as the main herbicide used for this purpose [22]. The second event was the introduction of genetically modified (GMO) cultivars, such as soybean, corn, and cotton, which were engineered to be glyphosate-resistant [23].

The lack of or insufficient rotation of herbicides with different modes of action, along with the failure to implement integrated weed management principles, led to excessive selection pressure for glyphosate on the weed community [24,25]. Over years of intensive glyphosate use, weed biotypes that are resistant to this herbicide were selected, spreading to the point that glyphosate-resistant populations are now found throughout Brazil [26,27]. Currently, there have been official reports of 20 cases of weeds with simple or multiple resistance to glyphosate [28]. Resistant biotypes identified in Brazil include the following species: *Lolium perenne*, *Conyza bonariensis*, *Conyza canadensis*, *Digitaria insularis*, *Conyza sumatrensis*, *Chloris elata*, *Amaranthus palmeri*, *Eleusine indica*, *Amaranthus hybridus*, *Euphorbia heterophylla*, *Echinochloa crus-galli*, and *Bidens subalternans* [29–32]. In addition to these species, a noteworthy challenge is the management of volunteer glyphosate-resistant corn plants, where grain losses during corn harvests have led to infestations in subsequent soybean crops, complicating chemical control [33,34].

The emergence and spread of resistant biotypes across various regions of Brazil necessitated the use of additional herbicides to complement glyphosate's action on previously controlled species [35]. Among the herbicides that began to be more intensively used as a result of the spread of resistant biotypes in Brazil's grain production areas are 2,4-D, triclopyr, diclosulam, flumioxazin, haloxyfop-methyl, clethodim, and glufosinate. Despite the increase in resistant populations, glyphosate remains the primary active ingredient used in Brazilian agriculture, with variations in formulations depending on the type of salt utilized.

Thus, the objective of this study was to present and discuss data on herbicide commercialization in Brazil, with the main hypothesis being that the increase in herbicide consumption is driven by the worsening cases of glyphosate-resistant weeds.

2. Materials and Methods

The primary data used for the analysis of pesticide commercialization and the number of formulated products per active ingredient were obtained from the *Instituto Brasileiro do Meio Ambiente e dos Recursos Naturais Renováveis* (IBAMA), the agency responsible for releasing commercialization data in Brazil. According to Brazilian legislation, manufacturing and importing companies must report their annual sales volume to IBAMA, which compiles this information to produce and publish reports on pesticide sales in the country. The data collection period from IBAMA covered from 2010 to 2020. After analyzing the primary database, it was necessary to make adjustments and standardize the historical series to align it with the guidelines of the Food and Agriculture Organization of the United Nations (FAO) [36] and the Organisation for Economic Co-operation and Development (OECD) [37].

To normalize and standardize the data from the IBAMA primary database, the following data were excluded: (1) The volume of non-agricultural pesticide (NA) sales in Brazil between 2010 and 2020. (2) The sales volume of products registered exclusively as adjuvants or surfactants (2010 to 2017), which were no longer classified as pesticides in Brazil following the *Ministério da Agricultura e Pecuária* (MAPA) act n° 104 of 20 November 2017, published in the Official Gazette of 21 November 2017. From 2018 onward, these products were no longer included in IBAMA's database, and by excluding data from 2010 to 2017, the historical series was standardized. (3) The sales volume of products classified as semiochemicals and microbiologicals (2010 to 2013). Since 2014, IBAMA began reporting these biopesticides separately from chemical, biochemical, and phytochemical products in its bulletins, and (4) sales of formulated pesticide products between industries in Brazil (2010 to 2020), because counting inter-company sales could result in double counting.

Additionally, after direct consultation with FAO and OECD technicians and statisticians, who also develop and disseminate international databases related to agricultural sustainability indicators, the following standardizations were implemented: (1) For the herbicides 2,4-D, glyphosate, picloram, MCPA, aminopyralid, florpyrauxifen-benzyl, fluroxypyr, haloxyfop-methyl, imazapic, imazaquin, imazapyr, imazethapyr, triclopyr, and cyhalofop-butyl, the data were recorded in acid equivalent (a.e.). These herbicides are formulated with different derivatives to ensure formulation stability, where standardizing the data in a.e. allows normalization regardless of the derivative used in the formulation. (2) For the herbicides diquat and paraquat, the data were recorded in tons of ion equivalent. Similar to the herbicides mentioned in Item 1, these compounds are also formulated with derivatives, and data standardization ensures consistent reporting. (3) For the fungicides copper hydroxide, copper oxychloride, copper oxide, and copper sulfate, the data were recorded in tons of metallic copper equivalent to maintain consistency, and (4) for all other pesticides, the data were recorded in tons of active ingredient (a.i.).

The increase in herbicide sales in Brazil from 2010 to 2020 was compared with the growth of the Brazilian agricultural area (perennial and temporary crops) and soybean area, using data from the SIDRA system, provided by the *Instituto Brasileiro de Geografia e Estatística* (IBGE) [38]. Additionally, it was compared with the expansion of non-degraded pasture areas (pastures that effectively use agricultural inputs, such as herbicides), calculated using data from the Image Processing and Geoprocessing Laboratory of the *Universidade Federal de Goiás* (LAPIG/UFG) [39].

3. Results and Discussion

3.1. Pesticide Sales in Brazil (2010 to 2020)

The data presented in Table 1 show the evolution of pesticide sales in Brazil, divided by usage classes, from 2010 to 2020. During this period, herbicides accounted for the highest sales volume among pesticides used in Brazilian agriculture. In 2010, 157,512 tons of herbicide active ingredients were sold in Brazil, representing 53.34% of all pesticides (chemical, biochemical, and phytochemical) commercialized that year. However, by 2020, herbicide sales had reached 329,697 tons of active ingredients, marking a 128.1[illegible]% increase over 11 years and accounting for 58.89% of total pesticide sales in Brazil. For comparison on a global scale, in 2019 herbicides represented approximately 52% of the total pesticide market share [40], which is slightly lower than the Brazilian percentage. Additionally, the global increase in herbicide consumption from 2010 to 2020 was 29.92% [36], significantly lower than the increase recorded in Brazil.

Two reasons can be given to justify the use of higher quantities of herbicides compared to other pesticide categories. The first reason is that fungicides and insecticides are typically applied during the crop cycle (from sowing to harvest), whereas herbicides are used before, during, and after the crop cycle [41]. Their use is common before crop establishment (pre-sowing burndown) and after harvest (burndown of spontaneous vegetation during the fallow period). The second reason is the limited availability of bioherbicides in the Brazilian market, unlike bioinsecticides and biofungicides, which are already well-established in Brazilian agriculture. This has led to the exclusive use of chemical molecules for weed control in agricultural areas. Bioherbicides can be formulated with microorganisms or phytotoxins derived from microorganisms, insects, or plant extracts that act as natural agents for weed control [42].

Table 1. Sales data and representative percentage of each pesticide class (2010 to 2020). Grouping standard proposed by OECD [37].

Year	Total	Insecticides [1]	Fungicides [2]	Herbicides	Others [3]
	Tons of Active Ingredient [4]				
2010	295,287	72,856	63,365	157,512	1554
2011	314,862	77,950	61,298	173,171	2443
2012	383,999	75,146	55,117	251,094	2642
2013	407,792	96,135	53,541	256,004	2112
2014	420,467	104,172	61,028	252,765	2502
2015	445,684	95,636	74,354	272,852	2842
2016	451,276	86,627	85,605	276,147	2897
2017	453,111	86,174	90,882	273,502	2553
2018	490,495	83,711	110,698	293,053	3033
2019	564,635	98,562	132,357	329,697	4019
2020	610,098	101,783	144,241	359,308	4766
	Percentage (%) of total sales in the year				
2010	100.00	24.67	21.46	53.34	0.53
2011	100.00	24.75	19.47	55.00	0.78
2012	100.00	19.57	14.35	65.39	0.69
2013	100.00	23.57	13.13	62.78	0.52
2014	100.00	24.78	14.50	60.12	0.60
2015	100.00	21.46	16.68	61.22	0.64
2016	100.00	19.20	18.97	61.19	0.64
2017	100.00	19.02	20.06	60.36	0.56
2018	100.00	17.07	22.57	59.74	0.62
2019	100.00	17.46	23.44	58.39	0.71
2020	100.00	16.68	23.64	58.89	0.79

[1] Includes insecticides, acaricides, molluscicides, and nematicides. [2] Includes fungicides, bactericides, and pesticides used in seed treatment. [3] Includes fumigants, growth regulators, and seed protectants. [4] For the herbicides 2,4-D, glyphosate, picloram, MCPA, aminopyralid, florpyrauxifen-benzyl, fluroxypyr, haloxyfop-methyl, imazapic, imazaquin, imazapyr, imazethapyr, triclopyr, and cyhalofop-butyl, data are reported in tons of acid equivalent. For the herbicides diquat and paraquat, data are reported in tons of ion equivalent. For fungicides such as copper hydroxide, copper oxychloride, copper oxide, and copper sulfate, data are reported in tons of metallic copper equivalent. All other pesticides are reported in tons of active ingredients.

Regarding pesticide sales data, the introduction of *Bt* technology in soybean, corn, and cotton cultivars, combined with the increased use of bioinsecticides (both microbiological and macrobiological), likely contributed to the reduced share of chemical insecticides in the Brazilian pesticide market [43,44]. On the other hand, the growing share of fungicides in Brazil can be attributed mainly to the following factors: (1) weakening of some breeding programs focused on disease resistance/tolerance, resulting in the release of cultivars susceptible to various diseases and requiring more chemical control intervention; and (2) increased pathogen resistance to fungicides, leading to a significant rise in the use of multi-site fungicides in preventive applications, particularly in crops like soybeans [45,46]. It is worth noting that the recommended dose per area for multi-site fungicides is significantly higher than for more recently introduced site-specific fungicides.

Several factors explain the increased use of herbicides during the studied period (2010 to 2020) in Brazil: (1) expansion of no-till farming areas [47,48], where herbicide applications for pre-sowing burndown are common, with higher doses often used due to the more advanced growth stages of weeds. Brazilian farmers have become more aware of soil conservation [49], leading to the replacement of mechanical weed control practices, such as plowing, harrowing, and scarifying, which expose the soil to erosion; (2) growth in second-crop corn areas [50], where farmers use herbicides for pre-harvest soybean desiccation to advance corn sowing, in addition to increased herbicide use in corn crops themselves; (3) reduction in rural labor availability, which hinders the use of manual weed control methods [51]. Data from the World Bank Group [52] show that the rural population in Brazil decreased by 31.58% between the years 2000 and 2020; (4) lack of bioherbicides or microbiological herbicides on the Brazilian market, unlike the growing availability of bioinsecticides and biofungicides; and (5) greater availability of products formulated from the same herbicide active ingredient, increasing competition among manufacturers/importers, which may have driven prices down for some herbicides. Examples of this price reduction can be found in the database of the *Companhia Nacional de Abastecimento* (CONAB) [53]: The herbicide clethodim (same formulated product and same active ingredient concentration) was sold for USD 51.41 per liter in 2010 and was marketed for USD 8.45 per liter in 2020 (based on data from Paraná State, Brazil). The herbicide haloxyfop (same formulated product and same active ingredient concentration) was sold for USD 32.82 per liter in 2010 and was marketed for USD 9.92 per liter in 2020 (based on data from São Paulo State, Brazil). (6) Increased areas infested with herbicide-resistant weed biotypes, particularly for glyphosate [54–56].

Herbicide-resistant weeds are the most significant factor behind the increased herbicide sales and are the central theme of the present material. It is important to note that the rise in herbicide sales does not necessarily indicate abusive behavior by farmers. Unlike other pesticides, there is often resistance among agricultural experts to increasing herbicide doses, as this could increase the risk of crop injuries, prolong residual herbicide activity, and cause carryover problems for subsequent crops. Additionally, higher herbicide doses raise production costs.

In addition to evaluating the commercialization of herbicides, it is relevant to investigate whether the environmental hazard potential profile of these products changed during the analyzed period. In Brazil, IBAMA is the institution responsible for assessing pesticides based on the environmental risks associated with their use, categorizing them into four classes: Class I: Highly hazardous to the environment; Class II: Very hazardous to the environment; Class III: Hazardous to the environment; and Class IV: Slightly hazardous to the environment To determine these classifications, IBAMA evaluates various factors, such as bioaccumulation, toxicity to non-target organisms, biodegradation, and soil mobility. In 2010, 0.001% of the herbicides sold in Brazil were classified as Class I, while 18.90% belonged to Class II and 81.10% to Class III. These proportions changed minimally by 2020, with 0.02% of the herbicides classified as Class I, 20.63% as Class II, and 79.35% as Class III. Notably, during this time frame, no herbicides classified as Class IV were sold in Brazil.

Regarding the human toxicity potential of herbicides sold in Brazil, it is important to highlight that between 2010 and 2020, no herbicides classified by the World Health Organization [57] as Ia (extremely hazardous) or Ib (highly hazardous) were marketed. An integrative analysis of the consolidated database showed that not all herbicide active ingredients experienced increased sales in Brazil from 2010 to 2020. Therefore, this study presents and discusses the sales results for herbicides that saw the highest sales growth during this period and are recommended for use in grain production areas.

3.2. Evolution of Herbicide Sales in Brazil (from 2010 to 2020)

The evolution of 2,4-D herbicide sales (tons of acid equivalent) is represented in Figure 1A, where a nearly linear increase from 2010 to 2020 can be observed. The herbicide 2,4-D is a synthetic auxin, characterized by causing various metabolic disturbances in sensitive plants, such as abnormal growth, death of root tissues, epinasty, and obstruction of phloem [58]. In Brazil, the herbicide 2,4-D is primarily marketed in the form of amine salt and is indicated for the control of broadleaf weeds [59]. In Brazil, 15,323 tons of this herbicide's active ingredient were sold in 2010, while in 2020, this number rose to 51,149 tons, representing an increase of 233.8%. The observed increase in usage exceeds the increase recorded during this 11-year period concerning the area of agriculture designated for harvesting (annual and perennial crops) (24.3%), the soybean area (59.4%), and the area of pastures without degradation (16.0%), which can be classified as the most demanding agricultural activities for 2,4-D in the country (Figure 1B).

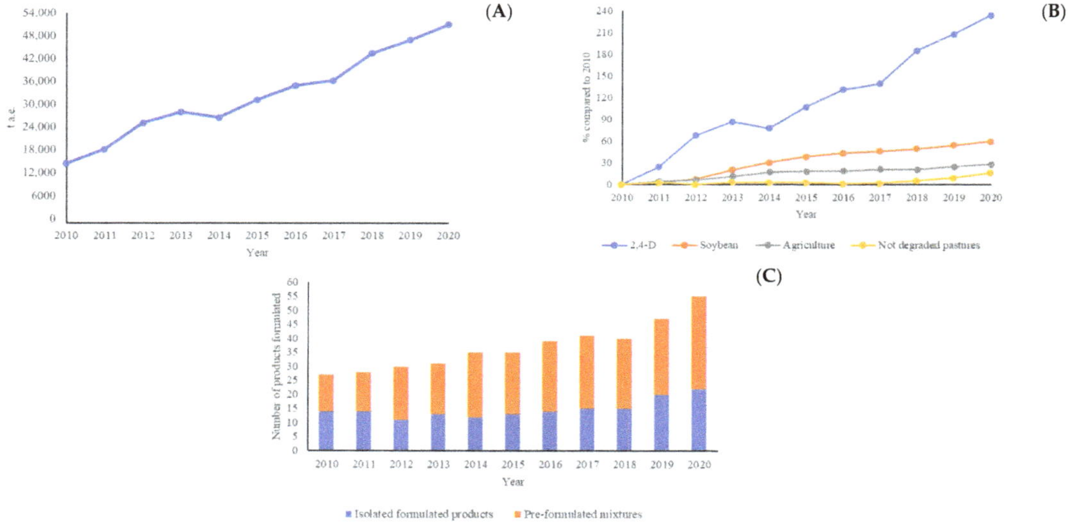

Figure 1. Sales of 2,4-D (tons of a.e.) from 2010 to 2020 (**A**). Evolution of herbicide sales in relation to the year 2010, compared to the evolution of soybean area (IBGE), the area of agriculture designated for harvesting (annual and perennial) (IBGE), and the area cultivated with pastures without degradation (LAPIG/UFG) (**B**). Number of formulated products based on 2,4-D sold in Brazil (**C**) (period from 2010 to 2020).

The main reason for the increased use of 2,4-D during this period is the worsening cases of glyphosate-resistant weed biotypes, specifically the increasing infestation of biotypes from different *Conyza* spp. species and areas infested with resistant *Amaranthus* spp. biotypes [27,60]. As a result, glyphosate, which previously provided effective control of these weed species, began requiring complementary herbicides during pre-sowing burndown to help manage resistant biotypes, driving the sales of 2,4-D in Brazil. It is important to note that the Enlist™ technology, which enables the selective use of 2,4-D, glyphosate, and glufosinate in crops like soybean and corn [61], was not available in the Brazilian market during the evaluated period (2010 to 2020) and, therefore, cannot be considered one of the factors responsible for the observed increase in use.

Other factors may have also contributed to this rise in 2,4-D consumption in Brazil, including: (1) increased use of herbicides in pastures between 2010 and 2020 [62]; (2) expansion of no-till farming in Brazil, as the main agricultural use of this herbicide is linked to pre-sowing burndown, especially before soybean crops; (3) increased availability of 2,4-D products in the Brazilian market. In 2010, 27 formulated products containing the active ingredient 2,4-D were sold, 14 containing only 2,4-D (as a single active ingredient) and 13 pre-formulated mixtures with other active ingredients. By 2020, the number of formulated products containing 2,4-D had increased to 55 (22 single active ingredient products and 33 pre-formulated mixtures), representing a 103.7% increase in product availability (Figure 1C); (4) expansion of agricultural areas infested with glyphosate-tolerant weeds, such as *Commelina benghalensis*, *Spermacoce latifolia*, *Ipomoea* spp., and *Tridax procumbens* [63,64]; and (5) shorter plantback intervals for soybean sowing compared to other auxin herbicides.

Another auxin herbicide, triclopyr, also experienced an increase in sales between 2010 and 2020 (Figure 2A). In 2010, 228 tons of triclopyr (acid equivalent) were sold, while in 2020, this volume increased to 2405 tons, representing a 953.5% increase in sales. This rise far exceeds the increase in soybean area (59.4%) and the area of pastures without degradation (16.0%), two agricultural activities where triclopyr is potentially used (Figure 2B). The main causes for the increase in triclopyr sales from 2010 to 2020 are as follows: (1) worsening cases of glyphosate-resistant weed biotypes, particularly the increased infestation of *Conyza* spp. biotypes, including those resistant to 2,4-D (rapid necrosis) [65], where triclopyr became a tool for management [66], and (2) expansion of areas infested with species from the genera *Spermacoce*, *Borreria*, and *Mitracarpus*, with triclopyr becoming one of the go-to herbicides for controlling these species during the off-season. Other less impactful factors may have also contributed to the rise in triclopyr sales, such as (1) increased availability of triclopyr-based products in the Brazilian market, especially products formulated with triclopyr as the sole active ingredient (a 400% increase in the availability of single-active ingredient formulations from 2010 to 2020) (Figure 2C), and (2) lower antagonism when mixed with ACCase-inhibiting herbicides compared to 2,4-D [67].

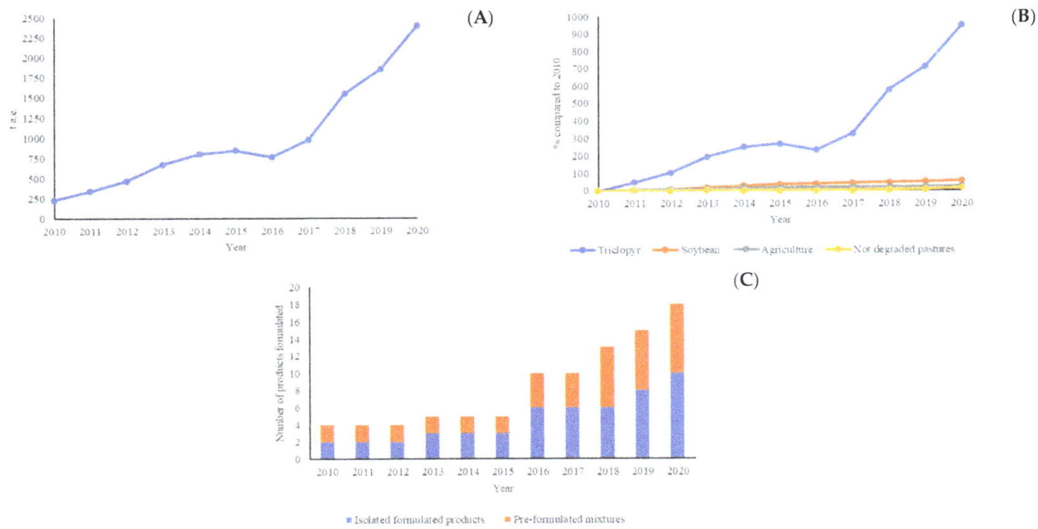

Figure 2. Sales of triclopyr (tons of a.e.) from 2010 to 2020 (**A**). Evolution of the commercialization of herbicide in relation to the year 2010, compared to the evolution of the soybean area (IBGE), the area of agriculture intended for harvest (temporary and permanent) (IBGE), and the area cultivated with pastures showing no degradation (LAPIG/UFG) (**B**). Number of formulated products based on triclopyr marketed in Brazil (**C**) (period from 2010 to 2020).

Haloxyfop-methyl and clethodim are herbicides used for post-emergence control of grasses, being selective for broadleaf crops such as soybean [68]. Both herbicides inhibit the enzyme acetyl-CoA carboxylase (ACCase), blocking lipid synthesis and, consequently, the formation of cell walls [69]. These two graminicides had increases in sales in Brazil between 2010 and 2020 (Figure 3A). In 2010, 111.3 tons of haloxyfop-methyl and 244.5 tons of clethodim (active ingredient/acid equivalent) were sold. By 2020, these amounts rose to 1109.8 tons of haloxyfop-methyl and 6779.6 tons of clethodim, representing an 896.9% increase in haloxyfop-methyl consumption and a 2672.8% increase for clethodim (Figure 3B). These increases far exceed the growth in agricultural area (24.3%) and soybean-planted area (59.4%) during the same period.

Among the main reasons for the sharp increase in the consumption of haloxyfop-methyl and clethodim herbicides are: (1) a significant rise in grain-growing areas infested with glyphosate-resistant biotypes of *Digitaria insularis* [26]. This situation created the need for supplementary control of this species, both in pre-sowing burndown operations and in post-emergence soybean management, substantially increasing the demand for these herbicides. More recently, another grass species resistant to glyphosate that could contribute to the increased use of haloxyfop-methyl and clethodim is *Eleusine indica*. However, this trend has not been observed with the same intensity, as multiple-resistant biotypes (EPSPS and ACCase) of this species have spread, limiting the effectiveness of these herbicides [70]; and (2) the increasing challenge posed by volunteer glyphosate-resistant corn plants. With the growing adoption of glyphosate-resistant corn hybrids in Brazil, controlling volunteer corn plants, resulting from mechanical harvesting failures, has become more problematic. This created the need to introduce new herbicides for controlling these plants before soybean sowing or even in established soybean fields. Among the effective herbicides for this purpose are ACCase inhibitors, such as haloxyfop-methyl and clethodim [71].

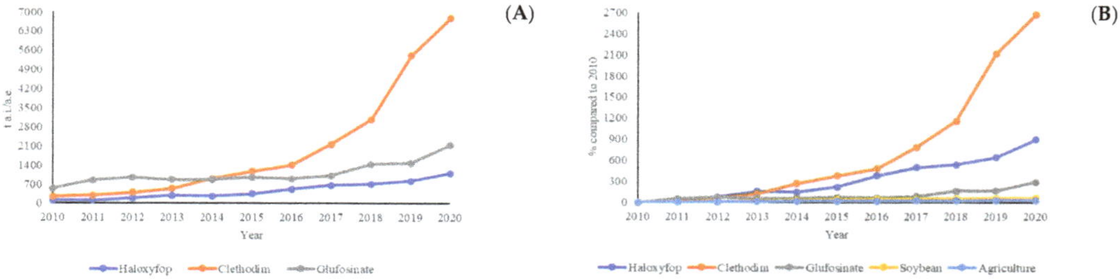

Figure 3. Sales of haloxyfop-methyl, clethodim, and glufosinate (tons of a.i./a.e.) (**A**). Evolution of the commercialization of herbicides in relation to the year 2010, compared to the evolution of the soybean area (IBGE) and the area of agriculture intended for harvest (temporary and permanent) (IBGE) (**B**). Sales data for haloxyfop-methyl are presented in effective ingredients.

The increased availability of haloxyfop-methyl and especially clethodim products may have also contributed to the rise in herbicide sales. In 2010, only one formulated product containing the isolated active ingredient haloxyfop-methyl was available in the Brazilian market. By 2020, this number had grown to nine products, including seven with haloxyfop-methyl as a single active ingredient and two pre-formulated mixtures (Figure 4A), representing an 800% increase in haloxyfop-methyl-based product availability over 11 years. For clethodim, the increase in formulated products during the same period was 600%, rising from two products in 2010 (one isolated and one pre-formulated mixture) to 14 products in 2020 (11 isolated and three pre-formulated mixtures) (Figure 4B).

The higher increase in clethodim consumption compared to haloxyfop-methyl may be related to the lower cost per hectare of clethodim application. It is important to note that the rise in glyphosate-resistant *Digitaria insularis* biotypes and volunteer corn plants (Roundup Ready™—RR™) was a key driver encouraging pesticide manufacturers and importers to register and market generic products based on haloxyfop-methyl and clethodim in Brazil, as these resistance cases created substantial demand for this group of herbicides. In summary, the increased availability of formulated products can be seen as a response to the worsening weed resistance to glyphosate in Brazil.

Three key factors can be identified as predominant drivers of the increase in the commercialization of the herbicide glufosinate in Brazil (Figures 3 and 4C): (1) increase in areas infested with glyphosate-resistant weeds, such as *Conyza* spp., *Digitaria insularis*, *Eleusine indica*, and *Amaranthus* spp. [72]. This is the most important factor driving the rise in glufosinate sales in Brazil; (2) introduction of transgenic soybean, corn, and cotton cultivars resistant to post-emergence applications of glufosinate in the Brazilian market; and (3) growth in the availability of glufosinate-based formulated products, especially after 2017 (one formulated product in 2010; nine in 2020). This increase is largely due to the sharp rise in glyphosate-resistant weeds in Brazil's grain crops, which has spurred companies' interest in marketing this active ingredient. It is also important to note that the ban on paraquat use in Brazil had a significant impact on glufosinate sales. Glufosinate shares some similar agronomic characteristics with paraquat, such as a broad spectrum of control, contact action, and the absence of residual soil activity [73]. However, the impact of this ban became more evident only in the later years of the analyzed period.

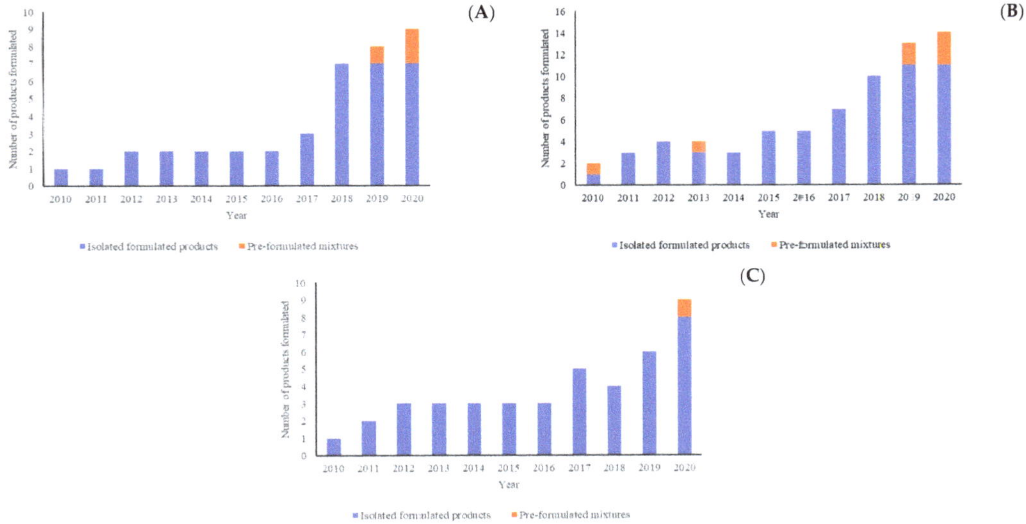

Figure 4. Number of formulated products based on haloxyfop-methyl (**A**), clethodim (**B**), and glufosinate (**C**) marketed in Brazil, subdivided by products containing only the active ingredient (isolated) or pre-formulated mixtures containing the respective herbicides (period from 2010 to 2020).

In the context of glyphosate-resistant weeds, the use of herbicides with residual soil activity has re-emerged in recent years in grain production areas. Looking ahead, there are expectations for the increased adoption of these herbicides in agricultural production systems, as they can control weed populations in the pre-emergence stage of crops [74]. Figures 5 and 6 show the sales data (tons of active ingredient and percentage growth) and the number of formulated products for the herbicides diclosulam and flumioxazin, which experienced sales growth between 2010 and 2020. Although flumioxazin had higher sales volumes, comparing the application rates per hectare for the crops registered for diclosulam suggests that diclosulam usage also increased substantially and is comparable to flumioxazin (Figure 5A). Moreover, considering only the percentage increase in sales relative to the first year of analysis (2010), both diclosulam and flumioxazin saw growth, with increases exceeding 520.0% (Figure 5B).

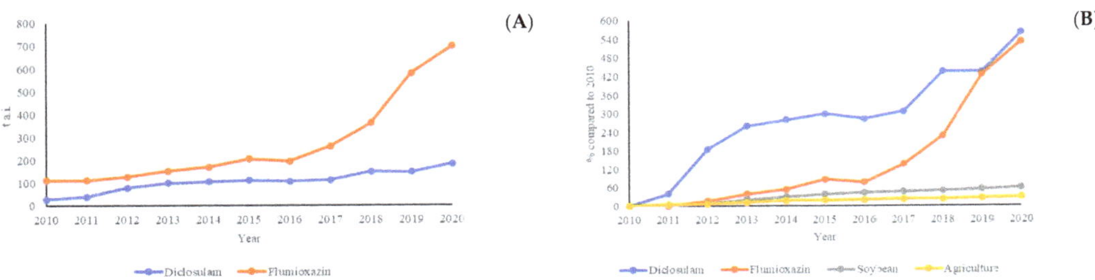

Figure 5. Sales of diclosulam and flumioxazin (tons of a.i.) (**A**). Evolution of herbicide sales compared to the year 2010, alongside the evolution of soybean area (IBGE) and the area of agriculture designated for harvest (temporary and permanent) (IBGE) (**B**).

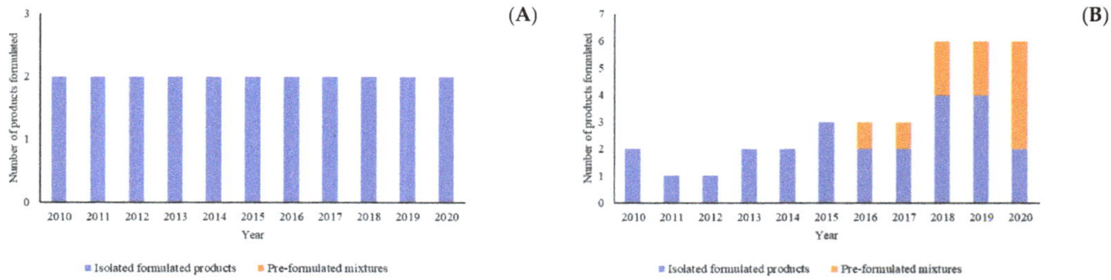

Figure 6. Number of formulated products based on diclosulam (**A**) and flumioxazin (**B**) marketed in Brazil, subdivided by products containing only the active ingredient (isolated) or pre-formulated mixtures containing the respective herbicides (period from 2010 to 2020).

Several factors can explain the increase in the commercialization of diclosulam and flumioxazin in the Brazilian market, among which the following stand out: (1) increased adoption of herbicides used in soybean pre-emergence applications, as certain products containing these active ingredients rank among the most widely used for this purpose [75]; (2) greater occurrence of glyphosate-resistant weeds, including *Conyza* spp. and the *Amaranthus* spp. complex (*A. hybridus* and *A. palmeri*), which diclosulam and flumioxazin effectively control in pre-emergence [60]; and (3) increased use of herbicides with residual soil activity, such as diclosulam and flumioxazin, in pre-sowing burndown applications for soybean crops.

Regarding the evolution of formulated products containing diclosulam and flumioxazin in the Brazilian agrochemical market, distinct behaviors can be observed depending on the active ingredient (Figure 6). For diclosulam, only two formulated products containing this active ingredient, one registered for sugarcane and another for soybean, were marketed during the period analyzed in this study (Figure 6A). Given the herbicide's high efficacy in controlling the main weeds infesting the crops for which diclosulam is registered [76], future prospects suggest a potential increase in the number of formulated products based on this active ingredient.

For flumioxazin, the number of formulated products marketed in Brazil increased, particularly from 2015 onward (Figure 6B). This reflects one of the unique characteristics of flumioxazin's weed control action: unlike other herbicides, flumioxazin exhibits residual soil activity (pre-emergence control), offering the added benefit of effective weed control in early post-emergence stages through contact action [77,78]. This is especially true when flumioxazin is used in combination with other herbicides. Additionally, due to its lower acquisition cost compared to other pre-emergent herbicides, along with these agronomic advantages, the number of formulated products containing flumioxazin has increased in recent years.

Despite the issues of weed resistance and tolerance discussed throughout this manuscript, data presented in Table 2 show that the commercialization of glyphosate in Brazil increased from 101,385 tons of active ingredient in 2010 to 209,345 tons in 2020, representing a 106.5% increase over the period. Furthermore, in 2010, glyphosate sales accounted for 64.37% of total herbicide sales in Brazil, whereas in 2020 this percentage dropped to 58.26%, indicating that other herbicides experienced higher growth rates in sales compared to glyphosate. This shift may reflect the widespread occurrence of glyphosate-resistant biotypes in agricultural areas across the country. Notably, glyphosate accounted for 34.33% of all pesticide sales in Brazil in 2010, a scenario that remained virtually unchanged in 2020 (34.31%). Several factors may explain the increased use of glyphosate in Brazil, including: (1) expansion of no-till farming; (2) greater adoption of GMO cultivars resistant to glyphosate, such as soybean, corn, and cotton; (3) increased use of cover crops (e.g., *Brachiaria* and *Panicum*), either in monoculture or intercropped systems, where glyphosate is the main tool for pre-sowing burndown [79]; (4) additive or synergistic effect when mixed with other herbicides with different modes of action, enhancing weed control [80]; and

(5) increased availability of glyphosate-based products. In 2010, 36 glyphosate-formulated products were sold in Brazil, including 35 single-ingredient products and only one pre-mix. By 2020, this number rose to 47 products, a 30.6% increase, with 45 single-ingredient products and two pre-mixes (Table 2).

Table 2. Information on glyphosate commercialization in Brazil in 2010 and 2020.

Information	Year	
	2010	2020
yphosate commercialization (tons of active ingredients)	101,385	209,345
of total herbicide commercialization	64.37	58.26
of total pesticide commercialization	34.33	34.31
umber of Formulated products	36	47
umber of isolated formulated products	35	45
umber of pre-formulated mixtures with other active ingredients	1	2
lt formulation	% of total glyphosate sales	
opropylamine	51.59	12.55
mmonium	21.49	27.06
otassium	26.92	35.88
methylammonium	0.00	0.75
iammonium	0.00	9.26
sopropylamine + potassium]	0.00	14.50

The type of salt formed from glyphosate also changed between 2010 and 2020. Chemically, glyphosate is formulated as ammonium, diammonium, dimethylammonium, potassium, or isopropylamine salts [81]. In 2010, isopropylamine salt accounted for 51.59% of glyphosate sales in Brazil. However, by 2020, its share dropped to 12.55%, with potassium salt becoming the most sold, representing 35.88% of total glyphosate sales (Table 2) Changes in the type of glyphosate salt used in commercial products can affect the speed of weed control [22] and influence environmental factors after application, such as absorption speed and the impact of rainfall following application (rainfastness).

The data indicate disproportionate growth in the sale of herbicides used to manage glyphosate-resistant weeds, including both pre- and post-emergence products. This suggests that the resistance of weeds to glyphosate is a primary factor driving increased herbicide consumption. As a result, the quantitative data clearly highlight the negative impacts of glyphosate resistance, leading to agronomic and economic losses and significantly increasing the need for additional herbicides.

Despite the spread of resistant biotypes, glyphosate has not been widely replaced by other herbicides. Instead, it is often supplemented with additional molecules [82], as discussed in this study, contributing to the rising consumption of herbicides in Brazil. This increase in herbicide use is primarily due to a greater number of active ingredients applied per season, not higher doses. It is important to note that herbicide selectivity for crops is based on dose limits [83], since exceeding these limits can cause significant yield losses. Before resistance became widespread, glyphosate isolated was considered sufficient to control weeds across many agricultural areas [84]. Today, however, most soybean fields require two, three, or more active ingredients to complement glyphosate's action, both before sowing (burndown) and during the crop cycle (pre- and post-emergence).

Given this scenario, it is essential to implement integrated weed management strategies, including preventive control (e.g., cleaning machinery and acquiring certified, weed-free seeds), and cultural control (e.g., crop rotation, no-till farming with surface mulch, narrower row spacing, selecting fast-growing cultivars, and using high-vigor seeds). Additionally, whenever possible, herbicide rotation with different modes of action should be practiced, along with the use of sprayers equipped with weed-detecting sensors for post-emergence applications. Research investments aimed at identifying biological control agents to develop bioherbicides, as well as improving physical weed control methods (e.g., robotics, laser beams, electromagnetic waves, or boiling water with sealing foam), are also strategic and sustainable approaches. These combined efforts are crucial to preventing a future collapse in chemical weed control effectiveness and avoiding unsustainable herbicide loads in agricultural environments.

4. Conclusions

The commercialization of herbicides in Brazil saw an increase between 2010 and 2020. Some herbicides, such as clethodim, haloxyfop-methyl, triclopyr, glufosinate, 2,4-D, diclosulam, and flumioxazin, experienced increases in sales of 2672.8%, 896.9%, 953.5%, 290.2%, 233.8%, 561.3%, and 531.6%, respectively, far exceeding the growth rate of Brazil's agricultural area. Several factors may have contributed to this substantial rise in herbicide consumption; however, the primary cause was the worsening cases of glyphosate-resistant weed biotypes in Brazilian agricultural areas, particularly among species such as *Conyza* spp., *Amaranthus* spp., *Digitaria insularis*, and *Eleusine indica*. This situation created the need to supplement glyphosate with other herbicides to achieve effective chemical control of these species.

The data highlight that weed resistance to glyphosate has driven a substantial increase in the commercialization of herbicides with different profiles for both pre- and post-emergence applications, resulting in agronomic and economic impacts. This scenario led to the need for glyphosate to be supplemented with other molecules rather than replaced, contributing to an increase in the number of active ingredients used per season. This situation underscores the importance of integrated practices, such as rotating herbicide modes of action and implementing preventive and cultural controls, to ensure the effectiveness of chemical control and reduce dependency on herbicides. Investments in bioherbicides, nanotechnology related to herbicide formulation, and physical methods such as robotics and laser emissions are becoming essential to tackle future challenges and offer innovative and sustainable alternatives in weed management.

Author Contributions: Conceptualization, S.d.O.P. and R.R.M.B.; methodology, S.d.O.P., R.R.M.B. and R.A.A.P.; software, R.R.M.B. and R.A.A.P.; validation, S.d.O.P., R.R.M.B., R.A.A.P. and M.A.B.M.; formal analysis, S.d.O.P., R.R.M.B. and R.A.A.P.; investigation, S.d.O.P., R.R.M.B. and R.A.A.P.; resources, S.d.O.P. and R.R.M.B.; data curation, S.d.O.P. and R.A.A.P.; writing—original draft preparation, S.d.O.P., R.R.M.B. and G.B.P.B.; writing—review and editing, S.d.O.P., R.R.M.B., R.A.A.P. and G.B.P.B.; visualization, S.d.O.P., R.R.M.B., R.A.A.P. and G.B.P.B.; supervision, S.d.O.P., R.R.M.B., R.A.A.P. and M.A.B.M.; project administration, R.R.M.B.; funding acquisition, R.R.M.B. All authors have read and agreed to the published version of the manuscript.

Funding: This research received funding from the *Ministério da Agricultura e Pecuária* (MAPA) through a resource administered by the *Empresa Brasileira de Pesquisa Agropecuária* (EMBRAPA) (Project number: 10.22.10.018.00.00), and the APC was funded by MAPA.

Institutional Review Board Statement: Not applicable.

Data Availability Statement: Data are available from the corresponding author.

Acknowledgments: Special thanks to EMBRAPA *Meio Ambiente* analyst Renan Milagres Lage Novaes for his valuable contributions in reviewing the present material.

Conflicts of Interest: The authors declare that the research was conducted in the absence of any commercial or financial relationships that could be construed as potential conflicts of interest.

References

1. Abouziena, H.F.; Haggag, W.M. Weed control in clean agriculture: A review. *Planta Daninha* **2016**, *34*, 377–392. [CrossRef]
2. Parven, A.; Meftaul, I.M.; Venkateswarlu, K.; Megharaj, M. Herbicides in modern sustainable agriculture: Environmental fa ecological implications, and human health concerns. *Int. J. Environ. Sci. Technol.* **2024**. [CrossRef]
3. Wang, A.; Zhang, W.; Wei, X. A review on weed detection using ground-based machine vision and image processing techniqu *Comput. Electron. Agric.* **2019**, *158*, 226–240. [CrossRef]
4. Pinto, P.H.G.; Lima, S.F.; Andrade, M.G.O.; Contardi, L.M.; Ávila, J.; Reis, B.O.; Bernardo, V.F.; Vendrusculo, E.P. Weeds soybean cultivation with different predecessor cover crops. *Rev. Agric. Neotrop.* **2021**, *8*, e5890. [CrossRef]

Zhang, H.; Cao, D.; Zhou, W.; Currie, K. Laser and optical radiation weed control: A critical review. *Precis. Agric.* **2024**, *25*, 2033–2057. [CrossRef]

Alptekin, H.; Ozkan, A.; Gurbuz, R.; Kulak, M. Management of weeds in maize by sequential or individual applications of pre- and post-emergence herbicides. *Agriculture* **2023**, *13*, 421. [CrossRef]

Nath, C.P.; Singh, R.G.; Choudhary, V.K.; Datta, D.; Nandan, R.; Singh, S.S. Challenges and alternatives of herbicide-based weed management. *Agronomy* **2024**, *14*, 126. [CrossRef]

Samim, M.; Haqmal, M.; Afghan, A.; Khaleeq, K.; Ahmadi, A. Response of soybean to nitrogen levels and weed management on growth, yield and economic efficiency. *J. Res. Appl. Sci. Biotechnol.* **2023**, *2*, 139–145. [CrossRef]

Kumar, S.; Rana, S.S. Weed management strategies in soybean (*Glycine max*)—A review. *Indian J. Agric. Sci.* **2022**, *92*, 438–444. [CrossRef]

Gorayeb, E.S.; Nascimento, S.C.; Santos, A.N.M.R.; Batalhon, L.; Albuquerque, M.R.M.; Oliveira, V.G.F.; Souza, V.B.; Bogo, A.; Silva, F.N. Survey of viruses and vectors in tomato plants, alternative hosts, and weeds in the state of Santa Catarina, Brazil. *Plant Pathol.* **2024**, *73*, 444–454. [CrossRef]

Scavo, A.; Mauromicale, G. Integrated weed management in herbaceous field crops. *Agronomy* **2020**, *10*, 466. [CrossRef]

Choudhary, C.S.; Behera, B.; Raza, M.B.; Mrunalini, K.; Bhoi, T.K.; Lal, M.K.; Nongmaithem, D.; Pradhan, S.; Song, B.; Das, T.K. Mechanisms of allelopathic interactions for sustainable weed management. *Rhizosphere* **2023**, *25*, 100667. [CrossRef]

Thompson, M.; Chauhan, B.S.; Thierfelder, C.; Esser, A. Weed control in conservation agriculture in southern Africa. *Sustainability* **2021**, *13*, 5950. [CrossRef]

Gianessi, L.P. The increasing importance of herbicides in worldwide crop production. *Pest Manag. Sci.* **2013**, *69*, 1099–1105. [CrossRef]

Kanatas, P.; Travlos, I.; Papastylianou, P.; Gazoulis, I.; Kakabouki, I.; Tsekoura, A. Yield, quality and weed control in soybean crop as affected by several cultural and weed management practices. *Not. Bot. Horti Agrobot. Cluj-Napoca* **2020**, *48*, 329–341. [CrossRef]

Rivas-Garcia, T.; Espinosa-Calderón, A.; Hernández-Vázquez, B.; Schwentesius-Rindermann, R. Overview of environmental and health effects related to glyphosate usage. *Sustainability* **2022**, *14*, 6868. [CrossRef]

Procópio, S.O.; Pires, F.R.; Menezes, C.C.E.; Barroso, A.L.L.; Moraes, R.V.; Silva, M.V.V.; Queiroz, R.G.; Carmo, M.L. Effects of burndown herbicides in weed control in soybean crop. *Planta Daninha* **2006**, *24*, 193–197. [CrossRef]

Toni, L.R.M.; Santana, H.; Zai, D.A.M. Adsorption of glyphosate on soils and minerals. *Quim. Nova* **2006**, *29*, 829–833. [CrossRef]

Ringselle, B.; Bakken, A.K.; Höglind, M.; Jørgensen, M.; Tørresen, K.S. Effects of integrated grassland renewal strategies on annual and perennial weeds in the sowing year and subsequent production years. *Eur. J. Agron.* **2023**, *146*, 126799. [CrossRef]

Dalcin, L.S.; Terra, T.G.R.; Leal, T.C.A.B.; Terra, M.A.; Chaves, C.C. Efeito residual de herbicidas em dessecação de pré-plantio na cultura do sorgo granífero. *J. Biotechnol. Biodivers.* **2019**, *7*, 363–373. [CrossRef]

Beckie, H.J.; Flower, K.C.; Ashworth, M.B. Farming without glyphosate? *Plants* **2020**, *9*, 96. [CrossRef] [PubMed]

Pereira, B.C.S.; Braz, G.B.P.; Souza, M.F.; Reginaldo, L.T.R.T.; Ferreira, C.J.B. Performance of glyphosate-based products applied alone and in combination with herbicides in burndown. *Rev. Caatinga* **2023**, *36*, 765–774. [CrossRef]

Braz, G.B.P.; Freire, E.S.; Pereira, B.C.S.; Farnese, F.S.; Souza, M.F.; Loram-Lourenço, L.; Sousa, L.F. Agronomic performance of RR® soybean submitted to glyphosate application associated with a product based on *Bacillus subtilis*. *Agronomy* **2022**, *12*, 2940. [CrossRef]

Arsenijevic, N.; Dewerff, R.; Conley, S.; Ruark, M.; Werle, R. Influence of integrated agronomic and weed management practices on soybean canopy development and yield. *Weed Technol.* **2022**, *36*, 73–78. [CrossRef]

Amaral, G.S.; Alcántara-De la Cruz, R.; Martinelli, R.; Rufino Junior, L.R.; Carvalho, L.B.; Azevedo, F.A.; Silva, M.F.G.F. Occurrence of multiple glyphosate-resistant weeds in Brazilian citrus orchards. *AgriEngineering* **2023**, *5*, 1068–1078. [CrossRef]

Ovejero, R.F.L.; Takano, H.K.; Nicolai, M.; Ferreira, A.; Melo, M.S.C.; Cavenaghi, A.L.; Christoffoleti, P.J.; Oliveira, R.S., Jr. Frequency and dispersal of glyphosate-resistant sourgrass (*Digitaria insularis*) populations across Brazilian agricultural production areas. *Weed Sci.* **2017**, *65*, 285–294. [CrossRef]

Mendes, R.R.; Takano, H.K.; Gonçalves Netto, A.; Picoli Junior, G.J.; Cavenaghi, A.L.; Silva, V.F.V.; Nicolai, M.; Christoffoleti, P.J.; Oliveira, R.S., Jr.; Melo, M.S.C.; et al. Monitoring glyphosate- and chlorimuron-resistant *Conyza* spp. populations in Brazil. *An. Acad. Bras. Ciênc.* **2021**, *93*, e20190425. [CrossRef]

Heap, I. Herbicide Resistant Weeds in Brazil. 2024. Available online: https://www.weedscience.com/Summary/Country.aspx?CountryID=5 (accessed on 5 August 2024).

Moreira, M.S.; Nicolai, M.; Carvalho, S.J.P.; Christoffoleti, P.J. Glyphosate-resistance in *Conyza canadensis* and *C. Bonariensis*. *Planta Daninha* **2008**, *25*, 157–164. [CrossRef]

Araújo, L.S.; Correia, N.M.; Tornisielo, V.L.; Labate, M.T.V.; Tsai, S.M.; Carbonari, C.A.; Victória Filho, R. Goosegrass (*Eleusine indica*) resistant to multiple herbicide modes of action in Brazil. *Weed Sci.* **2023**, *71*, 189–197. [CrossRef]

Correia, N.M. Chemical and cultural management strategies for glyphosate-resistant sourgrass in central Brazil. *Pesqui. Agropecu. Bras.* **2023**, *58*, e02900. [CrossRef]

Sulzbach, E.; Turra, G.M.; Cutti, L.; Kroth, L.V.E.; Tranel, P.J.; Merotto Júnior, A.; Markus, C. Smooth pigweed (*Amaranthus hybridus*) and unresolved *Amaranthus* spp. from Brazil resistant to glyphosate exhibit the EPSPS TAP-IVS substitution. *Weed Sci.* **2024**, *72*, 48–58. [CrossRef]

33. Marca, V.; Procópio, S.O.; Silva, A.G.; Volf, M. Chemical control of glyphosate-resistant volunteer maize. *Rev. Bras. Herbic.* **201** *14*, 103–110. [CrossRef]
34. Buchling, C.; Braz, G.B.P.; Procópio, S.O.; Ferreira, C.J.B.; Silva, A.G.; Coradin, J. Pre-emergence control and interference voluntary maize plants on a soybean crop in Brazilian *Cerrado*. *Acta Sci. Agron.* **2022**, *44*, e54544. [CrossRef]
35. Pereira, B.C.S.; Ferreira, C.J.B.; Braz, G.B.P.; Souza, M.F.; Tavares, R.L.M.; Rosa, M.; Carmo, E.L.; Vian, G.H.; Silva, A.P.S.; Machado, F Can soil compaction alter morphophysiological responses and soybean yield under application of selective herbicides? *Appl. Envir Soil Sci.* **2023**, *2023*, 5518677. [CrossRef]
36. FAOSTAT. Pesticide Use. 2024. Available online: https://www.fao.org/faostat/en/#data/RP (accessed on 22 September 202
37. OECD. *OECD Compendium of Agri-Environmental Indicators*; OECD Publishing: Paris, France, 2013. Available online: https://ww oecd-ilibrary.org/agriculture-and-food/oecd-compendium-of-agri-environmental-indicators_9789264186217-en (accessed c 26 February 2024).
38. IBGE (Instituto Brasileiro de Geografia e Estatística). Sistema IBGE de Recuperação Automática—SIDRA. 2024. Available onlin https://sidra.ibge.gov.br/home/pms/brasil (accessed on 18 April 2024).
39. LAPIG/UFG (Image Processing and Geoprocessing Laboratory of Universidade Federal de Goiás). Atlas das Pastagens Brasileir 2024. Available online: https://lapig.iesa.ufg.br/p/38972-atlas-das-pastagens (accessed on 14 April 2024).
40. Qu, R.Y.; He, B.; Yang, J.F.; Lin, H.Y.; Yang, W.C.; Wu, Q.Y.; Li, Q.X.; Yang, G.F. Where are the new herbicides? *Pest Manag. S* **2021**, *77*, 2620–2625. [CrossRef] [PubMed]
41. Chauhan, B.S.; Singh, R.G.; Mahajan, G. Ecology and management of weeds under conservation agriculture: A review. *Crop Pr* **2012**, *38*, 57–65. [CrossRef]
42. Hasan, M.; Ahmad-Hamdani, M.S.; Rosli, A.M.; Hamdan, H. Bioherbicides: An eco-friendly tool for sustainable weed manag ment. *Plants* **2021**, *10*, 1212. [CrossRef]
43. Horikoshi, R.J.; Dourado, P.M.; Berger, G.U.; Fernandes, D.S.; Omoto, C.; Willse, A.; Martinelli, S.; Head, G.P.; Corrêa, A. Large-scale assessment of lepidopteran soybean pests and efficacy of Cry1Ac soybean in Brazil. *Sci. Rep.* **2021**, *11*, 1595 [CrossRef]
44. Bortoloti, G.; Sampaio, R.M. Demandas tecnológicas: Os bioinsumos para controle biológico no Brasil. *Cad. Ciênc. Tecnol.* **2022**, *3* e26927. [CrossRef]
45. Zambolim, L.; Juliatti, F.C.; Guerra, W. How to cope with the vulnerability of site specific fungicides on the control of Asia soybean rust. *Int. J. Res. Agron.* **2021**, *4*, 14–25. [CrossRef]
46. Machado, F.J.; Barro, J.P.; Godoy, C.V.; Dias, A.R.; Forcelini, C.A.; Utiamada, C.M.; Andrade Júnior, E.R.; Juliatti, F.C Grigolli, J.F.J.; Campos, H.D.; et al. Is tank mixing site-specific premixes and multi-site fungicides effective and economi for managing soybean rust? A meta-analysis. *Crop Prot.* **2022**, *151*, 105839. [CrossRef]
47. Fuentes-Llanillo, R.; Telles, T.S.; Soares Junior, D.; Melo, T.R.; Friedrich, T.; Kassam, A. Expansion of no-tillage practice i conservation agriculture in Brazil. *Soil Tillage Res.* **2021**, *208*, 104877. [CrossRef]
48. Possamai, E.J.; Conceição, P.C.; Amadori, C.; Bartz, M.L.C.; Ralisch, R.; Vicensi, M.; Marx, E.F. Adoption of the no-tillage systen in Paraná State: A review. *Rev. Bras. Cienc. Solo* **2022**, *46*, e0210104. [CrossRef]
49. Tavares Filho, J.; Rinschede, M. Vision of the farmers, professional and students of the agronomic area in respect of soil and wate conservation in Londrina, PR. *Semin. Cienc. Agrar.* **2009**, *30*, 1195–1202. [CrossRef]
50. Ulrich, A.; Sousa, T.L.M.; Guimarães, A.G.; Thompson, W.; Lobo Junior, A.R.; Evaristo, A.B. Performance of second-crop maiz hybrids in different population densities. *Rev. Agrogeoambiental* **2023**, *15*, e20231740. [CrossRef]
51. Santiago, A.D.; Procópio, S.O.; Braz, G.B.P.; Ferreira, C.J.B. The use of pre-emergence herbicides in cassava decreases the need c manual weeding. *Rev. Ceres* **2020**, *67*, 223–230. [CrossRef]
52. World Bank Group. Urban Population (% of Total Population)—Brazil. 2024. Available online: https://data.worldbank.org indicator/SP.URB.TOTL.IN.ZS?locations=BR (accessed on 4 December 2024).
53. CONAB. Companhia Nacional de Abastecimento. Insumos Agropecuários. 2024. Available online: https://consultaweb.conal gov.br/consultas/consultaInsumo.do?method=acaoCarregarConsulta (accessed on 4 December 2024).
54. Krzyzaniak, F.; Albrecht, L.P.; Albrecht, A.J.P. *Digitaria insularis*: Cross-resistance between ACCase inhibitors and multipl resistance to glyphosate. *Aust. J. Crop Sci.* **2023**, *17*, 556–562. [CrossRef]
55. Guidugli, T.B.; Bacha, A.L.; Martins, H.L.; Novello, B.D.; Alves, P.L.C.A. Glyphosate-resistant *Digitaria insularis* effect o eucalyptus initial growth. *New For.* **2024**, *55*, 1209–1229. [CrossRef]
56. Kalsing, A.; Velini, E.D.; Merotto, A.; Carbonari, C.A. The Population genomics of *Conyza* spp. in soybean macroregions sugges the spread of herbicide resistance through intraspecific and interspecific gene flow. *Sci. Rep.* **2024**, *14*, 19536. [CrossRef]
57. World Health Organization. *The WHO Recommended Classification of Pesticides by Hazard and Guidelines to Classification 2019* World Health Organization: Geneva, Switzerland, 2020.
58. Marcinkowska, K.; Praczyk, T.; Gawlak, M.; Niemczak, M.; Pernak, J. Efficacy of herbicidal ionic liquids and choline salt based o 2,4-D. *Crop Prot.* **2017**, *98*, 85–93. [CrossRef]
59. Carini, A.J.; Galon, L.; Rossetto, E.R.O.; Silva, F.A.C.; Tonin, R.J.; Perin, G.F. Seletividade de sais de 2,4-D aplicados em diferente estádios fenológicos de cultivares de trigo. *Rev. Bras. Ciênc. Agrár.* **2024**, *19*, e3749. [CrossRef]
60. Braz, G.B.P.; Takano, H.K. Chemical control of multiple herbicide-resistant *Amaranthus*: A review. *Adv. Weed Sci.* **2022**, *40* e0202200062. [CrossRef]

Oliveira, T.S.; Martins, H.L.; Carrega, W.C.; Barroso, A.A.M.; Alves, P.L.C.A. Selectivity of 2,4-D choline salt, glyphosate, glufosinate, and their mixtures for Enlist E3™ soybeans. *Afr. J. Agric. Res.* **2024**, *20*, 693–701. [CrossRef]
Merotto Júnior, A.; Gazziero, D.L.P.; Oliveira, M.C.; Scursoni, J.; Garcia, M.A.; Figueroa, R.; Turra, G.M. Herbicide use history and perspective in South America. *Adv. Weed Sci.* **2022**, *40*, e020220050. [CrossRef]
Kalsing, A.; Rossi, C.V.S.; Lucio, F.R.; Minozzi, G.B.; Gonçalves, F.P.; Valeriano, R. Efficacy of control of glyphosate-tolerant species of the *Rubiaceae* family through double-knockdown applications. *Planta Daninha* **2020**, *38*, e020190700. [CrossRef]
Perissato, M.M.; Albrecht, A.J.P.; Albrecht, L.P.; Rosa, W.B.; Perissato, S.M.; Larini, W.F. Efficacy of herbicides applied to *Commelina benghalensis* in the west of the state of Paraná. *Rev. Agric. Neotrop.* **2023**, *10*, e7243. [CrossRef]
Queiroz, A.R.S.; Delatorre, C.A.; Lucio, F.R.; Rossi, C.V.S.; Zobiole, L.H.S.; Merotto Júnior, A. Rapid necrosis: A novel plant resistance mechanism to 2,4-D. *Weed Sci.* **2020**, *68*, 6–18. [CrossRef]
Cantu, R.M.; Albrecht, L.P.; Albrecht, A.J.P.; Silva, A.F.M.; Danilussi, M.T.Y.; Lorenzetti, J.B. Herbicide alternative for *Conyza sumatrensis* control in pre-planting in no-till soybeans. *Adv. Weed Sci.* **2021**, *39*, e2021000025. [CrossRef]
Pedrollo, N.T.; Peripolli, M.; Cassol, J.C.; Dornelles, S.H.B. Clethodim and triclopir in association in glyphosate resistant voluntary corn control. *Vivências* **2022**, *18*, 263–274. [CrossRef]
Lee, S.; Payne, C.; Rees, S.; Ahrens, H.; Arve, L.; Asmus, E.; Bojack, G.; Arsequell, E.L.B.; Gatzweiler, E.; Helmke, H.; et al. Investigation of Acetyl-CoA Carboxylase-inhibiting herbicides that exhibit soybean crop selectivity. *Pest Manag. Sci.* **2024**. [CrossRef]
Barroso, A.L.L.; Dan, H.A.; Procópio, S.O.; Toledo, R.E.B.; Sandaniel, C.R.; Braz, G.B.P.; Cruvinel, K.L. Efficacy of ACCase-inhibiting herbicides in controlling grass weeds in soybean crops. *Planta Daninha* **2010**, *28*, 149–157. [CrossRef]
Correia, N.M.; Araújo, L.S.; Bueno Júnior, R.A. First report of multiple resistance of goosegrass to herbicides in Brazil. *Adv. Weed Sci.* **2022**, *40*, e020220007. [CrossRef] [PubMed]
Braz, L.B.P.; Braz, G.B.P.; Procópio, S.O.; Silva, A.G.; Braz, A.J.B.P.; Ferreira, C.J.B. Chemical control in different glyphosate resistant maize hybrids. *Rev. Bras. Milho Sorgo* **2018**, *17*, 535–547. [CrossRef]
Adegas, F.S.; Correia, N.M.; Silva, A.F.; Concênço, G.; Gazziero, D.L.P.; Dalazen, G. Glyphosate-resistant (GR) soybean and corn in Brazil: Past, present, and future. *Adv. Weed Sci.* **2022**, *40*, e0202200102. [CrossRef]
Takano, H.K.; Dayan, F.D. Glufosinate-ammonium: A review of the current state of knowledge. *Pest Manag. Sci.* **2020**, *76*, 3911–3925. [CrossRef]
Schelter, M.L.; Prates, A.A.; Fruet, D.L.; Souza, M.P.; Guerra, N.; Oliveira Neto, A.M. Response of soybean cultivars with different maturation times to pre-emergence herbicides. *Semin. Cienc. Agrar.* **2023**, *44*, 841–858. [CrossRef]
Coradin, J.; Braz, G.B.P.; Machado, F.G.; Silva, A.G.; Sousa, J.V.A. Herbicidas aplicados em pré-emergência para o controle de milho voluntário e capim-amargoso. *Rev. Cientif. Rural.* **2019**, *21*, 28–38. [CrossRef]
Braz, G.B.P.; Oliveira Júnior, R.S.; Zobiole, L.H.S.; Rubin, R.S.; Voglewede, C.; Constantin, J.; Takano, H.K. Sumatran Fleabane (*Conyza sumatrensis*) control in no-tillage soybean with diclosulam plus halauxifen-methyl. *Weed Technol.* **2017**, *31*, 184–192. [CrossRef]
Stephenson, D.O.; Spivey, T.A.; Deliberto, M.A.; Blouin, D.C.; Woolam, B.C.; Buck, T.B. Effects of low-dose flumioxazin and metribuzin postemergence applications on soybean. *Weed Technol.* **2019**, *33*, 87–94. [CrossRef]
Ferrier, J.; Soltani, N.; Hooker, D.C.; Robinson, D.E.; Sikkema, P.H. Biologically effective dose of flumioxazin and pyroxasulfone for control of multiple herbicide–resistant waterhemp (*Amaranthus tuberculatus*) in soybean. *Weed Sci.* **2022**, *70*, 243–248. [CrossRef]
Santos, F.L.S.; Silva, W.T.; Calil, F.N.; Cunha, P.P.; Costa, R.B.; Ximenes, P.A. Desiccation of forage plants from *Urochloa* genus using glyphosate herbicide. *Rev. Agric. Neotrop.* **2022**, *9*, e6772. [CrossRef]
Lima, D.B.C.; Silva, A.G.; Procópio, S.O.; Barroso, A.A.L.; Dan, H.A.; Costa, E.C.; Braz, A.J.B.P. Herbicides selection for controlling soybean volunteer plants resistant to glyphosate. *Rev. Bras. Herbic.* **2011**, *10*, 1–12. [CrossRef]
Benbrook, C.M. Trends in glyphosate herbicide use in the United States and globally. *Environ. Sci. Eur.* **2016**, *28*, 3. [CrossRef] [PubMed]
Riley, E.B.; Bradley, K.W. Influence of application timing and glyphosate tank-mix combinations on the survival of glyphosate-resistant giant ragweed (*Ambrosia trifida*) in soybean. *Weed Technol.* **2014**, *28*, 1–9. [CrossRef]
Kudsk, P.; Streibig, J.C. Herbicides—A two-edged sword. *Weed Res.* **2003**, *43*, 90–102. [CrossRef]
Okada, M.; Hanson, B.D.; Hembree, K.J.; Peng, Y.; Shrestha, A.; Stewart, C.N.; Wright, S.D.; Jasieniuk, M. Evolution and spread of glyphosate resistance in *Conyza bonariensis* in California and a comparison with closely related *Conyza canadensis*. *Weed Res.* **2015**, *55*, 173–184. [CrossRef]

agriculture

MDPI

Article

Assessment of Fumonisin, Deoxynivalenol, and Zearalenone Levels and the Occurrence of Mycotoxigenic *Fusarium* Species in Cereal Grains from Muscat, Sultanate of Oman

Fatma Khuseib Hamed Al-Rashdi [1,*], Abdullah Mohammed Al-Sadi [1,2], Mostafa Ibrahim Waly [3], Shah Hussain and Rethinasamy Velazhahan [1,*]

1 Department of Plant Sciences, College of Agricultural and Marine Sciences, Sultan Qaboos University, Al-Khoud, Muscat 123, Oman; alsadi@squ.edu.om (A.M.A.-S.); shahpk85@gmail.com (S.H.)
2 College of Agriculture, University of Al Dhaid, Sharjah P.O. Box 27272, United Arab Emirates
3 Department of Food Science and Nutrition, College of Agricultural and Marine Sciences, Sultan Qaboos University, Al-Khoud, Muscat 123, Oman; mostafa@squ.edu.om
4 Center for Environmental Studies and Research (CESAR), Sultan Qaboos University, Al-Khoud, Muscat 123, Oman
* Correspondence: s117805@student.squ.edu.om (F.K.H.A.-R.); velazhahan@squ.edu.om (R.V.)

Abstract: Mycotoxin contamination in agricultural goods is a major global problem due to its negative impact on human and animal health. The principal mycotoxin producers are fungal species from the genera *Fusarium*, *Aspergillus*, *Alternaria*, and *Penicillium*. The toxigenic fungal species produce the mycotoxins as secondary metabolites when they invade agricultural commodities during crop cultivation in the field (preharvest) or after harvesting or during transport and storage. This study was designed to investigate the levels of *Fusarium* mycotoxins, viz., fumonisin (FUM), zearalenone (ZEN), and deoxynivalenol (DON) in cereal grain samples collected from Muscat, Sultanate of Oman during 2023-24. A total of 90 cereal grain (wheat, corn, rice, barley) samples from local markets at Muscat, the Plant Quarantine Department, Oman, and Oman Flour Mills Company were analyzed using competitive enzyme immunoassay kits. Furthermore, *Fusarium* spp. associated with the contaminated grain samples were isolated, and their mycotoxin-producing potential was assessed. The results indicated that FUM, ZEN, and DON levels were below the detection limit (LOD) in 81%, 97%, and 44% of the samples, respectively. Two out of fifteen corn samples and one out of thirty-seven wheat samples tested exceeded the maximum permissible limit for FUM and ZEN, respectively, as set by the European Commission. A total of 19 *Fusarium* spp. associated with the contaminated grain samples were isolated and identified through molecular techniques. Sixteen isolates of *F. verticillioides*, one isolate of *F. thapsinum*, and two new *Fusarium* species were identified based on nuclear ribosomal DNA internal transcribed spacer and elongation factor 1-alpha sequences. Two isolates of *F. verticillioides* (FQD-1 and FQD-20) produced FUM levels exceeding 2000 $\mu g\ kg^{-1}$. The maximum ZEN concentration was observed in *F. verticillioides* FQD-20 (9.2 $\mu g\ kg^{-1}$), followed by *F. verticillioides* FQD-2 (2.8 $\mu g\ kg^{-1}$) and *Fusarium* sp. FOFMC-26 (2.5 $\mu g\ kg^{-1}$). All tested *Fusarium* strains produced DON, with levels ranging from 25.6 to 213 $\mu g\ kg^{-1}$, with *F. thapsinum* FQD-4 producing the highest level (213 $\mu g\ kg^{-1}$). To our knowledge, this is the first report on the occurrence of *Fusarium* mycotoxins and mycotoxigenic *Fusarium* spp. in food commodities in Oman.

Keywords: cereals; enzyme-linked immunosorbent assay; food safety; *Fusarium* mycotoxins; molecular characterization; toxigenic *Fusarium*

1. Introduction

Food security is a serious issue among the global communities as the world population anticipated to expand to 9.7 billion by 2050 [1]. Threats to food security include limitations in the supply of nutritious and safe foods to consumers [2,3]. Mold contamination in agricultural commodities is considered as a major food safety hazard, as several mold species secrete harmful toxic secondary metabolites known as "mycotoxins" on various food matrices. Reports indicate that 5–10% of agricultural products worldwide are spoiled by mold contamination, rendering them unfit for human or animal consumption. Several mold species from the genera *Fusarium, Aspergillus, Alternaria*, and *Penicillium* produce mycotoxins on the substrates during their growth under favorable conditions [4]. This contamination can occur at various stages, including crop cultivation in the field, harvest, storage, and processing [5]. Consuming food that is contaminated with mycotoxins might lead to mycotoxicosis and food poisoning, potentially causing death in both humans and animals [6]. There are around 300 different mycotoxins, but only about 20 are known to cause toxic effects on humans when consumed through contaminated food. These include aflatoxin, sterigmatocystin, ochratoxin (A, B, C), trichothecenes (deoxynivalenol, diacetoxyscirpenol, T2 toxin, HT-2 toxin, nivalenol), fumonisin (B1, B2), moniliformin, zearalenone, cyclopiazonic acid, patulin, alternariol monomethyl ether, alternariol, tenuazonic acid, and citrinin [7]. Among them, fumonisins, aflatoxins, zearalenone, and deoxynivalenol are frequently found in cereal grains. Each mycotoxin is associated with specific health risks in humans and animals. Aflatoxins are considered the most carcinogenic mycotoxin and are classified in Group I. Fumonisins and ochratoxins are classified in Group 2B, while trichothecenes and zearalenone are categorized as non-carcinogenic and placed in Group 3 [8]. These mycotoxins are exceptionally stable compounds that can withstand high temperatures, storage, and processing conditions [9]. For example, DON remains stable at 120 °C, showing moderate stability at 180 °C and partial stability at 210 °C [10]. ZEN can be inactivated after 60 min at 175 °C [11], whereas fumonisins remain stable even at temperatures as high as 250 °C [12].

Fusarium spp. produce three major classes of mycotoxins as secondary metabolites: fumonisins, trichothecenes, and zearalenone [13]. Fumonisin (FUM) contamination is common in maize and maize-based products. Fumonisins are primarily produced by *F. verticillioides* and *F. proliferatum* as secondary metabolites [14,15]. Other *Fusarium* species such as *F. oxysporum, F. napiforme, F. nygamai, F. anthophilum*, and *F. dlamini* are also known to produce fumonisins [16,17]. About 15 types of fumonisin (fumonisin A, B, C, and P) have been characterized [18]. The primary food contaminants are fumonisin B1 (FB1), fumonisin B2 (FB2), and fumonisin B3 (FB3), with FB1 being the most dangerous [19]. *F. verticillioides* is an important pathogen in corn, causing seedling blight, ear rot, seed rot, and stalk rot. The pathogen produces fumonisins mainly during the pre-harvesting stage, and under improper storage conditions, the pathogen continues to produce the toxin during the post-harvesting stage [20]. Besides corn, fumonisins have been detected in wheat, barley, rye, oat, rice, and other millets [17,21]. Other food products reported to contain FB1 include dried figs, garlic, asparagus, beers, and milk [17]. Fumonisins have been connected with esophageal cancer [17] and harmful effects on the liver and kidneys in animals [22]. Fumonisin consumption has been associated with pulmonary edema syndrome in pigs [23], leukoencephalomalacia in horses [24], hepatotoxicity and nephrotoxicity in rats [25], and apoptosis in several different types of animal cells [26].

Zearalenone (ZEN; F-2 toxin; $C_{18}H_{22}O_5$) is a non-steroidal estrogenic compound primarily produced by *F. graminearum, F. oxysporum, F. verticillioides, F. culmorum, F. cerealis, F. semitectum, F. equiseti, F. acuminatum, F. sporotrichioides*, and *F. crookwellense* [27]. Maize, wheat, barley, oats, rice, rye, and sorghum are highly susceptible to ZEN contamination [28]. ZEN is a xenoestrogen biosynthesized via polyketide pathway [27]. The chemical structure of ZEN mimics natural estrogens like 17β-estradiol, allowing it to bind to estrogen receptor sites and disrupt hormonal balance, leading to various reproductive system diseases [29]. Furthermore, ZEN is rapidly absorbed in the body of mammals and metabolized into highly toxic compounds [30]. ZEN exposure has been proven to diminish the progesterone and serum testosterone levels in the bloodstream, resulting in sterility and decreased conception rates in animals such as rats, pigs, and cows [31,32].

Deoxynivalenol (DON; Vomitoxin; $C_{15}H_{20}O_6$) is a trichothecene mycotoxin produced by many *Fusarium* species such as *F. graminearum*, *F. oxysporum*, *F. avenae*, *F. asiaticum*, and *F. culmorum* [33,34]. Corn, wheat, barley, rye, and oats are commonly contaminated by DON following infection by the toxigenic fungi in the field and during storage [35,36]. DON is a highly stable mycotoxin in both storage and processing [37,38]. Acute poisoning from DON causes emesis, while chronic low-dose exposure leads to anorexia, growth retardation, immunotoxicity, and reproduction impairment in experimental animal models [39–41]. In addition, DON is phytotoxic and is regarded as a virulence factor in fungal pathogenesis [42–44]. In many cases, these mycotoxins can co-exist in agricultural commodities, leading to synergistic, additive, or antagonistic toxic effect on the host [45].

According to the European Commission regulatory standards, the maximum tolerable limits of fumonisins (FB1 + FB2) in unprocessed corn; deoxynivalenol in unprocessed durum wheat, oats, and maize; and zearalenone in unprocessed corn are 2000, 1750, and 200 $\mu g\ kg^{-1}$, respectively [46]. Oman sources most of its cereal grains from international markets [47]. Though there are a few reports on the occurrence of aflatoxins in food commodities [48–50], studies on the level of *Fusarium* toxins in imported and marketed agricultural products in Oman is limited. This study was conducted to quantify the levels of fumonisin, deoxynivalenol, and zearalenone in cereal grains (wheat, corn, rice, and barley) collected from Muscat, Oman. The occurrence of toxigenic strains of *Fusarium* in the mycotoxin-contaminated samples was also investigated.

2. Materials and Methods

2.1. Sample Collection

Ninety cereal grain samples consisting of wheat (37 samples), corn (15 samples), rice (5 samples), and barley (33 samples) (100 g to 1 kg) were collected from the local markets in Muscat, the Plant Quarantine Department of the Ministry of Agriculture, Fisheries and Water Resources (MAFWR), Oman, and Oman Flour Mills Company (S.A.O.G), Muscat in 2023 and 2024. The collected samples were analyzed for grain contamination with FUM, ZEN, and DON.

2.2. Mycotoxin Estimation

A quantitative analysis of *Fusarium* mycotoxins in the cereal grain samples was performed using RIDASCREEN® Fumonisin ECO (Art. No. R3411), RIDASCREEN® Zearalenon (Art. No. R1401), and RIDASCREEN® DON (Art. No. R5906) competitive enzyme immunoassay kits (R-Biopharm AG, Darmstadt, Germany), following the manufacturer's instructions. These kits had detection limits of 30 $\mu g\ kg^{-1}$ for Fumonisin and 1.75 $\mu g\ kg^{-1}$ for Zearalenone and 18.5 $\mu g\ kg^{-1}$ for DON. The software RIDASOFT Win.NET Food & Feed Version 1.5.2 (Art. No. Z9996FF; R-Biopharm AG, Darmstadt, Germany) was used for determination of mycotoxin levels.

2.3. Isolation of *Fusarium* spp.

The cereal grains (wheat, corn, rice, and barley) were surface-sterilized with 1% NaOCl for 2 min. To eliminate NaOCl residues, the samples were washed three times with sterilized distilled water (SDW) before being plated on a potato dextrose agar (PDA) medium (Oxoid Ltd., Basingstoke, UK). The plates were then incubated for 3–5 days at 27 °C. Pure cultures of the fungi were obtained by using the hyphal tip isolation method.

2.4. DNA Extraction, Amplification, and Sequencing

Lee and Taylor's [51] method was followed to recover the DNA from the fungal mycelium. The fungal mycelium (~80 mg) was taken in a sterile 2 mL centrifuge tube from a 7-day-old PDA culture with a sterile scalpel and ground in the presence of acid-washed sand and suspended in 600 µL extraction buffer (50 mM Tris-HCl pH 7.6, 50 mM EDTA, 3% SDS, 1% 2-mercaptoethanol). The sample was extracted with an equal volume of phenol:chloroform:isoamyl alcohol (25:24:1, *v*/*v*; Sigma-Aldrich, St. Louis, MO, USA). DNA was precipitated with 1/10 volume of 3 M sodium acetate and 0.6 volume of isopropanol, washed with 70% ethanol, and resuspended in 50 µL SDW. A NanoDrop 1000 Spectrophotometer (Thermo Fisher Scientific, Waltham, MA, USA) was used to evaluate the quality and amount of isolated DNA.

The primer combinations ITS5/ITS4 [52] and EF-1α-F/EF-1α-R [53] were used to amplify two DNA regions: internal transcribed spacer (ITS) and translation elongation factor (EF-1α) (Table 1). The reaction mixture (25 µL) consisted of 1 µL of DNA template (100 ng), 1 µL of each primer (10 µmol), a puReTaq Ready-to-Go PCR bead (GE Healthcare, Buckinghamshire, UK), and 22 µL of nuclease free water [54]. For ITS amplification, the PCR conditions described by Hussain et al. [55] were used. At Macrogen Inc. (Seoul, Republic of Korea), the PCR products were purified and subsequently sequenced in both directions using the same primers.

Table 1. Primers used for identification of fungi in this study.

Gene	Primer	Direction	Sequence (5′→3′)
ITS	ITS5	Forward	GGAAGTAAAAGTCGTAACAAGG
	ITS4	Reverse	TCCTCCGCTTATTGATATGC
EF-1α	EF-1α-F	Forward	ATGGGTAAGGAAGACAAGAC
	EF-1α-R	Reverse	GGAAGTACCAGTGATCATGTT

2.5. Sequence Alignment and Phylogenetic Analyses

BioEdit v. 7.0.9 was used to align and transform the forward and reverse primer readings of the ITS and EF-1α regions into consensus sequences [56]. Sequence similarity was checked against the GenBank sequences using BLAST tools (https://blast.ncbi.nlm.nih.gov/Blast.cgi) (accessed on 3 September 2024). The most similar species in BLAST results were *F. verticillioides*, *F. thapsinum*, and *F. chlamydosporum*. It is obvious from the literature that both *F. verticillioides* and *F. thapsinum* belonged to the *F. fujikuroi* species complex [57]. Therefore, we constructed a combined ITS-EF-1α dataset comprised of the *F. fujikuroi* species complex and the *F. chlamydosporum* species complex along with representative species from other species complexes [58]. The dataset consisted of 52 specimens, including *Fusicolla aquaeductuum* (CBS 734.79) as an outgroup taxon. The data matrix was aligned using Mafft v. 7 (https://mafft.cbrc.jp/alignment/server/) (accessed on 3 September 2024) [59]. A maximum likelihood (ML) method was used for the phylogenetic analyses, with the ML phylogeny generated using RAxML-HPC BlackBox on the Cipres Science Gateway [60,61]. According to jModelTest2, the best model (GTR + F + I + G4) was chosen [62]. In total, 1000 bootstrap repeats were used to evaluate branch support for the ML phylogeny, and a bootstrap (BT) percentage of >50 was deemed significant. The phylogenetic tree was visualized using FigTree v.1.4.2 [63] and annotated with Adobe Illustrator CC2019.

2.6. Mycotoxigenic Potential of Fusarium spp.

The mycotoxin-producing potential of *Fusarium* strains on the maize grain substrate was assessed following a method described by Shi et al. [64]. Briefly, 25 mL of distilled water was added to 50 g of mycotoxin-free maize and left overnight. The medium was then sterilized by autoclaving at 121 °C for 20 min. A 6 mm diameter agar disk cut from a 7-day-old PDA culture of the fungus was transferred to the sterilized maize grain medium and incubated at 25 °C for 21 days. After that, the medium containing a fungal mycelial mat was dried in an oven at 40–50 °C until constant weight was achieved, then ground into powders and analyzed for mycotoxins (FUM, DON, ZEN) by using R-Biopharm RIDASCREEN competitive enzyme immunoassay kits as described earlier. The uninoculated maize grain medium prepared in the same manner served as the control. Each treatment was replicated thrice.

3. Results

3.1. Occurrence of Fumonisin

Fumonisin levels in 73 out of 90 grain samples (81%) were below the detection limit (Table 2). Twelve corn samples, three wheat samples, and two barley samples tested positive for fumonisin. In general, corn samples exhibited higher levels of fumonisin, with two out of the fifteen tested samples exceeding the EC's maximum tolerance level of 2000 μg kg^{-1} for fumonisin. All five rice samples tested were below the detection limit for fumonisin.

Table 2. The levels of fumonisin in cereal grain samples collected from local markets, Oman flour mills company, and quarantine department, in Muscat, Oman.

Food Items	Analyzed Samples	Sample Count and Range of Fumonisin Concentrations (μg kg^{-1})			
		<LOD (<30)	31–1000	1001–2000	>2000
Corn	15	3	9 (38.7–954.3)	1 (1096.2)	2 (2011.3–2808.5)
Wheat	37	34	3 (34.9–95.9)	0	0
Rice	5	5	0	0	0
Barley	33	31	2 (45.1–712.9)	0	0
Total	90	73	14	1	2

3.2. Occurrence of Zearalenone

The results indicated that 87 out of 90 grain samples tested (97%) were below the detection limit (Table 3). One corn sample, one barley sample, and one wheat sample tested positive for zearalenone, with the wheat sample exceeding the EC's maximum tolerance level of 200 μg kg^{-1} for zearalenone. All five rice samples were below the detection level for zearalenone.

Table 3. The levels of zearalenone in cereal grain samples collected from local markets, Oman flour mills company, and quarantine department, in Muscat, Oman.

Food Items	Analyzed Samples	Sample Count and Range of Zearalenone Concentrations (μg kg^{-1})			
		<LOD (<1.75)	1.76–100	101–200	>200
Corn	15	14	1 (2.7)	0	0
Wheat	37	36	0	0	1 (293.4)
Rice	5	5	0	0	0
Barley	33	32	1 (40.7)	0	0
Total	90	87	2	0	1

3.3. Occurrence of DON

Out of ninety cereal grain samples tested, forty samples (44%) were below the detection level for DON, and none exceeded the EC's maximum tolerance level of 1750 μg kg^{-1} for unprocessed durum wheat, oats, or maize (Table 4). Nine of the fifteen corn samples had DON levels ranging from 21 to 398 μg kg^{-1}, with one sample reaching 519 μg kg^{-1}. Among the thirty-seven wheat samples, nineteen had DON levels ranging from 22 to 311 μg kg^{-1}, with one sample reaching 559 μg kg^{-1}. The DON levels in three out of five rice samples ranged from 20 to 56 μg kg^{-1}, while seventeen out of thirty-three barley samples had levels between 19 and 67 μg kg^{-1}, respectively.

Table 4. The levels of deoxynivalenol (DON) in cereal grain samples collected from local markets, Oman flour mills company, and quarantine department, in Muscat, Oman.

Food Items	Analyzed Samples	Sample Count and Range of Deoxynivalenol Concentrations (μg kg^{-1})				
		<LOD (<18.5)	18.6–500	501–1000	1001–1500	>1500
Corn	15	5	9 (20.7–398.3)	1 (519.3)	0	0
Wheat	37	17	19 (22.4–311.5)	1 (559.4)	0	0
Rice	5	2	3 (20.1–56.0)	0	0	0
Barley	33	16	17 (19.2–66.7)	0	0	0
Total	90	40	48	2	0	0

3.4. Isolation of *Fusarium* spp. from Cereal Grains

A total of 19 *Fusarium* species associated with the contaminated cereal grain samples were isolated on the PDA medium. Initial BLAST analysis of the ITS sequences confirmed that the isolates belonged to the *Fusarium* genus. Further identification of species within the *Fusarium* genus was conducted using combined ITS and EF-1α-sequence alignment.

3.5. Phylogenetic Analysis

The final ITS-EF-1α dataset was 2104 characters long, including 1557 constant sites, 390 parsimony informative sites, and 157 uninformative sites. The ML phylogeny is presented in Figure 1, where the species of *Fusarium* recovered in two clades. Clade-I corresponding to the *F. fujikuroi* species complex (FFSC) and Clade-II to the *F. chlamydosporum* species complex (FCSC), *F. tricinctum* species complex (FTSC), *F. citricola* species complex (FCCSC), and *F. incarnatum-equiseti* species complex (FIESC), respectively. There were sixteen isolates of *F. verticillioides* and one isolate of *F. thapsinum* (FQD-4) recovered in this study. Both *F. verticillioides* and *F. thapsinum* fall in FFSC. The two isolates (FOFMC-26, FOFMC-27), representing an undescribed species in *Fusarium*, recovered with an independent lineage in Clade-II. The GenBank accession numbers of the ITS and EF-1α gene sequences of the *Fusarium* spp. isolated in this study are given in Table 5.

Table 5. *Fusarium* species isolated from mycotoxin-contaminated cereal grains and GenBank accession numbers for the internal transcribed spacer (ITS) and translation elongation factor 1-alpha (EF-1α) sequences.

Isolate	Source	Identified Fungus	GenBank Accessions	
			ITS	EF-1α
FQD-1	Wheat	*Fusarium verticillioides*	PQ325649	PQ349965
FQD-2	Wheat	*Fusarium verticillioides*	PQ325650	PQ349966
FQD-3	Corn	*Fusarium verticillioides*	PQ325651	PQ349967
FQD-9	Barley	*Fusarium verticillioides*	PQ325652	PQ362702
FQD-10	Barley	*Fusarium verticillioides*	PQ325653	PQ604778
FQD-11	Barley	*Fusarium verticillioides*	PQ325654	PQ369625
FQD-12	Barley	*Fusarium verticillioides*	PQ325655	PQ349968
FQD-13	Corn	*Fusarium verticillioides*	PQ325656	PQ349973
FQD-16	Corn	*Fusarium verticillioides*	PQ325657	PQ409349
FQD-17	Corn	*Fusarium verticillioides*	PQ325658	PQ362701
FQD-18	Corn	*Fusarium verticillioides*	PQ325659	PQ362700
FQD-20	Corn	*Fusarium verticillioides*	PQ325660	PQ409350
FQD-21	Corn	*Fusarium verticillioides*	PQ325661	PQ349969
FQD-22	Corn	*Fusarium verticillioides*	PQ325662	PQ349970
FQD-23	Corn	*Fusarium verticillioides*	PQ325663	PQ349972
FQD-24	Corn	*Fusarium verticillioides*	PQ325664	PQ369624
FOFMC-26	Wheat	*Fusarium* sp.	PQ325667	PQ330259
FOFMC-27	Wheat	*Fusarium* sp.	PQ325668	PQ330260
FQD-4	Corn	*Fusarium thapsinum*	PQ325666	PQ330258

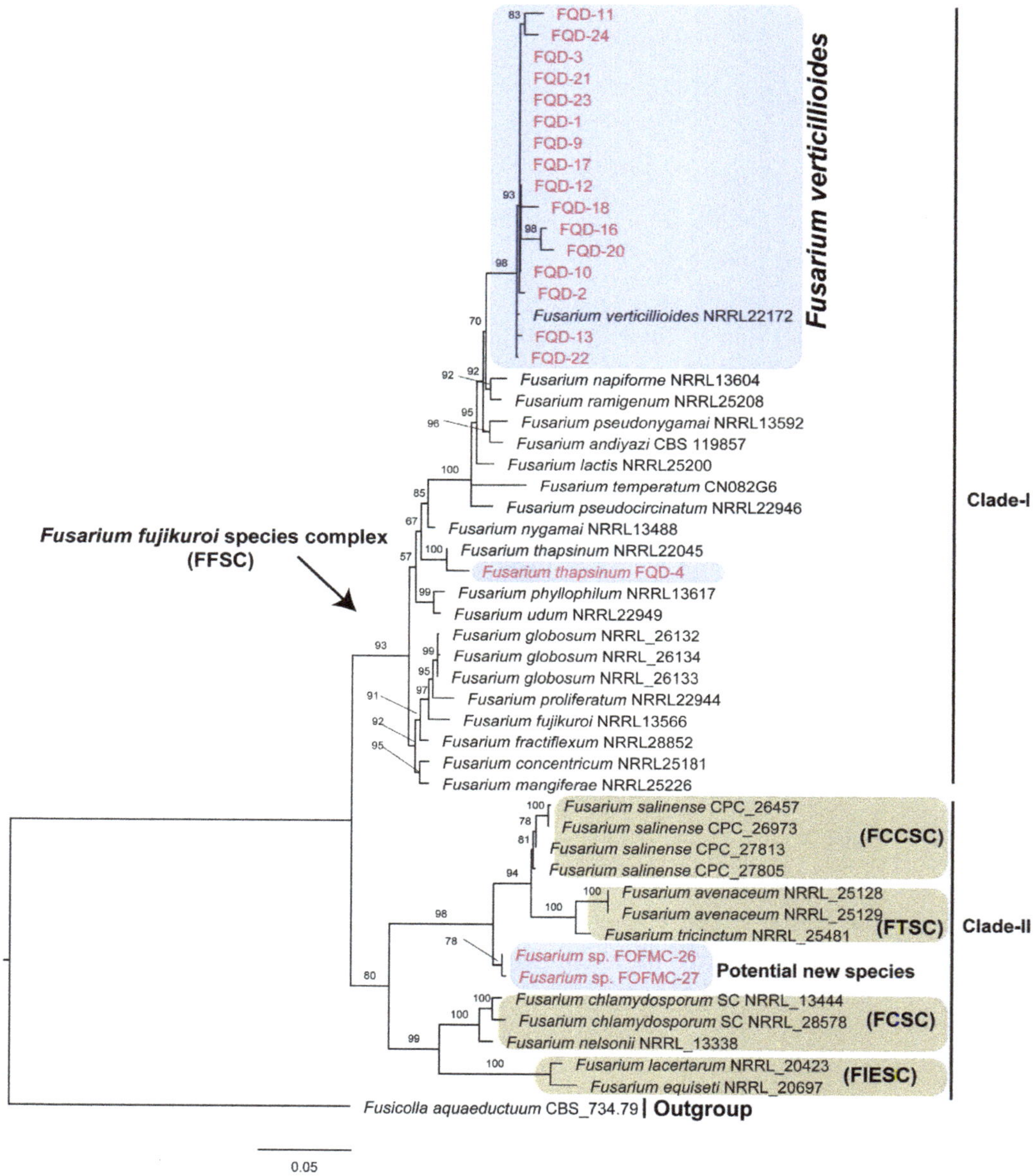

Figure 1. Maximum likelihood (ML) phylogeny of *Fusarium* species based on combined ITS-EF-1α sequences, with *Fusicolla aquaeductuum* as the outgroup taxon. Species of the genus are recovered in two clades, where Clade-I consists of species *F. fujikuroi* species complex (FFSC), and Clade-II with *F. citricola* species complex (FCCSC), *F. tricinctum* species complex (FTSC), *F. chlamydosporum* species complex (FCSC), and *F. incarnatum-equiseti* species complex (FIESC). There were 16 isolates of *F. verticillioides* recovered from corn (10 isolates), barley (4 isolates) and wheat (2 isolates), one isolate of *F. thapsinum* from corn, and two isolates of a potential new species of *Fusarium* from wheat. The bootstrap (BT) percentages above 50% are shown above nodes, all the newly generated sequences are in blue highlighted shade in red fonts.

3.6. Mycotoxin Production by Fusarium spp.

These *Fusarium* isolates produced varying levels of mycotoxins (Table 6). The *Fusarium* species produced FUM at levels ranging from 49.7 to 2489.2 μg kg^{-1}. *F. verticillioides* isolate FQD-1 produced the highest level of FUM (2489 μg kg^{-1}), followed by FQD-20 (2408 μg kg^{-1}), FQD-17 (1895 μg kg^{-1}), and FQD-13 (1044 μg kg^{-1}). No FUM production was detected in *F. verticillioides* isolates FQD-3, FQD-9, and FQD-10; *F. thapsinum* isolate FOD-4; and *Fusarium* sp. isolates FOFMC-26 and FOFMC-27. The production of ZEN by the *Fusarium* spp. ranged from 2.5 to 9.2 μg kg^{-1}. The maximum level of ZEN was produced by *F. verticillioides* FQD-20 (9.2 μg kg^{-1}), followed by *F. verticillioides* FQD-2 (2.8 μg kg^{-1}), and *Fusarium* sp. FOFMC-26 (2.5 μg kg^{-1}). All other isolates tested showed no ZEN production. All tested *Fusarium* strains produced DON, with levels ranging from 26 to 213 μg kg^{-1}. *F. thapsinum* FQD-4 produced the highest level (213 μg kg^{-1}) of DON, followed by F. *verticillioides* FQD-13 (192 μg kg^{-1}) and *Fusarium* sp. FOFMC-25 (188 μg kg^{-1}). The control samples (uninoculated corn substrate) showed no detectable levels of FUM, ZEN, and DON.

Table 6. Mycotoxin production by *Fusarium* species from cereal grains.

ungal Isolates	Deoxynivalenol (μg kg^{-1})	Fumonisin (μg kg^{-1})	Zearalenone (μg kg^{-1})
usarium verticillioides FQD-1	69.2 ± 4.5	2489.2 ± 27.3	nd
usarium verticillioides FQD-2	147.2 ± 13.2	336.5 ± 21.0	2.8 ± 0.1
usarium verticillioides FQD-3	108.2 ± 4.6	nd	nd
usarium verticillioides FQD-9	77.1 ± 8.6	nd	nd
usarium verticillioides FQD-10	161.0 ± 4.0	nd	nd
usarium verticillioides FQD-11	27.7 ± 0.9	379.6 ± 21.0	nd
usarium verticillioides FQD-12	54.1 ± 2.4	106.0 ± 1.7	nd
usarium verticillioides FQD-13	192.0 ± 9.9	1044.6 ± 45.1	nd
usarium verticillioides FQD-16	104.6 ± 12.7	173.9 ± 3.9	nd
usarium verticillioides FQD-17	67.7 ± 5.2	1895.5± 23.7	nd
usarium verticillioides FQD-18	133.3 ± 2.6	229.6 ± 60.9	nd
usarium verticillioides FQD-20	96.7 ± 11.8	2408.1 ± 32.9	9.2 ± 0.0
usarium verticillioides FQD-21	31.1 ± 0.7	137.8 ± 42.3	nd
usarium verticillioides FQD-22	25.6 ± 0.4	49.7 ± 5.1	nd
usarium verticillioides FQD-23	29.4 ± 0.9	101.4 ± 1.0	nd
usarium verticillioides FQD-24	149.0 ± 15.9	92.1 ± 13.2	nd
usarium sp. FOFMC-26	188.3 ± 23.6	nd	2.5 ± 0.1
usarium sp. FOFMC -27	68.5 ± 1.4	nd	nd
usarium thapsinum FQD-4	213.0 ± 9.8	nd	nd
ontrol	nd *	nd	nd

Data are means ± standard deviation from triplicate analysis. * Not detected, refers to values below quantification limit of assay.

Among the *Fusarium* isolates tested, *F. verticillioides* FQD-20 produced high levels of both FUM and ZEN. *F. thapsinum* FQD-4, which produced the highest level of DON, did not produce FUM and ZEN. Similarly, *F. verticillioides* FQD-1, which produced the highest level of FUM, produced no ZEN and low levels of DON.

4. Discussion

Food contamination by *Fusarium* toxins poses serious health and economic problems globally. For example, the natural occurrence of *Fusarium* toxins in wheat and corn varied significantly between China's high- and low-risk areas for esophageal cancer, according to Luo et al. [65]. Both the incidence and average levels of DON were higher in the high-risk areas compared to the low-risk areas. Hence, the occurrence of *Fusarium* toxins in agricultural commodities has been monitored worldwide [65–77]. In this study, out of 90 cereal grain samples consisting of wheat, corn, rice, and barley, only 3 samples (3.3%), one wheat and two corn, exceeded the EU's maximum permissible limit for ZEN

(200 μg kg^{-1}) and FUM (2000 μg kg^{-1}), respectively. DON levels remained within the permissible limits (1750 μg kg^{-1}). FUM, ZEN, and DON levels in 81%, 97%, and 44% of the samples, respectively, were below the detection limits.

Several studies reported the prevalence of *Fusarium* toxins in agricultural commodities [17,27,78,79]. Scudamore and Patel [71], while analyzing 82 consignments of maize imported into the UK for *Fusarium* toxins, reported that fumonisins were detected in almost every sample, and the maximum concentrations found for deoxynivalenol, nivalenol, zearalenone, and fumonisin B1 + fumonisin B2 were 444, 496, 165, and 5002 ppb, respectively. Logrieco et al. [66] reported that wheat and maize collected from Mediterranean countries were contaminated with DON, with levels reaching up to 8 mg kg^{-1} in wheat and 30 mg kg^{-1} in maize. Vrabcheva et al. [68] reported that DON and ZEA were the predominant toxins in wheat samples from Bulgaria, with contamination frequencies of 67% and 69%, respectively. The average levels in the contaminated samples were 180 μg kg^{-1} for DON and 17 μg kg^{-1} for ZEA, with maximum concentrations of 1800 μg kg^{-1} and 120 μg kg^{-1}, respectively. In Korea, Lee et al. [80] examined polished rice, brown rice, and two kinds of by-products, viz., discolored rice and blue-tinged (less-ripe) rice, for *Fusarium* mycotoxins. DON (59 to 1355 ng g^{-1}), Nivalenol (NIV) (66 to 4180 ng g^{-1}), and ZEA (25 to 3305 ng g^{-1}) were found in rice samples that had discolored. ZEA (26 to 3156 ng g^{-1}), NIV (50 to 3607 ng g^{-1}), and DON (86 to 630 ng g^{-1}) were detected in blue-tinged rice. The main contaminants found in brown rice samples were ZEA (47 to 235 ng g^{-1}) and NIV (52 to 569 ng g^{-1}). However, samples of polished rice were mostly devoid of mycotoxins. Birr et al. [81] analyzed forage maize samples collected from Northern Germany and reported high incidences of DON and ZEN (100 and 96%, respectively), with concentrations reaching up to 10,972 μg kg^{-1} for DON and 3910 μg kg^{-1} for ZEN. Deoxynivalenol-3-glucoside (DON3G) and 3- and 15-acetyl-deoxynivalenol (3 + 15-AcDON), two modified forms of DON, were also found in almost all samples (100 and 97%, respectively), with maximal values of 3038 μg kg^{-1} and 2237 μg kg^{-1}. Additionally, ZEN metabolites, α- and β-zearalenol (α-ZEL and β-ZEL), were found in lower incidences (59 and 32%), with concentrations up to 423 μg kg^{-1} and 203 μg kg^{-1}, respectively. The natural presence of *Fusarium* mycotoxins in barley and corn samples from Korea was examined by Kim et al. [82]. They discovered that the main contaminants in barley were DON, NIV, and ZEA, with mean concentrations of 170 ng g^{-1}, 1011 ng g^{-1}, and 287 ng g^{-1}, respectively. With mean concentrations of 310 ng g^{-1} and 297 ng g^{-1}, respectively, DON and 15-acetyldeoxynivalenol (15-ADON) were the main contaminants in corn. Shantika et al. [77] reported that more than 35% of maize food products collected from Indonesia markets contained both DON and FUM simultaneously. DON was found in 42.2% of the samples, but FUM was found in 84.4%. FUM and DON mean values varied from 0.12 μg kg^{-1} to 264.24 μg kg^{-1} and 2.62 μg kg^{-1} to 122.28 μg kg^{-1}, respectively. In their analysis of wheat grain samples gathered from Nakuru and Nyandarua, Kenya, Muthomi et al. [70] discovered that the majority of the samples were tainted with mycotoxins, with occurrence rates of up to 86% for T-2 toxin and 75% for deoxynivalenol (DON). Zearalenone and aflatoxin B1 were among the other mycotoxins found. Up to 35% of the samples had DON, T-2 toxin, and zearalenone co-occurring. Chilaka et al. [75] investigated maize, sorghum, millet, and *ogi* (akamu) samples from Nigerian markets for *Fusarium* mycotoxins contamination. Their findings revealed that fumonisins were the most common, particularly in maize and *ogi*, with occurrence rates of 65% and 93% and mean concentrations of 935 and 1128 μg kg^{-1}, respectively. In addition, multiple toxins were found to be present in 43% of the samples. Stanciu et al. [76] investigated the presence of *Fusarium* mycotoxins in Romanian wheat flour and grains. They reported that DON was detected in 14% of the samples, containing levels between 111 and 1787 μg kg^{-1}, while ZEN was present in 9% of the samples, with levels between 51 and 1135 μg kg^{-1}. The low levels of mycotoxin contamination in the cereal grain samples in this study might be due to strict import regulations on food commodities and monitoring and good storage practices followed by the traders.

In this study, nineteen *Fusarium* isolates were found in the tainted grain samples and identified by PCR. On the basis of combined analysis of ITS and EF-1α sequences, these fungal isolates were identified as *F. verticillioides* (16 isolates), *F. thapsinum* (1 isolate), and new *Fusarium* species (2 isolates). Chrpova et al. [83] detected the presence of *F. poae* and *F. graminearum* in winter wheat from the Czech Republic. They found that DON production by *F. graminearum* and *F. culmorum* was significantly elevated in instances of mixed infections with other species. *F. graminearum* was the most often observed species in wheat grown in the Netherlands, according to Van der Fels-Klerx et al. [72], followed by *F. avenaceum* and *Microdochium nivale*. The incidence and concentrations of DON were the highest, followed by ZEN and beauvericin. According to Lee et al. [80], the blue-tinged rice in Korea had the highest contamination rate with *F. graminearum* (3.8%), followed by discolored rice (2.4%), and brown rice (1.6%). According to Muthomi et al. [70], samples of freshly harvested wheat grain from Kenya's Nakuru and Nyandarua exhibited high levels of fungal contamination, particularly from *Epicoccum*, *Alternaria*, and *Fusarium* species. Isolates of *F. graminearum* demonstrated high virulence and significantly reducing kernel weight.

Numerous mycotoxins, such as FUM, ZEN, fusaric acid, T-2 toxin, HT-2 toxin, and DON, are known to be produced by *Fusarium* species [66,84–88]. Shi et al. [64] reported that *F. proliferatum*, *F. verticillioides*, *F. solani*, and *F. fujikuroi* produced fumonisins and fusaric acid; *F. oxysporum*, *F. temperatum*, *F. subglutinans*, *F. tricinctum*, *F. equiseti*, *F. concentricum*, *F. musae*, *F. andiyazi*, and *F. sacchari* produced only fusaric acid; *F. sporotrichioides*, *F. polyphialidicum*, and *F. langsethiae* produced type A trichothecenes; and *F. graminearum*, *F. poae*, *F. culmorum*, and *F. meridionale* produced type B trichothecenes. The *Fusarium* species isolated in this study produced FUM at levels ranging from 49.7 to 2489.2 μg kg^{-1}. *F. verticillioides* isolates exhibited the highest production of FUM. The production of ZEN by the *Fusarium* spp. ranged from 2.5 to 9.2 μg kg^{-1}. Among them, *F. verticillioides* FQD-20 yielded the highest concentration at 9.2 μg kg^{-1}. All *Fusarium* spp. tested produced DON, with concentrations ranging from 26 to 213 μg kg^{-1}. *F. thapsinum* FQD-4 recorded the highest DON level at 213 μg kg^{-1}, followed by *F. verticillioides* FQD-13 at 192 μg kg^{-1} and *Fusarium* sp. FCFMC-26 at 188 μg kg^{-1}. DON is primarily produced by *F. graminearum* and *F. culmorum* [89,90]. When examining the toxigenic potential of certain *Fusarium* isolates from Mediterranean cereals, Logrieco et al. [66] found that *F. culmorum* produced ZON, α- and β-ZOH, DON, and 3-AcDON; *F. graminearum* produced ZON, DON, and 15-AcDON; *F. crookwellense* produced ZON, α- and β-ZOH, FUS, and NIV; *F. equiseti* produced ZON and α-ZOH; and *F. avenaceum* produced moniliformin. None of the isolates of *F. moniliforme*, *F. proliferatum*, and *F. semitectum* were able to produce mycotoxins. However, Ramakrishna et al. [91] reported that *F. moniliforme* (syn: *F. verticillioides*) produced DON under optimum conditions, albeit at low concentrations (0.01–0.09 μg g^{-1} liquid medium). In this study, *F. thapsinum* FQD-4 isolated from corn produced the highest level of DON. *F. thapsinum* is a pathogen that causes stalk rot and grain mold of sorghum [92] and stalk rot of maize [93]. It is known to produce high levels of moniliformin and fusaric acid [94]. Leslie et al. [95] reported that *F. thapsinum* recovered from sorghum produced little to no fumonisins but substantial amounts of moniliformin. Considering the economic significance of corn in Oman and globally, *F. thapsinum* may be emerging as an important mycotoxigenic fungus.

5. Conclusions

This study revealed that *Fusarium* toxin levels in 97% of the cereal grain samples collected from Muscat, Oman, were within the permissible limits set by the European Commission. However, 13% of corn samples and 3% of wheat samples exceeded the maximum permissible limit for FUM and ZEN, respectively. The co-occurrence of these mycotoxins could pose health risks to local consumers due to their potential synergistic or additive effects. *Fusarium* species are a significant source of mycotoxins in food and feed. Furthermore, the presence of toxigenic *Fusarium* strains in certain grain samples could serve as a source of inoculum, potentially leading to disease outbreaks in wheat crops across various wilayats (provinces) in Oman, where wheat is cultivated. The mold growth and mycotoxin contamination could pose a threat to the quality of cereal grains. Oman imports most cereal grains from other countries, and hence, stricter quarantine regulations are essential to mitigate the risk of mycotoxin contamination. Enhancing consumer awareness about mycotoxins, ensuring proper storage conditions, and promoting food safety practices are essential for preventing *Fusarium* growth and toxin production in cereal grains. Further investigation is needed to ascertain the production of other mycotoxins by these *Fusarium* species. As far as we are aware, this is the first report documenting the presence of *Fusarium* spp. and their associated mycotoxins in cereal grains sold in Oman.

Author Contributions: F.K.H.A.-R., R.V., A.M.A.-S. and M.I.W. planned the study; F.K.H.A.-R. and S.H. conducted the laboratory experiments; F.K.H.A.-R., R.V., A.M.A.-S., M.I.W. and S.H. wrote the manuscript. All authors have read and agreed to the published version of the manuscript.

Funding: This research received no external funding.

Institutional Review Board Statement: Not applicable.

Data Availability Statement: This article includes all data generated or analyzed during this study.

Acknowledgments: The authors are grateful to Ali Al-Raeesi, Central Lab for Phytosanitary & Food Safety, United Integrated Laboratories LLC., Barka, Sultanate of Oman for his help in collection of samples. The first author is thankful to the Deanship of Postgraduate Studies, Sultan Qaboos University for the scholarship, research funds and facilities for this research.

Conflicts of Interest: The authors declare that they have no conflicts of interest.

References

1. Lal, R. Feeding 11 billion on 0.5 billion hectare of area under cereal crops. *Food Energy Secur.* **2016**, *5*, 239–251. [CrossRef]
2. Bazerghi, C.; McKay, F.H.; Dunn, M. The role of food banks in addressing food insecurity: A systematic review. *J. Communit Health* **2016**, *41*, 732–740. [CrossRef] [PubMed]
3. Vagsholm, I.; Arzoomand, N.S.; Boqvist, S. Food security, safety, and sustainability—Getting the trade-offs right. *Front. Sustai Food Syst.* **2020**, *4*, 16. [CrossRef]
4. Kabak, B.; Dobson, A.D.W.; Var, I.I.L. Strategies to prevent mycotoxin contamination of food and animal feed: A review. *Crit. Re Food Sci. Nutr.* **2006**, *46*, 593–619. [CrossRef]
5. Shephard, G.S. Impact of mycotoxins on human health in developing countries. *Food Addit. Contam.* **2008**, *25*, 146–151. [CrossRef
6. Wagacha, J.M.; Muthomi, J.W. Mycotoxin problem in Africa: Current status, implications to food safety and health and possibl management strategies. *Int. J. Food Microbiol.* **2008**, *124*, 1–12. [CrossRef]
7. Abdin, M.Z.; Ahmad, M.M.; Javed, S. Advances in molecular detection of *Aspergillus*: An update. *Arch. Microbiol.* **2010**, *19* 409–425. [CrossRef]
8. International Agency for Research on Cancer (IARC). Traditional herbal medicines, some mycotoxins, naphthalene and styren In *IARC Monographs on the Evaluation of Carcinogenic Risks to Humans*; International Agency for Research on Cancer: Lyon, Franc 2002; Volume 82, pp. 301–366.
9. Gallo, A.; Mosconi, M.; Trevisi, E.; Santos, R.R. Adverse effects of *Fusarium* toxins in ruminants: A review of in vivo and in vitr studies. *Dairy* **2022**, *3*, 474–499. [CrossRef]
10. Mishra, S.; Dixit, S.; Dwivedi, P.D.; Pandey, H.P.; Das, M. Influence of temperature and pH on the degradation of deoxynivaleno (DON) in aqueous medium: Comparative cytotoxicity of DON and degraded product. *Food Addit. Contam. Part A* **2014**, *3* 121–131. [CrossRef]
11. Ryu, D.; Hanna, M.A.; Eskridge, K.M.; Bullerman, L.B. Heat stability of zearalenone in an aqueous buffered model system *J. Agric. Food Chem.* **2003**, *51*, 1746–1748. [CrossRef]
12. Bryła, M.; Waśkiewicz, A.; Szymczyk, K.; Jędrzejczak, R. Effects of pH and temperature on the stability of fumonisins in maiz products. *Toxins* **2017**, *9*, 88. [CrossRef] [PubMed]
13. Ji, F.; He, D.; Olaniran, A.O.; Mokoena, M.P.; Xu, J.; Shi, J. Occurrence, toxicity, production and detection of *Fusarium* mycotoxi A review. *Food Prod. Process. Nutr.* **2019**, *1*, 1–14. [CrossRef]
14. Rheeder, J.P.; Marasas, W.F.; Vismer, H.F. Production of fumonisin analogs by *Fusarium* species. *Appl. Environ. Microbiol.* **2002**, *6* 2101–2105. [CrossRef]
15. Sanchez-Rangel, D.; SanJuan-Badillo, A.; Plasencia, J. Fumonisin production by *Fusarium verticillioides* strains isolated from maiz in Mexico and development of a polymerase chain reaction to detect potential toxigenic strains in grains. *J. Agric. Food Chem* **2005**, *53*, 8565–8571. [CrossRef] [PubMed]
16. Mogensen, J.M.; Frisvad, J.C.; Thrane, U.; Nielsen, K.F. Production of fumonisin B2 and B4 by *Aspergillus niger* on grapes and raisins. *J. Agric. Food Chem.* **2010**, *58*, 954–958. [CrossRef] [PubMed]

Kamle, M.; Mahato, D.K.; Devi, S.; Lee, K.E.; Kang, S.G.; Kumar, P. Fumonisins: Impact on agriculture, food, and human health and their management strategies. *Toxins* **2019**, *11*, 328. [CrossRef]
Braun, M.S.; Wink, M. Exposure, occurrence, and chemistry of fumonisins and their cryptic derivatives. *Compr. Rev. Food Sci. Food Saf.* **2018**, *17*, 769–791. [CrossRef]
Damiani, T.; Righetti, L.; Suman, M.; Galaverna, G.; Dall'Asta, C. Analytical issue related to fumonisins: A matter of sample comminution? *Food Control* **2019**, *95*, 1–5. [CrossRef]
Chulze, S.N. Strategies to reduce mycotoxin levels in maize during storage: A review. *Food Addit. Contam.* **2010**, *27*, 651–657. [CrossRef]
Cendoya, E.; Chiotta, M.L.; Zachetti, V.; Chulze, S.N.; Ramirez, M.L. Fumonisins and fumonisin-producing *Fusarium* occurrence in wheat and wheat by products: A review. *J. Cereal Sci.* **2018**, *80*, 158–166. [CrossRef]
Chu, F.S.; Li, G.Y. Simultaneous occurrence of fumonisin B1 and other mycotoxins in moldy corn collected from the People's Republic of China in regions with high incidences of esophageal cancer. *Appl. Environ. Microbiol.* **1994**, *60*, 847–852. [CrossRef] [PubMed]
Harrison, L.R.; Colvin, B.M.; Greene, J.T.; Newman, L.E.; Cole, J.R., Jr. Pulmonary edema and hydrothorax in swine produced by fumonisin B1, a toxic metabolite of *Fusarium moniliforme*. *J. Vet. Diagn. Investig.* **1990**, *2*, 217–221. [CrossRef] [PubMed]
Ross, P.F.; Rice, L.G.; Osweiler, G.D.; Nelson, P.E.; Richard, J.L.; Wilson, T.M. A review and update of animal toxicoses associated with fumonisin-contaminated feeds and production of fumonisins by *Fusarium* isolates. *Mycopathologia* **1992**, *117*, 109–114. [CrossRef] [PubMed]
Voss, K.A.; Plattner, R.D.; Riley, R.T.; Meredith, F.I.; Norred, W.P. In vivo effects of fumonisin B1-producing and fumonisin B1-nonproducing *Fusarium moniliforme* isolates are similar: Fumonisins B2 and B3 cause hepato-and nephrotoxicity in rats. *Mycopathologia* **1998**, *141*, 45–58. [CrossRef] [PubMed]
Jones, C.; Ciacci-Zanella, J.R.; Zhang, Y.; Henderson, G.; Dickman, M. Analysis of fumonisin B1-induced apoptosis. *Environ. Health Perspect.* **2001**, *109*, 315–320. [CrossRef]
Ropejko, K.; Twarużek, M. Zearalenone and its metabolites- General overview, occurrence, and toxicity. *Toxins* **2021**, *13*, 35. [CrossRef]
Golge, O.; Kabak, B. Occurrence of deoxynivalenol and zearalenone in cereals and cereal products from Turkey. *Food Control* **2020**, *110*, 106982. [CrossRef]
Kowalska, K.; Habrowska-Gorczynska, D.E.; Piastowska-Ciesielska, A.W. Zearalenone as an endocrine disruptor in humans. *Environ. Toxicol. Pharmacol.* **2016**, *48*, 141–149. [CrossRef]
Brodehl, A.; Möller, A.; Kunte, H.J.; Koch, M.; Maul, R. Biotransformation of the mycotoxin zearalenone by fungi of the genera *Rhizopus* and *Aspergillus*. *FEMS Microbiol. Lett.* **2014**, *359*, 124–130. [CrossRef]
Yang, J.Y.; Wang, G.X.; Liu, J.L.; Fan, J.J.; Cui, S. Toxic effects of zearalenone and its derivatives α-zearalenol on male reproductive system in mice. *Reprod. Toxicol.* **2007**, *24*, 381–387. [CrossRef]
Rai, A.; Das, M.; Tripathi, A. Occurrence and toxicity of a fusarium mycotoxin, zearalenone. *Crit. Rev. Food Sci. Nutr.* **2020**, *60*, 2710–2729. [CrossRef] [PubMed]
Khan, M.K.; Pandey, A.; Athar, T.; Choudhary, S.; Deval, R.; Gezgin, S.; Hamurcu, M.; Topal, A.; Atmaca, E.; Santos, P.A.; et al. Fusarium head blight in wheat: Contemporary status and molecular approaches. *3 Biotech* **2020**, *10*, 1–17. [CrossRef] [PubMed]
Sumarah, M.W. The deoxynivalenol challenge. *J. Agric. Food Chem.* **2022**, *70*, 9619–9624. [CrossRef] [PubMed]
European Food Safety Authority. Deoxynivalenol in food and feed: Occurrence and exposure. *EFSA J.* **2013**, *11*. 3379.
Yao, Y.; Long, M. The biological detoxification of deoxynivalenol: A review. *Food Chem. Toxicol.* **2020**, *145*, 111649. [CrossRef]
Lauren, D.R.; Smith, W.A. Stability of the *Fusarium* mycotoxins nivalenol, deoxynivalenol and zearalenone in ground maize under typical cooking environments. *Food Addit. Contam.* **2001**, *18*, 1011–1016. [CrossRef]
Ma, Y.Y.; Guo, H.W. Mini-review of studies on the carcinogenicity of deoxynivalenol. *Environ. Toxicol. Pharmacol.* **2008**, *25*, 1–9. [CrossRef]
Pestka, J.J.; Smolinski, A.T. Deoxynivalenol: Toxicology and potential effects on humans. *J. Toxicol. Environ. Health Part B Crit. Rev.* **2005**, *8*, 39–69. [CrossRef]
Pestka, J.J. Deoxynivalenol: Mechanisms of action, human exposure, and toxicological relevance. *Arch. Toxicol.* **2010**, *84*, 663–679. [CrossRef]
Pinto, A.C.S.M.; De Pierri, C.R.; Evangelista, A.G.; Gomes, A.S.D.L.P.B.; Luciano, F.B. Deoxynivalenol: Toxicology, degradation by bacteria, and phylogenetic analysis. *Toxins* **2022**, *14*, 90. [CrossRef]
Pestka, J.J. Deoxynivalenol: Toxicity, mechanisms and animal health risks. *Anim. Feed Sci. Technol.* **2007**, *137*, 283–298. [CrossRef]
Ferrigo, D.; Raiola, A.; Causin, R. *Fusarium* toxins in cereals: Occurrence, legislation, factors promoting the appearance and their management. *Molecules* **2016**, *21*, 627. [CrossRef] [PubMed]
Freire, L.; Sant'Ana, A.S. Modified mycotoxins: An updated review on their formation, detection, occurrence, and toxic effects. *Food Chem. Toxicol.* **2018**, *111*, 189–205. [CrossRef]
Speijers, G.J.A.; Speijers, M.H.M. Combined toxic effects of mycotoxins. *Toxicol. Lett.* **2004**, *153*, 91–98. [CrossRef] [PubMed]
European Commission. Commission Regulation (EC) No 1881/2006 of 19 December 2006 setting maximum levels for certain contaminants in foodstuffs. *Off. J. Eur. Union* **2006**, *L364*, 5–24.
Mbaga, M.D. Alternative mechanisms for achieving food security in Oman. *Agric. Food Secur.* **2013**, *2*, 3. [CrossRef]

48. Elshafie, A.E.; Al-Rashdi, T.A.; Al-Bahry, S.N.; Bakheit, C.S. Fungi and aflatoxins associated with spices in the Sultanate of On *Mycopathologia* **2002**, *155*, 155–160. [CrossRef]
49. Al-Alawi, A.K.S.; Al-Mandhari, A.A.S.; Al-Mahmooli, I.H.; Al-Harrasi, M.M.A.; Al-Bulushi, I.M.; Al-Sadi, A.M.; Velazhahar Assessment of aflatoxin B1 content and aflatoxigenic molds in imported food commodities in Muscat, Oman. *J. Agric. Mar.* **2023**, *28*, 1–6.
50. Akintola, A.; Al-Dairi, M.; Imtiaz, A.; Al-Bulushi, I.M.; Gibreel, T.; Al-Sadi, A.M.; Velazhahan, R. The extent of aflatoxin contamination in chili (*Capsicum annuum* L.) and consumer awareness and knowledge of aflatoxins in Oman. *Agriculture* **2** *14*, 1536. [CrossRef]
51. Lee, S.B.; Taylor, J.W. Isolation of DNA from Fungal Mycelia and Single Spores. In *PCR Protocols: A Guide to Methods Applications*; Innis, M.A., Gelfand, D.H., Sninsky, J.J., White, T.J., Eds.; Academic Press: San Diego, CA, USA, 1990; pp. 282–2
52. White, T.J.; Bruns, T.D.; Lee, S.B.; Taylor, J.W. Amplification and direct sequencing of fungal ribosomal RNA genes for phyloge ics. In *PCR Protocols: A Guide to Methods and Applications*; Academic Press, Inc.: New York, NY, USA, 1990; pp. 315–322.
53. O'Donnell, K.; Sarver, B.A.J.; Brandt, M.; Chang, D.C.; Noble-Wang, J.; Park, B.J.; Sutton, D.A.; Benjamin, L.; Lindsley, M.; Padh A.; et al. Phylogenetic diversity and microsphere array-based genotyping of human pathogenic Fusaria, including isolates fr the multistate contact lens-associated US keratitis outbreaks of 2005 and 2006. *J. Clin. Microbiol.* **2007**, *45*, 2235–2248. [CrossR
54. Al-Sadi, A.M.; Al-Ghaithi, A.G.; Al-Balushi, Z.M.; Al-Jabri, A.H. Analysis of diversity in *Pythium aphanidermatum* populati from a single greenhouse reveals phenotypic and genotypic changes over 2006 to 2011. *Plant Dis.* **2012**, *96*, 852–858. [CrossR [PubMed]
55. Hussain, S.; Al-Kharousi, M.; Al-Muharabi, M.A.; Al-Maqbali, D.A.; Al-Shabibi, Z.; Al-Balushi, A.H.; Al-Yahya'ei, M.N.; Al Saa N.; Abdel-Jalil, R.; Velazhahan, R.; et al. Phylogeny of *Agaricus* subgenus *Pseudochitonia* with the description of a new section a new species from Oman. *Mycol. Prog.* **2022**, *21*, 72. [CrossRef]
56. Hall, T.A. BioEdit: A user-friendly biological sequence alignment editor and analysis program for Windows 95/98/NT. *Nuc Acids Symp. Ser.* **1999**, *41*, 95–98.
57. Summerell, B.A.; Leslie, J.F. Fifty years of *Fusarium*: How could nine species have ever been enough? *Fungal Divers.* **2011**, 135–144. [CrossRef]
58. Sandoval-Denis, M.; Guarnaccia, V.; Polizzi, G.; Crous, P.W. Symptomatic citrus trees reveal a new pathogenic lineage in *Fusari* and two new *Neocosmospora* species. *Persoonia* **2018**, *40*, 1–25. [CrossRef]
59. Katoh, K.; Rozewicki, J.; Yamada, K.D. MAFFT online service: Multiple sequence alignment, interactive sequence choice a visualization. *Brief. Bioinform.* **2019**, *20*, 1160–1166. [CrossRef]
60. Miller, M.A.; Pfeiffer, W.; Schwartz, T. Creating the CIPRES Science Gateway for inference of large phylogenetic trees. Proceedings of the 2010 Gateway Computing Environments Workshop (GCE), New Orleans, LA, USA, 14 November 2010; IE Piscataway, NJ, USA, 2010; pp. 1–8.
61. Stamatakis, A. RAxML version 8: A tool for phylogenetic analysis and post-analysis of large phylogenies. *Bioinformatics* **2014**, 1312–1313. [CrossRef]
62. Darriba, D.; Taboada, G.L.; Doallo, R.; Posada, D. jModelTest 2: More models, new heuristics and high-performance computi *Nat. Methods* **2012**, *9*, 772. [CrossRef]
63. Rambaut, A. *FigTree, Tree Figure Drawing Tool Version 131*; Institute of Evolutionary 623 Biology, University of Edinbur Edinburgh, UK, 2012.
64. Shi, W.; Tan, Y.; Wang, S.; Gardiner, D.M.; De Saeger, S.; Liao, Y.; Wang, C.; Fan, Y.; Wang, Z.; Wu, A. Mycotoxigenic potentials *Fusarium* species in various culture matrices revealed by mycotoxin profiling. *Toxins* **2017**, *9*, 6. [CrossRef]
65. Luo, Y.; Yoshizawa, T.; Katayama, T. Comparative study on the natural occurrence of *Fusarium* mycotoxins (trichothecenes a zearalenone) in corn and wheat from high-and low-risk areas for human esophageal cancer in China. *Appl. Environ. Microb* **1990**, *56*, 3723–3726. [CrossRef]
66. Logrieco, A.; Bottalico, A.; Ricci, V. Occurrence of *Fusarium* species and their mycotoxins in cereal grains from some Mediterrane countries. *Phytopathol. Mediterr.* **1990**, *29*, 81–89.
67. Park, J.J.; Smalley, E.B.; Chu, F.S. Natural occurrence of *Fusarium* mycotoxins in field samples from the 1992 Wisconsin corn cr *Appl. Environ. Microbiol.* **1996**, *62*, 1642–1648. [CrossRef]
68. Vrabcheva, T.; Geßler, R.; Usleber, E.; Märtlbauer, E. First survey on the natural occurrence of *Fusarium* mycotoxins in Bulgari wheat. *Mycopathologia* **1996**, *136*, 47–52. [CrossRef] [PubMed]
69. Nikiema, P.N.; Worrillow, L.; Traore, A.S.; Wild, C.P.; Turner, P.C. Fumonisin contamination of maize in Burkina Faso, West Afri *Food Addit. Contam.* **2004**, *21*, 865–870. [CrossRef] [PubMed]
70. Muthomi, J.W.; Ndung'u, J.K.; Gathumbi, J.K.; Mutitu, E.W.; Wagacha, J.M. The occurrence of *Fusarium* species and mycotoxins Kenyan wheat. *Crop Prot.* **2008**, *27*, 1215–1219. [CrossRef]
71. Scudamore, K.A.; Patel, S. Occurrence of *Fusarium* mycotoxins in maize imported into the UK, 2004–2007. *Food Addit. Conta Part A* **2009**, *26*, 363–371. [CrossRef]
72. Van der Fels-Klerx, H.J.; De Rijk, T.C.; Booij, C.J.H.; Goedhart, P.W.; Boers, E.A.M.; Zhao, C.; Waalwijk, C.; Mol, H.G.J.; Van d Lee, T.A.J. Occurrence of *Fusarium* head blight species and *Fusarium* mycotoxins in winter wheat in the Netherlands in 2009. *Fo Addit. Contam. Part A* **2012**, *29*, 1716–1726. [CrossRef]

73. Juan, C.; Ritieni, A.; Mañes, J. Occurrence of *Fusarium* mycotoxins in Italian cereal and cereal products from organic farming. *Food Chem.* **2013**, *141*, 1747–1755. [CrossRef]
74. Rodríguez-Carrasco, Y.; Fattore, M.; Albrizio, S.; Berrada, H.; Mañes, J. Occurrence of *Fusarium* mycotoxins and their dietary intake through beer consumption by the European population. *Food Chem.* **2015**, *178*, 149–155. [CrossRef]
75. Chilaka, C.A.; De Boevre, M.; Atanda, O.O.; De Saeger, S. Occurrence of *Fusarium* mycotoxins in cereal crops and processed products (Ogi) from Nigeria. *Toxins* **2016**, *8*, 342. [CrossRef]
76. Stanciu, O.; Juan, C.; Miere, D.; Loghin, F.; Manes, J. Occurrence and co-occurrence of *Fusarium* mycotoxins in wheat grains and wheat flour from Romania. *Food Control* **2017**, *73*, 147–155. [CrossRef]
77. Shantika, Z.R.; Rahayu, W.P.; Lioe, H.N. Co-occurrence and exposure assessment of deoxynivalenol and fumonisins from maize and maize-based products in Indonesia. *Food Res.* **2024**, *8*, 413–426. [CrossRef] [PubMed]
78. Sohn, H.B. Co-occurrence of Fusarium mycotoxins in mouldy and healthy corn from Korea. *Food Addit. Contam.* **1999**, *16*, 153–158. [CrossRef] [PubMed]
79. Mahato, D.K.; Devi, S.; Pandhi, S.; Sharma, B.; Maurya, K.K.; Mishra, S.; Dhawan, K.; Selvakumar, R.; Kamle, M.; Mishra, A.K.; et al. Occurrence, impact on agriculture, human health, and management strategies of zearalenone in food and feed: A review. *Toxins* **2021**, *13*, 92. [CrossRef]
80. Lee, T.; Lee, S.H.; Lee, S.H.; Shin, J.Y.; Yun, J.C.; Lee, Y.W.; Ryu, J.G. Occurrence of *Fusarium* mycotoxins in rice and its milling by-products in Korea. *J. Food Prot.* **2011**, *74*, 1169–1174. [CrossRef]
81. Birr, T.; Jensen, T.; Preußke, N.; Sönnichsen, F.D.; De Boevre, M.; De Saeger, S.; Hasler, M.; Verreet, J.A.; Klink, H. Occurrence of *Fusarium* mycotoxins and their modified forms in forage maize cultivars. *Toxins* **2021**, *13*, 110. [CrossRef]
82. Kim, J.C.; Kang, H.J.; Lee, D.H.; Lee, Y.W.; Yoshizawa, T. Natural occurrence of *Fusarium* mycotoxins (trichothecenes and zearalenone) in barley and corn in Korea. *Appl. Environ. Microbiol.* **1993**, *59*, 3798–3802. [CrossRef]
83. Chrpova, J.; Sip, V.; Sumikova, T.; Salava, J.; Palicova, J.; Stockova, L.; Dzuman, Z.; Hajslova, J. Occurrence of *Fusarium* species and mycotoxins in wheat grain collected in the Czech Republic. *World Mycotoxin J.* **2016**, *9*, 317–327. [CrossRef]
84. Molto, G.A.; Gonzalez, H.H.L.; Resnik, S.L.; Gonzalez, A.P. Production of trichothecenes and zearalenone by isolates of *Fusarium* spp. from Argentinian maize. *Food Addit. Contam.* **1997**, *14*, 263–268. [CrossRef]
85. Bottalico, A.; Perrone, G. Toxigenic *Fusarium* species and mycotoxins associated with head blight in small-grain cereals in Europe. *Eur. J. Plant Pathol.* **2002**, *108*, 611–624. [CrossRef]
86. Quarta, A.; Mita, G.; Haidukowski, M.; Santino, A.; Mulè, G.; Visconti, A. Assessment of trichothecene chemotypes of *Fusarium culmorum* occurring in Europe. *Food Addit. Contam.* **2005**, *22*, 309–315. [CrossRef] [PubMed]
87. Schollenberger, M.; Muller, H.M.; Rufle, M.; Suchy, S.; Plank, S.; Drochner, W. Natural occurrence of 16 *Fusarium* toxins in grains and feedstuffs of plant origin from Germany. *Mycopathologia* **2006**, *161*, 43–52. [CrossRef] [PubMed]
88. Tian, Y.; Tan, Y.; Liu, N.; Liao, Y.; Sun, C.; Wang, S.; Wu, A. Functional agents to biologically control deoxynivalenol contamination in cereal grains. *Front. Microbiol.* **2016**, *7*, 395. [CrossRef] [PubMed]
89. Audenaert, K.; Vanheule, A.; Höfte, M.; Haesaert, G. Deoxynivalenol: A major player in the multifaceted response of *Fusarium* to its environment. *Toxins* **2013**, *6*, 1–19. [CrossRef] [PubMed]
90. Munkvold, G.P.; Proctor, R.H.; Moretti, A. Mycotoxin production in *Fusarium* according to contemporary species concepts. *Annu. Rev. Phytopathol.* **2021**, *59*, 373–402. [CrossRef]
91. Ramakrishna, Y.; Bhat, R.V.; Ravindranath, V. Production of deoxynivalenol by *Fusarium* isolates from samples of wheat associated with a human mycotoxicosis outbreak and from sorghum cultivars. *Appl. Environ. Microbiol.* **1989**, *55*, 2619–2620. [CrossRef]
92. Petrovic, T.; Walsh, J.L.; Burgess, L.W.; Summerell, B.A. *Fusarium* species associated with stalk rot of grain sorghum in the northern grain belt of eastern Australia. *Australas. Plant Pathol.* **2009**, *38*, 373–379. [CrossRef]
93. Tahir, A.; Khan, S.N.; Javaid, A.; Riaz, M. Morphological and molecular characterization of *Fusarium thapsinum*, causing stalk tot of maize in Punjab, Pakistan. *Mycopath* **2018**, *16*, 57–64.
94. Pena, G.A.; Sulyok, M.; Chulze, S.N. Effect of interacting conditions of water activity, temperature and incubation time on *Fusarium thapsinum* and *Fusarium andiyazi* growth and toxin production on sorghum grains. *Int. J. Food Microbiol.* **2020**, *318*, 108468. [CrossRef]
95. Leslie, J.F.; Zeller, K.A.; Lamprecht, S.C.; Rheeder, J.P.; Marasas, W.F. Toxicity, pathogenicity, and genetic differentiation of five species of Fusarium from sorghum and millet. *Phytopathology* **2005**, *95*, 275–283. [CrossRef]

agriculture

MDPI

Review

Progress in Research on Prevention and Control of Crop Fungal Diseases in the Context of Climate Change

Jien Zhou [1,2,3,†], Xueyan Zhang [1,2,†], Zheng Qu [1], Chenchen Zhang [1], Feng Wang [1,2], Tongguo Gao [4], Yanpo Yao [1,*] and Junfeng Liang [1,*]

1 Agro-Environmental Protection Institute, Ministry of Agriculture and Rural Affairs, Tianjin 300191, China; zjientongxue@163.com (J.Z.); 13571732795@163.com (X.Z.); quzhengwf@163.com (Z.Q.); chynacici0314@163.com (C.Z.); wangfeng_530@163.com (F.W.)
2 Dali Comprehensive Experimental Station of Environmental Protection Research and Monitoring Institute, Ministry of Agriculture and Rural Affairs (Dali Original Seed Farm), Dali 671004, China
3 College of Resources and Environment, Xinjiang Agricultural University, Urumqi 830052, China
4 College of Life Sciences, Hebei Agricultural University, Baoding 071001, China; gtgrxf@163.com
* Correspondence: yaoyanpo@caas.cn (Y.Y.); liangjunfeng@caas.cn (J.L.); Tel.: +86-22-23610159 (Y.Y.)
† These authors contributed equally to this work.

Abstract: With an advancement in global climate change, the frequency of extreme climatic events, such as high temperature, drought, and flooding, has increased. Meanwhile, outbreaks of crop fungal diseases are becoming more frequent and serious, and crop growth and food production are seriously threatened. This article focuses on the climate change-related aggravation of crop fungal diseases; summarizes the progress in research on the impact of climate change on soil-borne fungal diseases, air-borne fungal diseases, and seed-borne fungal diseases; and discusses the conventional methods for diagnosing crop fungal diseases. On the basis of comparative analysis, the concept of ecological control is proposed; ecological control can maintain the dynamic balance of crop–biology–soil, provide a good environment for the healthy growth of crops, and provide a new fungal disease control method in the context of climate change.

Keywords: climate change; crops; fungal diseases; advances

1. Introduction

By 2050, the world's population will exceed 9.7 billion, and human demand for food will continue to grow. However, crop yields globally have suffered severe declines due to disease, with 70–80% of crop diseases being caused by fungi [1]. There are approximately 1.5 million fungal species globally, of which *Fusarium* alone causes wilt and root rot in more than 100 crops [2], such as tomato wilt [3], corn rusts [4], and wheat fusarium head blight [5]. Indeed, most crops are susceptible to fungal diseases, with rice blast [6], wheat common smut, and wheat stem rust [7] causing 10–35% and 15–20% yield losses in rice [8] and wheat [9]. Although fungal diseases already pose a serious threat to global food security [10], the advancement of climate change has deepened this threat, highlighting the urgency of the challenge that this poses for humans.

It is widely recognized that climate change has the potential to decrease the productivity of food crops [11–14]. With global average temperatures having risen by 1.5 °C since the industrial revolution. The model predicts that an increase in temperature of 1.6 °C–3 °C will result in an 18–32% reduction in potato yields [15], which is particularly concerning given that the median global temperature is projected to rise by 2.6 °C–3.1 °C in 2100 [16,17]. As such, the reductions in crop yields seen to date may become far more severe in the future. In addition, climate change stimulates the growth of fungi that are pathogenic to crops and increases the incidence of fungal diseases. Higher temperatures would increase the relative abundance of most fungal pathogens globally [1], and fungal pathogens are more aggressive at higher temperatures (Figure 1) [18]. Flooding also promotes the spread of pathogenic fungi through soil runoff. For example, it has been shown that *Colletotrichum* rapidly infects and causes disease after reaching onion cultivations through rainfall washout transmission [19]. In addition, an increase in relative humidity due to flooding can trigger outbreaks of leaf spot or other fungal diseases in winter wheat [20]. It has also been reported that under drought and salt stress conditions, the rate of infection of broad bean root rot increases to 25–100% [21].

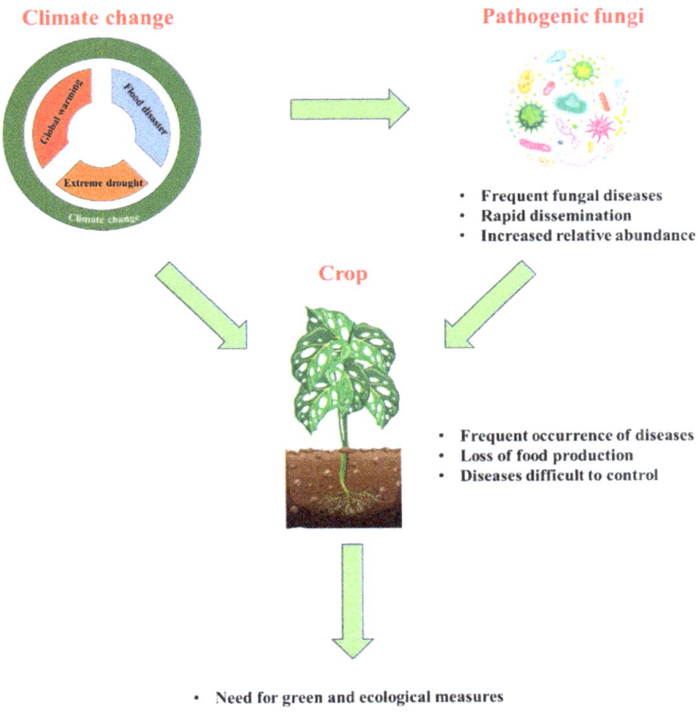

Figure 1. Relationship between climate, crops, and fungal pathogens.

Overall, in the context of global climate change, fungal diseases of crops become established more rapidly, spread over a wider area, and are more damaging [18]. Moreover, the chemical pesticides that are currently most commonly used are no longer adequate because they damage plants and lead to resistance of pathogenic fungi and environmental destruction [22,23]. This background has prompted researchers to rethink microbe–crop–soil interactions as a whole to achieve ecological prevention and control of fungal diseases. This article focuses on the climate change-related aggravation of crop fungal diseases; summarizes the progress of research on the impact of climate change on soil-borne, airborne, and seed-borne fungal diseases; discusses the methods for diagnosing crop fungal diseases; and proposes ecological prevention and control methods for maintaining crop health and preventing and controlling crop fungal diseases based on comparative analyses of such approaches.

2. Climate Change Exacerbates Crop Fungal Diseases

Increased frequency of extreme weather events due to climate change, such as droughts, floods, and high temperatures, has exacerbated the prevalence of crop fungal diseases at an alarming rate. This has in turn affected the growth cycle, yield, and quality of crops [24,25].

2.1. Drought Exacerbates Crop Fungal Diseases

Drought stress reduces crop fine roots and soil microbial diversity, allowing disease-causing fungi to infect crops more readily [26]. Studies have shown that drought exacerbates faba bean wilt [21], chickpea root rot [27], rice blast [28], and grapevine disease [29]. The lack of attention and efforts to prevent and control soil-borne pathogens in arid areas makes them more likely to cause significant damage [30]. It was reported that an annual drought in May–June significantly increases the incidence of rust in winter wheat in Poland [31]. Moreover, drought stress-induced soil salinization has been reported to increase the susceptibility of crops to pathogens and promote the spread of fungal crop diseases [32]. The interaction between crop salt stress and pathogen infection depends on the type of pathogen and the host crop species. For example, onions exposed to salt stress conditions suffer from more severe infestation with *Fusarium oxysporum* than those not exposed to salt stress [33,34]. Moreover, increased salinity of irrigation water significantly exacerbated cucumber damping-off disease caused by *Pythium aphanidermatum* [35].

2.2. Flooding and High Humidity Exacerbates Crop Fungal Diseases

Extreme precipitation can lead to frequent flooding and an increase in pathogenic fungi [31]. For example, precipitation or soil runoff can promote the spread of disease via spores of pathogenic fungi in the soil and in the air [36]. Prolonged rainfall makes it easier for *Fusarium* to attack wheat, which is more susceptible to tan spot disease [37]. Indeed, dramatic increases in the levels of pathogens such as *Phytophthora*, *Fusarium*, and *Pythium* have been reported after flooding [38]. In addition, the transport and agglomeration of fungal spores of *Epicoccum tritici* were found to be promoted by water condensation and evaporation [39]. Moreover, sugarcane is known to be more susceptible to false floral smut under conditions of high precipitation or high humidity [40], while the incidence of cashew leaf blight was shown to increase by 2% for every 1 mm increase in precipitation [41]. The number of spots in chickpea was also found to increase with increasing rainfall [42], while the incidence and severity of quinoa downy mildew increased under flooded conditions [43]. Finally, it was found that the likelihood of rice leaf blight and brown spot occurring simultaneously increases with increasing soil water content [44].

2.3. High Temperatures Exacerbate Crop Fungal Diseases

High temperatures are among the most direct manifestations of climate change, and prolonged high temperatures can increase the extent and severity of crop diseases [45]. For example, elevated temperatures can lead to pathogen spore germination, crop root damage, and an increased incidence of disease in susceptible crops. For example, the germination time of *Botryosphaeria* as the causative agent of grapevine trunk disease was found to be shorter at elevated temperatures [46], exacerbating this disease [47]. High temperatures were also shown to increase the incidence of root rot in celery, mainly due to the reduced expression of *Fusarium* resistance [48]. Soybean was also reported to suffer a higher incidence of *Fusarium solani*-induced root diseases at 30 °C [49], and soybean is more susceptible to sudden death syndrome at 37 °C, indicating that disease transmission is sensitive to temperature. Temperature changes also cause spot growth, increased infection efficiency, and spore formation of potato late blight, along with increases in infection efficiency, spot growth rate, and spore formation occurring under fluctuating temperature conditions compared with the rates at a constant temperature [50]. Moreover, elevated storage temperature also exacerbates crop fungal diseases, with a significant growth in the rice mold population being identified after 16 weeks of storage at 40 °C [51], showing incidence rates as high as 96.5%–99.1%. Moreover, while no decay occurred in carrots at a storage temperature of 10 °C, decay was greatest when the temperature rose to 35 °C [52]. It has also been found that increasing storage temperature from 2 °C to 25 °C increased the risk of infection of nectarines with brown rot by more than 60-fold [53] (Table 1).

Table 1. Prevalence of crop fungal diseases due to climate change.

Type of Climate Change	Crop	Disease	Pathogenic Fungus	Hazard	Reference
Desertification	Faba bean	Root rot	*Fusarium equiseti, Fusarium graminearum, Fusarium brachygibbosum*	Growth rate of pathogenic fungi increased by 30–44%	[21]
	Chickpea	Root rot	*Macrophomina phaseolina*	Chickpea production down 56.99%	[27]
	Rice	Rice blast	*Magnaporthe oryzae*	Plants under drought for longer have higher fungal abundance and poorer germination	[28]
Flooding and high humidity	Cucumber	Damping-off disease	*Pythium aphanidermatum*	Significant increase in morbidity from 40% to 93%	[35]
	Sugar cane	False floral smut	*Epicoccum andropogonis, Claviceps purpurea*	Heavy precipitation during the bloom period resulted in a higher incidence than in previous years	[40]
	Cashew	Leaf blight	*Cryptosporiopsis* spp.	2% increase in incidence per mm increase in rainfall	[41]
	Quinoa	Downy mildew	*Peronospora variabilis*	Decreased chlorophyll in quinoa after disease onset	[43]
High temperatures	Grapevine trunk	Grapevine trunk disease	*Lasiodiplodia theobromae*	Pathogenic fungi grow at 4 °C–40 °C and prefer higher temperatures	[46]
	Celery	Root rot	*Fusarium oxysporum*	*Fusarium acanthamoeba* levels and disease severity increased with increasing temperature	[48]
	Soybean	Root rot	*Fusarium solani*	Longer root lesion lengths and higher disease incidence of *Fusarium* at 30 °C	[49]

3. Impact of Climate Change on Crop Fungal Diseases with Different Routes of Transmission

Crop fungal diseases are mainly categorized into soil-borne fungal diseases, air-borne fungal diseases, and seed-borne fungal diseases according to the transmission route. Climate change can alter the sites of infestation, disease levels, and mechanisms by which fungi cause damage via different transmission routes.

3.1. Soil-Borne Fungal Diseases

Soil-borne fungal diseases occur when fungi residing in the soil alongside crop remnants invade crop roots or stems under favorable conditions. These diseases hinder root functionality and the absorption of water and nutrients, resulting in stunted growth and decreased crop yields [54]. Especially in cases where there are continuous cropping and soil degradation, outbreaks of soil-borne diseases can frequently occur [55]. For example, potato late blight causes annual losses of approximately EUR 6.1 billion globally [56], and potatoes are more susceptible to extreme summer temperatures and drought [57]. The ratio of mild evapotranspiration to precipitation limits crop diversity, and extreme summer temperatures and drought make it easier for pathogens to attack crops, with some soil-borne pathogens, such as *Pythium* spp., *Phytophthora* spp., *Fusarium* spp., and *Mycosphaerella* spp., leading to crop yield reductions of 50–75% [58]. Climate change causes soil acidification, which alters the physicochemical properties and microbial communities of soil, promoting the accumulation of pathogens and exacerbating soil-borne diseases [59]. Acidic soils are in turn associated with the greatest incidence of *Fusarium*-induced root rot diseases of soybeans. Extreme climate conditions predispose crops to injury, and such vulnerable crops are more susceptible to disease. It has been reported that temperatures above 30 °C trigger oxidative stress in grains, which directly damages enzymes and tissues [60], and fungi such as *Aspergillus* and *Fusarium oxysporum* often parasitize and proliferate through sites of damage in crops [61]. There is increasing evidence to suggest that soil-borne diseases are not solely caused by individual species, but rather by pathogenic complexes [62–65], and the complexity of soil-borne pathogen–environment interactions is highlighted by the existence of multifaceted and contrasting interactions, such as those with temperature and humidity [66]. This complexity is further aggravated by the unpredictable and fluctuating proportions of diverse pathogens, which vary significantly across seasons and regions [67].

3.2. Air-Borne Fungal Diseases

Air-borne fungal diseases involve the spread of crop pathogens via spores aided by air currents to cause crop disease epidemics and associated damage. Pathogenic fungi, such as *Cladosporium*, *Penicillium*, and *Aspergillus* (including the species *Aspergillus fumigatus*), are present in the air [68] and cause air-borne fungal diseases including gray mold, downy mildew, powdery mildew, and early blight [69]. Individuals or groups of pathogens that are rapidly transmitted through the air can cause a wide range of diseases, intense outbreaks, and serious disasters. The spread of yellow rust to Australia in 1978 is an example of the spread of an air-borne disease [70], where the original rust fungus originated in Europe [71,72]. Another case of this kind involved Australian wheat stem rust, which occurred for the first time in Ethiopia in the mid-1980s, followed by its gradual emergence in the Middle East and the Indian subcontinent [73,74]. Increased air-borne sporulation is driven by higher temperatures and increased humidity. Studies have shown that temperature is the most important variable for the spread of *Alternaria* spores in Spain [75]. Meanwhile, it has been reported that warm climatic conditions in France promote the rapid growth of *Cladosporium* fungal spores [76]. Spores are transported over long distances by diffusion into the surrounding area through droplets or air [77]. Cucurbit downy mildew caused by the oomycete Cuban pseudofungus spreads rapidly [78], and it has been shown that relative humidity above 90% and a temperature between 15 °C and 20 °C are the optimal conditions for downy mildew [79], leading to yield losses of up to 80% in European cucurbits [78,80]. Another example of an air-borne fungal disease is black spot caused by *Venturia inaequalis*, which is an important disease of apples worldwide [81] and can cause 70% yield loss [81–83]. Meanwhile, early blight caused by *Alternaria solani* is a common disease of potato [84,85] that causes great damage under high-temperature conditions [86,87]. It has been reported that early blight is likely to become more severe in European potato-cultivating areas in the future [88].

3.3. Seed-Borne Fungal Diseases

Seed-borne fungal diseases are those in which the seed produced by a crop carries a pathogenic fungal propagule that develops at a certain stage of growth after the seed has been sown when suitable climatic conditions prevail. The fungal pathogens can be present on the outside or inside of the seed and may cause seed septoria, rot, and necrosis as well as impairing seedling growth [89]. When a crop pathogenic fungus binds to the seed, it may cause significant economic losses [90]. For example, when healthy wheat seeds are infested with *Tilletia caries*, even when sown into uninfested soil, the seed germinates and the latent pathogen invades the wheat spike as it spreads through the wheat seed [56], resulting in reduced seed quality and toxic metabolite production [91]. *Aspergillus flavus*, the seed-borne fungal pathogen of mungbean, is a toxic pathogenic fungus whose pathogenicity is primarily related to aflatoxin. It has been found that the rate of *A. flavus* colonization and the aflatoxin concentration in peanuts increase at temperatures close to 35 °C [92]. Increased storage temperature and humidity can lead to seed ulceration. High relative humidity during storage conditions can lead to pathogenic fungal growth, which can negatively impact the germination and growth of rice seeds [93]. Meanwhile, with increasing water content or storage temperature of soybean seeds, pathogenic fungi produce chemicals that accelerate seed deterioration [94] (Table 2).

Table 2. Mechanisms by which climate change impacts fungal diseases with different transmission routes.

Transmission Route	Crop	Pathogenic Fungus	Mechanisms of Influence	Reference
Soil-borne disease	Tomato	*Fusarium solani, F. oxysporum* or *Ilyonectria destructans*	Crops are more susceptible to extreme summer temperatures and drought	[54,57–59]
	Maize	*Fusarium graminearum*	High temperatures trigger oxidative stress in grains that damages enzymes and tissues, making fungi more susceptible to infestation	[61]
Air-borne disease	Cucurbit	*Pseudoperonospora cubensis*	Relative humidity > 90% and temperature 15 °C–20 °C are optimal conditions for cucurbit downy mildew development	[78,79]
	Potato	*Alternaria solani*	*Streptomyces* levels exceed 50 spores/m^3 at average temperature > 18 °C and leaf humidity > 80%	[84,85]
Seed-borne disease	Wheat	*Tilletia caries*	Stored seeds are affected by temperature and humidity	[91]
	Peanut	*Aspergillus flavus*	Increase in the proportion of *Aspergillus flavus* colonizing peanuts and aflatoxin concentration with increasing temperature	[92]
	Rice	*Fusarium fujikuroi*	At >25 °C, mycelium grows most vigorously	[93]

4. Control of Fungal Diseases of Crops

The precise diagnosis of fungal diseases is a prerequisite for improving the effectiveness of prevention and control of such diseases in crops. Different crop diseases, different transmission routes, and diseases at different reproductive periods require the adoption of highly sensitive detection technologies and means. Disease prevention and control are currently divided into disease-resistant breeding, agronomic measures, chemical control, biological control, and ecological control.

4.1. Disease Detection

4.1.1. Methods for Diagnosing Fungal Diseases of Crops

The timely detection of crop diseases and their reliable diagnosis can make a major contribution to maintaining total food production globally [95]. Methods for diagnosing fungal diseases in crops can be divided into two categories: traditional and modern detection methods. Traditional detection methods mainly rely on morphological, microbiological, and biochemical identification, with fungal pathogens being detected by visual inspection methods and culture methods. The results of visual inspections are greatly influenced by the knowledge and experience of the inspectors, and have limited accuracy in the on-site diagnosis of fungal pathogens of crops [96]. Meanwhile, the culture method is time-consuming, with 3–5 days usually being required to cultivate the fungus [97]. In conclusion, traditional detection methods are time-consuming, ineffective, and unsuitable when efficient and rapid detection is required. Meanwhile, modern detection methods can be subdivided into direct and indirect detection methods, which are the main current methods for diagnosing fungal diseases.

Direct Detection Methods

Direct detection methods can be categorized into immunological methods and polymerase chain reaction (PCR) methods. The principle of immunological methods is based on the binding of antigen and antibody, and these methods include the immunodiffusion test, enzyme-linked immunosorbent assay, radioimmunosorbent assay (RISA), and dot immunobinding assay (DIBA) [98]. In terms of their advantages over traditional culture methods, immunological methods have a short detection time, but they have the disadvantages of low detection sensitivity and accuracy, as well as interference by contaminants. Meanwhile, PCR methods can detect specific target nucleic acids of the target pathogens and improve the specificity, sensitivity, and speed of detection [99]. However, they have the disadvantages of a lack of standardization, the need for additional preparation of samples, the difficulty of on-site operation, the need for special equipment, and the high cost of detection [100]. Overall, in the context of climate change, the above two methods are not suitable for diagnosing fungal pathogens in crops.

Indirect Detection Methods

Indirect detection methods identify crop diseases not by directly identifying the pathogen but by detecting the physiological response of the crop to the pathogen. The main indirect detection techniques include spectroscopy, biosensor detection, gas chromatography and mass spectrometry detection, and imaging techniques.

(1) Spectroscopy. Spectroscopy is used to assess the health of a crop by irradiating light of a specific wavelength onto the crop tissue and measuring the intensity of the light wavelengths reflected back. Examples of types of spectroscopy include fluorescence spectroscopy, visible spectroscopy, infrared spectroscopy, and nuclear magnetic resonance spectroscopy. Near-infrared spectroscopy can detect maize spot disease and olive leaf spot [101,102]. However, spectroscopic methods cannot detect diseases before they develop in crops and cannot detect multiple diseases occurring at the same time.
(2) Biosensors. A sensor is an analytical device that converts chemical, physical, or biological information into a useful analytical signal. A biosensor is "an integrated receptor–transducer device capable of providing selective quantitative or semiquantitative analytical information using biometric elements" [103]. In recent years, biosensors have attracted much attention due to their good results in detecting, classifying, diagnosing, and quantifying crop diseases. Biosensors can be categorized into optical biosensors, volatile biosensors, electrochemical biosensors, and mass-sensitive biosensors. Despite the ability of biosensors to rapidly and accurately detect disease-causing fungal pathogens, only a few devices are commercially available for detecting crop fungal pathogens in crops [104].
(3) Gas chromatography coupled with mass spectrometry (GC–MS). GC–MS is a technique for analyzing volatiles produced by fungal infections in crops [105]. It has the advantages of not destroying the crop and enabling continuous monitoring for a long time [106], but it has the disadvantages of environmental factors such as humidity potentially interfering with the sensor readings [107] and the need to collect volatile organic compound samples prior to the GC–MS analysis, which limits its application in the field.
(4) Imaging technology. Imaging technologies include thermal, hyperspectral, and red–green–blue (RGB) and fluorescence imaging, coupled with unmanned aerial vehicles (UAVs) that can monitor large farms. Using such technologies, diseases are diagnosed indirectly by detecting changes in the color, texture, and temperature of crop leaves. In this context, intelligent agricultural machines and robots can detect crop diseases at an early stage and monitor disease development remotely [108,109]. Meanwhile, aerial remote sensing (RS) using UAVs or unmanned aerial systems with intelligent vision systems is an efficient and inexpensive way of detecting diseases in crops [108–113]. It is reported that the integration of RGB, multispectral, hyperspectral, fluorescence, and

thermal infrared imaging sensors, coupled with efficient algorithms on unmanned aerial vehicles (UAVs), enables effective detection, differentiation, and quantification of the severity of symptoms induced by diverse pathogens under field conditions [114,115]. It has been shown that these methods can achieve reliable diagnosis in cereal crops, such as rice [116], corn [117], and wheat [118,119]; vegetables, such as grapevines [120], potatoes [121], soybeans [122], and tomatoes [123]; and forests, such as pine forests [108]. The combination of imaging obtained by drones with artificial intelligence algorithms, such as machine learning, is a promising new approach that not only enables early and accurate diagnosis of crop diseases but also improves crop yields while reducing the cost of disease treatment [124]. Utilizing RS technology to retrieve and gather data, encompassing meteorological information as well as the properties of the earth's surface and soil, and offering timely feedback, holds significant promise as an effective approach for diagnosing fungal crop diseases in the context of climate change [125] (Table 3).

Table 3. Methods for diagnosing crop fungal diseases.

Method Category	Diagnostic Method		Principle	Advantages and Disadvantages	Reference
Traditional detection methods	Visual inspection method		Isolation of pathogens and interpretation of visual symptoms of disease through microscopy	Highly subjective and error-prone	[96]
	Culture		Morphological characterization of pathogens by medium culture and microscopic observation	Cheap, simple, but accuracy and reliability depend on experience and skill, time-consuming	[97]
Modern testing methods	Direct detection methods	Immunological methods	Based on antigen–antibody binding	Short detection time, low sensitivity and accuracy	[98]
		Polymerase chain reaction (PCR)	Detection of target nucleic acids specific to the target pathogen	Fast detection, lack of standardization, additional sample preparation, difficult on-site operation, high detection cost	[99,100]
	Indirect detection methods	spectroscopy	Measurement of the intensity of reflected light wavelengths irradiated by a specific light source to assess the health of the crop	Inability to detect disease prior to onset, and inability to detect multiple diseases occurring at the same time	[101,102]
		Biosensor detection	Provides selective quantitative or semiquantitative analytical information using biometric elements	Quantitative information on crop diseases is available, but commercially available equipment is limited	[103,104]
		GC–MS	Assessment of changes in volatiles due to fungal infection of crops	No crop damage, continuous monitoring over long periods of time, environmental factors interfere with sensor readings, pre-collection of samples required	[105–107]
		Imaging technology	Indirect diagnosis of diseases by detecting changes in color, texture, or temperature of crop leaves	Efficient and inexpensive, suitable for climate change conditions	[108,109]

4.2. *Prevention and Control of Crop Fungal Diseases*

4.2.1. Breeding Measures for Disease Resistance

Disease resistance breeding is a method of breeding new crop varieties that develop strong resistance to diseases through targeted selection or alteration of certain genotypes. It is one of the most efficient and economical measures to cope with crop diseases. However, improving the disease resistance of crops can simultaneously affect their yield [126]. It has been shown that disease resistance breeding enhances *Phytophthora* blight resistance while simultaneously reducing *Rhizobia* colonization, resulting in reduced crop yields [127]. High-quality resistance results in the targeted selection of pathogens, while the speed and scale of pathogen spread increase under extreme climatic conditions. Moreover, research devoted to particular resistant varieties is often overtaken by the emergence of pathogens with greater pathogenicity [128].

4.2.2. Agronomic Measures

Agronomic measures involve the creation of environmental conditions that promote crop growth and the survival and reproduction of beneficial bacteria, but that inhibit the occurrence of pathogenic fungi. Such measures can involve practices such as fertilization, irrigation, and tillage. The application of fertilizer to the soil can change its physical and chemical properties and create an environment suitable for the growth of beneficial microorganisms. The application of chitin amendments to the soil was reported to contribute to soil carbon and nitrogen cycling [129]. Meanwhile, the application of cow dung after ammonia fumigation was found to reduce the abundance of *Fusarium spinosum* in soil and its infestation in bananas by approximately 55% [130]. Although irrigation can greatly reduce the negative effects of drought [131], there is a lack of effective agronomic measures to cope with disease outbreaks due to climatic extremes of rainfall and flooding; agronomic measures such as crop rotation and intercropping do not have effects that are rapid enough to control fungal disease in the current season [132], and although deep tillage may promote the onset of, for example, leaf spot disease in tea trees [133], the process of applying fertilizers is labor-intensive and time-consuming [134] and fertilizers pose the risk of nutrient leaching [135]. The greatest pressure on irrigation is the severe shortage of freshwater resources available for this purpose [136].

4.2.3. Chemical Control

Chemical control, which involves killing pathogenic bacteria using chemical pesticides to avoid crop damage, is a common method of managing fungal disease and is used for both disease prevention and treatment of diseased crops [137–139]. Chemicals play an important role in increasing crop yield and quality, improving food safety, reducing microbial toxins, and extending the shelf life of crops [140,141]. However, chemicals can have harmful effects on plant and animal health and nature [142–144]. The application of pesticides poisons organisms in the soil and water bodies and reduces soil biodiversity [145]. A previous study showed that higher spraying of fungicide reduces the diversity of *Xylaria* spp. isolates, which are beneficial to crops [146]. It has also been shown that penthiopyrad induces oxidative damage and lipid metabolism, causes mitochondrial dysfunction, and leads to apoptosis in fish [147]. The intensive use of chemicals has also been found to lead to increased pathogen resistance [148], and long-term pesticide application has significantly reduced the effectiveness of control of *Zymoseptoria tritici*, *Colletotrichum falcatum*, and *Venturia inaequalis* [149,150]. Moreover, with climate extremes, such as flooding, more than 90% of pesticides are lost in the field, resulting in serious risks to ecosystems. Some pesticides may even exert effects lasting for decades [151,152].

4.2.4. Biological Control

Biological control, which entails the utilization of organisms like crops, insects, and microorganisms, serves as a natural means to mitigate a diverse range of diseases triggered by pathogens [153], is considered a safe and environmentally friendly strategy that can reduce or eliminate the utilization of chemical pesticides [154]. The main forms of biological control involve the application of microbial or bio-organic fertilizers and bio-antagonistic fungicides to the soil. In terms of the microorganisms used for biological control, the main ones include bacteria such as *Pseudomonas* spp. and *Bacillus* spp. [155], fungi such as *Penicillium* spp. and *Trichoderma* spp., and actinomycetes such as *Streptomyces* spp. Regarding the mechanisms by which biocontrol bacteria exert their effects, these include the formation of biofilm by *Bacillus* spp. to promote crop growth, improvement of the microbial community in the soil, and inhibition of the growth of pathogenic bacteria [156]. In terms of how biocontrol fungi work, they can induce crop resistance, produce antimicrobial substances against pathogenic fungi (e.g., iron carriers, volatile organic compounds, solubilizing enzymes), and even occupy ecological niches in advance. For example, it was reported that *Chromolaena odorata* not only increases the height and shoot/root dry weights of melon plants to promote crop development, but also inhibits the wilt pathogen *Fusarium oxysporum* by 90% [157]. However, biological control has the shortcoming that the microorganisms used are susceptible to environmental influences, especially in extreme climatic conditions such as drought and high temperatures, leading to unstable prevention and control effects during the application of microbial agents.

4.2.5. Ecological Regulation

In the context of global climate change and the increasing frequency of extreme weather events, traditional singular control measures have become inadequate in effectively preventing and controlling fungal diseases in crops. At its core, ecological control involves a systematic approach to provide a favorable environment for healthy crop growth and maintain the dynamic balance among crop–biology–soil (Figure 2). Soil health is linked to crop health. Heavy metal-stressed vetiver fractionates heavy metals in the soil by secreting organic acids, while raising the level of organic carbon and the pH of soil [158]. Crops can influence the climate by regulating root secretions to reduce nitrogen loss and the carbon footprint, and also by recruiting beneficial microorganisms that enhance the ability to suppress pathogens [159]. The warming of the climate and N_2 and CO_2 also alter crop–soil interactions [160], while elevated CO_2 concentration indirectly affects the size of the population of microbes in the soil [161] and its diversity [162] through physicochemical properties such as soil temperature, moisture, and pH, as well as crop root morphology and root secretions [163]. Microorganisms can inhibit crop diseases by fixing nitrogen [164] and promoting crop growth [165,166], or even by changing the structure of the microbiome from a dysfunctional to a healthy state by altering environmental conditions [167]. The release of metabolites from maize roots affects the composition of root-associated microbiota [168], and the excretion of taxifolin by tomato plants serves as a catalyst for the attraction and recruitment of beneficial bacterial species [169]. The disadvantages of ecological regulation are the complexity of its application and the long time it takes to obtain results.

Effective responses to crop fungal diseases under global climate change conditions require a combination of control measures to realize the dynamic balance of crop–biology–soil, and the capacity of various control measures to enable adaptation to climate change should be fully exploited, such as improving the physicochemical properties of soil and reducing the relative abundance of soil pathogens by performing supplementation with soil conditioning agents, such as biochar, when microbial fungicides are applied [170]. Immobilization of microorganisms in materials can also reduce their exposure to extreme climatic conditions and improve environmental resilience [171]. Meanwhile, the intensive use of multidisciplinary tools, such as varietal performance prediction [172], can help develop new varieties that can adapt to climate change. Moreover, microbial pesticides or nanobiopesticides derived from microorganisms, such as bacteria, fungi, viruses, protozoa, and algae, are being intensively explored to control crop diseases. In this context, environmentally friendly and green control measures also have great potential, such as the application of biopesticides prepared from wood mold for the control of root rot of chickpea [173] and the application of biocontrol agents such as a *Trichoderma* and chitosan-polyethylene glycol (PEG) hybrid system for seed treatment [174]. Nanobiopesticides have beneficial physicochemical properties, such as a small particle size, easy reactivity, large surface area, clear biological interactions with crops, and clear transport and fate in the environment (Table 4).

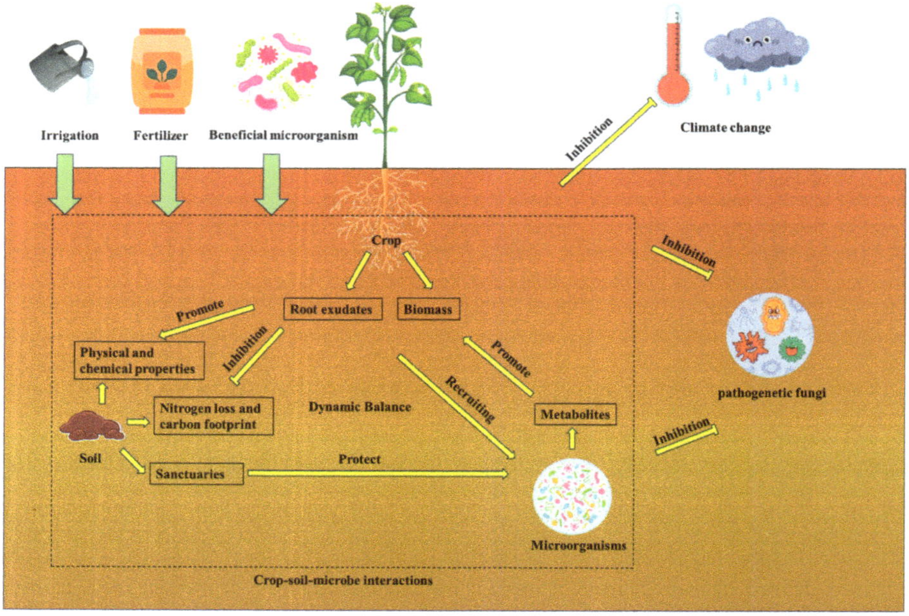

Figure 2. Dynamic equilibrium between crop–biology–soil.

Table 4. Control of crop fungal diseases under climate change.

Preventive and Curative Measures	Principles of Prevention and Treatment	Advantages	Disadvantages	Reference
Breeding for disease resistance	Targeted selection or alteration of certain genotypes to produce new varieties of crops that are resistant to disease	One of the most effective and economical measures	High-quality resistance results in the targeted selection of pathogens	[126–128]
Agronomic measures	Fertilization, irrigation, and so on to create environmental conditions suitable for the growth and development of vegetables and the survival and reproduction of beneficial organisms, but not conducive to the occurrence of pathogenic fungi	Improve the physical and chemical properties of the soil to provide a good growing environment for crops	High upfront costs, and labor-intensive and time-consuming process	[129–131,134–136]
Chemical control	Chemical pesticides kill pathogenic fungi	By far the most commonly used control method	Pathogen resistance, environmental risks	[137–152]
Biological control	Use of organisms such as crops, insects, and microorganisms to limit diseases caused by pathogens	A green, healthy, and promising approach	Colonization by free microorganisms is difficult, and the preventive effect is unstable in extreme climate conditions such as drought and high temperature	[153–157]
Ecological regulation	Provide a favorable environment for healthy crop growth and maintain crop–biology–soil dynamic balance	Maintaining a dynamic crop–biology–soil balance	the complexity of its application and long time to obtain results.	[158–170]

5. Summary and Outlook

With advancement in global climate change, extreme climatic events, such as high temperatures, droughts, and floods, are becoming more frequent and more severe. Meanwhile, outbreaks of fungal diseases of crops are becoming more frequent and serious and disrupting the dynamic balance among crops, organisms, and the soil. Intensifying climate change necessitates the development of new data acquisition technologies and diagnostic methods, such as RS, combined with artificial intelligence coupling technology, to collect and analyze disease data in an accurate, timely, and efficient manner on the basis of existing methods for monitoring and diagnosing crop fungal diseases. After determining the areas affected by disease, the ecological regulation can be used to organize the implementation of ecological control measures and traditional control measures, apply the control technology in the context of climate change, maintain the health of the soil and crop, protect the diversity of the overall community of microorganisms, and slow down the harm caused by crop fungal diseases in association with climate change. The ecological management of crop–biological–soil dynamics is considered to be an effective measure to combat fungal diseases in the future, and this new concept promotes the green development of agriculture and mitigates the damage caused to crops by fungal diseases under extreme climatic conditions.

Author Contributions: Conceptualization, J.Z. and X.Z.; methodology, J.Z., X.Z., and Z.Q.; validation, J.Z., X.Z., Z.Q., and C.Z.; investigation, J.Z., C.Z., F.W., and T.G.; resources, J.Z., F.W., Y.Y., and J.L.; data curation, J.Z. and Z.Q.; writing—original draft preparation, J.Z. and Z.Q.; visualization, J.Z.; supervision, J.Z., Z.Q., and F.W.; funding acquisition, J.Z. and F.W. All authors have read and agreed to the published version of the manuscript.

Funding: This research is financially supported by National Key Research and Development Program Projects (No. 2022YFD1901305-4), Yunnan Fundamental Research Projects (202101AT070002), Wang Feng Expert Primary-level Workstation, Yunnan Province.

Institutional Review Board Statement: Not applicable.

Informed Consent Statement: Not applicable.

Data Availability Statement: Not applicable.

Conflicts of Interest: The authors declare no conflicts of interest.

References

1. Delgado-Baquerizo, M.; Guerra, C.A.; Cano-Díaz, C.; Egidi, E.; Wang, J.; Eisenhauer, N.; Singh, B.K.; Maestre, F.T. The Proportion of Soil-Borne Pathogens Increases with Warming at the Global Scale. *Nat. Clim. Chang.* **2020**, *10*, 550–554. [CrossRef]
2. Li, E.; Ling, J.; Wang, G.; Xiao, J.; Yang, Y.; Mao, Z.; Wang, X.; Xie, B. Comparative Proteomics Analyses of Two Races of *Fusarium oxysporum* f. sp. Conglutinans that Differ in Pathogenicity. *Sci. Rep.* **2015**, *5*, 13663. [CrossRef]
3. Narayanasamy, P. *Microbial Plant Pathogens-Detection and Disease Diagnosis*; Springer: Berlin/Heidelberg, Germany, 2011. [CrossRef]
4. Brefort, T.; Doehlemann, G.; Mendoza-Mendoza, A.; Reissmann, S.; Djamei, A.; Kahmann, R. Ustilago Maydis as a Pathogen. *Annu. Rev. Phytopathol.* **2009**, *47*, 423–445. [CrossRef]
5. Nourozian, J.; Etebarian, H.R.; Khodakaramian, G. Biological Control of *Fusarium graminearum* on Wheat by Antagonistic Bacteria. *Songklanakarin J. Sci. Technol.* **2006**, *28* (Suppl. S1), 29–38.
6. Tamura, T.; Shinzato, N.; Ito, M.; Ueno, M. Microbial Secondary Metabolite Induction of Abnormal Appressoria Formation Mediates Control of Rice Blast Disease Caused Bymagnaporthe Oryzae. *J. Phytopathol.* **2019**, *167*, 156–162. [CrossRef]
7. Karelov, A.; Kozub, N.; Sozinova, O.; Pirko, Y.; Sozinov, I.; Yemets, A.; Blume, Y. Wheat Genes Associated with Different Types of Resistance against Stem Rust (*Puccinia graminis* Pers.). *Pathogens* **2022**, *11*, 1157. [CrossRef]
8. Asibi, A.E.; Chai, Q.; Coulter, J.A. Rice Blast: A Disease with Implications for Global Food Security. *Agronomy* **2019**, *9*, 451. [CrossRef]
9. Su, J.; Zhao, J.; Zhao, S.; Li, M.; Pang, S.; Kang, Z.; Zhen, W.; Chen, S.; Chen, F.; Wang, X. Genetics of Resistance to Common Root Rot (Spot Blotch), *Fusarium* Crown Rot, and Sharp Eyespot in Wheat. *Front. Genet.* **2021**, *12*, 699342. [CrossRef]
10. Pennisi, E. Armed and Dangerous. *Science* **2010**, *327*, 804–805. [CrossRef]
11. Lobell, D.; Burke, M.; Tebaldi, C.; Mastrandrea, M.D.; Falcon, W.; Naylor, R.L. Prioritizing Climate Change Adaptation Needs for Food Security in 2030. *Science* **2008**, *319*, 607–610. [CrossRef]
12. Godfray, H.C.; Beddington, J.R.; Crute, I.R.; Haddad, L.; Lawrence, D.; Muir, J.F.; Pretty, J.; Robinson, S.; Thomas, S.M.; Toulmin, C. Food Security: The Challenge of Feeding 9 Billion People. *Science* **2010**, *327*, 812–818. [CrossRef] [PubMed]
13. Alexandratos, N.; Bruinsma, J. *World Agriculture Towards 2030/2050: The 2012 Revision*; FAO: Rome, Italy, 2012.
14. Challinor, A.J.; Watson, J.E.M.; Lobell, D.; Howden, S.M.; Smith, D.R.; Chhetri, N. A Meta-Analysis of Crop Yield under Climate Change and Adaptation. *Nat. Clim. Chang.* **2014**, *4*, 287–291. [CrossRef]
15. Hijmans, R.J. The Effect of Climate Change on Global Potato Production. *Am. J. Potato Res.* **2003**, *80*, 271–279. [CrossRef]

16. Reisinger, A.; Kitching, R.; Chiew, F.; Hughes, L.; Newton, P.; Schuster, S.; Tait, A.; Whetton, P.; Barnett, J.; Becken, S.; et . *Climate Change 2014: Impacts, Adaptation, and Vulnerability. Working Group II Contribution to the Fifth Assessment Report of the IPC Australasia Intergovernmental Panel on Climate Change*; Cambridge University Press: Cambridge, UK, 2014.
17. Rogelj, J.; den Elzen, M.; Höhne, N.; Fransen, T.; Fekete, H.; Winkler, H.; Schaeffer, R.; Sha, F.; Riahi, K.; Meinshausen, M. Pa Agreement Climate Proposals Need a Boost to Keep Warming Well below 2 °C. *Nature* **2016**, *534*, 631–639. [CrossRef] [PubMe
18. Singh, N.; Anand, G.; Kapoor, R. Incidence and Severity of Fungal Diseases of Safflower in India. *Crop Prot.* **2019**, *125*, 1049([CrossRef]
19. Salunkhe, V.N.; Gedam, P.; Pradhan, A.; Gaikwad, B.; Kale, R.; Gawande, S. Concurrent Waterlogging and Anthracnose-Twist Disease in Rainy-Season Onions (*Allium cepa*): Impact and Management. *Front. Microbiol.* **2022**, *13*, 1063472. [CrossRef] [PubMe
20. Romero, F.; Cazzato, S.; Walder, F.; Vogelgsang, S.; Bender, S.F.; van der Heijden, M.G.A. Humidity and High Temperature A Important for Predicting Fungal Disease Outbreaks Worldwide. *New Phytol.* **2022**, *234*, 1553–1556. [CrossRef] [PubMed]
21. Haddoudi, I.; Mhadhbi, H.; Gargouri, M.; Barhoumi, F.; Romdhane, S.B.; Mrabet, M. Occurrence of Fungal Diseases in Faba Bea (*Vicia faba* L.) under Salt and Drought Stress. *Eur. J. Plant Pathol.* **2021**, *159*, 385–398. [CrossRef]
22. Feulefack, J.; Khan, A.; Forastiere, F.; Sergi, C.M. Parental Pesticide Exposure and Childhood Brain Cancer: A Systematic Revie and Meta-Analysis Confirming the IARC/WHO Monographs on Some Organophosphate Insecticides and Herbicides. *Childr* **2021**, *8*, 1096. [CrossRef] [PubMed]
23. Al Naggar, Y.; Wubet, T. Chronic Exposure to Pesticides Disrupts the Bacterial and Fungal Co-Existence and the Cross-Kingdo Network Characteristics of Honey Bee Gut Microbiome. *Sci. Total Environ.* **2024**, *906*, 167530. [CrossRef] [PubMed]
24. Alptekin, B.; Langridge, P.; Budak, H. Abiotic Stress Mirnomes in the *Triticeae*. *Funct. Integr. Genom.* **2017**, *17*, 145–170. [CrossRe [PubMed]
25. Naing, A.H.; Kim, C. Abiotic Stress-Induced Anthocyanins in Plants: Their Role in Tolerance to Abiotic Stresses. *Physiol. Plar* **2021**, *172*, 1711–1723. [CrossRef] [PubMed]
26. Schreiner, R.P.; Tarara, J.M.; Smithyman, R.P. Deficit Irrigation Promotes Arbuscular Colonization of Fine Roots by Mycorrhiz Fungi in Grapevines (*Vitis vinifera* L.) In an Arid Climate. *Mycorrhiza* **2007**, *17*, 551–562. [CrossRef] [PubMed]
27. Chilakala, A.R.; Mali, K.V.; Irulappan, V.; Patil, B.S.; Pandey, P.; Rangappa, K.; Ramegowda, V.; Kumar, M.N.; Puli, C.O.F Mohan-Raju, B.; et al. Combined Drought and Heat Stress Influences the Root Water Relation and Determine the Dry Root R Disease Development Under Field Conditions: A Study Using Contrasting Chickpea Genotypes. *Front. Plant Sci.* **2022**, *13*, 89055 [CrossRef] [PubMed]
28. Bidzinski, P.; Ballini, E.; Ducasse, A.; Michel, C.; Zuluaga, P.; Genga, A.; Chiozzotto, R.; Morel, J.B. Transcriptional Basis c Drought-Induced Susceptibility to the Rice Blast Fungus Magnaporthe oryzae. *Front. Plant Sci.* **2016**, *7*, 1558. [CrossRef] [PubMec
29. Choi, H.-K.; Iandolino, A.; Silva, F.; Cook, D. Water Deficit Modulates the Response of *Vitis vinifera* to the Pierce's Diseas Pathogen *Xylella fastidiosa*. *Mol. Plant-Microbe Interact. MPMI* **2013**, *26*, 643–657. [CrossRef] [PubMed]
30. Jat, R.A.; Wani, S.P.; Sahrawat, K.L. Conservation Agriculture in the Semi-Arid Tropics: Prospects and Problems. *Adv. Agro* **2012**, *117*, 191–273.
31. Radzikowski, P.; Jończyk, K.; Feledyn-Szewczyk, B.; Jóźwicki, T. Assessment of Resistance of Different Varieties of Winter Whea to Leaf Fungal Diseases in Organic Farming. *Agriculture* **2023**, *13*, 875. [CrossRef]
32. Pandey, P.; Irulappan, V.; Bagavathiannan, M.V.; Senthil-Kumar, M. Impact of Combined Abiotic and Biotic Stresses on Plar Growth and Avenues for Crop Improvement by Exploiting Physio-Morphological Traits. *Front. Plant Sci.* **2017**, *8*, 537. [CrossRe [PubMed]
33. Shoaib, A.; Meraj, S.; Nafisa, N.; Khan, K.; Javaid, A. Influence of Salinity and Fusarium Oxysporum as the Stress Factors o Morpho-Physiological and Yield Attributes in Onion. *Physiol. Mol. Biol. Plants* **2018**, *24*, 1093–1101. [CrossRef] [PubMed]
34. Chojak-Koźniewska, J.; Linkiewicz, A.M.; Sowa, S.; Radzioch, M.A.; Kuźniak, E. Interactive Effects of Salt Stress and *Pseudomona syringae* pv. *lachrymans* Infection in Cucumber: Involvement of Antioxidant Enzymes, Abscisic acid and Salicylic Acid. *Enviro Exp. Bot.* **2017**, *136*, 9–20.
35. Al-Sadi, A.M.; Al-masoudi, R.S.; Al-Habsi, N.; Al-Said, F.A.-J.; Al-Rawahy, S.A.; Ahmed, M.; Deadman, M.L. Effect of Salinity o Pythium Damping-Off of Cucumber and on the Tolerance of *Pythium aphanidermatum*. *Plant Pathol.* **2010**, *59*, 112–120. [CrossRef
36. Allard, S.M.; Ottesen, A.R.; Micallef, S.A. Rain Induces Temporary Shifts in Epiphytic Bacterial Communities of Cucumber an Tomato Fruit. *Sci. Rep.* **2020**, *10*, 1765. [CrossRef] [PubMed]
37. Czarnecka, D.; Czubacka, A.; Agacka-Mołdoch, M.; Trojak-Goluch, A.; Księżak, J. The Occurrence of Fungal Diseases in Maize i Organic Farming Versus an Integrated Management System. *Agronomy* **2022**, *12*, 558. [CrossRef]
38. Chávez-Arias, C.C.; Gómez-Caro, S.; Restrepo-Díaz, H. Physiological, Biochemical and Chlorophyll Fluorescence Parameters o *Physalis peruviana* L. Seedlings Exposed to Different Short-Term Waterlogging Periods and Fusarium Wilt Infection. *Agronom* **2019**, *9*, 213. [CrossRef]
39. Iliff, G.J.; Mukherjee, R.; Gruszewski, H.A.; Schmale Iii, D.G.; Jung, S.; Boreyko, J.B. Phase-Change-Mediated Transport an Agglomeration of Fungal Spores on Wheat Awns. *J. R. Soc. Interface* **2022**, *19*, 20210872. [CrossRef] [PubMed]
40. Gopi, R.; Chandran, K.; Ramesh Sundar, A.; Nisha, M.; Mahendran, B.; Keerthana; Jayaraman, S.; Viswanathan, R. Occurrence o False Floral Smut in Sugarcane Inflorescence and Associated Weather Variables. *SugarTech* **2023**, *25*, 1411–1418. [CrossRef]
41. Nene, W.; Kapinga, F.; Shomari, S.; Assenga, B. Cashew Leaf and Nut Blight Disease Outbreaks under Unimodal Rainfall Patter in Tanzania. *J. Plant Pathol.* **2022**, *104*, 929–938. [CrossRef]

Khaliq, I.; Moore, K.; Sparks, A.H. The Relationship between Natural Rain Intensity and Ascochyta Blight in Chickpea Development. *Eur. J. Plant Pathol.* **2022**, *164*, 313–323. [CrossRef] [PubMed]
Rollano-Peñaloza, O.M.; Palma-Encinas, V.; Widell, S.; Mollinedo, P.; Rasmusson, A.G. The Disease Progression and Molecular Defense Response in Chenopodium Quinoa Infected with Peronospora Variabilis, the Causal Agent of Quinoa Downy Mildew. *Plants* **2022**, *11*, 2946. [CrossRef] [PubMed]
Barro, M.; Kassankogno, A.I.; Wonni, I.; Sérémé, D.; Somda, I.; Kaboré, H.K.; Béna, G.; Brugidou, C.; Tharreau, D.; Tollenaere, C. Spatiotemporal Survey of Multiple Rice Diseases in Irrigated Areas Compared to Rainfed Lowlands in the Western Burkina Faso. *Plant Dis.* **2021**, *105*, 3889–3899. [CrossRef] [PubMed]
Desaint, H.; Aoun, N.; Deslandes, L.; Vailleau, F.; Roux, F.; Berthomé, R. Fight Hard or Die Trying: When Plants Face Pathogens under Heat Stress. *New Phytol.* **2021**, *229*, 712–734. [CrossRef] [PubMed]
Ji, T.; Salotti, I.; Altieri, V.; Li, M.; Rossi, V. Temperature-Dependent Growth and Spore Germination of Fungi Causing Grapevine Trunk Diseases: Quantitative Analysis of Literature Data. *Plant Dis.* **2023**, *107*, 1386–1398. [CrossRef] [PubMed]
Songy, A.; Fernandez, O.; Clément, C.; Larignon, P.; Fontaine, F. Grapevine Trunk Diseases under Thermal and Water Stresses. *Planta* **2019**, *249*, 1655–1679. [CrossRef] [PubMed]
Kaur, S.; Barakat, R.; Kaur, J.; Epstein, L. The Effect of Temperature on Disease Severity and Growth of *Fusarium oxysporum* f. sp. Apii Races 2 and 4 in Celery. *Phytopathology* **2022**, *112*, 364–372. [CrossRef] [PubMed]
Yan, H.; Nelson, B.D. Effect of Temperature on *Fusarium solani* and *F. Tricinctum* Growth and Disease Development in Soybean. *Can. J. Plant Pathol.* **2020**, *42*, 527–537. [CrossRef]
Shakya, S.K.; Goss, E.M.; Dufault, N.S.; van Bruggen, A.H. Potential Effects of Diurnal Temperature Oscillations on Potato Late Blight with Special Reference to Climate Change. *Phytopathology* **2015**, *105*, 230–238. [CrossRef] [PubMed]
Shafiekhani, S.; Wilson, S.A.; Atungulu, G.G. Impacts of Storage Temperature and Rice Moisture Content on Color Characteristics of Rice from Fields with Different Disease Management Practices. *J. Stored Prod. Res.* **2018**, *78*, 89–97. [CrossRef]
Odebode, A.C.; Unachukwu, N.E. Effect of Storage Environment on Carrot Root Rots and Biochemical Changes During Storage. *Z. Für Leb. Und-Forsch. A* **1997**, *205*, 277–281. [CrossRef]
Casagrande, E.; Génard, M.; Lurol, S.; Charles, F.; Bevacqua, D.; Martinetti, D.; Lescourret, F. Brown Rot Disease in Stored Nectarines: Modeling the Combined Effects of Preharvest and Storage Conditions. *Phytopathology* **2022**, *112*, 1575–1583. [CrossRef] [PubMed]
Meshram, S.; Adhikari, T.B. Microbiome-Mediated Strategies to Manage Major Soil-Borne Diseases of Tomato. *Plants* **2024**, *13*, 364. [CrossRef] [PubMed]
Mendes, L.W.; Raaijmakers, J.M.; de Hollander, M.; Mendes, R.; Tsai, S.M. Influence of Resistance Breeding in Common Bean on Rhizosphere Microbiome Composition and Function. *ISME J.* **2018**, *12*, 212–224. [CrossRef] [PubMed]
Adolf, B.; Andrade-Piedra, J.; Bittara Molina, F.; Przetakiewicz, J.; Hausladen, H.; Kromann, P.; Lees, A.; Lindqvist-Kreuze, H.; Perez, W.; Secor, G.A. Fungal, Oomycete, and Plasmodiophorid Diseases of Potato. In *The Potato Crop: Its Agricultural, Nutritional and Social Contribution to Humankind*; Campos, H., Ortiz, O., Eds.; Springer International Publishing: Cham, Switzerland, 2020; pp. 307–350. [CrossRef]
Guan, Z.; Biswas, T.; Wu, F. The U.S. Tomato Industry: An Overview of Production and Trade. *EDIS* **2018**, *2018*. [CrossRef]
Panth, M.; Hassler, S.C.; Baysal-Gurel, F. Methods for Management of Soilborne Diseases in Crop Production. *Agriculture* **2020**, *10*, 16. [CrossRef]
Zhang, Y.; Ye, C.; Su, Y.; Peng, W.; Lu, R.; Liu, Y.; Huang, H.; He, X.; Yang, M.; Zhu, S. Soil Acidification Caused by Excessive Application of Nitrogen Fertilizer Aggravates Soil-Borne Diseases: Evidence from Literature Review and Field Trials. *Agric. Ecosyst. Environ.* **2022**, *340*, 108176. [CrossRef]
Schauberger, B.; Archontoulis, S.; Arneth, A.; Balkovic, J.; Ciais, P.; Deryng, D.; Elliott, J.; Folberth, C.; Khabarov, N.; Müller, C.; et al. Consistent Negative Response of US Crops to High Temperatures in Observations and Crop Models. *Nat. Commun.* **2017**, *8*, 13931. [CrossRef] [PubMed]
Cheng, W.; Lin, M.; Qiu, M.; Kong, L.; Xu, Y.; Li, Y.; Wang, Y.; Ye, W.; Dong, S.; He, S.; et al. Chitin Synthase Is Involved in Vegetative Growth, Asexual Reproduction and Pathogenesis of *Phytophthora capsici* and *Phytophthora sojae*. *Environ. Microbiol.* **2019**, *21*, 4537–4547. [CrossRef] [PubMed]
You, M.P.; Lamichhane, J.R.; Aubertot, J.-N.; Barbetti, M.J. Understanding Why Effective Fungicides Against Individual Soilborne Pathogens Are Ineffective with Soilborne Pathogen Complexes. *Plant Dis.* **2019**, *104*, 904–920. [CrossRef] [PubMed]
Barbetti, M.J.; Khan, T.N.; Pritchard, I.; Lamichhane, J.R.; Aubertot, J.N.; Corrales, D.C.; You, M.P. Challenges With Managing Disease Complexes During Application of Different Measures Against Foliar Diseases of Field Pea. *Plant Dis.* **2021**, *105*, 616–627. [CrossRef] [PubMed]
Back, M.; Haydock, P.; Jenkinson, P. Disease Complexes Involving Plant Parasitic Nematodes and Soilborne Pathogens. *Plant Pathol.* **2002**, *51*, 683–697. [CrossRef]
Ahmadi, M.; Mirakhorli, N.; Erginbas-Orakci, G.; Ansari, O.; Braun, H.-J.; Paulitz, T.C.; Dababat, A. Interactions among Cereal cyst Nematode *Heterodera filipjevi*, Dryland crown rot *Fusarium culmorum*, and Drought on Grain Yield Components and Disease Severity in Bread wheat. *Can. J. Plant Pathol.* **2021**, *44*, 415–431. [CrossRef]

66. You, M.P.; Rensing, K.; Renton, M.; Barbetti, M.J. Modeling Effects of Temperature, Soil, Moisture, Nutrition and Variety Determinants of Severity of Pythium Damping-Off and Root Disease in Subterranean Clover. *Front. Microbiol.* **2017**, *8*, 2 [CrossRef] [PubMed]
67. Gibert, S.; Edel-Hermann, V.; Gautheron, E.; Gautheron, N.; Bernaud, E.; Sol, J.M.; Capelle, G.P.J.; Galland, R.; Bardon-Debats, Lambert, C.; et al. Identification, Pathogenicity and Community Dynamics of Fungi and Oomycetes Associated with Pea R Rot in Northern France. *Plant Pathol.* **2022**, *71*, 1550–1569. [CrossRef]
68. Haas, D.; Ilieva, M.; Fritz, T.; Galler, H.; Habib, J.; Kriso, A.; Kropsch, M.; Ofner-Kopeinig, P.; Reinthaler, F.F.; Strasser, A.; e Background Concentrations of Airborne, Culturable Fungi and Dust Particles in Urban, Rural and Mountain Regions. *Sci. T Environ.* **2023**, *892*, 164700. [CrossRef] [PubMed]
69. Macleod, A.; Pautasso, M.; Jeger, M.; Haines-Young, R. Evolution of the International Regulation of Plant Pest and Challenges Future Plant Health. *Food Secur.* **2010**, *2*, 49–70. [CrossRef]
70. Wellings, C.; Mcintosh, R.; Walker, J. *Puccinia striiformis* f.sp. Tritici in Eastern Australia Possible Means of Entry and Implicati for Plant Quarantine. *Plant Pathol.* **2007**, *36*, 239–241. [CrossRef]
71. Ding, Y.; Cuddy, W.S.; Wellings, C.R.; Zhang, P.; Thach, T.; Hovmøller, M.S.; Qutob, D.; Brar, G.S.; Kutcher, H.R.; Park, Incursions of Divergent Genotypes, Evolution of Virulence and Host Jumps Shape a Continental Clonal Population of the Str Rust Pathogen *Puccinia striiformis*. *Mol. Ecol.* **2021**, *30*, 6566–6584. [CrossRef] [PubMed]
72. Visser, B.; Meyer, M.; Park, R.F.; Gilligan, C.A.; Burgin, L.E.; Hort, M.C.; Hodson, D.P.; Pretorius, Z.A. Microsatellite Analysis a Urediniospore Dispersal Simulations Support the Movement of *Puccinia graminis* f. sp. Tritici from Southern Africa to Austr *Phytopathology* **2019**, *109*, 133–144. [CrossRef] [PubMed]
73. Sharma-Poudyal, D.; Chen, X.M.; Wan, A.M.; Zhan, G.M.; Kang, Z.S.; Cao, S.Q.; Jin, S.L.; Morgounov, A.; Akin, B.; Mert, Z.; e Virulence Characterization of International Collections of the Wheat Stripe Rust Pathogen, *Puccinia striiformis* f. sp. Tritici. *P Dis* **2013**, *97*, 379–386. [CrossRef] [PubMed]
74. Ali, S.; Gladieux, P.; Leconte, M.; Gautier, A.; Justesen, A.F.; Hovmøller, M.S.; Enjalbert, J.; de Vallavieille-Pope, C. Ori Migration Routes and Worldwide Population Genetic Structure of the Wheat Yellow Rust Pathogen *Puccinia striiformis* f. Tritici. *PLoS Pathog.* **2014**, *10*, e1003903. [CrossRef] [PubMed]
75. Picornell, A.; Rojo, J.; Trigo, M.M.; Ruiz-Mata, R.; Lara, B.; Romero-Morte, J.; Serrano-García, A.; Pérez-Badia, R.; Gutiér Bustillo, M.; Cervigón-Morales, P.; et al. Environmental Drivers of the Seasonal Exposure to Airborne *Alternaria spores* in Sp *Sci. Total Environ.* **2022**, *823*, 153596. [CrossRef] [PubMed]
76. Sindt, C.; Besancenot, J.-P.; Thibaudon, M. Airborne *Cladosporium* Fungal Spores and Climate Change in France. *Aerobiologia* 2(*32*, 53–68. [CrossRef]
77. Estève, R.S.; Baisnée, D.; Guinot, B.; Sodeau, J.; O'Connor, D.J.; Belmonte, J.; Besancenot, J.-P.; Petit, J.-E.; Thibaudon, M.; Oli G.; et al. Variability and Geographical Origin of Five Years Airborne Fungal Spore Concentrations Measured at Saclay, Fra from 2014 to 2018. *Remote Sens.* **2019**, *11*, 1671. [CrossRef]
78. Urban, J.; Lebeda, A. Variation of Fungicide Resistance in Czech Populations of *Pseudoperonospora cubensis*. *J. Phytopathol.* 2(*155*, 143–151. [CrossRef]
79. Zhanbin, S.; Yu, S.; Hu, Y.; Wen, Y. Biological Control of the Cucumber Downy Mildew Pathogen *Pseudoperonospora cuber Horticulturae* **2022**, *8*, 410. [CrossRef]
80. Lamichhane, J.R.; Bischoff-Schaefer, M.; Bluemel, S.; Dachbrodt-Saaydeh, S.; Dreux, L.; Jansen, J.P.; Kiss, J.; Köhl, J.; Kudsk Malausa, T.; et al. Identifying Obstacles and Ranking Common Biological Control Research Priorities for Europe to Manage M Economically Important Pests in Arable, Vegetable and Perennial Crops. *Pest Manag. Sci.* **2017**, *73*, 14–21. [CrossRef] [PubM
81. Bowen, J.K.; Mesarich, C.H.; Bus, V.G.; Beresford, R.M.; Plummer, K.M.; Templeton, M.D. Venturia Inaequalis: The Causal Ag of Apple Scab. *Mol. Plant Pathol.* **2011**, *12*, 105–122. [CrossRef] [PubMed]
82. González-Domínguez, E.; Armengol, J.; Rossi, V. Biology and Epidemiology of Venturia Species Affecting Fruit Crops: A Revi *Front. Plant Sci.* **2017**, *8*, 1496. [CrossRef]
83. Jha, G.; Thakur, K.; Thakur, P. The Venturia Apple Pathosystem: Pathogenicity Mechanisms and Plant Defense Responses *Biomed. Biotechnol.* **2009**, *2009*, 680160. [CrossRef] [PubMed]
84. Abuley, I.K.; Nielsen, B.J. Evaluation of Models to Control Potato Early Blight (*Alternaria solani*) in Denmark. *Crop Prot.* **2017**, 1 118–128. [CrossRef]
85. Meno, L.; Escuredo, O.; Rodríguez-Flores, M.S.; Seijo, M.C. Modification of the TOMCAST Model with Aerobiological Data Management of Potato Early Blight. *Agronomy* **2020**, *10*, 1872. [CrossRef]
86. Singh, V.; Shrivastava, A.; Jadon, S.; Wahi, N.; Singh, A.; Sharma, N. Alternaria Diseases of Vegetable Crops and Its Managem Control to Reduce the Low Production. *Int. J. Agric. Biol.* **2015**, *7*, 834.
87. Campos, H.; Ortiz, O. The Potato Crop: Its Agricultural, Nutritional and Social Contribution to Humankind. In *The Potato Cr* Springer: Berlin/Heidelberg, Germany, 2019.
88. Escuredo, O.; Seijo-Rodríguez, A.; Meno, L.; Rodríguez-Flores, M.S.; Seijo, M.C. Seasonal Dynamics of Alternaria during t Potato Growing Cycle and the Influence of Weather on the Early Blight Disease in North-West Spain. *Am. J. Potato Res.* **2019**, 532–540. [CrossRef]
89. Sarita; Buts, A.K.; Singh, R. Seed Borne Mycoflora of Mung Bean (*Phaseolus aureus* Roxb.) and Its Control by Fungicides. *A Appl. Sci. Res.* **2014**, *5*, 8–10.

91. Howell, C.R. Cotton Seedling Preemergence Damping-Off Incited by *Rhizopus oryzae* and *Pythium* spp. and Its Biological Control with *Trichoderma* spp. *Phytopathology* **2002**, *92*, 177–180. [CrossRef] [PubMed]
92. Pieczul, K.; Perek, A.; Kubiak, K. Detection of *Tilletia caries*, *Tilletia laevis* and *Tilletia controversa* Wheat Grain Contamination Using Loop-Mediated Isothermal DNA Amplification (LAMP). *J. Microbiol. Methods* **2018**, *154*, 141–146. [CrossRef] [PubMed]
93. Mkandawire, L.; Mhango, W.; Saka, V.W.; Juma, S.; Goodman, J.; Brandenburg, R.L.; Jordan, D.L. Influence of Plant Population and Harvest Date on Peanut (*Arachis hypogaea*) Yield and Aflatoxin Contamination. *Peanut Sci.* **2021**, *48*, 33–39. [CrossRef]
94. Pramunadipta, S.; Widiastuti, A.; Wibowo, A.; Priyatmojo, A. Rep-PCR Analysis of *Fusarium proliferatum* Causing Sheath Rot Disease and Its Relationship to Light, PH, Temperature and Rice Varieties. *Arch. Phytopathol. Plant Prot.* **2022**, *55*, 973–990. [CrossRef]
95. Rao, P.J.M.; Pallavi, M.; Bharathi, Y.; Priya, P.B.; Sujatha, P.; Prabhavathi, K. Insights into Mechanisms of Seed Longevity in Soybean: A Review. *Front. Plant Sci.* **2023**, *14*, 1206318. [CrossRef] [PubMed]
96. Farber, C.; Mahnke, M.; Sanchez, L.; Kurouski, D. Advanced Spectroscopic Techniques for Plant Disease Diagnostics. A Review. *TrAC Trends Anal. Chem.* **2019**, *118*, 43–49. [CrossRef]
97. Genaev, M.A.; Skolotneva, E.S.; Gultyaeva, E.I.; Orlova, E.A.; Bechtold, N.P.; Afonnikov, D.A. Image-Based Wheat Fungi Diseases Identification by Deep Learning. *Plants* **2021**, *10*, 1500. [CrossRef] [PubMed]
98. Kindo, A.J.; Tupaki-Sreepurna, A.; Yuvaraj, M. Banana Peel Culture as an Indigenous Medium for Easy Identification of Late-Sporulation Human Fungal Pathogens. *Indian J. Med. Microbiol.* **2016**, *34*, 457–461. [CrossRef] [PubMed]
99. Khater, M.; de la Escosura-Muñiz, A.; Merkoçi, A. Biosensors for Plant Pathogen Detection. *Biosens. Bioelectron.* **2017**, *93*, 72–86. [CrossRef] [PubMed]
100. Farber, C.; Bryan, R.; Paetzold, L.; Rush, C.; Kurouski, D. Non-Invasive Characterization of Single-, Double- and Triple-Viral Diseases of Wheat with a Hand-Held Raman Spectrometer. *Front. Plant Sci.* **2020**, *11*, 01300. [CrossRef] [PubMed]
101. Dyussembayev, K.; Sambasivam, P.; Bar, I.; Brownlie, J.C.; Shiddiky, M.J.A.; Ford, R. Biosensor Technologies for Early Detection and Quantification of Plant Pathogens. *Front. Chem.* **2021**, *9*, 636245. [CrossRef] [PubMed]
102. Abu-Khalaf, N.; Salman, M. Detecting Plant Diseases Using Visible/Near Infrared Spectroscopy. *NIR News* **2013**, *24*, 12–25. [CrossRef]
103. Larsolle, A.; Hamid Muhammed, H. Measuring Crop Status Using Multivariate Analysis of Hyperspectral Field Reflectance with Application to Disease Severity and Plant Density. *Precis. Agric.* **2007**, *8*, 37–47. [CrossRef]
104. Thévenot, D.R.; Toth, K.; Durst, R.A.; Wilson, G.S. Electrochemical Biosensors: Recommended Definitions and Classification. *Biosens. Bioelectron.* **2001**, *16*, 121–131. [CrossRef]
105. Neethirajan, S.; Ragavan, K.V.; Weng, X. Agro-Defense: Biosensors for Food from Healthy Crops and Animals. *Trends Food Sci. Technol.* **2018**, *73*, 25–44. [CrossRef]
106. Wilson, A.D. Noninvasive Early Disease Diagnosis by Electronic-Nose and Related VOC-Detection Devices. *Biosensors* **2020**, *10*, 73. [CrossRef] [PubMed]
107. Li, X.; Guo, L.; Guo, M.; Qi, D.; Zhou, X.; Li, F.; Wu, J. Three Highly Sensitive Monoclonal Antibody-Based Serological Assays for the Detection of Tomato Mottle Mosaic Virus. *Phytopathol. Res.* **2021**, *3*, 23. [CrossRef]
108. Gan, Z.; Zhou, Q.a.; Zheng, C.; Wang, J. Challenges and Applications of Volatile Organic Compounds Monitoring Technology in Plant Disease Diagnosis. *Biosens. Bioelectron.* **2023**, *237*, 115540. [CrossRef] [PubMed]
109. Qin, J.; Wang, B.; Wu, Y.; Lu, Q.; Zhu, H. Identifying Pine Wood Nematode Disease Using UAV Images and Deep Learning Algorithms. *Remote Sens.* **2021**, *13*, 162. [CrossRef]
110. Cui, S.; Ling, P.; Zhu, H.; Keener, H.M. Plant Pest Detection Using an Artificial Nose System: A Review. *Sensors* **2018**, *18*, 378. [CrossRef] [PubMed]
111. Martinelli, F.; Scalenghe, R.; Davino, S.; Panno, S.; Scuderi, G.; Ruisi, P.; Villa, P.; Stroppiana, D.; Boschetti, M.; Goulart, L.R.; et al. Advanced methods of plant disease detection. A review. *Agron. Sustain. Dev.* **2014**, *35*, 1–25. [CrossRef]
112. Ge, Y.; Thomasson, J.; Sui, R. Remote Sensing of Soil Properties in Precision Agriculture: A Review. *Front. Earth Sci.* **2011**, *5*, 229–238. [CrossRef]
113. Shi, Y.; Thomasson, J.A.; Murray, S.C.; Pugh, N.A.; Rooney, W.L.; Shafian, S.; Rajan, N.; Rouze, G.; Morgan, C.L.; Neely, H.L.; et al. Unmanned Aerial Vehicles for High-Throughput Phenotyping and Agronomic Research. *PLoS ONE* **2016**, *11*, e0159781. [CrossRef] [PubMed]
114. Herrmann, I.; Bdolach, E.; Montekyo, Y.; Rachmilevitch, S.; Townsend, P.; Karnieli, A. Assessment of Maize Yield and Phenology by Drone-Mounted Superspectral Camera. *Precis. Agric.* **2020**, *21*, 51–76. [CrossRef]
115. Bauriegel, E.; Herppich, W.B. Hyperspectral and Chlorophyll Fluorescence Imaging for Early Detection of Plant Diseases, with Special Reference to Fusarium Spec. Infections on Wheat. *Agriculture* **2014**, *4*, 32–57. [CrossRef]
116. Kuska, M.; Wahabzada, M.; Leucker, M.; Dehne, H.W.; Kersting, K.; Oerke, E.C.; Steiner, U.; Mahlein, A.K. Hyperspectral Phenotyping on the Microscopic Scale: Towards Automated Characterization of Plant-Pathogen Interactions. *Plant Methods* **2015**, *11*, 28. [CrossRef] [PubMed]
117. Zhang, D.; Zhou, X.; Zhang, J.; Lan, Y.; Xu, C.; Liang, D. Detection of Rice Sheath Blight Using an Unmanned Aerial System with High-Resolution Color and Multispectral Imaging. *PLoS ONE* **2018**, *13*, e0187470. [CrossRef]
118. Sun, Q.; Sun, L.; Shu, M.; Gu, X.; Yang, G.; Zhou, L. Monitoring Maize Lodging Grades via Unmanned Aerial Vehicle Multispectral Image. *Plant Phenomics* **2019**, *2019*, 5704154. [CrossRef]

118. Khot, L.R.; Sankaran, S.; Carter, A.H.; Johnson, D.A.; Cummings, T.F. UAS Imaging-Based Decision Tools for Arid Winter Whe and Irrigated Potato Production Management. *Int. J. Remote Sens.* **2016**, *37*, 125–137. [CrossRef]
119. Guo, A.; Huang, W.; Dong, Y.; Ye, H.; Ma, H.; Liu, B.; Wu, W.; Ren, Y.; Ruan, C.; Geng, Y. Wheat Yellow Rust Detection Usir UAV-Based Hyperspectral Technology. *Remote Sens.* **2021**, *13*, 123. [CrossRef]
120. Shahi, T.; Xu, C.-Y.; Neupane, A.; Guo, W. Recent Advances in Crop Disease Detection Using UAV and Deep Learning Techniqu *Remote Sens.* **2023**, *15*, 2450. [CrossRef]
121. Franceschini, M.H.D.; Bartholomeus, H.M.; Apeldoorn, D.v.; Suomalainen, J.M.; Kooistra, L. Assessing Changes in Potato Canop Caused by Late Blight in Organic Production Systems through UAV-Based Pushbroom Imaging Spectrometer. *ISPRS-Int. Arc Photogramm. Remote Sens. Spat. Inf. Sci.* **2017**, *42*, 109–112. [CrossRef]
122. Yamamoto, S.; Nomoto, S.; Hashimoto, N.; Maki, M.; Hongo, C.; Tatsuhiko, S.; Homma, K. Monitoring Spatial and Time-Seri Variations in Red Crown Rot Damage of Soybean in Farmer Fields Based on UAV Remote Sensing. *Plant Prod. Sci.* **2023**, *26*, 36–4 [CrossRef]
123. Abdulridha, J.N.; Ampatzidis, Y.; Kakarla, S.C.; Roberts, P.D. Detection of Target Spot and Bacterial Spot Diseases in Tomat Using UAV-Based and Benchtop-Based Hyperspectral Imaging Techniques. *Precis. Agric.* **2019**, *21*, 955–978. [CrossRef]
124. Tallapragada, P.; Ross, S.D.; Schmale, D.G. Lagrangian Coherent Structures Are Associated with Fluctuations in Airborn Microbial Populations. *Chaos* **2011**, *21*, 033122. [CrossRef] [PubMed]
125. Zhao, S.; Liu, M.; Tao, M.; Zhou, W.; Lu, X.; Xiong, Y.; Li, F.; Wang, Q. The Role of Satellite Remote Sensing in Mitigating an Adapting to Global Climate Change. *Sci. Total Environ.* **2023**, *904*, 166820. [CrossRef]
126. Wang, J.; Long, X.; Chern, M.; Chen, X. Understanding the Molecular Mechanisms of Trade-Offs between Plant Growth an Immunity. *Sci. China Life Sci.* **2021**, *64*, 234–241. [CrossRef] [PubMed]
127. Plett, J.M.; Plett, K.L.; Bithell, S.L.; Mitchell, C.; Moore, K.; Powell, J.R.; Anderson, I.C. Improved *Phytophthora* Resistance i Commercial Chickpea (*Cicer arietinum*) Varieties Negatively Impacts Symbiotic Gene Signalling and Symbiotic Potential in Som Varieties. *Plant Cell Environ.* **2016**, *39*, 1858–1869. [CrossRef] [PubMed]
128. Li, Q.; Wang, B.; Yu, J.; Dou, D. Pathogen-Informed Breeding for Crop Disease Resistance. *J. Integr. Plant Biol.* **2021**, *63*, 305–31 [CrossRef] [PubMed]
129. Hui, C.; Jiang, H.; Liu, B.; Wei, R.; Zhang, Y.; Zhang, Q.; Liang, Y.; Zhao, Y. Chitin Degradation and the Temporary Response c Bacterial Chitinolytic Communities to Chitin Amendment in Soil under Different Fertilization Regimes. *Sci. Total Environ.* **202** *705*, 136003. [CrossRef]
130. Shen, Z.; Xue, C.; Penton, C.R.; Thomashow, L.S.; Zhang, N.; Wang, B.; Ruan, Y.; Li, R.; Li, R.; Shen, Q. Suppression of Banan Panama Disease Induced by Soil Microbiome Reconstruction through an Integrated Agricultural Strategy. *Soil Biol. Biochem.* **201** *128*, 164–174. [CrossRef]
131. Troy, T.; Kipgen, C.; Pal, I. The Impact of Climate Extremes and Irrigation on US Crop Yields. *Environ. Res. Lett.* **2015**, *10*, 05401 [CrossRef]
132. Wheeler, T.A.; Bordovsky, J.P.; Keeling, J.W. The Effectiveness of Crop Rotation on Management of Verticillium Wilt over Tim *Crop Prot.* **2019**, *121*, 157–162. [CrossRef]
133. Kpu, K.A.; Annih, M.G.; Ambang, A.L.; Agwah, E.D. Influence of Tillage Systems and Sowing Dates on the Incidence of Lea Spot Disease in *Telfairia occidentalis* Caused by *Phoma sorghina* in Cameroon. *Sci. Rep.* **2022**, *12*, 19790. [CrossRef] [PubMed]
134. Shukla, M.; Sadhu, A.; Chinchmalatpure, A.; Prasad, I.; Kumar, S.; Camus, D. Fertigation-Modern Technique of Fertilize Application. *Indian Farmer* **2018**, *5*, 1062–1071.
135. Breda, C.C.; Soares, M.B.; Tavanti, R.F.R.; Viana, D.G.; Freddi, O.d.S.; Piedade, A.R.; Mahl, D.; Traballi, R.C.; Guerrini, I.A Successive Sewage Sludge Fertilization: Recycling for Sustainable Agriculture. *Waste Manag.* **2020**, *109*, 38–50. [CrossRef]
136. Saccon, P. Water for Agriculture, Irrigation Management. *Appl. Soil Ecol.* **2018**, *123*, 793–796. [CrossRef]
137. Collinge, D.B.; Jørgensen, H.J.L.; Latz, M.A.C.; Manzotti, A.; Ntana, F.; Rojas, E.C.; Jensen, B. Searching for Novel Funga Biological Control Agents for Plant Disease Control Among Endophytes. In *Endophytes for a Growing World*; Cambridge Universit Press: Cambridge, UK, 2019.
138. Lamichhane, J.R.; Dachbrodt-Saaydeh, S.; Kudsk, P.; Messéan, A. Toward a Reduced Reliance on Conventional Pesticides i European Agriculture. *Plant Dis.* **2016**, *100*, 10–24. [CrossRef] [PubMed]
139. Strange, R.N.; Scott, P.R. Plant Disease: A Threat to Global Food Security. *Annu. Rev. Phytopathol.* **2005**, *43*, 83–116. [CrossRef]
140. Cooper, J.; Dobson, H.M. The Benefits of Pesticides to Mankind and the Environment. *Crop Prot.* **2007**, *26*, 1337–1348. [CrossRef
141. Kawasaki, K.; Lichtenberg, E. Quality Versus Quantity Effects of Pesticides: Joint Estimation of Quality Grade and Crop Yield. In Proceedings of the 2015 AAEA & WAEA Joint Annual Meeting, San Francisco, CA, USA, 26–28 July 2015.
142. Carvalho, F. Agriculture, Pesticides, Food Security and Food Safety. *Environ. Sci. Policy* **2006**, *9*, 685–692. [CrossRef]
143. Damalas, C.A.; Eleftherohorinos, I.G. Pesticide Exposure, Safety Issues, and Risk Assessment Indicators. *Int. J. Environ. Re Public Health* **2011**, *8*, 1402–1419. [CrossRef] [PubMed]
144. Kim, K.H.; Kabir, E.; Jahan, S.A. Exposure to Pesticides and the Associated Human Health Effects. *Sci. Total Environ.* **2017**, *57* 525–535. [CrossRef] [PubMed]
145. Li, Z.; Sun, J.; Zhu, L. Organophosphorus Pesticides in Greenhouse and Open-Field Soils across China: Distribution Characteristi Polluted Pathway and Health Risk. *Sci. Total Environ.* **2021**, *765*, 142757. [CrossRef] [PubMed]

Carro-Huerga, G.; Mayo-Prieto, S.; Rodríguez-González, Á.; González-López, Ó.; Gutiérrez, S.; Casquero, P.A. Influence of Fungicide Application and Vine Age on Trichoderma Diversity as Source of Biological Control Agents. *Agronomy* **2021**, *11*, 446. [CrossRef]

Ren, B.; Zhao, T.; Li, Y.; Liang, H.; Zhao, Y.; Chen, H.; Li, L.; Liang, H. Enantioselective Bioaccumulation and Toxicity of the Novel Chiral Antifungal Agrochemical Penthiopyrad in Zebrafish (*Danio rerio*). *Ecotoxicol. Environ. Saf.* **2021**, *228*, 113010. [CrossRef] [PubMed]

Borel, B. When the Pesticides Run Out. *Nature* **2017**, *543*, 302–304. [CrossRef] [PubMed]

Bass, C.; Denholm, I.; Williamson, M.S.; Nauen, R. The Global Status of Insect Resistance to Neonicotinoid Insecticides. *Pestic. Biochem. Physiol.* **2015**, *121*, 78–87. [CrossRef] [PubMed]

Price, C.L.; Parker, J.E.; Warrilow, A.; Kelly, D.E.; Kelly, S.L. Azole Fungicides—Understanding Resistance Mechanisms in Agricultural Fungal Pathogens. *Pest Manag. Sci.* **2015**, *71*, 1054–1058. [CrossRef] [PubMed]

Li, L.; Xu, Z.; Kah, M.; Lin, D.; Filser, J. Nanopesticides: A Comprehensive Assessment of Environmental Risk Is Needed before Widespread Agricultural Application. *Environ. Sci. Technol.* **2019**, *53*, 7923–7924. [CrossRef] [PubMed]

Wang, M.; Hu, Z.; Yang, T.; Pei, H.; Zhang, F. A Dual Pesticide–Fertilizer Silicon-Base Nanocomposite to Synergistically Control Fungal Disease and Provide Nutrition. *Environ. Sci. Nano* **2023**, *10*, 3462–3475. [CrossRef]

Stenberg, J.A.; Sundh, I.; Becher, P.G.; Björkman, C.; Dubey, M.; Egan, P.A.; Friberg, H.; Gil, J.F.; Jensen, D.F.; Jonsson, M.; et al. When Is It Biological Control? A Framework of Definitions, Mechanisms, and Classifications. *J. Pest Sci.* **2021**, *94*, 665–676. [CrossRef]

Veras, F.F.; Silveira, R.D.; Welke, J.E. Bacillus spp. As a Strategy to Control Fungi and Mycotoxins in Food. *Curr. Opin. Food Sci.* **2023**, *52*, 101068. [CrossRef]

Glick, B.R. Bacteria with ACC Deaminase Can Promote Plant Growth and Help to Feed the World. *Microbiol. Res.* **2014**, *169*, 30–39. [CrossRef] [PubMed]

Dong, W.; Long, T.; Ma, J.; Wu, N.; Mo, W.; Zhou, Z.; Jin, J.; Zhou, H.; Ding, H. Effects of *Bacillus velezensis* GUAL210 Control on Edible Rose Black Spot Disease and Soil Fungal Community Structure. *Front. Microbiol.* **2023**, *14*, 1199024. [CrossRef] [PubMed]

Nuangmek, W.; Aiduang, W.; Kumla, J.; Lumyong, S.; Suwannarach, N. Evaluation of a Newly Identified Endophytic Fungus, *Trichoderma phayaoense* for Plant Growth Promotion and Biological Control of Gummy Stem Blight and Wilt of Muskmelon. *Front. Microbiol.* **2021**, *12*, 634772. [CrossRef] [PubMed]

Bhavani, P.; Suganya, K.; Parwin Banu, K.S.; Ramalakshmi, A.; Nalina, L. Effects of Root Exudates of Vetiver on Physicochemical Properties and Fractionation of Heavy Metals in Tannery Effluent Contaminated Soil. *Ecol. Environ. Conserv.* **2023**, *29*, 164–169. [CrossRef]

Berendsen, R.L.; Pieterse, C.M.J.; Bakker, P.A.H.M. The Rhizosphere Microbiome and Plant Health. *Trends Plant Sci.* **2012**, *17*, 478–486. [CrossRef] [PubMed]

Hasegawa, S.; Ryan, M.H.; Power, S.A. CO_2 Concentration and Water Availability Alter the Organic Acid Composition of Root Exudates in Native Australian Species. *Plant Soil* **2023**, *485*, 507–524. [CrossRef]

Wu, J.; Yu, S. Effect of Root Exudates of *Eucalyptus urophylla* and *Acacia mearnsii* on Soil Microbes under Simulated Warming Climate Conditions. *BMC Microbiol.* **2019**, *19*, 224. [CrossRef]

Climate Warming Threatens Soil Microbial Diversity. *Nat. Microbiol.* **2022**, *7*, 935–936. [CrossRef] [PubMed]

Li, X.; Dong, J.; Chu, W.; Chen, Y.; Duan, Z. The Relationship between Root Exudation Properties and Root Morphological Traits of Cucumber Grown under Different Nitrogen Supplies and Atmospheric CO_2 Concentrations. *Plant Soil* **2018**, *425*, 415–432. [CrossRef]

Vicente, E.J.; Dean, D.R. Keeping the Nitrogen-Fixation Dream Alive. *Proc. Natl. Acad. Sci. USA* **2017**, *114*, 3009–3011. [CrossRef] [PubMed]

Shah, A.; Nazari, M.; Antar, M.S.M.; Msimbira, L.A.; Naamala, J.; Lyu, D.; Rabileh, M.A.; Zajonc, J.; Smith, D.L. PGPR in Agriculture: A Sustainable Approach to Increasing Climate Change Resilience. *Front. Sustain. Food Syst.* **2021**, *5*, 667546. [CrossRef]

Zytynska, S.E.; Eicher, M.; Rothballer, M.; Weisser, W.W. Microbial-Mediated Plant Growth Promotion and Pest Suppression Varies Under Climate Change. *Front. Plant Sci.* **2020**, *11*, 573578. [CrossRef] [PubMed]

Berg, G.; Rybakova, D.; Fischer, D.; Cernava, T.; Vergès, M.C.; Charles, T.; Chen, X.; Cocolin, L.; Eversole, K.; Corral, G.H.; et al. Microbiome Definition Re-Visited: Old Concepts and New Challenges. *Microbiome* **2020**, *8*, 103. [CrossRef]

Zhu, S.; Vivanco, J.M.; Manter, D.K. Nitrogen Fertilizer Rate Affects Root Exudation, the Rhizosphere Microbiome and Nitrogen-Use-Efficiency of Maize. *Appl. Soil Ecol.* **2016**, *107*, 324–333. [CrossRef]

Zhou, X.; Zhang, J.; Khashi u Rahman, M.; Gao, D.; Wei, Z.; Wu, F.; Dini-Andreote, F. Interspecific Plant Interaction via Root Exudates Structures the Disease Suppressiveness of Rhizosphere Microbiomes. *Mol. Plant* **2023**, *16*, 849–864. [CrossRef] [PubMed]

Zhou, Y.; Liu, Y.; Li, S.; Yang, Q. The Combination of Biochar and *Bacillus subtilis* Biological Agent Reduced the Relative Abundance of Pathogenic Bacteria in the Rhizosphere Soil of Panax notoginseng. *Microorganisms* **2024**, *12*, 783. [CrossRef] [PubMed]

Chen, W.; Wu, Z.; Liu, C.; Zhang, Z.; Liu, X. Biochar Combined with *Bacillus subtilis* SL-44 as an Eco-Friendly Strategy to Improve Soil Fertility, Reduce Fusarium Wilt, and Promote Radish Growth. *Ecotoxicol. Environ. Saf.* **2023**, *251*, 114509. [CrossRef] [PubMed]

172. Xiong, W.; Reynolds, M.; Xu, Y. Climate Change Challenges Plant Breeding. *Curr. Opin. Plant Biol.* **2022**, *70*, 102308. [Cross [PubMed]
173. Samada, L.H.; Tambunan, U.S.F. Biopesticides as Promising Alternatives to Chemical Pesticides: A Review of Their Current Future Status. *J. Biol. Sci.* **2020**, *20*, 66–76. [CrossRef]
174. Chandrika, K.S.V.P.; Prasad, R.D.; Godbole, V. Development of Chitosan-PEG Blended Films using *Trichoderma*: Enhancemen Antimicrobial Activity and Seed Quality. *Int. J. Biol. Macromol.* **2019**, *126*, 282–290. [CrossRef] [PubMed]

agriculture

Article

Influence of Nitrogen Fertilisation Level and Weather Conditions on Yield and Quantitative Profile of Anti-Nutritional Compounds in Grain of Selected Rye Cultivars

Alicja Sułek [1], Grażyna Cacak-Pietrzak [2], Marcin Studnicki [3], Jerzy Grabiński [1], Anna Nieróbca [4], Marta Wyzińska [1] and Marcin Różewicz [1,*]

1 Department of Cereal Crop Production, Institute of Soil Science and Plant Cultivation—State Research Institute, 8 Czartoryskich Street, 24-100 Pulawy, Poland; sulek@iung.pulawy.pl (A.S.); jurek@iung.pulawy.pl (J.G.); mwyzinska@iung.pulawy.pl (M.W.)

2 Department of Food Technology and Assessment, Institute of Food Sciences, Warsaw University of Life Science, 159C Nowoursynowska Street, 02-787 Warsaw, Poland; grazyna_cacak_pietrzak@sggw.edu.pl

3 Department of Biometry, Institute of Agriculture, Warsaw University of Life Science, 159 Nowoursynowska Street, 02-776 Warsaw, Poland; marcin_studnicki@sggw.edu.pl

4 Department of Agrometeorology and Applied Informatics, Institute of Soil Science and Plant Cultivation—State Research Institute, 8 Czartoryskich Street, 24-100 Pulawy, Poland; anna.nierobca@iung.pulawy.pl

* Correspondence: mrozewicz@iung.pulawy.pl

Abstract: Cultivar, habitat conditions and agrotechnology have an influence on the yield and chemical composition of rye grain. The main anti-nutritional substances present in rye grain include alkylresorcinols, water-soluble pentosans and trypsin inhibitors. The aim of this study was to determine the variability in yield and the concentration of anti-nutritional compounds in the grain of selected winter rye cultivars in relation to nitrogen fertilisation levels and weather conditions. Field studies were conducted at the Experimental Station of IUNG-PIB in Osiny (Poland) in two growing seasons (2018/2019 and 2019/2020). The experiment was located on pseudo-polylic soil using the randomised sub-block method in three replications. The first factor of the experiment was the level of nitrogen fertilisation (0, 70 and 140 kg N·ha^{-1}) and the second was the population (Dańkowskie Skand, Piastowskie) and hybrid (KWS Vinetto, SU Performer) winter rye cultivars. The study showed that the yield of winter rye depended on the genotype and the level of nitrogen fertilisation. The hybrid cultivars yield ed 17.9% higher in relation to the population cultivars. The content of anti-nutritional compounds in rye grain depended significantly on genotype, level of nitrogen fertilisation and weather conditions. The reason for the higher synthesis of anti-nutrients in rye grain was the stressful weather conditions occurring in the 2019/2020 season. Nitrogen fertilisation influenced the content of alkylresorcinols, water-soluble pentosans and trypsin inhibitor activity in grain. The interaction of cultivar and fertilisation was also found to shape the content of the aforementioned anti-nutrients.

Keywords: alkylresorcinols; cultivars; grain yield; nitrogen fertilisation; pentosans; rye; trypsin inhibitors

1. Introduction

Rye (*Secale cereale* L.) is a cereal grown in Europe in a large area of the so-called "rye belt", which includes countries such as Germany, Poland, Ukraine and Belarus. In recent years, rye production in the European Union has been around 8 million tonnes per year [1]. In the food industry, rye grain serves mainly as a raw material for the production of different types of flour from which bread (rye, mixed) is made [2]. Products made from rye flour, especially whole-grain flour, are characterised by their high nutritional and health-promoting values [3]. Ongoing breeding work focused on improving grain yield and quality further increases the potential for the use of rye grain in both human and animal nutrition [4]. The breeding of new cultivars aims to increase grain yield and quality by using cultivars with stable yields under variable climatic conditions [5]. A particular direction of progress in breeding cultivars with high yield potential is to obtain hybrid cultivars of rye [6,7]. The heterosis effect occurring in hybrid cultivars allows for higher grain yields resulting from higher ear density [8–10]. Hybrid cultivars, as a result of transgressive segregation of alleles, may show a different grain chemical composition [11]. Higher grain yield and better grain quality influence the increased interest in the cultivation of hybrid cultivars. There is considerable interest in the cultivation of hybrid rye cultivars in Western European countries in particular, which account for 81% of total rye grain production, with population cultivars accounting for the remainder [12]. The main factors in favour of rye cultivation are its low soil requirements, resistance to low winter temperatures and relatively high grain yield. Yield and its quality are determined by weather conditions during the growing season and appropriately selected agrotechnics, including the level of nitrogen fertilisation, as well as by the cultivar [13,14].

A significant proportion of rye grain is used in the feeding of various animal species; mainly pigs and poultry [4]. However, a limitation to its wider use is its content of anti-nutritional substances that have a negative impact on animal health and development. These include alkylresorcinols, water-soluble pentosans and trypsin inhibitors. However, due to the varietal variation associated with the content of these substances, the grain of some modern rye cultivars may have a higher feed value [15]. The rye grain of hybrid cultivars can be used in poultry feed at a higher proportion than population cultivars [16–19].

The main anti-nutritional substances found in rye grain include alkylresorcinols, non-starch polysaccharides and trypsin inhibitors [19]. Alkylresorcinols are a group of phenolic lipids. Most of these compounds are located in the central parts of the fruit and seed coat. These compounds were found in the largest amount in rye grain (360–2180 mg·kg^{-1}), followed by triticale grain (294–1145 mg·kg^{-1}) and wheat (268–943 mg·kg^{-1}) [20]. Literature data indicate that alkylresorcinols interact with other compounds present in grain to reduce animal weight gain [21]. Feeding animals cereal grain from which alkylresorcinols have been extracted increases production effects [22]. Compared to other cereals, rye grain contains the highest amount of these compounds [23]. The high water-binding capacity of NSP and the resulting swelling hinders the penetration of digestive contents by enzymes that hydrolyse starch, proteins and fats. Compared to other cereals, rye grain contains the highest amount of these compounds, which swell in the digestive tract, causing reduced feed intake, poorer utilisation of the nutrients and energy contained in the feed, and ultimately lower body weight [24]. The only method to neutralise their anti-nutritional effect is to add xylanase enzymes to the feed [25]. Climate change and the resulting requirements to reduce the environmental impact of feed production are the main challenges in animal nutrition. In the case of wheat, the grain of which is widely used for the production of feed, rye uses a natural solution, primarily water and phosphorus, and its production produces a carbon footprint [26]. However, a large amount of non-starch polysaccharides used in rye grain may increase intestinal fermentation, which may affect the intestinal life, especially in young monogastric animals, such as piglets and chickens for fattening. Among the available anti-nutritional agents against the action of pentosans and alkylresorcinols are geese, in which hybrid rye grains are fed with a nutritional value comparable to oats [27]. In older animals, the anti-nutritional effect of alkylresorcinols and water-soluble pentosans is no longer as strong, and may even have an impact on health by acting prebiotically. The strongest anti-nutritional effect in animals, both young and adult, involves inhibiting trypsins. They inhibit one of the proteolytic enzymes (proteases) by use, which causes it to work more intensively and cause an increase in mass [28].

Higher trypsin inhibitor activities are found in rye grain than in the grains of other cereal types (wheat and triticale). These compounds are found in both endosperm and embryos. Their activity is associated with low molecular weight cereal proteins, which include albumin and globulins [29]. The adverse effects of inhibitors on the animal body are mainly due to a reduction in the activity of digestive enzymes and a decrease in digestibility, as well as the utilisation of nutrients, especially protein, leading to a reduction in animal weight gain [30].

The amount of anti-nutritional compounds in cereal grain depends mainly on the type, species and variety of cereal. However, it is also largely shaped by climatic, soil and agrotechnical conditions, as well as by the time and conditions of grain storage after harvest [31,32]. Among the agrotechnical factors, nitrogen fertilisation has the greatest impact on the yield and chemical composition of grain. The use of nitrogen by cereals depends on genetic and environmental factors [33]. Research carried out so far shows that rye varieties use increasing doses of nitrogen unequally [34]. Individual varieties differ significantly in terms of the rate of growth of vegetative mass and nitrogen accumulation, especially with abundant fertilisation with this element. The uptake and remobilization of nitrogen by cereal plants is largely influenced by weather conditions during the growing season. Climate changes, and especially droughts, cause cereal plants to less effective use nitrogen more, which leads to lower yields, and therefore the concentration of nitrates in the dry matter of grain is higher than in wet years [35,36]. The few available pieces of literature on this subject [17,37,38] indicate that nitrogen fertilisation affects the concentration of anti-nutritional compounds in cereal grains, such as alkylresorcinols, pentosans and the activity of trypsin inhibitors. Safar-Noori et al. [39] showed that, as the level of nitrogen fertilisation increases, the content of protein and pentosans in wheat grain increases. The literature on the subject lacks research results on the influence of the level of nitrogen fertilisation on the synthesis of anti-nutritional compounds in the grain of population and hybrid varieties of winter rye; however, the obtained research results will fill the gap regarding this issue. The research hypothesis assumed that both the nitrogen fertilisation dose and weather conditions during the growing season would differentiate the content of anti-nutritional compounds in the grain of the tested winter rye varieties.

The aim of the study was to determine the yield and variability in the concentration of anti-nutritional compounds in the grain of selected winter rye cultivars in relation to the level of nitrogen fertilisation and weather conditions.

2. Materials and Methods

2.1. Site Characteristic, Experimental Design, and Agronomic Practices

The field experiment was conducted in the growing seasons 2018/2019 and 2019/2020 at the Experimental Station in Osiny (51°27′ N; 22°2′ E), belonging to the Institute of Soil Science and Plant Cultivation—State Research Institute (IUNG–PIB) in Pulawy (Poland). The experiment was located on pseudo-loamy soil, good wheat complex, quality class IIIa and IIIb. The soil of pH 6.5 contained 181 mg K·kg^{-1} and 173 mg P·kg^{-1}. Before applying nitrogen fertilisation, the content of mineral nitrogen (NH_4^+, NO_3^-) in the soil was determined in early spring, which in 2019 was 39.6 kg N·ha^{-1} and 32.6 kg N·ha^{-1} in 2020, respectively. The two-factor experiment was set up using the randomised sub-block method in triplicate. The sowing of rye was carried out beetwen 20th–30th of September, and the crop was cultivated after winter oilseed rape in a crop rotation with 75% share of cereals in the sowing structure. The experiment was conducted with a ploughing system. After harvesting the forecrop, straw was shredded and ploughed to a depth of 8–10 cm, then harrowed with a heavy harrow, followed by pre-sowing ploughing to a depth of 20–22 cm.

The first factor of the experiment was the level of nitrogen fertilisation: 0, 70 and 140 kg N·ha^{-1}. A nitrogen dose of 70 kg N·ha^{-1} was applied at two dates: 40 kg N·ha^{-1} in spring at the start of vegetation and 30 kg N·ha^{-1} at the stalk shooting stage (BBCH 31). A nitrogen dose of 140 kg N·ha^{-1} was applied at three dates: 70 kg N·ha^{-1} in spring at the start of vegetation, 50 kg N·ha^{-1} at the stalk shooting stage (BBCH 31) and 20 kg N·ha^{-1} at the earing stage (BBCH 57). Nitrogen was applied in a form NH_4NO_3.

The second factor of the experiment was the winter rye cultivars. Two population cultivars were included, Dańkowskie Skand and Piastowskie, and two hybrid cultivars, KWS Vinetto and SU Performer.

Dańkowskie Skand—Hodowla Danko—cultivated since 2017 https://danko.pl/en/odmiany-type/dankowskie-skand-2/ (accesed on 21 February 2024).

Piastowskie—Poznańska Hodowla Roślin—cultivated since 2017 https://phr.pl/produkty/zboza-ozime/zyto-ozime-piastowskie/ (accessed on 21 February 2024).

KWS Vinetto—KWS cultivated since 2017—https://www.kws.com/pl/pl/produkty/zboza/zyto/przeglad-odmian-zyto/kws-vinetto/ (accessed on 21 February 2024).

SU Performer—Saaten Union cultivated since 2014 https://www.saaten-union.pl/zyto/zyto-ozime-hybrydowe/su-performer/ (accessed on 21 February 2024).

The area of the experimental plot to be harvested was 20.0 m^2. The population rye was sown at 2.5 million grains$\cdot$ha^{-1}, and hybrid rye at 2.0 million grains$\cdot$ha^{-1}. Pre-sowing mineral fertilisation was applied in the form of superphosphate at a rate of 60 kg $P_2O_5\cdot ha^{-1}$ and potassium salt at a rate of 90 kg $K_2O\cdot ha^{-1}$. Protection against diseases, pests and weeds was carried out according to the integrated method of reducing weeds and disease and pest pathogens after exceeding the threshold of harmfulness. Immediately after harvest, the grain yield per unit area was assessed.

2.2. Meteorological Conditions

The course of weather conditions was described on the basis of monthly mean values of air temperature and total precipitation in comparison with climatic norms and mean values for the period 1981–2010. Meteorological data were obtained from the Agro-meteorological Station located near the experimental fields at the Experimental Station IUNG–PIB in Osiny (φ = 51.469 N, λ = 22.052 E). The years in which the study was conducted differed in terms of thermal conditions and precipitation amounts. The 2018/2019 growing season was characterised by lower precipitation, while at the same time there were large temperature fluctuations (Figure 1a). During the autumn period (September to December), precipitation was average and temperatures were slightly higher compared to multi-year averages (1980–2010). An exceptionally warm period was February (3.4 °C) and March (4.5 °C), in these months the average temperature was 4.1 and 3.2 °C higher, respectively, compared to the multi-year period (1981–2010). In May, there was a change in the trend of thermal conditions. After an exceptionally warm spring, May was cool (12.9 °C) and characterised by high precipitation (86.1 mm). June saw another change in weather trends; it was a hot month with low rainfall. The temperature in June was 21.7 °C, 5.1 °C higher than the multi-year period (1981–2010), while precipitation was low at 38.7 mm, which was only 62% of normal.

The second growing season 2019/2020 from autumn to spring (September to March) was characterised by higher air temperatures compared to the 1981–2010 multi-year average (Figure 1b). A particularly outstanding period was winter (December to February), which was very warm with high precipitation. The temperature during this period was: in December 3.1 °C, in January 1.8 °C and in February 3.4 °C, compared to the multi-year average (1981–2010) it was higher by, respectively: 4.0, 4.1 and 4.9 °C. In addition, the period was characterised by high precipitation totalling 153 mm, which was 180% of the precipitation compared to the multi-year average (1981–2010). A change in the trend of weather conditions occurred in March and April. During this period, precipitation totals were very low, amounting to only 34 mm, which was 54% of the precipitation compared to the multi-year average (1980–2010). After a period with low precipitation, the next two months (May and June) were characterised by high precipitation. In May, the precipitation was 112 mm, while in June it was even higher, at 189 mm; compared to the multi-year period, precipitation was 101 and 206% higher, respectively. In contrast, thermal conditions in these months were different. May was cold (11.1 °C), while June was warm (18.4 °C). Compared to the multi-year period (1981–2010), the temperature in May was 2.8 °C lower, while in June it was 1.8 °C higher.

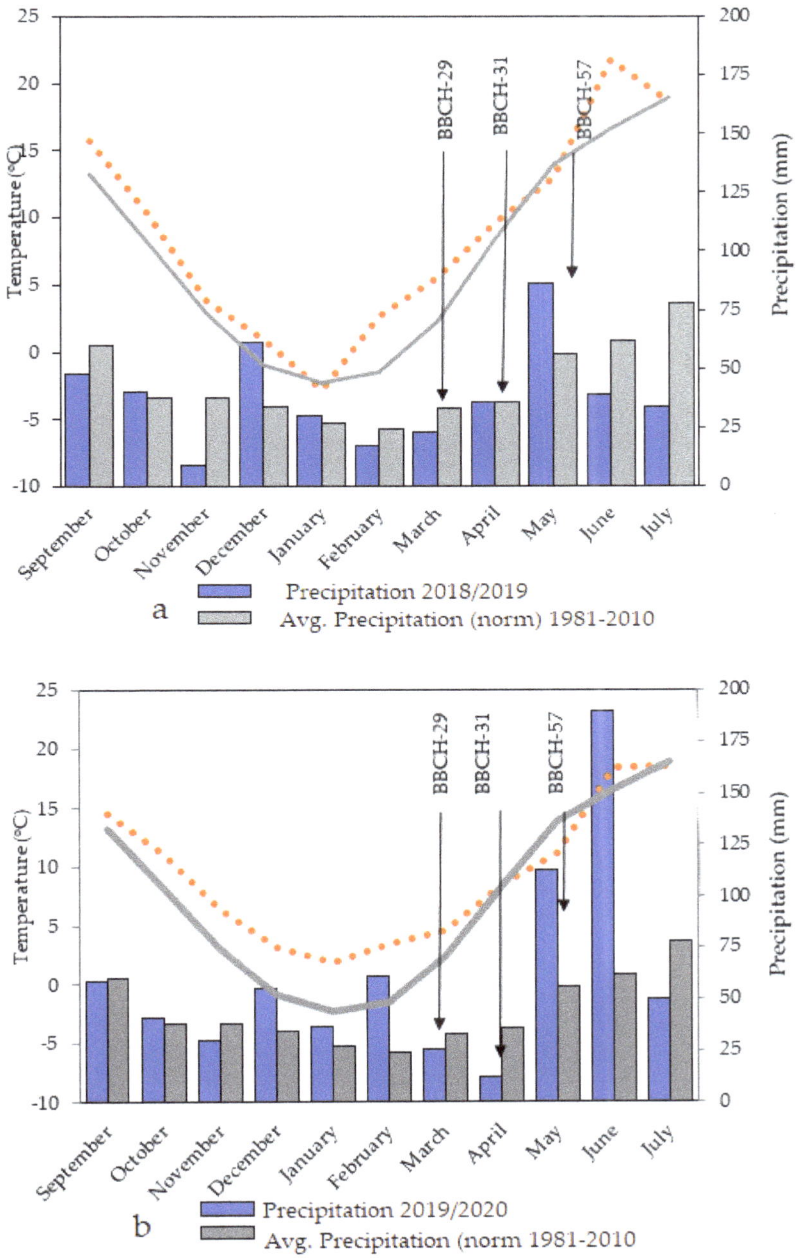

Figure 1. (**a**) Monthly air temperature (°C) and monthly of precipitation (mm) in years 2018/2019. The arrow in the graph indicates the date of nitrogen application in a given BBCH phase at the dose indicated in Section 2.1. (**b**) Monthly air temperature (°C) and monthly of precipitation (mm) in years 2019/2020. The arrow in the graph indicates the date of nitrogen application in a given BBCH phase at the dose indicated in Section 2.1.

2.3. Chemical Analyses

2.3.1. Determination of Alkylresorcinol Content

The analysis was performed by the spectrophotometer method [40]. Grains were ground to a particle size of <0.5 mm. Alkylresorcinols were extracted with acetone from the weighed amount of 1 g in the time of 3 h at 55 °C in a water bath. The content of alkylresorcinols was determined using acetone extracts after developing a colour reaction with the two-phase p-nitroaniline and measuring the absorbance of the coloured solution at 435 nm. For preparing the calibration curve, the orcinol was used.

2.3.2. Determination of Soluble Pentosans

The orcinol-hydrochloric acid method of Albaum and Umbreit [41] for pentose determination was used. It consisted of heating in boiling water for 30 min a solution containing 3 mL of pentose, 3 mL of 0.1% ferric chloride in concentrated hydrochloric acid, and 0.3 mL of 1.0% orcinol in 100% ethanol, cooling, and determining absorbance at 670 nm.

2.3.3. Determination of the Activity of Tripsin Inhibitors

Tripsin inhibitors were determined by the method of PN-EN ISO 14902 [42], which were extracted from the sample at pH 9.5. The residual trypsin activity was measured by adding benzoyl-L-arginine-p-nitroanilide (L-BABA) as a substrate. The amount of p-nitroanilide released was measured spectrometrically.

2.4. Statistical Analysis

Results recorded in the course of conducted chemical analyses were subjected to statistical analysis with the use of R 4.3.0 [43]. In order to compare the grain yield and contents of alkylresorcinol, soluble pentosans and tripsin inhibitors between the cultivars and N fertilisation effects, we used two-way ANOVA as a post hoc test, and Tukey's multiple comparison procedure was also used, with identical letters denoting a lack of differences at the significance level $\alpha = 0.05$. The assumption of normality of the layouts was checked using the QQ Plots for residuals and the homogeneity of variances using the Levene's test. Moreover, a principal component analysis (PCA) was used to evaluate the relationship between study traits and factors.

3. Results and Discussion

Winter rye yields depended on the cultivar and nitrogen fertilisation rate. In both 2019 and 2020, the highest grain yield was obtained from hybrid cultivars. On average, hybrid cultivars yielded 17.9% higher than population cultivars during the study years. In 2019, the hybrid cultivar KWS Vinetto had the highest yield (8.44 $t \cdot ha^{-1}$), and in 2020 the hybrid cultivar SU Performer (8.33 $t \cdot ha^{-1}$) had the highest yield (Figure 2). Literature data [44–46] also indicate that hybrid winter rye cultivars achieve higher grain yields compared to population cultivars. In a study by Latusek and Bujak [44], the grain yield of hybrid cultivars was 17.7% higher, while in a study by Wyzińska et al. [45] it was 11.2% higher in relation to population cultivars. Maciorowski et al. [34] found that hybrid cultivars yielded 9% higher compared to population cultivars. Also, a comparative study by Szuleta et al. [46] on 24 population and hybrid rye cultivars showed a yield advantage for the hybrid cultivars, especially in years with less favourable weather conditions. The higher grain yield of hybrid rye cultivars shown in both years of the study compared to population cultivars may have been due to the higher tolerance of these cultivars to lower rainfall, especially in March and April, which were lower than the multi-year average. Hybrid cultivars characterised by higher tillering and vegetative mass structure provided more stable yields than population cultivars [47]. The reduction in rye grain yields as a result of unstable weather conditions in spring could reach up to 17.1% [48,49].

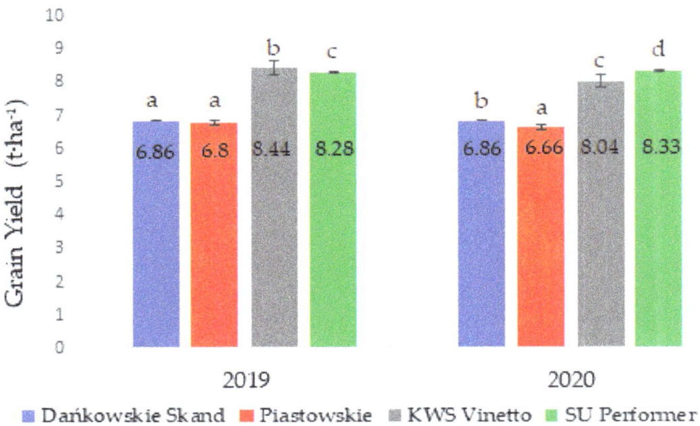

Figure 2. Grain yield of winter rye depending on the cultivar (average for a change). a, b, c, d—the same letters indicate no significant differences at significance level $\alpha = 0.05$.

A significant effect of nitrogen fertilisation level on winter rye yield was found. In 2019, a significant increase in grain yield occurred at a dose of 70 kg N·ha^{-1}, and further increases in nitrogen dose did not cause a significant increase in grain yield. In contrast, in 2020, grain yield increased significantly alongside the growing nitrogen fertilisation rates. The highest rye grain yield was obtained at the application of 140 kg N·ha^{-1}, and it was 4.6% higher with respect to the dose of 70 kg N·ha^{-1} (Figure 3). Latusek and Bujak [44] and Dopierała et al. [50], comparing rye yield at two agrotechnical levels, observed a positive response of both hybrid and population cultivars to higher levels of nitrogen fertilisation. In contrast, a study by Noland [51] showed that a significant increase in rye yield occurred only up to an application rate of 67 kg N·ha^{-1}. Population cultivars, however, showed a weaker yield response to nitrogen fertilisation than hybrid cultivars [52]. Szuleta et al. [53] found no significant differences in the grain yield of rye population cultivars as a result of doubling the nitrogen dose. The efficiency with which rye utilises the supplied nitrogen dose depends on weather conditions, especially the amount of rainfall during the growing season, but also the content of organic matter, which accumulates in the water reserve [48,54,55].

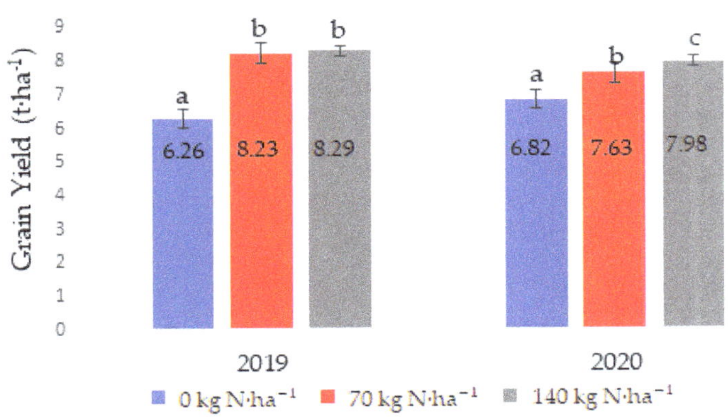

Figure 3. Grain yield of winter rye depending on the dose of nitrogen (average for nitrogen fertilisation). a, b, c—the same letters indicate no significant differences at significance level $\alpha = 0.05$.

It was found that the content of the tested anti-nutritional substances in rye grain was significantly influenced by the cultivar, nitrogen fertilisation rate and weather conditions during the period of field cultivation, i.e., 2018–2020.

The differences in the content of anti-nutritional compounds in rye grain from the different harvest years gave us an incentive to analyze the weather conditions during the different growing seasons (Figures 2 and 3). Significantly, the grain harvested in 2019 had the lowest content of anti-nutrients (Table 1). In rye grain from 2019, the average content of alkylresorcinols was 533.0 mg·kg^{-1}, soluble pentosans was 1.29% d.m. and trypsin inhibitor activity was 1.19 mg·g^{-1}. A higher content of anti-nutritional compounds was found in rye grain from 2020. The average content of alkylresorcinols was 658.2 mg·kg^{-1}, water-soluble pentosans was 1.46% of d.m. and trypsin inhibitor activity was 1.32 mg·g^{-1}. Increased synthesis of anti-nutritional compounds was ascribed to stress conditions during the growing season. In the 2019/2020 growing season, the stress was caused by high precipitation in the months of May and June, which was well above normal, resulting in high levels of fungal diseases in the rye plantation. Previous studies [17] have also shown that the content of anti-nutritional substances in rye grain (alkylresorcinols, water-soluble pentosans, trypsin inhibitors) is strongly influenced by weather conditions during the plant growing season. Increased synthesis of the aforementioned anti-nutritional compounds in the grain occurred when stress conditions for plant development occurred during the growing season. The highest levels of anti-nutritional substances were synthesised when plants experienced stress caused by a shortage of rainfall during their intensive growth period (at the stalk shooting stage), as well as too much rainfall in May and June (at the earing and flowering stages). On the other hand, Jaśkiewicz and Szczepanek [38], investigating the content of alkylresorcinols in winter triticale grain, found that lower accumulation of these compounds was favoured by limited precipitation at the stalk shooting stage and excess precipitation during grain maturation. Similarly, Bellato et al. [55] showed that durum wheat grain contained more alkylresorcinols when there was less rainfall during the growing years. Weather conditions during the growing season also affect the accumulation of water-soluble pentosans in rye grain. Less rainfall during the growing season favours higher content in rye grain [49,53]; in winter wheat, on the other hand, a higher accumulation of pentosans in the grain is favoured by a lower temperature at the grain pouring stage [56]. Marentes-Culma and Coy-Barrera [57] showed that grain of the same cultivar of triticale from cultivation under different environmental and weather conditions contains different amounts of alkylresorcinols.

The grain of the winter rye cultivars tested was characterised by different contents of anti-nutritional compounds. The highest content of alkylresorcinols in rye grain in both 2019 and 2020 was found in the grain of the KWS Vinetto hybrid cultivar, reaching 555.2 and 674.2 mg·kg^{-1}, respectively (Figure 4A). In 2019, the lowest content of alkylresorcinols was found in the grain of the rye population cultivars Dańkowskie Skand (510.3 mg·kg^{-1}) and Piastowskie (532.0 mg·g^{-1}). In 2020, the least amount of these compounds was contained in the grain of the hybrid cultivar SU Performer (643.3 mg·kg^{-1}). On average, over the years of the study, the grain of the KWS Vinetto hybrid cultivar had the highest content of alkylresorcinols (614.70 mg·kg^{-1}) (Table 1). Similarly, other authors [48,49,54] have shown that the content of alkylresorcinols in cereal grain is mainly determined by a genetic factor. Grabiński et al. [17] showed that, during the three-year study period, the highest alkylresorcinol content was found in grain of the hybrid cultivar Visello (mean 563 mg·kg^{-1}) and the lowest in the grain of the hybrid cultivar Brasetto (mean 535 mg·kg^{-1}). A study by Kulawianek et al. [58] showed that the highest accumulation of these compounds was in the grain of the hybrid cultivar Gradan (1152 mg·kg^{-1} d.m.) and the lowest in the grain of the population cultivar Amilo (1058 mg·kg^{-1} d.m.). Studies by Targońska-Karasek et al. [59] also showed the influence of the genetic factor on the synthesis of alkylresorcinols, while the lowest potential for alkylresorcinol synthesis was demonstrated by the population cultivar Dańkowskie Złote in comparison to the hybrid cultivar Daniello and the population cultivar L 318. Jaśkiewicz and Szczepanek [38] showed that, for both winter and spring triticale, there are varietal differences in the content of alkylresorcinols in grain. Similar relationships were found for winter wheat [57,59–61].

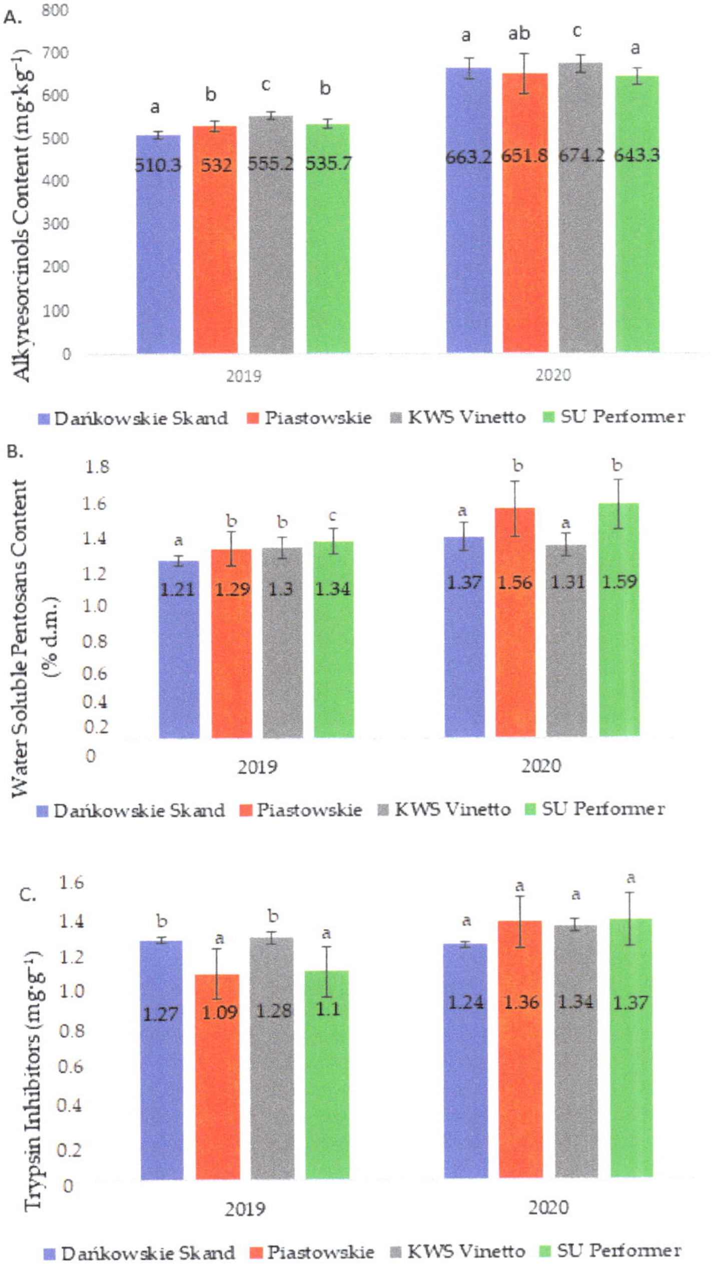

Figure 4. Content of selected anti-nutritive components: alkylresorcinols (**A**), water-soluble pentosans (**B**) and tripsin inhibitors (**C**) in grain cultivars in 2019–2020. a, b, c—the same letters indicate no significant differences at significance level $\alpha = 0.05$.

Table 1. Compounds of selected anti-nutritional substances in the grain of winter rye varieties depending on the growing season, variety and nitrogen fertilisation.

Specification	Alkylresorcinol Content ($mg{\cdot}kg^{-1}$)	Water-Soluble Pentosans Content (% d.m.)	Activity of Tripsin Inhibitors ($mg{\cdot}g^{-1}$)
Growing season			
2018/2019	533.29 ^{a}	1.28 ^{a}	1.23 ^{a}
2019/2020	661.79 ^{b}	1.46 ^{b}	1.46 ^{b}
Cultivar			
Dańkowskie Skand	586.75 ^{a}	1.29 ^{a}	1.26 ^{a}
Piastowskie	591.90 ^{ab}	1.42 ^{ab}	1.23 ^{a}
KWS Vinetto	614.70 ^{b}	1.31 ^{ab}	1.31 ^{a}
SU Performer	588.00 ^{ab}	1.47 ^{b}	1.24 ^{a}
Nitrogen Fertilisation ($kg{\cdot}ha^{-1}$)			
0	575.20 ^{a}	1.41 ^{a}	1.28 ^{a}
70	610.55 ^{b}	1.38 ^{a}	1.34 ^{b}
140	601.45 ^{a}	1.31 ^{a}	1.14 ^{a}

$^{a,\,b}$—the same letters indicate no significant differences at significance level $\alpha = 0.05$.

The content of water-soluble pentosans, like alkylresorcinols, was shaped by the genetic factor. In 2019, the content of water-soluble pentosans in rye grain ranged from 1.21 to 1.34% d.m., and in 2020 from 1.31 to 1.56% d.m. (Figure 4B). In 2019, the most water-soluble pentosans were contained in the grain of the hybrid cultivars KWS Vinetto and SU Performer. On the other hand, in 2020, the highest concentration of these compounds was found in the grain of the hybrid cultivar SU Performer and the population cultivar Piastowskie. On average, over the years of the study, the grain of the SU Performer hybrid variety had the highest content of soluble pentosans (1.47% d.m.) (Table 1). Studies by other authors [17,55,62] also showed that the content of soluble pentosans in rye grain depends on the cultivar. Buksa et al. [37] showed that there is a very high variation in the content of water-soluble pentosans in the grain of population cultivars of winter rye. According to Grabiński et al. [17], the most water-soluble pentosans are contained in the grain of hybrid rye cultivars. However, Kulichova et al. [62] showed no significant differences between rye cultivars in the content of water-soluble pentosans in grain. The varietal factor also has a strong influence on the content of pentosans in wheat and barley grain [50].

In 2019, the activity of the trypsin inhibitors in rye grain was shaped by a genetic factor. The grain of the population cultivar Dańkowskie Skand (1.27 $mg{\cdot}g^{-1}$) and the hybrid cultivar KWS Vinetto (1.27 $mg{\cdot}g^{-1}$) had significantly higher activity of these substances than the grain of the population cultivar Piastowskie (1.09 $mg{\cdot}g^{-1}$) and the hybrid cultivar SU Performer (1.10 $mg{\cdot}g^{-1}$) (Figure 4C). In 2020, there were no significant differences in the activity of trypsin inhibitors in the grain of the rye cultivars tested; there was only a trend towards lower activity in the grain of the population cultivar Dańkowskie Skand. Over the years, it was shown that the grain of the KWS Vinetto hybrid cultivar was characterized on average by the highest activity of trypsin inhibitors (1.31 $mg{\cdot}g^{-1}$) (Table 1). In a study by Grabiński et al. [17], it was shown that the activity of trypsin inhibitors depended on the genetic factor. The authors obtained the highest trypsin inhibitor activity (2.80 $mg{\cdot}g^{-1}$) in the grain of the hybrid cultivar Visello, and the lowest (2.01 $mg{\cdot}g^{-1}$) in the grain of the population cultivar Dańkowskie Diament. Kulichova et al. [62], in examining 19 rye cultivars, showed that most of them had similar trypsin inhibitor activity, but also found the occurrence of two cultivars with significantly lower activity of this parameter, indicating genetic variation. In addition, Simonetti et al. [63] showed that the activity of trypsin inhibitors in wheat grain depends on the species and cultivar, and also that environmental conditions also have a significant effect.

Our own research has shown the interaction of cultivar and nitrogen fertilisation rates in shaping the content of anti-nutrients in winter rye grain (Figure 5A–F). In 2019, a significant increase in the content of alkylresorcinols was obtained in the grain of the population cultivars Dańkowskie Skand and Piastowskie, as well as the hybrid cultivar KWS Vinetto, when a dose of 70 kg N·ha^{-1} was applied; further increasing the dose to 140 N·ha^{-1} did not result in an increase in the concentration of these substances in the grain (Figure 5A). In contrast, in the case of the hybrid cultivar SU Performer, grain with the highest alkylresorcinol content was obtained using a fertilisation rate of 140 kg N·ha^{-1}. In 2020, there was no significant effect of increasing the nitrogen fertilisation rate on the content of alkylresorcinols in the grain of the population cultivar Dańkowskie Skand (Figure 5B). In contrast, the population cultivar Piastowskie and the hybrid cultivar KWS Vinetto reacted with a significant increase in the content of these compounds when fertilized with 140 kg N·ha^{-1}. In the grain of the hybrid cultivar SU Performer, the highest content of alkylresorcinols was found after fertilisation at a dose of 70 kg N·ha^{-1}. Literature reports indicate that cereal cultivars show a differentiated response in alkylresorcinol content to increasing doses of nitrogen fertilisation [17,53,61]. Grabiński et al. [17] showed that an increase in the intensity of production technology, including nitrogen fertilisation, caused a significant increase in alkylresorcinol content by 7%. According to Jaśkiewicz and Ochman [64], nitrogen fertilisation at a dose of 100 kg·ha^{-1} favoured AR (alkylresorcinols) accumulation in spring triticale grain, especially under conditions of higher rainfall in April and May. On the other hand, under conditions of limited rainfall at the tillering stage and the beginning of stalk shooting, AR content in grain was similar at both fertilisation levels of 70 and 100 kg N·ha^{-1}. Takač et al. [65], found that spelt wheat grain contained more alkylresorcinols in organic cultivation compared to conventional cultivation where increased mineral nitrogen fertilisation is applied.

In 2019, the highest concentration of water-soluble pentosans was characterised by the grain of the population cultivar Piastowskie from the crop where the highest nitrogen fertilizer rate was applied (140 kg N·ha^{-1}). However, for the other rye cultivars, increasing nitrogen doses caused a decrease in the content of these compounds in the grain (Figure 5C). In 2020, as in 2019, the grain of the hybrid cultivars KWS Vinetto and SU Performer had the highest concentration of pentosans on the sites where no nitrogen fertilisation was applied, reaching 1.42 and 1.89% d.m., respectively. Population cultivars, on the other hand, responded with a significant increase in the content of water-soluble pentosans when the highest level of nitrogen fertilisation was applied, the content of these compounds being 1.67% d.m. (Figure 5D). Grabiński et al. [17] showed that higher nitrogen fertilisation increased water-soluble pentosans by 8.7%. A study by Noori [66] showed that both weather conditions, especially the amount of rainfall, as well as cultivar and nitrogen fertilisation, condition the pentosan content of wheat grain. Under drought conditions, despite higher fertilisation levels, the author obtained a lower pentosan content in grain than under higher nitrogen rates and optimum rainfall.

In 2019, grain of all rye cultivars tested showed the lowest trypsin inhibitor activity at the highest nitrogen fertilisation level applied in the experiment, of 140 N·ha^{-1} (Figure 5E). Grain of the hybrid cultivar KWS Vinetto had the highest trypsin inhibitor activity on the site without nitrogen fertilisation, as well as grain of the other cultivars after an application rate of 70 kg N·ha^{-1}. In 2020, with the exception of the hybrid cultivar KWS Vinetto, the grain with the lowest trypsin inhibitor activity was obtained similarly to 2019 after a fertilisation application of 140 N·ha^{-1} (Figure 5F). In the case of the population cultivars Dańkowskie Skand and Piastowskie, the highest trypsin inhibitor activity was shown by the grain after an application rate of 70 kg N·ha^{-1} in the case of the hybrid cultivar KWS Vinetto 140 N·ha^{-1}, and in the case of the hybrid cultivar SU Performer from a crop without nitrogen fertilisation. Studies on soybeans have shown that increasing rates of mineral nitrogen fertilisation result in a proportional decrease in trypsin inhibitor activity [67].

Xue et al. [68] found that dividing the nitrogen dose in relation to its single application resulted in a decrease in trypsin inhibitor activity in wheat grain. Grabiński et al. [17] showed that increasing the intensity of production technology, including nitrogen fertilisation, did not differentiate trypsin inhibitor activity in rye grain.

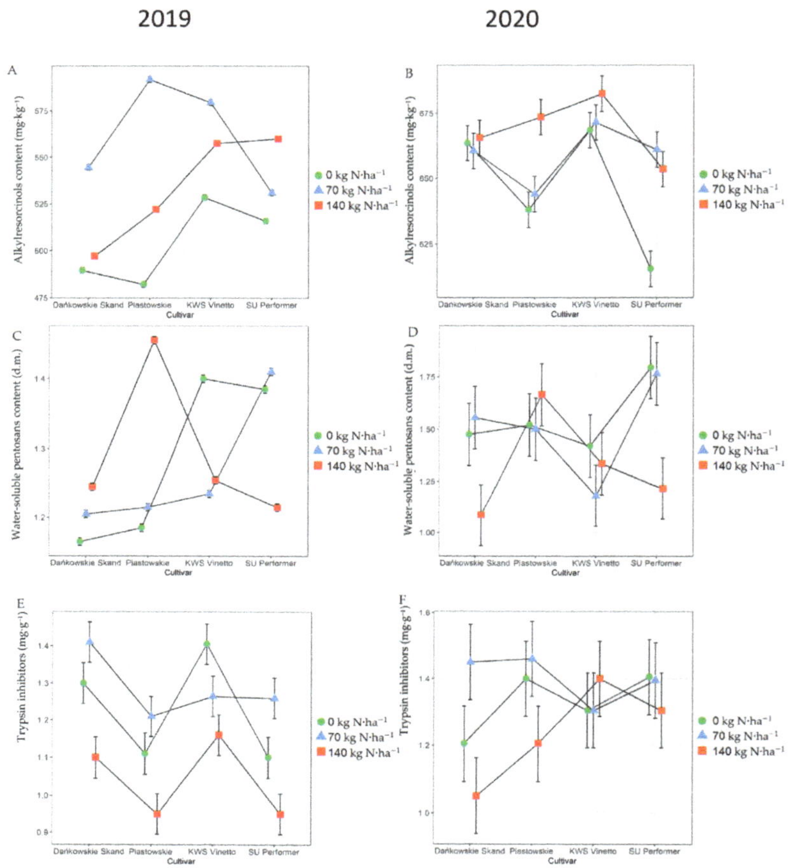

Figure 5. Interaction of nitrogen fertilisation and cultivar in the content of alkylresorcinols (**A**,**B**), water-soluble pentosans (**C**,**D**) and trypsin inhibitors (**E**,**F**) in rye grain.

Figure 6 shows the results of the PCA principal component analysis of the content of individual anti-nutrients in the grain of the rye cultivars tested in relation to the level of nitrogen fertilisation. We observe a strong positive correlation between yield and the content of alkylresorcinols in the grain. The hybrid cultivar KWS Vinetto showed the strongest correlation related to the application of a higher nitrogen dose and increased yield and alkylresorcinol content in grain. In contrast, the SU Performer hybrid variety showed the most-favourable response to fertilisation without significant changes in the content of water-soluble pentosans in the grain. We also observe a negative correlation between water-soluble pentosan content and trypsin inhibitor activity.

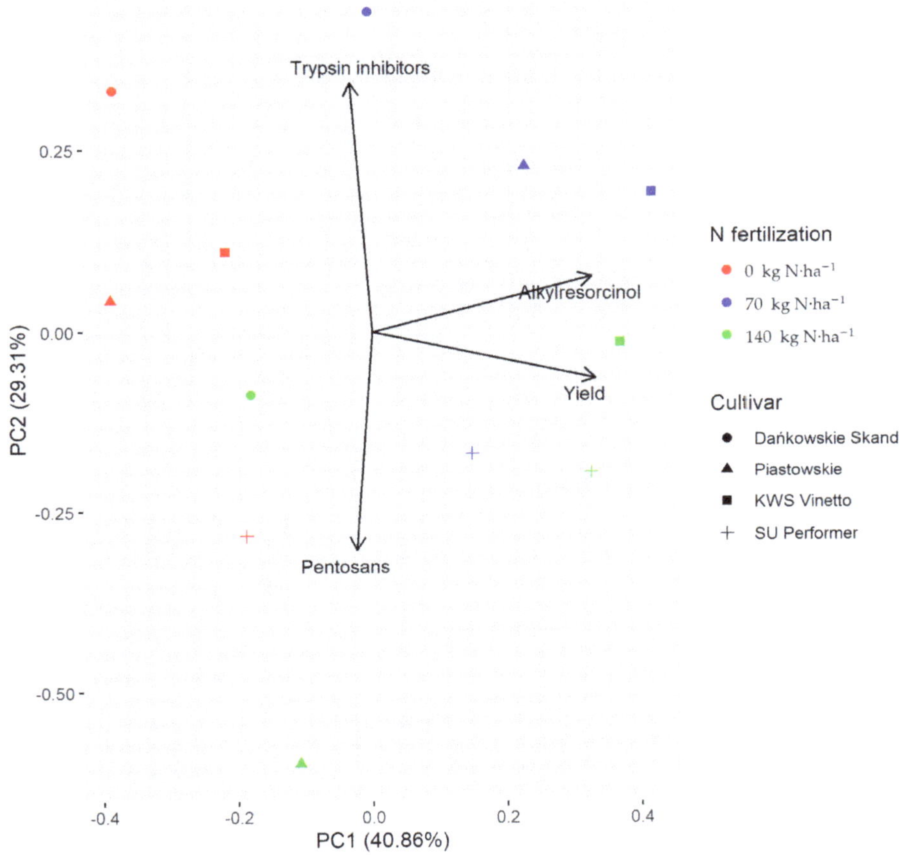

Figure 6. Biplot plot for PCA principal components for cultivar and nitrogen fertilisation as a function of alkylresorcinols, water-soluble pentosans and trypsin inhibitor activity and yield.

4. Conclusions

Our study showed that the contents of alkylresorcinols, water-soluble pentosans and trypsin inhibitors in rye grain depend on a cultivar. They are also affected by weather conditions and the level of nitrogen fertilisation. There was a correlation between cultivar and the rate of nitrogen fertilisation. A significantly greater synthesis of anti-nutritional substances in rye grain occurred in the growing season with high rainfall. The interaction of nitrogen fertilisation and cultivar affected the content of anti-nutritional substances in rye grain. The use of a higher nitrogen rate increased the content of alkylresorcinols to the largest degree at the KWS Vinetto hybrid cultivar. The Piastowskie population cultivar had the highest content of water-soluble pentosans in the grain at the highest nitrogen fertilisation rate used, while the Dańkowskie Skand population variety exhibited the highest activity of trypsin inhibitors in the grain at the low nitrogen fertilisation rate.

Author Contributions: Conceptualization, A.S. and J.G., methodology A.S. and M.W.; statistical analysis, A.S. and M.S., weather conditions analysis, A.N., Writing—original draft, A.S., G.C.-P. and M.R., Writing—revive and editing, A.S., J.G., G.C.-P. and M.W. All authors have read and agreed to the published version of the manuscript.

Funding: This research received no external funding.

Institutional Review Board Statement: Not applicable.

Data Availability Statement: Data available on request due to restrictions, e.g., privacy or ethical. The data presented in this study are available on request from the first author.

Conflicts of Interest: The authors declare no conflicts of interest.

References

1. AGRI Food. Available online: https://agridata.ec.europa.eu/ (accessed on 12 December 2023).
2. Stępniewska, S.; Słowik, E.; Cacak-Pietrzak, G.; Romankiewicz, D.; Szafrańska, A.; Dziki, D. Prediction of rye flour baking quali based on parameters of swelling curve. *Eur. Food Res. Technol.* **2018**, *244*, 989–997. Available online: https://link.springer.com article/10.1007/s00217-017-3014-z (accessed on 21 February 2024). [CrossRef]
3. Stępniewska, S.; Hassoon, W.H.; Szafrańska, A.; Cacak-Pietrzak, G.; Dziki, D. Procedures for Breadmaking Quality Assessment Rye Wholemeal Flour. *Foods* **2019**, *8*, 331. [CrossRef]
4. Brzozowski, L.J.; Szuleta, E.; Phillips, T.D.; Van Sanford, D.A.; Clark, A.J. Breeding cereal rye (*Secale cereale*) for quality traits. *Crc Sci.* **2023**, *63*, 1964–1987. [CrossRef]
5. Miedaner, T.; Korzun, V.; Wilde, P. Effective Pollen-Fertility Restoration Is the Basis of Hybrid Rye Production and Ergot Mitigatio *Plants* **2022**, *11*, 1115. [CrossRef] [PubMed]
6. Hackauf, B.; Siekmann, D.; Fromme, F.J. Improving Yield and Yield Stability in Winter Rye by Hybrid Breeding. *Plants* **202** *11*, 2666. [CrossRef] [PubMed]
7. Wilde, P.; Miedaner, T. Hybrid rye breeding. In *Compendium of Plant Genomes—Rye Genome*; Springer Publishing Hous Berlin/Heidelberg, Germany, 2021; pp. 13–41.
8. Galán, R.J.; Bernal-Vasquez, A.M.; Jebsen, C.; Piepho, H.P.; Thorwarth, P.; Steffan, P.; Miedaner, T. Integration of genotypi hyperspectral, and phenotypic data to improve biomass yield prediction in hybrid rye. *Theor. Appl. Genet.* **2020**, *133*, 3001–301 [CrossRef] [PubMed]
9. Pomortsev, A.V.; Dorofeev, N.V.; Zorina, S.Y.; Katysheva, N.B.; Sokolova, L.G.; Zhuravkova, A.S.; Mikhailova, E.V. Evaluation c Population and Hybrid Varieties of Winter Rye in the Conditions of Eastern Siberia. *Agronomy* **2023**, *13*, 1431. [CrossRef]
10. Mackay, I.J.; Cockram, J.; Howell, P.; Powell, W. Understanding the classics: The unifying concepts of transgressive segregatio inbreeding depression and heterosis and their central relevance for crop breeding. *Plant Biotechnol. J.* **2021**, *19.1*, 26–34. [CrossRef
11. Klimek-Kopyra, A.; Bacior, M.; Neugschandtner, R. Hybrid rye (*Secale cereale* L.) as a good crop component to enhance yiel stability in a winter cereal mixture. *Acta Agrobot.* **2023**, *76*, 172670. [CrossRef]
12. Macholdt, J.; Honermeier, B. Impact of highly varying seeding densities on grain yield and yield stability of winter rye cultivar under the influence of delayed sowing under sandy soil conditions. *Arch. Agron. Soil Sci.* **2017**, *63*, 1977–1992. [CrossRef]
13. Linina, A.; Kunkulberga, D.; Kronberga, A.; Locmele, I. Winter rye grain quality of hybrid and population cultivars. *Agron. Re* **2019**, *17*, 1380–1389. [CrossRef]
14. Schwarz, T.; Kuleta, W.; Turek, A.; Tuz, R.; Nowicki, J.; Rudzki, B.; Bartlewski, M. Assessing the efficiency of using a moder hybrid rye cultivar for pig fattening, with emphasis on production costs and carcass quality. *Anim. Prod. Sci.* **2014**, *55*, 467–47 [CrossRef]
15. Milczarek, A.; Osek, M.; Skrzypek, A. Effectiveness of using a hybrid rye cultivar in feeding broiler chickens. *Can. J. Anim. Sc* **2020**, *100*, 502–509. [CrossRef]
16. Janiszewski, P.; Lisiak, D.; Borzuta, K.; Grześkowiak, E.; Schwarz, T.; Siekierko, U.; Andres, K.; Świątkiewicz, S. The Effect o Feeding Chicken and Geese Broilers with Different Cereals on the Fatty Acids Profile in Meat. *Foods* **2021**, *10*, 2879. [CrossRef [PubMed]
17. Grabiński, J.; Sułek, A.; Wyzinska, M.; Stuper-Szablewska, K.; Cacak-Pietrzak, G.; Nieróbca, A.; Dziki, D. Impact of genotype weather conditions and production technology on the quantitative profile of anti-nutritive compounds in rye grains. *Agronom* **2021**, *11*, 151. [CrossRef]
18. Boros, D. Alkylresorcinols of cereal grains—Their importance in food and feed. *Biul. IHAR* **2015**, *277*, 7–20. (In Polish) [CrossRef
19. Ciurescu, G.; Vasilachi, A.; Lavinia, I.; Dumitru, M.; Reta, D. Assessing the efficiency of using a local hybrid of rye for broile chickens aged 1–42 d, with emphasis on performance and meat quality. *Arch Zootech.* **2022**, *25*, 5–21. [CrossRef]
20. Ismagilov, R.; Ayupov, D.; Nurlygayanov, R. Ways to reduce anti-nutritional substances in winter rye grain. *Physiol. Mol. Bio Plants* **2020**, *26*, 1067–1073. [CrossRef]
21. Skrzypek, A.; Makarska, E.; Kociuba, W.; Studziński, M. Antioxidant activity and content of resorcinol of lipids in hybrids strai of winter triticale. *Żywn. Nauka Technol. Jakość* **2007**, *2*, 51–59. (In Polish)
22. Dynkowska, W.M. Rye (*Secale cereale* L.) Phenolic compounds as health-related factors. *Plant Breed. Seed Sci.* **2019**, *79*, 9–24 [CrossRef]
23. Boros, D. Influence of R genome on the nutritional value of triticale for broiler chicks. *Anim. Feed Sci. Tech.* **1999**, *76*, 219–226 [CrossRef]

Lagaert, S.; Pollet, A.; Courtin, C.M.; Volckaert, G. β-Xylosidases and α-l-arabinofuranosidases: Accessory enzymes for arabinoxylan degradation. *Biotechnol. Adv.* **2014**, *32*, 316–332. [CrossRef] [PubMed]
Wilke, V.; Kamphues, J. Effects of substituting wheat by rye in diets for young fattening pigs on nutrient digestibility, performance, products of intestinal fermentation, and fecal characteristics. *Front. Vet. Sci.* **2023**, *10*, 1199505. [CrossRef]
Lisiak, D.; Janiszewski, P.; Grześkowiak, E.; Borzuta, K.; Lisiak, B.; Samardakiewicz, Ł.; Schwarz, T.; Powałowski, K.; Andres, K. Research on the Effects of Gender and Feeding Geese Oats and Hybrid Rye on Their Slaughter Traits and Meat Quality. *Animals* **2021**, *11*, 672. [CrossRef]
Aderibigbe, A.; Cowieson, A.J.; Sorbara, J.O.; Pappenberger, G.; Adeola, O. Growth performance and amino acid digestibility responses of broiler chickens fed diets containing purified soybean trypsin inhibitor and supplemented with a monocomponent protease. *Poult. Sci.* **2020**, *99*, 5007–5017. [CrossRef]
Lexhaller, B.; Colgrave, M.L.; Scherf, K.A. Characterization and relative quantitation of wheat, rye, and barley gluten protein types by liquid chromatography–tandem mass spectrometry. *Front. Plant Sci.* **2019**, *10*, 1530. [CrossRef] [PubMed]
Katoch, R. Nutritional and Anti-Nutritional Constituents in Forages. In *Nutritional Quality Management of Forages in the Himalayan Region*; Springer: Singapore, 2022. [CrossRef]
Piasecka-Kwiatkowska, D.; Warchlewski, J. The cereal protein inhibitors of hydrolytic enzymes and their role. Part I Protein inhibitors of alpha-amylase. *Żywn. Nauka Technol. Jakość* **2000**, *3*, 110–119. (In Polish)
Piasecka-Kwiatkowska, D.; Warchlewski, J. The cereal protein inhibitors of hydrolytic enzymes and their role. Part II Protein inhibitors of proteinases. *Żywn. Nauka Technol. Jakość* **2000**, *3*, 33–38. (In Polish)
Korbas, M.; Mrówczyński, M. *Integrated Production of Winter and Spring Wheat*; IOR-PIB: Poznań, Poland, 2009; pp. 1–166. (In Polish)
Sharma, S.; Kumar, T.; Foulkes, M.J.; Orfor, S.; Singh, A.M.; Wingen, L.U.; Karnam, V.; Nari, L.S.; Mandal Kumar, P.; Griffiths, S.; et al. Nitrogen uptake and remobilization from pre-and post-anthesis stages contribute towards grain yield and grown in limited nitrogen conditions. *CABI Agric. Biosci.* **2023**, *4*, 12. [CrossRef]
Maciorowski, R.; Stankowski, S.; Piech, M. Reaction of hybrid and open-pollinated rye cultivars to nitrogen fertilization and growth regulator. Part I. Grain yield, yield components and selected physiological traits. *Biul. IHAR* **2000**, *215*, 109–120. (In Polish)
Wojcieska, U. The physiological role of nitrogen in shaping plant yield. The influence of nitrogen on the yield of plants. *Postępy Nauka Rol.* **1994**, *1.94*, 115–127. (In Polish)
Faber, A.; Jarosz, Z.; Król, A. The impact of climate change on the efficiency of nitrogen use and its losses. *Probl. Rol. Swiat.* **2019**, *19*, 37–46. (In Polish)
Buksa, K.; Nowotna, A.; Gambuś, H.; Krawontka, J.; Sabat, R.; Noga, M. Technological evaluation and chemical composition of rye grain of selected varieties cultivated by 3 consecutive years. *Acta Agroph.* **2012**, *19*, 265–276. (In Polish)
Jaśkiewicz, B.; Szczepanek, M. Crop mangament and variety have influence on alkylresolcinol content in triticale grain. *Acta Agric. Scand.* **2016**, *66*, 570–574. [CrossRef]
Safar-Noori, M.; Dong, Q.; Saneoka, H. Improvement of grain yield, nutritional and anti-nutritional quality, and seed physiological performance of wheat by NPK fertilization. *Agric. Sci. Technol.* **2018**, *20*, 1467–1477. Available online: http://jast.modares.ac.ir/article-23-17204-en.html (accessed on 24 January 2024).
Tłuścik, F. Localization of the alkylresorcinols in rye and wheat caryopses. *Acta Soc. Bot. Pol.* **1978**. *47*, 211–2018. [CrossRef]
Album, H.G.; Umbereit, W.W. Differentiation between ribose-3phosphate and ribose-5 phosphate by means of the orcinol-pentose reaction. *J. Biol. Chem.* **1947**, *167*, 369–376. [CrossRef]
PN-EN ISO 14902:2001; Animal Feeding Stuffs—Determination of Trypsin Inhibitor Activity of Soya Products. International Organization for Standardization: Geneva, Switzerland, 2001.
R Core Team. *R: A Language and Environment for Statistical Computing*; R Foundation for Statistical Computing: Vienna, Austria, 2003.
Latusek, A.; Bujak, H. Genotype-environment interaction for yield of winter rye cultivars cultivated with the two levels of agricultural technology in climate conditions of Lower Silesia. *Biul. IHAR* **2012**, *265*, 47–57. (In Polish) [CrossRef]
Wyzińska, M.; Grabiński, J.; Sułek, A. Comparison on the profitability of cultivation of winter rye depending on the different production technology. *Agron. Sci.* **2022**, *LXXVII 1*, 45–52. [CrossRef]
Szuleta, E.; Phillips, T.; Knott, C.A.; Lee, C.D.; Van Sanford, D.A. Influence of Planting Date on Winter Rye Performance in Kentucky. *Agronomy* **2022**, *12*, 2887. [CrossRef]
Hadasch, S.; Laidig, F.; Macholdt, J.; Bönecke, E.; Piepho, H.P. Trends in mean performance and stability of winter wheat and winter rye yields in a long-term series of variety trials. *Field Crops Res.* **2020**, *252*, 107792. [CrossRef]
Miedaner, T.; Haffke, S.; Siekmann, D.; Fromme, F.J.; Roux, S.R.; Hackauf, B. Dynamic quantitative trait loci (QTL) for plant height predict biomass yield in hybrid rye (*Secale cereale* L.). *Biomass Bioenergy* **2018**, *115*, 10–18. [CrossRef]
Goncharov, A.A.; Safonov, T.A.; Malko, A.M.; Bocharov, G.A.; Goncharov, S.V. Climate change expected to increase yield of spring cereals and reduce yield of winter cereals in the Western Siberian grain belt. *Field Crops Res.* **2023**, *302*, 109038. [CrossRef]
Dopierała, P.; Bujak, H.; Kaczmarek, J.; Dopierała, A. Ocena interakcji genotypowo-środowiskowej plonu populacyjnych odmian żyta ozimego. *Biul. IHAR* **2003**, *230*, 243–253. (In Polish)
Noland, R.L. Rye grain response to nitrogen fertilizer and seeding rate. *Agrosyst. Geosci. Environ.* **2022**, *5*, e20239. [CrossRef]

52. Tupits, I.; Tamm, I.; Magistrali, A.; Rempelos, L.; Cakmak, I.; Leifert, C.; Grausgruber, H.; Wilkinson, A.; Butler, G.; Bilsborrow Evaluating the effect of nitrogen fertilizer rate and source on the performanceof open-pollinated rye (*Secale cereale* L.) cultiv incontrasting European environments. *Crop Sci.* **2022**, *62*, 928–946. [CrossRef]
53. Szuleta, E.; Shockley, J.M.; Knott, C.; Phillips, T.; Van Sanford, D.A. Influence of nitrogen rate on yield and profitability of grain production. *Crop Forage Turfgrass Manag.* **2023**, *9*, e20243. [CrossRef]
54. Studnicki, M.; Macholdt, J.; Macdonald, A.; Stępień, W. Effects of Fertilizers and Manures on Temporal Yield Variability of Wir Rye. *Agronomy* **2021**, *11*, 519. [CrossRef]
55. Bellato, S.; CIccoritti, R.; Del Frate, V.; Sqrulletta, D.M.; Carbone, K. Influence of genotype and environment on the content 5-n alkylresorcinols, total phenols and on the antiradical activity of whole durum wheat grains. *J. Cereal Sci.* **2013**, *57*, 162–1 [CrossRef]
56. Korge, M.; Alaru, M.; Keres, I.; Khaleghdoust, B.; Möll, K.; Altosaar, I.; Loit, E. The influence of cropping system, weat conditions and genotype on arabinoxylan content in wheat and barley grains. *J. Cereal Sci.* **2023**, *110*, 103650. [CrossRef]
57. Marentes-Culma, R.; Coy-Barrera, E. Effect of Soil Type on the Temporal and Spatial 5-n-alk (en) ylresorcinol Variation in *Triti* Grown Under Greenhouse Conditions. *J. Soil Sci. Plant Nutr.* **2022**, *22*, 4428–4437. [CrossRef]
58. Kulawianek, M.; Jaromin, A.; Kozubek, A.; Zarnowski, R. Alkylresorcinols in selected Polish rye and wheat cereals a whole-grain products. *J. Agric. Food Chem.* **2008**, *56*, 7236–7242. [CrossRef]
59. Targońska-Karasek, M.; Kwiatek, M.; Groszyk, J.; Walczewski, J.; Kowalczyk, M.; Pawelec, S.; Boczkowska, M.; Rucińska, Characteristic of the gene candidate SecARS encoding alkylresorcinol synthase in Secale. *Mol. Biol. Rep.* **2023**, *50*, 8373–83 [CrossRef]
60. Sun, T.; Zhang, Y.; Yang, O.; Ban, J.; Guo, B.; Zhang, B.; Zhao, H. Study on the spatio-temporal variation of arabinoxylan a alkylresorcinol in wheat grains. *J. Food Compos. Anal.* **2023**, *117*, 105103. [CrossRef]
61. Zhi, L.; Wang, F.; Xu, K.; Zhang, Z.; Zhu, Q.; Zhang, Y.; He, Z. Effects of nitrogen fertilization rate and environment on composition of alkylresorcinols in three elite winter wheat cultivars. *J. Cereal Sci.* **2023**, *112*, 103718. [CrossRef]
62. Kulichová, K.; Sokol, J.; Nemeček, P.; Maliarová, M.; Maliar, T.; Havrlentová, M.; Kraic, J. Phenolic compounds and biologi activities of rye (*Secale cereale* L.) grains. *Open Chem.* **2019**, *17*, 988–999. [CrossRef]
63. Simonetti, E.; Bosi, S.; Negri, L.; Dinelli, G. Amylase Trypsin Inhibitors (ATIs) in a Selection of Ancient and Modern Wheat: Eff of Genotype and Growing Environment on Inhibitory Activities. *Plants* **2022**, *11*, 3268. [CrossRef] [PubMed]
64. Jaśkiewicz, B.; Ochmian, I. Alkylresorcinol content in grains of spring triticale cultivars depending on the soil tillage system a nitrogen fertilization level. *Agron. Sci.* **2022**, *77*, 27–35. [CrossRef]
65. Takač, V.; Tóth, V.; Rakszegi, M.; Mikó, P.; Mikić, S.; Mirosavljević, M. The Influence of Farming Systems, Genotype and Th Interaction on Bioactive Compound, Protein and Starch Content of Bread and Spelt Wheat. *Foods* **2022**, *11*, 4028. [CrossR [PubMed]
66. Noori, M.S. Alleviation of drought stress in wheat (*Triticum aestivum* L.) by mineral fertilization. *J. Stress Physiol. Biochem.* **20** *17.1*, 82–93.
67. Perić, V.; Srebrić, M.; Jankuloski, L.; Jankulovska, M.; Žilić, S.; Kandić, V.; Mladenović Drinić, S. The effects of nitrogen on prote oil and trypsin inhibitor content of soybean. *Genetika* **2009**, *41*, 137–144. [CrossRef]
68. Xue, C.; Matros, A.; Mock, H.P.; Mühling, K.H. Protein composition and baking quality of wheat flour as affected by split nitrog application. *Front. Plant Sci.* **2019**, *10*, 642. [CrossRef] [PubMed]

agriculture

MDPI

rticle

\biotic and Biotic Factors Affecting Crop Growth and ʼroductivity: Unique Buckwheat Production in Egypt

Iohamed M. Hassona [1,2,*], Hala A. Abd El-Aal [1], Nahla M. Morsy [1] and Ahmed M. S. Hussein [3]

1 Department of Sustainable Development of Environment and Its Projects Management, Environmental Studies & Research Institute (ESRI), University of Sadat City (USC), Sadat City 32897, Menoufia, Egypt; hala.ahmed@esri.usc.edu.eg (H.A.A.E.-A.); nahla.morsy@esri.usc.edu.eg (N.M.M.)

2 Qur'anic Botanic Garden, Hamad bin Khalifa University (HBKU), Doha 34110, Qatar

3 Food Technology Department, National Research Center, Dokki 12622, Cairo, Egypt; a_said22220@yahoo.com

* Correspondence: mhassona@hbku.edu.qa or mmamh83@gmail.com

Abstract: Egypt did not previously grow buckwheat, due to the belief that the environment does not meet the factors of growth, development, and productivity in an arid or semiarid region. The present study investigated two species of buckwheat, *Fagopyrum tataricum* (FT) and *Fagopyrum esculentum* (FE), which were planted in two different soil, weather, and water property sites, including the first in Belbies City and the second in Sadat City in the middle of January, November, and March for two successive seasons, 2018–2019 and 2019–2020. The study uniquely focuses on Egypt to investigate three interactions of location × species, location × sowing date, and species × sowing date on growth and productivity. The parameters measured included plant height (cm), number of branches, internodes, and leaves per plant, fresh weight (gm), number of grains per plant, grain weight of 1 m^2, and yield (kg/Ha). Our results indicated significant differences in all measured interactions. For the location × species interaction, FT planted in Belbies City consistently outperformed all other combinations, with a plant height = 97.704 cm in the 1st season and productivity = 859.38 kg/ha in the 1st season, while FE in Sadat showed the lowest growth and productivity. For the interaction of location × sowing date, Belbies × mid-March sowing achieved the highest plant heights of 84.89 cm and 75.44 cm, and productivity of 702.88 kg/ha and 708.21 kg/ha in consecutive seasons. Conversely, Sadat City × Mid-March sowing resulted in the lowest plant heights of 57.500 cm, and 57.667 cm, and productivity of 490.67 kg/ha, and 444.55 kg/ha. The species × sowing date interaction further emphasized the superiority of FT sown in mid-March, which led to the best plant height growth of 95.78 cm in the 1st season and the highest productivity of 837.55 kg/ha in the 1st season. In contrast, FE sown in mid-March exhibited the poorest outcomes. The study provides an understanding of the two-way interactions affecting buckwheat cultivation in Egypt. Our results indicated its viability with appropriate species selection and sowing dates, contributing to agricultural diversity and sustainability.

Keywords: buckwheat; Egyptian environment; *Fagopyrum tataricum*; *Fagopyrum esculentum*; interaction effect; productivity

1. Introduction

Agriculture, at the heart of global sustainability and food security, faces unprecedented challenges in the era of climate change [1]. Arid and semiarid regions, such as those predominant in Egypt, are particularly vulnerable, necessitating innovative agricultural practices and crop diversification. Among the various crops, buckwheat stands out for its exceptional adaptability, nutritional benefits, and potential to thrive under water-limited conditions, presenting an untapped resource for regions experiencing climatic adversity [2]. However, the exploration of underutilized crops such as buckwheat represents a frontier in agricultural research driven by the need to address food security and sustainable farming practices in arid environments. Recent studies have begun to underscore the resilience of crops to poor soil conditions, drought, and low water availability, making them attractive options for enhancing agricultural diversity in challenging landscapes in Egypt [3,4]. The agricultural history of Egypt is marked by a reliance on water-intensive crops, an approach that is increasingly unsustainable given the country's water scarcity issues and the salinization of arable lands. The introduction of buckwheat, a crop previously overlooked due to misconceptions about its suitability, represents a paradigm shift toward sustainable agriculture, leveraging its minimal water requirement and adaptability [5]. However, this research challenges existing agricultural paradigms by demonstrating buckwheat's viability in Egypt, thereby contributing to the diversification of crops—a cornerstone of food security and ecological sustainability. By dissecting the interaction between abiotic and biotic factors, this study aimed to identify new agricultural practices that boost productivity while ensuring resilience against climate change [6,7]. Nevertheless, the interaction of genetic, environmental, and management factors is crucial in understanding crop growth and productivity, an area of agricultural science that has become increasingly relevant with the advent of climate change and the push for sustainable practices. Buckwheat has emerged as a crop of interest due to its adaptability to diverse climatic and soil conditions, underscoring the importance of this study within Egypt's unique agricultural patterns [8]. The variability in growth responses among buckwheat species to different environmental conditions highlights the significance of species selection, highlighting the complexity of genotype–environment interactions and their impact on crop performance across various settings [3,7,9,10]. Moreover, the influence of abiotic factors, such as soil conditions and external treatments, on buckwheat emphasizes the role of these factors in crop adaptability [6,11]. Environmental factors such as altitude and UV radiation also contribute to the complexity of buckwheat cultivation and have significant effects on growth and productivity [12,13]. Moreover, the impact of sowing date on buckwheat yield illustrates the significance of temporal factors in agricultural outcomes [5,14,15]. Challenges posed by seasonal transitions and specific soil conditions, such as salinity, necessitate a deeper understanding of optimal cultivation practices [4,16]. However, soil health and management practices play a critical role in determining crop productivity, highlighting the importance of soil conditions and agronomic practices in achieving sustainable yields [17–19]. The history of land use and previous cropping systems influences the success of current crops, revealing the intricate relationship between past and present agricultural practices [20,21].

Furthermore, the classification of *Fagopyrum* sp. involves distinguishing between *F. esculentum* and *F. tataricum*, which are species whose biological differences are emphasized. *F. esculentum*, commonly known as a common buckwheat species, is an annual Asian herb with small pinkish or white flowers and edible triangular seeds. In contrast, *F. tataricum*, which is also an erect annual herb, has a smaller seed size. Studies have shown that *F. esculentum* flowers earlier than *F. tataricum* and produces fewer nodes, branches, and inflorescences but more flowers per inflorescence [2,17]. Additionally, *F. tataricum* is more resistant to water stress and exhibits traits of drought tolerance, while *F. esculentum* has characteristics of drought avoidance [22]. Furthermore, there are differences in genome size between these species, with *F. esculentum* having a larger genome size than *F. tataricum* [7,18]. However, the influence of location, species choice, and sowing date on buckwheat growth and productivity, as well as the variability in certain buckwheat species in response to environmental conditions [3,9] and the role of genotype–environment interactions [7,10], and the effects of external treatments and soil conditions on buckwheat plants further expand our understanding of these interactions [6,11]. Additionally, the influence of altitude and UV radiation provides further context for understanding the environmental sensitivity of buckwheat [12,13], further to the impact of sowing date on buckwheat productivity [5,14,15]. Nevertheless, the transition between seasons [16] and the specific challenges of saline soils [4] further contribute to the understanding of the optimal conditions for buckwheat cultivation and the critical role of soil conditions in crop productivity. Previous crops and cultivation practices were influenced, along with insights into weed suppression and no-tillage systems [17–21,23–25].

This research investigated the interaction between biotic and abiotic factors in buckwheat cultivation in arid regions of Egypt, aiming to optimize growth conditions and unlock the potential for resilience and nutrition in sustainable farming systems. By exploring buckwheat's viability in these challenging climates, this study challenges traditional agricultural practices and offers insights into enhancing food security and sustainability amid environmental challenges. These findings suggest that selecting appropriate species and sowing dates, in addition to suitable farming practices, can significantly improve buckwheat productivity in Egypt, supporting the goals of agricultural diversification and sustainability in response to changing climate conditions.

2. Materials and Methods

Field experiments were conducted during the 2018–2019 and 2019–2020 seasons at two locations: the Belbies City site (BCS) and the experimental farm of the Environmental Studies & Research Institute (ESRI), University of Sadat City (SCS) in Sadat City, Monofiya. These experiments aimed to evaluate the growth and productivity of buckwheat plants under varying climatic conditions, focusing on the morphological characteristics and yield of grains from different buckwheat species and sowing dates.

The experimental treatments were as follows:

2.1. Materials

Locations of Cultivation: BCS Was Located at 30.4196° N, 31.5619° E, and SCS Was Located at 30.3594° N, 30.5327° E (Scheme 1).

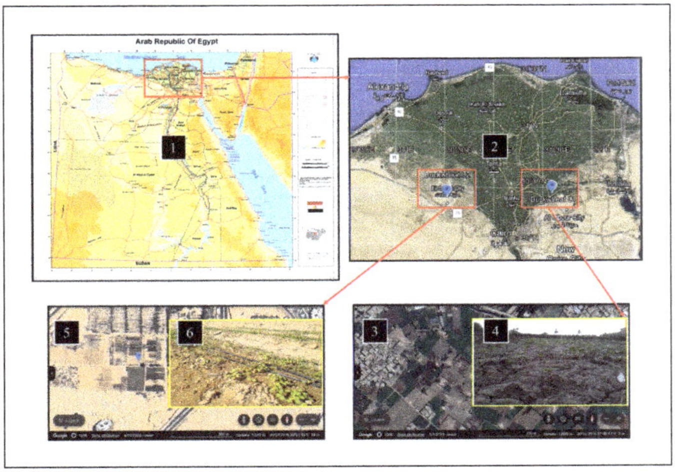

Scheme 1. The study locations where the typography of each planting site showed, where: 1 = Map of Egypt (sourced by CAMPASS, 2023). 2 = A focused satellite image of the regions of both study locations (by Google Earth Web, 2023). 3 & 4 are satellite images of the planting sites of Belbies City and Sadat City, respectively (sourced by Google Earth Web, 2023). 5 & 6 are real images of the planting sites at the early stages of growth in Belbies City and Sadat City, respectively. Photos edited and compiled by the authors.

Species: Fagopyrum esculentum, Represented by "Japanese"; Fagopyrum tataricum, by "Madawaska"; Both Cultivars were Purchased Online from a Commercial Seed Company in California, United States, and Shipped to Egypt following the force Laws of the State Ministry of Agriculture for Quarantining and Importing Seeds.

Sowing Dates

Season 1: Second week of November 2018, January 2019, and March 2019
Season 2: Second weeks of November 2019, January 2020, and March 2020

2.2. Methods

2.2.1. Seeds

The seeds were sealed in their original bags and then opened in the field for planting. The first batch was planted at the Belbies City site as afeer cultivation [14] (dry seeds in a dry land, a common planting method widespread in the cultivation of grain, e.g., wheat). The seeds were mixed with sand at a 1:2 v/v ratio to control the spread of the seeds on the designated plot and ensure that all the seeds were well sown with uniformity. The experiment used a rate of 160 seeds/m^2 for all 3 sowing dates at both locations, and the first planting occurred at the Belbies City site in mid-November 2018. A total of 6 plots were planted in each patch. Three plots were randomly distributed for *F. esculentum*, and three were allocated for *F. tataricum*; each plot was an area of 4.5 m^2 planted with ±720 seeds weighing approximately 21.800 g. In the second location, the seeds were sowed manually by clever farmers and then land scarfed via a handy fork to ensure the uniformity of the seeds and ensure that the soil was covered by a layer of soil at least 2 cm in length. However, due to the land structure, a raised bed is commonly used there, so the seeds were spread on the upper side of the raised bed where each plot was 6 m × 0.75 m = 4.5 m^2, with the same rate of seed and sand mixture used in the first location. Irrigation was applied to both sites directly after sowing. However, at the location of Sadat City, the raised beds were covered by plastic sheets to avoid bird attacks.

2.2.2. Experimental Site Analysis

At the beginning of each experimental season, three soil samples from the top layer (0–30 cm) were randomly collected from the experimental sites. Then, the samples were air dried, ground, and sieved through a 2 mm sieve and then bulked together before being subjected to soil analysis for classifying soil texture through particle size distribution analysis (mechanical analysis) as well as determining some soil chemical characteristics [7]. The physical and chemical properties of the experimental soil during both seasons of the study are presented in Table 1, and the water properties, including physical, chemical, and biological parameters, were analysed at both sites, as shown in Table 2. However, weather data for both study sites, including temperature, total precipitation, relative humidity, wind gusts, sunshine duration, and soil temperature [0–7 cm down], were obtained by the weather model Nonhydrostatic Meso-Scale Modelling Technology (NMM) of Meteoblue based in Basil, Switzerland, as shown in Table 3.

Table 1. Physical and chemical properties of the experimental soil samples.

Location	Parameter	Belbies City Site	Sadat City Site
Soil Properties			
pH	-	7.83	8.57
EC (dS/m)	Electrical Conductivity	1.48	1.11
SP %	Saturation Percentage	25	-
Soluble Anions (meq/L)			
CO_3^{2-}	Carbonate	-	-
HCO_3^-	Bicarbonate	0.25	1.9
Cl^-	Chloride	12.1	7.4
SO_4^{2-}	Sulfate	1.56	3.4

Table 1. *Cont.*

Location	Parameter	Belbies City Site	Sadat City Site
Soluble Cations (meq/L)			
Ca^{2+}	Calcium	3.7	1.2
Mg^{2+}	Magnesium	1.8	1.2
Na^{+}	Sodium	8	3.3
K^{+}	Potassium	0.5	0.18
Soil Composition			
$CaCO_3$%	Calcium Carbonate	3.9	1.7
OM %	Organic Matter	1.2	0.35
Available Nutrients (mg/kg)			
N	Nitrogen	175	-
P	Phosphorus	183.6	5.3
K	Potassium	7.96	508
Cu	Copper	0.75	1.43
Fe	Iron	5.25	1.11
Mn	Manganese	2.43	20
Zn	Zinc	2	2.75
Mechanical Analysis (%)			
Sand	-	83	83
Silt	-	15.5	10
Clay	-	1.2	7
Texture Grade	-	Clay sandy	Sandy loamy

Notes: EC (dS/m): Reflects soil salinity; lower values indicate less salinity. OM %: Indicates soil fertility; higher values suggest more organic matter and better soil health. Soluble anions and cations are expressed in milliequivalents per liter (meq/L), indicating the concentration of each ion in the soil solution Available levels of nutrients (P, K, Cu, Fe, Mn, Zn) are now correctly presented in milligrams per kilogram (mg/kg) for clarity, reflecting the concentration of each nutrient available in the soil for plant uptake. The "-" indicates data not available or not applicable.

Table 2. Irrigation water properties of the experimental soil samples.

Parameter	Unit	Belbies City Site	Sadat City Site
pH	-	7.41	7.98
EC	dS/m	0.64	1.26
Chemical Parameters			
Ca^{2+}	meq/L	1.70	4.20
Mg^{2+}	meq/L	1.00	2.60
Na^{+}	meq/L	3.20	5.50
K^{+}	meq/L	0.01	0.20
$CO_3{}^{2-}$	meq/L	0.00	0.00
$HCO_3{}^{-}$	meq/L	0.21	4.00
Cl^{-}	meq/L	4.70	7.00
$SO_4{}^{2-}$	meq/L	10.9	2.50
SAR	-	2.76	3.50
TDS	mg/L	410.00	896.00
Trace Elements			
Fe	ppm	0.27	0.11
Zn	ppm	0.01	0.18
Mn	ppm	0.20	<0.01
Cu	ppm	0.02	<0.01

Notes: EC (Electrical Conductivity) is measured in deciSiemens per meter (dS/m), indicating the water's salinity level. Lower values suggest lower salinity, which is generally better for irrigation purposes. SAR (Sodium Adsorption Ratio) is a measure of the suitability of water for use in agricultural irrigation, based on the sodium (Na^{+}) content relative to calcium (Ca^{2+}) and magnesium (Mg^{2+}) contents. TDS (Total Dissolved Solids) is measured in milligrams per liter (mg/L), reflecting the total concentration of dissolved substances in water. Higher TDS values indicate more dissolved minerals and salts. Trace elements like Fe (Iron), Zn (Zinc), Mn (Manganese), and Cu (Copper) are presented in parts per million (ppm).

Table 3. Weather data of the study locations.

Location / Parameter	Belbies City Site (BCS)		Sadat City Site (SCS)	
	Season 1 (2018–2019)	Season 2 (2019–2020)	Season 1 (2018–2019)	Season 2 (2019–2020)
Temperature [°C] Max	27.31–36.71	19.04–35.8	18.37–35.86	17.9–33.92
Temperature [°C] Min	8.51–24.48	9.75–21.27	7.97–23.87	9.57–20.64
Temperature [°C] Mean	13.65–30.37	14.18–28.24	12.92–29.33	13.61–26.81
Precipitation [mm] Total	0.00–0.47	0.00–1.42	0.00–2.00	0.00–1.12
Relative Humidity [%] Max	68.3–85.94	75.52–81.57	70.88–89.44	79.23–84.67
Relative Humidity [%] Min	19.63–33.3	22.22–35.56	21.26–37.89	24.26–42.92
Wind Gust [km/h] Max	27.53–40.62	30.6–39.4	28.36–41.82	30.7–41.12
Sun Duration [min] Total	492.23–777.84	413.8–804.67	514.65–788.43	402.65–802.5
Soil Temperature [°C] Max	28.01–40.08	19.7–38.58	20.32–43.27	20.31–40.71
Soil Temperature [°C] Min	11.27–28.89	12.57–26.29	9.00–26.96	10.87–23.56
Soil Temperature [°C] Mean	15.22–34.03	15.09–31.85	13.98–33.97	15.02–31.14

2.2.3. Experimental Site Preparation

The experimental soil was prepared by land plowing in two perpendicular directions, and all the experimental plots received organic manure as compost at a rate of 16 tons/hectare, phosphorus fertilizer at 80 units P_2O_5/hectare as calcium superphosphate (15.5% P_2O_5), and agricultural sulfur as agricultural sorghum. Superfine 98% S at 180 kg/hectare full doses of compost, superphosphate, and agricultural sulfur were applied once during the final preparation of the experimental soil and thoroughly mixed with the soil before the ridges were constructed. However, nitrogen fertilizer was applied at 180 units N/hectare in the form of ammonium sulfate (20.6% N). The nitrogen dose was split into 3 portions. The first portion was added 15 days after the seeding date, and the remaining portions were subsequently added at a rate of one portion every 15 days until grain setting. Nitrogen fertilizer was applied manually at the Belbies site and through injection in a drip irrigation system as a plant fertigation application at the Sadat site.

2.2.4. Data Recorded

Plant Vegetative Growth Parameters

After 70 days after sowing and after all the experimental treatments were carried out, a random and representative sample of 9 buckwheat plants from each experimental subplot was taken by cutting at ground level (at 2–3 cm above the ground surface) and subsequently transferred to the laboratory of the Environmental Studies and Research Institute to determine the following vegetative growth parameters.

Plant Height (cm)

The average plant height was recorded in centimeters (cm) from the point of contact of the plant stem with the soil up to the highest point of the inflorescence on the main stem of the buckwheat plant samples in each experimental subplot.

Number of Branches per Plant

The average number of lateral branches per plant was counted for the buckwheat plant samples in each experimental subplot, and the average number of branches per plant was calculated.

Whole-Plant Fresh Weight (gm)

The average fresh weight of the whole buckwheat plant (stem, lateral branches, and leaves) was recorded in grams (gm) by using a 3-decimal digital electronic balance directly after being taken from the fields of the study.

Harvesting of Buckwheat Grain Samples

At the harvesting stage, ±75 days after the sowing date, a representative sample of 9 individual buckwheat plants (group A plants), in addition to the plant samples harvested from one square meter from the middle of each experimental subplot (group B plants), was randomly taken by cutting at ground level (at 2–3 cm above the ground surface). Then, all the plant samples were left to air dry for at least 7–10 days for the plants planted in mid-November and January and for 3–5 days for the plants planted in mid-March. Afterward, each of the 10 dried plant samples was threshed by hand to obtain the grains from the dried main and lateral panicles per group of plants (group A plants) or per all plant samples harvested from each experimental subplot (group B plants). Moreover, the buckwheat grains were manually ground to clean and remove any impurities, foreign matter, or broken or immature grains. The cleaned grains of group B plants were used directly to determine the number of grains/plant, total grain yield per square meter (kg), and grain yield (kg/Ha), where Ha = 10,000 square meters.

Number of Seeds per Plant

The average number of seeds per plant in each experimental subplot was determined by manually counting the grains extracted from 9 plants in the dried main and lateral panicles of individual plants in group A. The total number of seeds was subsequently divided by 9 to obtain the grain count per plant.

Total Grain Weight per Square Meter (kg)

The average total grain yield per square meter was calculated by quantifying the number of harvested plants (group B) within one square meter. Then, the dried plants were threshed by hand to obtain the grains from the dried main and lateral panicles, and a 3-decimal-digit electronic balance was used to obtain the average grain yield/meter^2 in kilograms (kg) units.

Total Grain Yield per Hectare (kg/ha)

The average total grain yield per hectare was calculated by multiplying the average grain yield/square meter (g) by the average number of buckwheat plants per hectare (a plant density of approximately 122 thousand plants per hectare) to obtain the average grain yield/hectare in kilograms per hectare (kg/ha).

2.3. The Statistical Analysis

A randomized complete block design incorporating three factors was used for the data analysis, with each parameter being replicated thrice. The treatment means were subjected to pairwise comparison using the least significant difference (LSD) test, as described by

Snedecor and Cochran (1994) [26]. The data analysis was performed utilizing the Assistat software program, version 1.0.

3. Results

3.1. Evaluation of the 2-Way Interaction Effect of Biotic and Abiotic Factors on the Growth of Buckwheat

3.1.1. Interaction between Location and Species

Through our results shown in Table 4 and Figure 1, we observed interaction effects between location (Belbies City vs. Sadat City) and species (*Fagopyrum tataricum* vs. *Fagopyrum esculentum*) on various growth characteristics of buckwheat over two seasons. For plant height, in both seasons, *Fagopyrum tataricum* plants grown in Belbies City had the highest average plant height, significantly surpassing those of all the other combinations. The least significant difference (LSD) at the 0.05 level highlights the statistical significance of these differences, clearly indicating the superior growth performance of *Fagopyrum tataricum* in Belbies City. For the branch number per plant, similar patterns were observed for the number of branches, with *Fagopyrum tataricum* in Belbies City producing the most branches compared to the other groups. The differences between the groups were statistically significant, as confirmed by the L.S.D., underscoring the combined influence of species and location on this growth characteristic. For the internode number per plant, the trend continues with the number of internodes, where *Fagopyrum tataricum* in Belbies City exhibited the highest count, significantly outperforming the buckwheat grown under other conditions. These results are consistent across both seasons, with the L.S.D. values affirming the significance of these findings. For the number of leaves per plant, *Fagopyrum tataricum* in Belbies City also had a significantly greater number of leaves than the other plants, indicating that a robust growth pattern was influenced by both the choice of species and the cultivation location. These differences were validated by the corresponding L.S.D. values. Finally, for the fresh weight per plant, to reflect the trends in the aforementioned parameters, *Fagopyrum tataricum* plants grown in Belbies City also had the highest fresh weight per plant, indicating that this combination was the most productive in terms of biomass accumulation. The statistical analysis, supported by the L.S.D. values, confirmed the impact of species and location on this critical growth metric.

Table 4. Interaction between location and species and its effect on buckwheat growth characteristics.

Parameters	Treatments	*Fagopyrum tataricum*		*Fagopyrum esculentum*		L.S.D. at 0.05
		Belbies City	**Sadat City**	**Belbies City**	**Sadat City**	
Plant height cm per plant	*1st Season*	97.704 a	85.667 b	75.296 c	35.463 d	6.6183
	2nd Season	90.741 a	78.333 b	63.185 c	41.704 d	3.6779
Branch number per plant	*1st Season*	12.000 a	10.111 b	8.000 c	6.185 d	0.476
	2nd Season	10.704 a	9.445 b	7.778 c	6.815 d	0.334
Internode number per plant	*1st Season*	14.111 a	11.037 b	8.889 c	7.185 d	0.597
	2nd Season	12.259 a	10.889 b	9.037 c	7.296 d	0.3517
Number of leaves per plant	*1st Season*	46.852 a	27.889 b	18.778 c	11.037 d	2.6427
	2nd Season	31.778 a	19.630 b	14.370 c	8.556 d	2.872
Fresh weight per plant gm	*1st Season*	36.188 a	22.187 b	14.330 c	10.171 d	1.6143
	2nd Season	39.356 a	24.771 b	16.013 c	11.31 d	1.881

where L.S.D. = least significant difference. Different letters within the same row indicate significant differences ($p \leq 0.05$).

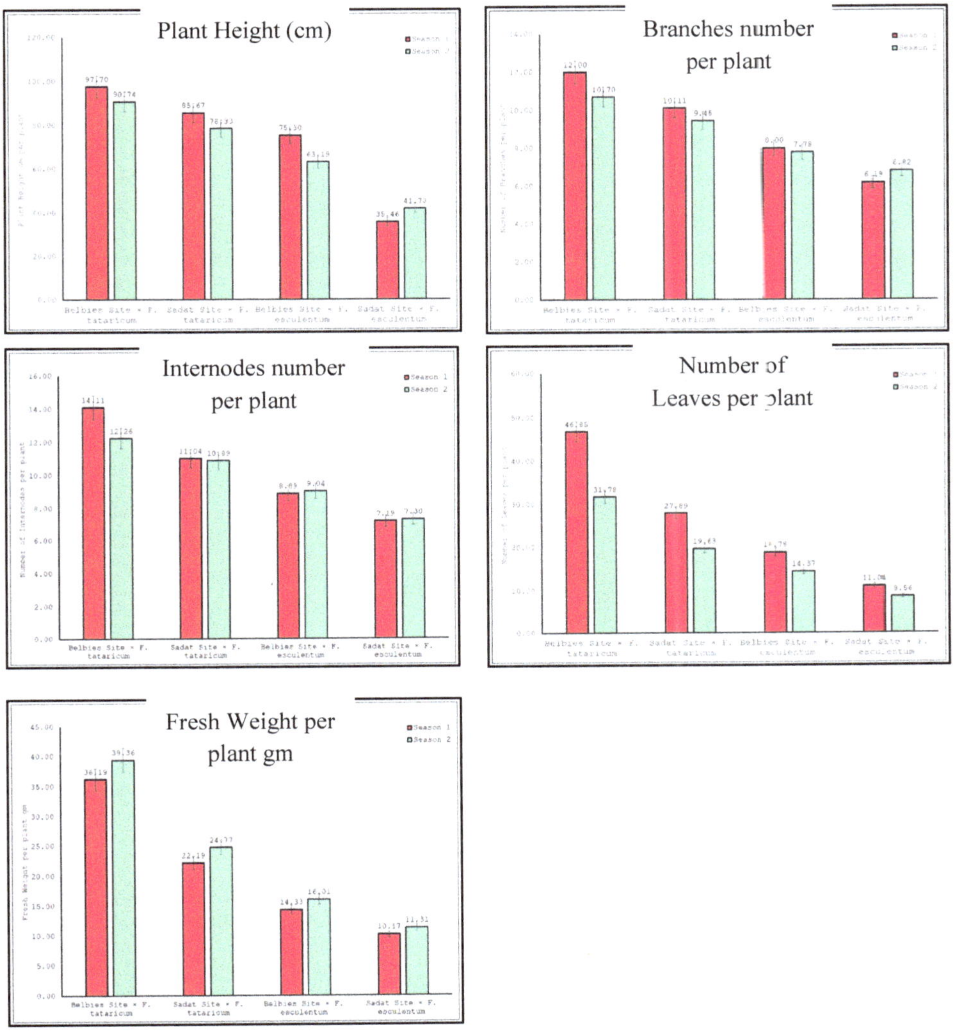

Figure 1. Evaluation of location × species on the growth of buckwheat.

3.1.2. Interaction of Location × Planting Date

The results in Table 5 and Figure 2 show the interaction effects of location and sowing time on the growth of buckwheat plants across two parameters: physical growth characteristics and fresh weight per plant. The study differentiates results by location (Belbies City site vs. Sadat City site) and by sowing time (mid-January, mid-November, mid-March), presenting findings for two growing seasons. For plant height, in both the first and second seasons, plant height was consistently greater in Belbies City than in Sadat City across all sowing times. The least significant difference (L.S.D.) at the 0.05 level for plant height indicates significant differences between locations, with no significant variation among sowing times within the same location. For instance, in the first season, plant heights in Belbies City ranged from 84.889 cm to 88.222 cm, which was significantly greater than those in Sadat City, where heights ranged from 57.500 cm to 62.806 cm ($p \leq 0.05$). For the branch and internode number, the number of branches and internodes per plant followed a similar pattern to that of plant height, with plants in Belbies City exhibiting more branches and internodes than those in Sadat City, regardless of sowing time. The L.S.D. values confirmed these differences as statistically significant ($p \leq 0.05$), demonstrating the impact of location over sowing time on these growth parameters. The number of leaves per plant showed notable variability, with significant differences observed not only between locations but also among sowing times within the same location, especially in the first season. For example, in Belbies City, during the first season, sowing in mid-March resulted in the highest number of leaves (35.778), which was significantly greater than that of the other sowing times, indicating that both location and sowing time influence leaf production. In terms of fresh weight per plant, fresh weight per plant demonstrated significant variation influenced by both location and sowing time, with higher weights generally observed in Belbies City across all sowing times. In the second season, sowing in mid-March in Belbies City resulted in the highest fresh weight (28.857 g), which was significantly greater than that of all the measurements in Sadat City, where the maximum observed weight was 20.921 g for mid-March sowing.

Table 5. Interaction effect of location × sowing time on the growth of buckwheat plants.

Parameters	Treatments	Belbies City Site			Sadat City Site			L.S.D. at 0.05
		Mid-Jan	Mid-Nov	Mid-Mar	Mid-Jan	Mid-Nov	Mid-Mar	
Plant height cm per plant	*1st Season*	88.222 a	86.389 a	84.889 a	62.806 b	61.389 b	57.500 b	8.106
	2nd Season	77.667 a	77.778 a	75.444 a	60.444 b	61.945 b	57.667 b	4.504
Branch number per plant	*1st Season*	10.056 a	9.944 a	10.000 a	8.222 b	8.111 b	8.111 b	0.583
	2nd Season	9.223 a	9.111 a	9.389 a	8.111 b	8.222 b	8.056 b	0.409
Internode number per plant	*1st Season*	11.611 a	11.444 a	11.445 a	9.000 b	9.056 b	9.278 b	0.731
	2nd Season	10.778 a	10.778 a	10.389 a	9.222 b	9.222 b	8.833 b	0.431
Number of leaves per plant	*1st Season*	32.500 a	30.167 b	35.778 b	17.722 c	19.444 cd	21.222 d	3.237
	2nd Season	21.889 b	21.889 b	25.444 a	13.834 c	14.500 c	13.944 c	3.517
Fresh weight per plant gm	*1st Season*	24.817 a	24.323 ab	26.637 b	15.433 c	15.239 d	17.810 d	1.977
	2nd Season	26.591	27.604 a	28.857 a	16.222 c	16.979 c	20.921 b	2.304

where L.S.D. = least significant difference. Different letters within the same row indicate significant differences ($p \leq 0.05$).

3.1.3. Interaction Effect of Species × Sowing Time

Our results in Table 6 and Figure 3 detail the interaction effects of species (*Fagopyrum tataricum* × *Fagopyrum esculentum*) and planting time (mid-March, mid-January, mid-November) on the growth of buckwheat across various growth parameters over two growing seasons. For plant height, the data showed that, compared with *Fagopyrum esculentum*, *Fagopyrum tataricum* consistently achieved greater plant heights across all planting times, with significant differences noted ($p \leq 0.05$). For instance, in the first season, the height reached 95.778 cm for *Fagopyrum tataricum* sown in mid-March, which was significantly taller than that of *Fagopyrum esculentum* in the same period, which reached only 46.611 cm. An L.S.D. of 0.05 validates the statistical significance of these differences. In terms of the number of branches, similar trends were observed for the number of branches, where *Fagopyrum tataricum* plants consistently had more branches than *Fagopyrum esculentum* plants across all planting times. The first season showed a significant difference with *Fagopyrum tataricum* sown in Mid-March having 11.722 branches, contrasting with 6.389 branches for *Fagopyrum esculentum* sown in the same period, supported by a statistically significant L.S.D. The number of internodes followed a pattern akin to plant height and branch number, with *Fagopyrum tataricum* exhibiting a higher number of internodes than *Fagopyrum esculentum*. This effect was consistent across all planting times and both seasons, demonstrating the influence of the species on this growth parameter. However, for the number of leaves, *Fagopyrum tataricum* plants also had a significantly greater number of leaves than *Fagopyrum esculentum* plants, with the greatest disparity observed in the first season. The difference in leaf count between the species was statistically significant, as indicated by the L.S.D. values. Finally, for fresh weight per plant, the fresh weight of *Fagopyrum tataricum* plants was significantly greater than that of *Fagopyrum esculentum* plants across all planting times and seasons. This parameter clearly illustrates the superior growth performance of *Fagopyrum tataricum* under the conditions tested.

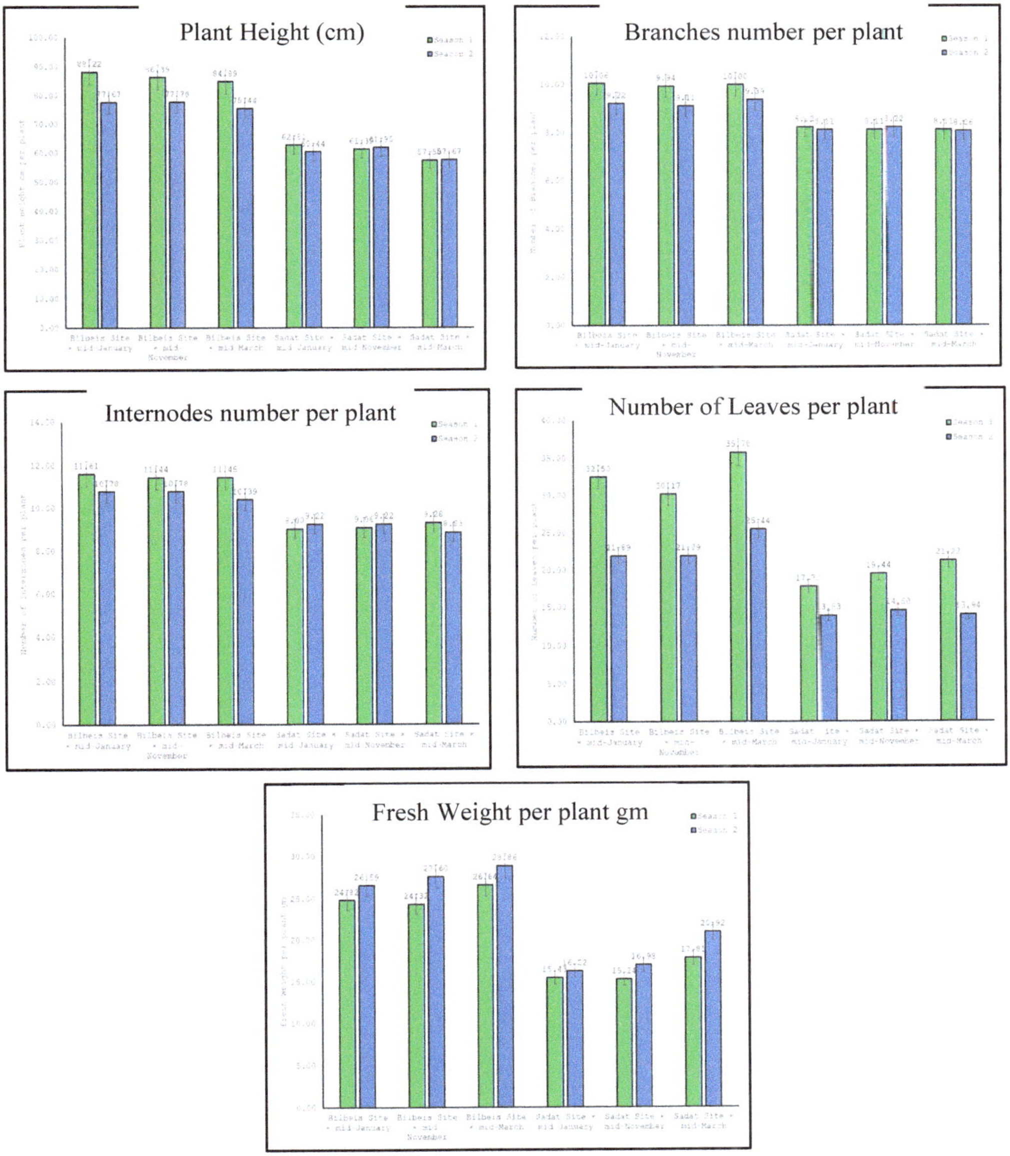

Figure 2. Evaluation of location × sowing date on the growth of buckwheat.

Table 6. Interaction effect of species × planting times on the growth of buckwheat.

Parameters	Treatments	*Fagopyrum tataricum* Mid-Mar	*Fagopyrum tataricum* Mid-Jan	*Fagopyrum tataricum* Mid-Nov	*Fagopyrum esculentum* Mid-Jan	*Fagopyrum esculentum* Mid-Nov	*Fagopyrum esculentum* Mid-Mar	L.S.D. at 0.05
Plant height cm per plant	*1st Season*	95.778 a	90.833 a	88.444 a	60.195 b	59.333 b	46.611 b	8.1057
	2nd Season	89.389 a	81.056 b	83.167 b	57.056 c	56.556 c	43.722 d	4.5045
Branch number per plant	*1st Season*	11.722 a	10.833 b	10.611 b	7.445 c	7.445 c	6.389 d	0.583
	2nd Season	10.500 a	9.778 b	9.945 b	7.556 c	7.389 c	6.945 d	0.4091
Internode number per plant	*1st Season*	13.334 a	12.278 b	12.111 b	8.334 c	8.389 c	7.389 d	0.7312
	2nd Season	11.944 a	11.389 b	11.389 b	8.611 c	8.611 c	7.278 d	0.4307
Leaf number per plant	*1st Season*	44.833 a	33.722 b	33.556 b	16.500 c	16.055 c	12.166 d	3.3267
	2nd Season	30.444 a	23.278 b	23.389 b	12.445 c	13.000 cd	8.944 d	3.5174
Fresh weight per plant gm	*1st Season*	33.495 a	27.483 b	26.584 b	12.767 c	13.032 cd	10.952 d	1.977
	2nd Season	37.697 a	28.418 b	30.075 b	14.395 c	14.508 c	12.081 d	2.3037

where L.S.D. = least significant difference. Different letters within the same row indicate significant differences ($p \leq 0.05$).

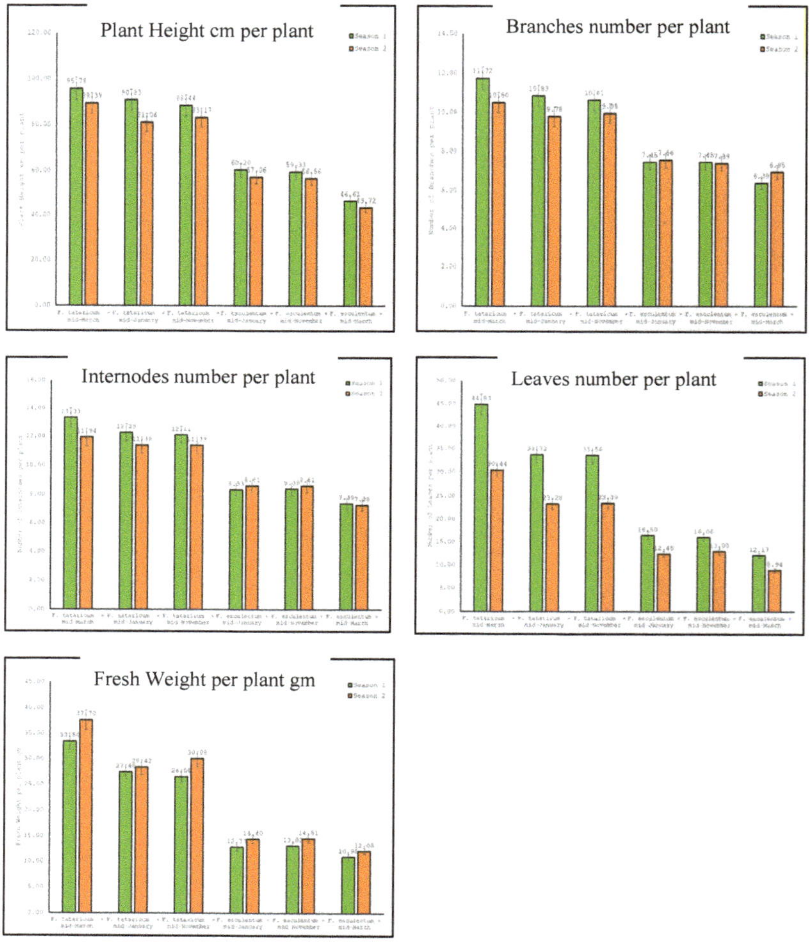

Figure 3. Evaluation of species × sowing date on the growth of buckwheat.

3.2. Evaluation of the 2-Way Interaction between Biotic and Abiotic Factors on the Productivity of Buckwheat

3.2.1. Interaction between Location and Species

The obtained results in Table 7 and Figure 4 show the interaction effects of location (belbies vs. Sadat) and species (*Fagopyrum tataricum* vs. *Fagopyrum esculentum*) on the productivity of buckwheat plants, examining parameters such as the number of seeds per plant, grains per square meter, and productivity rate per hectare over two growing seasons. For the number of seeds per plant, *Fagopyrum tataricum* in Belbies City exhibited the highest number of seeds per plant in both seasons (48.07 and 49.63, respectively), significantly outperforming all the other treatment combinations. *Fagopyrum esculentum* in Sadat City had the lowest seed count (13.67 and 14.63, respectively). A least significant difference (LSD) of 0.05 was used to confirm these differences as statistically significant. Similarly, *Fagopyrum tataricum* grown in Belbies City achieved the highest grain yield per square meter (0.0886 kg and 0.0867 kg, respectively, in the first and second seasons), markedly higher than the lowest yields observed for *Fagopyrum esculentum* in Sadat City (0.0314 kg and 0.032 kg, respectively). The LSD test results validated the significant impact of both location and species on grain yield. Additionally, for the kg/Hectare productivity rate, which reflects trends in the other parameters, *Fagopyrum tataricum* in Belbies City also led in terms of productivity per hectare, with 859.38 kg and 832.27 kg in the first and second seasons, respectively. In contrast, *Fagopyrum esculentum* in Sadat City had the lowest productivity (316.32 kg and 319.12 kg, respectively). The statistical significance of these differences was confirmed by the LSD test at 0.05.

Table 7. Interaction effect of location × species on the productivity of buckwheat.

Treatment	Parameter	Number of Seeds Per Plant		Grains of 1 Meter^2 (kg)		Productivity Rate kg/Hectare	
		1st Season	2nd Season	1st Season	2nd Season	1st Season	2nd Season
Fagopyrum tataricum	Belbies	48.07 a	49.63 a	0.088.6 a	0.0867 a	859.38 a	832.27 a
	Sadat	25.04 b	26.74 b	0.0663 b	0.0596 b	661.43 b	594.43 b
Fagopyrum esculentum	Belbies	17.74 c	18.59 c	0.0499 c	0.0467 c	498.02 c	467.72 c
	Sadat	13.67 d	14.63 d	0.0314 d	0.032 d	316.32 d	319.12 d
LSD at 0.05		3.407	3.1064	3.85×10^{-3}	5.82×10^{-3}	39.714	60.144

where L.S.D. = least significant difference. Different letters within the same row indicate significant differences ($p \leq 0.05$).

These results delineate a clear pattern of superior performance by *Fagopyrum tataricum*, especially when cultivated in Belbies City, across all the measured productivity parameters. Conversely, *Fagopyrum esculentum* in Sadat City consistently exhibited the lowest productivity. The statistically significant differences, as indicated by the LSD tests, affirm that both the choice of species and the cultivation location critically influence buckwheat productivity.

3.2.2. Interaction of Location × Sowing Time

Our findings in Table 8 and Figure 5 explore the interaction effect of location (Belbies City site vs. Sadat City site) and sowing time (mid-January, mid-November, and mid-March) on the productivity of buckwheat, focusing on the number of seeds per plant, grains per square meter, and the productivity rate per hectare across two growing seasons. For the number of seeds per plant, the data clearly illustrate the influence of both sowing time and location on the number of seeds per plant. In both seasons, sowing at the Belbies City site in mid-March resulted in the highest number of seeds (37.72 and 37.83, respectively), significantly surpassing the counts in Sadat city at any sowing time. A least significant difference (LSD) of 0.05 was used to confirm the statistical significance of these observations. Similarly, the highest grain yield per square meter was observed for buckwheat sown in mid-March at the Belbies City site (0.0723 kg and 0.073 kg in the first and second seasons, respectively), indicating optimal productivity when sown during this period. Again, the yields were significantly lower in Sadat City, regardless of the sowing time, with the LSD values substantiating the significance of the interaction between location and sowing time. However, for the kg/Hectare productivity rate, which reflects the trends in number of seeds per plant and grains per square meter, the highest productivity rates per hectare were achieved with mid-March sowing at the Belbies City site (702.88 kg and 708.21 kg in the first and second seasons, respectively). In contrast, all the productivity metrics in Sadat City were significantly lower, underscoring the influence of both sowing date and location on overall productivity. The differences across these groups were statistically significant, as indicated by the LSD at 0.05.

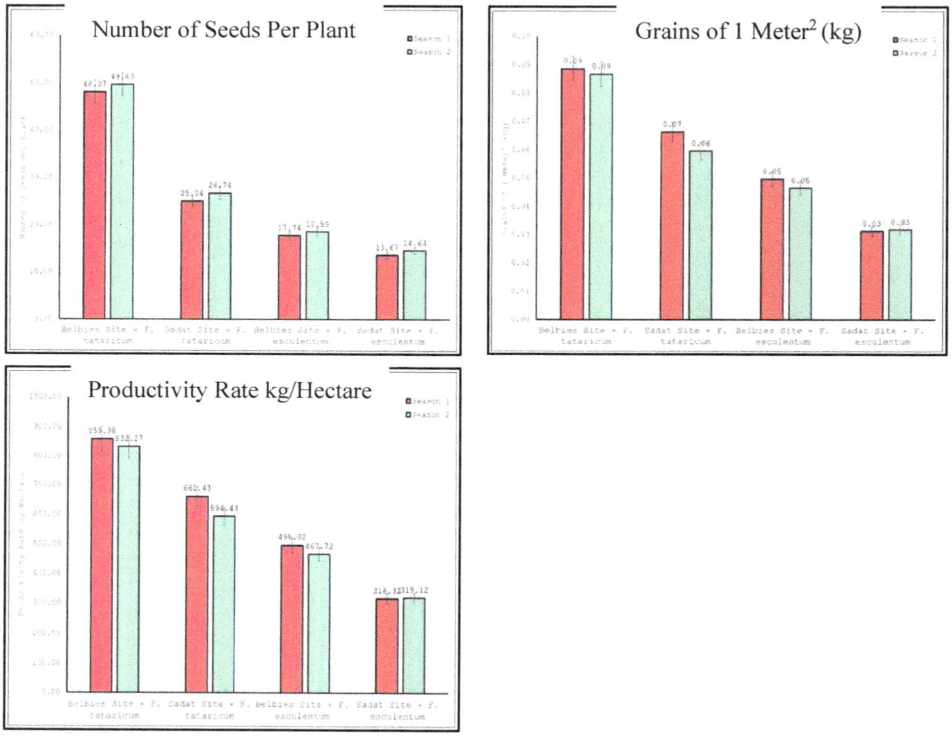

Figure 4. Evaluation of location × species on the productivity of buckwheat.

Table 8. Interaction effect of location × sowing time on the productivity of buckwheat.

Treatment	Parameter	Number of Seeds Per Plant		Grains of 1 Meter^2 (kg)		Productivity Rate kg/Hectare	
		1st Season	2nd Season	1st Season	2nd Season	1st Season	2nd Season
Belbies City Site	Mid-Jan	29.67 b	32.78 b	0.0673 b	0.0655 b	648.76 b	627.3 b
	Mid-Nov	31.33 b	31.72 b	0.0680 ab	0.0615 b	684.46 ab	614.48 b
	Mid-Mar	37.72 a	37.83 a	0.0723 a	0.073 a	702.88 a	708.21 a
Sadat City Site	Mid-Jan	17.72 c	19.61 cd	0.0495 c	0.0472 c	481.99 c	448.23 c
	Mid-Nov	18.61 c	19.22 d	0.0482 c	0.0457 c	493.96 c	477.56 c
	Mid-Mar	21.72 c	23.22 c	0.0490 c	0.0445 c	490.67 c	444.55 c
LSD at 0.05		4.1727	3.8045	4.71×10^{-3}	7.13×10^{-3}	48.639	73.662

Different letters within the same columns indicate statistically significant differences ($p \leq 0.05$).

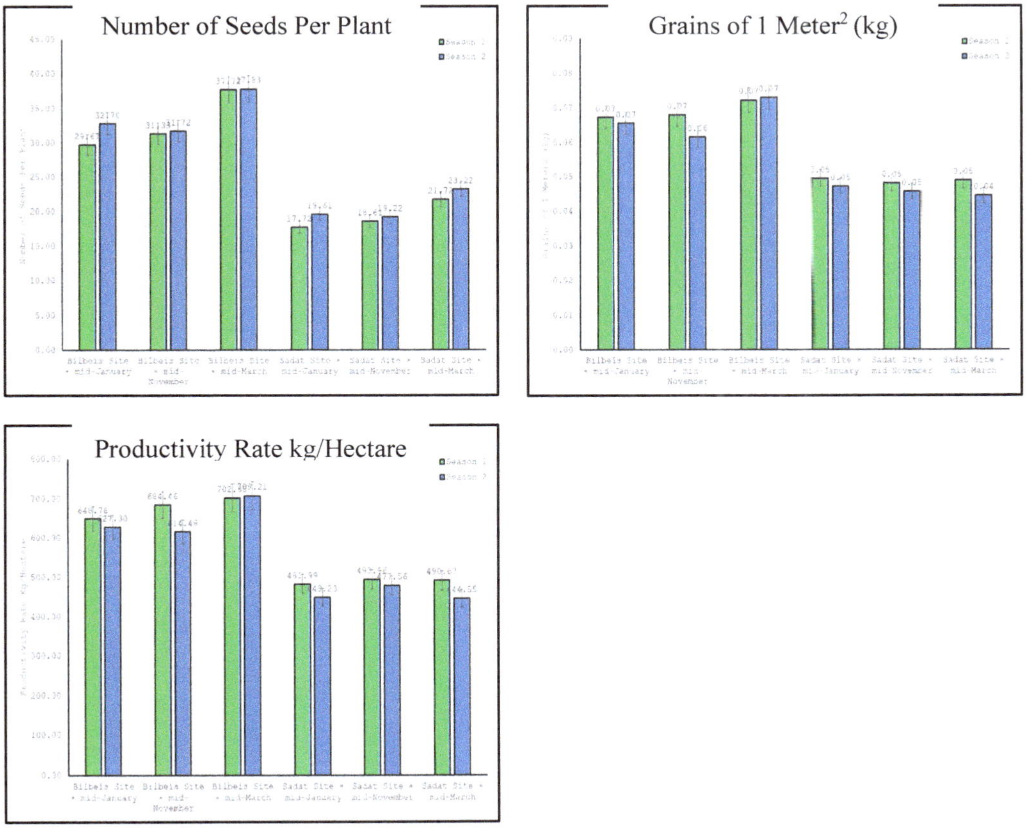

Figure 5. Evaluation of location × sowing date on the productivity of buckwheat.

These results demonstrated the significant effects of sowing time and location on the productivity of buckwheat, with mid-March sowing occurring at the Belbies City site consistently yielding the highest number of seeds, highest grain weight per square meter, and greatest productivity per hectare. Conversely, the Sadat City site had the lowest productivity across all sowing times, and the statistical analysis confirmed the robustness of these findings.

3.2.3. Interaction of Species × Sowing Date

Our obtained results in Table 9 and Figure 6 present the interaction effects of species (*Fagopyrum tataricum* vs. *Fagopyrum esculentum*) and sowing time (mid-March, mid-January, mid-November) on the productivity of buckwheat, with a focus on the number of seeds per plant, grains per square meter, and the productivity rate per hectare across two growing seasons. This analysis is structured to highlight the significant differences and trends in buckwheat productivity, providing insights into how these factors influence buckwheat productivity. For the number of seeds per plant, *Fagopyrum tataricum* produced significantly greater numbers of seeds per plant across all sowing times than did *Fagopyrum esculentum*, with the highest counts observed for mid-March sowing in both seasons (45.17 and 45.72, respectively). This indicates a clear advantage of *Fagopyrum tataricum* in terms of seed production, especially when sown in mid-March. A least significant difference (LSD) of 0.05 underscores the statistical significance of these differences. However, for grains per square meter, *Fagopyrum tataricum* sown in mid-March also yielded the highest grain weight per square meter (0.0858 kg and 0.084 kg in the first and second seasons, respectively), significantly outperforming *Fagopyrum esculentum* sown during the same period. This finding suggested that not only the choice of species but also the timing of sowing significantly affect grain production. The LSD test results confirmed the statistical significance of these observations. However, according to the results in seeds per plant and grains per square meter, *Fagopyrum tataricum* sown in mid-March achieved the highest productivity rates per hectare (837.55 kg and 818.36 kg in the first and second seasons, respectively). In contrast, *Fagopyrum esculentum*, especially when sown in mid-March, had the lowest productivity, highlighting the substantial impact of both species and sowing time on overall productivity. The differences across these groups were statistically significant, as indicated by the LSD at 0.05. These results demonstrate the superior performance of *Fagopyrum tataricum* over *Fagopyrum esculentum* in terms of productivity, as sowing in mid-March further enhanced its productivity.

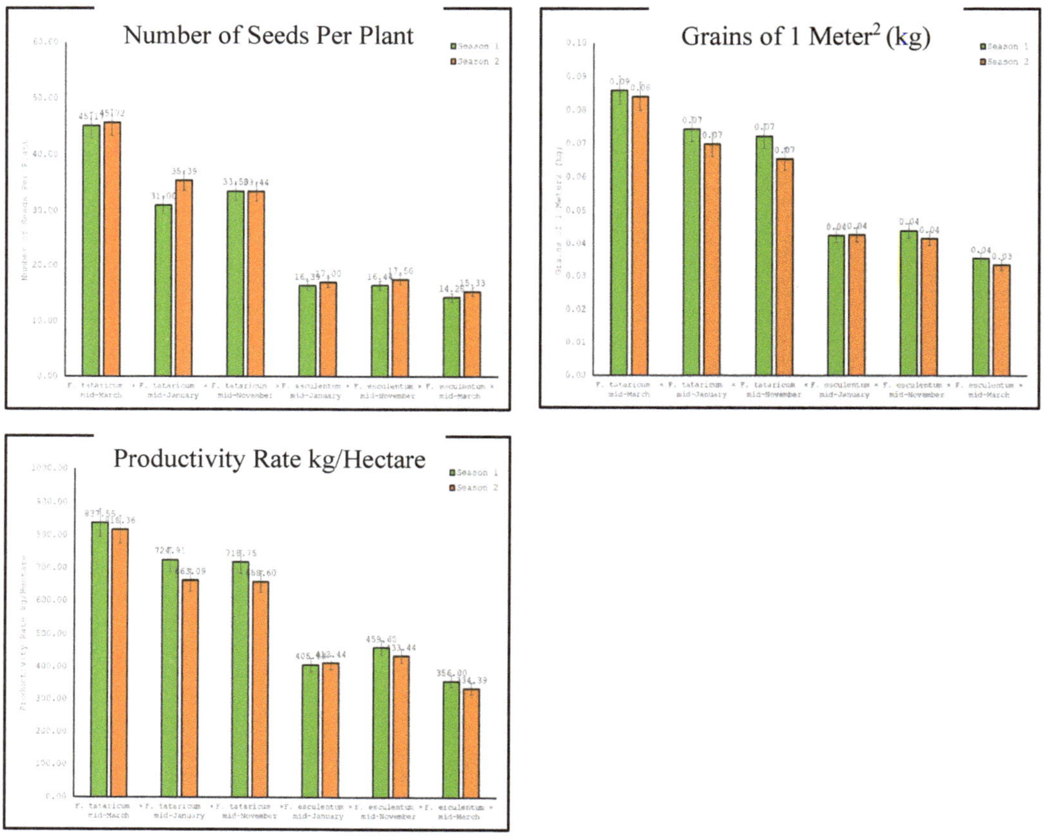

Figure 6. Evaluation of species × sowing date on the productivity of buckwheat.

Table 9. Interaction effect of species × sowing time on the productivity of buckwheat.

Treatment	Parameter	Number of Seeds Per Plant		Grains of 1 Meter^2 (kg)		Productivity Rate kg/Hectare	
		1st Season	2nd Season	1st Season	2nd Season	1st Season	2nd Season
Fagopyrum tataricum	Mid-Mar	45.17 a	45.72 a	0.0858 a	0.084 a	837.55 a	818.36 a
	Mid-Jan	31.00 b	35.39 b	0.0743 b	0.0698 b	724.91 b	663.09 b
	Mid-Nov	33.50 b	33.44 b	0.0722 b	0.0655 b	718.75 b	658.60 b
Fagopyrum esculentum	Mid-Jan	16.39 c	17.00 c	0.0425 c	0.0428 c	405.83 d	412.44 c
	Mid-Nov	16.44 c	17.50 c	0.0440 c	0.0417 c	459.68 c	433.44 c
	Mid-Mar	14.28 c	15.33 c	0.0355 d	0.0335 d	356.00 e	334.39 d
LSD at 0.05		4.1727	3.8045	4.71×10^{-3}	7.13×10^{-3}	48.639	73.662

where L.S.D. = least significant difference. Different letters within the same row indicate significant differences ($p \leq 0.05$).

4. Discussion

4.1. Evaluation of the 2-Way Interaction Effect of Biotic and Abiotic Factors on the Growth of Buckwheat

The observed variations in buckwheat growth and productivity across different locations, species, and sowing dates, as indicated in the experimental outcomes, can be attributed to complex interactions among genetic, environmental, and management factors [1,6,23]. The superior vegetative growth characteristics exhibited by *Fagopyrum tataricum* (FT) in Belbies (BCS) compared to those of the Sadat City site (SCS) and *Fagopyrum esculentum* (FE) underscore the influence of both location and species on buckwheat growth. This advantage is likely due to the genetic predisposition of FT to adapt better to the environmental conditions present in Belbies, such as soil type, pH, and microclimate, which may be more conducive to its growth than the conditions in Sadat city. Furthermore, the soil and water analyses (Tables 1 and 2) provide insights into these environmental conditions, revealing differences in pH, electrical conductivity (EC), and nutrient availability between the two sites, which could significantly affect plant growth [15]. The variations in temperature, precipitation, and relative humidity (Table 3) between the locations across the seasons also influenced these outcomes. Moreover, the impact of sowing time on growth parameters highlights the significance of temporal factors on agricultural outcomes, as demonstrated by the optimal performance of Belbies plants sown in mid-March, which could be attributed to more favorable weather conditions during critical growth stages. These findings are supported by weather data, indicating that seasonal transitions and specific climatic conditions during the growth period significantly influence vegetative development [17,24]. Additionally, the interaction effect between species and sowing time, particularly the superior performance of FT over FE across all sowing times, can be explained by the inherent genetic differences between the two species. The better adaptation of FT to water stress and its drought tolerance [10] may have contributed to its greater resilience and superior growth under varying environmental conditions. The differences in genome size between these species [21,22] may also play a role in their adaptability, with FT possibly possessing genetic traits that confer an advantage under the specific conditions experienced during the experimental periods. The methodologies employed in this study, including detailed site preparation, soil and water analysis, and careful selection of sowing times and species, provided a robust framework for evaluating the interactions between genetic, environmental, and management factors in buckwheat cultivation. The use of organic manure, phosphorus fertilizer, and split applications of nitrogen fertilizer reflects contemporary agricultural practices aimed at enhancing crop productivity while considering sustainability [8,15]. The differential performance of buckwheat species across locations and sowing times is a clear indication of the complexity of genotype–environment interactions. The adaptability of *Fagopyrum tataricum* (FT) to conditions at the Belbies City site, compared to that of *Fagopyrum esculentum* (FE), whose disparity in performance at the Sadat City site mirrors previous findings [6,27], highlighted species-specific responses to environmental conditions. This suggests that species selection cannot be universal but rather should be tailored to the specific environmental conditions of the cultivation site, taking into account the local soil, water, and climatic characteristics. The disparities in soil pH, EC, and nutrient availability between the two sites likely played a significant role in influencing plant growth and development. For instance, the relatively higher pH and EC values at the Sadat City site might have imposed additional stress on FE, which, coupled with its inherent drought avoidance characteristics, could explain its underperformance compared to that of FT, which exhibits traits of drought tolerance. These findings align with the physiological and genetic makeup of buckwheat species, which significantly influences their response to abiotic stresses [10]. The experimental design, which meticulously accounted for the variations in planting times and employed rigorous soil preparation and fertilization regimes, underscores the importance of management practices in optimizing crop yields. The application of organic manure, phosphorus, and nitrogen fertilizers, as described, likely enhanced the soil's fertility, thereby supporting the vegetative and reproductive growth of buckwheat. However, the critical role of soil health and management practices in sustainable crop production has been noted [8,15]. Furthermore, weather data analysis, which offers insights into temperature, precipitation, and relative humidity variations, provides a contextual background for interpreting the experimental results. The optimal growth and productivity observed for buckwheat sown in mid-March at the Belbies site could be attributed to the alignment of critical growth stages with favorable weather conditions, highlighting the temporal dimension of environmental suitability for crop cultivation. In summary, the findings of this study contribute to a deeper understanding of how genetic, environmental, and management factors converge to influence crop growth and productivity.

4.2. Evaluation of the 2-Way Interaction between Biotic and Abiotic Factors on the Productivity of Buckwheat

The differences in buckwheat productivity observed across various experimental conditions, including location, species selection, and sowing time, can be attributed primarily to the complex interplay of genetic, environmental, and agronomic factors. However, the significant impact of the location × species interaction particularly underscores the importance of selecting the appropriate species for specific environmental conditions, as *Fagopyrum tataricum* exhibited greater productivity in Belbies than in Sadat due to its genetic adaptability to local environmental and soil conditions [5]. Moreover, its inherent genetic traits, such as increased tolerance to local abiotic stresses indicated by soil pH, EC, and soil texture characteristics, further contributed to its superior performance in Belbies. Conversely, *Fagopyrum esculentum* showed reduced productivity across both locations, indicating its possibly lesser adaptability to the given environmental conditions or its specific genetic makeup, rendering it less efficient under the tested conditions [6,27]. Furthermore, the interaction between location and sowing date highlights the critical role of sowing timing in optimizing buckwheat productivity, with mid-March sowing in Belbies achieving the highest productivity due to the optimal alignment of the sowing date with favorable environmental conditions, such as temperature and humidity, conducive to buckwheat growth. This is supported by weather data analysis, where temperature and precipitation patterns likely favored the mid-March sowing period, resulting in enhanced growth and productivity [12,20]. Additionally, the interaction between species and sowing date provides insights into how genetic factors combined with agronomic practices influence crop outcomes, with *Fagopyrum tataricum* exhibiting superior performance after mid-March sowing, emphasizing its adaptability to the climatic conditions that prevailed during this period. This adaptability, possibly due to its genetic traits favouring growth under specific environmental conditions experienced in the spring season in Egypt, further accentuates the need for species-specific management practices to optimize productivity. The observed variances in productivity related to sowing time and species selection underscore the complex genotype-by-environment interactions that influence crop performance [8]. Moreover, comprehensive soil and water analyses were conducted at the beginning of each experimental season, and the physical and chemical properties of the soil and irrigation water quality were recorded; these analyses directly impacted plant growth, development, and ultimately productivity. The suitability of the soil and water properties in Belbies for buckwheat cultivation, especially for *F. tataricum*, may have contributed to the higher productivity observed at this location. Additionally, the methodological approach, including the use of organic manure, phosphorus fertilizer, and nitrogen fertilizer, likely provided an optimal nutrient regime that supported the growth and productivity of buckwheat plants under the tested conditions. The strategic application of fertilizers, based on the specific growth stages of buckwheat, enhances nutrient availability during critical periods of growth and development, contributing to the observed differences in productivity across different sowing times and species. In conclusion, the results of this study elucidate the intricate relationships among genetic, environmental, and management factors in determining crop productivity. The adaptability of *Fagopyrum tataricum* to the specific conditions in Belbies, especially when sown in mid-March, highlights the potential for optimizing buckwheat cultivation through careful selection of species and sowing times that align with local environmental conditions. These findings contribute to the broader understanding of sustainable agricultural practices that can enhance crop resilience and productivity in the face of changing climatic conditions, providing valuable insights into the development of sustainable agricultural practices that leverage genetic diversity and environmental management to optimize crop productivity [22,25,28].

5. Conclusions

This research highlights critical findings on buckwheat cultivation in Egypt from three interactions: location × species, location × sowing date, and species × sowing date. The study demonstrated that *Fagopyrum tataricum* significantly outperformed *Fagopyrum esculentum*, particularly when sown in mid-March. It also showed a significant increase by planting it in the Belbies City site, with productivity reaching up to 859 kg/ha. This underscores the adaptability of this species and the importance of choosing the right sowing dates. Additionally, the interaction between location and sowing date revealed that mid-March sowing in Belbies yielded the highest productivity, emphasizing the impact of environmental conditions on crop growth. The study's insights into these interactions offer valuable guidance for optimizing buckwheat cultivation in Egypt, contributing to the development of sustainable agricultural practices in arid and semiarid regions.

Author Contributions: Conceptualization, M.M.H. and H.A.A.E.-A.; methodology, M.M.H.; software, N.M.M.; validation, M.M.H., H.A.A.E.-A. and A.M.S.H.; formal analysis, N.M.M.; investigation, A.M.S.H.; resources, H.A.A.E.-A.; data curation, N.M.M.; writing—original draft preparation, M.M.H.; writing—review and editing, H.A.A.E.-A. and M.M.H.; visualization, N.M.M.; supervision, H.A.A.E.-A.; project administration, H.A.A.E.-A. and N.M.M.; funding acquisition, M.M.H. All authors have read and agreed to the published version of the manuscript.

Funding: This research was funded in part by Open Access funding provided by the Qatar National Library (QNL).

Data Availability Statement: The data that support the findings of this study are available on request from the corresponding author, Hassona, M, M.

Conflicts of Interest: The authors declare no conflict of interest.

eferences

Barrett, C.B. Overcoming global food security challenges through science and solidarity. *Am. J. Agric. Econ.* **2021**, *103*, 422–447. [CrossRef]

Keler, V.V.; Demeneva, A.A.; Martynova, O.V. Influence of various elements from cultivation technology on the buckwheat yield. *IOP Conf. Ser. Earth Environ. Sci.* **2021**, *659*, 12061. [CrossRef]

3. Siracusa, L.; Gresta, F.; Sperlinga, E.; Ruberto, G. Effect of sowing time and soil water content on grain yield and phenolic profi of four buckwheat (Fagopyrum esculentum Moench.) varieties in a Mediterranean environment. *J. Food Compos. Anal.* **2017**, *6* 1–7. [CrossRef]
4. Song, J.N.; Liu, X.H.; Wang, Y.Q.; Yang, H.B. Transcriptome analysis reveals salinity responses in four Tartary buckwheat cultiva *J. Plant Biochem. Biotechnol.* **2021**, *30*, 564–578. [CrossRef]
5. Sowmya, S.A.; Vijayakumar, B.; Vishwanath, Y.C.; Ganiger, V.M.; Hegde, L.; Lokesh, M.S. Evaluation of buckwheat (Fagopyru esculentum Moench.) genotypes under northern dry zone of Karnataka. *Pharma Innov. J.* **2021**, *10*, 1974–1979.
6. Zhang, K.; He, M.; Fan, Y.; Zhao, H.; Gao, B.; Yang, K.; Zhou, M. Resequencing of global Tartary buckwheat accessions revea multiple domestication events and key loci associated with agronomic traits. *Genome Biol.* **2021**, *22*, 23. [CrossRef] [PubMed]
7. Tandon, H.L.S. *Methods of Analysis of Soils, Plants, Waters, Fertilisers & Organic Manures*; Fertiliser Development and Consultati Organisation: New Delhi, India, 2005.
8. Calegari, A.; Tiecher, T.; Wutke, E.B.; Canalli, L.D.S.; Bunch, R.; Rheinheimer, D.D.S. The role and management of soil mulch ar cover crops in Conservation Agriculture systems. In *Advances in Conservation Agriculture*; Burleigh Dodds Science Publishir Cambridge, UK, 2020; pp. 179–248.
9. Germ, M.; Gaberščik, A. The effect of environmental factors on buckwheat. In *Molecular Breeding and Nutritional Aspects Buckwheat*; Academic Press: Cambridge, MA, USA, 2016; pp. 273–281.
10. Fesenko, N.N.; Glazova, Z.I.; Fesenko, I.N. Cold stress at seedlings stage of buckwheat optimizes development of both roots ar aboveground biomass and limits the excessive vegetative growth interfering with seed formation (an analytical review). *Ac Agric. Slov.* **2020**, *116*, 5–10. [CrossRef]
11. Hassona, M.M.; Hussein, A.S.; Morsy, N.; AAbd El-Aal, H. Chemical, rheological, sensorial and functional properties buckwhe semolina flour composite pasta. *Egypt. J. Chem.* **2023**, *66*, 467–478. [CrossRef]
12. Mariotti, M.; Macchia, M.; Cerri, D.; Gatta, D.; Arduini, I.; Saccomanni, G. Rutin content in the forage and grain of comm buckwheat (Fagopyrum esculentum) as affected by sowing time and irrigation in a Mediterranean environment. *Crop Pasture S* **2020**, *71*, 171–182. [CrossRef]
13. Salehi, A.; Fallah, S.; Neugschwandtner, R.W.; Mehdi, B.; Kaul, H.P. Growth analysis and land equivalent ratio of fenugree buckwheat intercrops at different fertilizer types. *Die Bodenkult. J. Land Manag. Food Environ.* **2018**, *69*, 105–119. [CrossRef]
14. Khazaei, H.; Link, W.; Street, K.; Stoddard, F.L. ILB 938, a valuable faba bean (*Vicia faba* L.) accession. *Plant Genet. Resour.* **2018**, 478–482. [CrossRef]
15. Oomah, B.D.; Mazza, G. Flavonoids and antioxidative activities in buckwheat. *J. Agric. Food Chem.* **1996**, *44*, 1746–1750. [CrossRe
16. Jung, G.H.; Kim, S.L.; Kim, M.J.; Kim, S.K.; Park, J.H.; Kim, C.G.; Heu, S. Effect of sowing time on buckwheat (Fagopyru esculentum Moench) growth and yield in central Korea. *J. Crop Sci. Biotechnol.* **2015**, *18*, 285–291. [CrossRef]
17. Maletić, R.O.; Jevđović, R. The influence of meteorological conditions on major quantitative and qualitative traits of buckwhe (Fagopyrum esculentum Moench). *J. Agric. Sci.* **2003**, *48*, 11–19. [CrossRef]
18. Facho, Z.H.; Khan, F.; Tao, W.; Ali, S. Species divergence and diversity in buckwheat landraces collected from the weste himalayan region of pakistan. *Pak. J. Bot* **2019**, *51*, 2215–2224. [CrossRef] [PubMed]
19. Brunori, A.; Végvári, G. Rutin content of the grain of buckwheat (Fagopyrum esculentum Moench. and *Fagopyrum tataricu* Gaertn.) varieties growtn in southern Italy. *Acta Agron. Hung.* **2007**, *55*, 265–272. [CrossRef]
20. Larney, F.J.; Blackshaw, R.E. Weed seed viability in composted beef cattle feedlot manure. *J. Environ. Qual.* **2003**, *32*, 1105–11 [CrossRef]
21. Guo, X.D.; Ma, Y.J.; Parry, J.; Gao, J.M.; Yu, L.L.; Wang, M. Phenolics content and antioxidant activity of tartary buckwheat fro different locations. *Molecules* **2011**, *16*, 9850–9867. [CrossRef]
22. JoJohansson, E.; Kuktaite, R.; Labuschagne, M.; Lama, S.; Lan, Y.; Nakimbugwe, D.; Repo-Carrasco-Valencia, R.; Tafesse, Tesfaye, K.; Vazquez, D. Adaptation to abiotic stress factors and their effects on cereal and pseudocereal grain quality. *Developing Sustainable and Health Promoting Cereals and Pseudocereals*; Academic Press: Cambridge, MA, USA, 2023; pp. 339–35
23. Boglaienko, D.; Soti, P.; Shetty, K.G.; Jayachandran, K. Buckwheat as a cover crop in Florida: Mycorrhizal Status and soil analys *Agroecol. Sustain. Food Syst.* **2014**, *38*, 1033–1046. [CrossRef]
24. Omidbaigi, R.; Mastro, G.D.E. Influence of sowing time on the biological behaviour, biomass production, and rutin content buckwheat (Fagopyrum esculentum Moench). *Ital. J. Agron.* **2004**, *8*, 47–50.
25. Björkman, T.; Shail, J.W. Using a buckwheat cover crop for maximum weed suppression after early vegetables. *HortTechnolo* **2013**, *23*, 575–580. [CrossRef]
26. Snedecor, G.W.; Cochran, W.G. *Statistical Methods*, 8th ed.; Iowa State University Press: Ames, IA, USA, 1994.
27. Bulan, M.T.S.; Stoltenberg, D.E.; Posner, J.L. Buckwheat species as summer cover crops for weed suppression in no-tilla vegetable cropping systems. *Weed Sci.* **2015**, *63*, 690–702. [CrossRef]
28. Mariotti, M.; Masoni, A.; Arduini, I. Forage and grain yield of common buckwheat in Mediterranean conditions: Response sowing time and irrigation. *Crop Pasture Sci.* **2016**, *67*, 1000–1008. [CrossRef]

agriculture

MDPI

Article

Factors Influencing the Accumulation of Free Asparagine in Wheat Grain and the Acrylamide Formation in Bread

Anna Szafrańska [1,*], Grażyna Podolska [2], Olga Świder [1], Danuta Kotyrba [1], Edyta Aleksandrowicz [2], Agnieszka Podolska-Charlery [3] and Marek Roszko [1]

1 Institute of Agricultural and Food Biotechnology—State Research Institute, 36 Rakowiecka St., 02-532 Warsaw, Poland; olga.swider@ibprs.pl (O.Ś.); marek.roszko@ibprs.pl (M.R.)
2 Institute of Soil Science and Plant Cultivation—State Research Institute, Czartoryskich 8, 24-100 Puławy, Poland; grazyna.podolska@iung.pulawy.pl (G.P.)
3 National Food Institute, Technical University of Denmark, Kemitorvet Building 201, 2800 Kongens Lyngby, Denmark
* Correspondence: anna.szafranska@ibprs.pl

Abstract: Asparagine is one of the precursors of acrylamide that can form during bread production. The aim of this work was to determine the effect of genotype, environment, sulfur fertilization, and the interaction of those factors on the asparagine content, technological value of wheat, and acrylamide level in bread. The research material consisted of five wheat cultivars grown in two locations in Poland with nitrogen fertilization of 110 kg ha^{-1} and sulfur fertilization of 30 kg ha^{-1}. The standard ISO method for analyzing the milling and baking properties of wheat was used. The UHPLC-MS/MS method for analyzing the amino acids and the GC/MS method for acrylamide in bread were implemented. The analysis of variance results indicated that the location influenced the total variance in the measured asparagine content and quality of wheat the most, followed by the cultivar and then by the interaction between the environment and cultivar. Sulfur fertilization had no significant effect on the asparagine content, but slightly lowered the gluten quality and loaf volume of bread. However, sulfur fertilization in connection with the cultivar characterized by low starch damage had a positive effect on lowering the acrylamide in bread. Asparagine content in wheat and acrylamide in bread varies mostly depending on cultivar and environment.

Keywords: asparagine; acrylamide; baking value; starch damage; sulfur fertilization; wheat cultivars

1. Introduction

Wheat is known for being one of the most important crops grown in the world. In 2021, the total global production equaled 770 million tons, with an average yield of 3.6 t ha^{-1} and a harvested area of about 220 million ha [1]. Bread is a staple food for a large proportion of the world's population. Wheat bread, without a doubt, is an integral part of the daily diet and delivers not only carbohydrates and protein but also crucial health-promoting ingredients, especially when consumed as whole grain [2]. However, in the process of baking bread, acrylamide is formed, which may be a health risk, because acrylamide has been classified as being probably cancerogenic to humans (Group 2A). It has also been shown to have neurological and reproductive effects at high doses in rodent toxicology studies. Acrylamide forms mainly from free asparagine and carbonyl sources as part of the Maillard reaction [3]. Numerous food processing parameters, including temperature, formulation, and cooking time, affect acrylamide formation from asparagine. Acrylamide can be found in most cereal products, including bread and breakfast cereals, however, in particularly high concentrations in crispbread, cookies, and gingerbread [4].

A number of investigations focused on strategies to minimize the levels and the formation of acrylamide in wheat products. Based on the results, the reduction in heat and the addition of specific ingredients—for example, cysteine or asparaginase—alongside prolonged fermentation time is proposed [5]. However, all technological efforts to reduce acrylamide might be in vain if unsuitable raw material is used. Amrein et al. [6] proved that in French fries and potato chips, acrylamide content is dependent on the potato cultivar. Additionally, several studies showed that cultivars play a very important role in acrylamide content in baking products. Claus et al. [7] described that acrylamide content in wheat cultivars varied in the range of 8.7 to 24.9 mg/100 g. Also, Taeymans et al. [8] reported a range of 7.4 to 66 mg/100 g. Claus et al. [7] stated that, generally, wheat varieties belonging to the good quality category of wheat—E and A—show higher acrylamide levels than cultivars belonging to the low category—B and Ck. However, these authors also note that, among high-quality category A cultivars, it is possible to find cultivars with a high crude protein content and relatively low acrylamide concentration. The differences result from different amounts of asparagine in the grain of wheat [7,9]. Rapp et al. [10] showed that the mean asparagine content ranged from 143.25 to 392.75 mg/kg for the different wheat cultivars. Malunga et al. [11] found that the asparagine content of Canadian wheat varied from 302 to 965 and 116 to 336 mg kg^{-1} for whole meal and white flour, respectively. Genotype was a factor that had a significant effect on the asparagine content, suggesting that breeding strategies are an interesting avenue to pursue in order to produce cultivars with lower levels of this amino acid.

For this reason, focusing studies on the reduction of the precursors of acrylamide in wheat grains by lowering the asparagine content seems like the best way forward. Asparagine accumulates under stress conditions as a biological response to the restriction of protein synthesis. Stressful conditions can be caused by drought, salt in the soil, pathogen attack, toxicity, solar radiation, or nutrient deficiencies [9,12,13]. Asparagine often plays a pivotal role in the active site of enzymes [14,15]. Interestingly, the biosynthesis of all essential amino acids can be achieved from asparagine as a donor of fixed nitrogen. High free asparagine concentrations in grain indicate a poor efficiency of nitrogen utilization, defined as the conversion of assimilated nitrogen to grain protein, resulting in lower-quality grain. Asparagine content in flour and dough significantly affected acrylamide content [7]. Furthermore, it was found that asparagine concentration in grain was affected by both genotype and production environment [7,9,11,13]. In the Claus et al. [7] study, both asparagine and acrylamide content were shown to be dependent on weather conditions during the growing season, especially temperature.

Nitrogen is a vital component of amino acids; therefore, nitrogen fertilization has a strong impact on crude protein and free asparagine contents in flour [16,17]. Higher doses of nitrogen influenced the increase of asparagine in wheat grains [7,18]. In addition, sulfur was shown to have an important role in acrylamide formation and its precursors [16,17]. Curties and Halford [16] demonstrated a significant effect of sulfur fertilization on asparagine content in wheat grain. In conditions of sulfur deficiency in the soil, the amount of asparagine in grain was 26 mmol/kg (3435 mg/kg), while under sufficient sulfur conditions, the amount was 1.6 mmol/kg (211 mg/kg). Halford et al. [17] showed that even flours from wheat grown in the field with an intermediate level of sulfur application (10 kg sulfur per hectare) contained more acrylamide after processing than flours from wheat grown with the application of 40 kg sulfur per hectare.

The use of wheat for consumption requires appropriate quality characteristics that are important from a bakery's point of view. Wheat baking quality parameters depend on cultivars, weather conditions, and fertility. Sulfur and nitrogen fertilizer application on winter wheat can have many benefits. Nitrogen is an essential nutrient important for wheat growth and protein biosynthesis. Low nitrogen availability can limit crop growth, yield, and quality [19]. Adequate sulfur is also required for the proper growth and development of wheat, sulfur being essential for the biosynthesis of amino acids, proteins, and chlorophyll, in particular [20]. Sulfur is a key element of enzymes involved in nitrogen metabolism [21–23]. Moreover, sulfur deficiency can decrease nitrogen assimilation [24]. Without an adequate supply of sulfur, wheat is unable to reach its full yield potential and cannot efficiently utilize nitrogen for protein biosynthesis [25]. Therefore, nitrogen use efficiency and crop yields are increased when nitrogen and sulfur are applied together [24,26,27] Two critically important amino acids, methionine and cysteine, contain sulfur. Methionine is the initiating amino acid in the synthesis of all proteins, and cysteine plays a pivotal function in protein folding [28]. Nitrogen and sulfur deficiency are detrimental to baking quality due to their influence on the formation of disulfide bonds formed from the sulfhydryl groups of cysteine. Such bonds are important because they influence the viscoelasticity of dough [29].

Sulfur can be deposited in the soil in a variety of ways. According to Ref. [30], the US Federal Clean Air Act revisions introduced in 1990 have strongly improved air quality by regulating power plant air emissions. As a result, sulfur emissions have decreased by 80%, resulting in the reduction of sulfur deposition along with lower sulfur deposition. Lower deposition combined with a lack of sulfur fertilization results in soil sulfur deficiency throughout the world [31]. Although the topic of asparagine and acrylamide has been discussed for several years, the number of publications on the interactions between genotype, growing location, and sulfur, as well as their impact on the asparagine content and baking quality of wheat, is still limited and requires supplementation. We hypothesize that the use of sulfur will reduce the amount of free asparagine and affect the baking quality of winter wheat, but the effect of sulfur on reducing asparagine will depend on the cultivar.

Another factor that we chose to consider in this research is the influence of starch damage in wheat flour on the formation of acrylamide in bread. Damaged starch formation occurs during the milling of grain due to the mechanical shear between the mill rolls [32]. Among the factors affecting the amount of damaged starch in the flour are wheat grain hardness, wheat tempering conditions, and the setting of the mill (i.e., roll diameter and speed ratio, pressure, and gap size between the reduction rolls, milling time, and roll temperature) [32,33]. Starch damage strongly affects the behavior of dough during processing (e.g., water absorption, gelatinization, and rheological properties), as well as the quality of the finished product (e.g., color, shelf life) [32]. However, levels of damaged starch in the work of Wang et al. [34] and Mulla et al. [35] for wheat flours showed a strong positive correlation with acrylamide formation in bread. An increase in starch damage was also correlated with reduced sugar content in flour [34,35]. Another hypothesis of our work was to prove that starch damage might influence the acrylamide content in bread.

Therefore, the goal of this study was to evaluate the influence of wheat cultivars, sulfur fertilization, and growing location impact on the asparagine and acrylamide content and baking quality of winter wheat.

2. Materials and Methods

Plant material. This research work concerned five winter wheat cultivars representing two technological groups approved by the Polish National List of Agricultural Plant Varieties issued yearly by COBORU (The Research Center for Variety Testing). Group A comprises quality bread cultivars, namely 'Hondia' and 'Pokusa'. Group B comprises bread cultivars 'Belissa', 'Hybery F_1', and 'Bonanza'.

Field trials. Field trials were conducted during the 2020–2021 and 2021–2022 growing seasons at the Experimental Stations in Wielichowo (52°07′ N 16°21′ E) and Werbkowice (50°45′ N 23°45′ E), Institute of Soil Science and Plant Cultivation—State Research Institute (IUNG-PIB) Puławy, Poland. The soil in Werbkowice was typical chernozem, class I soil categorized as a very good wheat soil complex that is characteristic of the region. In Wielichowo, the soil was pseudopodsolic, class Iva, categorized as a very good rye complex. Winter rapeseed was the fore crop. The experiment was established using the long strip method. The first factor was the cultivar ('Hondia', 'Pokusa', 'Belissa', 'Hybery F_1', and 'Bonanza'), and the second factor was sulfur fertilization. Plots with no sulfur fertilization were used as control objects. Before the start of the trial, the soil was analyzed down to a depth of 30 cm. In Werbkowice, the soil was characterized by 7.9 pH_{KCl}, and the soil fertility indicators were 19.5 mg of K_2O, 10.2 mg of P_2O_5, and 0.16 mg of S per 100 g of soil, and 12.5 mg of Mg per 100 g of soil. In Wielichowo, the soil was characterized by 6.2 pH_{KCl}, and the soil fertility indicators were 19.2 mg of K_2O, 25.3 mg of P_2O_5, and 0.78 mg of S per

100 g of soil, and 6.4 mg of Mg per 100 g of soil. In both experimental stations, the nitrogen in the planned total dose of 110 kg ha^{-1} was applied in divided doses. The first dose of 60 kg ha^{-1} was applied after resumption of spring vegetation and the second dose of 50 kg ha^{-1} was applied at the stem elongation stage. One dose of sulfur was 30 kg ha^{-1} and was applied after the resumption of spring vegetation together with nitrogen fertilization. Winter wheat was sown on 5 October (Werbkowice) and 28 September (Wielichowo) in an optimal sowing term at a seed rate of 450 germinable kernels per m^2. The area of the harvested plot was 0.7 ha. Pre-sowing practices, pre-sowing plowing, and mineral fertilization (NPK) were accomplished before sowing. A 350 kg ha^{-1} dose of Polifoska 6-NPK (6 20 30) fertilizer was applied, which was at a concentration equivalent to 21 kg of N, 70 kg of P_2O_5, and 105 kg of K_2O per ha. All winter wheat cultivars were chemically protected against pests. Weed control consisted of 1.0 L ha^{-1} of Kantor Forte 195SE and 0.5 L ha^{-1} Puma Universal 0.69 EW. Disease control consisted of 1.0 L ha^{-1} of Tilt Turbo at the flag leaf stage and 2.0 L ha^{-1} of Adaxar Plus after blossoming. Pest control consisted of 0.3 L ha^{-1} of Fury 100 EW at the flag leaf stage. The crop was harvested at full maturity on 2 August 2021 and on 28 July 2022.

Weather conditions during the growing season of winter wheat were assessed on the basis of the Hydrothermal Coefficient of Selyaninov (HTC). During the growing seasons from February to August, the HTC was 1.04 (2021) and 1.14 (2022) in Werbkowice and 1.32 (2021) and 1.28 (2022) in Wielichowo. This indicates that in Werbkowice, the year was dry, while in Wielichowo, it was optimal. Particularly large differences occurred in July during the grain ripening period. In Werbkowice, the HTC was 0.54 (2021) and 0.62 (2022), while the HTC in Wielichowo was 1.43 (2021) and 1.32 (2022). Therefore, during the grain ripening period, there was a drought in Werbkowice, while in Wielichowo, there were optimal weather conditions.

Because of a comparable impact of factors (cultivar and sulfur fertilization) in both years and taking into account the large amount of work needed to determine milling tests and the assessment of rheological characteristics of dough and baking tests, the cultivar samples from both years were mixed appropriately for further research.

Technological quality analysis. Test weight (ISO 7971-3:2019) [36], protein content (ISO 20483:2013) [37], gluten quantity (ISO 21415-2:2015) [38], Zeleny sedimentation index (ISO 5529:2007) [39], and ash content (ISO 2171:2007) [40] were evaluated to determine the baking value of five wheat cultivars. Grain samples were ground to whole meal flours using hammer mill FN 3100 (Perten InstrumentsAB, Hägersten, Sweden) for protein and gluten content analyses. The Sedimat laboratory mill (Brabender GmbH & Co. KG, Duisburg, Germany) was used to prepare samples for the Zeleny sedimentation test.

Amino acid analysis. Sample preparation. Free amino acid content was analyzed based on the method described by Świder et al. [41] with slight modifications. In brief, 2 g of a sample was placed in a centrifuge tube and spiked with internal standard (1.7—diaminoheptane solution, Merck, Darmstadt, Germany). In order to perform extraction, 40 mL of trichloroacetic acid (5% solution, Avantor Performance Materials, Gliwice, Poland) was added and the sample was shaken thoroughly. After centrifugation (10,000× g for 10 min.), 100 μL of the supernatant was mixed in a 15 mL volume centrifuge tube with 2.5 mL of di-sodium tetraborate (3% water solution, J.T. Baker, Pol-Aura, Gliwice, Poland) and 2.5 mL of dansyl-chloride (20 mMol in acetonitrile, abcr GmbH, Karlsruhe, Germany) to derivatize the analyzed compounds. Derivatization was performed by shaking in a water bath at 40 °C for 1 h. Then, 10 μL of formic acid (98–100% purity, Avantor Performance Materials, Gliwice, Poland) was added to stop derivatization. The sample was filtered into a chromatographic vial using a syringe filter (pore size: 0.22 μm, Captiva Econofilters, Agilent, Santa Clara, CA, USA). Amino acid standards were supplied by Merck (Darmstadt, Germany) and their assays were as follows: asparagine, arginine, aspartic acid, threonine, valine, methionine, phenylalanine, leucine, isoleucine, lysine, tyrosine, histidine ≥98%, glutamine, proline, ornithine, glutamic acid, and serine ≥99%.

Chromatographic analysis. UHPLC-MS/MS (ultra-high performance liquid chromatography coupled with a high-resolution mass spectrometer) (Q Exactive Orbitrap Focus MS Thermo Fisher Scientific, Waltham, MA, USA) analysis was performed to determine free amino acid concentration. The compounds were separated on the Cortecs UPLC C18 2.1 × 100 mm, 1.6 μm column (Waters, Milford, MA, USA). Phase A consisted of water:ACN (90:10)/0,1% FA/5 mM ammonium formate, and phase B contained can:water (90:10)/0.1% FA/5 mM ammonium formate. LC-MS-grade water and LC-MS-grade acetonitrile were supplied by Witko (Łódź, Poland), and formic acid 98–100% and ammonium formate ≥97% for LC-MS were supplied by Chem-Lab (Zedelgem, Belgium). The following liquid phase gradient at the flow rate 0.3 mL/min was applied: A:B (%)0–2 min.—90:10—waste, 2–22 min.—0:100, 22–25 min.—0:100, 25–26 min.—90:10, 26–28 min.—90:10. MS was equipped with heated electrospray ionization (HESI) source. Analysis was performed in positive polarization mode with an injection volume of 2.5 μL. The detailed MS parameters were set as follows: spray voltage: 3 kV, capillary temperature: 256 °C, sheath gas flow rate: 48, auxiliary gas flow rate: 11, sweep gas flow rate: 2, probe heater temperature: 413 °C, S-lens RF level: 50, resolution: 70,000 in simultaneous scan and 35,000 in all ion fragmentation. Acquisition and analysis of the data were performed using Xcalibure 4.2.47 software.

Milling and baking quality analysis. Wheat grain samples weighing approx. 2.5 kg each were moistened to 16% with water. After resting for 16 ± 1 h, the wheat samples were milled using an MLU-202 laboratory flour mill (Bűhler AG, Uzwil, Switzerland) to obtain flour for the measurement of the rheological characteristics of dough by mixolab acc. to ISO 17718:2013 [42] (KPM Analytics, Chopin Technologies, Villeneuve-la-Garenne, France) and to perform laboratory baking test. Flour yield was in the range of 68.1 to 75.4%. Damage starch of flour was analyzed according to ISO 17715:2013 [43] by Chopin SDmatic (KPM Analytics, Chopin Technologies, Villeneuve-la-Garenne, France).

Baking test. The wheat flour samples were evaluated by a standard baking test for pan bread [44]. The dough was prepared by one-stage method at 28–30 °C by mixing flour (100%), water (acc. To water absorption specified by Mixolab + 3%), yeast (3%), and salt (1%) using a laboratory mixer (KitchenAid, Benton Harbor, MI, USA). The dough in bulk was fermented for 60 min in a proofing chamber at 30 °C/75% RH. Then, after 30 min, kneading by hand was performed. The dough was next divided into three dough pieces with a mass of 250 g each that were molded round and placed in baking pans in a proofing chamber for 38–47 min (the time needed for optimal dough development). The loaves were baked in an oven (230 °C, 30 min) (Piccolo Wachtel Winkler, Wachtel ABT GmbH, Pulsnitz, Germany). The bread samples were assessed after 20 ± 1 h of cooling for loaf volume, bread crumb hardness (Instron 1140, Instron, Norwood, MA, USA), and acrylamide content using GC/MS after bromination [45]. Anion exchange solid phase extraction combined with octadecyl silica was used to perform sample cleanup. Based on validation experiment results, it was confirmed that the method performs well at low acrylamide concentration levels with LOQ values at 10 μg kg^{-1}, and recovery relative standard deviation below 6%. The validation process was performed according to ISO/IEC 17025:2017 [46] requirements. Detailed statistical parameters of the described method have been reported by Roszko et al. [45].

Statistical analysis. The results were statistically analyzed using the three-way analysis of variance (ANOVA) with subsequent Tukey's HSD test with the significance level of $p < 0.05$. The three main factors were localization, sulfur fertilization applied, and the wheat cultivar used. A principal component analysis (PCA) was carried out on the average results of each cultivar. Data were analyzed using Statistica v. 13 software (Tibco Inc. Palo Alto, CA, USA).

3. Results and Discussion

3.1. Technological Value of Wheat Samples

The tested wheat samples were varied in terms of basic quality characteristics used in assessing suitability for processing grain for consumption purposes. The analysis of variance results (Table 1 and Table S1) indicated that the localization had the biggest influence on the total variance in the measured technological parameters of wheat grain, followed by cultivar, and then by the interaction between the environment and cultivar.

Table 1. F values calculated in the three-way ANOVA made for the qualitative parameters of grain of the examined wheat samples.

	Sources of Variation						
Parameter	**Localization (A)**	**S Fertilization (B)**	**Cultivar (C)**	**A × B**	**A × C**	**B × C**	**A × B × C**
Test weight	205.30 **	0.22	28.85 **	2.33	12.03 **	0.85	2.82
Protein content	2695.88 **	1.02	78.34 **	2.61	162.79**	3.09 *	6.92 **
Gluten content	2160.08 **	12.51 **	92.71 **	2.52	84.61 **	0.60	1.56
Zeleny index	84.70 **	0.42	9.97 **	0.58	14.52 **	0.91	0.74
Ash content	15.80 **	0.00	1.54	0.10	1.71	0.20	0.42
Asparagine	3418.61 **	1.50	43.33 **	27.35 **	17.04 **	2.82	4.88 **

Significant at: * $p < 0.05$, ** $p < 0.01$.

Wheat samples grown on location in Wielichowo were characterized by the highest average value of the test weight (74.0 kg hl^{-1}), which is an important predictor of the flour extraction rate for wheat (Table 2). However, these samples showed lower protein and gluten content (13.8% and 27.8%, respectively), with lower gluten quality (Zeleny index) and ash content than wheat grown in the Werbkowice location. Wheat grain samples from the Werbkowice location were also characterized by the highest value of asparagine content (525.1 mg kg^{-1}). Field location significantly influences the protein content and asparagine content. The analysis of variance results (Table 1) indicated that localization predominated (91.1%) the total variance in the measured free asparagine concentration in tested whole meal samples. The genotype, as well as the interaction localization and cultivar, contributed only 4.6 and 1.8%, respectively. For the wheat cultivars tested in this study, fertilization had the least impact on free asparagine content, contributing only 0.4% to the variance in free asparagine concentration. Similar to the result of Xie et al. [47], the interaction between genotype and fertilization and the interaction between localization and fertilization contributed less than 2% to the total variance.

The protein content of grain in the Werbkowice location was, on average, 2.5 p.% (percent points) higher than in Wielichowo. Additionally, we found interaction cultivar × localization in protein content (Table 1). In Wielichowo, sulfur fertilization did not cause significant changes in protein content, while in Werbkowice, sulfur fertilization caused an increase in protein content in the 'Hybery F_1' and 'Bonanza' cultivars (Table S2). For the cv. 'Pokusa' grown in Werbkowice, sulfur fertilization resulted in a reduction of the asparagine content in the grain. However, in Wielichowo, sulfur caused a decrease in the asparagine content in the seeds of the 'Belissa' and 'Bonanza' cultivars. Previous research also investigated environmental factors and agronomic differences that significantly influenced the free asparagine content in wheat grain [48]. They attempted to find answers to a question of how different environmental conditions and certain agronomic practices impact asparagine content in grain. Research from Malunga et al. [49] also showed that the asparagine concentration was significantly influenced by location and was responsible for 80% of the variation compared to 13% for the cultivar factor. Tafuri et al. [50] proved that several Italian cultivars seemed to be highly influenced by the environment, whereas others indicated relative stability in free asparagine content across locations and years. Xie et al. [47] stated that cultivating wheat in suitable environments, together with the selection of wheat genotypes with lower potential to form free asparagine content, is the most effective proposed action to control free asparagine content in Canadian wheat.

Table 2. Technological value of grain of wheat cultivars tested in the experiment.

Factor	Test Weight (kg hl^{-1})	Protein Content (N × 5.7) (%d.m.)	Gluten Content (%)	Zeleny Index (cm^3)	Ash Content (% d.m.)	Asparagine Content (mg kg^{-1})
Localization (A)						
Wielichowo	74.0 ^{a}	13.8 ^{b}	27.8 ^{b}	55 ^{b}	1.59 ^{b}	276.2 ^{b}
Werbkowice	71.5 ^{b}	16.3 ^{a}	34.2 ^{a}	63 ^{a}	1.68 ^{a}	525.1 ^{a}
Fertilization (B)						
N + 0S	72.7	15.0	30.7 ^{b}	59	1.63	392.4
N + 30S	72.8	15.1	31.2 ^{a}	59	1.63	408.9
Cultivar (C)						
Hondia	73.6 ^{ab}	14.6 ^{d}	29.9 ^{d}	56 ^{b}	1.60	378.3 ^{cd}
Pokusa	72.9 ^{bc}	15.8 ^{a}	33.4 ^{a}	63 ^{a}	1.65	397.1 ^{c}
Belissa	71.3 ^{d}	15.3 ^{b}	31.4 ^{b}	60 ^{ab}	1.59	426.8 ^{b}
Hybery F_1	73.8 ^{a}	14.6 ^{d}	29.7 ^{d}	56 ^{b}	1.66	363.5 ^{d}
Bonanza	72.2 ^{c}	15.0 ^{c}	30.7 ^{c}	58 ^{b}	1.64	437.4 ^{a}

a, b, c, d—values marked with the same letters do not differ significantly at $p < 0.05$ according to Tukey's multiple test.

In our study, the applied form of sulfur fertilization had no significant effect on the mentioned technological grain parameters, except gluten content (Table 1). Similar to the results of Klikocka et al. [51], the application of sulfur caused a slight positive effect on the increase in gluten content. Sulfur addition had no negative effect on the grain characteristics for processing for baking purposes. In our results, the asparagine content was higher in low-sulfur soil (Werbkowice) compared with high-sulfur soils (Wielichowo). Our results are consistent with the research of Wilson et al. [52] and Muttucumaru et al [53]. In the research of Wilson et al. [52] and Muttucumaru et al. [53], free asparagine content in grain cultivated in low-sulfur soils was strikingly high—up to 30 times more than that cultivated in soil with normal sulfur levels. Wilson et al. [52] showed that low nitrogen use efficiency genotypes produced grain with greater asparagine content than high nitrogen use efficiency genotypes under sulfur-deficient conditions, compared to other research. Despite that, sulfur application reduces asparagine content to baseline levels. According to Xie et al. [47], fertilization had the least impact on asparagine content. Sulfur fertilization strategies for reducing free asparagine concentration in wheat were not always effective. In the research conducted by Stockmann et al. [15] in non-sulfate-deficient soils, sulfur fertilization within a conventional farming system did not influence the protein content and free asparagine amount significantly. Sufficient sulfur accumulation in the grains is presumably supplied from the available sulfur in the soil during the grain-filling period [15,54]. However, different accumulation of asparagine content in different cultivars, as a result of sulfur fertilization, was observed in both locations. The asparagine content in Werbkowice was found to be lower by 16.3 mg kg^{-1} for cv. 'Pokusa' and higher for cv. 'Hondia', 'Belissa', 'Hybery F_1', and 'Bonanza' (on average, 52 mg kg^{-1}) for grain with N + 30S treatment compared to N treatment. For the Wielichowo location, the differences were from −34 mg kg^{-1} for cv. 'Belissa' to +9 mg kg^{-1} for cv. 'Pokusa') (Table S2).

Tested wheat cultivars were characterized by high protein and gluten content with good potential for bread making. Among analyzed wheat cultivars, cv. 'Pokusa' was characterized by the highest value of wheat for baking purposes with the highest value of protein and gluten content and the best gluten quality determined by the Zeleny index (Table 1). The 'Hondia' and 'Hybery F_1' cultivars, despite the highest test weight, showed a slightly lower baking value.

The free asparagine content in wheat varied widely both within cropping systems and cultivars [15]. This was also confirmed in this study as the asparagine content in tested wheat samples ranged from 235.6 (cv. 'Hondia', N fertilization) to 587.9 mg kg^{-1} (cv. 'Bonanza', N + 30S fertilization) (Table S2). Therefore, wheat grown in Poland has similar asparagine levels to wheat cultivated in other parts of the world, i.e., in Canada where asparagine content ranges from 302.2 to 700.3 mg kg^{-1} for whole meal flour [11], and was lower than those of whole meal obtained from hard red spring cultivated in North Dakota and Nebraska, USA (357–1037 mg kg^{-1} [55] and 200–110 mg kg^{-1} [13], respectively). The asparagine content of wheat grown in Europe was in a wider range (320–1560 mg kg^{-1}) [56], while that of wheat cultivated in Australia was relatively lower (137–437 mg kg^{-1}) [57]. Free asparagine content in a total of 54 bread wheat cultivars for the Italian market ranged from 0.55 to 2.84 mmol kg^{-1} dry matter (73 to 375 mg kg^{-1}) [50]. Malunga et al. [49] suggested that breeding strategies should aim to identify cultivars that are low asparagine accumulating and are stable across different growing environments. These observations suggest that five tested wheat cultivars are not stable in diverse environments. However, since only two crop years and two growing locations were used in this research, multiple years, locations, and cultivars are required to understand the differences in polish conditions. In an ideal scenario, cultivars in which the influence of localization on the asparagine content is not significant would be found.

Asparagine content was negatively correlated with test weight (r = −0.831) and positively correlated with protein content (r = 0.855), Zeleny index (r = 0.706), and ash content (r = 0.614) (Table S3). Higher protein content may increase levels of free asparagine. Stockmann et al. [15] reported a poor relationship between protein and free asparagine for nitrogen trials, whereas the conventional sulfur trial showed a good correlation for both traits. Ohm et al. [55] found significant but low simple linear correlations between free asparagine content and test weight, as well as wheat protein content and bread volume, suggesting that high asparagine content might be associated with good bread quality.

The significant interaction between localization and fertilization was stated only for asparagine content (Table 1). The interaction between localization and cultivar had a significant impact on test weight, protein content, gluten content, Zeleny index, and asparagine, contributing 11.8, 17.5, 11.6, 27.6, and 1.8% of the total variance, respectively (Table S1). The interaction between fertilization and cultivar was found only for protein content, whereas interactions between all three tested factors—localization × fertilization × cultivar—were found only for protein content and asparagine content and contributed only 0.7 and 0.5% of the total variance, respectively.

3.2. Quality Characteristics of Wheat Flour Samples Obtained from Tested Cultivars

Flour samples from wheat cultivars obtained in laboratory milling were evaluated in terms of quality parameters describing the properties of both protein and starch complexes. Both indirect and direct methods were used for conducting laboratory baking tests.

The ANOVA results (Tables 3 and S4) indicated that both localization and cultivar, as well as the interaction between these two factors, had the greatest share in the total variance in the measured quality parameters that characterized the protein complex of flour. Localization had the greatest share in the total variance of such quality parameters as protein content (63.8% in the total variance) and protein weakening in points C2 (78.9%) and C_{18} (81.3%), respectively (Table S4). For the wheat samples tested in this study, the cultivar had the second strongest impact on such parameters as water absorption, development time T1, gluten index, gluten content, and stability, contributing to 58.5%, 47.0%, 43.3%, 40.6%, and 30.7% of the variance in their concentration, respectively.

Results of the quality characteristics of the protein and starch complex of flour are presented in Table 4. Flour samples obtained from grain from the Werbkowice localization were characterized by significantly higher protein and gluten content than samples obtained from the Wielichowo location. Even though the quality of gluten measured by the gluten index was slightly lower, it still met the requirements for good baking value of flour for bread production [58]. Protein quantity and quality influenced the water absorption of flour and the technological properties of dough measured by mixolab. Flours from grain samples from the Werbkowice location, which were also characterized by higher ash content, showed higher water absorption values (on average 57.3%) compared to

Wielichowo location (54.4%). However, the lowest quality of gluten samples from the Werbkowice location caused significantly lower values of dough stability and torque in points C2 and C_{18}.

Table 3. F values calculated in the three-way ANOVA made for the qualitative parameters of the flour obtained from the grain of the examined wheat samples.

	Sources of Variation					
Parameter	**Localization (A)**	**S Fertilization (B)**	**Cultivar (C)**	**A × B**	**A × C**	**B × C**
Protein content	530.45 **	1.25	23.05 **	1.25	50.07 **	0.62
Gluten content	93.06 **	0.45	34.22 **	2.82	24.06 **	0.95
Gluten index	135.16 **	0.02	63.21 **	6.45	44.00 **	2.43
Ash content	102.38 **	0.03	7.68 *	1.44	6.79 *	1.35
Water absorption	232.96 **	2.14	127.96 **	0.70	30.58 **	0.26
Development time, T1	2.78	28.44 **	214.14 **	1.00	224.03 **	8.44 *
Stability	13.06 *	7.60	5.45	1.98	4.23	1.42
C2	770.06 **	27.56 **	14.44 *	5.06	24.44 **	3.50
C_{18}	586.51 **	10.51 *	7.87 *	8.18 *	16.78 **	3.51
Falling number	868.97 **	0.86	21.85 **	10.85 *	20.81 **	1.62
Starch damage	2.24	0.31	17.47 **	0.00	14.40 *	0.76
C3	2653.80 **	21.76 **	108.00 **	13.49*	62.60 **	4.55
C4	1263.83 **	0.03	40.29 **	5.98	13.93 *	0.42
C5	2311.20 **	0.04	52.44 **	15.96 *	14.06 *	1.31
Final temp. of gelatinization D3	0.60	11.43 *	1.49	1.20	7.31 *	2.67

Significant at: * $p < 0.05$, ** $p < 0.01$.

Table 4. Quality characteristics of protein complex of flour obtained from the tested cultivars.

Factor	Protein Content (N × 5.7) (%d.m.)	Gluten Content (%)	Gluten Index (%)	Ash Content (% d.m.)	Water Absorption (%)	Development Time, T1 (min)	Stability (min)	C2 (Nm)	C_{18} (Nm)
Localization (A)									
Wielichowo	12.3 ^{b}	31.2 ^{b}	91 ^{a}	0.50 ^{b}	54.4 ^{b}	3.4	10.0 ^{a}	0.52 ^{a}	0.54 ^{a}
Werbkowice	14.3 ^{a}	35.4 ^{a}	82 ^{b}	0.56 ^{a}	57.3 ^{a}	3.5	9.4 ^{b}	0.40 ^{b}	0.41 ^{b}
Fertilization (B)									
N + 0S	13.3	33.4	86	0.53	55.7	3.6 ^{a}	9.9	0.47 ^{a}	0.48 ^{a}
N + 30S	13.4	33.1	86	0.53	56.0	3.3 ^{b}	9.5	0.45 ^{b}	0.46 ^{b}
Cultivar (C)									
Hondia	13.0 ^{cd}	31.8 ^{b}	90 ^{bc}	0.50 ^{b}	53.2 ^{d}	3.4 ^{b}	10.2	0.48 ^{a}	0.49 ^{a}
Pokusa	14.0 ^{b}	35.9 ^{a}	84 ^{c}	0.53 ^{ab}	57.6 ^{b}	4.5 ^{a}	9.5	0.46 ^{a}	0.48 ^{ab}
Belissa	13.5 ^{bc}	36.6 ^{a}	76 ^{d}	0.52 ^{ab}	59.0 ^{a}	4.2 ^{a}	9.1	0.44 ^{b}	0.45 ^{b}
Hybery F_1	12.7 ^{d}	29.8 ^{b}	88 ^{bc}	0.54 ^{ab}	54.0 ^{d}	3.1 ^{b}	9.7	0.48 ^{a}	0.48 ^{ab}
Bonanza	13.4 ^{bc}	32.4 ^{b}	94 ^{a}	0.56 ^{a}	55.4 ^{c}	2.0 ^{c}	10.0	0.45 ^{ab}	0.46 ^{ab}

a, b, c, d—values marked with the same letters do not differ significantly at $p < 0.05$ according to Tukey's multiple test.

The addition of sulfur fertilization had no effect on flour protein and gluten content, and neither had any effect on gluten quality, ash content, and water absorption of flours (Table 4). However, we observed lower values of development time and torque in points C2 and C_{18}, which indicates that dough is at risk of becoming less stable and weaker during processing [59]. The differences were statistically significant, but not from a technological point of view. Sulfur application in the research of Wilson et al. [52] caused increased average water absorption of flour, development time, and dough stability determined by farinograph.

Flour obtained from tested wheat cultivars was differentiated in terms of protein properties. Cultivar 'Pokusa' was characterized by the highest baking value—with the highest protein and gluten content (14.0% and 35.9%, respectively) (Table 4). The lowest baking value was found for cv. 'Hondia' and 'Hybery F_1', which were also characterized by the lowest water absorption values (53.2 and 54.0%, respectively). Cultivar 'Belissa', with one of the greatest protein and gluten contents and the lowest gluten index, was characterized to be the most favorable in terms of water absorption of flour (59.0%) compared to the rest of the cultivars. All tested cultivars showed protein weakening—a dough torque in point C2 below 0.5 Nm, which indicates the appropriate quality to produce bread [60]. High values of dough torque measured 18 min after the mixolab test started (C_{18}) indicate a better tolerance to the weakening of the gluten structure during mixing and gradually increasing the temperature of the dough [60].

The interaction between localization and fertilization was stated only for the dough torque measured 18 min after the test started (C_{18}) (Table 3). The interaction between localization and cultivar was found for protein content, gluten content, gluten index, ash content, water absorption, dough development time, C2, and C_{18}, whereas the interaction between fertilization and cultivar was found only for development time.

Asparagine concentration was positively correlated with factors affecting bread-making quality: gluten content (r = 0.597) and water absorption of flour (r = 0.616) (Table S3). With the increase of asparagine content in flour, the gluten index decreased (r = −0.514), as well as the dough torque in point C_{18} (r = −0.952, respectively). No significant correlation between the asparagine content of wheat and protein or gluten content was found in the research of Malunga et al. [11]. However, weak correlations were found between asparagine content and wheat protein [13], gluten content, gluten index, the Zeleny sedimentation index, and water absorption [56].

The importance of the level of alpha-amylase activity is crucial for the dough fermentation process and for obtaining high-quality bread. Alpha-amylase activity and starch properties of tested flours were characterized by several parameters. The ANOVA results (Tables 3 and S4) indicated that localization had the greatest share in the total variance in the measured quality parameters characterizing the starch complex of flour, such as starch retrogradation C5 (88.8% in the total variance), stability of hot starch paste C4 (84.7%), starch gelatinization C3 (78.2%), and falling number (81.8%) (Table S4). For the wheat samples tested in this study, the cultivar was the second most impactful factor on parameters such as starch damage, contributing to 51.0% of the variance in their concentration.

Tested flours were characterized by average and low alpha-amylase activity, measured as falling number (in the range of 187 to 372 s). Flours obtained from grain cultivated in the Wielichowo localization were characterized by significantly higher falling numbers compared to the Werbkowice localization (Table 5). This lower alpha-amylase activity of grain samples from the Wielichowo localization directly affected the higher values of dough torque measured in points C3, C4, and C5 (2.15, 2.17, and 3.46 Nm, respectively) than from Werbkowice (Table 5).

Additional sulfur fertilization had no negative effect on the starch properties of tested flours. It influenced only the lower value of dough torque in point C3 measures of starch gelatinization and the lower final temperature of gelatinization (1.87 Nm and 74.8 °C, respectively) (Table 5). According to Klikocka et al. [51], sulfur addition had no effect on the starch content of tested wheat cultivars.

Alpha-amylase activity determined by falling numbers varied between the tested cultivars. Cultivar 'Belissa' was characterized by the highest value of falling number (on average, 317 s) compared to other cultivars (Table 5). The lowest falling number was stated for cv. 'Bonanza' (266 s). Despite the highest falling number value, cv. 'Belissa', along with cv. 'Pokusa' and cv. 'Bonanza', was characterized by the significantly lowest dough torque in points C3 (starch gelatinization), C4 (stability of hot starch paste), and C5 (starch retrogradation) of the mixolab curves. The highest values were stated for cv. 'Hondia' and 'Hybery F_1'.

Table 5. Quality characteristics of starch complex of flour obtained from the tested cultivars.

Factor	Falling Number (s)	Starch Damage (UCD)	C3 (Nm)	C4 (Nm)	C5 (Nm)	Final Temperature of Gelatinization, D3 (°C)
Localization (A)						
Wielichowo	342 a	18.2	2.15 a	2.17 a	3.46 a	75.7
Werbkowice	237 b	18.8	1.63 b	1.16 b	1.69 b	75.4
Fertilization (B)						
N + 0S	291	18.6	1.92 a	1.67	2.58	76.3 a
N + 30S	288	18.4	1.87 b	1.66	2.57	74.8 b
Cultivar (C)						
Hondia	288 bc	16.3 c	2.00 a	1.85 a	2.84 a	75.2
Pokusa	282 bc	20.3 a	1.79 b	1.51 b	2.36 b	74.8
Belissa	317 a	19.7 ab	1.81 b	1.62 b	2.47 b	76.3
Hybery F_1	296 ab	17.2 bc	2.04 a	1.90 a	2.95 a	75.6
Bonanza	266 c	18.9 ab	1.82 b	1.46 b	2.28 b	75.8

a, b, c—values marked with the same letters do not differ significantly at $p < 0.05$ according to Tukey's multiple test.

Starch damage is one of the most important criteria for assessing the quality of baking flour. It affects the water absorption of flour, the rheological properties of dough, the fermentation of the dough, and the structure of the bread crumb [61]. The starch damage of tested flours varied from 14.0 to 21.3 UCD and differed depending on the cultivar (Table 5). The highest average starch damage was found for cv. 'Pokusa', whereas the lowest value was found for cv. 'Hondia'. A wide range of starch damage indicates the possibility of using these flours in the production of many assortments of baking and pastry products. Wheat flours intended to produce bread and confectionery products are characterized by starch damage, usually in the wide range from 14 to 24 UCD [62]. According to Ma et al. [63], an increase in the starch damage of flour significantly decreased the falling number and increased the alpha-amylase activity. One of its benefits is greater susceptibility to amylolytic enzymes, which results in increased maltose and dextrin production during dough fermentation [32,34]. With the increase in the amount of fermentable carbohydrates in the dough, the activity of yeast and bacteria is stimulated [64]. In addition, the increase in the level of starch damage may also favor the increased reduction in carbohydrates, which are also involved in acrylamide formation. The conversion ratio of asparagine to acrylamide could be enhanced when starch damage is increased [34].

Mulla et al. [35] found a correlation between damaged starch content in the flour and acrylamide content in bread. The extent of milling is known to affect the contents by both reducing sugars [65] and amino acids [66]. This observation indicated that the mitigation of acrylamide in bread can be obtained by reducing starch damage in wheat flour [34]. To clearly unravel genotypic differences and their interaction with environmental factors and, especially, nitrogen and sulfur fertilization, further research is needed.

Similar to the research of Malunga et al. [11], no significant correlation was found with asparagine content. Asparagine content was negatively correlated with the falling number (r = −0.905) (Table S3). Doughs from flours with higher asparagine content were also characterized by lower consistency in points C3, C4, and C5 (r = −0.932, r = −0.964, and r = −0.977, respectively).

The interaction between localization and fertilization was found for the following quality parameters: falling number and dough torque in points C3 and C5 (Table 3). The interaction between localization and cultivar was found for falling number; starch damage; dough torque in points C3, C4, C5; and final temperature of gelatinization, whereas no interaction between fertilization and cultivar was detected.

3.3. Bread Quality Properties

The ANOVA results (Tables 6 and S5) indicated that the localization had the greatest share in the total variance of the measured acrylamide content in bread (55.8% in the total variance) and the color of the bread crumb (85.5%—parameter a^*). The wheat cultivar had the second most significant impact on parameters such as loaf volume (62.5% of the total variance), acrylamide content (13.5%), and bread crumb color (L 38.7% and b^* 55.2%).

Table 6. F values calculated in the three-way ANOVA made for the qualitative parameters for the bread obtained in the baking trial from the grain of the examined wheat samples.

Parameter	Sources of Variation						
	Localization (A)	S Fertilization (B)	Cultivar (C)	A × B	A × C	B × C	A × B × C
Loaf volume	12.22 **	18.03 **	68.75 **	1.47	20.02 **	1.74	1.51
Bread crumb hardness	119.35 **	25.87 **	4.33 **	3.08	6.25 **	29.19 **	37.75 **
Acrylamide content	1518.09 **	359.17 **	92.03 **	9.21 **	8.15 *	65.03 *	37.92 *
*L*crumb	404.92 **	0.61	85.12 **	1.59	11.63 **	0.84	0.81
*a**crumb	3059.32 **	12.69 **	62.99 **	0.61	25.49 **	4.68 **	12.79 **
*b**crumb	248.16 **	4.36 *	187.01 **	0.00	51.10 **	4.91 **	12.92 **
*L*crust	36.33 **	24.89 **	2.08	0.41	13.96 **	2.03	1.17
*a**crust	0.60	2.85	4.03 *	6.67 *	9.71 **	1.43	0.40
*b**crust	39.13 **	33.20 **	3.48 *	0.03	7.25 **	1.96	2.14

Significant at: * $p < 0.05$, ** $p < 0.01$.

The data on bread's physical characteristics are shown in Table 7. Bread obtained in laboratory baking was characterized by a proper appearance with the shape of a well-risen loaf and the right color of the crust, as well as a relatively uniform porosity of the crumb (Figures 1–4). The most favorable in terms of baking value and the possibility of using it for baking purposes was bread obtained from grain cultivated in Wielichowo with the highest bread volume (322 cm^3) and the lowest acrylamide content (11.9 µg 100^{-1} g) compared to samples from Werbkowice (Table 7).

Table 7. Quality characteristics of bread obtained in laboratory baking from tested cultivars.

Factor	Loaf Volume (cm^3/100 g)	Bread Crumb Hardness (N)	Acrylamide Content (µg 100^{-1} g)	Breadcrumb Color			Bread Crust Color		
				L	*a**	*b**	*L*	*a**	*b**
Localization (A)									
Wielichowo	322 a	14.9 a	11.9 b	74.2 a	−1.07 b	18.44 a	53.36 a	13.97	30.45 a
Werbkowice	315 b	11.9 b	17.9 a	70.1 b	−0.23 a	17.60 b	49.73 b	13.83	27.88 b
Fertilization (B)									
N + 0S	323 a	12.7 b	16.4 a	72.2	−0.68 b	17.97 b	50.04 b	14.06	27.98 a
N + 30S	315 b	14.1 a	13.3 b	72.1	−0.62 a	18.08 a	53.04 a	13.70	30.35 a
Cultivar (C)									
Hondia	334 a	12.7 b	14.1 b	75.1 a	−0.58 b	17.25 c	52.34	14.23 a	30.02 ab
Pokusa	303 c	13.8 ab	14.2 b	69.2 d	−0.45 a	18.26 b	50.35	14.28 a	28.47 ab
Belissa	297 c	14.3 a	17.8 a	72.7 b	−0.72 c	18.81 a	51.64	13.95 ab	28.97 ab
Hybery F_1	337 a	12.9 b	13.9 b	72.3 bc	−0.75 c	17.08 c	52.60	13.24 c	30.10 a
Bonanza	323 b	13.5 ab	14.4 b	71.5 c	−0.75 c	18.72 a	50.78	13.81 ab	28.27 b

L, lightness, (L = 0 is black; L = 100 is white); *a**, green-red opponent colors ($-a$ = green; $+a$ = red); *b**, blue-yellow opponent colors ($-b$ = blue; $+b$ = yellow). a, b, c, d—values marked with the same letters do not differ significantly at $p < 0.05$ according to Tukey's multiple test.

(a) (b) (c) (d) (e)

Figure 1. Cross-sectional view of bread from the tested wheat cultivars from Wielichowo localization with N + 0S fertilization: (**a**) Hondia; (**b**) Pokusa; (**c**) Belissa; (**d**) Hybery F_1; (**e**) Bonanza.

(a) (b) (c) (d) (e)

Figure 2. Cross-sectional view of bread from the tested wheat cultivars from Wielichowo localization with N + 30S fertilization: (**a**) Hondia; (**b**) Pokusa; (**c**) Belissa; (**d**) Hybery F_1; (**e**) Bonanza.

(a) (b) (c) (d) (e)

Figure 3. Cross-sectional view of bread from the tested wheat cultivars from Werbkowice localization with N + 0S fertilization: (**a**) Hondia; (**b**) Pokusa; (**c**) Belissa; (**d**) Hybery F_1; (**e**) Bonanza.

(a) (b) (c) (d) (e)

Figure 4. Cross-sectional view of bread from the tested wheat cultivars from Werbkowice localization with N + 30S fertilization: (**a**) Hondia; (**b**) Pokusa; (**c**) Belissa; (**d**) Hybery F_1; (**e**) Bonanza.

The addition of sulfur fertilization had a negative effect on the bread volume by lowering it from 323 cm^3 to 315 cm^3. Additionally, the hardness of the bread crumb was increased to 14.1 N (Table 7). Shahsavani and Gholami [67] and Li et al. [68] found that wheat grown at an intermediate sulfur fertilization rate yielded the highest loaf volume. Asparagine reductions achieved by increasing sulfur fertilization partially coincided with a loss in the functionality of bread.

An undoubtedly positive health effect of sulfur fertilization was lowering the acrylamide content in bread, on average, from 16.4 to 13.3 μg 100^{-1} g (Table 7). Considering the results of individual samples in a given location, the highest reduction in acrylamide content was found for grain samples cultivated in Wielichowo (on average, 3.6 μg 100^{-1} g lower) than in Werbkowice (on average, 2.5 μg 100^{-1} g lower) (Table S2). Bread obtained from cv. 'Hondia' and 'Belissa' cultivated with sulfur fertilization was characterized by the greatest reduction in acrylamide content in both localizations (on average, 3.7 and 6.9 μg 100^{-1} g lower).

Unlike the results of Claus et al. [7], bread obtained from wheat cultivars belongs to Class A—good category wheat (cv. 'Hondia', 'Pokusa')—and shows a lower value of acrylamide than from other bread category cultivars ('Belissa', Hybery F_1', 'Bonanza'). It is possible to indicate cultivars with good baking value, the bread obtained from which will be characterized by a low acrylamide content.

Among tested cultivars, cv. 'Hondia' and cv. 'Hybery F_1' with the lowest starch damage of flour were characterized by the highest bread volume, the lowest bread crumb hardness, and one of the lowest acrylamide contents. Cultivar 'Belissa', with the second highest degree of starch damage, was characterized significantly by the lowest bread volume with the highest bread crumb hardness and the highest acrylamide content (296.9 cm^3, 14.3 N and 17.8 μg 100^{-1} g, respectively), and was rated as the least favorable.

Browning products obtained from the Maillard reaction influence sensory perception. Bread crust obtained from flours with higher protein and amino acid contents was darker. Localization and fertilization, but not cultivar, were factors that significantly influenced the lightness L of the tested bread crusts (Table 7). The highest value of brightness L of the crust was obtained for the bread samples obtained from Wielichowo (with also a lower value of protein content, asparagine content in grain, and acrylamide content in bread), and with N + 30S treatment (with lower acrylamide content) (Table 7). Li et al. [68] indicated that the environmental factor and cultivar, but not sulfur fertilization, influence the brightness of the bread crust. Bread obtained from grain cultivated with sulfur was also characterized by a smaller share of green and a greater share of red in the color of the crumb, higher brightness of the crust, and a higher share of red tone.

Asparagine content influenced the acrylamide content in bread (r = 0.768) (Table S3). Bread from flours with higher asparagine content was also characterized by darker crumb color and a greater share of red color (a^*crumb) (r = −0.733 and r = 0.897, respectively). In this work, we did not observe the influence of starch damage on the acrylamide content in bread when the results of all cultivars, fertilization, and location were compared. However, bread obtained from flour with higher starch damage was characterized by darker crust color, with a greater share of red, and a smaller share of yellow (correlation with L^* value r = −0.607, a^* r = 0.548, and b^* r = −0.545, respectively).

The results of a principal component analysis (PCA) demonstrated that the first two principal components (PC1 and PC2) accounted for 8.3% of the variation (Figure 5). PC1 explained 51.4% and PC2 explained 30.9% of the variation. Figure 5 shows that cultivars 'Hondia' and 'Hybery F_1' with the lowest asparagine content, starch damage, water absorption of flour, bread crumb hardness, acrylamide content, and the highest loaf volume belong to one group, while the remaining cultivars form three separate groups. Comparing the position of the cases on the chart with the forms of components and factor loadings, it can be concluded that the cv. 'Hondia' is negatively correlated with PC1 and positively with PC2, while the cv. 'Hybery F_1' is negatively correlated with PC1 and

PC2. However, the cultivars 'Bonanza', 'Belissa', and 'Pokusa' were positively correlated with PC1.

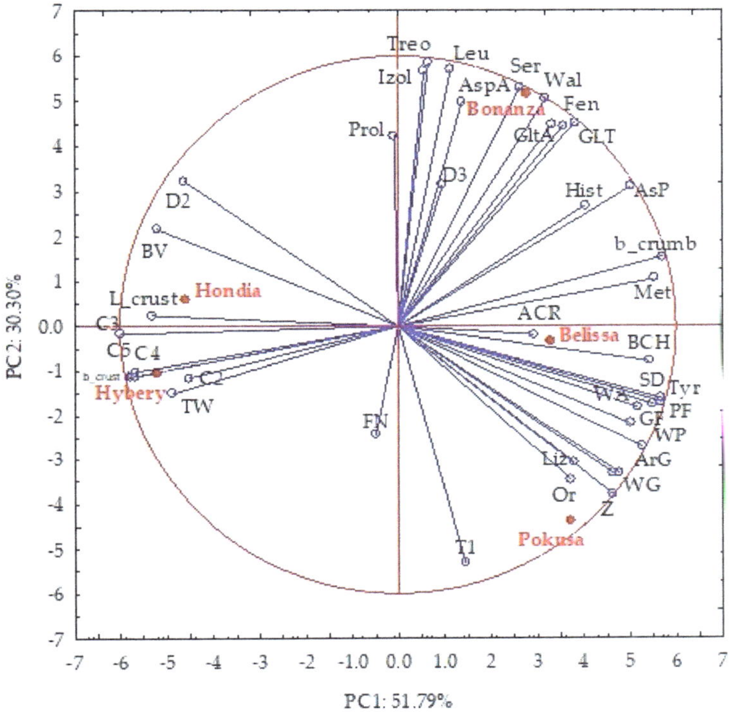

Figure 5. Variable graph of PCA showing the quality parameters and cultivars of wheat. Explanations: TW—test weight, FN—falling number, Z—Zeleny index, WP—wheat protein content, PF—flour protein content, WG—wheat gluten content, SD—starch damage, WA—water absorption, T1—development time, BV—bread volume, BCH—bread crumb hardness, ACR—acrylamide content, Treo—threonine, Leu—leucine, Ser—, Wal—valine, Fen—phenylalanine, Izol—isoleucine, Prol—proline, GltA—glutamic acid, GLT—glutamine, Hist—histidine, AsP—asparagine, AspA—aspartic acid, Met—methionine, Tyr—tyrosine, ArG—arginine, Liz—lysine, Or—ornithine.

Figure 5 shows that cultivars 'Hondia' and 'Hybery F_1' are classified in the left section of the chart, which indicates relatively higher values of bread volume (BV), brightness of crust (*L*crust), starch gelatinization (C3), amylolytic activity (C4), retrogradation (C5), and protein weakening (C2), while featuring low values of amino acids, total protein, and gluten content. However, cultivars 'Bonanza', 'Belissa', and 'Pokusa' scored low values of parameters such as the starch–amylolytic complex and higher amino acid values than the previously mentioned cultivars. The cv. 'Bonanza' is characterized by the highest value of aspartic acid, threonine, leucine, serine, and valine among the three cultivars mentioned. The cv. 'Pokusa' is characterized by the highest protein content both in grain and flour regarding gluten content and arginine, and the most favorable features characterizing the quality of gluten proteins, i.e., the Zeleny sedimentation index and water absorption of flour. The obtained results indicate a positive correlation of the starch damage (SD) with water absorption (WA) and bread crumb hardness (BCH), and a negative correlation with the brightness of a crust (*L*crust) and starch retrogradation (C5). Cultivars 'Belissa', 'Bonanza', and 'Pokusa' were characterized by higher average values of starch damage (SD), water absorption (WA), and bread crumb hardness (BCH). Bread obtained from cultivars 'Hondia' and 'Hybery F_1' had the lowest acrylamide content and bread from cultivars 'Bonanza', 'Belissa' and, 'Pokusa' had the highest acrylamide content due to the high protein content and amino acid content, including asparagine and aspartic acid, and higher access to maltose and dextrin products during fermentation as a result of high starch damage.

4. Conclusions

To our knowledge, this work presents the first comprehensive analysis related to the accumulation of asparagine in grain and acrylamide content in bread, including the assessment of the baking value of five winter wheat varieties cultivated in Poland.

Our research showed a significant influence of localization (environment) and cultivar and their interaction on the asparagine and acrylamide content and technological parameters of grain. The influence of sulfur application on the asparagine content in grain and on the baking value has not been confirmed. However, we found a negative effect of sulfur fertilization on the acrylamide content in bread. The obtained results indicate that, on the one hand, there is no effect of sulfur on the asparagine content in grain and, on the other hand, a decrease in acrylamide in bread may be surprising because the highly significant positive correlation between these features was confirmed in our research. Probably, the effectiveness of sulfur in reducing the asparagine in the endosperm depends on the wheat cultivar and the location of the crop, which requires further research and confirmation.

The food industry requires cultivars with reduced free asparagine content and favorable baking properties, and which show stable characteristics regardless of weather conditions. The obtained localization × cultivar interactions indicate different responses of wheat cultivars depending on habitat conditions. The interaction was confirmed in the asparagine and acrylamide content and bread-making quality. In our research, the asparagine content in tested cultivars ranged from 235 to 588 mg kg^{-1}, the acrylamide content from 8.0 to 23.4 µg 100^{-1} g, and the protein content in wheat grain from 12.3 to 17.6% d.m. (Table S2). Considering the quality requirements of the baking industry (high protein content) and food safety requirements (low asparagine and acrylamide content), the most valuable cultivar is 'Pokusa'. Cv. 'Pokusa' is characterized by high protein content in grain (15.8% d.m.) and one of the lowest asparagine contents in grain (397 mg kg^{-1}) and acrylamide content in bread (14.2 µg 100^{-1} g). From the point of view of the health of the raw material, we can also recommend the 'Hybery F_1' cultivar. Bread obtained from this cultivar has the lowest acrylamide content (13.9 µg 100^{-1} g), but the grain is characterized by the lowest protein content (14.6% d.m.) and gluten content (29.7%) among tested cultivars.

A novel aspect of our research is the analysis of the influence of the amylolytic complex and starch damage on the acrylamide content in bread. Our results indicate a significant negative correlation of starch quality parameters with asparagine and acrylamide content. The results also indicate differences in cultivars. Among others, cv. 'Belissa', with the highest falling number and starch damage values, is characterized by a higher asparagine content in grain and acrylamide content in bread. Further research should therefore consider the aspect of starch properties on the formation of asparagine in grain and acrylamide in bread.

Supplementary Materials: The following supporting information can be downloaded at: https://www.mdpi.com/article/10.3390/agriculture14020207/s1. Table S1: The ratio of the factorial SS to the total sum of squares illustrating the percentage share of the total variance for the qualitative parameters of grain of the examined wheat samples. Table S2: Protein and asparagine content in wheat, starch damage of flour, and acrylamide content of bread depend on cultivar, localization, and fertilization. Table S3: Correlation between asparagine concentration, acrylamide content, and quality parameters. Table S4: The ratio of the factorial SS to the total sum of squares illustrating the percentage share of the total variance for the qualitative parameters of the flour obtained from the grain of the examined wheat samples. Table S5: The ratio of the factorial SS to the total sum of squares illustrating the percentage share of the total variance for the qualitative parameters for the bread obtained in the baking trial.

Author Contributions: Conceptualization, A.S. and G.P.; methodology, A.S., G.P. and M.R.; validation, A.S. and O.Ś.; formal analysis A.S.; investigation, A.S., E.A. and O.Ś.; resources, A.S. and M.R.; data curation, A.S. and D.K.; writing—original draft preparation, A.S.; writing—review and editing, A.P.-C. and M.R.; visualization, A.S.; supervision, A.S.; project administration, A.S.; funding acquisition, A.S. and G.P. All authors have read and agreed to the published version of the manuscript.

Funding: This research is supported by Institute of Agricultural and Food Biotechnology—State Research Institute (statutory research activity in 2022–2023—grant number ZZ-107-01) supported by the Polish Ministry of Education and Science.

Institutional Review Board Statement: Not applicable.

Data Availability Statement: Data are contained within the article or supplementary material. The data presented in this study are available in the supplementary material. Table S1: The ratio of the factorial SS to the total sum of squares illustrating the percentage share of the total variance for the qualitative parameters of grain of the examined wheat samples. Table S2: Protein and asparagine content in wheat, starch damage of flour, and acrylamide content of bread depend on cultivar, localization, and fertilization. Table S3: Correlation between asparagine concentration, acrylamide content, and quality parameters. Table S4: The ratio of the factorial SS to the total sum of squares illustrating the percentage share of the total variance for the qualitative parameters of the flour obtained from the grain of the examined wheat samples. Table S5: The ratio of the factorial SS to the total sum of squares illustrating the percentage share of the total variance for the qualitative parameters for the bread obtained in the baking trial.

Acknowledgments: This research was based upon the work carried out in the frame of COST Action ACRYRED—"Reducing Acrylamide Exposure of Consumers by a Cereals Supply-chain Approach Targeting Asparagine", grant number CA 21149, where the author Anna Szafrańska is a WG1 and WG2 Member, and Marek Roszko is a WG5 Member.

Conflicts of Interest: The authors declare no conflicts of interest.

References

1. FAOSTAT. *Crops and Livestock Products*; FAOSTAT: Rome, Italy, 2023. Available online: https://www.fao.org/faostat/en/#data/QCL (accessed on 10 December 2023).
2. Huang, T.; Xu, M.; Lee, A.; Cho, S.; Qi, L. Consumption of whole grains and cereal fiber and total and cause-specific mortality: Prospective analysis of 367,442 individuals. *BMC Med.* **2015**, *13*, 59, Erratum in *BMC Med.* **2015**, *13*, 85. [CrossRef]
3. Zyzak, D.V.; Sanders, R.A.; Stojanovic, M.; Tallmadge, D.H.; Eberhart, B.L.; Ewald, D.K.; Gruber, D.C.; Morsch, T.R.; Strothers, M.A.; Rizzi, G.P.; et al. Acrylamide formation mechanism in heated foods. *J. Agric. Food Chem.* **2003**, *51*, 4782–4787. [CrossRef]
4. Becalski, A.; Lau, B.P.; Lewis, D.; Seaman, S.W. Acrylamide in foods: Occurrence, sources, and modeling. *J. Agric. Food Chem.* **2003**, *51*, 802–808. [CrossRef]
5. Claus, A.; Carle, R.; Schieber, A. Acrylamide in cereal products: A review. *J. Cereal Sci.* **2008**, *47*, 118–133. [CrossRef]
6. Amrein, T.M.; Schönbächler, B.; Rohner, F.; Lukac, H.; Schneider, H.; Keiser, A.; Escher, F.; Amadò, R. Potential for acrylamide formation in potatoes: Data from the 2003 harvest. *Eur. Food Res. Technol.* **2004**, *219*, 572–578. [CrossRef]
7. Claus, A.; Schreiter, P.; Weber, A.; Graeff, S.; Herrmann, W.; Claupein, W.; Schieber, A.; Carle, R. Influence of Agronomic Factors and Extraction Rate on the Acrylamide Contents in Yeast-Leavened Breads. *J. Agric. Food Chem.* **2006**, *54*, 8968–8976. [CrossRef] [PubMed]
8. Taeymans, D.; Wood, J.; Ashby, P.; Blank, I.; Studer, A.; Stadler, R.H.; Gondé, P.; Van Eijck, P.; Lalljie, S.; Lingnert, H.; et al. A review of acrylamide: An industry perspective on research, analysis, formation, and control. *Crit. Rev. Food Sci. Nutr.* **2004**, *44*, 323–347. [CrossRef] [PubMed]
9. Curtis, T.Y.; Powers, S.J.; Wanga, R.; Halford, N.G. Effects of variety, year of cultivation and sulphur supply on the accumulation of free asparagine in the grain of commercial wheat varieties. *Food Chem.* **2018**, *239*, 304–313. [CrossRef] [PubMed]
10. Rapp, M.; Schwadorf, K.; Leiser, W.L.; Würschum, T.; Longin, C.F.H. Assessing the variation and genetic architecture of asparagine content in wheat: What can plant breeding contribute to a reduction in the acrylamide precursor? *Theor. Appl. Genet.* **2018**, *131*, 2427–2437. [CrossRef] [PubMed]
11. Malunga, L.N.; Ames, N.P.; Masatcioglu, M.T.; Khorshidi, A.S.; Thandapilly, S.J.; Cuthbert, R.D.; Sopiwnyk, E.; Scanlon, M.G. Free asparagine concentrations in Canadian hard red spring wheat cultivars. *Can. J. Plant Sci.* **2019**, *99*, 338–347. [CrossRef]
12. Lea, P.J.; Sodek, L.; Parry, M.A.J.; Shewry, P.R.; Halford, N.G. Asparagine in plants. *Ann. Appl. Biol.* **2007**, *150*, 1–26. [CrossRef]

13. Navrotskyi, S.; Baenziger, P.S.; Regassa, T.; Guttieri, M.J.; Rose, D.J. Variation in asparagine concentration in Nebraska wheat *Cereal Chem.* **2018**, *95*, 264–273. [CrossRef]
14. Mansfield, J.; Gebauer, S.; Dathe, K.; Ulbrich-Hofmann, R. Secretory phospholipase A2 from *Arabidopsis thaliana*: Insights into the three-dimensional structure and the amino acids involved in catalysis. *Biochemistry* **2006**, *45*, 5687–5694. [CrossRef]
15. Stockmann, F.; Weber, E.A.; Schreiter, P.; Merkt, N.; Claupein, W.; Graeff-Hönninger, S. Impact of Nitrogen and Sulphur Supply on the Potential of Acrylamide Formation in Organically and Conventionally Grown Winter Wheat. *Agronomy* **2018**, *8*, 284. [CrossRef]
16. Curtis, T.Y.; Halford, N.G. Reducing the acrylamide-forming potential of wheat. *Food Energy Secur.* **2016**, *5*, 153–164. [CrossRef]
17. Halford, N.G.; Muttucumaru, N.; Curtis, T.Y.; Parry, M.A.J. Genetic and agronomic approaches to decreasing acrylamide precursors in crop plants. *Food Addit. Contam.* **2007**, *24* (Suppl. S1), 26–36. [CrossRef] [PubMed]
18. Lerner, S.E.; Seghezzo, M.L.; Molfese, E.R.; Ponzio, N.R.; Cogliatti, M.; Rogers, W.J. N- and S-fertiliser effects on grain composition industrial quality and end-use in durum wheat. *J. Cereal Sci.* **2006**, *44*, 2–11. [CrossRef]
19. Sinclair, T.R.; Horie, T. Leaf Nitrogen, Photosynthesis, and Crop Use Efficiency: A Review. *Crop Sci.* **1989**, *29*, 90–97. [CrossRef]
20. Jamal, A.; Moon, Y.-S.; Abdin, M.Z. Sulphur–A general overview and interaction with nitrogen. *Aust. J. Crop Sci.* **2010**, *4*, 523–529
21. Cambell, W.H. Nitrate reductase structure, function and regulation: Bridging the Gap between Biochemistry and Physiology *Annu. Rev. Plant Physiol. Plant Mol. Biol.* **1999**, *50*, 277–303. [CrossRef]
22. Mandal, U.K.; Singh, G.; Victor, U.S.; Sharma, K.L. Green manuring: Its effect on soil properties and crop growth under rice–wheat cropping system. *Eur. J. Agron.* **2003**, *19*, 225–237. [CrossRef]
23. Swamy, U.; Wang, M.; Tripathy, J.N.; Kim, S.-K.; Hirasawa, M.; Knaff, D.B.; Allen, J.P. Structure of Spinach Nitrite Reductase Implications for Multi-electron Reactions by the Iron−Sulphur:Siroheme Cofactor. *Biochemistry* **2005**, *44*, 16054–16063. [CrossRef] [PubMed]
24. Salvagiotti, F.; Castellarin, J.M.; Miralles, D.J.; Pedrol, H.M. Sulphur fertilization improves nitrogen use efficiency in wheat by increasing nitrogen uptake. *Field Crops Res.* **2009**, *113*, 170–177. [CrossRef]
25. Yu, Z.; Juhasz, A.; Islam, S.; Diepeveen, D.; Shang, J.; Wang, P.; Ma, W. Impact of mid-season sulphur defciency on wheat nitrogen metabolism and biosynthesis of grain protein. *Sci. Rep.* **2018**, *8*, 2499. [CrossRef] [PubMed]
26. De Ruiterr, J.M.; Martinn, R.J. Management of nitrogen and sulphur fertilizer for improved bread wheat (*Triticum aestivum*) quality. *N. Z. J. Crop Hortic. Sci.* **2001**, *29*, 287–299. [CrossRef]
27. Aulakh, M.S.; Malhi, S.S. Interaction of Nitrogen with Other Nutrients and Water: Effects on Crop Yield and Quality, Nutrient Use Efficiency, Carbon Sequestration and Environmental Pollution. *Adv. Agron.* **2004**, *86*, 341–409. [CrossRef]
28. Brosnan, J.T.; Brosnan, M.E. The sulphur-containing amino acids: An overview. *J Nutr.* **2006**, *136* (Suppl. 6), 1636S–1640S. [CrossRef]
29. Győri, Z. Sulphur Content of Winter Wheat Grain in Long Term Field Experiments. *Commun. Soil Sci. Plant Anal.* **2005**, *36* 373–382. [CrossRef]
30. *National Atmospheric Deposition Program*; NADP Program Office: Leading Ridge, PA, USA; Illinois State Water Survey, University of Illinois: Champaign, IL, USA, 2015.
31. David, M.B.; Gentry, L.E.; Mitchell, C.A. Riverine Response of Sulfate to Declining Atmospheric Sulphur Deposition in Agricultural Watersheds. *J. Environ. Qual.* **2016**, *45*, 1313–1319. [CrossRef]
32. Jukić, M.; Koceva Komlenić, D.; Mastanjević, K.; Mastanjević, K.; Lučan, M.; Popovici, C.; Nakov, G.; Lukinac, J. Influence of damage starch on the quality parameters of wheat dough and bread. *Ukr. Food J.* **2019**, *8*, 512–521. [CrossRef]
33. Putri, N.A.; Shalihah, I.M.; Widyasna, A.F.; Damayanti, R.P. A review of starch damage on physicochemical properties of flour *Food Scien Tech J.* **2020**, *2*, 1. [CrossRef]
34. Wang, S.; Yu, J.; Xin, Q.; Wang, S.; Copeland, L. Effects of starch damage and yeast fermentation on acrylamide formation in bread. *Food Control* **2017**, *73 Pt B*, 230–236. [CrossRef]
35. Mulla, M.Z.; Bharadwaj, V.R.; Annapure, U.S.; Singhal, R.S. Effect of damaged starch on acrylamide formation in whole wheat flour based Indian traditional staples, *chapattis* and *pooris*. *Food Chem.* **2010**, *120*, 805–809. [CrossRef]
36. *ISO 7971-3:2019*; Cereals Determination of Bulk Density, Called Mass per Hectolitre Part 3: Routine Method. International Organization for Standardization: Geneva, Switzerland, 2019.
37. *ISO 20483:2013*; Cereals and Pulses. Determination of the Nitrogen Content and Calculation of the Crude Protein Content Kjeldahl Method. International Organization for Standardization:: Geneva, Switzerland, 2013.
38. *ISO 21415-2:2015*; Wheat and Wheat Flour Gluten Content Part 2: Determination of Wet Gluten and Gluten Index by Mechanical Means. International Organization for Standardization: Geneva, Switzerland, 2015.
39. *ISO 5529:2007*; Wheat Determination of the Sedimentation Index Zeleny Test. International Organization for Standardization Geneva, Switzerland, 2007.
40. *ISO 2171:2007*; Cereals, Pulses and By-Products Determination of Ash Yield by Incineration. International Organization for Standardization: Geneva, Switzerland, 2007.
41. Świder, O.; Roszko, M.Ł.; Wójcicki, M.; Szymczyk, K. Biogenic Amines and Free Amino Acids in Traditional Fermented Vegetables-Dietary Risk Evaluation. *J. Agric. Food Chem.* **2020**, *68*, 856–868. [CrossRef]
42. *ISO 17718:2013*; Wholemeal and Flour from Wheat (*Triticum aestivum* L.) Determination of Rheological Behaviour as a Function of Mixing and Temperature Increase. International Organization for Standardization: Geneva, Switzerland, 2013.
43. *ISO 17715:2013*; Flour from Wheat (*Triticum aestivum* L.) Amperometric Method for Starch Damage Measurement. International Organization for Standardization: Geneva, Switzerland, 2013.

Suchowilska, E.; Szafrańska, A.; Słowik, E.; Wiwart, M. Flour from *Triticum polonicum* L. as a potential ingredient in bread production. *Cereal Chem.* **2019**, *96*, 554–563. [CrossRef]

Roszko, M.Ł.; Szczepańska, M.; Szymczyk, K.; Rzepkowska, M. Dietary risk evaluation of acrylamide intake with bread in Poland, determined by two comparable cleanup procedures. *Food Addit. Contam. Part B Surveill.* **2020**, *13*, 1–9. [CrossRef]

ISO/IEC 17025. General Requirements for the Competence of Testing and Calibrarion Laboratories. International Organization for Standardization: Geneva, Switzerland, 2017.

Xie, Y.; Malunga, L.N.; Ames, N.P.; Waterer, J.; Khorshidi, A.S.; Scanlon, M.G. Effects of growing environment, genotype, and commercial fertilization levels on free asparagine concentration in Western Canadian wheat. *Cereal Chem.* **2021**, *98*, 89–99. [CrossRef]

Oddy, J.; Elmore, J.S.; Halford, N.G. Accounting for environmental variation in the free asparagine content of wheat grain. *J. Food Compos. Anal.* **2023**, *120*, 105333. [CrossRef]

Malunga, L.N.; Ames, N.; Khorshidi, A.S.; Thandapilly, S.J.; Yan, W.; Dyck, A.; Waterer, J.; Malcolmson, L.; Cuthbert, R.; Sopiwnyk, E.; et al. Association of asparagine concentration in wheat with cultivar, location, fertilizer, and their interaction. *Food Chem.* **2021**, *344*, 128630. [CrossRef]

Tafuri, A.; Zuccaro, M.; Ravaglia, S.; Pirona, R.; Masci, S.; Sestili, F.; Lafiandra, D.; Ceriotti, A.; Baldoni, E. Exploring variability of free asparagine content in the grain of bread wheat (*Triticum aestivum* L.) varieties cultivated in Italy to reduce acrylamide-forming potential. *Plants* **2023**, *2*, 1349. [CrossRef]

Klikocka, H.; Cybulska, M.; Barczak, B.; Narolski, B.; Szostak, B.; Kobiałka, A.; Nowak, A.; Wójcik, E. The effect of sulphur and nitrogen fertilization on grain yield and technological quality of spring wheat. *Plant Soil Environ* **2016**, *62*, 230–236. [CrossRef]

Wilson, T.L.; Guttieri, M.J.; Nelson, N.O.; Fritz, A.; Tilley, M. Nitrogen and sulphur effects on hard winter wheat quality and asparagine concentration. *J. Cereal Sci.* **2020**, *93*, 102969. [CrossRef]

Muttucumaru, N.; Halford, N.G.; Elmore, J.S.; Dodson, A.T.; Parry, M.; Shewry, P.R.; Mottram, D.S. The formation of high levels of acrylamide during the processing of flour derived from sulfate-deprived wheat. *J. Agric. Food Chem.* **2006**, *54*, 8951–8955. [CrossRef]

Weber, E.A.; Koller, W.D.; Graeff, S.; Hermann, W.; Merkt, N.; Claupein, W. Impact of different nitrogen fertilizers and an additional sulfur supply on grain yield, quality, and the potential of acrylamide formation on winter wheat. *J. Plant Nutr. Soil Sci.* **2008**, *171*, 643–655. [CrossRef]

Ohm, J.-B.; Mergoum, M.; Simsek, S. Variation of free asparagine concentration and association with quality parameters for hard red spring wheat grown in North Dakota. *Cereal Chem.* **2017**, *94*, 712–716. [CrossRef]

Corol, D.I.; Ravel, C.; Rakszegi, M.; Charmet, G.; Bedo, Z.; Beale, M.H.; Shewry, P.R.; Ward, J.L. 1H-NMR screening for the high-throughput determination of genotype and environmental effects on the content of asparagine in wheat grain. *Plant Biotechnol. J.* **2016**, *14*, 128–139. [CrossRef]

Emebiri, L.C. Genetic variation and possible SNP markers for breeding wheat with low-grain asparagine, the major precursor for acrylamide formation in heat-processed products. *J. Sci. Food Agric.* **2014**, *94*, 1422–1429. [CrossRef]

Bonfil, D.J.; Posner, E.S. Can bread wheat quality be determined by gluten index? *J. Cereal Sci.* **2012**, *56*, 115–118. [CrossRef]

Dabija, A.; Codină, G.G.; Fradinho, P. Effect of yellow pea flour addition on wheat flour dough and bread quality. *Rom Biotechnol. Lett.* **2017**, *22*, 12888–12897.

Wiwart, M.; Szafrańska, A.; Suchowilska, E. Grain of Hybrids Between Spelt (*Triticum spelta* L.) and Bread Wheat (*Triticum aestivum* L.) as a New Raw Material for Breadmaking. *Pol. J. Food Nutr. Sci.* **2023**, *73*, 265–277. [CrossRef]

Litvyak, V. Size and morphological features of native starch granules of different botanical origin. *Ukr. Food J.* **2018**, *7*, 563–576. [CrossRef]

Liu, X.; Li, E.; Jiang, Y.; Tian, Y. Milling process of starch. In *Physical Modification of Starch*; Sui, Z., Kong, X., Eds.; Springer Nature: Singapore, 2018; pp. 155–174. [CrossRef]

Ma, S.; Wang, X.-X.; Zheng, X.-L.; Bian, K.; Bao, Q.-D. Effect of mechanically damaged starch from wheat flour on the quality of frozen dough and steamed bread. *Food Chem.* **2016**, *202*, 120–124. [CrossRef]

Codină, G.G.; Sarion, C.; Dabija, A. Effects of Dry Sourdough on Bread-Making Quality and Acrylamide Content. *Agronomy* **2021**, *11*, 1977. [CrossRef]

Sadd, P.A.; Hamlet, C.G.; Liang, L. Effectiveness of methods for reducing acrylamide in bakery products. *J. Agric. Food Chem.* **2008**, *56*, 6154–6161. [CrossRef]

Claeys, W.L.; De Vleeschouwer, K.; Hendrickx, M.E. Quantifying the formation of carcinogens during food processing: Acrylamide. *Trends Food Sci. Technol.* **2005**, *16*, 181–193. [CrossRef]

Shahsavani, S.; Gholami, A. Effect of Sulphur Fertilization on Breadmaking Quality of Three Winter Wheat Varieties. *Pak. J. Biol. Sci.* **2008**, *11*, 2134–2138. [CrossRef]

Li, X.; Tyl, C.E.; Kaiser, D.E.; Annor, G.A. Effect of sulphur fertilization rates on wheat (*Triticum aestivum* L.) functionality. *J. Cereal Sci.* **2019**, *87*, 292–300. [CrossRef]

agriculture

MDPI

Article

Impact of Biochar Dose and Origin on Winter Wheat Grain Quality and Quantity

Marta Wyzińska *, Adam Kleofas Berbeć and Jerzy Grabiński

Institute of Soil Science and Plant Cultivation-State Research Institute, 24-100 Pulawy, Poland; aberbec@iung.pulawy.pl (A.K.B.); jurek@iung.pulawy.pl (J.G.)

* Correspondence: mwyzinska@iung.pulawy.pl

Abstract: The agricultural application of biocarbons (biochar) derived from different biomass sources in the process of pyrolysis is a promising solution for crop productivity and quality, soil health improvement, and carbon sequestration. In a three-year study, the effects of low doses of biochar (1 t·ha^{-1} and 3 t·ha^{-1}) of different origins on winter wheat grain quantity and quality were tested. Six different biochar types were used: biochar derived from wheat husk (WHB), (2) extracted medical plant biomass biochar (MPB), (3) wood chip biochar (WCB), (4) wood sawdust biochar (SB), (5) biochar made from straw of rye (RSB), and (6) meat and bone biochar (MBMB). Higher doses of biocarbon had a positive effect only on wet gluten content. The use of different types of biochar showed a significant impact on grain parameters; however, the results were different in different years of this study. Among the tested biochars, SB (Saw Dust biochar) showed rather good results for most of the parameters tested (the highest grain yield in 2018, the highest weight of 1000 g in 2019, the lowest wet gluten content and gluten index in 2020, the lowest falling number in 2019, and the highest Zellenys index in 2019). MBMB biochar was one of the highest yielders in 2018, had the highest wet gluten content in 2018, and the highest gluten index in 2019 and 2020; the lowest Zelleny's sedimentation index in 2019; and one of the lowest in 2020. Those made SB and MBMB the most promising biochars tested in this study.

Keywords: biochar; winter wheat; cereal yields; grain quality

1. Introduction

The recent increase in the world's population means that the demand for food is also increasing. This demand, depending on the scenario, will have to increase by 35% to 56% between 2010 and 2050 [1]. Some estimates indicate that food production between 2005 and 2050 will have to increase even more, even up to 70% by 2050 [2]. The total earth's population is expected to reach 9.7 billion people in 2050 [3]. These projections are becoming realistic through the breeding of more productive and higher-yielding varieties, but also through agrotechnical advances and more efficient use of nutrients [4,5]. Another way to meet those growing demands is the introduction of products and substances that improve soil fertility and its properties, and thus soil productivity. One of the key aspects is defining soil properties: biological, physical, chemical, and microbiological. Thanks to such determinations, we can observe how agricultural activities affect the soil environment and ecosystem services provided by soil [6,7]. The impact of crop cultivation under intensive agriculture principles is rarely positive. Recently, one of the main issues and challenges of global agriculture has been coping with negative environmental impacts, such as the loss of soil organic matter (SOM) [8,9]. SOM is of unique importance for many biochemical processes, including climate change mitigation potential and agricultural water management. This is prompting a search for alternative sources of organic matter that can contribute to soil organic matter and humus concentration and thus improve soil fertility [10]. Significant changes in the quality and quantity of humus compounds can be caused by the addition of an external source of organic matter [11]. The challenge of increased sequestration of carbon in soil is also a challenge for the European Union's (EU) policymakers, as reflected in recent programs and strategies, of which the European Green Deal [12] is of the greatest importance. In the European Union (EU), current legislation allows the legal use of biocarbon as a soil improver for agricultural purposes, including organic farming, according to specific criteria for contaminant content, including heavy metals and Polycyclic aromatic hydrocarbons (PAHs) [13].

One of the external sources of organic matter for soil is biochar (also known as biocarbon), which has been one of the leading research topics in Europe in recent years. Industries such as biocarbon generation, its modifications, and the manufacture of biocarbon-based products are developing rapidly. The International Biochar Initiative defines biocarbon as a fine-grained carbonate with a high organic carbon content and low degradability, obtained by pyrolysis of biomass and biodegradable wastes [14].

Biocarbon is primarily used as a measure in environmental protection and agriculture. It is a solid renewable fuel extracted from various types of biomass by pyrolysis [15]. The material from which biocarbon is produced can be diverse, and almost any biomass can be the substrate for biocarbon generation (energy crops, forestry waste, rapeseed straw, sunflowers, maize cobs, but also sewage sludge, organic waste, or manure) [16–20]. Biochar can enhance soil fertility by providing a habitat for microorganisms and beneficial bacteria. It acts as a porous structure, offering a refuge for soil microbes and promoting their activities. This microbial activity helps in nutrient cycling and improves the availability of essential nutrients to plants. Biochar has a high cation exchange capacity (CEC), meaning it can retain and exchange essential nutrients such as nitrogen, phosphorus, and potassium. This can lead to better nutrient availability for plant uptake and reduce nutrient leaching, promoting more efficient fertilizer use [21]. Moreover, biochar's porous structure also improves water retention in soils. It can hold onto water during wet periods and release it during dry periods, helping to maintain more consistent moisture levels. Additionally, it improves drainage in clayey soils, preventing waterlogging and enhancing aeration. The water-holding potential of biochar is especially visible on coarse-textured soils [22]. Biochar can also help regulate and stabilize soil pH. It has a neutral to slightly alkaline pH, and when added to acidic soils, it can contribute to raising the pH, making the soil more suitable for a broader range of crops, and, again, contributing to nutrient availability for plants [23]. Some studies suggest that biochar can bind with heavy metals in the soil, reducing their availability for uptake by plants. This may be beneficial for grain quality by minimizing the accumulation of heavy metals in grains [24–26].

The above-mentioned properties of biocarbon indicate the great potential for the use of this material in cereal crop production. Among cereals in the European Union, wheat accounts for nearly 46% of grain production. It is also an important crop in Poland, as it takes first place in terms of its share in the sowing structure (28%). In the case of wheat, it is not only the volume of grain yield obtained that is important, but also its quality, which is determined by the requirements of the milling and baking industry [27].

The impact of biochar (biocarbon) on wheat yield can depend on various factors mentioned above, including the type of biochar used (determining the size of carbon particles and its chemical composition) and its application rate as main factors. The aim of this study was to determine the effect of the type (biochar origin) and dose of biocarbon on winter wheat yields and selected characteristics of the technological value of grain and flour.

2. Materials and Methods

In the three growing seasons of 2017/2018, 2018/2019, and 2019/2020, a two-factor pot experiment was conducted in six replicates at the Vegetation Experiment Hall of IUNG-PIB in Puławy, Poland. The first factor of the first order was the type of biochar (source material that was used during the pyrolysis process): (0) control object (no biochar added), (1) wheat husk biochar (WHB), (2) biochar made from extracted medical plant biomass

(MPB), (3) wood chip biochar (WCB), (4) wood sawdust biochar (SB), (5) biochar made from rye straw (RSB), and (6) meat and bone biochar (MBMB). WHB and MPB biocarbons were market products with the appropriate certificates (certifying their quality and composition) purchased in Germany. WCB biocarbon was a market product purchased in Poland (it was available on the market at the time but is now discontinued). The other three biocarbons (SB, RSB, and MBMB) were produced for the purposes of this experiment by an external entity. Pyrolysis temperature and time, as well as the chemical composition of the tested bicarbons, are given in Table 1. Subjectively, according to this study authors, MBMB biocarbon was the most fine (smallest particle size). In contrast, the RSB biocarbon was characterized by the largest particles (fragments of charred stalks—straws were visible). Other biocarbons tested had medium particle sizes. Charred seed husks were visible in WHA biocarbon, while pieces of charred plants have been visible in MPB biocarbon. Accurate particle size measurements were not made under this study. The second factor of the experiment was the rate (dose) of biochar: (1) 1 $t \cdot ha^{-1}$, (2) 3 $t \cdot ha^{-1}$. The experiment was conducted in a completely randomized design, in Micherlich pots filled with 7 kg of soil. Soil was taken from a field where the crop rotation was 100% cereals. The soil for this study was taken from the top layer (0–30 cm) of Haplic Luvisol, made of clay. The content of selected elements in the soil was: total carbon—0.90%; organic carbon—0.78%; total N—0.10%; P_2O_5—27.7 $mg \cdot 100\ g^{-1}$; K_2O—28.2 $mg \cdot 100\ g^{-1}$; pH—6.08. The soil for the experiments was taken after 3 years of research and stored frozen. The soil was replaced in each year of this study, and fertilizers and biocarbons were added in the same way in each year of this study. Thus, each year, the seeds were sown in new soil with the same physical and chemical parameters. Because of this design, it was impossible to track the addition of carbon over time, and the results of each year must be evaluated independently. The experimental plant was winter wheat, variety Hondia (10 plants per vase). Before filling the vases with soil, the appropriate type of biocarbon was added to the substrate along with the relevant fertilizers and then mixed with the soil. Nitrogen in the form of NH_4NO_3 was applied at a dose of 3.6 g $N \cdot vase^{-1}$: 1/2 at the beginning of vegetation (spring) and 1/2 dose at the BBCH 30–32 (*ger.* Biologische Bundesanstalt, Bundessortenamt und Chemische Industrie)—a stem shooting stage of wheat. Fertilization of the remaining components per pot was as follows: P_2O_5—2.52 g in the form of KH_2PO_4; K_2O—2.04 g in the form of K_2SO_4; Mg—0.5 g in the form of $MgSO_4$. In addition, iron (50 mg per pot), boron (5 mg per pot), manganese (3 mg per pot), and copper (3 mg per pot) were also added to the substrate in the forms of $Fe(NH_4)_2$, $C_3H_6N_6 \cdot xBH_3O_3$, $C_{10}H_{12}MnN_2O_8 \cdot 2Na$, and $Cu(NH_4)_2$, respectively. The moisture content of the substrate was maintained at 60% water capacity throughout the growing season. The characteristics of the biocarbon types are shown in Table 1.

Table 1. Chemical composition of the tested biochar.

Type of Biochar	Total Carbon (%)	N (%)	S (%)	P (%)	pH	Pyrolysis Temperature (°C)	Time of Pyrolysis (min)
WHB	70.8	3.9	0.074	0.320	8.50	550	nd [1]
MPB	63.5	1.8	0.890	0.095	10.00	600	nd
WCB	67.6	2.4	<0.100	0.330	6.79	500	4–7
SB	42.6	3.6	0.224	4.420	6.78	550	240
RSB	70.5	1.2	0.146	0.260	9.65	550	180
MBMB	75.6	1.9	0.136	0.120	8.32	550	240

[1] nd—no data (the producer did not provide it in the certificate).

When wheat reached the full maturity-growing stage, tested plants were harvested. After the harvest, the following were determined: grain yield per pot and selected characteristics of the technological value of grain and flour: weight of 1000 grains (PN-68/R-74017 [28]), falling number (PN-EN ISO 3093 [29]), wet gluten content (PN-A-74042 [30]), gluten index, and Zelleny sedimentation index (PN-EN ISO 5529 [31]).

The falling number was determined using the Hagberg-Perten method [29]. The grain sample was ground on a laboratory mill (grinder). An automatic equipment *Falling Number 1500* was used for the determination. A total of 7 g of flour with a moisture content of 15% was weighed for analysis in order to ensure a constant level of dry matter content. A sample of middlings in a viscometric tube, together with a stirrer, was placed in a boiling water bath. The use of equipment automatically measured the stirrer's rate of descent in the starch glue. The determination was performed in duplicate. The amount of wet gluten in wheat grain and its index were determined.

The amount of wet gluten in wheat grain and its index were determined according to PN-A-74042 [30]. Grain samples were ground on a laboratory mill (grinder). A *Gluten Index System* was used to determine the amount of wet gluten and its quality. The method involves kneading the dough (with 10 g of flour sample and 4.8 cm^3 of 2% NaCl) and washing out the starch with a 2% NaCl solution. The process was carried out automatically in a single sequence for 5 min. Centrifugation of the formed gluten was carried out on sieve cassettes at 6000 +/−5 rpm.

The Zeleny sedimentation index was determined according to PN-EN ISO 5529 [31] for wheat. The determination of sedimentation index and Zeleny Test was carried out on an apparatus consisting of a measuring panel and a SWD—89 Sadkiewicz type shaker (Sadkiewicz Instruments, Bydgoszcz, Poland).

Average monthly temperatures during subsequent years of this study are given in Table 2.

Table 2. Monthly temperatures during the subsequent year of the experiment and multi-year annual (1981–2010) (°C).

Month	2017/2018	2018/2019	2019/2020	1981–2010
IX	14.1	15.5	14.4	13.3
X	9.5	10.0	10.8	8.0
XI	4.5	4.2	6.4	2.7
XII	2.4	0.9	3.1	−1.4
I	0.6	−2.4	1.7	−3.3
II	−3.5	2.9	3.4	−2.3
III	0.3	5.7	4.7	1.6
IV	13.9	10.0	8.9	8.7
V	17.7	13.9	11.9	14.5
VI	19.6	22.7	19.1	17.2
VII	21.0	19.4	19.3	19.5
VIII	21.1	20.4	20.3	17.8

The results were statistically analyzed with the Tukey test at $p \leq 0.05$.

3. Results and Discussion

3.1. Winter Wheat Yields

Winter wheat grain yield did not depend on the biocarbon dose (Figure 1). However, in the first year of this study, the type of biocarbon had a significant effect on winter wheat grain yield. Both biocarbons from forestry waste (WCB and SB) had a positive effect on yields (Figure 2). It should also be noted that in the other years of this study, the differences were not statistically significant, but the introduction of biocarbon into the soil resulted in a trend towards higher grain yields in winter wheat (Figure 2), regardless of the type of biocarbon. On the other hand, recent studies showed that different types (origins) of biochar can have a significant effect on yields. For example, Lilli et al. [32] have proven that tomato yields can increase even up to 175% when treated with biochar of specific origin and with appropriate fertilization. A study by Gebremedhin et al. [33] indicates a beneficial effect of biocarbons on wheat yield and yield structure traits. The application of biocarbon at 4 t·ha^{-1} resulted in an increase in grain and straw yield compared to the control. Różyło et al. [24] found that amendment of sewage sludge with 5% biochar can significantly improve yield quantity, but also some grain quality traits. In a study by Khan et al. [34], an increase in wheat grain yield was also observed, but after the application of biocarbon at much higher doses than in the present study (20 t·ha^{-1} and urea at 150 kg N·ha^{-1}. A positive effect of biochar amendments on cereal yields, especially when combined with nitrogen fertilization, has been reported in multiple studies [35–38]. Zhu et al. [39] also found that biochar amendment to soil could also stimulate the growth of crop root systems, which can also be a factor that, over time, improves the utilization rate of nutrients available in soil.

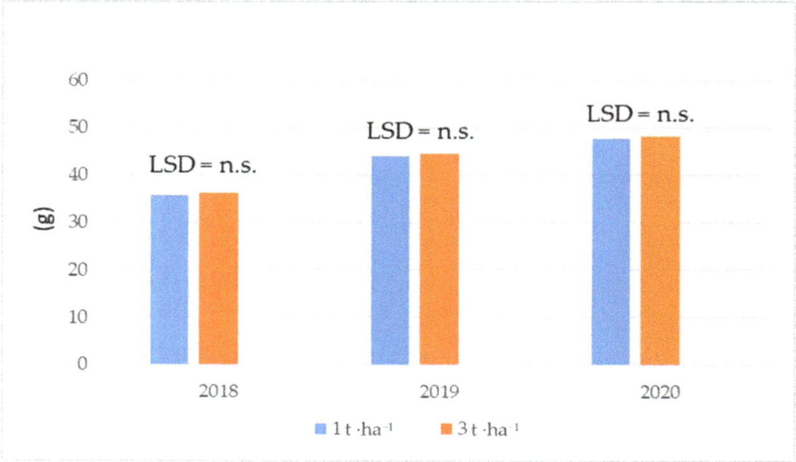

Figure 1. Winter wheat grain yield (g per pot) depends on the biochar rate (g per pot).

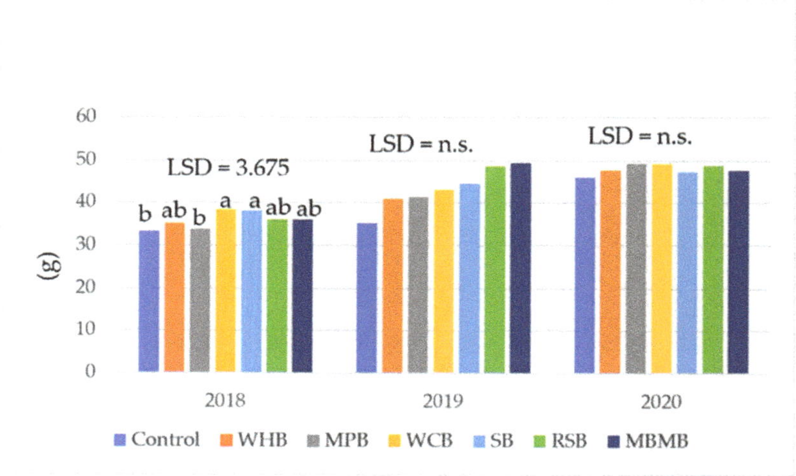

Figure 2. Winter wheat grain yield (g per pot) in the following years of the experiment, depending on the biochar type. Different lowercase letters indicate statistically significant differences between biochar types.

No statistical differences between tested biochar (and control) were found in the third year of this study, and all tested objects reached high yields (of about 45–50 g per pot). The yields for almost all objects (also for control) showed a trend to be the highest, and at a similar level, in the third year of this study. This may indicate that in the third year of this study, factors other than biochar amendment were mainly responsible for wheat yields. As the soil used in the experiment was removed after each year of the experiment, the increasing saturation of the soil sorption complex with nutrients in the following years could not be a factor. The possible answer is that biochar impact on plant yields was strongly affected by natural conditions, which were, naturally, slightly different in each year of this study.

3.2. Grain Characteristics

The grain of wheat utilized for consumption must be distinguished by its quality, which is determined by grain and flour quality parameters. In general, a distinction is made between the physical parameters of the grain (e.g., 1000 grain weight) and the qualitative parameters (e.g., falling number, Zelleny sedimentation index value, gluten content, gluten index). In our study, the assessment concerned selected physical and qualitative properties of the grain.

3.2.1. Mass of Thousand Grains

The mass of a thousand grains is a measure used in agriculture to assess the weight of a specific number of grains. It is an important parameter as it provides information about the size and weight of individual grains. It can be influenced by crop genotype, growing conditions, and agricultural practices. Farmers and researchers use this information to evaluate crop yield and seed quality and to make decisions about seed selection and planting practices.

In all years of this study, only a trend towards higher 1000 grain weight was found with the application of biocarbon at 3 t·ha^{-1}. However, the differences were not statistically proven (Figure 3). Billah et al. [40] found that an amendment of biochar can significantly improve the mass of a thousand grains. However, the strength of the effect is supported by the size of the biochar particles—the smaller the particles; the more visible the positive effect.

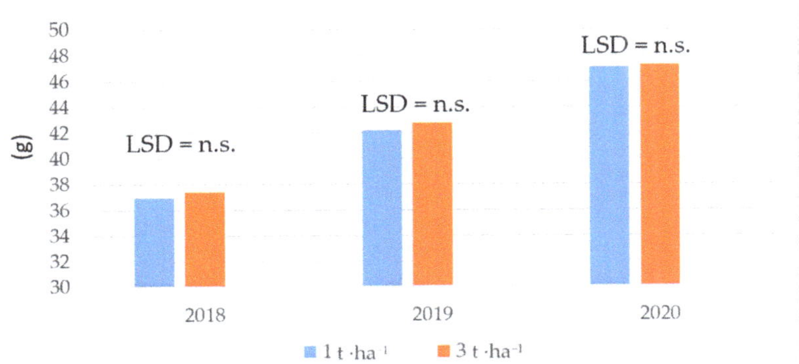

Figure 3. Winter wheat weight of 1000 grains (g) in the following years of the experiment, depending on the biochar dose.

Analysis of the effect of the type of biocarbon on the weight of 1000 grains showed that only in the second year of this study, amendments of biocarbon significantly increased the mass of thousand grains compared to the control object (no added biocarbon). However, none of the tested biochar types showed statistically better results than others in this matter. Amendment of biocarbon obtained from woody biomass (WCB and SB) again showed a trend to have the best potential of 1000 grain mass (similarly as for yield potential) (Figure 4).

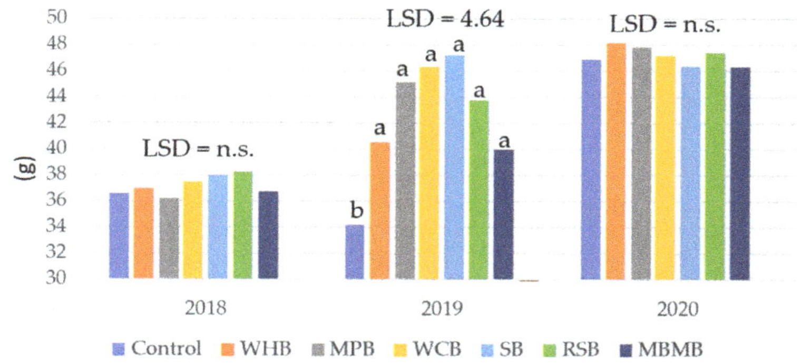

Figure 4. Winter wheat weight of 1000 grains (g) in the following years of the experiment, depending on the biochar type. Different lowercase letters indicate statistically significant differences between biochar types.

3.2.2. Wet Gluten Content and Gluten Index

Gluten is a complex mixture of proteins found in wheat and related grains like barley and rye. It plays a crucial role in the baking industry due to its unique properties that contribute to the texture, structure, and overall quality of baked goods. Since nitrogen fertilization is one of the key aspects influencing gluten content [41], the amendment of biocarbons, which can hold and release nitrogen, might have a significant impact on gluten content. In the presented study, the total dose of biocarbon significantly differentiated the amount of wet gluten only in the first year of this study (Figure 5). The application of biocarbon at 3 t·ha^{-1} resulted in a higher value for this parameter. In the following years of this study, only a trend towards increased wet gluten amounts with increased biocarbon doses was observed. According to Khan et al. [34], the application of biocarbon at much higher rates than in the presented study (20 t·ha^{-1}) resulted in an increase in the protein content of wheat grain and straw. This was also confirmed by Selivanovskaya [42] for biochar used at 30 t ha^{-1}. Also, Shahzad et al. [43] found an increase (by 6.8 g·kg^{-1}) in the protein content of wheat grain cultivated with biochar amendment. Kraska et al. [44] found that 20 t ·ha^{-1} of biochar amendment resulted in a higher protein content of rye grain than biochar applied at 30 t ·ha^{-1}. In the presented study, the type of biocarbon significantly shaped the amount of wet gluten in winter wheat grain in 2018 and 2020 (Figure 6). In 2018, the best value for this indicator was found on the site where meat and bone meal biocarbon was used. In 2020, MPB biochar had the highest content of wet gluten. WCB biochar showed one of the highest wet gluten contents in both 2018 and 2020.

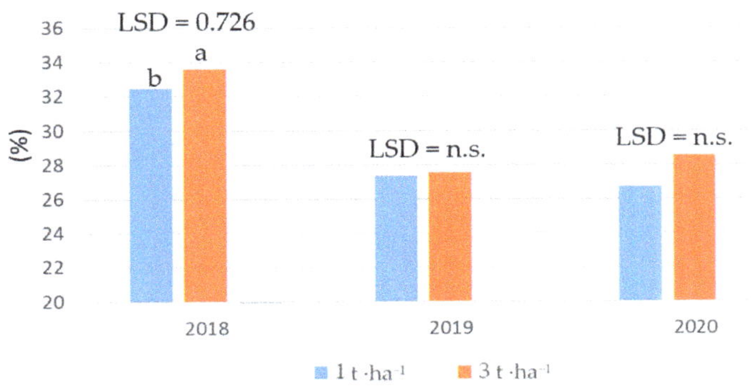

Figure 5. Winter wheat wet gluten content (%) in the following years of the experiment, depending on the biochar dose. Different lowercase letters indicate statistically significant differences between biochar doses.

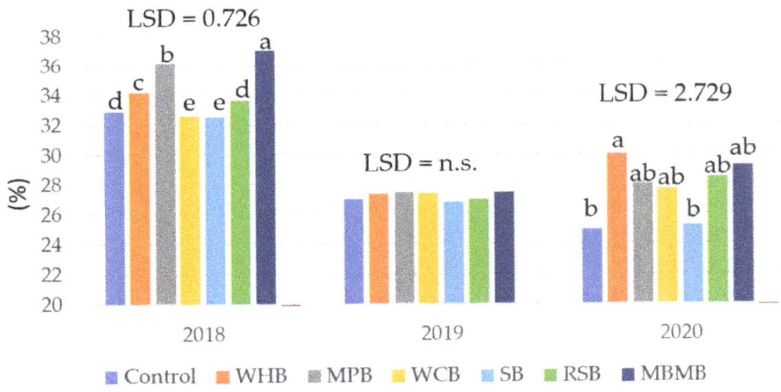

Figure 6. Winter wheat wet gluten content (%) in the following years of the experiment, depending on the biochar type. Different lowercase letters indicate statistically significant differences between biochar types.

The gluten index is a measure of the gluten strength in wheat flour. It provides information about the quality of gluten and its ability to form a stable and elastic network in dough. The gluten index is an important parameter in the baking industry, particularly in the production of bread and other baked goods. There was no significant effect of the biocarbon dose on the gluten index (Figure 7). There was only a trend toward better gluten quality with a higher biocarbon dose, but the differences were not statistically proven. In all years of this study, the type of biocarbon had an effect on the gluten index. However, the effect was not the same in all years. In 2018, the highest value of this parameter was found for the biocarbon amendment made from forestry waste (WCB). In 2019 and 2020, a significantly higher value of this index was found in the objects where biocarbon derived from meat and bone meal was used (MBMB) (Figure 8).

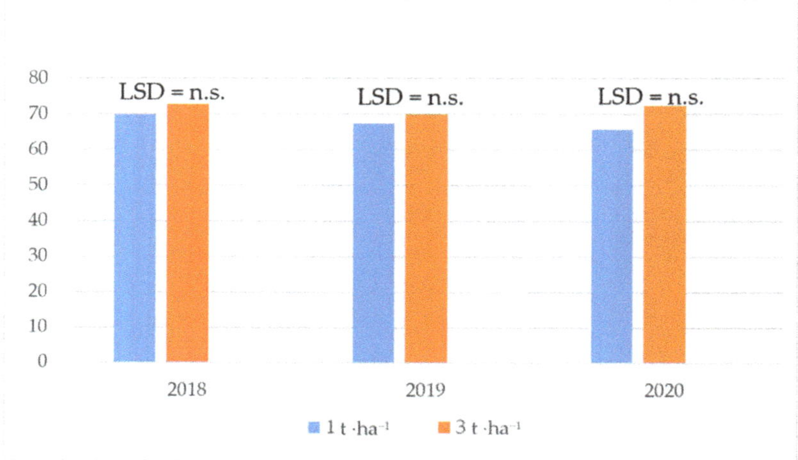

Figure 7. Winter wheat gluten index in the following years of the experiment, depending on the biochar dose.

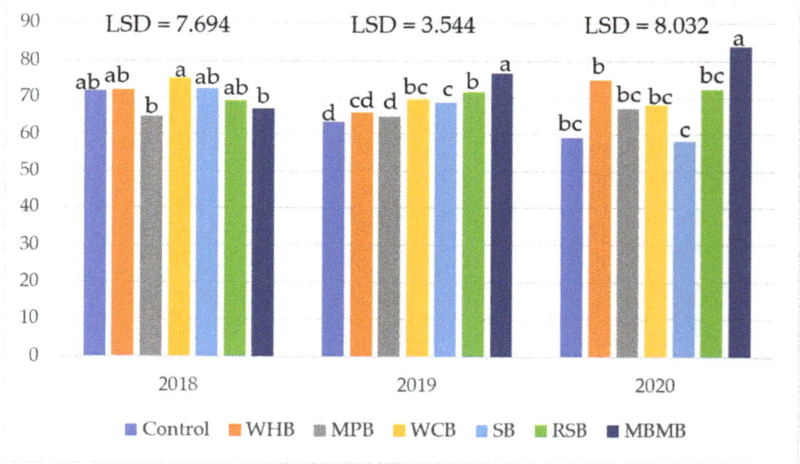

Figure 8. Winter wheat gluten index in the following years of the experiment, depending on the biochar type. Different lowercase letters indicate statistically significant differences between biochar types.

3.2.3. Grain Falling Number and Sedimentation Index

The falling number (FN) is a measure of alpha-amylase activity in grain and serves as an international standard of grain quality. It is determined through a test known as the Falling Number Test or Amylograph Test. This test is widely used in the flour milling and baking industries to assess the enzymatic activity that affects the quality of wheat and flour. High nitrogen fertilizer application can decrease the falling number due to increased alpha-amylase activity caused by the higher moisture of plant tissues. However, nitrogen application can also delay maturity, which may maintain a high falling number.

The relationship between nitrogen application and alpha-amylase activity is complex and context-dependent [45]. Excessive nitrogen, especially if applied late in the growing season, can increase the risk of pre-harvest sprouting and affect enzyme activity [46]. Moreover, grain falling can be affected by weather conditions. Barnard and Smith [47] found a negative impact of rainfall during the late stages of grain development on FN for some of the tested cultivars. The falling number, according to Hruskova et al. [48], can be strongly affected by weather conditions, especially higher temperatures, which can be positively correlated with FN. In the present study, FN showed a tendency to reach higher values in 2019, with temperatures in June being the highest of the three years. However, at the same time, the average temperature in April and May 2019 was lower than in 2018 and 2020 (Table 2). This might show that temperatures, especially at the last development stages, can be crucial for the development of quality parameters.

The Zelleny sedimentation index relates to both the quality and quantity of protein in the grain and therefore has an impact on the quality of the bread obtained and, in particular, its structure. Its higher values are desirable, which can be promoted by, for example, nitrogen fertilization [49]. A high sedimentation rate should be combined with a high content of glutein proteins, especially glutein itself, which is particularly important for the baking industry [50]. In the present study, there was no effect of biocarbon dose on the value of the falling number (Figure 9), however falling number was significantly influenced by biocarbon type in 2018 and 2019 (Figure 10). The value of the Zelleny sedimentation index was not influenced by biocarbon dose (Figure 11). However, the type of biocarbon used significantly differentiated the Zelleny sedimentation index values between tested objects (Figure 12). Differences in falling numbers between different biochar types were found for 2018 and 2019. In 2018, MBMB had the highest value of falling numbers, whereas in 2019, MBMB biochar had the lowest value of falling numbers. This indicates that this parameter was strongly affected by other factors than biochar amendment type, most likely weather conditions. The biochar type used had a significant impact on the Zelleny sedimentation index in all three years of this study. In 2018, all tested objects showed low values of the index; however, the highest values were observed for the control object and MBMB biochar. In 2019, SB biochar had the highest value of the Zelleny index (biochar with one of the highest concentrations of nitrogen). In 2019, the MBMB biochar showed the lowest value of this parameter. In 2020, RSB biochar had the highest Zellenys sedimentation index (biochar with the lowest nitrogen content), while SB biochar had one of the highest values of this parameter. The results showed one more time that the impact of biochar type on grain quality parameters is visible but highly variable and dependent on external parameters. Zelleny's sedimentation index, as a grain-falling number, can be strongly influenced by weather conditions [51]. Stepien and Wojtkowiak [52] found that higher N fertilization rates positively impact Zelleny's sedimentation index. However, as in our study, biochars with the highest concentration of nitrogen (SB and RSB) had different results in Zelleny's sedimentation index in different years, indicating weather as the most important driving factor for Zelleny's sedimentation index values. As wheater conditions (minmal, maximal temperatures, rainfall, and humidity) have a strong impact on both falling number and sedimentation index, we assume that those parameters of wheat's grain were strongly influenced by both weather conditions in the following years of this study, as well as nitrogen availability (modified by the type of biocarbon used), hence the large variations in this factor between years and between the types of biocarbon used.

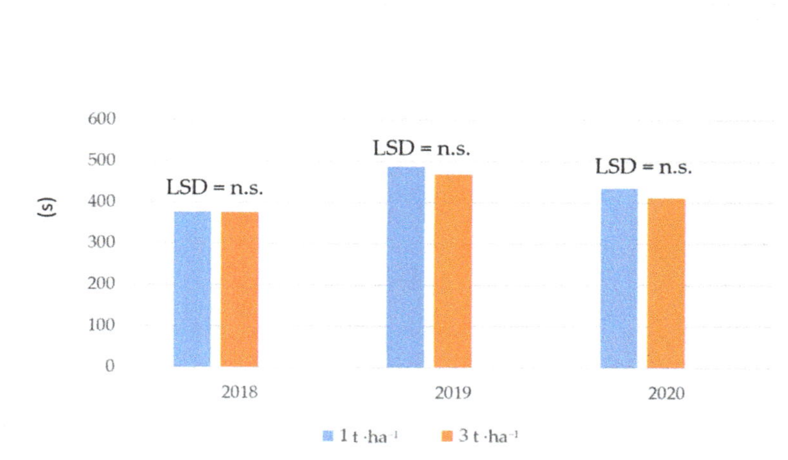

Figure 9. Winter wheat falling number (s) in the following years of the experiment, depending on the biochar dose.

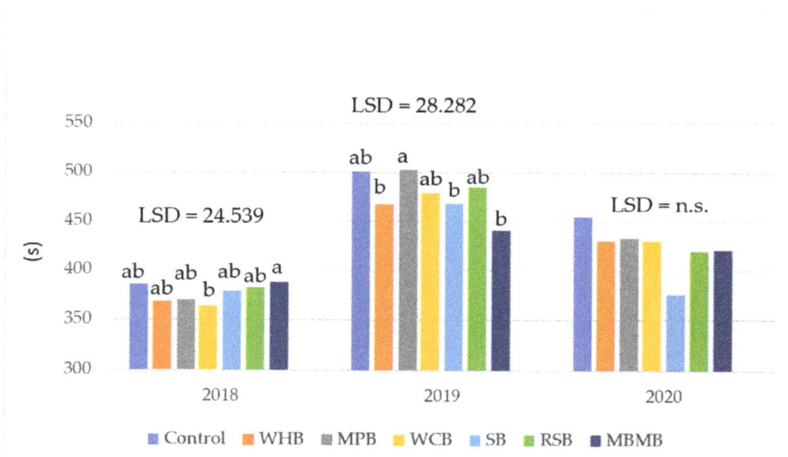

Figure 10. Winter wheat falling number (s) in the following years of the experiment, depending on the biochar type. Different lowercase letters indicate statistically significant differences between biochar types.

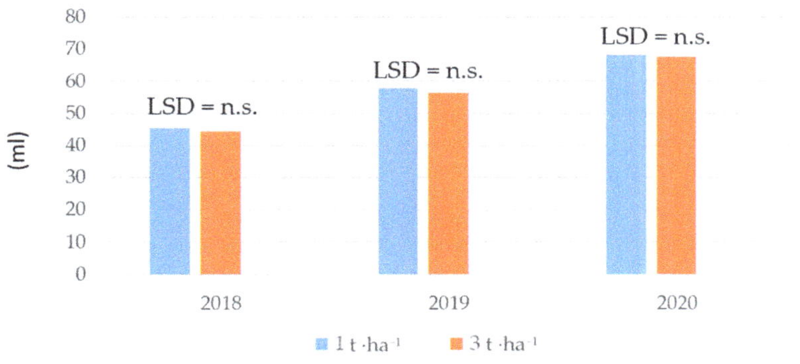

Figure 11. Winter wheat Zelleny sedimentation value (ml) in the following years of the experiment, depending on the biochar doses.

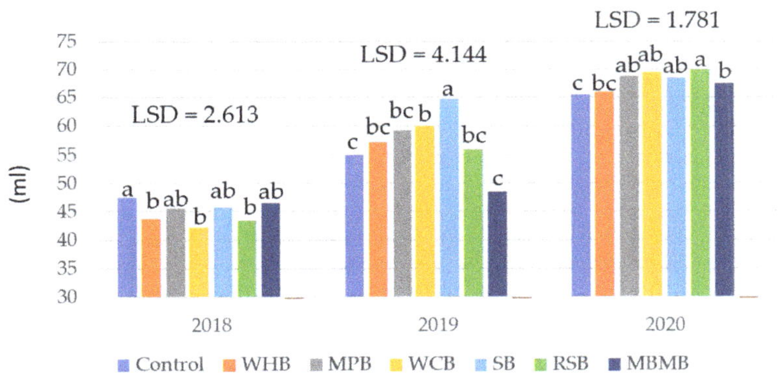

Figure 12. Winter wheat gluten value (ml) in the following years of the experiment, depending on the biochar type. Different lowercase letters indicate statistically significant differences between biochar types.

4. Conclusions

The dose of biocarbon had little effect on the quantity or quality of winter wheat grain. Of the parameters tested, only the gluten content showed a positive response to a higher dose of biocarbon. The lack of a significant response from the other factors tested may be indirectly due to the deliberate selection of low doses of biocarbon treatments. In the literature, tested biochar doses are often 20 tons or even 30 tons per hectare. These are doses at which the effect of biocarbon is pronounced, but the unit cost of applying this treatment makes this treatment not feasible for everyone (the cost of 1 tonne of biochar was estimated by Nematian et al. [53] at USD $450, even up to USD $1850). Lower doses seem to be more realistic and thus worth testing. The type of biocarbon used, which was inversely related to the dose, had a significant effect on many of the parameters tested. This could have been influenced by the characteristics of the raw material, its physical properties (particularly the size of biochar particles) or chemical properties (carbon content, macro- and micronutrient content), and pH. SB biocarbon (Saw Dust biochar) showed rather good results for most of the parameters tested. Also, MBMB biochar showed interesting results, especially in terms of gluten content. It also showed a tendency to promote high grain yields. Those two types of biocarbons had different properties, especially in terms of carbon content. In addition, the MBMB was the biocarbon that were physically the finest (had the smallest size of particles), according to the authors' data (no confirmation of this is in the table). In the present study, those two types of biochar can be recommended for further testing. Especially the low carbon content and high Nitrogen and Phosphorus content are the two parameters that could make SB biochar valuable for agricultural production. Tomczyk et al. [20] found that C content is higher in biochars produced at higher pyrolysis temperatures. Lower C content promotes higher ash and volatile solid contents. Moreover, sawdust is a waste that could possibly be obtained easily, in high amounts, and with a high rate of uniformity.

Author Contributions: M.W. was responsible for this study design and statistical analysis. A.K.B. and M.W. analyzed the data and wrote this paper. J.G. confirmed the methodology and coordinated this research. All authors participated in reviewing the manuscript and discussing the results. All authors have read and agreed to the published version of the manuscript.

Funding: This research received no external funding.

Institutional Review Board Statement: Not applicable.

Informed Consent Statement: Not applicable.

Data Availability Statement: The data presented in this study are available on request from the corresponding author. The data are not publicly available due to the internal rules of IUNG-PIB (the data owner).

Acknowledgments: This research was carried out as part of the independent, in-house scientific activity of IUNG-PIB.

Conflicts of Interest: The authors declare no conflicts of interest.

References

1. Van Dijk, M.; Morley, T.; Rau, M.L.; Saghai, Y. A meta-analysis of projected global food demand and population at risk of hung for the period 2010–2050. *Nat. Food* **2021**, *2*, 494–501. [CrossRef] [PubMed]
2. Noel, S. *Economics of Land Degradation Initiative: Report for Policy and Decision Makers_Reaping Economic and Environmental Benef from Sustainable Land Management*; ELD Initiative and Deutsche Gesellschaft für Internationale Zusammenarbeit (GIZ) Gmb Bonn, Germany, 2015; ISBN 978-92-808-6063-4.
3. Dorling, D. World population prospects at the UN: Our numbers are not our problem? In *The Struggle for Social Sustainabili* Policy Press: Bristol, UK, 2021; pp. 129–154. [CrossRef]
4. Ladha, J.; Pathak, H.; Krupnik, T.; Six, J.; van Kessel, C. Efficiency of fertilizer nitrogen in cereal production: Retrospects ar prospects. *Adv. Agron.* **2005**, *87*, 85–156. [CrossRef]
5. Anioł, A. Wpływ biotechnologii i procesów globalizacji w gospodarce na hodowlę roślin i wspierające ten sektor badan naukowe. *Biul. IHAR* **2010**, *256*, 3–13. [CrossRef]
6. Maron, P.A.; Lemanceau, P. Soil as a support of biodiversity and functions. In *Soil Carbon: Science, Management and Policy Multiple Benefits*; CABI: Wallingford, UK, 2015; pp. 141–153. [CrossRef]
7. Adhikari, K.; Hartemink, A.E. Linking soils to ecosystem services—A global review. *Geoderma* **2016**, *262*, 101–111. [CrossRef]
8. Alhameid, A.; Tobin, C.; Maiga, A.; Kumar, S.; Osborne, S.; Schumacher, T. Intensified agroecosystems and changes in soil carb dynamics. In *Soil Health and Intensification of Agroecosytems*; Academic Press: Cambridge, MA, USA, 2017; pp. 195–214. [CrossR
9. Garratt, M.P.; Bommarco, R.; Kleijn, D.; Martin, E.; Mortimer, S.R.; Redlich, S.; Senapathi, D.; Steffan-Dewenter, I.; Świtek, Takács, V.; et al. Enhancing soil organic matter as a route to the ecological intensification of European arable systems. *Ecosyste* **2018**, *21*, 1404–1415. [CrossRef]
10. Bahuguna, A.; Sharma, S.; Yadav, J. Effect of different organic sources on physical, chemical and biological properties of soil inceptisols of Varanasi. *Int. J. Plant Soil Sc.* **2021**, *33*, 41–52. [CrossRef]
11. Wyzinska, M.; Smreczak, B. The influence of type and rate of biochar on productivity of winter wheat. In Proceedings of the 20 International Conference "Engineering for Rural Development", Jelgava, Latvia, 22–24 May 2019; pp. 594–599. [CrossRef]

12. European Green Deal. Communication from the Commission to the European Parliament, the European Council, the Council, the European Economic and Social Committee and the Committee of the Regions the European Green Deal. Publications Office of the European Union; COM/2019/640 Final. Available online: https://eur-lex.europa.eu/legal-content/EN/TXT/?uri=COM:2019:640:FIN (accessed on 20 November 2023).
13. EBC. *European Biochar Certificate—Guidelines for a Sustainable Production of Biochar*; European Biochar Foundation (EBC): Arbaz, Switzerland, 2012–2022; Available online: http://european-biochar.org (accessed on 20 November 2023).
14. Malińska, K. Prawne i jakościowe aspekty dotyczące wymagań dla biowęgla. *Inżynieria Ochr. Śr.* **2015**, *18*, 359–371.
15. Bis, Z. Biowęgiel–powrót do przeszłości, szansa dla przyszłości. *Czysta Energia* **2012**, *6*, 28–31.
16. Sanchez, M.E.; Lindao, E.; Margaleff, D.; Martinez, O.; Moran, A. Pyrolysis of agricultural residues from rape and sunflower: Production and characterization of bio-fuels and biochar soil management. *J. Anal. Appl. Pyrolysis* **2009**, *85*, 142–144. [CrossRef]
17. Kwapinski, W.; Byrne, C.M.P.; Kryachko, E.; Wolfram, P.; Adley, C.; Leahy, J.J.; Novotny, E.H.; Hayes, M.H.B. Biochar from Biomass and Waste. *Waste Biomass Valorization* **2010**, *1*, 177–189. [CrossRef]
18. Ibarrola, R.; Shackely, S.; Hammond, J. Pyrolysis biochar systems for recovering biodegradable materials: A life cycle carbon assessment. *Waste Manag.* **2012**, *32*, 859–868. [CrossRef] [PubMed]
19. Song, W.; Guo, M. Quality variations of poultry litter biochar generated at different pyrolysis temperatures. *J. Anal. Appl. Pyrolysis* **2012**, *94*, 138–145. [CrossRef]
20. Tomczyk, A.; Sokołowska, Z.; Boguta, P. Biochar physicochemical properties: Pyrolysis temperature and feedstock kind effects. *Rev. Environ. Sci. Biotechnol.* **2020**, *19*, 191–215. [CrossRef]
21. Dai, Z.; Xiong, X.; Zhu, H.; Xu, H.; Leng, P.; Li, J.; Tang, C.; Xu, J. Association of biochar properties with changes in soil bacterial, fungal and fauna communities and nutrient cycling processes. *Biochar* **2021**, *3*, 239–254. [CrossRef]
22. Razzaghi, F.; Obour, P.B.; Arthur, E. Does biochar improve soil water retention? A systematic review and meta-analysis. *Geoderma* **2020**, *361*, 114055. [CrossRef]
23. Hossain, M.Z.; Bahar, M.M.; Sarkar, B.; Donne, S.W.; Ok, Y.S.; Palansooriya, K.N.; Kirkham, M.B.; Chowdhury, S.; Bolan, N. Biochar and its importance on nutrient dynamics in soil and plant. *Biochar* **2020**, *2*, 379–420. [CrossRef]
24. Różyło, K.; Świeca, M.; Gawlik-Dziki, U.; Stefaniuk, M.; Oleszczuk, P. The potential of biochar for reducing the negative effects of soil contamination on the phytochemical properties and heavy metal accumulation in wheat grain. *Agric. Food Sci.* **2017**, *26*, 34–46. [CrossRef]
25. Medyńska-Juraszek, A.; Rivier, P.A.; Rasse, D.; Joner, E.J. Biochar affects heavy metal uptake in plants through interactions in the rhizosphere. *Appl. Sci.* **2020**, *10*, 5105. [CrossRef]
26. Nkoh, J.N.; Ajibade, F.O.; Atakpa, E.O.; Abdulaha-Al Baquy, M.; Mia, S.; Odii, E.C.; Xu, R. Reduction of heavy metal uptake from polluted soils and associated health risks through biochar amendment: A critical synthesis. *J. Hazard. Mater. Adv.* **2022**, *6*, 100086. [CrossRef]
27. Różewicz, M.; Grabiński, J.; Wyzińska, M. Wpływ metody strip-till na plonowanie i jakośc ziarna pszenicy ozimej w zależności od odmiany i zakresu uprawy pożniwnej Część II. Jakość ziarna. *Agron. Sci.* **2023**, *78*, 19–28. [CrossRef]
28. *PN-68/R-74017*; Cereals and Pulses—Determinationa of the Mass Od 1000 Grains. Polish Committee for Standardization: Warsaw, Poland, 1968.
29. *PN-EN ISO 3093:2010*; Wheat, Rye and Flours Derived from Them—Durum Wheat and Semolina. Determination of Falling Number by the Hagberg-Perten Method. Polish Committee for Standardization: Warsaw, Poland, 2010.
30. *PN-A-74042-03:1993/Az1:1996*; Cereal Grains and Cereal Products—Determination of Wet Gluten by Mechanical Equipment. Wheat Flour. Polish Committee for Standardization: Warsaw, Poland, 1996.
31. *PN-EN ISO 5529:2010*; Wheat—Determination of Sedimentation Rate. Zeleny Test. Polish Committee for Standardization: Warsaw, Poland, 2010.
32. Lilli, M.A.; Paranychianakis, N.V.; Lionoudakis, K.; Saru, M.L.; Voutsadaki, S.; Kritikaki, A.; Komnitsas, K.; Nikolaidis, N.P. Characterization and Risk Assessment of Different-Origin Biochars Applied in Agricultural Experiments. *Sustainability* **2023**, *15*, 9036. [CrossRef]
33. Gebremedhin, G.H.; Halieselassie, B.; Berhe, D.; Belay, T. Effect of biochar on yield and yield components of wheat and post-harvest soil properties in Tigray, Ethiopia. *J. Fertil. Pestic.* **2015**, *6*, 2. [CrossRef]
34. Khan, M.A.; Basir, A.; Fahad, S.; Adnan, M.; Saleem, M.H.; Iqbal, A.; Al-Huqail, A.A.; Alosaimi, A.A.; Saud, S.; Liu, K.; et al. Biochar optimizes wheat quality, yield, and nitrogen acquisition in low fertile calcareous soil treated with organic and mineral nitrogen fertilizers. *Front. Plant Sci.* **2022**, *13*, 879788. [CrossRef] [PubMed]
35. Akhtar, S.S.; Andersen, M.N.; Liu, F. Residual effects of biochar on improving growth, physiology and yield of wheat under salt stress. *Agric. Water Manag.* **2015**, *158*, 61–68. [CrossRef]
36. Ali, K.; Wang, X.; Riaz, M.; Islam, B.; Khan, Z.H.; Shah, F.; Munsif, F.; Haq, S.I.U. Biochar: An eco-friendly approach to improve wheat yield and associated soil properties on sustainable basis. *Pak. J. Bot.* **2019**, *51*, 1255–1261. [CrossRef]
37. Dawar, K.; Fahad, S.; Alam, S.S.; Khan, S.A.; Dawar, A.; Younis, U.; Danish, S.; Datta, R.; Dick, R.P. Influence of variable biochar concentration on yield-scaled nitrous oxide emissions, Wheat yield and nitrogen use efficiency. *Sci. Rep.* **2021**, *11*, 16774. [CrossRef] [PubMed]

38. Khan, Z.; Rahman, M.H.U.; Haider, G.; Amir, R.; Ikram, R.M.; Ahmad, S.; Schofield, H.K.; Riaz, B.; Iqbal, R.; Fahad, S.; et al. Chemical and biological enhancement effects of biochar on wheat growth and yield under arid field conditions. *Sustainability* **2021**, *13*, 5890. [CrossRef]
39. Zhu, X.; Chen, L.; Kong, X.; Bao, S.; Wu, S.; Fang, L.; Shen, Y. Biochar alters the morphology of plant roots to enable optimized and reduced nitrogen fertilizer applications. *Plant Soil* **2023**, *492*, 655–673. [CrossRef]
40. Billah, M.M.; Ahmad, W.; Ali, M.; Khan, F. Effect of biochar particle size and biofertilizers on lentil (*Lense culinarous* M.) yield and available fractions of soil nutrients. *Soil Environ.* **2018**, *37*, 143–151.
41. Litke, L.; Gaile, Z.; Ruža, A. Effect of nitrogen fertilization on winter wheat yield and yield quality. *Agron. Res.* **2018**, *16*, 500–509. [CrossRef]
42. Selivanovskaya, S.; Galitskaya, P.; Mukharamova, S.; Gordeev, A.; Kanunnikov, K. Biochar From Chicken Manure Enriched With Iodine. In Proceedings of the 19th International Multidisciplinary Scientific GeoConference SGEM, Albena, Bulgaria, 28 June–6 July 2019; pp. 59–66. [CrossRef]
43. Shahzad, K.; Abid, M.; Sintim, H.Y.; Hussain, S.; Nasim, W. Tillage and biochar effects on wheat productivity under arid conditions. *Crop Sci.* **2019**, *59*, 1191–1199. [CrossRef]
44. Kraska, P.; Andruszczak, S.; Oleszczuk, P.; Świeca, M.; Kwiecińska-Poppe, E.; Gierasimiuk, P.; Różyło, K.; Pałys, E. The content of elements and quality parameters of winter rye grain as influenced by biochar-amended soil. *Zemdirbyste-Agric.* **2018**, *105*, 11–20. [CrossRef]
45. Gooding, M.J.; Kettlewell, P.S.; Davies, W.P.; Hocking, T.J. Effects of spring nitrogen fertilizer on the Hagberg falling number of grain from breadmaking varieties of winter wheat. *J. Agric. Sci.* **1986**, *107*, 475–477. [CrossRef]
46. Silva, R.R.; Zucareli, C.; Fonseca IC, D.B.; Riede, C.R.; Gazola, D. Nitrogen management, cultivars and growing environments on wheat grain quality. *Rev. Bras. Eng. Agrícola Ambient.* **2019**, *23*, 826–832. [CrossRef]
47. Barnard, A.; Smith, M.F. Determination of the influence of climate on falling number of winter wheat in the dryland production areas of the Free State Province of South Africa. *Euphytica* **2012**, *188*, 15–24. [CrossRef]
48. Hrušková, M.; Škodová, V.; Blažek, J. Wheat sedimentation values and falling number. *Czech J. Food Sci.* **2004**, *22*, 51. [CrossRef]
49. Knapowski, T.; Ralcewicz, M.; Spychaj-Fabisiak, E.; Ložek, O. Ocena jakości ziarna pszenicy ozimej uprawianej w warunkach zróżnicowanego nawożenia azotem. *Fragm. Agron.* **2010**, *27*, 73–80.
50. Murawska, B.; Spychaj-Fabisiak, E.; Keutgen, A.; Wszelaczyńske, E.; Pobereży, J. Cechy technologiczne badanych odmian ziarna pszenicy ozimej uprawianych w warunkach Polski i Wielkiej Brytanii. *Inżynieria Apar. Chem.* **2014**, *2*, 96–98.
51. Liana, A.; Alda, S.; Fora, C.; Diana, M.; Gogoasa, I.; Bordean, D.; Cârciu, G.; Cristea, T. Climatic conditions influence on the variation of quality indicators of some Romanian and foreign winter wheat cultivars. *J. Hortic. For. Biotechnol.* **2012**, *16*, 68–72.
52. Stepien, A.; Wojtkowiak, K. Evaluation of the effect of different levels of nitrogen and manganese fertiliser on the yield macronutrient content and technological properties of winter wheat. *J. Elementol.* **2019**, *24*, 661–675. [CrossRef]
53. Nematian, M.; Keske, C.; Ng'ombe, J.N. A techno-economic analysis of biochar production and the bioeconomy for orchard biomass. *Waste Manag.* **2021**, *135*, 467–477. [CrossRef]

agriculture

MDPI

eview

'oliar Fertilization of Crop Plants in Polish Agriculture

afał Januszkiewicz [1], Grzegorz Kulczycki [1,*] and Mateusz Samoraj [2]

[1] Institute of Soil Science Plant Nutrition and Environmental Protection, Wrocław University of Environmental and Life Sciences, Grunwaldzka Str. 53, 50-363 Wrocław, Poland; rafal.januszkiewicz@upwr.edu.pl

[2] Department of Advanced Material Technologies, Faculty of Chemistry, Wrocław University of Science and Technology, Smoluchowskiego 25, 50-372 Wrocław, Poland; mateusz.samoraj@pwr.edu.pl

* Correspondence: grzegorz.kulczycki@upwr.edu.pl; Tel.: +48-71-320-5654

Abstract: Foliar fertilization makes it possible to quickly supply plants with deficient nutrients, in case of both their deficiency in the soil and hindered uptake. Crops are characterized by varying nutrient requirements for micronutrients, as well as varying sensitivity to their deficiency. The paper presents practical aspects of the foliar feeding of plants with micronutrients using foliar fertilizers, and their general classification and characteristics. The requirements of basic crops (cereals, rapeseed and corn) for the application of micronutrient fertilization and the degree of their sensitivity to micronutrient deficiency were characterized. The market of foliar fertilizers and the directions of its development were evaluated. The possibilities of foliar fertilizers containing amino acids and silicon, and the possibility of using them for biofortification are presented. It was found that foliar fertilization is one of the most popular and effective methods for the biofortification of plants, as it allows the delivery of the right amount of specific elements in a specific stage of plant development and is thus an economical and environmentally safe procedure. In conclusion, the analysis of the foliar fertilizer market shows that its development is very dynamic, and foliar fertilization is becoming one of the basic elements in effective crop production. Further expansion of the range of foliar fertilizers produced should be expected in accordance with the growing expectations of agricultural producers.

Keywords: foliar fertilization; micronutrients; biofortification; cereals; corn; rapeseed

1. Introduction

The systematic growth of the world's population requires providing it with sufficient food [1]. With the shrinking in areas devoted to crop cultivation, the only possible method for meeting the nutritional needs of such a large population is to increase the yield per unit area [2]. This increase is associated with the depletion of nutrients contained in soils, by way of uptake by plants and dispersal in the environment, resulting in severe nutrient deficiencies. The way to make up for these deficiencies is to use fertilizers, especially mineral fertilizers, in which the components are most often found in readily available forms. The use of mineral fertilizers in a soil-based manner is linked to their transformation in the soil, which affects their availability, resulting in the excessive or insufficient uptake of nutrients by plants. These phenomena can affect the quantity and quality of the yield obtained and the loss of nutrients due to their leaching into the soil and their excessive volatilization into the atmosphere, causing increasing environmental pollution [3].

The aforementioned aspects and problems associated with the use of mineral fertilizers inspire the search for global solutions to increase the efficiency of the use of nutrients from fertilizers, enabling the production of sufficient quantities of good-quality food while maintaining a healthy environment. These goals are served by, among others, the UN Sustainable Development Strategy, which plans to eliminate hunger and all forms of malnutrition worldwide by 2030 [4], as well as the European Green Deal strategy [5], which additionally indicates that by 2030, nutrient losses should be reduced by at least 50% and fertilizer use should be reduced by a minimum of 20%.

Achieving these goals is possible, but it means developing more efficient and effective fertilizer systems. Their goal should be to increase agricultural productivity with less fertilizer inputs [6]. Of the fertilization methods currently in use, such expectations are largely met by the foliar delivery of nutrients, the effectiveness of which can be up to several times greater than that of soil application [7].

The delivery of nutrients with foliar fertilizers generally involves (1) the application of an aqueous solution to the plant surface in the form of droplets, (2) their retention on the leaf surface, (3) the absorption of the nutrient into living plant cells and (4) the translocation of the nutrient to areas where it can be used by the plant in its life processes. The efficiency of foliar fertilization largely depends on the absorption mechanisms of foliar-applied molecules, as they are affected by many environmental factors. There are review papers in the literature that focus on the mechanisms of penetration of foliar-applied nutrient solutions through the leaf surface [8–10].

Foliar fertilization is an increasingly common way in plant nutrition to provide nutrients to plants, and nowadays, it is also aimed at the following goals [10]: (1) increasing the concentration of components in plant foods with biofortification (enrichment), using fertilizers containing readily available forms of deficient elements (selenium, iodine, zinc, iron), the deficiency of which occurs in the diet of animals and humans and affects about 25% of the population; (2) improving the utilization of the supplied elements by introducing substances that increase their uptake, water uptake and resistance to stress caused by abiotic factors. These include biostimulants that activate plant metabolic processes, reduce stress and alleviate nutrient deficiencies. These are substances that are said to completely change crop production in the near future.

The growing demand for quality food products, resulting from a growing world population, requires the use of optimal solutions to provide plants with optimal conditions for their growth and development [11]. While the weather conditions that affect the course of vegetation and crop yields are difficult to predict and control [12], in the process of feeding plants, we have the opportunity to take measures to increase the availability of nutrients for plants.

Proper foliar fertilization is one of the indispensable elements of agrotechnology to achieve high yields. This method of fertilization makes it possible to provide plants with all the necessary nutrients in every stage of development. It is also an effective way to stimulate and realize the potential of a given plant. It is an excellent way to support the root system of plants during periods of intensive growth. It makes it possible to provide plants, in a quick and effective way, with missing components during periodic shortages, resulting from the inability to absorb these components from the soil. Through foliar fertilization, we can also selectively supply micronutrients to sensitive plant species, so we are able to optimally meet the nutritional needs of plants. Foliar fertilizers provide high efficiency of fertilization and thus allow us to reduce the amount of components we introduce into the environment.

It is assumed that foliar fertilizers, depending on the crop, the level of agrotechnique, soil quality, weather conditions and, very importantly, the quality of the products, give the possibility of obtaining a yield increase of several to several tens of percentage points [13–17]. By supplying the plant with the missing essential nutrients, we can influence the yield and its quality parameters, which was stated as early as 1972, more than 50 years ago [18]. Thanks to foliar fertilization, we can also enrich plants in selected mineral components [19].

The market of foliar fertilizers in Poland is growing dynamically, gaining importance. It is estimated that more than 2000 different fertilizer products are currently available, and there is a 10-fold increase in their quantity compared with 2009 [7]. The wide range of foliar fertilizers on the market allows agricultural producers to choose products tailored to their current needs and field situation. However, the effective use of these fertilizers requires expert knowledge of the possibilities and advisability of their application. Agricultural producers expect products that are simple to use, characterized by full solubility, easy to dose and well miscible with chemicals, if such mixtures are approved for use [20].

Nowadays, in addition to the high nutrient content of foliar fertilizers, fertilizer manufacturers enrich their composition with various substances of a stimulating nature or supporting plant metabolism (plant extracts, algae, vitamins or amino acids) [21–23]. This makes it possible to increase the plant's resistance to stress factors, diseases and pests, or to support plant regeneration after a stress factor. Preininger et al. [24] emphasize the positive effect of using bacteria, fungi or viruses on foliar fertilization of plants.

Currently, the rules for the introduction of foliar fertilizers to the Polish market are regulated by Polish regulations [25] and EU regulations on, inter alia, making fertilizer products available on the EU market [26] According to the regulations, in order for a fertilizer to be approved for marketing, it must be made from raw materials that meet the requirements of one of the component material categories (CMCs) in Annex 2 [26]. So, among other things, what is important is the ingredient itself, its origin or its purity, while it is not important, for example, the particle size or chemical form of the element in question. In practice, most manufacturers aim to introduce and make their solutions available in accordance with the EU regulations (mainly due to the freedom of circulation in member countries), and if this is not possible, national regulations are used.

2. Discussion

2.1. Nutritional Demands of Cereals, Rapeseed and Corn

The nutritional requirements of cultivated crop species vary widely, and these needs are further influenced by the direction of production, habitat conditions, and weather. Significant differences in nutritional requirements also exist for varieties within a single species [27]. By using proper fertilization, we can also influence quality parameters, such as protein quality and content, gluten or sedimentation rate [28,29]. Therefore, it is very important that the size of fertilizer doses is adapted to the nutritional requirements of the plant species grown and the selected variety and that the current abundance of soils is taken into account, so that fertilization, which is intended to improve quantitative and qualitative parameters, does not have the opposite effect, i.e., a decrease in yield [30].

The contents of the forms of available nutrients in soils vary widely and depend on the species of soil, among other factors. Heavier soils, which contain more clay fractions, are characterized by a higher content of micronutrients compared with lighter, sandy soils [31]. Nutrient compactness can also be affected by tillage intensity, crop rotation and applied fertilization. In planning nutrient rates, it is important to know the content of nutrients in the soil in soluble (available) forms for plants. An example is the determination of plant doses of phosphorus, as it may turn out that despite the high content of the total form of this nutrient in the soil, up to 80% of it may be in a form that is not available to plants [32].

On the basis of long-term studies in Poland carried out by chemical and agricultural stations in cooperation with the Institute of Crop and Soil Sciences (IUNG) [31], a large share of soils with low abundance of micronutrients was found, especially for boron and copper (Table 1).

Table 1. Percentage of soils low in available forms of micronutrients in Poland.

Microelement	1987–1993	1994–1999	2000–2012	2016–2017	
				Wheat	Rapeseed
Boron (B)	75	79	74	19	45
Copper (Cu)	37	34	34	30	14
Iron (Fe)	-	28	21	21	10
Manganese (Mn)	11	7	3	16	15
Molybdenum (Mo)	23	-	-	-	-
Zinc (Zn)	14	13	17	20	14

- data not available.

Soil pH, which strongly influences the effectiveness of agrotechnical treatments, also has an impact on limiting the availability of components [33]. In 2016, about 72% of soils in Poland were characterized by very acidic or acidic pH (41% very acidic soils, 31% acidic soils); 19%, slightly acidic pH; and 9%, neutral pH [34]. Low soil pH contributes to an increase in the toxicity of aluminum ions to the root system, a reduction in the development of beneficial microorganisms, or the hydration and leaching of nutrients deep into the soil profile.

Fertilization, both in soil and foliar forms, should ensure that plants have access to nutrients in an amount that covers their nutritional requirements, especially in critical stages of their development. Stanislawska-Glubiak and Korzeniowska [31] indicate that crop plants differ significantly in their sensitivity to micronutrient deficiency (Table 2).

Table 2. Sensitivity of crop plants to micronutrient deficiencies.

Crop Plant	B	Cu	Mn	Mo	Zn
Wheat	1	3	3	0	1
Barley	0	3	2	1	0
Rye	0	0	2	0	0
Triticale	0	1	1	0	0
Oats	0	3	3	1	0
Rape	3	1	2	2	0
Sugar beet	3	2	3	2	1
Corn	2	2	2	0	3

0—none; 1—small; 2—medium; 3—large.

Among the basic nutrients, six macronutrients and six micronutrients can be specifically distinguished in terms of their indispensability to crop plants. From among the basic macronutrients such as nitrogen, phosphorus, potassium, magnesium, sulphur and calcium, plants take up the most nitrogen and potassium and this can be as much as over 200 kg per hectare. On the other hand, among the micronutrients boron, copper, iron, manganese, molybdenum and zinc, there is a wide variation in their requirements for plants. These needs range from a few to several hundred grams per hectare. It clearly follows that the main source of macronutrients for plants must be soil fertilization, while in the case of micronutrients, only foliar application can fully meet the nutritional needs of the plant [7].

2.2. Cereals

Cereals are among the most popular crops in Poland and also dominate global production. In Poland, in 2021, the area sown with cereals was 7.45 million hectares [35], and the popularity of this group of crops is due to the possibility of their versatile use. Cereal grains are used in the food and feed industries but can also be used in the energy or pharmaceutical industries [35]. The area of cereal crops in Poland has fluctuated over the past 50 years. The 1965–1980 period saw a decline in the area under cultivation, followed by an increase in interest in this group of crops between 1980 and 2000, and another decline since 2001. The 2000 season saw the largest area of cereal planting in Poland—8.81 million hectares [36].

Cereal cultivation is dominated by winter varieties, which have higher yields and better economic efficiency of production (Table 3). The estimated average yield of winter cereals in Poland in the 2021 season was 46 dt ha^{-1}, while that of spring cereals was 35 dt ha^{-1} [34].

Table 3. Cereal yield in Poland in 2010–2021.

Species of Cereal	2010	2015	2016	2017	2018	2019	2020	2021
	dt ha^{-1}							
Basic Cereals with Cereal Mixtures	**35.1**	**36.7**	**37.5**	**40.0**	**32.3**	**35.2**	**44.8**	**42.6**
Winter wheat	45.7	47.6	47.2	51.1	43.0	46.4	54.2	51.8
Spring wheat	34.3	33.5	38.3	38.5	31.5	32.6	41.7	39.6
Rye	26.9	27.8	28.9	30.6	24.2	27.2	35.1	33.1
Winter barley	40.7	41.3	44.6	47.1	37.8	43.0	51.1	47.7
Spring barley	33.0	33.0	35.8	38.0	29.5	32.1	40.0	37.8
Oats	26.4	26.5	28.4	29.8	23.5	24.9	33.2	31.4
Winter triticale	35.2	36.3	37.1	40.4	32.8	35.9	45.0	43.1
Spring triticale	28.4	28.4	31.7	32.9	25.1	27.5	36.4	33.7
Winter cereal mixes	30.9	30.9	32.4	34.4	28.2	30.6	38.1	36.6
Spring cereal mixes	30.5	27.2	29.8	32.2	25.0	26.2	34.5	33.7

Different cereal species are characterized by different nutrient requirements. Winter wheat requires 22–26 kg of nitrogen (N), 8 kg of phosphorus (P_2O_5), 5 kg of potassium (K_2O), 2 kg of magnesium (MgO) and 1 kg of calcium (CaO) to produce 1 t of grain. In turn, this species' micronutrient requirements in grams per hectare are as follows: boron (B), 115 g; copper (Cu), 120 g; manganese (Mn), 500 g; molybdenum (Mo), 7 g; and zinc (Zn), 350 g [28].

2.3. Rapeseed

In recent years, there has been a steady increase in interest in the cultivation of rapeseed. In the 2021 season, the area under cultivation of this crop in Poland amounted to 0.99 million hectares [35]. In comparison, in 1947, the cultivation area was only 51 thousand hectares, and in 2002, 439 thousand hectares. With the increase in the area of cultivation of this plant, an increase in the yield of production per 1 ha is also noted. Over the past 10 years, the average yield per 1 ha has increased from 23.6 dt to 32.1 dt ha^{-1}, or as much as 36%. The increase in the yield of this crop is due to the introduction of new, improved varieties and improved agrotechnology, including fertilization. The high popularity of rapeseed, as in the case of cereals, is the result of significant market demand for this type of product. Among other uses, rapeseed is used in the production of cooking oils and in the production of feedstuffs, as well as biofuels [37].

Rapeseed is characterized by very high nutrient requirements. To produce 30 dt of rapeseed, it is necessary to provide 213 kg of nitrogen (N), 89 kg of phosphorus (P_2O_5), 287 kg of potassium (K_2O), 157 kg of calcium (CaO), 70 kg of nitrogen (MgO) and 75 kg of sulfur (S) [38]. Of the micronutrients, rapeseed shows the greatest sensitivity to and high demand for boron and manganese. The demand according to various authors ranges from 80 g ha^{-1} of boron and 100 g ha^{-1} of manganese [39] to 300 g of boron and 500 g of manganese for the assumed yield of 5 t ha^{-1}. Regarding other micronutrients, rapeseed needs about 50–200 g of copper, 300–750 g of zinc and 5–10 g of molybdenum for the assumed yield of 5 t ha^{-1} [40].

2.4. Corn

Corn is a crop of major economic importance worldwide [41]. The volume of yield and the acreage devoted to its cultivation place it among the three most important crops, next to wheat and rice [42]. It is also of great importance in our country, due to its multiple uses, mainly for grain and silage, but it can also be used in the production of biogas or biofuels. The last few decades have seen a significant increase in the area under cultivation. In Poland, in 1995, the species occupied an area of 181 tys. ha, while in 2009, the cultivation area increased to 695 tys. ha [43]. In 2019, the area of corn cultivated for grain was 665 tys, and that for silage, 600 tys. ha, giving us a total of 1.26 million ha and placing it second, in the area of sown crops in Poland, after cereals and just ahead of rapeseed [35].

Table 4 summarizes the average yield of corn in Poland in the period from 2010 to 2020, with average grain yield ranging from 47.1 to 71.5 dt ha^{-1}, and green forage, from 357 to 493 dt ha^{-1} (Table 4) [35].

Table 4. Corn yield in Poland.

Use of the Corn Crop	2010	2015	2016	2017	2018	2019	2020
	dt ha^{-1}						
Grain	59.7	47.1	72.9	71.5	59.9	56.2	56.2
Forage	437	357	493	487	426	406	459

Corn is characterized by a sizable demand for nitrogen; it is assumed to require 25 kg of this element to produce one ton of grain and a corresponding weight of straw [44]. It is also characterized by a high demand for phosphorus and potassium. To produce 1 ton of grain with adequate straw, it requires about 10 kg of phosphorus (P_2O_5) and 30 kg of potassium (K_2O). Thus, for the yield of 10.0 tons of grain ha^{-1}, the nutritional needs are 250 kg of nitrogen (N), 100 kg of phosphorus (P_2O_5) and 300 kg of potassium (K_2O). Good sulfur supply for corn plays a key role in nitrogen utilization and conversion [45], enabling higher yield with less nitrogen fertilization. The beneficial effect of sulfur on increased plant uptake and utilization of the applied nutrient has also been reported [46]. Balanced plant nutrition should take into account the supply of both macronutrients and micronutrients to plants, which affect the efficiency of the uptake of the supplied components [47] but also contribute to the increase in the obtained yield themselves [48].

Corn is very sensitive to zinc and boron deficiencies, while it is less sensitive to deficiencies of copper, manganese and molybdenum. Fertilization with micronutrients is particularly important because corn is mainly grown on light soils, which are characterized by a much lower content of micronutrients compared with heavy soils. The requirement for corn to produce one ton of grain in relation to micronutrients is as follows: 20 g of boron (B), 12 g of copper (Cu), 45 g of iron (Fe), 35 g of manganese (Mn), 1 g of molybdenum (Mo) and 50 g of zinc (Zn). The use of micronutrients in corn production contributes to higher yield, but the effectiveness of micronutrient fertilization treatment depends, among other things, on the availability of macronutrients [48], indicating that the combined application of macro- and micronutrients produces better yield-forming effects [49].

2.5. Types of Foliar Fertilizers

The number of foliar fertilizers available on the Polish market has been steadily increasing for several years, mainly due to the growing demand of agricultural producers for dedicated, specialized products adapted to the requirements of individual crops. Scientific research confirming the applicability of the products in agricultural practice contributes to the increase. We can divide the available products into several different groups, depending on the selected criterion:

1. Physical form: (a) liquid fertilizers—in this group, we can distinguish among liquid fertilizers, fertilizers in suspension and gel; (b) loose fertilizers—soluble in water, they are in the form of powders and crystals of various shapes.

2. Purpose: (a) universal—fertilizers that can be used in any crop; (b) dedicated—tailored to the nutritional needs of selected plants and crops.

3. Amount of components in the fertilizer: (a) monocomponent—dominant, high content of one macro- or micronutrient (N, B, Cu, Zn, Mn); (b) bicomponent—high content of two components, whether macronutrients (N + Ca, P + K, S + Mg, N + K, etc.), micronutrients (B + Mo, B + Zn, B + Mn, Mn + Zn) or a mix of macro- and micronutrients (N + Mn, N + Mo, P + B, P + Zn); (c) multicomponent—containing a minimum of three or more nutrients (N + P + K, N + P + K+ micro, P + K + Mg, Zn + B + Mg).

4. Type of components: (a) primary macronutrients with high content—N, P, K; (b) secondary macronutrients with high content—Ca, Mg, Na, S; (c) micronutrients with high content—B, Cu, Fe, Zn, Mn, Mo; (d) mixed fertilizers—fertilizer mixtures with increased content of selected macro- and micronutrients.

5. Forms of nutrients: (a) "pure" ionic forms; (b) complexed—with complexing agents, such as lignosulfonic acid (LS), glutamine hydroxamate (HGA), organic acids; (c) chelated—chelating agents, such as ethylenediaminetetraacetic acid (EDTA), 2-hydroxyethylethylenediaminetriacetic acid (HEEDTA), diethylenetriaminepentaacetic acid (DTPA), ethylenediamine-N,N′-bis(2-hydroxyphenylacetic) acid (EDDHA), ethylenediamine-di (o-hydroxy-o-methylphenylacetic) acid (EDDHMA), ethylenediamine di-(2-carboxy-5-hydroxyphenylacetic) acid (EDDCHA), ethylenediamine-N-N=bis(2-hydroxy-5-sulfophenylacetic) acid (EDDHSA), N-(1,2-dicarboxyethyl)-D,L-aspartic acid (IDHA), N,N-di(2-hydroxybenzyl)ethylenediamine-N,N-diacetic acid (HBED) and ethylenediaminedisuccinic acid (EDDS).

6. Homogeneity: (a) complex fertilizers—fertilizers containing at least two nutrients, characterized by the fact that in the solid phase, each granule has exactly the same content of each of the declared components; (b) blended fertilizers (blends)—fertilizers resulting from the physical mixing of at least two other fertilizers, without chemical reactions.

7. The content of additional components: (a) deficient components—silicon (Si), iodine (I), chromium (Cr); (b) adjuvants—supportive agents, improving the effectiveness of foliar fertilization treatments; (c) anti-stress substances—substances that increase resistance to stress factors(amino acids, plant extracts, vitamins); (d) stimulants—substances and chemical compounds that stimulate the plant for intensive development (e.g., amino acids, hormones, humic substances).

2.6. Foliar Fertilizer Market

The foliar fertilizer market in Poland is currently estimated at around PLN 300 million. The main suppliers of the products are Polish companies, but many solutions from Europe and further corners of the world are also available. We can count, among the leaders of the domestic market, the companies ADOB®, EKOPLON® and INTERMAG®. These are companies that have been engaged in the production of foliar fertilizers for more than 30 years, supplying their products both to the Polish market and to many other countries around the world. Among foreign manufacturers, the Polish market is dominated by European companies coming from Italy, France, Spain, Great Britain, Belgium and Turkey, but there are also companies from the United States of America, China or Japan. Every year, new domestic and foreign manufacturers of foliar fertilizers and biostimulants appear on the Polish market. Such a large number of suppliers allows Polish agricultural producers to benefit from the latest solutions in the field of foliar fertilization and biostimulation.

The market of foliar fertilizers has undergone quite a transformation over the past 30 years. In the 1990s, liquid fertilizers containing basic macro- and micronutrients dominated the market. Some of these products are still available on the market and do not promise to disappear in the near future. An example of such a product is liquid boron fertilizers containing boron in organic form—boroethanolamine—which account for about 70–80% of all boron fertilizers used on farms. The remaining 20–30% is in bulk, soluble forms based on boric acid and sodium borates. Blends of boron fertilizers with macro- and micronutrients are also available.

Between 2000 and 2010, loose soluble fertilizers began to appear and gain importance on the market. These were products mainly containing nitrogen, phosphorus and potassium, along with micronutrients. Four products were standard in the offer of manufacturers: balanced fertilizer, containing nitrogen, phosphorus, potassium and micronutrients at the same level; fertilizer with increased nitrogen content; fertilizer with increased phosphorus content; and fertilizer with increased potassium content. Compared with liquid fertilizers, bulk fertilizers require slightly longer preparation time due to the need for pre-dissolution. However, in many cases, they have a much higher content and concentration of nutrients. In addition, they are cheaper to transport and more resistant to changing weather conditions during storage.

The biggest boom in the market of foliar fertilizers occurred after 2010, at which time, macro- and micronutrient products, enriched with various additives, began to appear. Currently, on the market, we have many innovative solutions tailored to the current specifics of production, plant requirements and changing environmental conditions.

In the tables below (Tables 5–7) is a comparison of selected fertilizers dedicated to the cultivation of cereals, rapeseed and corn, as well as popular directions in which producers and science are directing their research and development efforts.

Table 5. Selected foliar fertilizers used in cereal crops.

Foliar Fertilizer	Chemical Composition											Type of Complexes/Chelates
	N	P_2O_5	K_2O	MgO	SO_3	B	Cu	Fe	Mn	Mo	Zn	
Liquid fertilizers												
Ekolist cereals	127	38.1	38.1	-	-	0.13	11.4	8.9	10.1	0.06	8.9	EDTA organic complexes
Plonvit cereals	195	-	-	26	59	0.18	11.7	10.4	14.3	0.06	13	EDTA organic complexes
Vital cereals	198	-	-	66	-	1.32	3.3	6.6	11.8	0.16	13.2	No
Sarplon cereals	245	-	34.3	10	-	1.32	6.6	13.2	19.8	0.26	4	EDTA/DTPA
Suplofol Micro Z	204	-	-	27	68	2	6.8	13.6	25.9	0.2	13.6	No
Crystalline fertilizers												
Maximus amino micro cereals	11	-	70	-	-	3.4	50	20	40	0.4	20	Glicyna
Adob Micro cereals	100	-	50	-	310	-	15	3	30	0.2	5	EDTA
Amino Ultra cereals	-	-	-	20	-	1.6	16	65	65	0.7	49	Glicyna
Cereals forte	50	150	150	78	20	0.2	10	1	10	0.01	0.04	EDTA/DTPA
Dr Green cereals	-	-	-	-	-	5	50	60	80	0.5	20	Micro Activ
Opti cereals	140	160	160	30	180	-	3	1.5	5	0.4	1.5	EDTA/DTPA
Suspension fertilizers												
Yaravita Gramitrel	64	-	-	250	-	-	50	-	150	-	80	Oxide form

Table 6. Selected foliar fertilizers used in rapeseed cultivation.

Foliar Fertilizer	Chemical Composition											Type of Complexes/Chelates
	N	P_2O_5	K_2O	MgO	SO_3	B	Cu	Fe	Mn	Mo	Zn	
Liquid fertilizers												
Ekolist rape	150	50	37.5	-	-	8.7	0.1	8.7	8.7	0.06	0.12	Technology ACTIVE
Plonvit rape	186	-	-	31	31	6.2	1.2	6.2	6.2	0.06	6.2	Technology INT
Vital rape	188	-	-	40	-	6.9	1.2	3.1	4.4	0.09	3.7	No
Sarplon rape	285	-	30.4	12.4	1.45	5.3	2.38	2.38	16.8	0.53	2.4	EDTA/DTPA
Suplofol micro BR	195	-	-	26	65	6.2	0.85	1	23.4	0.2	9.5	No

Table 6. *Cont.*

Foliar Fertilizer	Chemical Composition											Type of Complexes/Chelates
	N	P_2O_5	K_2O	MgO	SO_3	B	Cu	Fe	Mn	Mo	Zn	
Crystalline fertilizers												
Maximus Amino Micro rape	110	-	70	-	-	20	15	30	40	0.4	15	Glicyna
Adob Micro rape	47	-	-	-	135	100	5	3	15	1	3	EDTA
Rapeforte	50	150	150	46	147	30	0.03	1.5	10	0.01	0.04	EDTA/DTPA
Dr Green rape	-	-	-	-	-	100	2	25	50	0.5	20	Micro Activ
OPTI rape	110	150	210	20	190	15	1	1.5	2	0.4	1.5	EDTA
Suspension fertilizers												
Yaravita Gramitrel	64	-	-	250	-	-	50	-	150	-	80	Oxide form

Table 7. Selected foliar fertilizers used in corn cultivation.

Foliar Fertilizer	Chemical Composition											Type of Complexes/Chelates
	N	P_2O_5	K_2O	MgO	SO_3	B	Cu	Fe	Mn	Mo	Zn	
Liquid fertilizers												
Ekolist corn	75.6	126	37.8	-	-	6.3	1.2	7.5	2.5	0.06	11.3	Technologia ACTIVE
Plonvit corn	195	-	-	26	54.6	5.2	7.8	9.1	9.1	0.065	14.3	Technologia INT
Vital corn	203	-	-	69	-	1.6	4.1	6.7	13.5	0.13	17.6	Brak
Sarplon corn	271	-	11.9	18.5	-	4	1.3	1.3	6.6	0.46	19.8	EDTA
Suplofol Micro K	188	-	-	25	63	2.5	1.25	3.75	5	0.38	20	No
Crystalline fertilizers												
Maximus Amino Micro corn	-	110	70	-	-	20	20	20	30	0.4	50	Glicyna
Adob Mikro corn	70	20	-	30	100	20	1	2	5	0.1	40	EDTA
Corn forte	50	200	150	42	220	15	0.07	1	0.1	0.01	15	EDTA. DTPA
Dr Green corn	-	-	-	-	-	5	2	60	70	0.5	80	Micro Activ
OPTI corn	100	210	140	30	140	5	2	1	0.3	3	1	EDTA
Suspension fertilizers												
YaraVita Zeatrel	-	440	75	67	-	-	-	-	-	-	46	Oxide form

From the compilation of fertilizers in the tables (Tables 5–7), each product, even if dedicated to the same crop, is characterized by a different composition. On the market, we have products that contain all basic macro- and micronutrients, as well as those that only contain selected nutrients. Individual products also differ in the content of additives, the main purpose of which is to improve the efficiency of fertilization and/or support the plant in case of biotic and abiotic stresses. Large discrepancies can also be seen in the recommended dosages and prices of individual fertilizers. The above comparison shows that when choosing a particular solution, many factors should be taken into account, including composition, forms of components and their availability to the plant, as well as dosage, the price of the fertilizer or the type of chelating, complexing substances, which should be selected to maximize the effectiveness of foliar fertilization under specific conditions in the field.

2.7. Foliar Fertilizers Containing Amino Acids

The last 10 years have seen a significant increase in interest in amino acids used in plant fertilization. Currently, on the market, we have many products containing amino acids. Depending on the product, the content of these components can vary from less than 1% to more than 40% of the total weight of the fertilizer [23]. The various solutions also differ in the source of amino acids, e.g., plant or animal origin, and in the content of individual amino acids. All amino acids except glycine can differ in optical activity. They can exist in L- and D-forms, and very importantly, both forms can be taken up and utilized by plants [50]. There have been many publications on the effectiveness of amino acids, with authors confirming the effectiveness of fertilizers containing amino acids of both plant and animal origin [51]. The effectiveness of amino acids is due, among other things, to their nitrogen content, which is necessary for optimal plant growth and development, as well as their effect on the efficiency of uptake and utilization of other nutrients [23]. They can also influence the content of chlorophyll or carotenoids in plants, key substances involved in photosynthesis [52,53].

2.8. Foliar Fertilizers Containing Nano-Elements

Among the products used in foliar fertilization are those that contain nanocomponents. Their effectiveness can depend on the timing of application, concentration or particle size [54]. These fertilizers are gaining importance due to their ability to reduce the negative impact of fertilizers on the environment and their effectiveness. A positive effect of nano-iron on fruit yield and quality parameters was shown by [55], while in an experiment with rapeseed, a yield increase of 1298 kg ha^{-1} was reported [56]. Vishekaii et al. [57] studied the effect of boron nano-chelates on fruit and olive oil yield. Among the available literature, one can also find information on the positive effects on growth and yield of nano-molybdenum [58], nano-zinc [54] or nano-silicon [59], nano-copper, and nano-silver [60]. Meena et al. [61] studied the effects of nano-phosphorus, potassium and zinc on wheat cultivation, showing positive effects of nanoparticles on growth and yield at levels ranging from a few to tens of percentage points, depending on the combination.

2.9. Foliar Fertilizers Containing Silicon

For several years, the foliar fertilizer market has seen an increase in interest in foliar fertilizers containing silicon. Among other things, the use of these products contributes to the strengthening of plant cell walls, thereby increasing resistance to damage [62]. Foliar application of silicon also has a positive effect on reducing water loss through the leaves [63]. The positive effect of silicon on the response of plants to high temperatures was demonstrated in their experiment by Basirat and Mousavi [64]. The experiment was conducted in a greenhouse under controlled, high-temperature conditions (36 °C). Foliar application of silicon resulted in a 36.1% increase in total cucumber yield and a 40.3% increase in marketable yield. The positive effect of silicon applied in the form of nanoparticles was also demonstrated by Shalaby et al. [59]. In their study, they showed that the effectiveness of silicon fertilizers depended on the number of fertilization treatments and habitat conditions, and that yield gains could range from 6.4% (one treatment) to 12.9% (three treatments) in years with optimal rainfall and from 12.2% to 17.6% in dry years [63].

2.10. Foliar Fertilizers for Biofortification

The word biofortification is becoming more and more popular every year. There is more and more talk about deficiencies of selected elements in the diet of humans and animals and thus the need to develop effective methods for providing these elements [65]. One of the most effective ways is plant biofortification, that is, enriching plants with specific elements and improving their availability. The process of biofortification can take place via fertilization (topdressing and foliar) or breeding varieties that are able to accumulate increased amounts of selected elements. Foliar fertilization is one of the most popular and effective methods of plant biofortification, as it allows the delivery of the right amount of specific elements in a specific developmental stage of the plant and is thus an economical and environmentally safe procedure.

Most often, iodine and selenium are mentioned in biofortification, but the topic can also apply to other nutrients. In the case of iodine, the minimum human requirement is 1 μg kg^{-1} body weight day^{-1}, and the optimal dose is 95–150 μg day^{-1}. The human body has a similar requirement for selenium, where the optimal level for one person is 50–200 μg day^{-1} [19]. Currently, there are no foliar fertilizers for biofortification on the domestic market, but scientific research has been conducted for more than 10 years to determine the optimal doses and chemical forms of iodine or selenium. The effects of foliar and soil application of iodine and selenium in their experiments were studied by [66–69]. Biofortification using agrotechnical methods is a direction that will become increasingly important in the coming years, as the problem of nutrient deficiency in the diet of humans and animals affects not only our country but also the entire globe.

3. Conclusions

The analysis of the foliar fertilizer market in Poland indicates that its development is very dynamic. High competition among manufacturers of foliar fertilizers contributes to the appearance of many new innovative products on the market. Currently, products enriched with natural or synthetic additives to improve the efficiency of fertilization are expected to become the standard.

Author Contributions: Conceptualization, R.J., G.K. and M.S., data curation—compilation and analysis of the results, R.J. and G.K.; writing–original draft preparation, R.J. and G.K.; writing–review and editing, R.J., G.K. and M.S. All authors have read and agreed to the published version of the manuscript.

Funding: The APC/BPC is financed/co-financed by Wrocław University of Environmental and Life Sciences.

Institutional Review Board Statement: Not applicable.

Data Availability Statement: Not applicable.

Conflicts of Interest: The authors declare that they have no conflict of interest.

References

1. FAO. (Ed.) Repurposing Food and Agricultural Policies to Make Healthy Diets More Affordable In *The State of Food Security and Nutrition in the World*; FAO: Rome, Italy, 2022; ISBN 978-92-5-136499-4.
2. Hemathilake, D.M.K.S.; Gunathilake, D.M.C.C. Chapter 31—Agricultural Productivity and Food Supply to Meet Increased Demands. In *Future Foods*; Bhat, R., Ed.; Academic Press: Cambridge, MA, USA, 2022; pp. 539–553. ISBN 978-0-323-91001-9.
3. Nair, K.P. Soil Fertility and Nutrient Management. In *Intelligent Soil Management for Sustainable Agriculture: The Nutrient Buffer Power Concept*; Nair, K.P., Ed.; Springer International Publishing: Cham, Switzerland, 2019; pp. 165–189. ISBN 978-3-030-15530-8.
4. United Nations. Sustainable Development. Available online: https://www.un.org/sustainabledevelopment/sustainable-development-goals/ (accessed on 26 February 2023).
5. European Commission. A Green Deal. Available online: https://commission.europa.eu/strategy-and-policy/priorities-2019-2024/european-green-deal_en (accessed on 23 May 2023).
6. EPRS. Rolnictwo Precyzyjne a Przyszłość Rolnictwa w Europie. In *EPRS | Biuro Analiz Parlamentu Europejskiego, Dział prognoz Naukowych (STOA) PE 581.892*; EPRS: Brussels, Belgium, 2016; ISBN 978-92-846-1033-4.
7. Szewczuk, C.; Sugier, D. Ogólna Charakterystyka i Podział Nawozów Dolistnych Oferowanych Na Polskim Rynku. *Ann. Univ. Mariae–Curie–Skłodowska Lub.–Pol.* **2009**, *LXIV*, 29–36. [CrossRef]
8. Fernández, V.; Eichert, T. Uptake of Hydrophilic Solutes Through Plant Leaves: Current State of Knowledge and Perspectives of Foliar Fertilization. *Crit. Rev. Plant Sci.* **2009**, *28*, 36–68. [CrossRef]
9. Fernández, V.; Sotiropoulos, T.; Brown, P. *Foliar Fertilization: Scientific Principles and Field Pratices*; International Fertilizer Industry Association: Paris, France, 2013; ISBN 979-10-92366-00-6.
10. Ishfaq, M.; Kiran, A.; ur Rehman, H.; Farooq, M.; Ijaz, N.H.; Nadeem, F.; Azeem, I.; Li, X.; Wakeel, A. Foliar Nutrition: Potential and Challenges under Multifaceted Agriculture. *Environ. Exp. Bot.* **2022**, *200*, 104909. [CrossRef]
11. Calicioglu, O.; Flammini, A.; Bracco, S.; Bellù, L.; Sims, R. The Future Challenges of Food and Agriculture: An Integrated Analysis of Trends and Solutions. *Sustainability* **2019**, *11*, 222. [CrossRef]

12. Arora, N.K. Impact of Climate Change on Agriculture Production and Its Sustainable Solutions. *Environ. Sustain.* **2019**, *2*, 95–96. [CrossRef]
13. Szewczuk, C.; Michalojc, Z. Praktyczne Aspekty Dolistnego Dokarmiania Roslin. *Acta Agrophysica* **2003**, *85*, 19–29.
14. Jarecki, W.; Bobrecka-Jamro, D. Wpływ Dolistnie Stosowanego Mocznika z Mikrokomplexem Na Wielkość I Jakość Plonu Nasion Rzepaku Jarego (Influence Of Used on Leaves Urea with Microcomplex on Size and Quality). *Zesz. Nauk. Uniw. Przyr. We Wrocławiu Rol.* **2010**, *97*, 267–274.
15. Liszewski, M.; Błażewicz, J. Wpływ Nawożenia Dolistnego Miedzią i Manganem Na Przydatność Słodowniczą Ziarna Jęczmienia (Badania Wstępne). *Pol. J. Agron.* **2015**, *23*, 18–23.
16. Tobiasz-Salach, R.; Krochmal-Marczak, B.; Bobrecka-Jarmo, D. Ocena Wpływu Nawożenia Dolistnego Na Plonowanie i Skład Chemiczny Nasion Gryki (Fagopyrum Esculentum Moench). *Fragm Agron* **2018**, *35*, 106–114.
17. Oleksy, A.; Staron, J.; Kolodziejczyk, M.; Kulig, B.; Brodowicz, T. Wpływ Dolistnego Nawożenia Mikro-i Makroelementowego Na Plonowanie Oraz Zawartość Tłuszczu w Nasionach Rzepaku. *Fragm. Agron.* **2019**, *36*, 54–66.
18. Byszewski, W.; Moldovany, K.; Sadowska, A. Dolistne Żywienie Roślin. *Postępy Nauk Rol. 172 R* **1972**, 75–94.
19. Dziennik UstawLiszka-Skoczylas, M.; Żmudziński, D.; Rudnik, D. Biofortyfikacja Roślin Uprawnych Jako Metoda Walki z Deficytem Składników Mineralnych w Diecie Człowieka. In *Składniki bioaktywne surowców i produktów roślinnych*; Oddział Małopolski Polskiego Towarzystwa Technologów Żywności: Kraków, Poland, 2014; pp. 58–65.
20. Grzebisz, W.; Korbas, M.; Rybacki, P.; Szczepaniak, W.; Wolna-Maruwka, A.; Woźnica, Z. *Nawożenie Dolistne Roślin Uprawnych*; Polskie Wydawnictwo Rolnicze: Poznań, Poland, 2019.
21. Chojnacka, K.; Michalak, I.; Dmytryk, A.; Gramza, M.; Słowiński, A.; Górecki, H. Algal Extracts as Plant Growth Biostimulants. In *Marine Algae Extracts*; John Wiley & Sons, Ltd.: Hoboken, NJ, USA, 2015; pp. 189–212. ISBN 978-3-527-67957-7.
22. Soppelsa, S.; Kelderer, M.; Casera, C.; Bassi, M.; Robatscher, P.; Matteazzi, A.; Andreotti, C. Foliar Applications of Biostimulants Promote Growth, Yield and Fruit Quality of Strawberry Plants Grown under Nutrient Limitation. *Agronomy* **2019**, *9*, 483. [CrossRef]
23. Souri, M.K. Aminochelate Fertilizers: The New Approach to the Old Problem; a Review. *Open Agric.* **2016**, *1*, 118–123. [CrossRef]
24. Preininger, C.; Sauer, U.; Bejarano, A.; Berninger, T. Concepts and Applications of Foliar Spray for Microbial Inoculants. *Appl. Microbiol. Biotechnol.* **2018**, *102*, 7265–7282. [CrossRef]
25. Dziennik Ustaw Ustawa z Dnia 10 Lipca 2007 r. o Nawozach i Nawożeniu, Nr 147, Poz. 1033. 14 August 2007. Available online: https://isap.sejm.gov.pl/isap.nsf/DocDetails.xsp?id=wdu20071471033 (accessed on 25 August 2023).
26. Rozporządzenie Parlamentu Europejskiego i Rady (UE). 2019/1009 z Dnia 5 Czerwca 2019 r. Ustanawiające Przepisy Dotyczące Udostępniania Na Rynku Produktów Nawozowych UE, Zmieniające Rozporządzenia (WE) Nr 1069/2009 i (WE) Nr 1107/2009 Oraz Uchylające Rozporządzenie (WE) Nr 2003/2003. *Parlamentu Europejskiego i Rady UE*. 2019. Available online: https://eur-lex.europa.eu/eli/reg/2019/1009/oj (accessed on 25 August 2023).
27. Korzeniowska, J. Response of Ten Winter Wheat Cultivars to Boron Foliar Applicatio. *Ann. UMCS Agric.* **2009**, *63*, 72–77.
28. Błaszczyk, K. Wymagania Siedliskowe i Pokarmowe Pszenicy Ozimej. In *Problematyka Nauk Przyrodniczych i Technicznych*; Uniwersytet Przyrodniczy we Wrocławiu: Wroclaw, Poland, 2019; Volume 3, pp. 14–27.
29. Niu, J.; Liu, C.; Huang, M.; Liu, K.; Yan, D. Effects of Foliar Fertilization: A Review of Current Status and Future Perspectives. *J. Soil Sci. Plant Nutr.* **2021**, *21*, 104–118. [CrossRef]
30. Krysztoforski, M. Ekonomiczne i Środowiskowe Efekty Racjonalnego Nawożenia. In *Ograniczenie Zanieczyszczenia Azotem Pochodzenia Rolniczego Metodą Poprawy Jakości Wód*; S-PRINT: Warszawa, Poland, 2018; p. 3. ISBN 978-83-952011-0-3.
31. Stanisławska-Glubiak, E.; Korzeniowska, J. Analiza Rynku Nawozów Mikroelementowych Na Tle Potrzeb Nawożenia Mikroelementami w Polsce. *Stud. Rap. IUNG-PIB* **2020**, *63*, 145–161. [CrossRef]
32. Podleśna, A. Czynniki Kształtujące Pobieranie i Wykorzystanie Fosforu Przez Rośliny Oraz Jego Straty z Gleb Uprawnych. *Stud. I Rap. IUNG-PIB* **2019**, *56*, 59–76. [CrossRef]
33. Tkaczyk, P.; Rutkowska, A. Zmiany Odczynu i Zasobności w Składniki Pokarmowe Gleb Uprawnych Lubelszczyzny. *Ann UMCS Sect. E Agric.* **2020**, *75*, 3. [CrossRef]
34. Pietrzak, S.; Juszkowska, D.; Nawalany, P. Zmiany Odczynu i Zasobności Gleb Użytków Zielonych w Polsce Między 2008 a 2016 Rokiem. *Zagadnienia Doradz. Rol.* **2019**, *95*, 50–71.
35. GUS. *Rocznik Statystyczny Rolnictwa*; Główny Urząd Statystyczny: Warsaw, Poland, 2022.
36. Rachoń, L.; Kawczyńska, M. Changes in the Structure of Sowing Area, Yields and Harvests of Cereal Crops in Poland in the Years 1965–2015. *Ann. UMCS Sect. E Agric.* **2018**, *73*, 4. [CrossRef]
37. Raboanatahiry, N.; Li, H.; Yu, L.; Li, M. Rapeseed (Brassica Napus): Processing, Utilization, and Genetic Improvement. *Agronomy* **2021**, *11*, 1776. [CrossRef]
38. Wielebski, F. Rola Siarki w Kształtowaniu Ilości i Jakości Plonu Rzepaku Ozimego. *Rośliny Oleiste-Oilseed Crops* **2015**, *36*, 39–59.
39. Szulc, K. Strategie Nawożenia Rzepaku Mikroelementami. *Farmer.pl* 2015, 12 October 2015. Available online: https://www.farmer.pl/produkcja-roslinna/rosliny-oleiste/strategie-nawozenia-rzepaku-mikroelementami,59632.html (accessed on 25 August 2023).
40. Kobus, A. Priorytetowe Mikroelementy w Rzepaku. *Farmer.pl* 2015, 13 April 2015. Available online: https://www.farmer.pl/produkcja-roslinna/rosliny-oleiste/priorytetowe-mikroelementy-w-rzepaku,55637.html (accessed on 25 August 2023).
41. Erenstein, O.; Jaleta, M.; Sonder, K.; Mottaleb, K.; Prasanna, B.M. Global Maize Production, Consumption and Trade: Trends and R&D Implications. *Food Secur.* **2022**, *14*, 1295–1319. [CrossRef]

2. Lepiarczyk, A.; Filipek-Mazur, B.; Tabak, M.; Joniec, A. Wpływ Nawożenia Azotem i Siarką Na Plonowanie i Skład Chemiczny Ziarna Kukurydzy. Część I. Wielkość i Komponenty Plonu Ziarna Kukurydzy. *Fragm. Agron.* **2013**, *30*, 115–122.
3. Księżak, J.; Bojarszczuk, J.; Staniak, M. Produkcyjność Kukurydzy i Sorga w Zależności Od Poziomu Nawożenia Azotem. *Pol. J. Agron.* **2012**, *8*, 20–28.
4. Subedi, K.; Ma, B. Corn Crop Production: Growth, Fertilization and Yield. In *Corn Crop Production: Growth, Fertilization and Yield*; Danforth, A.T., Ed.; Nova Science Publishers, Inc.: Hauppauge, NY, USA, 2011; pp. 1–85.
5. Rehm, G.W. Sulfur in a Fertilizer Program for Corn. In *Sulfur: A Missing Link between Soils, Crops, and Nutrition*; Jez, J., Ed.; ASA-CSSA-SSSA: Madison, WI, USA, 2008; pp. 143–152.
6. Podleśna, A.; Podleśny, J.; Klikocka, H. Wpływ Nawożenia Siarką i Azotem Na Azotowo-Fosforową Gospodarkę Kukurydzy. *Przem. Chem.* **2017**, *96*, 1374–1377. [CrossRef]
7. Barczak, B.; Murawska, B.; Spychaj-Fabisiak, E. Zawartość Siarki i Azotu w Ziarnie Kukurydzy w Zależności Od Typu Gleby i Zastosowanego Nawożenia. *Fragm. Agron.* **2011**, *28*, 7–14.
8. Korzeniowska, J.; Gembarzewski, H. Potrzeby Nawożenia Mikroelementami Kukurydzy Uprawianej Na Kiszonkę. *Rocz. Glebozn.* **1999**, *50*, 79–84.
9. Świtkowski, M.; Kozera, W.; Barczak, B. Ocena Efektywności Dolistnego Nawożenia Zbóż. *Nauka Niejedno Ma Imię* **2015**, *T.3*, 131–139.
0. Pipiak, P.; Skwarek, M. Zastosowanie Nawozów Aminokwasowych w Rolnictwie. *Technol. Jakość Wyr.* **2020**, *65*, 144–157.
1. Wang, D.; Deng, X.; Wang, B.; Zhang, N.; Zhu, C.; Jiao, Z.; Li, R.; Shen, Q. Effects of Foliar Application of Amino Acid Liquid Fertilizers, with or without Bacillus Amyloliquefaciens SQR9, on Cowpea Yield and Leaf Microbiota. *PLoS ONE* **2019**, *14*, e0222048. [CrossRef]
2. Noroozlo, Y.A.; Souri, M.K.; Delshad, M. Stimulation Effects of Foliar Applied Glycine and Glutamine Amino Acids on Lettuce Growth. *Open Agric.* **2019**, *4*, 164–172. [CrossRef]
3. Miri Nargesi, M.; Sedaghathoor, S.; Hashemabadi, D. Effect of Foliar Application of Amino Acid, Humic Acid and Fulvic Acid on the Oil Content and Quality of Olive. *Saudi J. Biol. Sci.* **2022**, *29*, 3473–3481. [CrossRef]
4. Janmohammadi, M.; Amanzadeh, T.; Sabaghnia, N.; Dashti, S. Impact of Foliar Application of Nano Micronutrient Fertilizers and Titanium Dioxide Nanoparticles on the Growth and Yield Components of Barley under Supplemental Irrigation. *Acta Agric. Slov.* **2016**, *107*, 265–276. [CrossRef]
5. Bozorgi, H.R. Effects of Foliar Spraying with Marine Plant Ascophyllum Nodosum Extract and Nano Iron Chelate Fertilizer on Fruit Yield and Several Attributes of Eggplant (*Solanum Melongena* L.). *ARPN J. Agric. Biol. Sci.* **2012**, *7*, 357–362.
6. Mahdi, H.H.; Mutlag, L.A.; Mouhamad, R.S. Study the Effect of Khazra Iron Nano Chelate Fertilizer Foliar Application on Two Rapeseed Varieties. *Rev. Bionatura* **2019**, *4*, 841–845. [CrossRef]
7. Vishekaii, Z.R.; Soleimani, A.; Fallahi, E.; Ghasemnezhad, M.; Hasani, A. The Impact of Foliar Application of Boron Nano-Chelated Fertilizer and Boric Acid on Fruit Yield, Oil Content, and Quality Attributes in Olive (*Olea Europaea* L.). *Sci. Hortic.* **2019**, *257*, 108689. [CrossRef]
8. Manjili, M.J.; Bidarigh, S.; Amiri, E. Study the Effect of Foliar Application of Nano Chelate Molybdenum Fertilizer on the Yield and Yield Components of Peanut. *Egypt. Acad. J. Biol. Sci. H Bot.* **2014**, *5*, 67–71. [CrossRef]
9. Shalaby, T.A.; Abd-Alkarim, E.; El-Aidy, F.; Hamed, E.-S.; Sharaf-Eldin, M.; Taha, N.; El-Ramady, H.; Bayoumi, Y.; Dos Reis, A.R. Nano-Selenium, Silicon and H2O2 Boost Growth and Productivity of Cucumber under Combined Salinity and Heat Stress. *Ecotoxicol. Environ. Saf.* **2021**, *212*, 111962. [CrossRef] [PubMed]
0. Jaskulska, I.; Jaskulski, D. Efekty Stosowania Nanocząstek Srebra i Miedzi w Nawozach Dolistnych. *Przem. Chem.* **2020**, *99*, 250–253. [CrossRef]
1. Meena, R.; Jat, G.; Jain, D. Impact of Foliar Application of Different Nano-Fertilizers on Soil Microbial Properties and Yield of Wheat. *J. Environ. Biol.* **2021**, *42*, 302–308. [CrossRef]
2. Puppe, D.; Sommer, M. Chapter One—Experiments, Uptake Mechanisms, and Functioning of Silicon Foliar Fertilization—A Review Focusing on Maize, Rice, and Wheat. In *Advances in Agronomy*; Sparks, D.L., Ed.; Academic Press: Cambridge, MA, USA, 2018; Volume 152, pp. 1–49.
3. Trawczyński, C. Ocena Plonowania i Jakości Bulw Po Aplikacji Dolistnej Krzemu i Mikroelementów. *Agron. Sci.* **2021**, *76*, 1. [CrossRef]
4. Basirat, M.; Mousavi, S.M. Effect of Foliar Application of Silicon and Salicylic Acid on Regulation of Yield and Nutritional Responses of Greenhouse Cucumber Under High Temperature. *J. Plant Growth Regul.* **2022**, *41*, 1978–1988 [CrossRef]
5. Cakmak, I.; Kutman, U.B. Agronomic Biofortification of Cereals with Zinc: A Review. *Eur. J. Soil Sci.* **2018**, *69*, 172–180. [CrossRef]
6. Smoleń, S.; Ledwożyw-Smoleń, I.; Strzetelski, P.; Sady, W.; Rożek, S. Wpływ Nawożenia Jodem i Azotem Na Efektywność Biofortyfikacji w Jod Oraz Na Jakość Biologiczną Marchwi. The Effect of Iodine and Nitrogen Fertilization on Efficiency Biofortification in Iodine as Well as on Biological Quality of Carrot. *Ochr. Śr. Zasobów Nat.* **2009**, *40*, 313–320.
7. Rożek, S.; Smoleń, S.; Ledwożyw, I.; Strzetelski, P. Wstępna Ocena Wpływu Nawożenia i Dokarmiania Dolistnego Jodem Na Efektywność Biofortyfikacji Salaty w Jod Oraz Na Jej Sklad Mineralny. *J. Elem.* **2010**, *15*, 78–79.

68. Janečka, L.; Jůzl, M.; Elzner, P.; Drápal, K.; Mareček, V. Impact of the Aplication of Foliar Fertilizers Containing Selenium on th Yield Indicators in Potatoes. *Mendelnet* **2011**, 50–56.
69. Darecki, A.; Saeid, A.; Górecki, H. Selen w Perspektywie Fortyfikacji Roślin o Znaczeniu Gospodarczym Dla Polski. *Wiad. Chem* **2015**, *69*, 1067–1081.

agriculture

MDPI

Article

Perennial Ryegrass (*Lolium perenne* L.) Response to Different Forms of Sulfur Fertilizers

Grzegorz Kulczycki [1,*], Elżbieta Sacała [1], Anna Koszelnik-Leszek [2] and Łukasz Milo [3]

1 Institute of Soil Science Plant Nutrition and Environmental Protection, Wrocław University of Environmental and Life Sciences, Grunwaldzka Str. 53, 50-363 Wrocław, Poland; elzbieta.sacala@upwr.edu.pl
2 Department of Botany and Plant Ecology, Wrocław University of Environmental and Life Sciences, Grunwaldzki Sq. 24A, 50-363 Wrocław, Poland; anna.koszelnik-leszek@upwr.edu.pl
3 Chemical Plants "Siarkopol" TARNOBRZEG Ltd., Chemiczna Str. 3, 39-400 Tarnobrzeg, Poland; lmilo@zchsiarkopol.pl
* Correspondence: grzegorz.kulczycki@upwr.edu.pl; Tel.: +48-71-320-5654

Abstract: The aim of the study was to compare sulfate fertilizers and mixtures of elemental sulfur (S^0) and sulfate in terms of yield and nitrogen (N) and sulfur (S) status in perennial ryegrass. Mixtures of sulfate and S^0 can reduce the consumption of sulfate alone. The plants were grown in soil cultures. The plants were supplemented with S^0, K_2SO_4, $MgSO_4$, and $(NH_4)SO_4$ or a mixture of these salts with S^o. Two sulfur doses were applied and the ryegrass was harvested three times. Fresh and dry weights of each swath, the N and S content, and their uptake were determined. The total fresh yield of sulfur-fertilized plants was 25 to 94% higher compared to unfertilized plants. The increases in dry matter were even more significant. Fertilizers, being a mixture of S^0 and sulfate, showed the same efficiency as those containing sulfate alone. Sulfur fertilization resulted in a higher S content and its uptake, lowered N concentration in second and third swatch, and a decrease in total N uptake. In conclusion, to achieve high crop yields, soil sulfur deficiency should be corrected and fertilizers that are the mixture of elemental sulfur and sulfate are a beneficial and effective approach.

Keywords: elemental sulfur; sulfate; N:S ratio; ryegrass

1. Introduction

Sulfur (S), a macronutrient found in plants in the smallest amounts relative to other essential macronutrients, is currently recognized as one of the most important yield-forming elements [1,2]. In recent years, a shortage of plant-available sulfur in soils almost all over the world has been [3–6]. This adverse phenomenon is due to a drastic reduction in air pollution by sulfur, and consequently, a reduction in its deposition in the soil [6,7]. High-yielding crops grown today take up large amounts of sulfur from the soil and exacerbate this problem. It is expected that this phenomenon will intensify in the coming years [6,8] and in consequence, many agricultural areas will require fertilization with sulfur to maintain high yields and quality.

Therefore, research concerning sulfur application in agricultural plant production that will indicate an environmentally favorable, cheap for farmers, and efficient for plants solution of this problem is an urgent need. For proper growth and development, plants require sulfur at a level of 0.1–1.0% on a dry weight basis, and the average concentration of S in plant tissues ranges from 0.2 to 0.5% [9]. The main inorganic form of S directly available to plants is sulfate (SO_4^{2-}), which in mineral fertilizers, occurs as calcium, magnesium, potassium, or ammonium sulfate. Unfortunately, sulfate can be easily leached from the root zone, particularly from sandy soil. Furthermore, under sulfur limitation, plants absorb SO_4^- very rapidly, which results in the formation of a sulfate depletion zone. On the other hand, sulfur is classified as an immobile mineral element in plants, which is not readily remobilized to younger leaves during deficiency. Its deficiency at any stage of plant growth negatively affects plant metabolism and growth, and ultimately results in reduced yields [10]. This is due to the fact that sulfur is a multifunctional element which is not merely a component of certain amino acids (cysteine, methionine) and consequently proteins, but also builds other important cellular compounds [9,11]. They include coenzymes, lipids, and secondary metabolites that are involved in cellular metabolism, plant reaction to environmental stresses, and interactions with animals and pathogens [12]. Protein synthesis requires an adequate amount of two key macro-elements, nitrogen and sulfur; hence, a correct balance of these nutrients is particularly important [13].

Literature data indicate that sulfur and nitrogen metabolism are linked, and sulfur availability improves nitrogen uptake by plants, thereby affecting their quality and optimizing the N:S ratio [14,15]. The N:S ratio is considered to be a responsive indicator of sulfur supply to plants [16–19]. Fertilization with sulfur increases the total content of sulfur and sulfates in plants and deficit of this element results in the accumulation of non-protein nitrogen compounds that reduce the biological value of crop plants [15,20].

Fertilizers that are a mixture of sulfate and elemental sulfur (S^0) can be a good way to optimize sulfur supply to plants and improve its bioavailability in the long term. At the same time, they are cheaper and more environmentally friendly than sulfates alone. Elemental sulfur is chemically inert, difficult to leach from the soil compared to the anionic sulfate form, and available for longer in the soil. S^0 is a suitable source of this element for plants but must first be oxidized to the sulfate form by soil microorganisms. Oxidation of elemental sulfur and mineralization of organic sulfur, and thus the amount of available sulfur depends on the microbial activity of the soil. The temperature and moisture of the soil and its physico-chemical properties have a significant impact on this activity [11]. Sulfate is actively taken up through the roots and translocated to the shoot, and can be stored in vacuoles [9]. Plants are able to take up sulfate from the soil over a wide range of concentrations through the use of high-affinity and low-affinity transporters [11].

Literature data indicate that the application of sulfur increases the yield of crop and pasture plants and the magnitude of the response is dependent on the type of plant [21–23]. Some crops, such as oilseed rape and mustard, respond very well, while others respond much less well. There is little information on the response of meadow grasses; however, it is known that multi-cut grasses are more susceptible to sulfur deficiency than other crops [24]. In general, sulfur fertilization is expected to increase yields by an average of 25% under conditions of severe sulfur deficiency [25,26].

Perennial ryegrass—a species of the grass family—grows wild almost throughout Europe, northern Africa, the temperate zone of Asia, and was artificially introduced in North America and Australia. It is an excellent pasture grass that forms the basis of productive pastures for cattle. It can also be used to stabilize soil and prevent soil erosion, as well as to create hardy turf for lawns and golf courses. It has produced many varieties. Ryegrass is a highly productive plant and can be harvested multiple times during the growing season.

There is a relatively small number of studies that examine the sulfur fertilization needs of grasslands, especially in the temperate zone and over the past decade under conditions of very limited sulfur input from the atmosphere. Aspel et al. [27] in lysimeter experiments on perennial ryegrass showed that sulfur fertilization increased crop yields, nitrogen recovery from fertilizers, and significantly reduced nitrate leaching.

The aim of the study was to compare sulfate fertilizers and a mixture of elemental sulfur (S^0) and sulfate in terms of yield and N and S status in perennial ryegrass.

2. Materials and Methods

2.1. Pot Experiment

Plants were grown in pots filled with 2.5 kg of a sandy Arenosol whose granulometric composition was as follows: sand 86% with dominant medium and fine fractions, silt 12%, and clay 2%. The total content of carbon determined by the Dumas dry combustion method [28] and total content of sulfur determined by the Butters–Chenery method [29] were very low (Table 1). The plant-available phosphorus, potassium, magnesium, and sulfate sulfur form content in the soil are given in Table 1. The soil had a very acidic pH prior to liming, a high content of phosphorus, and a low content of potassium and magnesium. Before sowing, the soil acidity was adjusted to pHKCl 6.62 by applying lime at a rate $CaCO_3$ (2.12 g kg^{-1}). The level of total sulfur content determined in the soil, along with $S\text{-}SO_4$, classifies this soil type as low sulfur soil. Macronutrients were applied to the soil before sowing in the following doses: N 104 mg kg^{-1}, P 124 mg kg^{-1}, K 293 mg kg^{-1}, and Mg 92 mg kg^{-1}. The amounts of applied doses of macronutrients (N, P, K, Mg) were balanced taking into account the amount added with examined sulfates so that they were even and the same in all pots.

Table 1. Physical-chemical soil properties before the experiment.

Agronomic Category/Soil Texture	pH	C_{total}	N_{total}	S_{total}	P	K	Mg	$S\text{-}SO_4$
					Soluble Forms			
	1M KCl dm^{-3}	g kg^{-1} soil		mg kg^{-1}	mg kg^{-1} Soil			
Light/Loamy sand	3.9	6.18	0.57	115	74	78	13	6.82

2.2. Schedule of Experience

Two sulfur doses of 60 and 120 mg S kg^{-1} soil were applied in the form of elemental sulfur (S^0), sulfates (K_2SO_4, $MgSO_4$, $(NH_4)_2SO_4$), and a mixture of elemental sulfur and sulfates. The following treatments were established: no S—control soil without sulfur fertilization; S^0—elemental sulfur application in the form of Wigor S fertilizer ("Siarkopol" Tarnobrzeg Co. Ltd., Poland); S^0 + K_2SO_4; K_2SO_4 alone; S^0 + $MgSO_4$; $MgSO_4$; S^0 + $(NH_4)_2SO_4$; $(NH_4)_2SO_4$). Each treatment included four repetitions (pots).

2.3. Cultivated Plant and Its Vegetation

In the pot experiment, perennial ryegrass (*Lolium perenne* L. variety Solen) was cultivated. About 100 seeds (0.3 g) of perennial ryegrass were sown in 20 holes and covered with ~0.5 cm of soil. Soil moisture was maintained at 60% water-holding capacity by adding deionized water. Plants were grown in a growth chamber under controlled conditions (photoperiod 16 h/8 h light/dark and temperature 26–28 °C/16–18 °C day/night). The grass was cut after 30, 45, and 60 days of vegetation.

2.4. Sample Preparation and Methods for Chemical Analysis

Representative soil samples were taken before and after vegetation experiments. Soil pH was determined by the potentiometric method using 1 mol dm^{-3} KCl, the total C and N content by analyzer (LECO) [28], total S content by the Butters–Chenery method [29], sulfate(VI) content by the Bardsley and Lancaster method [30], soluble forms of phosphorus and potassium by the Egner–Riehm method [31] and magnesium by the method in [32].

Plants were collected after each grass cutting and the fresh mass of ryegrass was determined. Then, plants were dried at 105 °C for one hour (to kill plant tissues and avoid dry matter loss through respiration) and then at 60 °C to constant weight. Total nitrogen level was determined by the combustion method [28], and total S content by the Butters–Chenery method [29].

2.5. Statistical Analysis

All results obtained were subjected to one-way analysis of variance. Prior to performing the analysis of variance, a test for homogeneity of variance within groups was performed using Levene's test and the Shapiro–Wilk test of variables' conformity to normal distribution. The significance of differences between the averages was assessed using the Tukey's post hoc test with a significance level of $p < 0.05$. For all statistical analyses, the statistical program R was used [33].

3. Results and Discussion

3.1. Fresh and Dry Mass

Under the influence of 60 mg sulfur, the highest fresh and dry total mass was found for ammonium sulfate, and the observed increases were 94 and 139%, respectively, compared to the control (without S fertilization) (Figure 1). Similar increases were observed for the mixtures of elemental sulfur and sulfate (S^0 + K_2SO_4, S^0 + $MgSO_4$). Remaining S fertilizers also resulted in a statistically significant stimulation of ryegrass growth but not so spectacular. Elemental sulfur alone increased the total fresh and dry mass of ryegrass by 25 and 33%, respectively.

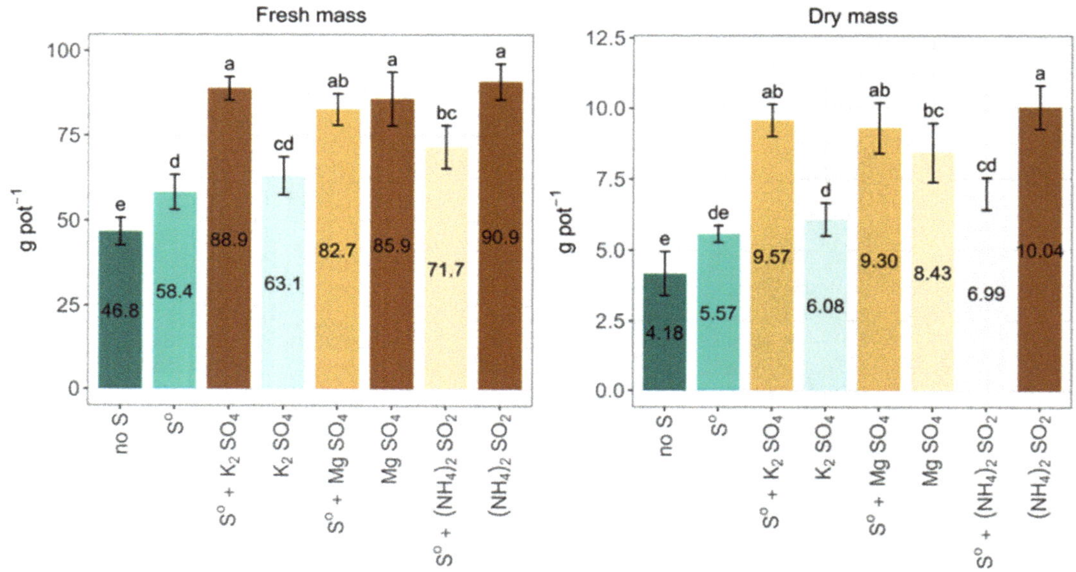

Figure 1. Total fresh and dry mass of ryegrass depending on sulfur fertilizer at a rate of S 60 mg/kg soil. Values labeled with the same letter are not significantly different ($p < 0.05$).

The response of ryegrass to the higher dose of sulfur (120 mg) was similar to the response to the lower dose and the least effective fertilizers were elemental sulfur and a combination of S^0 and $(NH_4)_2SO_4$ (Figure 2). These results show that ryegrass reacts very well to sulfur fertilization and a mixture of S^0 and sulfate gives significantly better results than the use of elemental sulfur alone and comparable results with the use of sulfates alone.

Figure 3 shows the dry matter yield of ryegrass obtained on the three harvest dates depending on the form of sulfur fertilizer used and the dosage rate of this element. In all the terms of ryegrass harvest, the dry matter yield obtained on fertilizers containing S^0 + S-SO_4 was similar to the objects where only sulfate forms were fertilized. The exception was the object with S^0 + $(NH_4)_2SO_4$ at the first harvest date, where significantly lower yields of ryegrass were obtained. As the vegetation period lengthened, the differences in ryegrass dry matter yields between the fertilizers applied decreased.

In order to visualize the differences between the fertilizers used, they were grouped according to the form of sulfur they contained (Figure 4). In all harvest dates, the combination of the S^0 + S-SO_4 form allowed dry matter yields to be obtained that did not differ from those obtained when fertilizing with the sulfate form alone.

Dry matter yields of ryegrass were also compared in relation to the sulfur doses applied (Figure 5). It was found that there were no significant differences between the sulfur doses used, either in the total dry matter yield or in the yields obtained at different harvest periods.

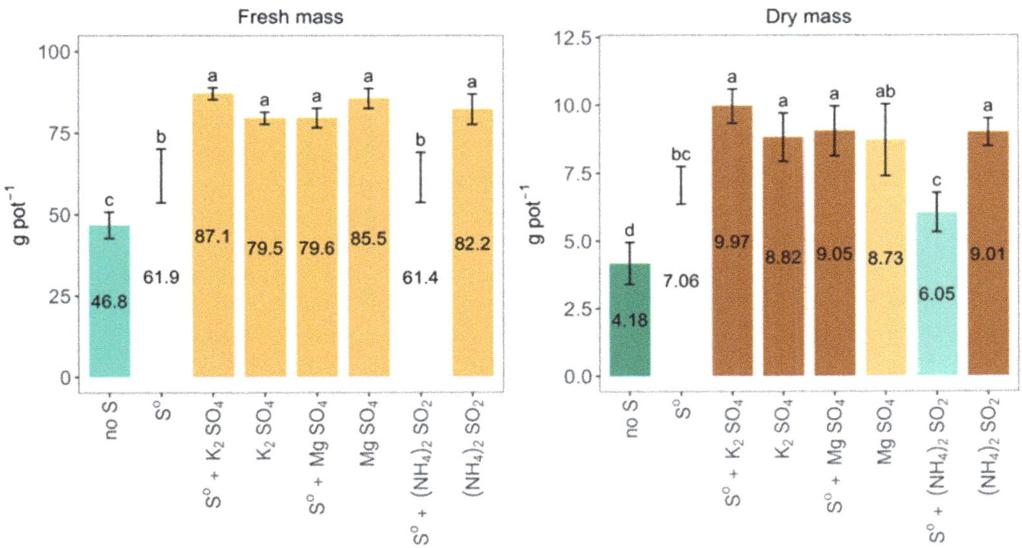

Figure 2. Total fresh and dry mass of ryegrass depending on S fertilizer at a rate of S 120 mg/kg soil. Values labeled with the same letter are not significantly different ($p < 0.05$).

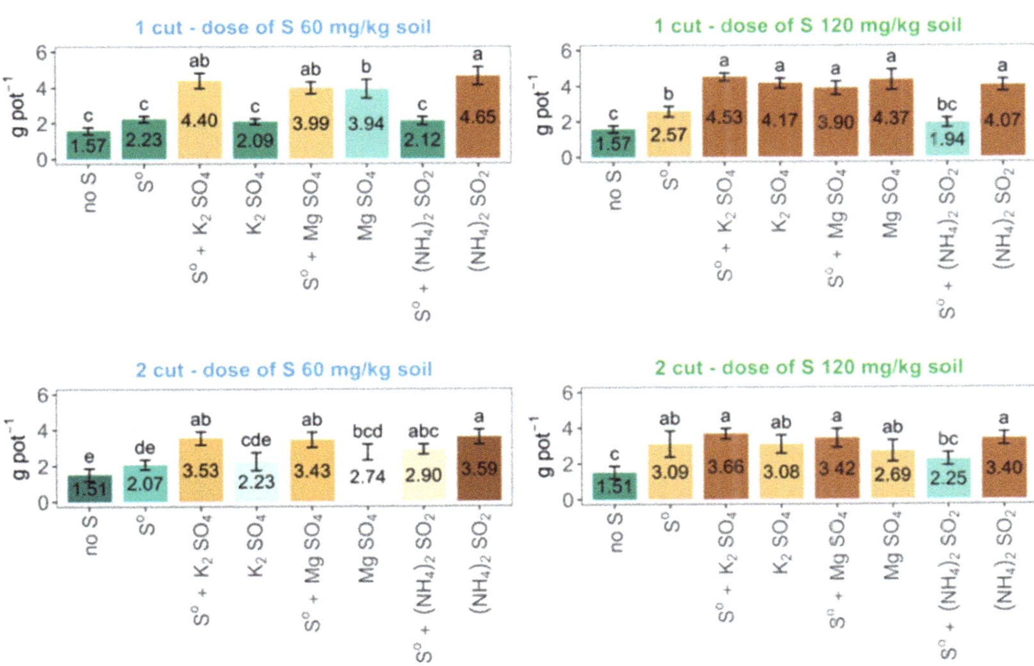

Figure 3. *Cont.*

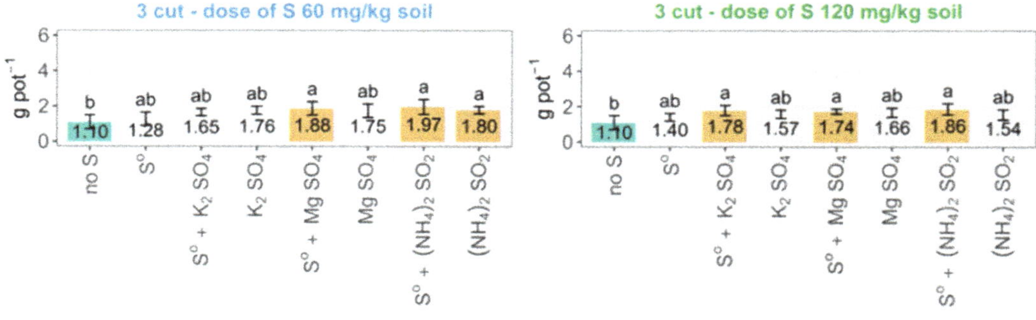

Figure 3. Dry mass of ryegrass depending on S fertilizer. Values labeled with the same letter are not significantly different ($p < 0.05$).

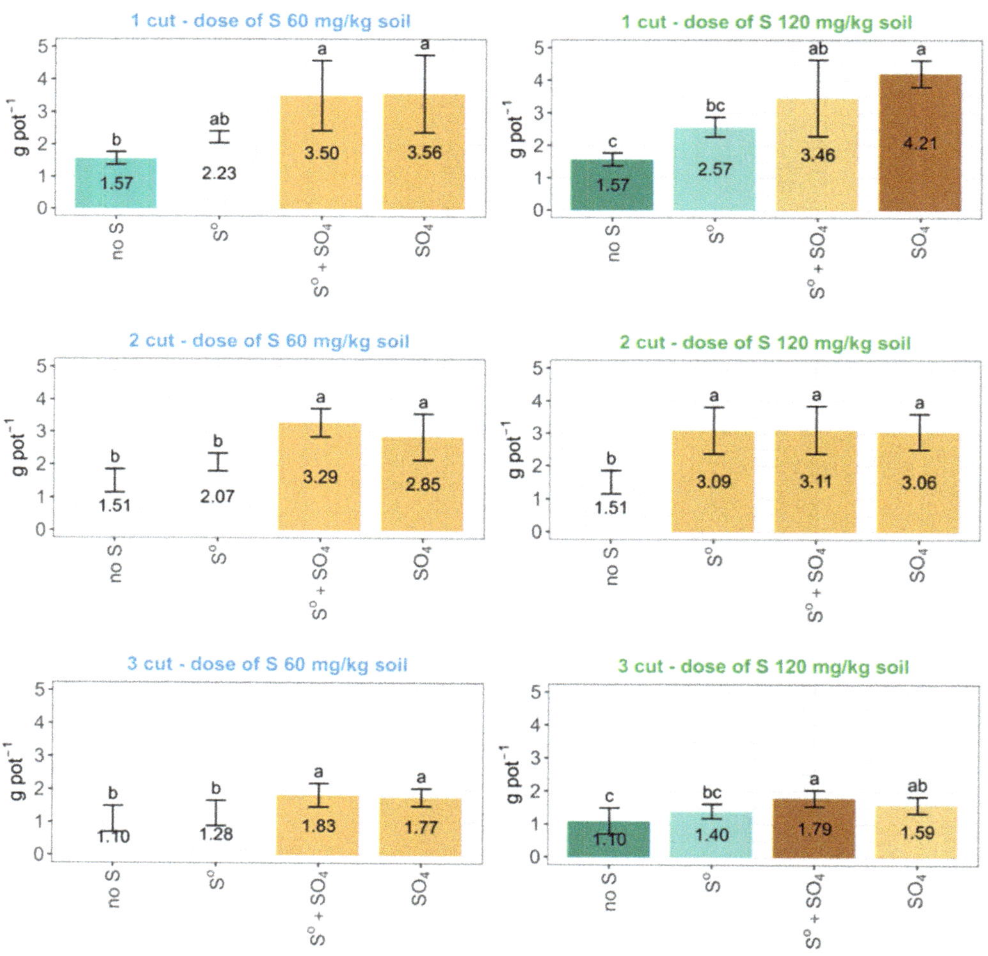

Figure 4. Comparison of dry mass of ryegrass according to forms of S. Values labeled with the same letter are not significantly different ($p < 0.05$).

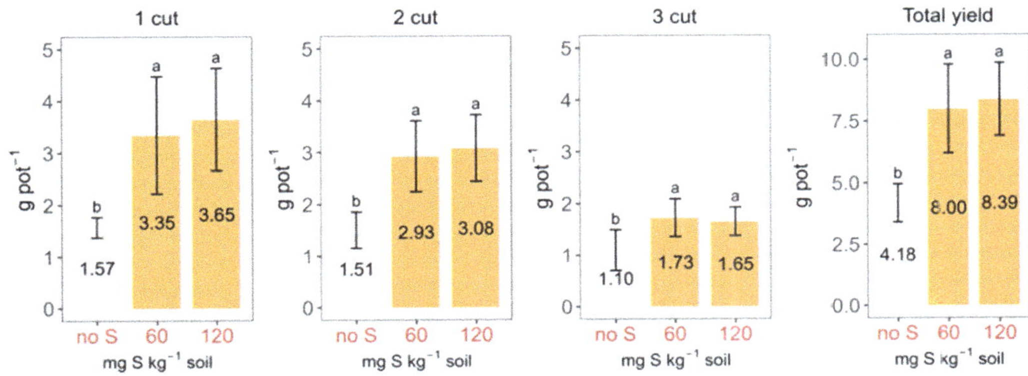

Figure 5. Dry mass of ryegrass in relation to S dose. Values labeled with the same letter are not significantly different ($p < 0.05$).

Degryse et al. [34] claim that both elemental sulfur and S^0- enriched sulfate fertilizers sustain plant sulfur requirements over a longer period than SO_4-S alone. Of the three swaths made in our experiment, the last one was significantly smaller than the earlier ones. This may have been due to the depletion of mineral nutrients in the soil as a result of their intensive uptake by the fast-growing grass. In the early stages of growth, the mineral requirements of plants are very high. Taube et al. [35] also found a varying response of successive regrowths of ryegrass to sulfur and nitrogen fertilization, with the greatest effect in the first regrowth. The results suggest that improved sulfur availability as a result of fertilization increases plant vigor, improves plant metabolism and photosynthetic activity, and ultimately results in better growth, higher stress tolerance, and improved nitrogen use efficiency. Our results are consistent with other studies that report a significant yield response of different crops to sulfur fertilization [25,27,34,36]. However, there are also reports indicating that plant growth response to sulfur addition is not always so positive and unequivocal [17,37]. Under field conditions, the response of plants to sulfur application is complex (among other immobilization, mineralization, leaching processes may have great impact) and, despite numerous studies, the exact requirements for sulfur fertilization are not known [34].

3.2. Nitrogen Content and Uptake

Nitrogen is one of the most important yield forming factors in agriculture. Literature data indicate that plant nitrogen content is significantly modified by the plant-available sulfur in the soil [27,37]. The results showed that the nitrogen content varied from one swath to the next, with the smallest differences (mostly statistically insignificant) occurring in the first swath (Table 2).

Under all applied treatments (form and dose of sulfur), the N concentration in the plants from the first swath ranged from 40.1 to 44.7 g kg^{-1} dry matter. Plants from the first and second swaths contained significantly more nitrogen compared to the third swath. In the latter, under sulfur fertilization conditions, N concentration ranged from 19 g kg^{-1} dry matter (for 120 mg S^0 + K_2SO_4) to 36.8 g kg^{-1} dry matter (for 120 mg S^0) and the response to sulfur fertilization was similar for both sulfur doses. In plants not fertilized with sulfur, N concentrations were 42, 55.3, and 45.9 g kg^{-1} dry mass in the first, second, and third swaths, respectively. Thus, nitrogen concentration in plants grown without sulfur addition was significantly higher than in plants fertilized with sulfur. The observed decrease was probably due to a dilution effect caused by the strong stimulation of ryegrass growth by sulfur fertilizers. As previously mentioned, the application of sulfur resulted in a significant increase in fresh and dry mass of ryegrass shoots (Figures 1 and 2). A decrease in the N content under sulfur fertilization was also observed in other plants [38]. Under

S^0 treatment, the average N content of the three swaths was similar to that recorded in the unfertilized plants and not significantly different from the sulfate-fertilized plants.

Table 2. Nitrogen content in perennial ryegrass.

Treatments	N Content (g kg^{-1} Dry Mass)							
	1 Cut		2 Cut		3 Cut		Mean in Cuts	
Sulfur dose 60 mg kg^{-1} soil								
no S	42.0	*ab*	55.3	*a*	45.9	*a*	47.7	*a*
S^0	42.5	*ab*	54.8	*a*	34.5	*b*	43.9	*ab*
$S^0 + K_2SO_4$	43.1	*ab*	39.1	*d*	19.8	*d*	34.0	*e*
K_2SO_4	42.4	*ab*	50.1	*abc*	29.8	*b*	40.8	*bc*
$S^0 + MgSO_4$	40.1	*b*	45.3	*bcd*	20.2	*cd*	35.2	*de*
$MgSO_4$	43.9	*ab*	53.3	*a*	29.0	*bc*	42.1	*bc*
$S^0 + (NH_4)_2SO_4$	45.9	*a*	52.0	*ab*	29.5	*b*	42.5	*bc*
$(NH_4)_2SO_4$	42.8	*ab*	43.5	*cd*	30.4	*b*	38.9	*cd*
Comparison of sulfur forms for dose 60 mg kg^{-1} soil								
no S	42.0	*a*	55.3	*a*	45.9	*a*	47.7	*a*
S^0	42.5	*a*	54.8	*a*	34.5	*b*	43.9	*ab*
$S^0 + SO_4$	43.0	*a*	45.5	*b*	23.2	*c*	37.2	*c*
SO_4	43.0	*a*	49.0	*ab*	29.7	*b*	40.6	*b*
Sulfur dose 120 mg kg^{-1} soil								
no S	42.0	*a*	55.3	*a*	45.9	*a*	47.7	*a*
S^0	41.5	*a*	50.6	*a*	36.8	*b*	43.0	*ab*
$S^0 + K_2SO_4$	40.9	*a*	33.3	*c*	19.0	*d*	31.1	*c*
K_2SO_4	42.5	*a*	50.6	*a*	28.2	*c*	40.4	*b*
$S^0 + MgSO_4$	40.6	*a*	37.5	*bc*	21.7	*cd*	33.3	*c*
$MgSO_4$	43.3	*a*	46.8	*ab*	28.2	*c*	39.4	*b*
$S^0 + (NH_4)_2SO_4$	44.7	*a*	50.1	*a*	36.4	*b*	43.7	*ab*
$(NH_4)_2SO_4$	44.5	*a*	48.1	*a*	29.5	*bc*	40.7	*b*
Comparison of sulfur forms for dose 120 mg kg^{-1} soil								
no S	42.0	*a*	55.3	*a*	45.9	*a*	47.7	*a*
S^0	41.5	*a*	50.6	*a*	36.8	*ab*	43.0	*ab*
S^0+SO_4	42.1	*a*	40.3	*b*	25.7	*c*	36.0	*c*
SO_4	43.4	*a*	48.5	*a*	28.6	*bc*	40.2	*b*
Comparison of sulfur doses: 60 and 120 mg kg^{-1} soil								
no S	42.0	*a*	55.3	*a*	45.9	*a*	47.7	*a*
60 mg kg^{-1} soil	43.0	*a*	48.3	*ab*	27.6	*b*	39.6	*b*
120 mg kg^{-1} soil	42.6	*a*	45.3	*b*	28.5	*b*	38.8	*b*

Values labeled with the same letter are not significantly different ($p < 0.05$).

The uptake of nitrogen by plants from individual swaths showed a similar pattern to that observed in nitrogen concentration (Table 3). The lowest nitrogen uptake was recorded in the third swath, and this amount was several times lower than in the first swath.

This downward effect was probably due to the depletion of readily available nitrogen in the soil. The highest total uptake (sum of three swaths) was recorded with $(NH_4)_2SO_4$ application (both S doses) and the lowest with no sulfur fertilization. Sulfur fertilization markedly increased nitrogen uptake by ryegrass plants, which in turn contributed to a significant stimulation of plant growth. In general, the effect of both sulfur doses was similar, with the exception of the S^0 fertilization where the total nitrogen uptake was significantly greater under the higher S dose, and counted 303 mg pot^{-1} against 245 for the lower dose. Reports on the interaction between N and S suggest that they affect each other, from the soil uptake, transport in the plant, and assimilation in cells [14]. Sulfur-deficient crops utilize nitrogen inefficiently, which results in increased losses of nitrogen to the environment [39,40].

Table 3. Nitrogen uptake of perennial ryegrass.

Treatments	N Uptake (mg pot^{-1})							
	1 Cut		2 Cut		3 Cut		Total Uptake	
Sulfur dose 60 mg kg^{-1} soil								
no S	66.0	*c*	82.8	*d*	50.4	*a*	199	*d*
S^0	94.6	*c*	113	*bcd*	44.2	*a*	245	*cd*
$S^0 + K_2SO_4$	189	*ab*	138	*abc*	32.5	*a*	325	*b*
K_2SO_4	88.6	*c*	110	*cd*	53.3	*a*	248	*cd*
$S^0 + MgSO_4$	160	*b*	155	*a*	36.6	*a*	327	*ab*
$MgSO_4$	174	*ab*	147	*abc*	51.1	*a*	356	*ab*
$S^0 + (NH_4)_2SO_4$	97.3	*c*	151	*ab*	56.5	*a*	296	*bc*
$(NH_4)_2SO_4$	199	*a*	156	*a*	54.3	*a*	339	*a*
Comparison of sulfur forms for dose 60 mg kg^{-1} soil								
no S	66.0	*c*	82.8	*c*	50.4	*a*	199	*b*
S^0	94.6	*bc*	113	*bc*	44.2	*a*	252	*b*
$S^0 + SO_4$	149	*ab*	148	*a*	41.9	*a*	339	*a*
SO_4	154	*a*	138	*ab*	52.9	*a*	344	*a*
Sulfur dose 120 mg kg^{-1} soil								
no S	66.0	*d*	82.8	*c*	50.4	*ab*	199	*d*
S^0	107	*c*	157	*ab*	51.3	*ab*	303	*bc*
$S^0 + K_2SO_4$	185	*ab*	121	*abc*	33.6	*b*	308	*abc*
K_2SO_4	178	*ab*	154	*ab*	44.1	*b*	355	*ab*
$S^0 + MgSO_4$	158	*b*	125	*abc*	37.5	*b*	299	*bc*
$MgSO_4$	189	*a*	126	*abc*	46.2	*ab*	342	*ab*
$S^0 + (NH_4)_2SO_4$	86.8	*cd*	113	*bc*	67.7	*a*	265	*c*
$(NH_4)_2SO_4$	181	*ab*	164	*a*	45.6	*ab*	367	*a*
Comparison of sulfur forms for dose 120 mg kg^{-1} soil								
no S	66.0	*c*	82.8	*c*	50.4	*a*	199	*c*
S^0	107	*bc*	157	*a*	51.3	*a*	315	*b*
$S^0 + SO_4$	143	*b*	120	*b*	46.3	*a*	309	*b*
SO_4	183	*a*	148	*a*	45.3	*a*	376	*a*
Comparison of sulfur doses: 60 and 120 mg kg^{-1} soil								
no S	66	*b*	83	*b*	50.4	*a*	199	*b*
60 mg kg^{-1} soil	143	*a*	139	*a*	46.9	*a*	329	*a*
120 mg kg^{-1} soil	155	*a*	137	*a*	46.6	*a*	339	*a*

Values labeled with the same letter are not significantly different ($p < 0.05$).

3.3. Sulfur Content and Uptake

Overall, as expected, sulfur fertilization resulted in a better supply of sulfur to the ryegrass, and in some cases, the effect was more pronounced with a higher dose of sulfur. This was particularly evident after the application of fertilizers containing only sulfates. Under these conditions, the average sulfur content of the three swaths was 5.66 and 7.0 g kg^{-1} dry mass for the 60 and 120 mg S kg^{-1} soil doses, respectively (Table 4).

The highest sulfur concentration (as an average of three swaths) was recorded for plants fertilized with potassium sulfate (120 mg kg^{-1} soil) and this value (8.02 g kg^{-1} dry mass) was 70% higher than that of unfertilized plants. Fertilizing the plants with the mixture of elemental sulfur with potassium or magnesium sulfate did not increase the sulfur content in ryegrass shoots. Some positive effects were found only in the first swath. However, these were sufficient amounts for proper growth and function. Our results also indicate that ryegrass is capable of maintaining relatively high sulfur concentrations in shoots (4.74 g kg^{-1} dry mass) despite low soil content. The recorded concentration is within the range expected for most crops (0.2 to 0.5% dry matter) [9]. However, a negative consequence was a strong inhibition of plant growth compared to sulfur-fertilized plants. In terms of sulfur requirements, multiple-cut grasses and Brassica crops are more prone to sulfur deficiency than other plants [20]. The sulfur demand is dependent not only on the plant species but also on its developmental stage [41]. During vegetative growth, sulfur uptake is optimized for growth. All plants regulate sulfur uptake and are able to adjust to very variable (low and high soil sulfur levels) and time-varying sulfur supply [42]. The uptake of sulfur by unfertilized plants remained at a similar level in the next three swaths (6.41, 7.27, and 5.71 mg pot^{-1}, respectively; Table 5). As expected, sulfur-fertilized plants took up significantly larger amounts of sulfur and, in general, the values were higher at the increased S dose. Analyzing the dynamics of sulfur uptake by S-fertilized plants, it can be stated that it was greatest during the first two periods of growth (Table 5).

Table 4. Sulfur content in perennial ryegrass.

Treatments	S Content (g kg^{-1} Dry Mass) in Perennial Ryegrass							
	1 Cut		2 Cut		3 Cut		Mean in Cuts	
Sulfur dose 60 mg kg^{-1} soil								
no S	4.09	*b*	4.84	*d*	5.28	*bc*	4.74	*d*
S^0	4.88	*a*	5.62	*bc*	6.27	*ab*	5.59	*abc*
$S^0 + K_2SO_4$	4.72	*ab*	5.40	*bcd*	4.16	*cd*	4.76	*d*
K_2SO_4	4.94	*a*	6.01	*ab*	7.33	*a*	6.09	*a*
$S^0 + MgSO_4$	4.79	*a*	5.58	*bc*	3.96	*d*	4.78	*d*
$MgSO_4$	4.55	*ab*	5.40	*bcd*	5.45	*b*	5.13	*cd*
$S^0 + (NH_4)_2SO_4$	4.93	*a*	5.25	*cd*	5.86	*b*	5.35	*bc*
$(NH_4)_2SO_4$	4.93	*a*	6.30	*a*	6.04	*b*	5.76	*ab*
Comparison of sulfur forms for dose 60 mg kg^{-1} soil								
no S	4.09	*b*	4.84	*c*	5.28	*ab*	4.74	*b*
S^0	4.88	*a*	5.62	*ab*	6.27	*a*	5.59	*a*
$S^0 + SO_4$	4.81	*a*	5.41	*b*	4.66	*b*	4.96	*b*
SO_4	4.80	*a*	5.90	*a*	6.27	*a*	5.66	*a*
Sulfur dose 120 mg kg^{-1} soil								
no S	4.09	*e*	4.84	*cd*	5.28	*de*	4.74	*d*
S^0	5.37	*c*	5.53	*bc*	6.77	*c*	5.89	*c*
$S^0 + K_2SO_4$	4.96	*cd*	4.50	*d*	4.71	*e*	4.72	*d*
K_2SO_4	6.88	*a*	6.00	*b*	11.2	*a*	8.02	*a*
$S^0 + MgSO_4$	4.78	*d*	5.34	*bc*	4.43	*e*	4.85	*d*
$MgSO_4$	5.15	*cd*	5.77	*b*	5.63	*cde*	5.52	*c*
$S^0 + (NH_4)_2SO_4$	5.48	*c*	5.30	*bcd*	6.06	*cd*	5.61	*c*
$(NH_4)_2SO_4$	6.21	*b*	7.58	*a*	8.56	*b*	7.45	*b*
Comparison of sulfur forms for dose 120 mg kg^{-1} soil								
no S	4.09	*c*	4.84	*b*	5.28	*b*	4.74	*b*
S^0	5.37	*ab*	5.53	*ab*	6.77	*ab*	5.89	*b*
$S^0 + SO_4$	5.07	*b*	5.04	*b*	5.07	*b*	5.06	*b*
SO_4	6.08	*a*	6.45	*a*	8.46	*a*	7.00	*a*
Comparison of Sulfur doses: 60 and 120 mg kg^{-1} soil								
no S	4.09	*c*	4.84	*b*	5.28	*b*	4.74	*b*
60 mg kg^{-1} soil	4.82	*b*	5.65	*ab*	5.58	*b*	5.35	*b*
120 mg kg^{-1} soil	5.55	*a*	5.72	*a*	6.76	*a*	6.01	*a*

Values labeled with the same letter are not significantly different ($p < 0.05$).

For sulfur fertilization at 60 mg S kg^{-1} soil, the highest total uptake (56.4 mg pot^{-1}, almost 3 times higher compared to the unfertilized plants) was observed in plants fertilized with ammonium sulfate. In plants exposed to 120 mg S, the highest sulfur uptake was found in the presence of potassium sulfate and ammonium sulfate.

The assimilation of sulfur and nitrogen are strongly linked, so a proper balance of these nutrients is very important. The N/S ratio is used to diagnose sulfur deficiency in plants [43–45]. Empirically, it has been established that for every 15 parts of N in a protein, there is 1 part of S. This means that the N/S ratio is fixed within a very narrow range of 15:1. Sulfur deficiency can limit protein synthesis, even if a large amount of N is available. This relationship has very important implications for human and animal nutrition.

Our results showed that for all fertilization variants, the N/S ratio was below the critical value of 15:1 (Table 6). For the averages of the three swaths, the ratio ranged from 5.82 to 10.3 and this highest value relates to non-fertilized plants. It may be considered that the recorded N/S ratios (all below 15:1) were satisfactory. Similar trends were observed for both sulfur doses. The reduced N/S ratio indicates that the plants did not suffer from sulfur deficiency (based on the value of the N/S ratio), and that plants not fertilized with sulfur were able to take up sulfur despite the low soil content. However, the growth of plants was significantly affected.

Table 5. Sulfur uptake of perennial ryegrass.

Treatments	S Uptake (mg pot^{-1})							
	1 Cut		2 Cut		3 Cut		Total Uptake	
Sulfur dose 60 mg kg^{-1} soil								
no S	6.41	*d*	7.27	*d*	5.71	*e*	19.4	*e*
S^0	10.9	*c*	11.6	*cd*	7.90	*cde*	30.4	*d*
$S^0 + K_2SO_4$	20.7	*ab*	19.0	*ab*	6.85	*de*	46.5	*b*
K_2SO_4	10.4	*cd*	13.5	*c*	12.9	*a*	36.7	*cd*
$S^0 + MgSO_4$	19.2	*ab*	19.2	*ab*	7.21	*de*	45.5	*b*
$MgSO_4$	17.9	*b*	14.8	*bc*	9.40	*bcd*	42.1	*bc*
$S^0 + (NH_4)_2SO_4$	10.4	*c*	15.2	*bc*	11.3	*ab*	36.9	*cd*
$(NH_4)_2SO_4$	22.9	*a*	22.6	*a*	10.9	*abc*	56.4	*a*
Comparison of sulfur forms for dose 60 mg kg^{-1} soil								
no S	6.41	*b*	7.27	*b*	5.71	*b*	19.4	*b*
S^0	10.9	*ab*	11.6	*b*	7.90	*b*	30.4	*b*
$S^0 + SO_4$	16.8	*a*	17.8	*a*	8.46	*b*	43.0	*a*
SO_4	17.1	*a*	17.0	*a*	11.0	*a*	45.1	*a*
Sulfur dose 120 mg kg^{-1} soil								
no S	6.41	*e*	7.27	*d*	5.71	*d*	19.4	*d*
S^0	13.8	*d*	17.0	*bc*	9.38	*bcd*	40.2	*bc*
$S^0 + K_2SO_4$	22.4	*bc*	16.4	*bc*	8.49	*cd*	47.4	*b*
K_2SO_4	28.7	*a*	18.5	*b*	17.5	*a*	64.8	*a*
$S^0 + MgSO_4$	18.5	*c*	18.0	*b*	7.67	*cd*	44.2	*b*
$MgSO_4$	22.5	*bc*	15.5	*bc*	9.21	*bcd*	47.3	*b*
$S^0 + (NH_4)_2SO_4$	10.6	*de*	11.9	*cd*	11.3	*bc*	33.8	*c*
$(NH_4)_2SO_4$	25.3	*ab*	25.7	*a*	13.4	*ab*	64.4	*a*
Comparison of sulfur forms for dose 120 mg kg^{-1} soil								
no S	6.41	*c*	7.27	*c*	5.71	*b*	19.4	*c*
S^0	13.8	*b*	17.0	*ab*	9.38	*ab*	40.2	*b*
$S^0 + SO_4$	17.2	*b*	15.4	*b*	9.15	*b*	41.8	*b*
SO_4	25.5	*a*	19.9	*a*	13.4	*a*	58.8	*a*
Comparison of sulfur doses: 60 and 120 mg kg^{-1} soil								
no S	6.41	*c*	7.27	*b*	5.71	*b*	19.4	*c*
60 mg kg^{-1} soil	16.1	*b*	16.5	*a*	9.49	*a*	42.1	*b*
120 mg kg^{-1} soil	20.3	*a*	17.6	*a*	11.0	*a*	48.9	*a*

Values labeled with the same letter are not significantly different ($p < 0.05$).

The lowest N/S ratio value was found in the last swath, probably due to a reduced pool of plant-available nitrogen in the soil and its limited uptake (Table 5). In the third swath, N/S values ranged between 2.52 (K_2SO_4 in a higher dose) and 8.74 (unfertilized plants). Analyzing the results, it can be concluded that an N/S ratio below 9.0 provides an optimum yield of perennial ryegrass. Ref. [27], in determining the potential of sulfur fertilization to increase grassland yields and N use efficiency, found that the N/S ratio was variable during the growing season. Its values ranged from 5:1 to 20:1 depending on the harvest date and the applied sulfur fertilization. Ref. [46] claims that S supplementation is more necessary at higher N doses.

Table 6. N:S ratio in perennial ryegrass.

Treatments	N:S in Perennial Ryegrass							
	1 Cut		2 Cut		3 Cut		Mean in Cuts	
Sulfur dose 60 mg kg^{-1} soil								
no S	10.3	*a*	11.4	*a*	8.74	*a*	10.3	*a*
S^0	8.76	*bc*	9.78	*b*	5.52	*b*	8.30	*bc*
$S^0 + K_2SO_4$	9.15	*abc*	7.26	*c*	4.81	*b*	7.73	*cd*
K_2SO_4	8.62	*bc*	8.39	*bc*	4.10	*b*	6.92	*d*
$S^0 + MgSO_4$	8.42	*c*	8.11	*c*	5.10	*b*	7.75	*cd*
$MgSO_4$	9.66	*ab*	9.91	*ab*	5.42	*b*	8.81	*b*
$S^0 + (NH_4)_2SO_4$	9.32	*abc*	9.92	*ab*	5.02	*b*	8.25	*bc*
$(NH_4)_2SO_4$	8.69	*bc*	6.91	*c*	5.01	*b*	7.25	*d*
Comparison of sulfur forms for dose 60 mg kg^{-1} soil								
no S	10.3	*a*	11.4	*a*	8.74	*a*	10.3	*a*
S^0	8.76	*b*	9.78	*ab*	5.52	*b*	8.30	*b*
$S^0 + SO_4$	8.96	*b*	8.43	*b*	4.98	*b*	7.91	*b*
SO_4	8.99	*b*	8.40	*b*	4.84	*b*	7.66	*b*
Sulfur dose 120 mg kg^{-1} soil								
no S	10.3	*a*	11.4	*a*	8.74	*a*	10.3	*a*
S^0	7.75	*bc*	9.18	*bc*	5.49	*bc*	7.81	*b*
$S^0 + K_2SO_4$	8.26	*bc*	7.38	*cde*	4.10	*cd*	7.18	*b*
K_2SO_4	6.17	*d*	8.52	*bcd*	2.52	*e*	5.82	*c*
$S^0 + MgSO_4$	8.53	*b*	6.99	*de*	4.97	*bc*	7.27	*b*
$MgSO_4$	8.41	*b*	8.15	*bcde*	5.02	*bc*	7.66	*b*
$S^0 + (NH_4)_2SO_4$	8.19	*bc*	9.47	*b*	6.00	*b*	7.91	*b*
$(NH_4)_2SO_4$	7.19	*cd*	6.36	*e*	3.47	*de*	6.05	*c*
Comparison of sulfur forms for dose 120 mg kg^{-1} soil								
no S	10.3	*a*	11.4	*a*	8.74	*a*	10.3	*a*
S^0	7.75	*bc*	9.18	*b*	5.49	*b*	7.81	*b*
$S^0 + SO_4$	8.33	*b*	7.95	*b*	5.02	*b*	7.45	*b*
SO_4	7.26	*c*	7.68	*b*	3.67	*c*	6.51	*c*
Comparison of sulfur doses: 60 and 120 mg kg^{-1} soil								
no S^0	10.3	*a*	11.4	*a*	8.74	*a*	10.3	*a*
60 mg kg^{-1} soil	8.95	*b*	8.61	*b*	5.00	*b*	7.52	*b*
120 mg kg^{-1} soil	7.79	*c*	8.01	*b*	4.51	*b*	6.77	*c*

Values labeled with the same letter are not significantly different ($p < 0.05$).

3.4. Soil Properties after Experiment

Fertilization with sulfur in both elemental and sulfate forms significantly decreased the soil reaction as compared to the object where this nutrient was not applied (Table 7). It was also found that the size of the sulfur dose had a significant effect on the decrease in soil reaction. It is often emphasized in the literature that the application of sulfur can affect the acidity of soils, especially when applied at higher doses [47–50].

In our study, sulfur fertilization also contributed to a significant decrease in both soil carbon and nitrogen content. The greatest decrease in soil carbon and nitrogen content was observed when a higher dose of sulfur was applied in its elemental form. The availability of sulfur to plants is related to the transformation of organic matter in the soil, which mainly depends on soil microbial activity. In turn, soil microbial activity is dependent on temperature, moisture, pH, and substrate availability [51–53]. Ikoyi et al. [54] found that short-term sulfate fertilization promotes perennial ryegrass growth by outweighing negative feedback from some of the soil biota, while Magnucka et al. [55] found that the application of sulfur fertilizers with the mineral NPKMg promotes soil fertility due to aggregate stabilization and a decrease in water-soluble organic compounds. Nitrogen mineralization from soil organic matter can occur more rapidly compared to the release of sulfur [56,57].

Table 7. Physical-chemical soil properties after experiment.

Treatments	pH		$C_{organic}$		N_{total}		S_{total}		S-SO_4		S-SO_4/S_{total}	
	1M KCl dm^{-3}		(g kg^{-1} Soil)				(mg kg^{-1} Soil)				%	
Sulfur dose 60 mg kg^{-1} soil												
no S	6.59	a	8.58	a	0.610	a	94	e	3.99	f	4.23	c
S^0	6.47	b	6.83	bc	0.544	abc	106	d	12.7	a	11.9	a
S^0 + K_2SO_4	6.48	b	7.27	b	0.522	bcd	136	a	10.4	bc	7.65	b
K_2SO_4	6.43	b	6.86	bc	0.468	d	104	d	6.02	ef	6.35	b
S^0 + $MgSO_4$	6.43	b	5.84	c	0.568	ab	126	ab	7.65	de	6.06	bc
$MgSO_4$	6.48	b	7.83	ab	0.532	bcd	110	cd	7.06	de	6.43	b
S^0 + $NH_4)_2SO_4$	6.38	b	6.96	b	0.496	cd	117	bc	11.7	ab	10.1	a
$(NH_4)_2SO_4$	6.44	b	7.74	ab	0.554	abc	133	a	8.58	cd	6.47	b
Comparison of sulfur forms for dose 60 mg kg^{-1} soil												
no S	6.59	a	8.58	a	0.610	a	94	c	3.99	d	4.23	d
S^0	6.47	b	6.83	bc	0.544	ab	106	b	12.7	a	11.9	a
S^0 + SO_4	6.43	b	6.69	c	0.529	b	126	a	9.92	b	7.93	b
SO_4	6.45	b	7.48	b	0.518	b	116	ab	7.22	c	6.42	c
Sulfur dose 120 mg kg^{-1} soil												
no S	6.59	a	8.58	a	0.610	a	94	e	3.99	c	4.23	e
S^0	6.37	b	5.39	c	0.466	c	147	cd	23.0	a	15.7	a
S^0 + K_2SO_4	6.36	b	7.62	ab	0.518	bc	158	b	21.8	a	13.8	ab
K_2SO_4	6.36	b	7.34	b	0.498	bc	144	d	13.5	b	9.37	cd
S^0 + $MgSO_4$	6.36	b	6.84	b	0.536	bc	159	ab	19.7	a	12.4	bc
$MgSO_4$	6.41	b	7.73	ab	0.526	bc	153	bc	10.7	b	7.00	de
S^0 + $(NH_4)_2SO_4$	6.40	b	7.20	b	0.500	bc	146	cd	21.4	a	14.6	ab
$(NH_4)_2SO_4$	6.37	b	7.57	b	0.556	ab	167	a	10.8	b	6.52	de
Comparison of sulfur forms for dose 120 mg kg^{-1} soil												
no S	6.59	a	8.58	a	0.610	a	94	b	3.99	c	4.23	c
S^0	6.37	b	5.39	c	0.466	c	147	a	23.0	a	15.7	a
S^0 + SO_4	6.37	b	7.22	b	0.518	bc	154	a	21.0	a	13.6	a
SO_4	6.38	b	7.55	b	0.527	b	155	a	11.7	b	7.63	b
Comparison of sulfur doses: 60 and 120 mg kg^{-1} soil												
no S^0	6.59	a	8.58	a	0.610	a	94	c	3.99	c	4.23	c
60 mg kg^{-1} soil	6.44	b	7.05	b	0.526	b	119	b	9.15	b	7.78	b
120 mg kg^{-1} soil	6.38	c	7.10	b	0.514	b	153	a	17.3	a	11.3	a

Values labeled with the same letter are not significantly different ($p < 0.05$).

Under the influence of sulfur fertilization with all analyzed forms of sulfur, the content of total sulfur and sulfate sulfur in the soil increased significantly compared to the control object. Similar studies comparing sulfur fertilization in the sulfate and elementary form also found similar relationships in the content of this element in soil [49]. This may be due to the fact that sulfur applied in the sulfate form is directly available to plants and a significantly higher overall uptake of sulfur by ryegrass was observed in comparison with objects fertilized with the elementary form. This confirms that fertilization with elemental forms of sulfur has a greater subsequent effect of sulfur delivery from the soil compared to sulfate forms.

Maintaining the optimal soil abundance of available forms of sulfur can have both agronomic and environmental benefits [27].

Increasing doses of sulfur also had a significant effect on the content of this element in the soil. Both lower and higher doses of sulfur were found to have the highest sulfate content in the soil when fertilized with the elemental form of sulfur, indicating a slower release of the available sulfur form into the soil during plant growth. This was also confirmed by the highest proportion of the total sulfate form in the soil when fertilizing with elemental sulfur. The combined use of elemental sulfur and sulfate sulfur also significantly increased the sulfate sulfur content of the soil compared with the sulfate form of sulfur. It is safer to use the elemental form of sulfur at higher sulfur rates applied to the grassland due to sulfate leaching into the soil profile [58].

4. Conclusions

The results indicate that fertilizers that are a mixture of S^0 and sulfate are as effective as those containing only sulfate in terms of stimulating perennial ryegrass growth. Both applied doses of sulfur were equally effective. Fertilization with all forms of sulfur increased the soil's content of this nutrient, making it possible to compensate for deficiencies of this element in the soil. Economic and environmental considerations indicate that the use of fertilizers that are a mixture of elemental and sulfate sulfur is a good approach in agriculture.

Author Contributions: Conceptualization, G.K., Ł.M. and E.S; methodology, G.K., A.K.-L. and E.S.; investigation, G.K. and E.S.; data curation—compiled and analyzed the results, G.K., E.S. and A.K.-L.; writing—original draft preparation, E.S. and G.K.; writing—review and editing, G.K., A.K.-L. and E.S. All authors have read and agreed to the published version of the manuscript.

Funding: These studies were co-financed by the National Center for Research and Development, Poland, under contract No. POIR.01.01.02-00-0145/16-00.

Data Availability Statement: The data presented in this manuscript are available from the authors upon reasonable request.

Conflicts of Interest: The authors declare no conflict of interest.

References

1. Zenda, T.; Liu, S.; Dong, A.; Duan, H. Revisiting Sulphur—The Once Neglected Nutrient: It's Roles in Plant Growth, Metabolism, Stress Tolerance and Crop Production. *Agriculture* **2021**, *11*, 626. [CrossRef]
2. Kopriva, S.; Malagoli, M.; Takahashi, H. Sulfur Nutrition: Impacts on Plant Development, Metabolism, and Stress Responses. *J. Exp. Bot.* **2019**, *70*, 4069–4073. [CrossRef] [PubMed]
3. Haider, K.; Schäffer, A. Turnover of Nitrogen, Phosphorus and Sulfur in Soils and Sediments. In *Soil Biochemistry*; Science Publishers: Enfield, NH, USA, 2009; pp. 49–75.
4. Messick, D. *The Sulphur Outlook, TFI Fertilizer Outlook Presentations*; The World Sulphur Institute: Tampa, FL, USA, 2013; Available online: http://www.firt.org (accessed on 20 August 2023).
5. Webb, J.; Jephcote, C.; Fraser, A.; Wiltshire, J.; Aston, S.; Rose, R.; Vincent, K.; Roth, B. Do UK Crops and Grassland Require Greater Inputs of Sulphur Fertilizer in Response to Recent and Forecast Reductions in Sulphur Emissions and Deposition? *Soil Use Manag.* **2016**, *32*, 3–16. [CrossRef]
6. Feinberg, A.; Stenke, A.; Peter, T.; Hinckley, E.-L.S.; Driscoll, C.T.; Winkel, L.H.E. Reductions in the Deposition of Sulfur and Selenium to Agricultural Soils Pose Risk of Future Nutrient Deficiencies. *Commun. Earth Environ.* **2021**, *2*, 101. [CrossRef]
7. Hartmann, K.; Lilienthal, H.; Schnug, E. Risk Mapping of Potential Sulphur Deficiency in Agriculture under Actual and Future Climate Scenarios in Germany. In *Effects of Climate Change on Plants: Implications for Agriculture*; Halford, N.G., Jones, H.D., Lawlor, D.W., Eds.; Aspects of Applied Biology; Association of Applied Biologists: Wellesbourne, UK, 2008; Volume 88, pp. 113–120.
8. Engardt, M.; Simpson, D.; Schwikowski, M.; Granat, L. Deposition of Sulphur and Nitrogen in Europe 1900–2050. Model Calculations and Comparison to Historical Observations. *Tellus B Chem. Phys. Meteorol.* **2017**, *69*, 1328945. [CrossRef]
9. Rengel, Z.; Cakmak, I.; White, P.J. *Marschner's Mineral Nutrition of Plants*, 4th ed.; Elsevier Ltd.: Oxford, UK, 2023.
10. Taiz, L.; Zeiger, E.; Møller, I.M.; Murphy, A.S. (Eds.) *Plant Physiology and Development*, 6th ed.; Sinauer Associates, Inc.: Sunderland, MA, USA, 2015.
11. Narayan, O.P.; Kumar, P.; Yadav, B.; Dua, M.; Johri, A.K. Sulfur Nutrition and Its Role in Plant Growth and Development. *Plant Signal. Behav.* **2022**, 2030082. [CrossRef]
12. Castro, V.; Carpena, M.; Fraga-Corral, M.; Lopez-Soria, A.; Garcia-Perez, P.; Barral-Martinez, M.; Perez-Gregorio, R.; Cao, H.; Simal-Gandara, J.; Prieto, M.A. Sulfur-Containing Compounds from Plants. In *Natural Secondary Metabolites: From Nature, Through Science, to Industry*; Carocho, M., Heleno, S.A., Barros, L., Eds.; Springer International Publishing: Cham, Switzerland, 2023; pp. 363–402. ISBN 978-3-031-18587-8.
13. Shah, S.H.; Islam, S.; Mohammad, F. Sulphur as a Dynamic Mineral Element for Plants: A Review. *J. Soil Sci. Plant Nutr.* **2022**, *22*, 2118–2143. [CrossRef]
14. Fageria, V.D. Nutrient Interactions in Crop Plants. *J. Plant Nutr.* **2001**, *24*, 1269–1290. [CrossRef]
15. Skwierawska, M.; Skwierawska, M.; Skwierawski, A.; Benedycka, Z.; Jankowski, K. Sulphur as a Fertiliser Component Determining Crop Yield and Quality. *J. Elem.* **2016**, *21*, 609–623. [CrossRef]
16. Stevens, R.J.; Watson, C.J. The Response of Grass for Silage to Sulphur Application at 20 Sites in Northern Ireland. *J. Agric. Sci.* **1986**, *107*, 565–571. [CrossRef]
17. Hahtonen, M.; Saarela, I. The Effects of Sulphur Application on Yield, Sulphur Content and N/S-Ratio of Grasses for Silage at Six Sites in Finland. *Acta Agric. Scand. Sect. B-Soil Plant Sci.* **1995**, *45*, 104–111. [CrossRef]

18. Zhao, J.F.; Withers, A.P.J.; Evans, J.E.; Monaghan, J.; Salmon, E.S.; Shewry, R.P.; McGrath, P.S. Sulphur Nutrition: An Important Factor for the Quality of Wheat and Rapeseed. *Soil Sci. Plant Nutr.* **1997**, *43*, 1137–1142. [CrossRef]
19. Grzebisz, W.; Przygocka-Cyna, K. Spring Malt Barley Response to Elemental Sulphur– the Prognostic Value of N and S Concentrations in Malt Barley Leaves. *Plant Soil Environ.* **2007**, *53*, 388–394. [CrossRef]
20. Zhao, F.J.; Tausz, M.; De Kok, L.J. Role of Sulfur for Plant Production in Agricultural and Natural Ecosystems. In *Sulfur Metabolism in Phototrophic Organisms*; Hell, R., Dahl, C., Knaff, D., Leustek, T., Eds.; Springer: Berlin/Heidelberg, Germany, 2008; pp. 417–435.
21. De Bona, F.D.; Monteiro, F.A. Nitrogen and Sulfur Fertilization and Dynamics in a Brazilian Entisol under Pasture. *Soil Sci. Soc. Am. J.* **2010**, *74*, 1248–1258. [CrossRef]
22. Rathore, S.S.; Shekhawat, K.; Kandpal, B.K.; Premi, O.P. Improving Nutrient Use Efficiency in Oilseeds Brassica. In *Nutrient Use Efficiency: From Basics to Advances*; Rakshit, A., Singh, H.B., Sen, A., Eds.; Springer: New Delhi, India, 2015; pp. 317–327, ISBN 978-81-322-2169-2.
23. Singh, D.P.; Gulpadiya, V.K.; Chauhan, R.S.; Singh, S.P. Effect of Sulphur on Productivity, Economics and Nutrient Uptake in Spinach. *Ann. Plant Soil Res.* **2015**, *17*, 29–32.
24. Mathot, M.; Thélier-Huché, L.; Lambert, R. Sulphur and Nitrogen Content as Sulphur Deficiency Indicator for Grasses. *Eur. J. Agron.* **2009**, *30*, 172–176. [CrossRef]
25. Mathot, M.; Mertens, J.; Verlinden, G.; Lambert, R. Positive Effects of Sulphur Fertilisation on Grasslands Yields and Quality in Belgium. *Eur. J. Agron.* **2008**, *28*, 655–658. [CrossRef]
26. Edmeades, D.C.; Thorrold, B.S.; Roberts, A.H.C. The Diagnosis and Correction of Sulfur Deficiency and the Management of Sulfur Requirements in New Zealand Pastures: A Review. *Aust. J. Exp. Agric.* **2005**, *45*, 1205–1223. [CrossRef]
27. Aspel, C.; Murphy, P.N.C.; McLaughlin, M.J.; Forrestal, P.J. Sulfur Fertilization Strategy Affects Grass Yield, Nitrogen Uptake, and Nitrate Leaching: A Field Lysimeter Study#. *J. Plant Nutr. Soil Sci.* **2022**, *185*, 209–220. [CrossRef]
28. FAO. *Standard Operating Procedure for Soil Total Nitrogen-Dumas Dry Combustion Method*; FAO: Rome, Italy, 202[illegible].
29. Butters, B.; Chenery, E.M. A Rapid Method for the Determination of Total Sulphur in Soils and Plants. *Analyst* **1959**, *84*, 239–245. [CrossRef]
30. Bardsley, C.E.; Lancaster, J.D. Determination of Reserve Sulfur and Soluble Sulfates in Soils. *Soil Sci. Soc. Am. J.* **1959**, *24*, 265–268. [CrossRef]
31. Egner, H.; Riehm, H. Doppellaktatmethode. In *Methodenbuch Band I. Die Untersuchung von Boden*; Thun, R., Hersemann, R., Knickmann, E., Eds.; Neumann Verlag: Radebeul/Berlin, Germany, 1955; pp. 110–125.
32. Schachtschabel, P. Das Pflanzenverfügbare Magnesium Des Boden Und Seine Bestimmung. *Z. Pflanzenernährung Düngung Bodenkd.* **1954**, *67*, 9–23.
33. R Core Team. R: A Language and Environment for Statistical Computing. In *R Foundation for Statistical Computing*; R Core Team: Vienna, Austria, 2023; Available online: https://www.R-project.org (accessed on 6 June 2023).
34. Degryse, F.; da Silva, R.C.; Baird, R.; Beyrer, T.; Below, F.; McLaughlin, M.J. Uptake of Elemental or Sulfate-S from Fall- or Spring-Applied Co-Granulated Fertilizer by Corn—A Stable Isotope and Modeling Study. *Field Crops Res.* **2018**, *221*, 322–332. [CrossRef]
35. Taube, F.; Jahns, U.; Wulfes, R.; Südekum, K.-H. Einfluß Der Schwefelversorgung Auf Ertrag Und Inhaltsstoffe von Deutschem Weidelgras (*Lolium perenne* L.). *Pflanzenbauwissenschaften* **2000**, *4*, 42–51.
36. Grant, C.A.; Mahli, S.S.; Karamanos, R.E. Sulfur Management for Rapeseed. *Field Crops Res.* **2012**, *128*, 119–128. [CrossRef]
37. Grygierzec, B.; Luty, L.; Musiał, K. The Efficiency of Nitrogen and Sulphur Fertilization on Yields and Value of N:S Ratio for Lolium x Boucheanum. *Plant Soil Environ.* **2015**, *61*, 137–143. [CrossRef]
38. Jankowski, K.J.; Kijewski, Ł.; Groth, D.; Skwierawska, M.; Budzyński, W.S. The Effect of Sulfur Fertilization on Macronutrient Concentrations in the Post-Harvest Biomass of Rapeseed (*Brassica napus* L. Ssp. Oleifera Metzg). *J. Elem.* **2015**, *3*, 585–597. [CrossRef]
39. Brown, L.; Scholefield, D.; Jewkes, E.C.; Preedy, N.; Wadge, K.; Butler, M. The Effect of Sulphur Application on the Efficiency of Nitrogen Use in Two Contrasting Grassland Soils. *J. Agric. Sci.* **2000**, *135*, 131–138. [CrossRef]
40. De Kok, L.J.; Tausz, M.; Hawkesford, M.J.; Hoefgen, R.; McManus, M.T.; Norton, R.M.; Rennenberg, H.; Saito, K.; Schnug, E.; Tabe, L. *Sulfur Metabolism in Plants: Mechanisms and Applications to Food Security and Responses to Climate Change*; Springer: Dordrecht, The Netherlands, 2012.
41. Hawkesford, M.J.; De Kok, L.J. Managing Sulphur Metabolism in Plants. *Plant Cell Environ.* **2006**, *29*, 382–395. [CrossRef]
42. Hawkesford, M.J. Plant Responses to Sulphur Deficiency and the Genetic Manipulation of Sulphate Transporters to Improve S-Utilization Efficiency. *J. Exp. Bot.* **2000**, *51*, 131–138. [CrossRef]
43. Dijkshoorn, W.; van Wijk, A.L. The Sulphur Requirements of Plants as Evidenced by the Sulphur-Nitrogen Ratio in the Organic Matter a Review of Published Data. *Plant Soil* **1967**, *26*, 129–157. [CrossRef]
44. Mathot, M.; Lambert, R.; Toussaint, B.; Peeters, A. Total Sulphur Content and N:S Ratio as Indicators for S Deficiency in Grasses. *Int. Grassl. Congr. Proc.* **2005**, 976.
45. Tallec, T.; Diquélou, S.; Lemauviel, S.; Cliquet, J.B.; Lesuffleur, F.; Ourry, A. Nitrogen:Sulphur Ratio Alters Competition between Trifolium Repens and Lolium Perenne under Cutting: Production and Competitive Abilities. *Eur. J. Agron.* **2008**, *29*, 94–101. [CrossRef]

46. Murphy, M.D.; O'Donnell, T. Sulphur Deficiency in Herbage in Ireland: 2. Sulphur Fertilisation and Its Effect on Yield and Quality of Herbage. *Ir. J. Agric. Res.* **1989**, *28*, 79–90.
47. Boswell, C.; Friesen, D. Elemental Sulfur Fertilizers and Their Use on Crops and Pastures. *Nutr. Cycl. Agroecosystems* **1993**, *35*, 127–149. [CrossRef]
48. Deubel, A.; Braune, H.; Tanneberg, H.; Merbach, W. Conversion and Acidifying Effect of Elemental Sulphur in an Alkaline Loess Soil. *Arch. Agron. Soil Sci.* **2007**, *53*, 161–171. [CrossRef]
49. Skwierawska, M.; Zawartka, L.; Zawadzki, B. The Effect of Different Rates and Forms of Sulphur Applied on Changes of Soil Agrochemical Properties. *Plant Soil Environ.* **2008**, *54*, 171–177. [CrossRef]
50. Kulczycki, G. The Effect of Elemental Sulfur Fertilization on Plant Yields and Soil Properties. In *Advances in Agronomy*; Academic Press: Cambridge, MA, USA, 2021; ISBN 0065-2113.
51. Zhao, F.J.; Wu, J.; McGrath, S.P. Soil Organic Sulphur and Its Turnover. In *Humic Substances in Terrestrial Ecosystems*; Alessandro P., Ed.; Elsevier Science B.V.: Amsterdam, The Netherlands, 1996; pp. 467–506. ISBN 978-0-444-81516-3.
52. Jaggi, A.F.F.; Aulakh, M.S.; Sharma, R. Temperature Effects on Soil Organic Sulphur Mineralization and Elemental Sulphur Oxidation in Subtropical Soils of Varying PH. *Nutr. Cycl. Agroecosyst.* **1999**, *54*, 175–182. [CrossRef]
53. Haque, M.A.; Mclaren, R.G.; Condron, L.M.; Williams, P.H.; Francis, G.S. Organic Sulphur and Nitrogen in Pasture-Cereal Mixed Cropping Soils and Related Factors—A Review . *J. Soil Nat.* **2007**, *1*, 39–54.
54. Ikoyi, I.; Fowler, A.; Storey, S.; Doyle, E.; Schmalenberger, A. Sulfate Fertilization Supports Growth of Ryegrass in Soil Columns but Changes Microbial Community Structures and Reduces Abundances of Nematodes and Arbuscular Mycorrhiza. *Sci. Total Environ.* **2020**, *704*, 135315. [CrossRef]
55. Magnucka, E.G.; Kulczycki, G.; Oksińska, M.P.; Kucińska, J.; Pawęska, K.; Milo, Ł.; Pietr, S.J. The Effect of Various Forms of Sulfur on Soil Organic Matter Fractions and Microorganisms in a Pot Experiment with Perennial Ryegrass (*Lolium perenne* L.). *Plants* **2023**, *12*, 2649. [CrossRef] [PubMed]
56. Freney, J.; Spencer, K. Soil Sulphate Changes in the Presence and Absence of Growing Plants. *Aust. J. Agric. Res.* **1960**, *11*, 339–345. [CrossRef]
57. Barrow, N.J. Studies on Mineralization of Sulphur from Soil Organic Matter. *Aust. J. Agric. Res.* **1961**, *12*, 306–319. [CrossRef]
58. Gilbert, M.; Shaw, K. Residual Effects of Sulfur Fertilizers on Cut Swards of a Stylosanthes Guianensis and Native Grass Pasture on a Euchrozem Soil in North Queensland. *Aust. J. Exp. Agric.* **1981**, *21*, 334–342. [CrossRef]

agriculture

MDPI

Review

A Crop of High Nutritional Quality and Health Maintenance Value: The Importance of Tartary Buckwheat Breeding

Ivan Kreft [1,2], Aleksandra Golob [1] and Mateja Germ [1,*]

[1] Biotechnical Faculty, University of Ljubljana, Jamnikarjeva 101, SI-1000 Ljubljana, Slovenia; ivan.kreft@guest.arnes.si (I.K.); aleksandra.golob@bf.uni-lj.si (A.G.)
[2] Nutrition Institute, Tržaška 40, SI-1000 Ljubljana, Slovenia
* Correspondence: mateja.germ@bf.uni-lj.si

Abstract: Tartary buckwheat (*Fagopyrum tataricum* (L.) Gaertn.), originating in the Himalayan area, is cultivated in central Asia and northern, central, and eastern Europe. Tartary buckwheat grain and sprouts are rich in flavonoid metabolites rutin and quercetin. The synthesis of flavonoids in plants is accelerated by UV-B solar radiation to protect the plants against radiation damage. During Tartary buckwheat food processing, a part of rutin is enzymatically converted to quercetin. Rutin and quercetin are able to pass the blood–brain barrier. Studies have investigated the effects of rutin and quercetin on blood flow to the brain, consequently bringing more nutrients and oxygen to the brain, and causing improved brain function. In addition to the impact on blood flow, rutin and quercetin have been shown to have antioxidative properties. The goals of breeding Tartary buckwheat are mainly to maintain and enhance the high nutritional quality. The goals could be reached via the breeding of Tartray buckwheat for larger cotyledons. Other main breeding efforts should be concentrated on the easy husking of the grain, the prevention of seed shattering, and the improvement in growth habits to obtain uniformity in grain ripening and a stable and high yield.

Keywords: *Fagopyrum tataricum*; flavonoids; rutin; quercetin; food; nutrition; breeding

1. Introduction

The place of origin of Tartary buckwheat (*Fagopyrum tataricum* (L.) Gaertn.) is the mountain area of the Himalayas. Tartary buckwheat cultivation takes place at high altitudes around the Himalayas, in China, Nepal, Bhutan, India, and Pakistan, as well in some other countries: Korea, Japan, Kazakhstan, Russia, Ukraine, Belarus, Sweden, Poland, Italy, Luxembourg, and Slovenia [1,2] (Figure 1a–d). In Bosnia and Herzegovina, Tartary buckwheat is grown mainly as a mixed crop with common buckwheat (*Fagopyrum esculentum* Moench) (Figure 1d) [1].

Tartary buckwheat seeds are protected by a thick husk and phenolic substances, and may remain alive but dormant in the soil for many years. Under suitable environmental conditions, they can grow again. Tartary buckwheat survives under strong ultraviolet radiation, which takes place at high altitudes. During their evolution, the plants survived while gradually accumulating genes for substances that allow the plants to reproduce and survive in the stony areas of the high Himalayas, exposed to intense UV-B radiation. The main protective substances that will be discussed in the present paper are flavonoids and other phenolic substances. There are many steps in the synthesis of flavonoid substances and phenolic acids based on phenylalanine; synthesis is enabled by genes and enzymes, active under the impact of UV-B radiation (Figure 2) [2]. The important function of Tartary buckwheat phenolic substances is thus the protection of plants against UV-B radiation.

Many genes involved in the biosynthesis of phenolic substances and their regulation were established in buckwheat [3]. The feasibility of Tartary buckwheat to be able to survive in conditions of high levels of abiotic stress is attributed to the development of several complexes of genes influencing the transduction of signals and the regulation of genes [3]. Such changes in the genetic layout cause the adaptation of plants to harsh ecological conditions. The significance of the genes for the biosynthetic pathway of rutin and the relevant MYB transcription factors (Figure 2) [3–5] is well-known.

Figure 1. (**a**) Flowering and ripening plant of Tartary buckwheat. (**b**) Inflorescence with ripening Tartary buckwheat grain. (**c**) Field of Tartary buckwheat in Wermland, Sweden. (**d**) Fields growing a mixture of Tartary and common buckwheat in the central part of Bosnia and Herzegovina.

Buckwheat metabolites with benzene rings included protect buckwheat plants in several ways. The protection of buckwheat plants from fungal attacks is important for fungi that form mycotoxins in the grain [6–8]. In food, however, the plant-protecting substances of Tartary buckwheat are important for protecting human health.

The flavonoid rutin is present in the grain of Tartary buckwheat, mainly in the cotyledons [9–11]. In addition to rutin, there are also enzymes in the grain which transform rutin into quercetin (Figure 3). Rutin and quercetin molecules are similar, with the difference that rutin has two sugar molecules attached to the aglycone part of the molecule. The enzyme rutinasidase enables the splitting apart of the sugar part of the aglycone and the transformation of rutin into quercetin. In the intact grain, rutin is separated from the enzyme and thus is protected from the transformation. After crushing Tartary buckwheat seeds, the rutin-degrading enzyme is mixed with rutin and during the preparation of dough, after moistening, the conditions are suitable for the decomposition of rutin molecules and the appearance of quercetin. As the consequence, the bitter substance quercetin appears [12–14].

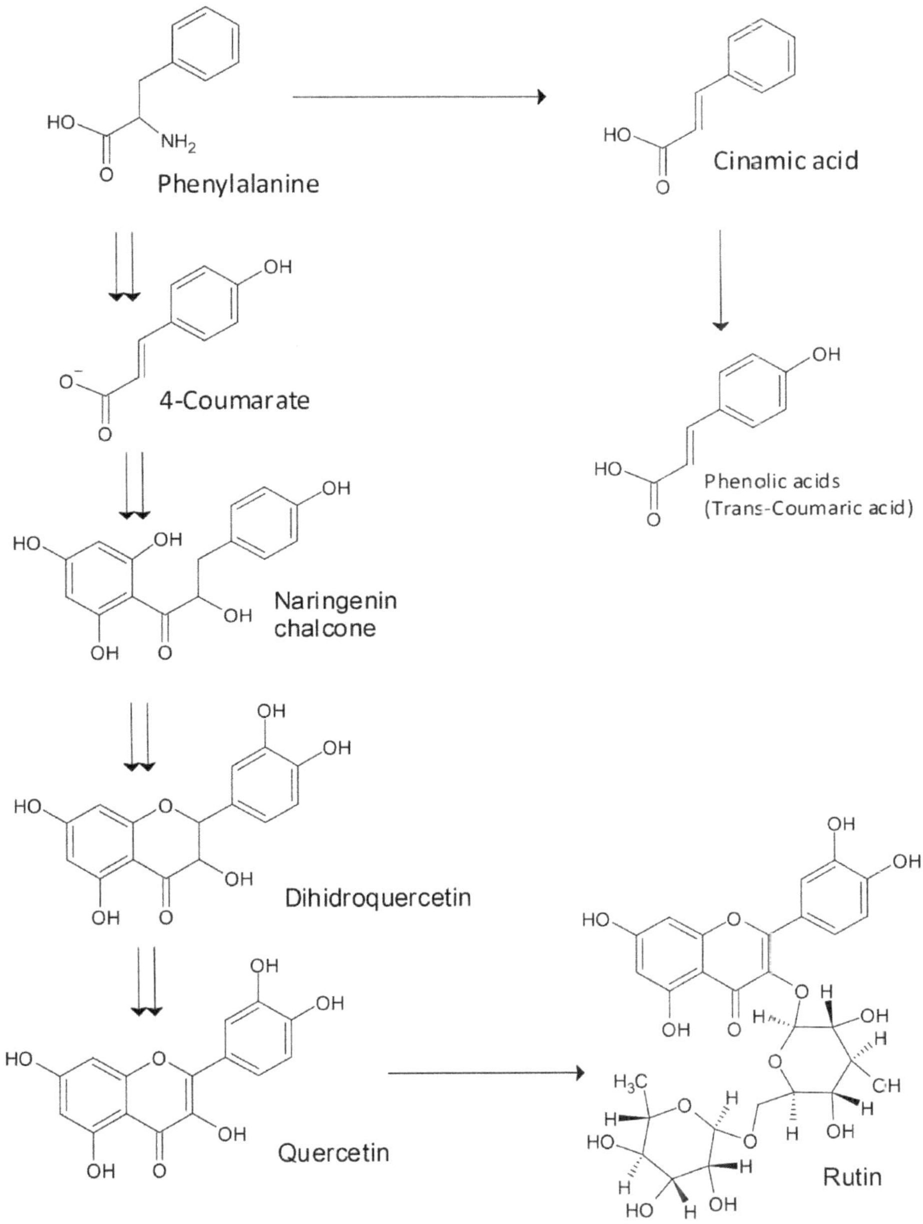

Figure 2. Main steps in the biosynthesis of rutin from phenylalanine in buckwheat plants, each supported by respective genes. A side branch of the synthetic pathway leads to phenolic acids.

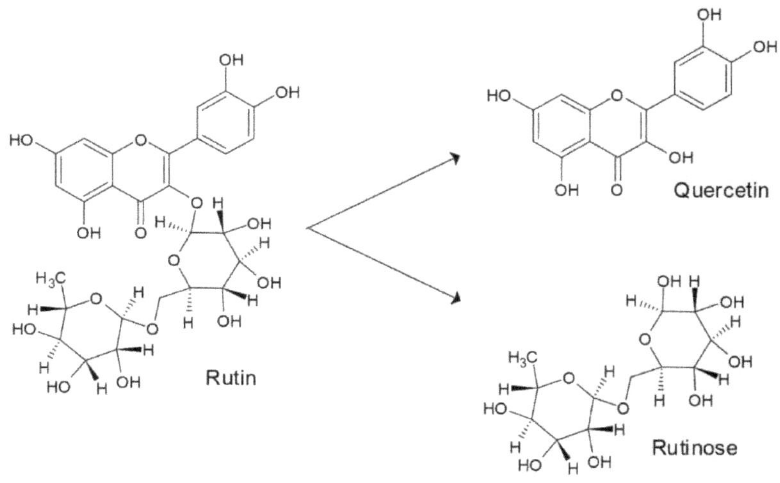

Figure 3. The enzymatic degradation of rutin to flavonoid quercetin and sugar rutinose.

Tartary buckwheat has, in comparison to cereals, a high nutritional value. It has good fatty acid composition and excellently balanced amino acid composition, with favorable vitamin B content [15–17], and a very high rutin level [18–20]. Rutin is known for its ability to strengthen blood vessels, supporting the impact of vitamin C and aiding in reducing blood clots and cholesterol levels [21–23]. Consuming dishes with Tartary buckwheat is important for the prevention of gallstone formation, hypertension, cardiovascular problems, and obesity [24–26]. The effects are mainly due to slowly digestible proteins, digestion-resistant starch, phenolic substances, and their interactions with other grain constituents [27]. From a comparison of the amounts of flavonoids rutin and quercetin in the collection of Tartary buckwheat grain and sprout samples from Europe and Asia, it was reported that the Nepal grain samples had the highest rutin content (13.3 g/kg in dry matter). The concentration of rutin in the sprouts from the same Himalayan area was up to 54.4 g/kg in dry matter. The concentration of quercetin in sprouts was 10–90-fold higher in comparison to that established in the grain [27,28]. Tartary buckwheat grain and sprouts have important potential for providing flavonoids in the human diet and for the production of functional dishes, rich in flavonoids.

When comparing common buckwheat and Tartary buckwheat, there is a considerable difference in the metabolic layout of the two species. It was found that 61 flavonoids and 94 non-flavonoid substances had significantly higher content (at least double) in Tartary buckwheat in comparison to common buckwheat. Tartary buckwheat flour has a yellow color, compared to the light gray color of common buckwheat flour (Figure 4a). The dough and bread made from Tartary buckwheat flour also express a yellow-green hue (Figure 4b–d). Tartary and common buckwheat grains are rich in secondary metabolites beneficial to human health and among them, non-flavonoid metabolites are important. They contribute high health-supporting effects in Tartary buckwheat compared to common buckwheat [29,30]. Zhang et al. [31] reported the sequencing and assembly of Tartary buckwheat [31–33]. Milling the grain of Tartary buckwheat and mixing the flour with water results in the formation of quercetin, which is a degradation product of rutin after rutinosidase activity [14,32,33].

Figure 4. (**a**) Tartary buckwheat flour (**left**) compared to common buckwheat flour (**right**) at the farmers' market in Xichang, Liangshan Yi Autonomous Prefecture, Sichuan, China. (**b**) Dough made from Tartary buckwheat flour, expressing a yellow-green hue due to Tartary buckwheat's secondary metabolites. (**c**) Pretzels made from a mixture of Tartary buckwheat (40%) and wheat flour, expressing a green hue due to the Tartary buckwheat's secondary metabolites content. (**d**) Slovenian olive-oil-seasoned vegan soup, thickened with boiled husked common buckwheat grain (to the left) and husked Tartary buckwheat grain.

Many novel dishes have recently been developed based on traditional Tartary and common buckwheat grain materials, rich in rutin and quercetin (Figure 4c,d) [11,34,35]. Due to the protein content, with suitable amino acid composition, innovative foods based on Tartary buckwheat are a source of high quality proteins in the diet of vegans and other people who do not consume animal proteins. In Korea and China, Tartary buckwheat sprouts are developed as a food material, rich in flavonoids rutin and quercetin [36].

2. Bioactivity of Flavonoids Rutin and Quercetin

In Tartary buckwheat, the complexation of quercetin with starch has been reported [37]. Enzyme molecules hinder the in vitro digestion of starch complexed with quercetin [37,38]. This causes the transformation of native starch to the resistant one and changes the physical and chemical characteristics of the Tartary buckwheat starch [39]. The effects of such a quercetin–starch complexation are that foods made from Tartary buckwheat will have limited digestibility. Quercetin in Tartary buckwheat can reduce the concentration of low-density lipoprotein, serum triacylglycerols, and body weight. In experimental animals, a diet with 0.1% quercetin was shown to significantly impact the lowering of low-density lipoprotein concentrations in serum, without any effect on high-density lipoprotein. Tartary buckwheat has also prevented an increase in body weight and fat deposition during high-fat intake in rats [40]. A buckwheat diet can improve glucose intolerance in patients by reducing the insulin response [41]. The reduced postprandial glycemic responses in common buckwheat were shown [41].

Phenolic substances are often transformed in the gut before their absorption. The microbiota in the colon enable this process [42]. Small-sized phenolic molecules, which result from the colon's microbial conversion, are more easily absorbed than large-sized phenolic metabolites [43–45].

Radiation-induced brain injury is a serious adverse effect of brain radiotherapy in oncology. Flavonoid quercetin has a wide range of biological activities, including the ability to regulate gut microbiota. After the oral submission of quercetin preparations in experimental mice, the spontaneous activity behavior, short-term memory ability, and anxiety level were improved [46]. The long-term administration of buckwheat whole meal flour suppresses cognitive decline by increasing hippocampal brain-derived neurotrophic factor production in experimental mice [47].

Important bioactivities have been found for quercetin and the related molecules, not just in the gastrointestinal system but also in blood vessels, muscles, and the brain [48]. The blood–brain barrier is a highly selective semipermeable border, consisting of endothelial cells. It prevents solutes from circulating blood from non-selectively crossing into the extracellular space of the central nervous system [49–51]. Ingested rutin and quercetin can accumulate in the brain tissue after crossing the blood–brain barrier [52]. The challenge in treating Alzheimer's disease and some other neurological conditions is the inability of medical substances to enter the brain due to poor solubility and the blood–brain barrier [53]. Quercetin could cross the blood–brain barrier and exerts neuroprotective effects in many neurological disease situations [54]. In vitro and in vivo experiments have shown that molecules, including quercetin, and quercetin derivatives, can penetrate the blood–brain barrier without toxicity; quercetin could be delivered to the ischemic area of the brain, ensuring the targeted delivery and antioxidative impact. The mitochondrial targeting of damaged neurons is also achieved in such a way. Substances with quercetin as part of the molecules may have better neuroprotective ability than quercetin itself [55].

Taile et al. [56] established that polyphenols are important dietary antioxidants with anti-inflammatory action, able to improve the effects of stroke and other cerebrovascular problems. The protective impact of quercetin and some other polyphenolic substances is connected with their bioavailability. It is important to apply polyphenol-based strategies for the improvement of the clinical picture of stroke.

Blood–brain barriers hinder not only the endogenous or exogenous toxicants, but also compounds with therapeutic properties. A group of chemo-sensing receptors was identified in the blood–cerebrospinal fluid barrier. Chemosensing bitter taste receptors are promising as potential targets of drugs. Quercetin, resveratrol and other metabolites with neuroprotective activity as ligands for transporting molecules can potentially counteract drug resistance in the delivery to the brain for the treatment of central nervous system disorders [57,58]. Extracts from buckwheat leaves and flowers have an impact on the antioxidant situation of the liver and brain of mice. Such extracts were given orally to mice for 21 days. The effects in the mouse brain were the following: the amount of glutathione and malondialdehyde was reduced, superoxide dismutase activity was significantly decreased, and catalase activity significantly increased [59].

Choi et al. [60] studied the impact of rutin and the n-butanol fraction extracted from Tartary buckwheat on learning and memory deficits in a mouse model of amyloid beta (A beta)-induced Alzheimer's disease. The impaired cognition and memory of experimental animals were attenuated by the oral submission of an n-butanol fraction and rutin extracted from plants of Tartary buckwheat. According to the results from Choi et al. [60], the n-butanol fraction and rutin extracted from Tartary buckwheat had protective effects, and therapeutic applications for the treatment of Alzheimer's disease are thus suggested.

3. Precautions for Possible Adverse Effects of Tartary Buckwheat Metabolites

One of buckwheat's secondary metabolites, fagopyrin, has a health threat when the green parts of common or Tartary buckwheat are eaten [61–63]. But the ingestion of buckwheat grain products seems to be safe due to the low concentration of fagopyrin in the grain [63–65]. A buckwheat allergy is a very rare event [66].

Suzuki et al. [12] investigated the potential toxicity of rutin-rich dough made from Tartary buckwheat through acute and subacute toxicity studies in experiments with laboratory rats. The concentration of rutin in Tartary buckwheat material was 1570 mg/100 g. In the experiment, no toxic or other non-regular symptoms were detected. The resulting body weight was not significantly different among the groups of experimental animals. The conclusion of the experiment was that Tartary buckwheat at a given dose was without noticeable adverse effects. The results of Vogrinčič et al. [67] and Suzuki et al. [68] also established that no genotoxic effects were expressed in Tartary buckwheat grain materials.

Quercetin is genotoxic to salmonella, but it was approved to be safe for human application [69]. The study of Cunningham et al. [70] confirmed the safety of quercetin for use in mice when applied daily at 12.5, 25, or 50 mg/kg of body weight for 98 days. In any case, more studies have to be performed to determine any possible quercetin toxicity effects after the chronic ingestion of Tartary buckwheat or other quercetin-rich food [70]

4. Main Goals and Methods for Breeding Tartary Buckwheat

Tartary buckwheat is a semi-wild plant [1]. It was not long ago when people started to cultivate it. Even now, wild forms of Tartary buckwheat grow outside the fields on the rocky slopes of the Himalayas. Tartary buckwheat is found as a weed in common buckwheat crops or other crops. The shattered seeds may return from the fields to less friendly areas, as wild plants, occupying poorly fertile soil. For the propagation of wild plants, the shattering is a very important trait. In newly domesticated crop-like Tartary buckwheat, seed shattering is a detrimental characteristic for growing in fields and consistent efforts must be made to eliminate this trait during breeding. When Tartary buckwheat is cultivated, an important part of the yield could be lost by the shattering of seeds. Therefore, breeding against shattering and obtaining the uniform ripening of Tartary buckwheat seeds are important goals in Tartary buckwheat breeding. In practical terms, the selection against shattering could be performed by selecting thicker peduncles supporting individual seeds. Other selection methods against seed shattering could be through the vibrating of plants and scoring which portion of seeds remain on the plants after the vibration impact. This selection method is not very effective because of the non-uniform ripening of the seeds on Tartary buckwheat plants. An important issue in obtaining less shattering is a uniform time for ripening of all seeds on the plant. In common buckwheat, the gene for the determinate growth habit is known [71–73]. Determinate buckwheat plants have more uniform ripening and stronger side branches, which supports the prevention of seed shattering. But until now, no report on determinate growth habits in Tartary buckwheat is known. It should be desirable to find Tartary buckwheat with determinate growth habits. One way to reach this goal could be a mass screening of the Tartary buckwheat population for this trait, or mutation breeding. Determinate common buckwheat plants are lower, so they could be hidden in the canopy. Determinate plants in the tight canopy condition have a lower possibility of surviving and reproducing seeds. Therefore, in canopies, selection pressure is working against plants with determinate growth habits, unless plants grow in conditions with less competition from neighboring plants [71,74,75]. Why determinate Tartary buckwheat plants have not yet been found among the wildly growing populations is unknown. Determinate Tartary buckwheat would be interesting because of its expected resistance to lodging, the simultaneous and uniform seed set and the resistance to seed shattering. One of the main problems with the production of Tartary buckwheat is its indeterminant growth. As a result, the seeds ripen very unevenly and the first ripe seeds can fall off before the last formed and filled seeds mature.

Mutation breeding in Tartary buckwheat should not be too complicated as it is a self-fertile plant and homozygotes for the mutated gene could already be expressed and scored in early generations after treatment, with mutagenic impact.

The lodging of plants does not seem to be a serious problem in Tartary buckwheat. If lodged, plants are very soon recovered. If plants can be created by breeding the determinate Tartary buckwheat, they are expected to be more resistant to lodging than non-determinate plants. For example, in common buckwheat, determinate plants are much better resistant to lodging in comparison to non-determinate plants [71].

As Tartary buckwheat grains have coarse and hard husks, it is very difficult to dehusk them to obtain groats. Only a few producers can make husked Tartary buckwheat [1]. Several easily husked or "Rice-Tartary" buckwheat forms are known. Rice-Tartary is a special type of Tartary buckwheat, with seeds with a loose husk that make dehusking easier [76,77]. It was established that by reciprocally crossing Tartary and rice-Tartary buckwheats, and backcrossing rice-Tartary-type progeny with Tartary buckwheat, the non-adhering husk trait depends on a single recessive gene [78]. However, according to the new results from Duan et al. [79], three genes regulate the trait of easily husked Tartary buckwheat.

In crossing, hot water emasculation was used. The hybrid rice-Tartary buckwheat cultivar Mikuqiao18 was obtained in China through the pedigree selection of crossbreeding 'Miqiao' with 'Jingqiaomai2' [77]. The hybrid had a lower yield than parent varieties because the grain mass was lower due to the thinner husk.

In Tartary buckwheat grain, the most valuable substances for good nutritional value and the maintenance of human health are cotyledons. The increased size of cotyledons is important for a high content of proteins and polyphenols. It is not yet clear how to select plants to achieve this goal. One possibility would be to evaluate the size of cotyledons through the transverse sections of grain. Another possibility would be to estimate the size of cotyledons through the evaluation of milling fractions. Contrary to the situation in common buckwheat, in Tartary buckwheat all the grain on a given Tartary buckwheat plant is genetically uniform because of self-fertilization. So, a certain grain from a Tartary buckwheat plant would be, genetically, a representative sample for other seeds from the same plant.

It has been established that the expression levels of genes for phenylalanine ammonia lyase (PAL) and 4-coumaric acid coenzyme A ligase (4CL) were positively correlated with the content of flavonoids in Tartary buckwheat [80]. This finding suggests the possibility for breeding Tartary buckwheat with stable or enhanced flavonoid content, which is very important for the utilization value of Tartary buckwheat.

Other important quality traits are plant height, leaf blade width, stem color (green or red), number of primary branches, inflorescence length, flower color (greenish-yellow, white, pink, red), seed anthocyanin color (green, brown, black), seed shape (ovate, with sharp edges or with growths) and seed weight. These accessions can be of vital significance for future buckwheat breeding programs.

Traditional breeding and selection methods are chiefly concentrated on visible properties, including plant morphology and yield. The changes in metabolic profile cannot be observed by the methods usually used. Research on changes in the metabolite profile during domestication and breeding efforts is very important. Metabolome profiling in Tartary buckwheat will make genetic improvements of traits possible in Tartary buckwheat, which is important for medical use and resistance against diseases [81].

Tartary buckwheat is mainly tolerant to pests and plant diseases. In any case, strains of *Rhizoctonia solani*, a soil-born pathogen, may damage the plants [82]. Using a multiomic approach, it is feasible to identify genes related to resistance against this pathogen in Tartary buckwheat. This finding could accelerate the molecular breeding of *Rhizoctonia*-resistant cultivars in Tartary buckwheat [82]. Regarding climatic changes, obtaining Tartary buckwheat cultivars that are tolerant to drought is essential. Relevant genes and regulatory systems connected to drought tolerance have been reported in Tartary buckwheat [83].

The transformation and genome editing of common and Tartary buckwheat have some restrictions in their application [84]. In several European countries (for example, Austria, Czech Republic, and Italy), common buckwheat and Tartary buckwheat are grown and used in nutrition as ecological crops. According to the official demands for ecological crops, the use of transgenic cultivars is not allowed. Moreover, transgenic cultivars should not be cultivated close to ecological crops, to prevent the pollination of ecologically grown plants with the pollen of transgenic plants [84].

5. Future Perspectives

Due to its excellent nutritional value and its adaptability to adverse climatic and soil conditions, Tartary buckwheat undoubtedly has a future. Above all, it is necessary to increase the yield. Care must be taken to ensure that entire yield is harvested on the fields, i.e., it is necessary to improve the resistance of Tartary buckwheat to weather incidents, to prevent lodging and losing seeds through shattering. The starting point for the successful breeding of Tartary buckwheat is genetic material that must be collected and used in breeding.

The genetic material of Tartary buckwheat is available from diverse sources. The starting material for breeding could be the domestic populations of Tartary buckwheat grown by farmers, especially in areas of the Himalayas. The starting genotypes for successful Tartary buckwheat breeding could also be wild plants, and weedy Tartary buckwheat plants growing in the fields among the plants of common buckwheat. Weedy Tartary buckwheat plants are already adapted to grow in cultivated fields. Within Tartary buckwheat populations there is a great diversity and polymorphism, so selection could be very effective.

To increase the diversity and accelerate breeding, induced mutations are feasible, as well as obtaining tetraploids or crossing Tatary buckwheat plants with plants of other, related species.

When breeding for increased yield, care must be taken to maintain and improve the resistance of plants against diseases, pests and weather disasters. Care should also be taken to preserve the excellent nutritional value of Tatary buckwheat, especially its high content of flavonoids, other polyphenolic substances, proteins and fiber.

6. Conclusions

Tartary buckwheat grain and sprouts are rich in flavonoids rutin and quercetin. After crushing Tartary buckwheat grain and mixing the obtained material with water, the rutin-degrading enzyme of the grain starts its activity of decomposing the rutin molecules, as the result the concentration of rutin becomes lower and the concentration of quercetin becomes higher.

Studies have shown the effects of flavonoids rutin and quercetin on the human body, including the brain, producing promising results. Rutin and quercetin have been shown to have anti-inflammatory effects, which are believed to play a role in various neurological conditions, including depression, anxiety, and Alzheimer's disease. Despite the promising results, further research is needed to completely understand the effects of rutin and quercetin and to determine the optimal dose for long-term use. Tartary buckwheat is a valuable food source of flavonoids rutin and quercetin and other phenolic metabolites. However, maintaining a balanced diet is most important, and relying on a single food as the main source of nutrients should be avoided.

Tartary buckwheat is a self-fertile plant, so mutation breeding is feasible. Few generations after mutagenic treatment, homozygotes for mutated genes could appear and mutated traits would be visible for selection. Another breeding method is screening Tartary buckwheat varieties and wild populations for desirable traits, and using hybridization to introduce the desirable traits to the target cultivars.

The main goal of Tartary buckwheat breeding is suggested to be maintaining and enhancing the excellent properties of flavonoid content and the high nutritional value of proteins.

Author Contributions: Conceptualization I.K., A.G. and M.G.; data curation A.G.; validation, writing original draft preparation, review and editing, all authors equally responsible; visualization, A.G.; project funding acquisition and administration, I.K. and M.G.; supervision, I.K. All authors have read and agreed to the published version of the manuscript.

Funding: This work was the result of a study financed by the Slovenian Research Agency, through the programs P1-0212 "Biology of Plants" and P3-0395 "Nutrition and Public Health", projects J1-3014, J4-3091, and the applied project L4-9305, co-financed by the Ministry of Agriculture, Forestry and Food, Republic of Slovenia.

Institutional Review Board Statement: Not applicable.

Informed Consent Statement: Not applicable.

Data Availability Statement: Not applicable.

Acknowledgments: Preparing samples for photographing is acknowledged to Blanka Vombergar, Stanko Vorih and Marija Horvat.

Conflicts of Interest: The authors declare that they have no conflict of interest.

References

1. Kreft, I. (Ed.) *Bitter Seed Tartary Buckwheat*; Slovenian Academy of Sciences and Arts; Maribor: Fagopyrum—Slovenian Association for Buckwheat Promotion: Ljubljana, Slovenia, 2022.
2. Kreft, I.; Vollmannová, A.; Lidiková, J.; Musilová, J.; Germ, M.; Golob, A.; Vombergar, B.; Kocjan Ačko, D.; Luthar, Z. Molecular Shield for Protection of Buckwheat Plants from UV-B Radiation. *Molecules* **2022**, *27*, 5577. [CrossRef] [PubMed]
3. Zhang, L.; Li, X.; Ma, B.; Gao, Q.; Du, H.; Han, Y.; Li, Y.; Cao, Y.; Qi, M.; Zhu, Y.; et al. The Tartary Buckwheat Genome Provides Insights into Rutin Biosynthesis and Abiotic Stress Tolerance. *Mol. Plant* **2017**, *10*, 1224–1237. [CrossRef]
4. Zhou, M.; Sun, Z.; Ding, M.; Logacheva, M.D.; Kreft, I.; Wang, D.; Yan, M.; Shao, J.; Tang, Y.; Wu, Y.; et al. FtSAD2 and FtJAZ1 Regulate Activity of the FtMYB11 Transcription Repressor of the Phenylpropanoid Pathway in Fagopyrum Tataricum. *New Phytol.* **2017**, *216*, 814–828. [CrossRef]
5. Chitarrini, G.; Nobili, C.; Pinzari, F.; Antonini, A.; De Rossi, P.; Del Fiore, A.; Procacci, S.; Tolaini, V.; Scala, V.; Scarpari, M.; et al. Buckwheat Achenes Antioxidant Profile Modulates Aspergillus Flavus Growth and Aflatoxin Production. *Int. J. Food Microbiol.* **2014**, *189*, 1–10. [CrossRef] [PubMed]
6. Gauthier, L.; Bonnin-Verdal, M.N.; Marchegay, G.; Pinson-Gadais, L.; Ducos, C.; Richard-Forget, F.; Atanasova-Penichon, V. Fungal Biotransformation of Chlorogenic and Caffeic Acids by Fusarium Graminearum: New Insights in the Contribution of Phenolic Acids to Resistance to Deoxynivalenol Accumulation in Cereals. *Int. J. Food Microbiol.* **2016**, *221*, 61–68. [CrossRef]
7. Li, Z.; Li, Z.; Huang, Y.; Jiang, Y.; Liu, Y.; Wen, W.; Li, H.; Shao, J.; Wang, C.; Zhu, X. Antioxidant Capacity, Metal Contents, and Their Health Risk Assessment of Tartary Buckwheat Teas. *ACS Omega* **2020**, *5*, 9724–9732. [CrossRef] [PubMed]
8. Nobili, C.; De Acutis, A.; Reverberi, M.; Bello, C.; Leone, G.P.; Palumbo, D.; Natella, F.; Procacci, S.; Zjalic, S.; Brunori, A. Buckwheat Hull Extracts Inhibit Aspergillus Flavus Growth and AFB1 Biosynthesis. *Front. Microbiol.* **2019**, *10*, 1997. [CrossRef]
9. Vombergar, B.; Luthar, Z. Starting Points for the Study of the Effects of Flavonoids, Tannins and Crude Proteins in Grain Fractions of Common Buckwheat (*Fagopyrum esculentum* Moench) and Tartary Buckwheat (*Fagopyrum tataricum* Gaertn.). *Folia Biol. Geol.* **2018**, *59*, 101. [CrossRef]
10. Vombergar, B. Rutin and Quercetin in Common Buckwheat and Tartary Buckwheat Flour. *Folia Biol. Geol.* **2020**, *61*, 257–280. [CrossRef]
11. Vombergar, B. Rutin and Quercetin in Common and Tartary Buckwheat Flour and Dough. *Fagopyrum* **2021**, *38*, 43–53. [CrossRef]
12. Suzuki, T.; Morishita, T.; Noda, T.; Ishiguro, K. Acute and Subacute Toxicity Studies on Rutin-Rich Tartary Buckwheat Dough in Experimental Animals. *J. Nutr. Sci. Vitaminol.* **2015**, *61*, 175–181. [CrossRef]
13. Suzuki, T.; Morishita, T.; Takigawa, S.; Noda, T.; Ishiguro, K. Characterization of Rutin-Rich Bread Made with 'Manten-Kirari', a Trace-Rutinosidase Variety of Tartary Buckwheat (*Fagopyrum tataricum* Gaertn.). *Food Sci. Technol. Res.* **2015**, *21*, 733–738. [CrossRef]
14. Suzuki, T.; Morishita, T.; Takigawa, S.; Noda, T.; Ishiguro, K.; Otsuka, S. Development of Novel Detection Method for Rutinosidase in Tartary Buckwheat (*Fagopyrum tataricum* Gaertn.). *Plants* **2022**, *11*, 320. [CrossRef] [PubMed]
15. Bonafaccia, G.; Marocchini, M.; Kreft, I. Composition and Technological Properties of the Flour and Bran from Common and Tartary Buckwheat. *Food Chem.* **2003**, *80*, 9–15. [CrossRef]
16. Kreft, M. Buckwheat Phenolic Metabolites in Health and Disease. *Nutr. Res. Rev.* **2016**, *29*, 30–39. [CrossRef]
17. Sytar, O.; Brestic, M.; Zivcak, M.; Phan Tran, L.-S. The Contribution of Buckwheat Genetic Resources to Health and Dietary Diversity. *Curr. Genom.* **2016**, *17*, 193–206. [CrossRef]
18. Fabjan, N.; Rode, J.; Košir, I.J.; Wang, Z.; Zhang, Z.; Kreft, I. Tartary Buckwheat (*Fagopyrum tataricum* Gaertn.) as a Source of Dietary Rutin and Quercitrin. *J. Agric. Food Chem.* **2003**, *51*, 6452–6455. [CrossRef]

19. Sytar, O.; Kosyan, A.; Taran, N.; Smetanska, I. Anthocyanin's as Marker for Selection of Buckwheat Plants with High Rutin Content. *Gesunde Pflanz.* **2014**, *66*, 165–169. [CrossRef]
20. Kuwabara, T.; Han, K.H.; Hashimoto, N.; Yamauchi, H.; Shimada, K.I.; Sekikawa, M.; Fukushima, M. Tartary Buckwheat Sprout Powder Lowers Plasma Cholesterol Level in Rats. *J. Nutr. Sci. Vitaminol.* **2007**, *53*, 501–507. [CrossRef]
21. Nishimura, M.; Ohkawara, T.; Sato, Y.; Satoh, H.; Suzuki, T.; Ishiguro, K.; Noda, T.; Morishita, T.; Nishihira, J. Effectiveness of Rutin-Rich Tartary Buckwheat (*Fagopyrum tataricum* Gaertn.) 'Manten-Kirari' in Body Weight Reduction Related to Its Antioxidant Properties: A Randomised, Double-Blind, Placebo-Controlled Study. *J. Funct. Foods* **2016**, *26*, 460–469. [CrossRef]
22. Suzuki, T.; Morishita, T.; Mukasa, Y.; Takigawa, S.; Yokota, S.; Ishiguro, K.; Noda, T. Breeding of 'Manten-Kirari', a Non-Bitter and Trace-Rutinosidase Variety of Tartary Buckwheat (*Fagopyrum tataricum* Gaertn.). *Breed. Sci.* **2014**, *64*, 344–350. [CrossRef]
23. Li, L.; Lietz, G.; Seal, C. Buckwheat and CVD Risk Markers: A Systematic Review and Meta-Analysis. *Nutrients* **2018**, *10*, 619. [CrossRef] [PubMed]
24. Tomotake, H.; Shimaoka, I.; Kayashita, J.; Yokoyama, F.; Nakajoh, M.; Kato, N. A Buckwheat Protein Product Suppresses Gallstone Formation and Plasma Cholesterol More Strongly than Soy Protein Isolate in Hamsters. *J. Nutr.* **2000**, *130*, 1670–1674. [CrossRef] [PubMed]
25. Tomotake, H.; Yamamoto, N.; Yanaka, N.; Ohinata, H.; Yamazaki, R.; Kayashita, J.; Kato, N. High Protein Buckwheat Flour Suppresses Hypercholesterolemia in Rats and Gallstone Formation in Mice by Hypercholesterolemic Diet and Body Fat in Rats Because of Its Low Protein Digestibility. *Nutrition* **2006**, *22*, 166–173. [CrossRef]
26. Tomotake, H.; Yamamoto, N.; Kitabayashi, H.; Kawakami, A.; Kayashita, J.; Ohinata, H.; Karasawa, H.; Kato, N. Preparation of Tartary Buckwheat Protein Product and Its Improving Effect on Cholesterol Metabolism in Rats and Mice Fed Cholesterol-Enriched Diet. *J. Food Sci.* **2007**, *72*, S528–S533. [CrossRef]
27. Luthar, Z.; Golob, A.; Germ, M.; Vombergar, B.; Kreft, I. Tartary Buckwheat in Human Nutrition. *Plants* **2021**, *10*, 700. [CrossRef] [PubMed]
28. Yu, J.H.; Kwon, S.J.; Choi, J.Y.; Ju, Y.H.; Roy, S.K.; Lee, D.-G.; Park, C.H.; Woo, S.-H. Variation of Rutin and Quercetin Contents in Tartary Buckwheat Germplasm. *Fagopyrum* **2019**, *36*, 51–65. [CrossRef]
29. Li, H.; Lv, Q.; Liu, A.; Wang, J.; Sun, X.; Deng, J.; Chen, Q.; Wu, Q. Comparative Metabolomics Study of Tartary (*Fagopyrum tataricum* (L.) Gaertn) and Common (*Fagopyrum esculentum* Moench) Buckwheat Seeds. *Food Chem.* **2022**, *371*, 131125. [CrossRef] [PubMed]
30. Huang, Y.; Zhang, K.; Guo, W.; Zhang, C.; Chen, H.; Xu, T.; Lu, Y.; Wu, Q.; Li, Y.; Chen, Y. Aspergillus Niger Fermented Tartary Buckwheat Ameliorates Obesity and Gut Microbiota Dysbiosis through the NLRP3/Caspase-1 Signaling Pathway in High-Fat Diet Mice. *J. Funct. Foods* **2022**, *95*, 105171. [CrossRef]
31. Zhang, C.; Zhang, R.; Li, Y.M.; Liang, N.; Zhao, Y.; Zhu, H.; He, Z.; Liu, J.; Hao, W.; Jiao, R.; et al. Cholesterol-Lowering Activity of Tartary Buckwheat Protein. *J. Agric. Food Chem.* **2017**, *65*, 1900–1906. [CrossRef]
32. Germ, M.; Árvay, J.; Vollmannová, A.; Tóth, T.; Golob, A.; Luthar, Z.; Kreft, I. The Temperature Threshold for the Transformation of Rutin to Quercetin in Tartary Buckwheat Dough. *Food Chem.* **2019**, *283*, 28–31. [CrossRef] [PubMed]
33. Fujita, K.; Yoshihashi, T. Heat-Treatment of Tartary Buckwheat (*Fagopyrum tataricum* Gaertn.) Provides Dehulled and Gelatinized Product with Denatured Rutinosidase. *Food Sci. Technol. Res.* **2019**, *25*, 613–618. [CrossRef]
34. Asami, Y.; Ikeda, S.; Ikeda, K. Leaving Buckwheat Noodles after Their Making and Subsequent Cooking Leads to Remarkable Changes in Mechanical Characteristics. *Fagopyrum* **2022**, *39*, 5–11. [CrossRef]
35. Park, C.H.; Kim, Y.B.; Choi, Y.S.; Heo, K.; Kim, S.L.; Lee, K.C.; Chang, K.J.; Lee, H.B. Rutin Content in Food Products Processed from Groats, Leaves, and Flowers of Buckwheat. *Fagopyrum* **2000**, *17*, 63–66.
36. Park, M.O.; Kim, H.J.; Choi, I.Y.; Park, C.H. Development and Utilization of Buckwheat Sprouts in Korea. *Fagopyrum* **2022**, *39*, 19–26. [CrossRef]
37. Ikeda, K.; Ishida, Y.; Ikeda, S.; Asami, Y.; Lin, R. Tartary, but Not Common, Buckwheat Inhibits α-Glucosidase Activity: Its Nutritional Implications. *Fagopyrum* **2017**, *34*, 13–18. [CrossRef]
38. Luo, K.; Zhou, X.; Zhang, G. The Impact of Tartary Buckwheat Extract on the Nutritional Property of Starch in a Whole Grain Context. *J. Cereal Sci.* **2019**, *89*, 102798. [CrossRef]
39. Li, Y.; Gao, S.; Ji, X.; Liu, H.; Liu, N.; Yang, J.; Lu, M.; Han, L.; Wang, M. Evaluation Studies on Effects of Quercetin with Different Concentrations on the Physicochemical Properties and in Vitro Digestibility of Tartary Buckwheat Starch. *Int. J. Biol. Macromol.* **2020**, *163*, 1729–1737. [CrossRef]
40. Peng, L.; Zhang, Q.; Zhang, Y.; Yao, Z.; Song, P.; Wei, L.; Zhao, G.; Yan, Z. Effect of Tartary Buckwheat, Rutin, and Quercetin on Lipid Metabolism in Rats during High Dietary Fat Intake. *Food Sci. Nutr.* **2020**, *8*, 199–213. [CrossRef] [PubMed]
41. Skrabanja, V.; Liljeberg Elmståhl, H.G.M.; Kreft, I.; Björck, I.M.E. Nutritional Properties of Starch in Buckwheat Products: Studies in Vitro and In Vivo. *J. Agric. Food Chem.* **2001**, *49*, 490–496. [CrossRef]
42. Selma, M.V.; Espín, J.C.; Tomás-Barberán, F.A. Interaction between Phenolics and Gut Microbiota: Role in Human Health. *J. Agric. Food Chem.* **2009**, *57*, 6485–6501. [CrossRef]
43. Lineva, A.; Benković, E.T.; Kreft, S.; Kienzle, E. Remarkable Frequency of a History of Liver Disease in Dogs Fed Homemade Diets with Buckwheat. *Tierärztliche Prax. Ausg. K Kleintiere/Heimtiere* **2019**, *47*, 242–246. [CrossRef]
44. Liu, J.; Song, Y.; Zhao, Q.; Wang, Y.; Li, C.; Zou, L.; Hu, Y. Effects of Tartary Buckwheat Protein on Gut Microbiome and Plasma Metabolite in Rats with High-Fat Diet. *Foods* **2021**, *10*, 2457. [CrossRef]

45. Valido, E.; Stoyanov, J.; Gorreja, F.; Stojic, S.; Niehot, C.; Kiefte-de Jong, J.; Llanaj, E.; Muka, T.; Glisic, M. Systematic Review of Human and Animal Evidence on the Role of Buckwheat Consumption on Gastrointestinal Health. *Nutrients* **2023**, *15*, 1. [CrossRef] [PubMed]
46. Hu, J.; Jiao, W.; Tang, Z.; Wang, C.; Li, Q.; Wei, M.; Song, S.; Du, L.; Jin, Y. Quercetin Inclusion Complex Gels Ameliorate Radiation-Induced Brain Injury by Regulating Gut Microbiota. *Biomed. Pharmacother.* **2023**, *158*, 114142. [CrossRef] [PubMed]
47. Katayama, S.; Okahata, C.; Onozato, M.; Minami, T.; Maeshima, M.; Ogihara, K.; Yamazaki, S.; Takahashi, Y.; Nakamura, S. Buckwheat Flour and Its Starch Prevent Age-Related Cognitive Decline by Increasing Hippocampal BDNF Production in Senescence-Accelerated Mouse Prone 8 Mice. *Nutrients* **2022**, *14*, 2708. [CrossRef]
48. Ballabh, P.; Braun, A.; Nedergaard, M. The Blood–Brain Barrier: An Overview: Structure, Regulation, and Clinical Implications. *Neurobiol. Dis.* **2004**, *16*, 1–13. [CrossRef] [PubMed]
49. Daneman, R.; Prat, A. The Blood–Brain Barrier. *Cold Spring Harb. Perspect. Biol.* **2015**, *7*, a020412. [CrossRef]
50. Wolff, A.; Antfolk, M.; Brodin, B.; Tenje, M. In Vitro Blood–Brain Barrier Models—An Overview of Established Models and New Microfluidic Approaches. *J. Pharm. Sci.* **2015**, *104*, 2727–2746. [CrossRef]
51. Gupta, S.; Dhanda, S.; Sandhir, R. Anatomy and Physiology of Blood-Brain Barrier. In *Brain Targeted Drug Delivery System*; Academic Press: Cambridge, MA, USA, 2019; pp. 7–31. [CrossRef]
52. Kawabata, K.; Mukai, R.; Ishisaka, A. Quercetin and Related Polyphenols: New Insights and Implications for Their Bioactivity and Bioavailability. *Food Funct.* **2015**, *6*, 1399–1417. [CrossRef]
53. Woon, C.K.; Hui, W.K.; Abas, R.; Haron, M.H.; Das, S.; Lin, T.S. Natural Product-Based Nanomedicine: Recent Advances and Issues for the Treatment of Alzheimer's Disease. *Curr. Neuropharmacol.* **2022**, *20*, 1498–1518. [CrossRef]
54. Wu, Y.; Wei, H.; Li, P.; Zhao, H.; Li, R.; Yang, F. Quercetin Administration Following Hypoxia-Induced Neonatal Brain Damage Attenuates Later-Life Seizure Susceptibility and Anxiety-Related Behavior: Modulating Inflammatory Response. *Front. Pediatr.* **2022**, *10*, 73. [CrossRef]
55. Cen, J.; Zhang, R.; Zhao, T.; Zhang, X.; Zhang, C.; Cui, J.; Zhao, K.; Duan, S.; Guo, Y. A Water-Soluble Quercetin Conjugate with Triple Targeting Exerts Neuron-Protective Effect on Cerebral Ischemia by Mitophagy Activation. *Adv. Healthc. Mater.* **2022**, *11*, 2200817. [CrossRef] [PubMed]
56. Taïlé, J.; Bringart, M.; Planesse, C.; Patché, J.; Rondeau, P.; Veeren, B.; Clerc, P.; Gauvin-Bialecki, A.; Bourane, S.; Meilhac, O.; et al. Antioxidant Polyphenols of Antirhea Borbonica Medicinal Plant and Caffeic Acid Reduce Cerebrovascular, Inflammatory and Metabolic Disorders Aggravated by High-Fat Diet-Induced Obesity in a Mouse Model of Stroke. *Antioxidants* **2022**, *11*, 858 [CrossRef] [PubMed]
57. Dogra, N.; Jakhmola-Mani, R.; Potshangbam, A.M.; Buch, S.; Pande Katare, D. CXCR4 as Possible Druggable Target Linking Inflammatory Bowel Disease and Parkinson's Disease. *Metab. Brain Dis.* **2023**, *1*, 1079–1096. [CrossRef]
58. Duarte, A.C.; Costa, A.R.; Gonçalves, I.; Quintela, T.; Preissner, R.; Santos, C.R.A. The Druggability of Bitter Taste Receptors for the Treatment of Neurodegenerative Disorders. *Biochem. Pharmacol.* **2022**, *197*, 114915. [CrossRef] [PubMed]
59. Sadauskiene, I.; Liekis, A.; Bernotiene, R.; Sulinskiene, J.; Kasauskas, A.; Zekonis, G. The Effects of Buckwheat Leaf and Flower Extracts on Antioxidant Status in Mouse Organs. *Oxid. Med. Cell. Longev.* **2018**, *2018*, 6712407. [CrossRef]
60. Choi, J.Y.; Lee, J.M.; Lee, D.G.; Cho, S.; Yoon, Y.H.; Cho, E.J.; Lee, S. The N-Butanol Fraction and Rutin from Tartary Buckwheat Improve Cognition and Memory in an In Vivo Model of Amyloid-β-Induced Alzheimer's Disease. *J. Med. Food* **2015**, *18*, 631–641 [CrossRef]
61. Eguchi, K.; Anase, T.; Osuga, H. Plant Production Science Development of a High-Performance Liquid Chromatography Method to Determine the Fagopyrin Content of Tartary Buckwheat (*Fagopyrum tartaricum* Gaertn.) and Common Buckwheat (*F. esculentum* Moench). *Plant Prod. Sci.* **2009**, *12*, 475–480. [CrossRef]
62. Benković, E.T.; Žigon, D.; Friedrich, M.; Plavec, J.; Kreft, S. Isolation, Analysis and Structures of Phototoxic Fagopyrins from Buckwheat. *Food Chem.* **2014**, *143*, 432–439. [CrossRef]
63. Kočevar Glavač, N.; Stojilkovski, K.; Kreft, S.; Park, C.H.; Kreft, I. Determination of Fagopyrins, Rutin, and Quercetin in Tartary Buckwheat Products. *LWT—Food Sci. Technol.* **2017**, *79*, 423–427. [CrossRef]
64. Kim, J.; Hwang, K.T. Fagopyrins in Different Parts of Common Buckwheat (*Fagopyrum esculentum*) and Tartary Buckwheat (*F. tataricum*) during Growth. *J. Food Compos. Anal.* **2020**, *86*, 103354. [CrossRef]
65. Szymański, S.; Majerz, I. Theoretical Studies on the Structure and Intramolecular Interactions of Fagopyrins—Natural Photosensi tizers of Fagopyrum. *Molecules* **2022**, *27*, 3689. [CrossRef] [PubMed]
66. Norbäck, D.; Wieslander, G. A Review on Epidemiological and Clinical Studies on Buckwheat Allergy. *Plants* **2021**, *10*, 607 [CrossRef]
67. Vogrinčič, M.; Kreft, I.; Filipič, M.; Žegura, B. Antigenotoxic Effect of Tartary (*Fagopyrum tataricum*) and Common (*Fagopyrum esculentum*) Buckwheat Flour. *J. Med. Food* **2013**, *16*, 944–952. [CrossRef]
68. Suzuki, T.; Morishita, T.; Takigawa, S.; Noda, T.; Ishiguro, K. Evaluation of the Mutagenicity Potential of Trace-Rutinosidase Variety of Tartary Buckwheat (*Fagopyrum tataricum* Gaertn.) Using the Ames Test. *J. Agric. Chem. Environ.* **2016**, *5*, 100–105 [CrossRef]
69. Okamoto, T. Safety of Quercetin for Clinical Application (Review). *Int. J. Mol. Med.* **2005**, *16*, 275–278. [CrossRef] [PubMed]

70. Cunningham, P.; Patton, E.; VanderVeen, B.N.; Unger, C.; Aladhami, A.; Enos, R.T.; Madero, S.; Chatzistamou, I.; Fan, D.; Murphy, E.A.; et al. Sub-Chronic Oral Toxicity Screening of Quercetin in Mice. *BMC Complement. Med. Ther.* **2022**, *22*, 279. [CrossRef] [PubMed]
71. Kreft, I. Breeding of Determinate Buckwheat. *Fagopyrum* **1989**, *9*, 57–59.
72. Fesenko, N.V. A Genetic Factor Responsible for the Determinate Type of Plants in Buckwheat. *Genetica* **1968**, *4*, 165–166.
73. Fesenko, N.V.; Martinenko, G.E. Contemporary Buckwheat Breeding Work in Russia. In *Current Advances in Buckwheat Research, Proceedings of the 6th International Symposium Buckwheat, Ina, Japan, 24–29 August 1995*; Shinshu University Press: Nagano, Japan, 1995; pp. 269–275.
74. Ohsawa, R. Invited Review Current Status and Prospects of Common Buckwheat Breeding in Japan. *Breed. Sci.* **2020**, *70*, 3–12. [CrossRef] [PubMed]
75. Funatsuki, H.; Suvorova, G.; Sekimura, K. Determinate Type Variants in Japanese Buckwheat Lines. *Breed. Sci.* **1996**, *46*, 275–277. [CrossRef]
76. Wang, Y.; Campbell, C.G. Tartary Buckwheat Breeding (*Fagopyrum tataricum* L. Gaertn.) through Hybridization with Its Rice-Tartary Type. *Euphytica* **2007**, *156*, 399–405. [CrossRef]
77. Wang, Y.; Guan, Z.; Liang, C.; Liao, K.; Xiang, D.; Huang, J.; Wei, C.; Shi, T.; Chen, Q. Agronomic and Metabolomics Analysis of Rice-Tartary Buckwheat (*Fagopyrum tataricum* Gaertn) Bred by Hybridization. *Sci. Rep.* **2022**, *12*, 11986. [CrossRef] [PubMed]
78. Mukasa, Y.; Suzuki, T.; Honda, Y. Suitability of Rice-Tartary Buckwheat for Crossbreeding and for Utilization of Rutin. *Japan Agric. Res. Q. JARQ* **2009**, *43*, 199–206. [CrossRef]
79. Duan, Y.; Yin, G.; He, R.; Yang, X.; Cai, S.; Wang, Y.; Lu, W.; Sun, D.; Wang, L.; Wang, Y.; et al. Identification of Candidate Genes for Easily-Shelled Traits in Tartary Buckwheat Based on BSA-Seq and RNA-Seq Methods. *Euphytica* **2022**, *218*, 91. [CrossRef]
80. Wang, D.; Yang, T.; Li, Y.; Deng, F.; Dong, S.; Li, W.; He, Y.; Zhang, J.; Zou, L. Light Intensity—A Key Factor Affecting Flavonoid Content and Expression of Key Enzyme Genes of Flavonoid Synthesis in Tartary Buckwheat. *Plants* **2022**, *11*, 2165. [CrossRef]
81. Zhao, H.; He, Y.; Zhang, K.; Li, S.; Chen, Y.; He, M.; He, F.; Gao, B.; Yang, D.; Fan, Y.; et al. Rewiring of the Seed Metabolome during Tartary Buckwheat Domestication. *Plant Biotechnol. J.* **2023**, *21*, 150–164. [CrossRef]
82. He, Y.; Zhang, K.; Li, S.; Lu, X.; Zhao, H.; Guan, C.; Huang, X.; Shi, Y.; Kang, Z.; Fan, Y.; et al. Multiomics Analysis Reveals the Molecular Mechanisms Underlying Virulence in Rhizoctonia and Jasmonic Acid–Mediated Resistance in Tartary Buckwheat (*Fagopyrum tataricum*). *Plant Cell* **2023**, *35*, 2773–2798. [CrossRef]
83. Meng, H.L.; Sun, P.Y.; Wang, J.R.; Sun, X.Q.; Zheng, C.Z.; Fan, T.; Chen, Q.F.; Li, H.Y. Comparative Physiological, Transcriptomic, and WGCNA Analyses Reveal the Key Genes and Regulatory Pathways Associated with Drought Tolerance in Tartary Buckwheat. *Front. Plant Sci.* **2022**, *13*, 985088. [CrossRef]
84. Luthar, Z.; Fabjan, P.; Mlinarič, K. Biotechnological Methods for Buckwheat Breeding. *Plants* **2021**, *10*, 1547. [CrossRef] [PubMed]

MDPI AG
Grosspeteranlage 5
4052 Basel
Switzerland
Tel.: +41 61 683 77 34

Agriculture Editorial Office
E-mail: agriculture@mdpi.com
www.mdpi.com/journal/agriculture

www.ingramcontent.com/pod-product-compliance
Lightning Source LLC
Chambersburg PA
CBHW061434210926
51850CB00008B/1511
9783725858316